10/87

QK
731
.P595
1987

D0849529

PLANT HORMONES AND THEIR ROLE IN PLANT GROWTH AND
DEVELOPMENT

The cover drawing shows four genetic lines of pea differing in internode length and gibberellin (GA) content in their vegetative tissue. Left to right: Nana (*na*): an ultradwarf containing no detectable GAs; Dwarf (*Na le*) containing GA_{20}; Tall (*Na le*) containing GA_1; Slender (*la cry*[s]): an ultratall also with no detectable GAs (see chapter E4). The inset formula is that for GA_1.

(Drawing by Barbara Bernstein from a photograph by Peter J. Davies)

Plant Hormones and their Role in Plant Growth and Development

Edited by

PETER J. DAVIES
Section of Plant Biology
New York State College of Agriculture and Life Sciences
Cornell University, Ithaca, New York, USA

D. HIDEN RAMSEY LIBRARY
U.N.C. AT ASHEVILLE
ASHEVILLE, N. C. 28814

1987 **MARTINUS NIJHOFF PUBLISHERS**
a member of the KLUWER ACADEMIC PUBLISHERS GROUP
DORDRECHT / BOSTON / LANCASTER

Distributors

for the United States and Canada: Kluwer Academic Publishers, P.O. Box 358, Accord Station, Hingham, MA 02018-0358, USA
for the UK and Ireland: Kluwer Academic Publishers, MTP Press Limited, Falcon House, Queen Square, Lancaster LA1 1RN, UK
for all other countries: Kluwer Academic Publishers Group, Distribution Center, P.O. Box 322, 3300 AH Dordrecht, The Netherlands

Library of Congress Cataloging in Publication Data


```
Plant hormones and their role in plant growth and
   development.

   Includes bibliographies and index.
   1. Plant hormones.  2. Growth (Plants)  3. Plants--
Development.  I. Davies, Peter J.
QK731.P595  1987        581.3'1         87-5669
```


ISBN 90-247-3497-5 (Hardback)
ISBN 90-247-3498-3 (Paperback)

Book information

Set in type by the Media Services Department, New York State College of Agriculture and Life Sciences, Cornell University, Ithaca, New York 14853, USA.

Copyright

© 1987 by Martinus Nijhoff Publishers, Dordrecht.

All rights reserved. No part of this publication may be reproduced, stored in a retrieval system, or transmitted in any form or by any means, mechanical, photocopying, recording, or otherwise, without the prior written permission of the publishers,
Martinus Nijhoff Publishers, P.O. Box 163, 3300 AD Dordrecht,
The Netherlands.

PRINTED IN THE NETHERLANDS

PREFACE

Plant hormones play a crucial role in controlling the way in which plants grow and develop. While metabolism provides the power and building blocks for plant life it is the hormones that regulate the speed of growth of the individual parts and integrate these parts to produce the form that we recognize as a plant. In addition, they play a controlling role in the processes of reproduction. This book is a description of these natural chemicals: how they are synthesized and metabolized; how they work; how we measure them; and a description of some of the roles they play in regulating plant growth and development. This is not a conference proceedings but a selected collection of newly written, integrated, illustrated reviews describing our knowledge of plant hormones and the experimental work which is the foundation of this knowledge.

The information in these pages is directed at advanced students and professionals in the plant sciences: botanists, biochemists, molecular biologists, or those in the horticultural, agricultural and forestry sciences. It is intended that the book should serve as a text and guide to the literature for graduate level courses in the plant hormones, or as a part of courses in plant or comparative development. Scientists in other disciplines who wish to know more about the plant hormones and their role in plants should also find this volume invaluable. It is hoped that anyone with a reasonable scientific background can find valuable information in this book expounded in an understandable fashion.

Gone are the days when one person could write a comprehensive book in an area such as plant hormones. I have thus drawn together a team of forty-one experts who have individually or jointly written about their own area. At my direction they have attempted to tell a story in a way that will be both informative and interesting. Their styles and approaches vary, because they each have a tale to tell from their own perspective. The choice of topics has been my own. Within each topic the coverage and approach has been decided by the authors. While the opinions expressed by the authors are their own, they are, in general, also mine because I knew their perspective before I invited them to join the project.

A volume such as this cannot be encyclopedic. For this sort of coverage the reader is refered to the Encyclopedia of Plant Physiology (new series, volumes 9, 10, and 11). Nevertheless, we have covered the majority of topics in which active research is taking place. The author of each chapter has provided a set of references which will guide the reader to more detailed recent reviews, as well as classical papers and the latest advances in that area. The number of references has, however, been limited so as not to disrupt the narrative excessively. The subject matter ranges from basic biochemistry and molecular biology to the use of natural and synthetic plant growth regulators in agriculture and horticulture, with a preview of the potential manipulation of plant hormone content via prokaryotes. It is most noticeable as one progresses through the chapters that while we know a lot, though certainly not everything, about the action of plant hormones at the molecular and cellular level, our knowledge of hormones in the whole plant functions of importance to agriculture and horticulture, such in flowering, tuberization and dormancy, is at a very superficial level. Only when such systems are fully understood can we hope to manipulate plant growth to human advantage.

I would like to thank the authors who made this volume possible. The production of the final copy was accomplished from author-submitted computer discs by Marilyn Kelly, Donna Vantine, Anne Mac Arthur, and Parrish Kelley at Cornell University Media Services, whose efforts are gratefully acknowledged. I would also like to thank Linda DeNoyer for the translation of several computer discs and for assistance in the preparation of the index and Barbara Bernstein for preparing some of the artwork.

Peter J. Davies
Ithaca, New York
December 1986

PLANT HORMONES
AND THEIR ROLE IN PLANT GROWTH
AND DEVELOPMENT

Contents

E. THE FUNCTIONING OF HORMONES IN PLANT GROWTH AND DEVELOPMENT

A.   INTRODUCTION

A1. The Plant Hormones: Their Nature, Occurrence, and Functions

Peter J. Davies

Section of Plant Biology, Cornell University, Ithaca, New York 14853, USA.

INTRODUCTION

The Meaning of a Plant Hormone

Plant hormones are a group of naturally occurring, organic substances which influence physiological processes at low concentrations. The processes influenced consist mainly of growth, differentiation and development, though other processes, such as stomatal movement, may also be affected. Plant hormones have also been refered to as 'phytohormones' though this term is seldom used.

The term "hormone" comes originally from the Greek and is used in animal physiology to denote a chemical messenger. Its original use in plant physiology was derived from the mammalian concept of a hormone. This involves a localized site of synthesis, transport in the bloodstream to a target tissue, and the control of a physiological response in the target tissue via the concentration of the hormone (14). Auxin was similarly thought to produce a growth response at a distance from its site of synthesis, and thus to fit the definition of a *transported* chemical messenger. It is now clear that plant hormones do not fulfil the requirements of a hormone in the mammalian sense. In a controversial article that has caused a rethinking of much dogma in the area, Trewavas (14) has argued that plant physiologists have been so concerned with making plant hormones fit the same characteristics as animal hormones that we have overlooked their unique characteristics in controlling growth and development.   The synthesis of plant hormones may be localized (as occurs for animal hormones) but may also occur in a wide range of tissues, or cells within tissues. While they may be transported and have their action at a distance this is not always the case. They may also act in the tissue in which they are synthesized or even within the same cell. At one

1

extreme we find the transport of cytokinins from roots to leaves where they prevent senescence and maintain metabolic activity, while at the other extreme the production of the gas ethylene may bring about changes within the same tissue, or within the same cell, where it is synthesized. Thus, we must abandon the concept of *transport* as being an *essential* property of a plant hormone. Trewavas has also strongly argued that control is not by concentration but by a change in sensitivity of the tissue to the compound. This issue is more hotly debated and will be discussed later. However, as a result of the lack of direct rigid parallels with animal hormones he has argued that we should abandon the term "plant hormone". In a rebuttal Cleland (15) has characterized this argument as "semantic quibbling". Trewavas suggests we should replace the term hormone with "plant growth substance".[1] The disadvantage of this name, besides being cumbersome and containing the rather vague term "substance," is that it does not describe fully what these natural regulators do. Growth is only one of the many processes influenced. Changing the name to "plant growth and development substances" becomes even more awkward and still may not cover all cases. While the term plant growth regulator is a little more precise this term has been (unfortunately?) usurped by the agrichemical industry to denote synthetic plant growth regulators (see chapter E19) as distinct from endogenous growth regulators. Thus the term plant growth regulator is largely unused in reference to endogenous regulators of growth and differentiation. One could invent an entirely new single word name for this group of compounds but, rather like the international language Esperanto, that would seem to be unlikely to catch on in the face of over fifty years of habit. Thus we are left with the imperfect term "plant hormone" and all that it implies. We must break with the characteristics expected of animal hormones: plant hormones are a unique set of compounds, with unique metabolism and properties, that form the subject of this book. Their only universal characteristics are that they are natural compounds in plants with an ability to affect physiological processes at concentrations far below those where either nutrients or vitamins would affect these processes. In fact, this notion of a plant hormone is much closer to the meaning of the Greek origin of the word (*to set in motion* or *to stimulate*) than is the current meaning of the word *hormone* used in the context of animal physiology. The Greek origin of the word does not have in it the implicit idea of either transport or action at a distance. The term *plant hormones* has, therefore, been retained in this book.

----

[1] The international society for the study of plant hormones is named the "International Plant Growth Substance Association" (IPGSA).

The Discovery, Identification and Quantitation of Plant Hormones.

The concept of plant hormones derives from Darwin's experiments on the phototropism of coleoptiles, which indicated the presence of a transported signal. This led to Went's elucidaton of auxin in 1928 and its subsequent identification as indoleacetic acid (IAA) (5). Other lines of investigation led to the discovery of the other hormones: research in plant pathogenesis led to gibberellins (GA); efforts to culture tissues led to cytokinins (CK); the control of abscission and dormancy led to abscisic acid (ABA); and the effects of illuminating gas and smoke led to ethylene. These accounts are told in virtually every elementary plant physiology textbook, and further elaborated in either personal accounts (5,13) or advanced treatises devoted to individual hormones (2,10) so that they need not be repeated here. It is interesting to note that, of all the plant hormones, only the chemical identification of abscisic acid was made from higher plant tissue. The original identification of the others came from extracts that produced hormone-like effects in plants: auxin from urine, gibberellins from fungal culture filtrates, cytokinins from autoclaved herring sperm DNA, and ethylene from illuminating gas. Today we have at our disposal methods of purification (such as high performance liquid chromatography:HPLC) and characterization (gas chromatography-mass spectrometry:GC-MS) that can operate at levels undreamed of by early investigators (see Chapter D1). Thus while early purifications from plant material utilized tens or even hundreds of kilograms of tissues modern analyses can be performed on a gram or less of tissue, making the characterization of hormone levels in individual leaves, buds, or even from tissues within the organs much more feasible. The advent of immunoassay (Chapter D2) promises to enable the localization of the hormones within cells. Only when the exact level and location of the hormones within the tissues are known will their precise roles and modes of action in a process be fully elucidated.

THE NATURE, OCCURRENCE, AND EFFECTS OF THE PLANT HORMONES

Before we become involved in the various subsequent chapters covering aspects of hormone biochemistry and action it is necessary to review what hormones do. In subsequent chapters some or most of these effects will be described in more detail while others will not be referred to again. It is impossible to give detailed coverage of every hormonal effect. The choice of topics for subsequent chapters has been determined largely by whether there is active research in progress in that area. Thus, while the mode of action of auxin and some effects of auxin are subjects of subsequent chapters, most of the current work on cytokinins is on their synthesis and metabolism (which is described), but so little progress has been made on how they act that there is little to add to the information below. The effects produced by each hormone have been elucidated largely from exogenous applications. In some cases we have evidence that the endogenous hormone also fulfills that role, while in other cases it has not been conclusively proved that the endogenous hormone functions in the same manner. The nature, occurrence, transport and effects of each hormone (or hormone group) are given below. It should, however, be emphasized that hormones do not act alone but in conjunction, or in opposition, to each other such that the final condition of growth or development represents the net effect of a hormonal balance (7).

Auxin

Nature

Indole-3-acetic acid (IAA) is the main auxin in most plants.

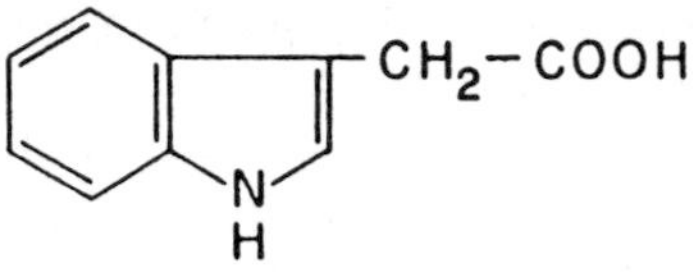

INDOLEACETIC ACID

Compounds which serve as IAA precursors may also have auxin activity (e.g., indoleacetaldehyde). Some plants contain other compounds that display weak auxin activity (e.g., phenylacetic acid) (16). IAA may also be present as various conjugates such as indoleacetyl aspartate (Chapter B1)). 4-chloro-IAA has also been reported in several species(11), though it is not clear to what extent the endogenous auxin activity in plants can be accounted for by 4-Cl-IAA. Several synthetic auxins are also used in commercial applications (Chapter E19).

Sites of biosynthesis

IAA is synthesized from tryptophan (Chapter B1) primarily in leaf primordia and young leaves, and in developing seeds.

Transport

IAA transport is cell to cell (Chapter E5 and E6). Transport to the root probably also involves the phloem.

Effects
• Cell enlargement—auxin stimulates cell enlargement and stem growth (Chapter C1).

- Cell division—auxin stimulates cell division in the cambium and, in combination with cytokinin, in tissue culture (Chapters E6 and E18).
- Vascular tissue differentiation—auxin stimulates differentiation of phloem and xylem (Chapter E6).
- Root initiation—auxin stimulates root initiation on stem cuttings, and also the development of branch roots and the differentiation of roots in tissue culture (Chapter E18).
- Tropistic responses—auxin mediates the tropistic (bending) response of shoots and roots to gravity and light (Chapter E7). It should be noted here that an endogenous role for auxin is controversial. Auxin may be involved in some instances and not in others.
- Apical dominance—the auxin supply from the apical bud represses the growth of lateral buds (Chapter E8).
- Leaf senescence—auxin delays leaf senescence (Chapter E16).
- Leaf and fruit abscission—auxin may inhibit or promote (via ethylene) leaf and fruit abscission depending on the timing and position of the source (Chapters E1, E8 and E19).
- Fruit setting and growth—auxin induces these processes in some fruit (Chapter E19)
- Assimilate partitioning—assimilate movement is enhanced towards an auxin source possibly by an effect on phloem transport (Chapter E12).
- Fruit ripening—auxin delays ripening (Chapters E1 and E17).
- Flowering—auxin promotes flowering in Bromeliads (Chapter E10).
- Growth of flower parts—stimulated by auxin (Chapter E1).
- Promotes femaleness in dioecious flowers (via ethylene) (Chapters E1 and E10).

In several systems (e.g., root growth) auxin, particularly at high concentrations, is inhibitory. Almost invariably this has been shown to be mediated by auxin-produced ethylene (3,6) (Chapter E1). If the ethylene synthesis is prevented by various ethylene synthesis inhibitors, the ethylene removed by hypobaric conditions, or the action of ethylene opposed by silver salts (Ag^+), then auxin is no longer inhibitory.

Gibberellins (GAs).

Nature

The gibberellins (GAs) are a family of compounds based on the *ent*-gibberellane structure (Chapter B2). While the most widely available compound is GA_3 or gibberellic acid, which is a fungal product, the most important GA in plants is probably GA_1, which is the GA primarily responsible for stem elongation (Chapters B2, E3 and E4). Many of the other GAs are precursors of the growth active GAs such as GA_1 and GA_{20}.

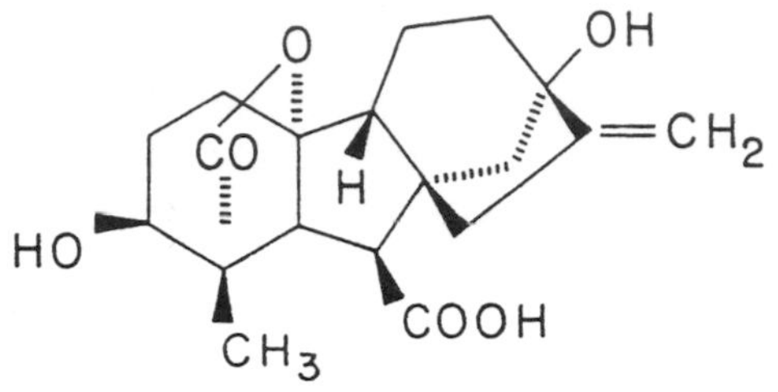

GA₁

Sites of biosynthesis.
GAs are synthesized from mevalonic acid (Chapter B2) in young tissues of the shoot (exact location uncertain) and developing seed. It is uncertain whether synthesis also occurs in roots (12).

Transport
GAs are probably transported in the phloem and xylem.

Effects
- Stem growth—GA_1 causes hyperelongation of stems by stimulating both cell division and cell elongation (Chapters E3 and E4). This produces tall, as opposed to dwarf, plants.
- Bolting in long day plants—GAs cause stem elongation in response to long days (Chapter E10).
- Induction of seed germination—GAs can cause seed germination in seeds that normally require cold (stratification) or light to induce germination (Chapter E15).
- Enzyme production during germination—GA stimulates the production of numerous enzymes, notably α-amylase, in germinating cereal grains (Chapter C2).
- Fruit setting and growth—This can be induced by exogenous applications in some fruit (e.g., grapes) (Chapter E19). The endogenous role is uncertain.
- Induction of maleness in dioecious flowers (Chapter E10).

Cytokinins (CKs)

Nature
CKs are adenine derivatives characterized by an ability to induce cell division in tissue culture (in the presence of auxin). The most common cytokinin base in plants is zeatin. Cytokinins also occur as ribosides and ribotides (Chapter B3).

Sites of biosynthesis
CK biosynthesis is through the biochemical modification of adenine (Chapter B3). It occurs in root tips and developing seeds.

6

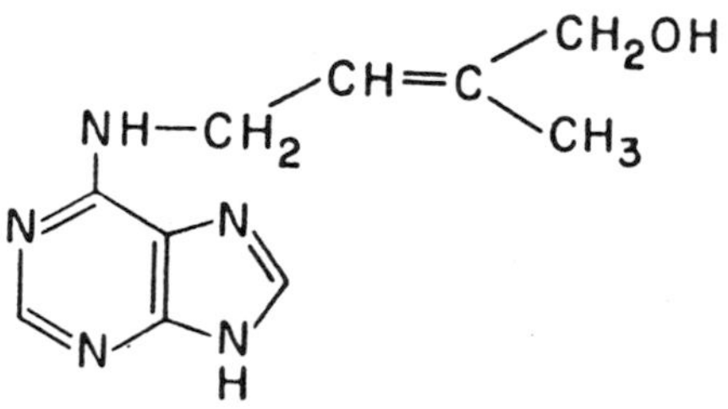

$$NH-CH_2-CH=C\underset{CH_3}{\overset{CH_2OH}{<}}$$

ZEATIN

Transport

CK transport is via the xylem from roots to shoots.

Effects
- Cell division—exogenous applications of CKs induce cell division in tissue culture in the presence of auxin (Chapter E18). This also occurs endogenously in crown gall tumors on plants (Chapter E20). The presence of CKs in tissues with actively dividing cells (eg. fruits, shoot tips) indicates that CKs may naturally perform this function in the plant.
- Morphogenesis—in tissue culture (Chapter E18) and crown gall (Chapter E20) CKs promote shoot initiation. In moss, CKs induce bud formation (Chapter E4).
- Growth of lateral buds—CK applications can cause the release of lateral buds from apical dominance (Chapter E8).
- Leaf expansion (8)—resulting solely from cell enlargement. This is probably the mechanism by which the total leaf area is adjusted to compensate for the extent of root growth, as the amount of CKs reaching the shoot will reflect the extent of the root system.
- CKs delay leaf senescence (Chapter E16).
- CKs may enhance stomatal opening in some species (Chapter E9).
- Chloroplast development—the application of CK leads to an accumulation of chlorophyll and promotes the conversion of etioplasts into chloroplasts (9).

Mode of Action

CKs are the only hormone in this book for which there is no chapter on the mode of action in any system. This is because the action of CKs is poorly understood and insufficient evidence exists to conclusively identify any biochemical point of action (4).

Ethylene

Nature

The gas ethylene (C_2H_4) is synthesized from methionine (Chapter B4) in many tissues in response to stress. It does not seem to be essential for normal vegetative growth (Chapter E1). It is the only hydrocarbon with a pronounced effect on plants.

Sites of synthesis

Ethylene is synthesized by most tissues in response to stress. In particular, it is synthesized in tissues undergoing senescence or ripening (Chapters E1 and E16).

Transport

Being a gas, ethylene moves by diffusion from its site of synthesis. A crucial intermediate in its production, 1-aminocyclopropane-1-carboxylic acid (ACC) can, however, be transported and may account for ethylene effects at a distance from the causal stimulus (Chapter E1).

Effects

The effects of ethylene are fully described in Chapter E1. They include:

- Release from dormancy.
- Shoot and root growth and differentiation.
- Adventitious root formation.
- Leaf and fruit abscission.
- Flower induction in some plants (Chapter E10).
- Induction of femaleness in dioecious flowers (Chapter E10).
- Flower opening.
- Flower and leaf senescence (Chapters E1 and E16).
- Fruit ripening.

Abscisic acid (ABA)

Nature

Abscisic acid is a single compound with the following formula:

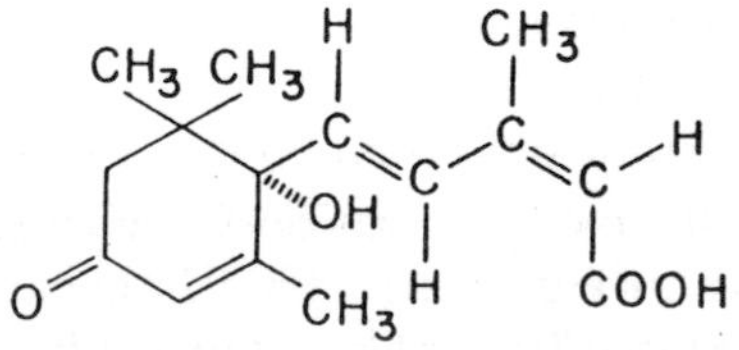

ABSCISIC ACID

Its name is rather unfortunate. The first name given was "abscisin II" because it was thought to control the abscission of cotton bolls. At almost the same time another group named it "dormin" for a purported role in bud dormancy. By a compromise the name abscisic acid was coined (1). It now appears to have little role in either function (Chapters E1 and E15) but we are stuck with this name. As a result of the original association with abscission and dormancy, ABA has become thought of as an inhibitor. While exogenous applications can inhibit growth in the plant, ABA appears to act as much as a promoter (e.g., storage protein synthesis in seeds—Chapter E13) as an inhibitor, and a more open attitude towards its overall role in plant development is warranted.

Sites of synthesis

ABA is synthesized from mevalonic acid (Chapter B5) in mature leaves particularly in response to water stress (Chapters E9 and E12). Seeds are also rich in ABA which may be imported from the leaves or synthesized *in situ* (Chapter E12).

Transport

ABA is exported from leaves in the phloem. There is some evidence that ABA may circulate to the roots in the phloem and then return to the shoots in the xylem (Chapter E12).

Effects

- Stomatal closure—water shortage brings about an increase in ABA which leads to stomatal closure (Chapter E9).
- ABA induces transport of photosynthate towards developing seeds, and its subsequent uptake by growing embryos (Chapter E12).
- ABA induces storage protein synthesis in seeds (Chapter E13).
- ABA counteracts the effect of gibberellin on α-amylase synthesis in germinating cereal grains (Chapter C3).
- ABA may affect the induction and maintenance of dormancy in seeds and buds. This role of ABA is currently uncertain. If it is involved it is certainly not the only factor in dormancy (Chapters E13 and E15).

Polyamines

There is some controversy as to whether these compounds (fully described in Chapter E2) should be classified as hormones, even within our rather broad current definition. They were first tentatively accepted by their inclusion (in the form of a specific session) at the International Conference on Plant Growth Substances in 1982. Galston (personal communication) justifies their classification as hormones on the following grounds:
- They are widespread in all cells and can exert regulatory control over growth and development at micromolar concentrations.
- In plants where the content of polyamines is genetically altered, development is affected. (E.g., in tissue cultures of carrot or *Vigna* when the polyamine level is low only callus growth occurs; when

polyamines are high, embryoid formation occurs. In tobacco plants, which are overproducers of spermidine, anthers are produced in place of ovaries.)

Such developmental control is more characteristic of hormonal compounds than nutrients such as amino acids or vitamins.

Polyamines have a wide range of effects on plants and appear to be essential for plant growth, particularly cell division and normal morphologies. At present it is not possible to make an easy, distinct list of their effects as for the other hormones. A variety of cellular and organismal effects is discussed in Chapter E2. It appears that polyamines are present in all cells rather than having a specific site of synthesis.

Whether in the long run we classify polyamines as hormones or not is irrelevant. Hormones are a human classification and organisms care naught for human classifications. Chemical compounds affect growth and development in various ways, or they do not do so. Polyamines clearly fall under the first of the two groups.

References

1. Addicott, F.T., Carns, H.R., Cornforth, J.W., Lyon, J.L., Milborrow, B.V., Ohkuma, K., Ryback, G., Smith, O.E., Thiessen, W.E., Wareing, P.F. (1968) Abscisic acid: a proposal for the redesignation of abscisin II (dormin). *In* Biochemistry and Physiology of Plant Growth Substances, pp. 1527-1529, Wightman, F., Setterfield, G., ed. Runge Press, Ottawa.
2. Addicott, F.T., Carns, H.R. (1983) History and Introduction. *In* Abscisic acid, pp. 1-21, Addicott, F.T., ed. Praeger, New York.
3. Burg, S.P., Burg, E.A. (1966) Interaction between auxin and ethylene and its role in plant growth. Proc. Natl. Acad. Sci. USA. 55, 262-269.
4. Horgan, R. (1984) Cytokinins. *In* Advanced Plant Physiology, pp. 53- 75, Wilkins, M.B., ed. Pitman, London.
5. Jacobs, W.P. (1979) Plant hormones and plant development. Cambridge University Press, Cambridge, New York. 339 pp.
6. Mulkey, T.J., Kuzmanoff, K.M., Evans, M.L. (1982) Promotion of growth and hydrogen ion efflux by auxin in roots of maize pretreated with ethylene biosynthesis inhibitors. Plant Physiol. 70, 186-188.
7. Leopold, A.C. (1980) Hormonal regulating systems in plants. *In* Recent Developments in Plant Sciences, pp. 33-41, Sen S.P., ed. Today and Tomorrow Publishers, New Delhi.
8. Letham, D.S. (1971) Regulators of cell division in plant tissues. XII. A cytokinin bioassay using excised radish cotyledons. Physiol. Plant 25, 391-396.
9. Parthier, B. (1979) Phytohormones and chloroplast development. Biochem. Physiol. Pflanzen 174, 173-214.
10. Phinney, B.O. (1983) The history of gibberellins. *In* The Biochemistry and Physiology of Gibberellins, Vol 1, pp. 19-52, Crozier. A., ed. Praeger, New York.
11. Pless, T., Bottger, M., Hedden, P., Graebe, J. (1984) Occurrence of 4-Cl-indoleacetic acid in broad beans and correlation of its levels with seed development. Plant Physiol. 74, 320-323.
12. Stoddart, J.L. (1983) Sites of gibberellin biosynthesis and action. *In* The Biochemistry and Physiology of Gibberellins, Vol. 2, pp. 1- 55, Crozier, A., ed. Praeger, New York.
13. Thimann, K.V. (1977) Hormone action in the whole life of plants. Univ. of Massachusetts Press, Amherst. 448 pp.

14. Trewavas, A. (1981) How do plant growth substances act? Plant Cell Environment 4, 203-228.
15. Trewavas, A.J., Cleland, R.E. (1983) Is plant development regulated by changes in the concentration of growth substances or by changes in the sensitivity to growth substances? Trends in Biochem. Sci. 8, 354-357.
16. Wightman, F., Lighty, D.G. (1982) Identification of phenylacetic acid as a natural auxin in the shoots of higher plants. Physiol. Plant 55, 17-24.

A2. The Plant Hormone Concept: Transport, Concentration, and Sensitivity

Peter J. Davies

Section of Plant Biology, Cornell University, Ithaca, New York 14853, USA.

THE ROLE OF TRANSPORT AND REDISTRIBUTION IN PLANT HORMONE FUNCTION.

The idea that transport was an essential part of the role of plant hormones originally came from experiments on the phototropic control of coleoptile growth. The hypothesis was that the IAA was synthesized in the tip, transported basipetally and was then redistributed laterally to give differential growth and bending. We now know that most of the IAA coming from the coleoptile tip is not synthesized *in situ* from tryptophan but comes from an IAA-inositol source in the endosperm of the grain (see Chapter B1), and is transported, as free IAA or as the IAA-conjugate, to the tip of the coleoptile where conjugate hydrolysis occurs. While so far we have only a minor deviation from the original concept, many studies have failed to show a redistribution of IAA in response to tropistic stimuli. In addition, studies by Firn and Digby (7) show that, at least in some cases, growth begins all along the stem at the same time, and faster than auxin can be transported from the tip. Thus, if auxin is involved in tropistic responses of stems it must already be in, or synthesized in, the responding tissue rather than being redirected in transit from the tip. If a stem is cut longitudinally in half and then laid on its side, the bottom half will grow faster than the top half, so that redistribution across the entire stem is not necesary for differential growth. This is not to say that auxin is not involved but rather that, if redistribution is important, then it must be on a smaller scale than the entire stem (see Chapter E7).

While auxin is undoubtedly transported in a very specific manner (Chapter E5), the auxin undergoing transport may be more involved in vascular differentiation (Chapter E6), or in lateral root initiation, rather than in the control of directional growth. In stems the action of auxin in elongation growth is primarily in the epidermal and/or sub-epidermal cells. If a stem section is surrounded by an auxin solution, growth occurs with a lag of about 15 minutes (see Chapter C1). If, however, the auxin is supplied to the apical end of a vertical stem section, and allowed to enter the stem via polar transport of the auxin, there is very little growth

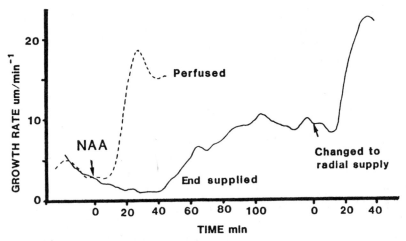

Fig. 1. The effect on growth of auxin (10 μM naphthalene acetic acid) supplied either only to the cut end, or to the whole section of sunflower hypocotyl sections. Auxin transporting into the segments from the ends has little influence on growth. From (7).

response even over a time period several times that needed for the auxin to transport through the section (8) (Fig. 1). In addition, auxin transport inhibitors have no effect on growth. We must, therefore, conclude that auxin in the transport system in stems stays in that system and does not move laterally to influence the growth-response of epidermal cells, at least over short time periods. It is unknown whether the auxin comes out of the transport channel on geotropic stimulation, and moves downward to influence the lower epidermal cells, or whether some other mechanism is involved.

A reinvestigation as to whether transported auxin has a role in the phototropic curvature of maize coleoptiles has, however, seemingly answered the question in the affirmative as there is a basipetal migration of the response from the tip, and this occurs at the same rate as the growth stimulation caused by exogenous auxin (2). Thus at the present time there is evidence both for and against a role of transported auxin in growth. Almost certainly the role of auxin varies with the experimental system, though what actually occurs in the non-manipulated growing plant is uncertain.

While we must now abandon transport as an essential part of the definition of a plant hormone this does not mean that transport plays no part in hormone functioning. A prime function of hormones in plants, which lack a nervous system, is to convey information from one part of the plant to another. The role of cytokinins in transmitting information on the status of the root system to the shoots is an obvious example: the leaves of newly transplanted trees are often much smaller than established trees so that the transpiring surface is reduced to match the supply of water from the roots. Abscisic acid appears to be a medium of communication between leaves and developing fruits (see Chapter E12), while auxin moving basipetally from the shoot tip induces vascular differentiation (Chapter E6) and, in coleoptiles, seems to be involved in phototropic curvature (Chapter E7). Trewavas has, however, argued (18) that transport is unimportant in the action of "plant growth substances". As a prime example he claimed that grafts between tall plants and short plants show that tallness is not transmitted. We now know, from the elegant work of Reid and co-workers on peas (Chapter E4), that the results obtained depend on the tissue. The control of tallness resides in GA_1, which is found only in the youngest internodes and thus will not be transmitted through a graft. GA_1 is, however, the final product of a synthetic pathway which produces the biologically active compound, and the transmission of tallness *can* be seen if the correct system is chosen. The genotype *Na le* is dwarf as it has the ability to produce GA_{20} (through the gene *Na*), but it lacks the ability to convert GA_{20} to GA_1, because it lacks the dominant gene *Le*. The genotype *na Le* is ultradwarf (nana) because it lacks the ability to synthesize GAs (because of *na*), but it has the ability to convert GA_{20} to GA_1 (because of *Le*), though normally it cannot do so because GA_{20} is lacking. Now if a dwarf scion is grafted onto a nana stock the resulting plant is tall. The stock synthesizes the GA_{20} and passes it to the scion which converts the GA_{20} to GA_1, giving tall growth (13). This shows that a thorough knowledge of the system in question is necessary before conclusions can be accurately drawn. Transport may be important in the role of plant hormones in some, and possibly most, systems.

CONCENTRATION VERSUS SENSITIVITY AS THE CONTROLLING ASPECT OF HORMONE ACTION.

The concept of control by changing concentrations is crucial to the original concept of hormones in mammals. A great stir was created amongst biologists working with plant hormones by the suggestion of Trewavas (18) that there is no evidence that plant hormones act via changes in the amount or concentration of the hormone, and that all change in response must be attributed to changes in the sensitivity of the tissue. The reason for this suggestion is the frequent lack of correlation between hormone concentrations measured in a tissue and the response of the tissue. In addition Trewavas points out that, in most plants, growth is proportional to the logarithm of the applied hormone concentration, such that there may be an increasing response over three orders of magnitude in concentration. However, changes in the *endogenous* concentration in tissues are usually far smaller than would be expected to produce the vast changes in growth or development observed; (plant hormone workers tend to regard a doubling in concentration as a large change!). According to Trewavas, as concentration changes cannot account for the differences in growth and development then something else must be responsible, and tissue sensitivity is the only other logical alternative that comes to mind. There is in fact relatively little known about tissue sensitivity. Our knowledge of hormone binding (Chapter C4) is still rather rudimentary. Nonetheless, it is probable that sensitivity of tissues certainly varies. Our best examples of changing sensitivity are to ethylene. Immature flower or fruit tissue show no response to ethylene, but mature tissue responds to ethylene with ripening or senescence (see Chapter E1). However, an important question is what we mean by the rather vague term "sensitivity". As Firn (6) has pointed out, a change in sensitivity simply refers to an observation that the response to a given amount of hormone has changed. This could be caused by a change in the number of receptors (*receptivity*), a change in receptor affinity (*affinity*), or a change in the subsequent chain of events (*response capacity*). A change in the response to a given amount of hormone could also be caused by a change in the level of other endogenous substances that enhance, or inhibit, the response to the hormone. When dealing with exogenously applied hormones *uptake efficiency* must also be taken into account. A change in each of these would give a very different response curve to changing hormone concentrations (Fig. 2) (6), and even the response of a single system has not been examined in detail to determine which, if any, of these factors might be operating.

Does concentration play any role? It seems that Trewavas has, in reality, overstated the case against concentration. Even if sensitivity changed there would still be a response to a change in concentration, though of a different magnitude (6). In addition we should ask what are we measuring when we calculate "concentration," and where we are measuring it, in relation to the tissues or cells that respond? If the accuracy of measurement or localization is vague then so is our knowledge of hormone concentration at the active site. Below I will list some of the pitfalls that can occur in measurement of these parameters.

Correlations of Growth with Hormonal Concentration.

One of Trewavas's principal examples to support his case was the seeming lack (at that time) of any correlation between the endogenous GA concentration and tallness (in tall versus dwarf plants), despite the fact that applications of GA_3 to dwarf plants produced tall phenotypes (18). This has now completely been refuted by the work of Reid and co-workers in peas (Chapter E4) (and by Phinney and co-workers in maize (16) though the hormonal story in maize does not resolve quite so neatly). They have shown that tall plants have GA_1 while dwarf plants do not. The tallness

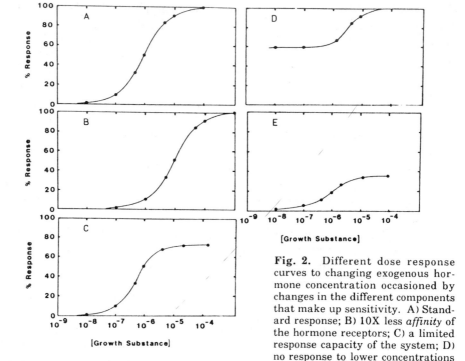

Fig. 2. Different dose response curves to changing exogenous hormone concentration occasioned by changes in the different components that make up sensitivity. A) Standard response; B) 10X less *affinity* of the hormone receptors; C) a limited response capacity of the system; D) no response to lower concentrations because of a high endogenous concentration. The response only occurs when the endogenous concentration is exceeded; E) A decrease in the number of hormone receptors (*receptivity*). Note change in slope of main part of response curve. From (6).

gene in peas, *Le*, controls the conversion (by 3-hydroxylation) of GA_{20} to GA_1. In the presence of *le* GA_{20} builds up and is not converted to GA_1 (9). If the gibberellin content is measured by bioassay, without prior chromatographic separation, then tall and dwarf plants are found to contain the same amount of "gibberellin activity." This is because the vast majority of bioassays will show no difference between GA_{20} and GA_1, since the bioassay plant has the ability to convert GA_{20} to the tall-active GA_1. (One newer bioassay, Waito-C rice, has the ability to distinguish 3C-OH GAs and on such a bioassay the difference can be seen.) It should also be noted that the GA_1 is only present in the youngest internodes of the stem. Often the extracts used in comparing tall and dwarf plants have come from whole shoots, mature internodes, or even seeds, which do not display any difference between tall and dwarf.

While it is relatively hard to find cases where precise variations in concentration correlate with parallel variations in growth or development, there are numerous cases where the presence of the hormone correlates with a distinct change in growth and development

(19). Amongst these could be cited the effect of auxin coming from a leaf primordium on vascular development (Chapter E6) or the hyperelongation of the submerged stem of deep water rice caused by the build-up of ethylene (Chapter E1). A developmental continuum from crown gall tumors with roots, to undifferentiated tumors, to tumors with shoots is shown to be correlated with increases in the cytokinin/auxin ratio upon infection with different strains of *Agrobacterium tumefaciens* bacteria (Fig. 4 in Chapter E20). Here we can see a distinct relationship with hormone concentration; sensitivity cannot be a factor as the infected tissue was the same in all cases.

A response to a wide (logarithmic) concentration range is not typical of all tissues. In maize coleoptiles the range is more linear, and over only about a two fold range of applied IAA concentration (2) (Fig. 3). It has also been pointed out (2) that the results indicating a response to the logarithm of applied IAA concentration have been obtained with isolated segments incubated in the solution, a possibly artifactual situation.

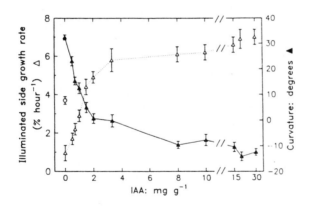

Fig. 3. The growth stimulation (Δ) and curvature repression (▲) induced by the application of different concentrations of IAA in lanolin (50 µg spot) to the irradiated side of phototropically stimulated maize coleoptiles. The single open circle shows the growth rate of the controls. Note that the growth response to IAA is approximately linear over the range of 0.5-2 mg/g and auxin totally overcomes the curvature at 2 mg/g of lanolin (a higher concentration than in aqueous solutions because of the limited diffusion in lanolin and the limited total quantity). The endogenous growth rate corresponds to that obtained by an application of IAA of about 1 mg/g or 50% of the concentration giving the maximum growth response. From (2).

Hormone Quantitation and Its Interpretation

One of the difficulties in correlating endogenous hormone concentration with differences in growth and development is the problem of what is measured when the hormone is assayed, even if this is done by highly accurate physico-chemical means (see Chapter D1). Quantitation is normally done by measuring total extractable hormone. Often this is of little relevance because the total hormonal amount tells us little about the hormonal concentration in the tissue in question, let alone in the cell, cell compartment, or at the hormone receptor site.

Many hormonal extractions are done of whole shoots, roots or fruits. While this may enable a first approximation it can be equated with analyzing the hormonal content of, say, the pituitary gland of a mammal by grinding up and assaying the whole animal. Studies investigating hormone metabolism in seeds have indicated that there are specific qualitative and quantitative differences between adjacent tissues. The embryo of pea has a different ABA and gibberellin content and metabolism than the seed coat (see Chapter E12). Applied GA_{20} is metabolized to GA_{29} in the embryo and this GA_{29} is then further metabolized to GA_{29} catabolite in the seed coat (17). We do not know what further subdivisions may exist within the embryo. Differences probably also exist within seemingly uniform tissues. We often analyze the whole tissue in the case of phototropism or geotropism, or, at best, half the stem or coleoptile. However, there may be more subtle differences within the half tissue, even though there is only a small measurable difference between the two halves (3) (Fig. 4), and this may lead to the observed differential growth. Experiments on the gradient between the upper and lower epidermis of dicotyledonous shoots indicate that the difference may be two to four times the gradient recorded between the upper and lower halves. In maize coleoptiles a two fold difference between the IAA content of the two sides is sufficient to account for the observed phototropic curvature (2).

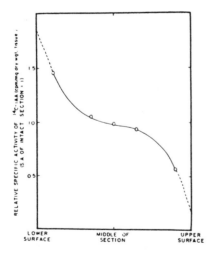

Fig. 4. The distribution of radioactivity from applied IAA-^{14}C across the diameter of a horizontally positioned pea stem section split horizontally into quarters. The extrapolations (dashed lines) indicate that the differences in IAA in the epidermal tissues, which are primarily responsible for growth, are greater than determined from measurements on half stems. From (3).

In general, we have little idea of cell to cell variation. We know even less of the differences in hormonal contents within cells. It was noted many years ago that growth of stem or coleoptile sections correlates with the amount of auxin that will diffuse out of the tissue rather than the total extractable auxin (15). This tells us that much of the hormone in the tissue is probably compartmentalized away from the growth active site, possibly within an organelle.

A demonstration of this compartmentation can be seen if a stem segment is loaded with radiolabelled auxin and then washed over an extended period to remove the auxin effluxing from the tissue. Growth initially increases in response to the applied IAA, but then declines to a very low rate over 3- 4 hours, even though about 70% of the radiolabel is still in the tissue and can be chromatographically identified as IAA. If the tissue is then put in an anaerobic environment, under which growth ceases, the rate of auxin efflux increases. A return to aerobic conditions is accompanied by a burst in the growth rate such that the total growth more than makes up for the loss of growth during anaerobiosis (termed *emergent growth*), and the rate of auxin efflux gradually returns to its previous condition (10) (Fig. 5). We interpret this as an indication that the IAA is sequestered within an inner membrane-bound compartment,

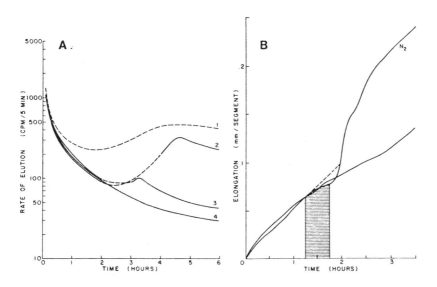

Fig. 5. A) The elution of label from [¹⁴C] IAA-treated pea stem segments as influenced by anaerobiosis and subsequent oxygenation. Under aerobic condition (bubbled with oxygen-- solid line) the rate of efflux steadily decreases (curve 4) but if the sections become anaerobic (bubbled with nitrogen) as indicated by the dashed line (curves 2 and 3), the rate of efflux increases. The increase can subsequently be reversed by the return of aerobic conditions. Segments in curve 1 were anaerobic throughout. B) In segments that become anaerobic for a short period, and then return to aerobic conditions, growth is stimulated beyond that of those that remain in aerobic conditions. From (10).

19

from which it leaks at a slower rate than when it leaves the cell through the plasmalemma. The growth active site, with which auxin interacts, is exterior to the cellular compartment in which it is contained. Once the auxin originally in the cytoplasm leaves the cell, the auxin concentration at the growth active site remains very low (under normal aerobic conditions) because the auxin exits through the plasmalemma faster than it leaves the internal compartment. Under anaerobic conditions the compartment membrane becomes leaky, possibly because an inwardly-directed, ATP-requiring IAA carrier becomes inactivated, allowing outward passive diffusion to predominate. Alternatively the compartment interior becomes more acid, thus increasing the diffusibility of the IAA, because a greater proportion of the IAA becomes protonated. This allows more IAA to leak into the rest of the cell and to the exterior. On the return of oxygen the high concentration of auxin at the growth active site allows ATP-requiring growth to take place, while the compartment membrane ceases being leaky. We do not know the location of the IAA-storage compartment, nor the location of the growth active site (see Chapters C1 and C4 for the latter). One possibility for the storage compartment is the chloroplast (or proplastid) which, because of its relatively high pH, would tend to trap acidic molecules (see Chapter E5). The vacuole seems unlikely because its acidic pH would dictate an auxin concentration only about 1/10th that in the cytoplasm, though its vast bulk in parenchymatous cells might make it a hormone reservoir. (The vacuole could possibly function as a reservoir of hormone conjugates that might be later hydrolyzed to release free hormone.) The development of hormone immunoassay (Chapter D2) promises a method by which we may be able to localize hormones within tissues and cells provided a way can be found to immobilize the highly soluble hormone molecules within the cells.

The Sensitivity Factor.

There is little doubt that changes in hormone concentration at the hormone-responsive active site can account for many of the hormone-related changes in growth and development in plants. Nonetheless, it is likely that sensitivity to the hormone plays a role in at least some cases. The increasing responsiveness of flowers and fruit to ethylene as they mature (Chapter E1) may represent either: a) a shift in the hormonal balance so that an inhibitor of ethylene action, such as IAA (20), may decrease with advancing maturity; b) a genuine change in the number or affinity of the ethylene responding sites, or c) the lack of inducibility of the subsequent enzymatic changes in immature tissue for reasons unknown. Only when hormone binding and receptor sites are fully understood can we hope to fully illucidate which of the above mechanisms might be operating.

Our best opportunities for studying hormone sensitivity lie in genetic mutants (Chapter E4) which lack sensitivity to the hormone, or appear to already be "turned on". In peas two stem-length mutants have been found with these characteristics. The slender mutant, determined by the gene combination *la cry*s, has extremely long internodes (12). It looks and behaves as if it is saturated with GA, yet the GA level is either normal, or deficient, depending on the genetic background. In the presence of the *na* allele the plants are free of detectable gibberellin, yet the plants still grow ultratall. Either the gibberellin receptor, or a subsequent receptor-controlled step, is turned on in an abnormal fashion. On the other side of the coin, a dwarf pea containing the allele *lk* is non-responsive to applied GA (14). The nature of the lesion is unknown. Since *lk* is largely epistatic to (inhibits the expression of) the *la cry*s gene combination its action would appear to be after that of *la cry*s.

CONCLUSION

The stimulating ideas of Trewavas (18) have had a great impact in revitalizing thinking within the world of plant hormones. However, the reasoned arguments presented by Trewavas have become totally misinterpreted in some places. For example, an article appeared in the New Scientist in 1984 under the headline "Do plants really have hormones?" which claimed that botanists were reevaluating the existence of hormones (22). In reply, Wang (21) points out the bias of the presented argument and counters that "by any definition these compounds are biologically active. Just as plant physiologists should not elevate hypotheses into dogma without sufficient data...we should not elevate normal scientific discussion into controversy on the same basis."

Nonetheless, we should ask whether hormones *do* have any controlling role. Canny has argued that there must be other controllers as the "hormones" are not specific or numerous enough to satisfy "Ashby's law of requisite variety: A situation can only be controlled by a controller that matches the variety of the situation" (4). He claims that plant hormones are simple compounds that cannot match the variety of developmental directions. By contrast, animal hormones, being compounds like proteins, have a high information content and can match the required variety of controlled reactions. In plants we do have the variability provided by concentration and interaction between different hormones. Canny suggests, however, that these features are still not enough and that either the plant body is autonomous or that other messages, such as oligosaccharides[1], await discovery. In rebuttal to Canny, Firn notes that hormones do not carry information because they don't have to (5). They only provide a "turn on" or "turn off" signal with the *information* being provided by the cell. Vertebrate hormones are complex because they need the *specificity* supplied by polypeptides. Concentration dependence could then provide some degree of control for the magnitude of the response. By comparison Ca^{2+} appears to be a very important regulator of a wide variety of processes, yet calcium is very simple; the specificity for the effects of Ca^{2+} depend on the cell.

In general we can answer whether a designated plant hormone is active in controlling any process by invoking the PESIGS rules provided by Jacobs (10). These state that:

- **Presence**—the chemical must be present in an organism and parallel variation should exist between the amount of the chemical and the relative activation of the process. (We should, however, modify the latter requirement such that: a) the hormone level should be measured in the exact tissue, cells, or even subcellular compartment where the reponse is occurring; and b) the possible changes in sensitivity that may occur during development should be taken into account.)
- **Excision**—removal of the organ that is the source of the hormone should lead to a cessation of the process.
- **Substitution**—the substitution of the pure chemical for the source organ should lead to the restoration of the process.
- **Isolation**—when as much as possible of the reacting system is isolated then the effect of the chemical is the same as in the less isolated system.
- **Generality**—the chemical should be involved in all similar situations.
- **Specificity**—the chemical should be specific.

Despite being formulated about thirty years ago these rules still provide a good set of guidelines for determining hormonal involvement. We should probably add one more proviso now possible with modern genetics, and that is the principal of genetic control: a correlation should be shown between the presence, absence, or level of a process, and the corresponding presence, absence, or level of a hormone in genetic lines of plants differing in the process purportedly controlled by the hormone.

There can be no doubt that hormones are important control agents in plants. The subsequent chapters provide more detail of our knowledge of this group of compounds and their effects.

---

[1]Oligosaccharides (oligosaccharins) have been suggested as controllers of development in plants, but at the present time there is little evidence to support the concept of a wide role in plant growth and development other than as a response to invasion by disease-causing organisms. Oligosaccharins have, however, been shown to produce developmental changes, including flower induction in tobacco tissue in culture (1). This is an area of potential future development.

Acknowledgments

I would like to thank Douglas Hamilton, Carol Mapes, David Law, Frederick Behringer, and Linda DeNoyer for constructive criticism of the first two chapters and Lois Geesy for typing the manuscripts.

References

1. Albersheim, P., Darvill, A.G. (1985) Oligosaccharins. Scientific American 253, No. 3, 58-64.
2. Baskin, T.I., Briggs, W.R., and Iino, M. (1986) Can lateral redistribution of auxin account for phototropism of maize coleoptiles? Plant Physiol. 81, 306-309.
3. Burg, S.P., Burg, E.A. (1967) Lateral auxin transport in stems and roots. Plant Physiol. 42, 891-893.
4. Canny, M.J. (1985) Ashby's law and the pursuit of plant hormones: a critique of accepted dogmas, using the concept of variety. Aust. J. Plant Physiol. 12, 1-7.
5. Firn, R.D. (1985) Ashby's law of requisite variety and its applicability to hormonal control. Aust. J. Plant Physiol. 12, 685-687.
6. Firn, R.D. (1986) Growth substance sensitivity: the need for clearer ideas, precise terms and purposeful experiments. Physiol. Plant. 67, 267-272.
7. Firn, R.D., Digby, J. (1980) The establishment of tropic curvatures in plants. Ann. Rev. Plant Physiol. 31, 131-148.
8. Firn, R.D., Tamimi, S. (1986) Auxin transport and shoot tropisms-- the need for precise models. In Plant Growth Substances 1985, pp 236-240. Bopp, M., ed. Springer, Heidelberg.
9. Ingram, T.J., Reid, J.B., Murfet, I., Gaskin, P., Willis, C.L., MacMillan, J. (1984) Internode length in Pisum. The Le gene controls 3-hydroxylation of gibberellin A_{20} to gibberellin A_1. Planta 160, 455-463.
10. Jacobs, W.P. (1979) Plant hormones and plant development. Cambridge University Press, Cambridge, New York, 339 pp.
11. Parrish, D.J. and Davies, P.J. (1977) Emergent growth--an auxin- mediated response. Plant Physiol. 59, 745-749.
12. Potts, W.C., Reid, J.B., Murfet, I.C. (1985) Internode length in Pisum. Gibberellins and the slender phenotype. Physiol. Plant. 63, 357-364.
13. Reid, J.B., Murfet, I.C., Potts, W.C. (1983) Internode length in Pisum. II. Additional information on the relationship and action of loci Le, La, Cry, Na and Lm. Jour. Exp. Bot. 34, 349-364.
14. Reid, J.B., Potts, W.C. (1986) Internode length in Pisum. Two further mutants lh and lc, with reduced gibberellin synthesis, and a gibberellin insensitive mutant lk. Physiol. Plant. 66, 417-426.
15. Scott, T.K, Briggs, W.R. (1962) Recovery of native and applied auxin from the light-grown Alaska pea seedlings. Amer. Jour. Bot. 49, 1056-1063.
16. Spray, C., Phinney, B.O., Gaskin, P., Gilmour, S.I., MacMillan, J. (1984) Internode length in Zea mays L. The dwarf-1 mutant controls the 3-hydroxylation of gibberellin A_{20} to gibberellin A_1. Planta 160, 464-468.
17. Sponsel, V.M. 1983. The localization, metabolism and biological activity of gibberellins in maturing and germinating seeds of Pisum sativum cv. Progress No. 9. Planta 159, 454-468.
18. Trewavas, A. (1981) How do plant growth substances act? Plant Cell Environment 4, 203-228.
19. Trewavas, A.J., Cleland, R.E. (1983) Is plant development regulated by changes in the concentration of growth substances or by changes in the sensitivity to growth substances? Trends in Biochem. Sci. 8, 354-357.
20. Vendrell, M. (1985) Dual effect of 2,4-D on ethylene production and ripening of tomato fruit tissue. Physiol. Plant 64, 559-563.
21. Wang, T.L. (1984) Plant hormones. New Scientist 103, No. 1413, 40.
22. Wyers, J. (1984) Do plants really have hormones? New Scientist 102, No. 1410, 9-13.

B. HORMONE SYSTHESIS AND METABOLISM

B1. Auxin Biosynthesis and Metabolism

Dennis M. Reinecke and Robert S. Bandurski

Department of Botany and Plant Pathology, Michigan State University, East Lansing, Michigan 48824, USA.

INTRODUCTION

Hormones regulate growth and development in plants, but how is the amount of the hormone regulated? In this chapter, we attempt to bring together information relevant to this question. We will consider only the auxin, indole-3-acetic acid (IAA)* since it is for IAA that the greatest body of knowledge exists. Auxins such as phenylacetic acid and 4-chloro-indole-3-acetic acid have been considered elsewhere (53,54,55,56) and other plant hormones are reviewed in separate chapters of this volume.

At the outset it must be emphasized that there exists a hiatus in knowledge of this field. There is an enormous amount known concerning the response of plants to IAA application. There is a small amount known concerning the endogenous amounts of IAA in plant tissues. But there is even less knowledge of how the amount of IAA in the tissue is regulated and what the relationship is between the amount of endogenous IAA and the plant response. Indeed, we do not know whether there will be a correlation between the response of the plant and the *amount* of IAA or the *turnover* of IAA, or possibly even a relationship with numbers, or activity of IAA binding sites (60).

In the hope that we can begin to answer such questions this chapter focuses on the "inputs to", and "outputs from" the IAA pool. Some of the knowledge is qualitative and indicates only the existence of the pathway, but there is also a growing body of knowledge of the pool size of IAA and its metabolites, and some knowledge of the turnover of these pools. When this knowledge is more complete, it should be possible to account for the steady state level of IAA, *and,* to understand how environmental stimuli,

---

Abbreviations

IAA, indole-3-acetic acid; IAInos, indole-3-acetyl-*myo*-inositol; IAInosGal, indole-3-acetyl-*myo*-inositol-galactose; OxIAA, oxindole-3-acetic acid; DiOxIAA, dioxindole-3-acetic acid.

such as tropic stimuli, alter the level of IAA in the tissue. In addition to the data presented here, several recent reviews provide a more detailed analysis of particular aspects of auxin synthesis and metabolism (2,3,13,53,54,55,56). Other chapters in this volume are concerned with how hormones work, auxin transport, and the role of hormones in photosynthate partitioning.

INPUTS TO AND OUTPUTS FROM THE IAA POOL

The inputs to the IAA pool would include: 1) *de novo* synthesis, whether from tryptophan, or from non-indolylic precursors; 2) hydrolysis of both amide and ester IAA conjugates; and 3) transport from one site in the plant to another site.

The outputs from the IAA pool would include: 1) oxidative catabolism; 2) conjugate synthesis; and 3) "use" in growth. This latter output assumes some special mechanism of IAA destruction to assure that the same molecule is not used repeatedly, thus losing its control function. Thus, the outputs "1" and "3" may be the same or closely related.

In Fig. 1 we present a summary of the known metabolic reactions which would affect the steady-state amount of IAA in the tissue.

Fig. 2 presents an example of how knowledge of the structures and amounts of IAA and its conjugates can be used to provide knowledge of pool sizes and pool turnover.

INPUTS TO THE IAA POOL

De Novo Synthesis

From Tryptophan

An enormous body of knowledge demonstrates that IAA can be synthesized from tryptophan and a summary of some of the known reactions is presented in Fig. 3. These data have been reviewed (13,54,56) and here we add some new developments and possible desirable additions.

There are serious hindrances to progress in this field: 1) tryptophan is readily converted to IAA by micro-organisms (53); 2) radiolabeled tryptophan can be non-enzymatically converted to IAA, presumably by radio-chemical decomposition reactions (17); the pool size of tryptophan is 3, or more, orders of magnitude larger than that of IAA (17); 4) almost none of the published studies have attempted to calculate the *amount* of IAA synthesized; 5) there are no published studies for higher plants, showing that a mutant plant deficient in its capacity to produce IAA from tryptophan is, in fact, IAA deficient.

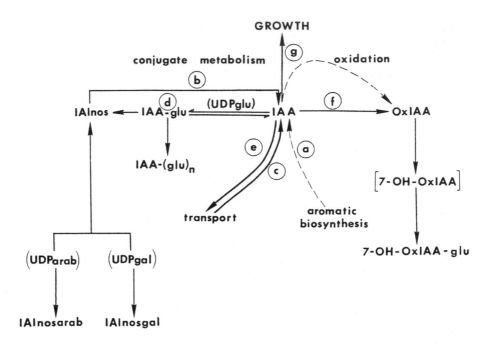

Fig. 1. The "inputs to" and the "outputs from" the IAA pool determine the steady state level of IAA in a particular plant tissue (e.g., corn). IAA inputs include: a) *de novo* synthesis; b) conjugate hydrolysis; and c) transport. IAA outputs include d) conjugate synthesis; f) catabolic oxidation; g) IAA "use" during the growth process; and e) transport.

Microbial and nonenzymatic IAA synthesis: Epstein et al., have described the conversion of 5-[³H]-tryptophan to labeled IAA in almost 30% yield by drying the tryptophan *in vacuo* in a glass tube (17). This oxidative conversion could be mediated by peroxides and free radicals which accumulate in the radioactive tryptophan solution. Precautions against this conversion have been described (17). The microbial conversion of tryptophan to IAA, probably occurs via indole-3-pyruvic acid, and may be avoided by the customary aseptic precautions. Owing to the enormous disparity in the relative amounts between tryptophan and IAA, even a minute microbial conversion of tryptophan to IAA would appear to be sufficient for the plant's IAA needs.

Pool size of tryptophan: Law and Hamilton (36) have described a system that may isolate the enormous tryptophan pool from the IAA pool. They observed that L-tryptophan plus gibberellin A_3, or D-tryptophan alone, were as effective as IAA in promoting elongation of pea (*Pisum sativum*) stem segments. They postulate that D-tryptophan is the immediate precursor of IAA and that the racemization of L to D tryptophan is GA controlled. This experiment, is difficult to control, owing to the disparity in pool sizes of D and L tryptophan but, if correct, the

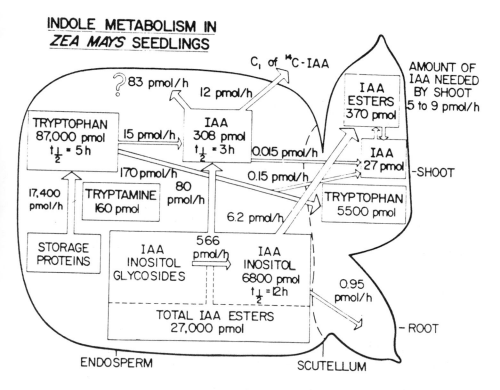

Fig. 2. Pool sizes and rates of turnover of the indolylic compounds in the kernels of *Zea mays* after 4 days of germination. Adapted from (17).

mechanism would provide a means of regulating the synthesis of the small IAA pool from the huge tryptophan pool.

The Pseudomonas savastanoi system: By far the best understood conversion of tryptophan to IAA is that catalyzed by the plant pathogen *Pseudomonas savastanoi*. IAA is a virulence factor for the pathogen in that it produces galls on the host plant (35). The pathway involves first the four electron oxidative decarboxylation of tryptophan to indole-3-acetamide by a tryptophan monooxygenase and then the hydrolysis of indole-3-acetamide to yield IAA (see Fig. 3). The genes for IAA production lie on a plasmid and are organized into an operon (19). Isolates from the oleander plant can further convert the IAA to IAA-lysine. Presumably this conjugate would play a role in the homeostatic control of IAA amount, perhaps in a manner similar to that indicated for higher plants (35).

De Novo *Synthesis of IAA from Non-indolylic Precursors*

Recent experiments from two laboratories in which young plants of *Zea mays* were grown on 30% deuterium oxide demonstrate that some of the IAA is made by a route resulting in the incorporation of, at least, one,

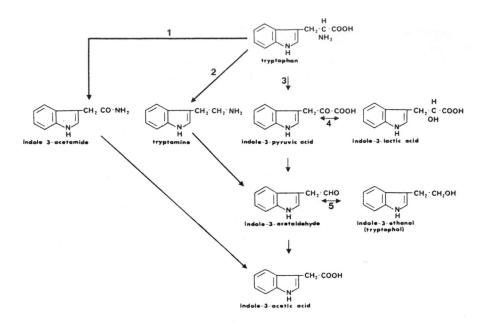

Fig. 3. pathways for the systhesis of IAA from tryptophan: 1) indole-3-acetamide pathway of *Pseudomonas savastanoi*; 2) tryptamine; and 3) indole-3-pyruvic acid. Indole-3-lactic acid (pathway 4), and indole-3-ethanol (pathway 5) may be side branches of these pathways for the regulation of IAA levels.

(51) or more (44) deuterium molecules into non-exchangeable positions of the indole ring of IAA. Since the degree of isotope discrimination is unknown, it is not possible to quantitatively evaluate these experiments and determine exactly what percent is made by *de novo* synthesis. However it is certain that some of the IAA is made *de novo* and the incorporation pattern suggests a precursor such as anthranilic acid, rather than preformed tryptophan.

Hydrolysis of IAA Conjugates

The Seed Auxin Precursor

Almost all knowledge of conjugate hydrolysis leading to the formation of free IAA has been derived from the study of seeds and young seedlings. Thus, we simply do not know whether mature plants are making and hydrolyzing conjugates, or whether conjugate synthesis and hydrolysis is primarily associated with seed maturation and germination. It will be an important extension of existing knowledge to determine what conjugates occur in the mature plant and whether they are being metabolized.

Seedlings of *Zea mays* (sweet corn) are the most thoroughly studied system (12). Some structures of IAA conjugates are shown in Fig. 4. The ester, indole-3-acetyl-*myo*-inositol (IAInos)* occurs in both the kernel and the shoot of corn (9). Five-[³H]-IAInos applied to the kernel is transported to the shoot and is in part hydrolyzed to yield free 5-[^{3}H]-IAA (10,43). Thus, in a qualitative sense, it is known that IAInos from the kernel supplies free IAA to the shoot.

In a quantitative sense, there is disagreement as to how much of the IAA of the shoot is derived from conjugates in the kernel (10,27). Studies with 5-[^{3}H]-IAInos indicate that between 2 to 6 pmol of IAA can be supplied to the shoot from the kernel and this would be a major source of the estimated 10 pmol required by each shoot in one hour (43). There are, however, difficulties in determining how much the 5-[^{3}H]-IAInos is diluted by endogenous IAInos (17) and thus the quantitative aspects of the source of IAA for the shoot require additional study.

5-[^{3}H]-indole-3-acetyl-*myo*-inositol-[^{14}C]-galactoside applied to the endosperm is also transported to the shoot, there to yield free IAA (34). In this case the double label permitted the conclusion that hydrolysis of the galactose from the conjugate occurred *after* the conjugate left the endosperm and *before* it entered the shoot—possibly in the scutellum— (34) (and personal communication).

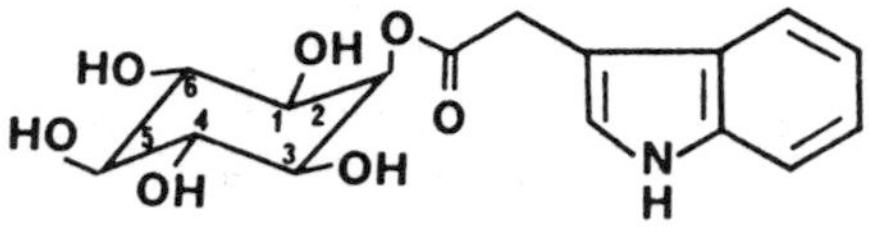

1

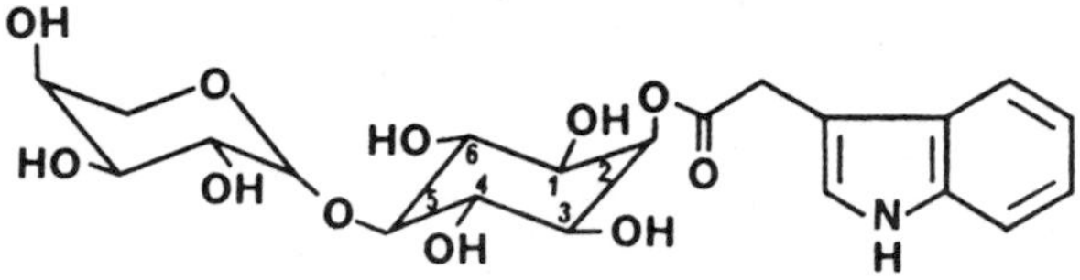

2

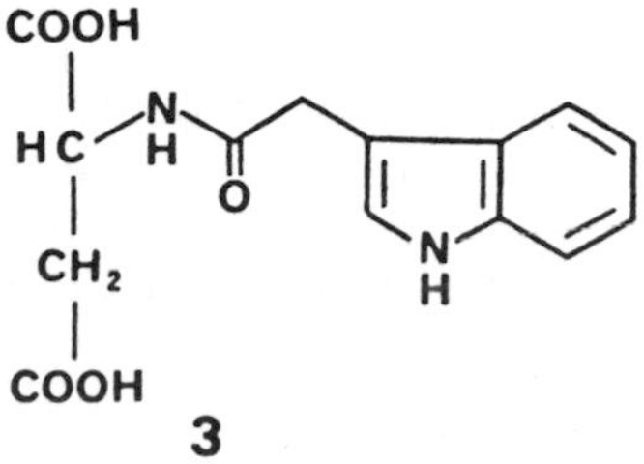

3

Fig. 4. Some examples of naturally occurring IAA conjugates. 1) indole-3-acetyl-*myo*-inositol; 2) indole-3-acetyl-*myo*-inositol-arabinoside; and 3) indole-3-acetyl-L-aspartate.

"Slow Release" Conjugates

Various conjugates of IAA, both ester and amide, have been used as "slow release" forms of IAA for tissue cultures (25). It would seem that IAA conjugates, differing in ease of hydrolysis by the plant's enzymes, and having conjugating moieties of varying degrees of lipophilicity could be used to "target" the IAA to a particular tissue or particular cell organelle with delivery of the hormone at the required rate. Thus, the conjugating moiety might be used as a "zip code," to bring the IAA to the desired location.

The studies of Hangarter and Good (25) indicated that the biological effect of the conjugate was related to its ability to release IAA when applied to the tissue. Additionally, the work of Bialek et al., (7) established that the differential growth induced by IAA-conjugates applied to stems of bean (*Phaseolus vulgaris*) was quantitatively related to the degree of hydrolysis of the conjugate by the tissue.

Enzymatic Hydrolysis of IAA-Conjugates

Hamilton et al., showed that extraction of corn shoots or roots with ether for 3 hr at 4°C yielded IAA whereas an extraction with 80% ethanol yielded no IAA (24). It was interpreted that the tissue was autolyzing in the wet ether with consequent enzymatic hydrolysis of IAA conjugates to yield detectable amounts of free IAA. Further, using water-soaked corn kernels, they demonstrated increased amounts of free IAA as a function of time of germination—again, most likely involving enzymatic hydrolysis of IAA conjugates in the kernels. The "bound" IAA of the seed could also be released by alkaline hydrolysis, as had been much earlier established (24). Control experiments with tryptophan and zein failed to yield IAA. It was, in fact, these early studies which led to studies of the chemical structure of the "bound IAA".

Despite the ease of observing conjugate hydrolysis in ether induced autolyzing tissues, it has been difficult to obtain *in vitro* enzymatic hydrolysis of IAA conjugates. Many commercial proteases and esterases that might be expected to hydrolyze IAA-amino acid conjugates, fail to do so. Hall and Bandurski have reported an enzyme preparation from corn which will hydrolyze IAA-*myo*-inositol (23) and Bialek et al., observed an enzyme which would hydrolyze indole-3-acetyl-L-alanine (8). In both cases these hydrolytic enzymes have proven difficult to extract and purify. In the case of the IAInos hydrolyzing enzyme from corn, the best preparations were prepared from acetone powders of the plant tissue. This requirement for acetone solubilization of the enzyme, together with the difficulties encountered in attempts at purification are suggestive of a regulated membrane-localized enzyme complex. Success in purifying one of the conjugate-hydrolyzing enzymes should be of interest in understanding the control of the amount of free IAA, and should also be of practical value in developing synthetic conjugates of varying degrees of ease of hydrolysis.

Transport of IAA from One Site to Another Site within the Plant

Transport from Seed to Shoot

Germinating seeds: Although not involving a net increase in IAA, or its conjugates, transport from one to another site within the plant can cause perturbations of the amount of IAA within a tissue or organ. Transport from seed to shoot or root would serve as an example. A second case, and one most probably involved in tropic stimulated asymmetric growth, involves movement of IAA from the vascular stele to the surrounding cortical tissues.

Transport from seed to shoot has been in part discussed above, under *Hydrolysis of IAA Conjugates.* These studies mainly involved application of labeled IAInos to the kernel and determination of IAA, IAInos, and other esters in the shoot tissue (10,43). Also Komoszynski and Bandurski (34) applied doubly labeled IAInosGal* to the endosperm. No intact IAInosGal could be detected in the shoot, although both labeled galactose and IAInos were present in the shoot. Further, the labeled galactose from IAInosGal was more efficiently transported from the endosperm into the shoot than was an equivalent amount of labeled free galactose applied to the endosperm. Several conclusions can be drawn concerning the transport process. First, it appears that IAInos is a transport form whereas IAInosGal is not. Secondly, it appears that the removal of the galactose moiety occurred after the IAInosGal had moved from the endosperm into the scutellum. Had hydrolysis occurred in the endosperm then free galactose should have been as effective in labeling the shoot as was the galactose from IAInosGal.

Polar transport of IAA from tip to base of coleoptile: The polar transport of IAA is discussed in chapter E5. A cardinal mystery remains, and that is to account for the greater speed of transport from top to bottom of the coleoptile than from bottom to top. A partial answer to this problem is provided in the finding by Jacobs (29) that a protein probably involved in IAA transport is localized primarily at the base of the cells involved in transport. Somewhat mysteriously this protein binds inhibitors of IAA transport, such as N-1-naphthylphthalamic acid, but does not bind IAA. It is possible that two adjacent proteins are involved (see chapter E5) or that the N-1-naphthylphthalamic acid binds to a "hinge point" of a protein transport system. Nonetheless the basal localization of a protein involved in transport could explain the polarity of transport.

Asymmetric transport of IAA during tropic stimulation: Among the earliest studies of IAA transport were those related to the problem of how an asymmetric stimulus could induce an asymmetric distribution of IAA. This phenomenon was described, but not explained, by the phrase, "lateral transport", as used in the Went-Cholodny theory for tropic responses (65). Several recent findings help to explain how lateral transport may occur. First, there is the finding that a geotropic stimulus results in an asymmetric distribution of both free and ester IAA in the mesocotyl of *Zea mays* seedlings (4,65). There is also a transient change in the ratios of free to conjugated IAA, as was earlier predicted, (65). However, since both free and ester IAA increased on the lower side of the gravity-stimulated mesocotyl, there must have been selective transport of IAA and/or its esters from the central vascular stele into the mesocotyl cortex (5).

The above data have permitted the formulation of a working theory for the transport of IAA in the maize mesocotyl. The postulates of the theory are that the IAA moves symplastically in the stele and also symplastically from the vascular stele into the cortical cells. A further postulate of the theory is that the plasmodesmatal connections between tissues are voltage "gated" in a manner analogous to the gating of animal gap junction cells. The gravity stimulus could thus alter the bioelectric potential of the IAA transporting cells and open or close the voltage-gated plasmodesmata, thus bringing about asymmetric movement of IAA from the central stele into the upper and lower cortical cells (4,65).

The implications of the above theory are considerable in that the proposed mechanism suggests that the movement of IAA within the plant symplast is under metabolic control, and thus selective movement of IAA within the plant and even from the plant into seeds or fruit could be controlled by metabolic gating of the plasmodesmatal connections of the plant symplast.

OUTPUTS FROM THE IAA POOL

Oxidative Catabolism

IAA catabolism is the chemical modification of the indole nucleus or side chain resulting in loss of auxin activity. The carbon skeleton may then re-enter the general metabolism of the plant. This discussion of IAA catabolism will include the oxidative decarboxylation of the side chain, and oxidation at the 2 and 3 positions of the indole nucleus without decarboxylation. IAA catabolism is of interest to plant scientists since it is the only irreversible output regulating IAA levels, and therefore may play an important role in regulating IAA mediated growth.

The interest in IAA catabolism predates identification of IAA as a ubiquitous auxin in plants. In 1934 (59) it was observed that water extracts of leaves inactivate the "growth promoting substance" of *Avena*. Later work showed that peroxidase catalyzes the oxidative decarboxyla-tion of IAA (18,30), and this reaction was generally assumed to be the endogenous route for oxidative catabolism. More recently a new pathway of IAA catabolism with the retention of the carboxyl side chain and oxidation of the indole nucleus has been identified (46,61) and convincingly shown to occur in *Zea mays*.

Decarboxylation Pathway

The decarboxylation pathway is catalyzed by peroxidases from numerous plant species (56), and often by several peroxidase isozymes per plant species. The *in vitro* reaction may be monitored by [^{14}C]-CO_2 evolution from 1-[^{14}C]-IAA, manometrically by O_2 uptake, colorimetrically by loss of Salkowski color, or by UV absorbance changes. In the literature, "IAA oxidase" and peroxidative oxidation of IAA with decarboxylation have been synonymous. This definition of IAA oxidation is now too narrow with the discovery of the oxidative non-decarboxylation pathways of IAA catabolism in *Zea mays* and *Vicia faba*.

The peroxidative decarboxylation of IAA may occur without added cofactors with purified horseradish peroxidase, the best studied "IAA oxidase" (26). Mn^{++} and monophenols increase reaction rate, and H_2O_2 reduces the lag period for IAA oxidation (54). Tomato peroxidases are dependent on H_2O_2 for the oxidation of IAA (33). The main products of peroxidase oxidation of IAA are 3-methyleneoxindole, 3-hydroxymethyloxindole (oxindole-3-carbinol), indole-3-aldehyde, and indole-3-methanol (26,40) (Fig. 5). The ratio of the various products depends on the enzyme/substrate ratio, cofactors, and pH of the reaction (21,48). A high enzyme/substrate ratio favors formation of indole-3-aldehyde and indole-3-methanol, and low amounts of the oxindoles. The addition of Mn^{++}, and 2,4-dichlorophenol also stimulates the formation of indole-3-methanol production at the expense of the other oxidation products. Indole-3-methanol is a precursor of indole-3-aldehyde since indole-3-methanol is converted by peroxidase to indole-3-aldehyde (21). [^{18}O] experiments (40) indicate that most of the oxygen incorporation into indole-3-methanol is

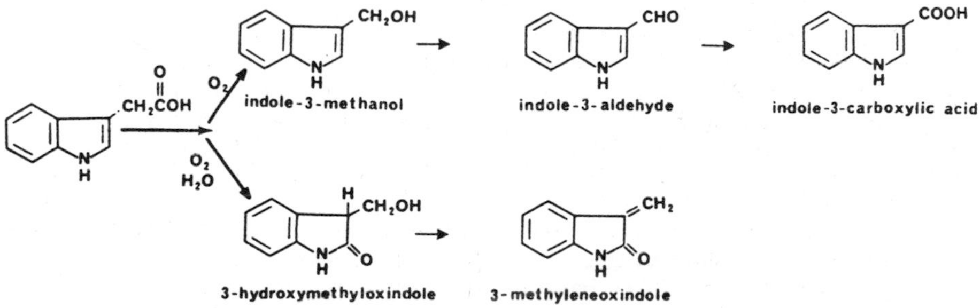

Fig. 5. Oxidative decarboxylation pathway. The peroxidase pathway is catalyzed by extracts from many plant species. The natural occurence of some ot these catabolites has been demonstrated.

from molecular oxygen and not water. Three-hydroxymethyloxindole is the immediate precursor of 3-methyleneoxindole via a nonenzymatic dehydration (26). [^{18}O] experiments (40) have shown that most of the oxygen in the 3-methyleneoxindole ring originates in water and not molecular oxygen. The additional information that labeled indole-3-methanol is not metabolized by horseradish peroxidase to 3-hydroxymethyloxindole demonstrates that oxindole and indole formation are separate branches of the peroxidase pathway. It is thought that peroxidase initially reacts with IAA by forming an IAA free radical which is subsequently attacked by oxygen (26,40). Following decarboxylation of the IAA, the reaction proceeds either by the indole-3-aldehyde route or the 3-methyleneoxindole route.

As mentioned previously cofactors for IAA oxidation may include Mn^{++}, and monophenols. Interestingly, Mn^{+++} will nonenzymatically oxidize IAA, and it is thought that peroxidase plus monophenol oxidizes Mn^{++} to Mn^{+++} (6). The Mn^{+++} may then oxidize IAA with the subsequent decarboxylation of IAA. Monophenols and m-diphenols stimulate IAA oxidation, while p-diphenols, o-diphenols, coumarins, and polyphenols inhibit the enzyme reaction (56). *In vivo* regulatory functions for these compounds in peroxidase oxidation of IAA have been suggested.

Non-decarboxylation pathway—Oxindole-3-acetic acid/Dioxindole-3-acetic acid pathway

This oxidation pathway has been observed in several plant species including rice, corn, and broad bean. Colorimetric assays indicate that oxindole-3-acetic acid (OxIAA)* may also occur in germinating seeds of *Brassica rapa*, and developing seeds of *Ribes rubrum* (32). The standard methods for monitoring the peroxidative decarboxylation of IAA may not distinguish between the two pathways in an *in vitro* system. For example, IAA oxidation to OxIAA would result in O_2 uptake, loss of Salkowski color, and the reduction in the 280 nm indole spectrum with an increase in the 247 nm oxindole spectrum—chracteristics also present in the peroxidase system. Measurement of $^{14}CO_2$ evolution from feeding 1-[^{14}C]-IAA is a clear indication of the peroxidase oxidative decarboxylation pathway. OxIAA has been shown not to be an intermediate or a substrate for the peroxidase pathway (26) so the pathways are independent. Careful chromatographic isolation and physicochemical identification of the catabolites is the best method to identify the IAA catabolic pathways occurring in a particular plant.

Oxindole-3-acetic acid and dioxindole-3-acetic acid (DiOxIAA)* were found to be synthesized by *Zea mays* and *Vicia faba* respectively following feeding 1-[^{14}C]-IAA (47,61) (Fig. 6). Isotope dilution experiments (46) showed that OxIAA was a naturally occurring compound in *Zea mays* endosperm and shoot tissues occurring in amounts of 357 pmol per endosperm and 47 pmol per shoot, or about the level of free IAA in these

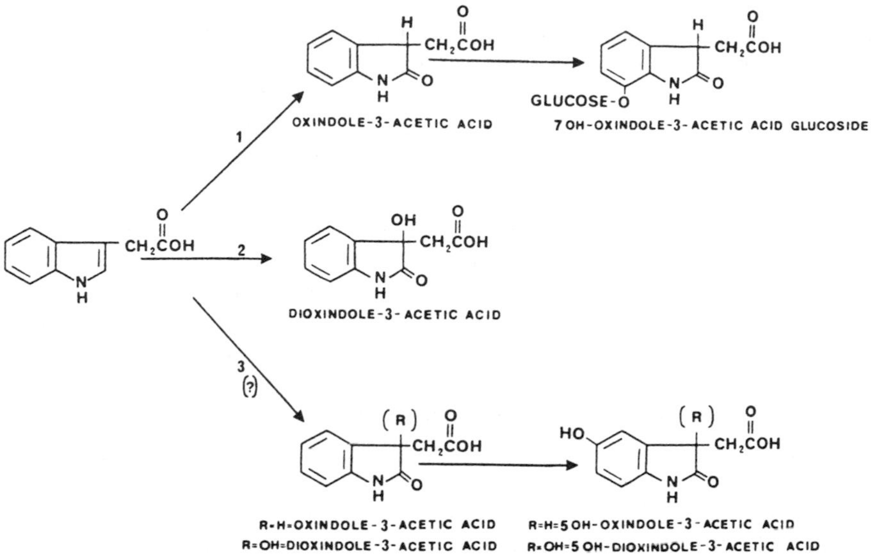

Fig. 6. Non-decarboxylation oxidative pathway. This is the IAA oxidation pathway in *Zea mays* (1), *Vicia faba* (2), and Rice (3).

tissues. In *Vicia faba*, DiOxIAA was estimated by UV measurements at 1 µmole per Kg fresh weight in root tissues. In both these experiments labeled IAA was fed to etiolated seedlings via endosperm tissue or cotyledons. The first report of IAA oxidation to OxIAA was in the basidiomycete *Hygrophorus conicus* (52). The conversion of tryptamine to IAA and then to OxIAA was unique to *Hygrophorus conicus* of the 12 basidiomycetes tested. *Oryza sativa*, rice, is interesting in that it is the only plant known to have OxIAA and DiOxIAA in the same plant (31). Rice was also shown to have the 5-hydroxy analogs of OxIAA and DiOxIAA. The oxindole-3-acetic acids were not isolated from vegetative tissues of rice, nor were radiolabel-feeding studies undertaken to show the precursor-product relationship with IAA. Whether OxIAA can be a precursor for DiOxIAA is also not resolved. In corn OxIAA is further metabolized by hydroxylation and glucose addition at the 7 position of the indole nucleus forming the 7-OH-OxIAA-glucoside (41). Isotope dilution assays have shown that the 7-OH-OxIAA-glucoside is a naturally occurring compound in corn in amounts of 62 pmol per shoot and 4.8 nmol per endosperm. *Vicia faba* is also reported to conjugate DiOxIAA to a glucose and aspartate conjugate of unknown linkage (61).

The enzymology of the OxIAA and DiOxIAA pathways is much less understood than the peroxidase pathway owing to the more recent discovery of the OxIAA/DiOxIAA pathway, and the commercial availability of horseradish peroxidase. In *Zea mays*, the rate of enzymatic activity for the oxidation of IAA to OxIAA has been measured in shoot,

35

root, scutellar, and endosperm tissues at 1-10 pmol.h^{-1}.mg protein^{-1} (45). The enzyme is soluble as evidenced by ultracentrifugation studies. Enzyme activity was reduced by 90 per cent when assayed under argon indicating an oxygen requirement for the reaction. However, [^{18}O] feeding studies have not yet positively identifed the source of oxygen in the oxindole ring. Enzyme activity was stimulated up to ten fold by extracting corn tissue with a nonionic detergent. A heat stable component of the Triton X-100 extracted tissue also increased enzyme activity when added to buffer extracted enzyme (45). Cofactors of mono-oxygenation reactions as well as peroxidase cofactors, are inactive in stimulating OxIAA formation. The identification of the heat stable, detergent-extracted factor may clarify the mechanism of OxIAA synthesis. The enzymology of IAA oxidation to DiOxIAA has not been investigated.

Physiological Occurrence of Pathways

Ideally the catabolic route for a plant hormone would be established by the feeding of isotopically-labeled compounds via a natural route and using physiological concentrations. Subsequently, the endogenous amount of the metabolite should be determined by an isotope dilution method (46). Finally the catabolites biological activity should be measured. (It should be noted that care must be taken with IAA metabolic studies since IAA may also be non-enzymatically degraded by light, acid, silica gel, etc. (54).)

Despite all the research on the *in vitro* peroxidase-catalyzed decarboxylation of IAA, only a few reports on the isolation and physicochemical identification of endogenous peroxidase catabolites have been reported. Indole-3-methanol, and indole-3-carboxylic acid have been identified in pine (50,57), and indole-3-aldehyde, indole-3-methanol, and indole-3-carboxylic acid were reported in pea sections fed IAA (38). In pine needles there was 2.3 ng per g fresh weight indole-3-carboxylic acid or about ten per cent of the level of free IAA (49). In etiolated pine shoots there was 19.7 ng per g fresh weight of indole-3-methanol. Whether indole-3-carboxylic acid is a natural metabolite of IAA is in question since in *Brassica* its occurrence can be an artifact of glucosinolate breakdown (57), and in pine [^{14}C]-IAA was not metabolized to indole-3-carboxylate (49). Three-hydroxymethyloxindole, and 3-methyleneoxindole are unstable compounds. However 3-hydroxymethyloxindole has been reported to occur in pea and wheat after feeding labeled IAA to sections (16,57). Indole-3-methanol glycoside, and indole-3-carboxylic acid were also reported to be catabolites of IAA in wheat sections (37). Generally, catabolites from radiolabel-feeding studies have been identified by thin-layer chromatography in several solvent systems; substantiation by physicochemical methods is needed.

There is no doubt that peroxidase can oxidize IAA *in vitro* but proof of the natural occurrence of peroxidase catabolites in untreated tissues is lacking. For example, an *in vitro* corn peroxidase system decarboxylated IAA to several compounds (6), but feeding labeled IAA to either corn endosperm or root segments resulted in rapid metabolism of IAA with minimal decarboxylation (42). Also, some of the studies measuring 1-[^{14}C]-IAA catabolism by $^{14}CO_2$ evolution may have overestimated peroxidase-catalyzed decarboxylation since some of this catabolism might be a cut surface phenomenon (63). For example, 75% of the "IAA oxidase" in pea segments was removed by a 4 minute buffer wash, and "IAA oxidase" activity was proportional to the number of pieces into which the sections were cut. These findings together with the fact that over 70% of the total enzyme activity was wall localized may indicate a wound response role for peroxidase-catalyzed oxidation of IAA.

Oxindole-3-acetic acid, and dioxindole-3-acetic acid have been demonstrated to be endogenous compounds in three plant species, although the precursor-product relationship has been shown only for broad bean and corn. The occurrence of the carboxyl-retention pathway of IAA catabolism in a monocotyledonous and a dicotyledonous plant indicates that the pathway may be more widely distributed. Since radiolabeled peroxidase metabolites and OxIAA/DiOxIAA metabolites can be synthesized chemically or enzymatically (47,57,61), the natural occurrence of these pathways in other plants may now be examined by an isotope dilution assay. A comprehensive determination of the biological activity of all the IAA catabolites from the peroxidase, and OxIAA/DiOxIAA pathways in several auxin bioassays has not yet been performed. However, those catabolites tested have been found to be inactive in promoting growth in auxin bioassays (46,56).

The answer to whether IAA is catabolized *in vivo* by the relatively non-specific peroxidase isozyme system, and/or specifically by another enzyme will clarify how and where IAA catabolism is involved in IAA mediated growth. Attempts have been made to correlate peroxidase levels and age of tissue with responsiveness of tissues to IAA. In some studies there was a positive correlation while other studies showed no correlation (56). The physicochemical measurement of IAA levels in an overproducing or underproducing peroxidase mutant versus the wild type would identify the *in vivo* role of peroxidases in regulating IAA levels.

In summary, there are two pathways of IAA catabolism: the peroxidase catalyzed oxidative decarboxylation of IAA, and the oxidation without decarboxylation to OxIAA or DiOxIAA. The peroxidase mechanism of oxidation is well understood with purified enzymes, but its physiological significance must still be elucidated. The OxIAA/DiOxIAA pathway has been identified in only two species; its occurrence in other species must be examined, and the mechanism of enzyme catalysis must be studied further. Future research will determine if IAA catabolism occurs in tissues non-responsive to IAA, or only in actively growing tissues where the oxidase reaction would destroy the hormone following the "growth promoting act". In either case, the elucidation of how and where IAA is catabolized in plants will help clarify how IAA levels are regulated during growth and development.

Conjugate Synthesis

Synthesis of conjugates from endogenous IAA occurs in developing seeds and also under conditions leading to a dramatic change in growth rate (12). In addition, the exogenous application of IAA leads to conjugate formation (43) and the application of one conjugate, for example IAInos, can lead to the formation of other conjugates (43). Lastly, in the case of the *Pseudomonad* system, the gene for synthesis of IAA-lysine has been cloned.

Enzymatic Synthesis of IAA-Conjugates of Zea Mays

The conjugate-synthesizing reactions which have been demonstrated *in vitro* are shown in Fig. 7. The first step in the synthesis of the IAA-*myo*-inositol family of conjugates is the synthesis of IAA-1-O-beta-D-glucose from IAA and UDPG (39). The IAA is then transacylated to *myo*-inositol to form IAInos. IAInos may then be glycosidated to form IAA-*myo*-inositol-galactoside or IAA-*myo*-inositol arabinoside by reaction with the appropriate uridine diphosphosugar (15). Thus, all of the low molecular weight conjugates of *Zea mays* have been synthesized with enzyme extracts (14).

Synthesis of IAA-Amino Acid Conjugates

IAA-L-aspartate (Fig. 4) was the first IAA amide conjugate to be chemically characterized (11). It is also widespread in nature (1) and is

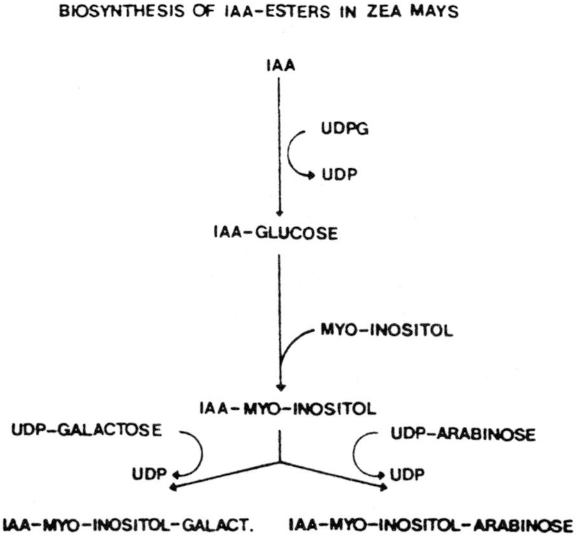

BIOSYNTHESIS OF IAA-ESTERS IN ZEA MAYS

IAA

UDPG
UDP

IAA-GLUCOSE

MYO-INOSITOL

IAA-MYO-INOSITOL

UDP-GALACTOSE   UDP-ARABINOSE

UDP   UDP

IAA-MYO-INOSITOL-GALACT.   IAA-MYO-INOSITOL-ARABINOSE

Fig. 7. The biosynthetic pathways for the low molecular weight esters indole-3-acetyl-*myo*-inositol, indole-3-acetyl-*myo*-inositol-galactose, and indole-3-acetyl-*myo*-inositol-arabinose as identified for *Zea mays*.

known to be an endogenous component of soybean seeds. Further, it is readily formed when IAA is applied to plant tissues (62) by, what appears to be, an adaptive enzyme system. Mysteriously, and despite numerous attempts, it has never been synthesized by an *in vitro* enzymatic system. This critical gap in knowledge must be remedied for improved understanding of how conjugate synthesis is regulated to homeostatically control endogenous IAA levels.

Recently, IAA-glutamate has also been demonstrated to occur naturally in some legumes, and cases are known where IAA is linked in amide linkage to a glycoprotein or to a small peptide (personal communication, J. Cohen and F. Percival).

Synthesis of Indole-3-acetyl-epsilon-L-lysine

In some strains of *Pseudomonas savastanoi* (35) the IAA produced from tryptophan represents only a transitory intermediate since the IAA is then rapidly conjugated to form IAA-lysine. The synthesis of IAA-lysine requires L-lysine and ATP, and the IAA-lysine formed is then further metabolized to an unknown conjugate (35). The IAA-lysine gene has been cloned into E. Coli and the IAA-lysine locus is some 2 kb upstream from the IAA operon.

Use of IAA in Growth

IAA promotes a cascade of reactions that lead to growth. The components of this cascade include rapid induction of mRNAs (22,58), membrane phenomenon such as permeability changes, media pH changes and enzyme modification (e.g., 8). Regardless of how IAA acts, a kinetic argument may be made that IAA is somehow destroyed at the moment it commits the growth promoting act. If the hormone is transported for example, from the tip to the growing region of a stem then the destruction or conjugation of the hormone is probably closely linked to the actual growth promoting reaction.

SUMMARY

The steady state levels of IAA may be modulated by the "inputs to" the IAA pool including synthesis, conjugate hydrolysis and transport, and by the "outputs from" the IAA pool including oxidative catabolism, conjugate synthesis, and IAA "use" during growth. Further work is needed to substantiate these processes in plants, so that ultimately we may understand how these processes are integrated optimally during IAA-induced growth and development of plants.

Acknowledgment

This work was supported by the U.S. National Science Foundation, PCM 79-04637 and the National Aeronautics and Space Administration, NAGW-97. Manuscript preparation was facilitated by Wendy Whitford.

References

1. Andreae, W.A., Good, N.E. (1955) The formation of indole- acetylaspartic acid in pea seedlings. Plant Physiol. *30*, 380-382.
2. Bandurski, R.S. (1984) Metabolism of indole-3-acetic acid. *In:* The Biosynthesis and Metabolism of Plant Hormones. Soc. for Exp. Biol. *23*, 183-200, Crozier, A., Hillman, J.R. eds. Cambridge Univ. Press, Cambridge.
3. Bandurski, R.S., Nonhebel, H.M. (1984) Auxins. *In:* Advanced Plant Physiol. 1-20. Wilkins, M.B. ed. Pitman, London.
4. Bandurski, R.S., Schulze, A., Reinecke, D. (1985). An attempt to localize and identify the gravity sensing mechanism of plants. The Phytologist *27*(S), 111-112.
5. Bandurski, R.S., Schulze, A. (1985) A working theory for the mechanism of the gravity-induced asymmetric distribution of IAA in the *Zea mays* mesocotyl. Plant Physiol. *77*(S), 57.
6. BeMiller, J.N., Colilla, W.(1972) Mechanism of corn indole-3-acetic acid oxidase *in vitro*. Phytochem. *11*, 3393-3402.
7. Bialek, K., Meudt, W.J., Cohen, J.D. (1983) Indole-3-acetic acid (IAA) and IAA conjugates applied to bean stem sections. IAA content and the growth response. Plant Physiol. *73*, 130-134.
8. Bialek, K., Cohen, J.D. (1984) Hydrolysis of an indole-3- acetic acid amino acid conjugate by an enzyme preparation from *Phaseolus vulgaris*. Plant Physiol. *75*(S), 108.
9. Chisnell, J.R. (1984) *Myo*-inositol esters of indole-3- acetic acid are endogenous components of *Zea mays* L. shoot tissue. Plant Physiol. *74*, 278-283.
10. Chisnell, J.R. (1984) The presence and translocation of indole-3-acetyl-*myo*-inositol in shoots of *Zea mays* L. Ph.D. Thesis, MSU.
11. Cohen, J.D. (1982) Identification and quantitative analysis of indole-3-acetyl-L-aspartate from seeds of *Glycine max*. Plant Physiol. *70*, 749-753.
12. Cohen, J.D., Bandurski, R.S. (1982) Chemistry and Physiology of the bound auxin. Ann. Rev. Plant Physiol. *33*, 403-430.
13. Cohen, J.D., Bialek, K. (1984) The biosynthesis of indole- 3-acetic acid in higher plants. *In:* The Biosynthesis and Metabolism of Plant Hormones. Soc. for Exp. Biol. 23, 165-182, Crozier, A., Hillman, J.R. eds. Cambridge Univ. Press, Cambridge.
14. Corcuera, L.J., Bandurski, R.S. (1982) Biosynthesis of indole-3-acetyl-*myo*-inositol arabinoside in kernels of *Zea mays* L. Plant Physiol. *70*, 1664-1666.
15. Corcuera, L.J., Michalczuk, L., Bandurski, R.S. (1982) Enzymatic synthesis of indol-3-yl-acetyl-*myo*-inositol galactoside. Biochem. J. *207*, 283-290.
16. Davies, P.J. (1972) The fate of exogenously applied indoleacetic acid in light grown stems. Physiol. Plant. *27*, 262-270.
17. Epstein, E., Cohen, J.D., Bandurski, R.S. (1980) Concentration and metabolic turnover of indoles in germinating kernels of *Zea mays* L. Plant Physiol. *65*, 415-421.
18. Galston, A.W., Bonner, J., Baker, R.S. (1953) Flavoprotein and peroxidase as components of the indoleacetic acid oxidase system of peas. Arch. Biochem. Biophys. *42*, 456-470.
19. Glass, M.L., Kosuge, T. (1985) Molecular cloning of the gene for the synthesis of IAA-lysine from *Pseudomonas syringae* pv *Savastanoi*. J. Cell. Biochem. Supplement *9* C, 217.

20. Goldsmith, M.H.M. (1977) The polar transport of auxin. Ann. Rev. Plant Physiol. 28, 439-478.
21. Grambow, H.J., Langenbeck-Schwich, B. (1983) The relationship between oxidase activity, peroxidase activity, H_2O_2, and phenolic compounds in the degradation of indole-3-acetic acid in vitro. Planta 157, 131-137.
22. Hagen, G., Guilfoyle, T.J. (1985) Rapid induction of selective transcription by auxins. Mol. Cell Biol. 5, 1197-1203.
23. Hall, P.J., Bandurski, R.S. (1986) [³H] Indole-3-acetyl-myo-inositol hydrolysis by extracts of Zea mays L. vegetative tissue. Plant Physiol. 80, 374-377.
24. Hamilton, R.H., Bandurski, R.S., Grigsby, B.H. (1961) Isolation of indole-3-acetic acid from corn kernels and etiolated corn seedlings. Plant Physiol. 36, 354-359.
25. Hangarter, R.P., Good, N.E. (1981) Evidence that IAA conjugates are slow-release sources of free IAA in plant tissues. Plant Physiol. 68, 1424-1427.
26. Hinman, R.L., Lang, J. (1965) Peroxidase-catalyzed oxidation of indole-3-acetic acid. Biochem. 4, 144-158.
27. Iino, M., Carr, D.J. (1982) Sources of free IAA in the mesocotyl of etiolated maize seedlings. Plant Physiol. 69, 1109-1112.
28. Jackson, D.L., McWha, J.A. (1983) Translocation and metabolism of endosperm applied 2-[14C] indoleacetic in etiolated Avena sativa L. seedlings. Plant Physiol. 73, 316-323.
29. Jacobs, M., Gilbert, S.F. (1983) Basal localization of the presumptive auxin transport carrier in pea stem cells. Science 220, 1297-1300.
30. Kenten, R.H. (1955) The oxidation of indolyl-3-acetic acid by waxpod bean root sap and peroxidase systems. Biochem. J. 59, 110.
31. Kinashi, H., Suzuki, Y., Takeuchi, S., Kawarada, A. (1976) Possible metabolic intermediates from IAA to B-Acid in rice bran. Agr. Biol. Chem. 40, 2465-2470.
32. Klämbt, H.D. (1959) Die 2-hydroxy-indol-3-essigsäure, ein pflanzliches indolderivat. Naturwiss 46, 649.
33. Kokkinakis, D.M., Brooks, J.L. (1979) Hydrogen peroxide- mediated oxidation of indole-3-acetic acid by tomato peroxidase and molecular oxygen. Plant Physiol. 64, 220-223.
34. Komoszynski, M., Bandurski, R.S. (1986) Transport and metabolism of indole-3-acetyl-myo-inositol-galactoside in seedlings of Zea mays. Plant Physiol. 80, 961-964.
35. Kosuge, T., Comai, L., Glass, N.L. (1983) Virulence determinants in plant pathogen interactions. In: Plant Molecular Biology, pp. 167-177, Goldberg, R.B., ed. Alan R. Liss, Inc. New York.
36. Law, D.M., Hamilton, R.H. (1985) Mechanism of GA-enhanced IAA biosynthesis. Plant Physiol. 77(S), 2.
37. Langenbeck-Schwich, B., Grambow, H.J. (1984) Metabolism of indole-3-acetic acid and indole-3-methanol in wheat leaf segments. Physiol. Plant. 61, 125-129.
38. Magnus V., Iskric, S., Kveder, S. (1971) Indole-3-methanol—A metabolite of indole-3-acetic acid in pea seedlings. Planta 97, 116-125.
39. Michalczyk, L., Bandurski, R.S. (1982) Enzymatic synthesis of indol-3-yl-acetyl-1-O-beta-D-glucose and indol-3-yl-acetyl-myo-inositol. Biochem. J. 207, 273-291.
40. Nakono, M., Kobayashi, S., Sugioka, K. (1982) Peroxidase- catalyzed oxidation of indole-3-acetic acid. In: Oxygenases and oxygen metabolism, pp. 245-254, Nozaki, M., ed. Academic Press, New York.
41. Nonhebel H., Bandurski, R.S. (1984) Oxidation of indole-3- acetic acid and oxindole-3-acetic acid to 2,3-dihydro-7-hydroxy-2-oxo-1H indole-3-acetic acid 7-O-B-D- glucopyranoside in Zea mays seedlings. Plant Physiol. 76, 979-983.
42. Nonhebel, H.M., Hillman, J.R., Crozier, A., Wilkins, M.B. (1985) Metabolism of [^{14}C] indole-3-acetic acid by coleoptiles of Zea mays L. J. Exp. Bot. 36, 99-109.
43. Nowacki, J., Bandurski, R.S. (1980) Myo-inositol esters of indole-3-acetic acid as seed auxin precursors of Zea mays L. Plant Physiol. 65, 422-427.
44. Pengelly, W.L., Bandurski, R.S. (1983) Analysis of indole- 3-acetic acid metabolism in Zea mays using deuterium oxide as tracer. Plant Physiol. 73, 445-449.

45. Reinecke, D.M., Bandurski, R.S (1985) Further characterization of the enzymatic oxidation of indole-3-acetic acid to oxindole-3-acetic acid. Plant Physiol.77(S), 3.
46. Reinecke, D.M., Bandurski, R.S. (1983) Oxindole-3-acetic acid, an indole-3-acetic acid catabolite in Zea mays. Plant Physiol. 71, 211-213.
47. Reinecke, D.M., Bandurski, R.S. (1981) Metabolic conversion of ^{14}C-indole-3-acetic acid to ^{14}C-oxindole-3-acetic acid. Biochem. and Biophys. Res. Commun. 103, 429-433.
48. Sabater, F., Sanchez-Bravo, J., Acosta, M. (1983) Effects of enzyme/substrate ratio and of cofactors on the oxidation of indole-3-acetic acid catalyzed by peroxidase. Revista Española de Fisiologia 39, 169-174.
49. Sandberg, G. (1984) Biosynthesis and metabolism of indole-3-ethanol and indole-3-acetic acid by Pinus sylvestris L. needles. Planta 161, 398-403.
50. Sandberg, G., Jensen, E., Crozier, A (1984) Analyses of indole-3-carboxylic acid in Pinus sylvestris needles. Phytochem. 23, 99-102.
51. Schärer, S., Ribant, J.M. (1983) Biosynthèse de l'AIA dans les plantules entières et les racines excisées de Zea mays. Experimental work completed under the supervision of L. Rivier, Université de Lausanne for the Plant Physiology Certificate.
52. Schuytema, E.C., Hargie, M.P., Merits, I., Schenck, J.R., Siehr, D.J., Smith, M.S., Varner, E.L. (1966) Isolation, characterization, and growth of basidiomycetes. Biotech. & Bioeng. 8, 275-286.
53. Schneider, E.A., Wightman, F. (1974) Metabolism of auxin in higher plants. Ann. Rev. Plant Physiol. 25, 487-513.
54. Schneider, E.A., Wightman, F. (1978) Auxins. In: Phytohormones and Related Compounds: A Comprehensive Treatise 1, 29-105, Letham, D.S., Goodwin, P.B., Higgins, T.J.V., eds. Elsevier/North-Holland, Amsterdam.
55. Sembdner G. (1974) Conjugates of plant hormones. In: Biochemistry and Chemistry of Plant Growth Regulators, pp.283-302, Schreiber, K., Schutte, H.R., Sembdner, G., eds. GRD: Inst. Plant Biochem, Halle.
56. Sembdner G, Gross, D., Liebisch, H-W., Schneider, G. (1981) Biosynthesis and metabolism of plant hormones. In: Hormonal Regulation of Development. I. Molecular Aspects of Plant Hormones (Encyclopedia of Plant Physiology Vol. 9), pp. 281-444, MacMillan, J., ed. Springer, Berlin.
57. Sundberg, B., Sandberg, R., Jensen, E. (1985) Identification of indole-3-methanol in etiolated seedlings of Scots Pine (Pinus sylvestris L.). Plant Physiol. 77, 952-955.
58. Theologis, A., Huynh, T.V., Davis, R.W (1985) Rapid induction of specific mRNAs by auxin in pea epicotyl tissue. J. Mol. Biol. 183, 53-68.
59. Thimann, K.V. (1934) Studies on the growth hormone of plants. VI. The distribution of the growth substance in plant tissues. J. Gen. Phys. 18, 23-34.
60. Trewavas, A.J. (1982) Growth substance sensitivity: the limiting factor in plant development. Physiol. Plant. 55, 60-72.
61. Tsurumi, S., Wada, S.(1980) Metabolism of indole-3- acetic acid and natural occurrence of dioxindole-3- acetic acid derivatives in Vicia roots. Plant & Cell Physiol. 21, 1515-1525.
62. Venis, M.A. (1972) Auxin-induced conjugation systems in peas. Plant Physiol. 49, 24-27.
63. Waldrum, J.D., Davies, E. (1981) Subcellular localization of IAA oxidase in peas. Plant Physiol. 68, 1303-1307.
64. Weiler, E.W., Wischnewski, S. (1984) The relationship between diffusible, extractable and conjugated (base-labile) forms of indole-3-acetic acid in isolated coleoptile tips of Zea mays L. Planta 162, 30-32.
65. Went, F.W., Thimann, K.V.(1978) Phytohormones (Reprinted) Allanheld, Osmun New York 1978.

B2. Gibberellin Biosynthesis and Metabolism

Valerie M. Sponsel
School of Chemistry, The University of Bristol, Bristol BS8 1TS, England.

INTRODUCTION

The gibberellins (GAs), unlike other groups of plant hormones, are defined by their chemical structure rather than by their biological activity. The GAs are all diterpenoid acids which have the same basic *ent-gibberellane* ring structure (structure 1). Within this basic ring system there are two main types of GAs, namely the C_{20}-GAs which have a full complement of 20 carbon atoms (structure 2), and the C_{19}-GAs in which the twentieth carbon atom has been lost by metabolism (structure 3). In all but one of the C_{19}-GAs the carboxylic acid at carbon-19 bonds to carbon-10 to give a lactone bridge.

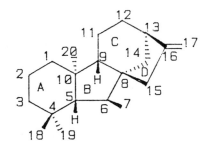

(1) ent-gibberellane

(2) GA_{12} (a C_{20}-GA)   (3) GA_9 (a C_{19}-GA)

There is a large array of structural diversification of the basic ring system—for instance there are different oxidative states of carbon-20, namely methyl (CH_3), hydroxymethyl (CH_2OH), aldehyde (CHO) or carboxylic acid (COOH). Additional functional groups can be added, especially to the C_{19}-GAs. These include the insertion of up to four hydroxyl (OH) groups, or of epoxide ($>$0) or ketone ($=$0) functions. The position and stereochemistry of these functional groups is very important. Thus the insertion of a hydroxyl group in α- or β-stereo-chemistry (designated by ‖‖‖‖‖‖‖‖‖ or ▶ bonds, respectively) may have different biochemical significance. Other structural modifications encountered are double bonds in ring A between C-1 and C-2, C-2 and C-3, and C-1 and C-10, and hydration of the double bond between C-16 and C-17.

The possible number of structural modifications within the basic ring pattern is very large. Thus there is a large and ever increasing number of known GAs. Each different GA which is found to be naturally occurring and whose structure has been chemically characterized is allocated an A number by MacMillan and Takahashi (37). Gibberellin A_1 (GA_1) through to GA_{72} are known to date. Their numerical order approximates to the chronological order to their discovery.

Gibberellins can also exist as conjugates. The most common naturally occurring GA-conjugates are those in which the GA is linked to a molecule of glucose, either by an ether or an ester linkage.

Gibberellins were first isolated from the fungus *Gibberella fujikuroi* in which they occur in large quantities as secondary metabolites (3). They are now known to be present in a diversity of vascular plants. In mono- and di-cotyledons, which have been investigated most thoroughly, GA levels are highest in reproductive tissues (up to 10 μg fresh weight^{-1}). Therefore many of the higher plant GAs were initially identified in developing seeds (3). Of the 72 known GAs, 25 have been identified in *Gibberella*, 61 in higher plants and 14 are present in both. Like *Gibberella*, individual angiosperms can contain many different GAs. For instance seeds of the cucurbit *Sechium edule* contain at least 20 GAs (1), and 16 known GAs have been identified in seeds of *Phaseolus* species (13, 30).

Gibberellic acid (GA_3), which is the end-product of GA metabolism in *G. fujikuroi*, has been commercially available for many years. Its application to dwarf or rosette plants, dormant buds, or dormant and germinating seeds can result in dramatic and diverse effects on growth (see Chapters C3 and E3). However, because of the vast array of naturally occurring GAs, the plant physiologist wishing to relate the content of *native* GAs to plant growth and development is faced with a very daunting task. Fortunately after many years of biosynthetic and metabolic studies, coupled with GA structure/activity investigations, it is now possible to make some rationalizations in this field.

SITES OF GA BIOSYNTHESIS

It is generally accepted that there are at least three sites of GA biosynthesis in higher plants, namely in developing fruits/seeds, in elongating shoot apical regions and in roots (22). There is incontrovertible evidence for GA biosynthesis occurring in developing fruits and seeds. Thus, enzymes which can convert mevalonic acid to a C_{19}-GA have been extracted from young fruits, seeds and seed parts (21,33). Clear evidence that GA biosynthesis occurs in shoots and roots is less easy to obtain, probably because the levels of GAs are comparatively low in vegetative tissues. For this reason most of the following discussion is concerned with GA biosynthesis in fruits and seeds.

There appear to be two main phases or flushes of GA biosynthesis in developing seeds (see 47). The first phase occurs shortly after anthesis and appears to be correlated with fruit growth. At this stage of development seeds are very small, and GA levels are similar to those in vegetative tissues. Thus GA biosynthesis during the early stages of seed development is difficult to study. The second phase of GA biosynthesis occurs as the maturing seeds are increasing in size, and it results in a large accumulation of GAs in the seeds. In most species chosen for study the seeds at this stage of development are large enough to handle with ease and can readily be separated into constituent parts.

Most studies on GA biosynthesis have utilized cell-free preparations from seed-parts. The use of cell-free systems obviates problems associated with substrate penetration into tissue, and in addition it provides an opportunity to study biochemical properties of the enzymes concerned. Preparations derived from the liquid endosperm of developing cucurbit seeds are particularly active. For example, endosperm preparations from both wild cucumber (*Marah macrocarpus*, formerly *Echinocystis macrocarpa*) and pumpkin (*Cucurbita maxima*) have been used extensively. Preparations derived from legume seeds, e.g. from cotyledons of garden pea (*Pisum sativum*) also yield active biosynthesising systems.

Vegetative tissues yield less active preparations, probably because the levels of enzymes for GA metabolism are very much lower than those in reproductive tissues. Nevertheless kaurene-synthesising preparations from pea shoot tips and from etiolated shoots of corn (*Zea mays*) have been described.

The elucidation of the GA biosynthetic pathway is now historical. The following sections provide an outline of the pathway up to GA_{12}-aldehyde, which is the same in *G. fujikuroi* and all higher plants examined so far. Details of the characterization of individual steps in individual systems are well covered in review articles (22,28,49).

THE GIBBERELLIN BIOSYNTHESIS PATHWAY

The pathway from acetate to geranylgeranyl pyrophosphate (GGPP) (Fig. 1).

In the terpenoid pathway C_5-building blocks are linked head-to-tail to give branched polymers of different chain length, which can then undergo cyclization and other changes. In this way natural products such as mono-terpenes (C_{10}), sesquiterpenes (C_{15}, from the Latin sesqui $= 1\frac{1}{2}$), diterpenes (C_{20}), triterpenes (C_{30}) etc. are formed. Each of these classes of compounds have carbon skeletons which can be 'dissected' into the appropriate number of C_5-isoprene or isopentane units.

The diterpenoid nature of GAs was recognized by Cross and associates (8). Early steps in the terpenoid pathway have been characterized in plant, animal and bacterial systems (4). Birch et al. (5) were the first to show that [14C]-labelled acetate was converted to [14C]GA$_3$ by cultures of G. fujikuroi, and mevalonic acid (MVA) was found to be a key intermediate (see Fig. 1). Mevalonate is formed by the reduction of hydroxymethylglutaryl coenzyme-A in a virtually irreversible reaction so that MVA is a very efficient precursor of terpenes. It is converted to mevalonate-5-pyrophosphate (MVAPP) in two steps by mevalonate kinases in the presence of ATP and Mg^{2+} or Mn^{2+}. This is then decarboxylated to isopentenyl pyrophosphate (IPP). Reversible isomerization of IPP gives dimethylallyl pyrophosphate (DMAPP). The enzyme IPP isomerase, which catalyses this conversion, has been isolated from pumpkin fruits (39). DMAPP is the starter unit for terpene biosynthesis. It condenses in a head-to-tail fashion with a molecule of IPP to give the C_{10}-intermediate geranyl pyrophosphate (GPP). This in turn condenses with another molecule of IPP to give farnesyl pyrophosphate (FPP). The pathway can branch at this point to sesquiterpenes, and this is a potential route to abscisic acid. Further condensation of FPP with a third molecule of IPP gives the C_{20}-intermediate geranylgeranyl pyrophosphate (GGPP). GGPP is the parent compound for all diterpenes. The three condensation reactions are catalysed by the same enzyme, namely GGPP synthetase, which belongs to a class of enzymes named prenyl transferases. GGPP synthetase has been isolated from a number of sources, including pumpkin fruits (40). It is a soluble enzyme requiring Mn^{2+} for high activity.

Cyclization of GGPP to ent-kaurene (Fig. 2).

At GGPP the terpenoid pathway diverges to give linear diterpenes (e.g., phytol), cyclic diterpenes (e.g., the kaurenoids) and tetraterpenes (e.g., the carotenoids). The cyclization of GGPP to *ent*-kaurene is a two-stage reaction catalysed by the soluble enzymes *ent*-kaurene synthetase A and *ent*-kaurene synthetase B. The A activity catalyses the conversion of

46

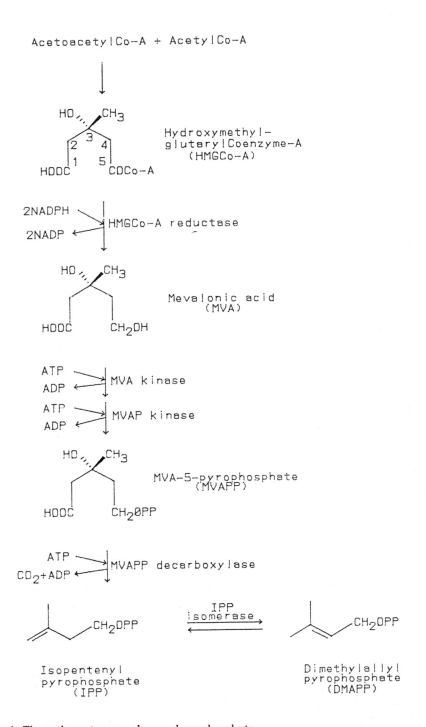

Fig. 1. The pathway to geranylgeranylpyrophosphate.

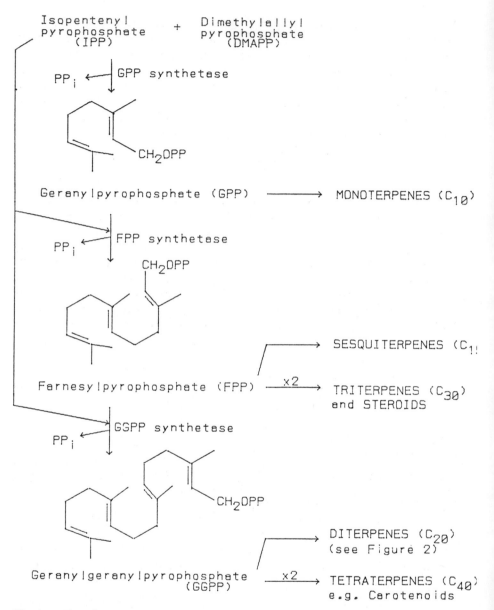

Fig. 1 continued

GGPP to a bicyclic intermediate copalyl pyrophosphate (CPP), whereas the B activity catalyses the further conversion of CPP to the tetracyclic diterpene ent-kaurene[a] (Fig. 2). The mechanism of these cyclization reactions has been discussed in detail (25,28).

The ent-kaurene synthetase enzymes have been studied extensively. The A and B activities have proved difficult to separate physically, although it has long been known that they possess different pH optima and different divalent cation requirements. Duncan and West (12) have resolved the two activities from *Marah* endosperm using a combination of separatory techniques to give two proteins each with a molecular weight of ca. 82 x 10³. It is known, however, that the two proteins associate during *ent*-kaurene synthesis because [¹⁴C]CPP generated *in situ* from [¹⁴C]GGPP is more readily converted to *ent*-kaurene than [¹³C]CPP which is added directly to the incubation mixture. Thus CPP appears to be channelled from the catalytic site of enzyme A to the catalytic site of enzyme B without equilibration with free CPP.

The conversion of GGPP to CPP by the A enzyme is a step of crucial importance in the biosynthesis of cyclic diterpenes. Unfortunately the A activity of *ent*-kaurene synthetase has been notoriously difficult to demonstrate in preparations from a variety of sources, due to its instability and susceptibility to natural and synthetic inhibitors (15,10). Thus some studies have been confined to the B activity. In pea shoots the B activity appears to reside in the chloroplast stroma, but the presence of plastid membranes is essential for maximal activity (42). The enzymes of the pathway up to and including the *ent*-kaurene synthetases are soluble, and those catalysing the next part are particulate. Thus the possible interaction of the *ent*-kaurene synthetases with a membrane component may serve to facilitate the unhindered progression from one phase of the GA biosynthetic pathway to the next. However indications are that enzymes catalysing reactions in the pathway beyond *ent*-kaurene are associated with endoplasmic reticulum not with plastids (19).

ent-Kaurene to GA₁₂-aldehyde (Fig. 3).

In the next part of the pathway the methyl group at carbon-19 of *ent*-kaurene is oxidised in the sequence $CH_3 \rightarrow CH_2OH \rightarrow CHO \rightarrow COOH$,

$$-CH_3 \longrightarrow -CH_2OH \longrightarrow -CH(OH)_2 \longrightarrow -C(OH)_3$$
$$H_2O \leftarrow\!\downarrow \qquad\qquad H_2O \leftarrow\!\downarrow$$
$$-CHO \qquad\qquad\qquad -COOH$$

to give *ent*-kaurenol, *ent*-kaurenal and *ent*-kaurenoic acid (Fig. 3). This sequence of oxidations is actually a series of successive hydroxylations so that all the reactions may be catalysed by a single enzyme active site. Evidence for and against this idea has been reviewed (25), leading to the

---

a Using IUPAC nomenclature the precursor of GAs is the enantiomeric form of (+) − kaurene, hence the term *ent*-kaurene. By convention α-substituents on *ent*-kaurene are termed *ent*-β, and β-substituents are termed *ent*-α.

49

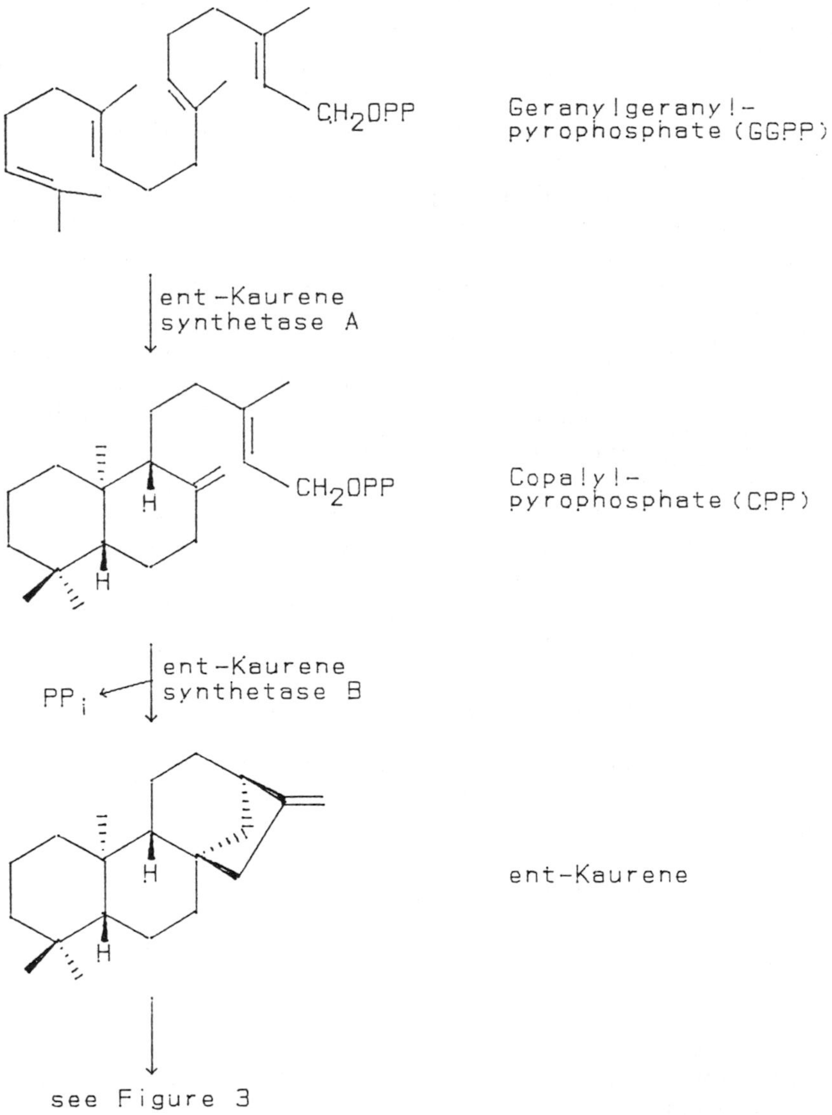

Geranylgeranyl-
pyrophosphate (GGPP)

ent–Kaurene
synthetase A

Copalyl-
pyrophosphate (CPP)

ent–Kaurene
synthetase B

ent–Kaurene

see Figure 3

Fig. 2. Cyclization of geranylgeranyl pyrophosphate to *ent*-kaurene.

conclusion that there are probably separate but similar catalytic sites for each substrate.

The enzymes involved in this series of oxidations have been studied in detail in several plants, and especially in *Marah* (23,24). The enzymes are microsomal and each oxidative step requires molecular oxygen and a reduced pyridine nucleotide (e.g., NADPH) for activity. These properties are characteristic of the mixed-function oxygenase or mono-oxygenase enzymes. This type of enzyme catalyses the insertion of one oxygen atom from molecular oxygen into an organic substrate: the second oxygen atom is reduced to water by electrons derived from a second substrate (e.g., NADPH). In *Marah*, inhibition of the oxidation of *ent*-kaurene to *ent*-kaurenol, and of *ent*-kaurenal to *ent*- kaurenoic acid by carbon monoxide can be reversed by light of wavelength around 450 nm. Thus in the *Marah* system cytochrome P450 is implicated as the electron acceptor at the active site of the enzyme.

The GA biosynthetic pathway proceeds with hydroxylation of *ent*-kaurenoic acid at the 7β-position to give *ent*-7α-hydroxykaurenoic acid. (Alternatively *ent*-kaurenoic acid can be directed towards the synthesis of kaurenolide derivatives which accumulate in *G. fujikuroi* and seeds of some higher plants).

At *ent*-7α-hydroxykaurenoic acid the pathway diverges once again, giving two products, GA_{12}-aldehyde and *ent*-6α,7α-dihydroxykaurenoic acid. This branch-point is of critical importance. GA_{12}-aldehyde is formed by the contraction of the B ring with extrusion of carbon-7. It is the first-formed GA in all systems and thus appears to be the precursor of all other GAs. In contrast, *ent*-6α,7α-dihydroxykaurenoic acid cannot be converted to GAs, nor has it any known function in plants. In pumpkin, in which the reaction has been studied most intensively, it has not been possible to resolve two enzyme activities since the formation of both products has the same pH and temperature optima and the same cofactor (NADPH and O_2) requirements. Detailed studies by Graebe and co-workers (18) have employed stereospecifically-labelled kaurenoic acid prepared by MacMillan's group. Results indicate that GA_{12}-aldehyde and *ent*-6α,7α-dihydroxykaurenoic acid are formed from a single common intermediate. The formation of this intermediate involves the removal of the *ent*-6α-hydrogen atom, which may be rate limiting.

POSSIBLE CONTROL MECHANISMS IN THE PATHWAY TO GA_{12}-ALDEHYDE.

Regulation of early steps in the terpenoid pathway in plants and animals has been reviewed (4). Feedback inhibition of MVA kinase by FPP, and to a lesser extent by GPP and GGPP has been shown in *Phaseolus vulgaris* shoots. MVAPP decarboxylase is inhibited in several

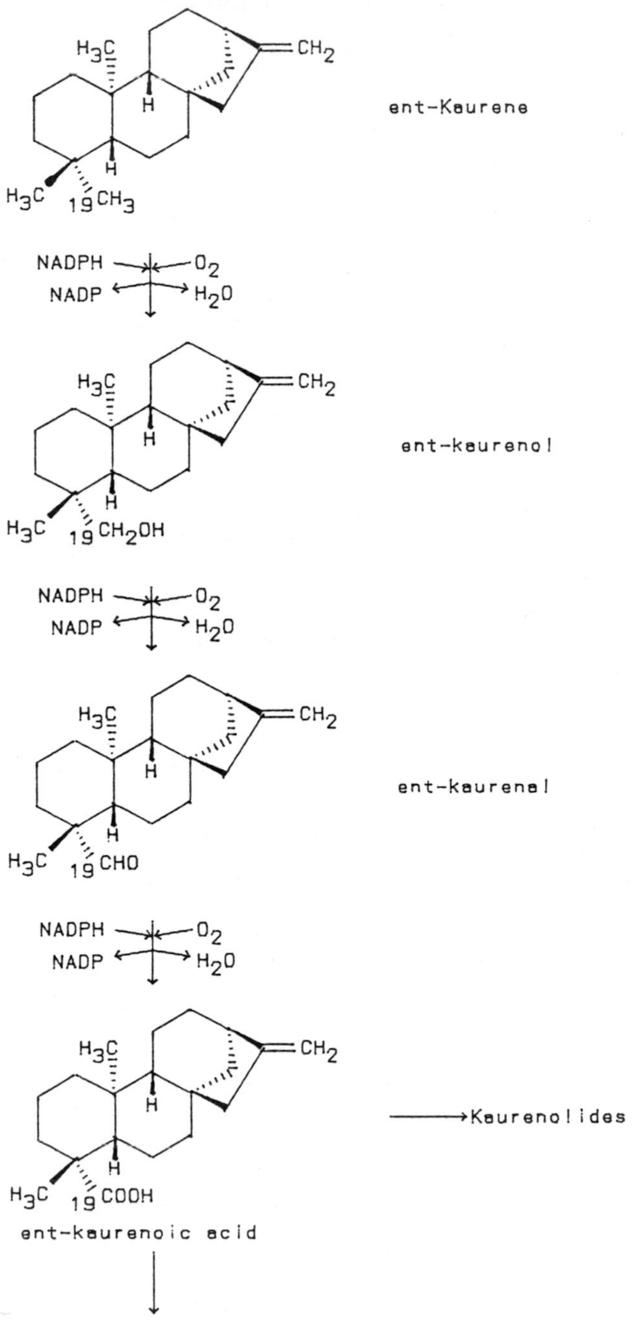

ent-Kaurene

NADPH → O2
NADP ← → H2O

ent-kaurenol

NADPH → O2
NADP ← → H2O

ent-kaurenal

NADPH → O2
NADP ← → H2O

⟶ Kaurenolides

ent-kaurenoic acid

Fig. 3. The pathway from *ent*-kaurene to GA$_{12}$-aldehyde.

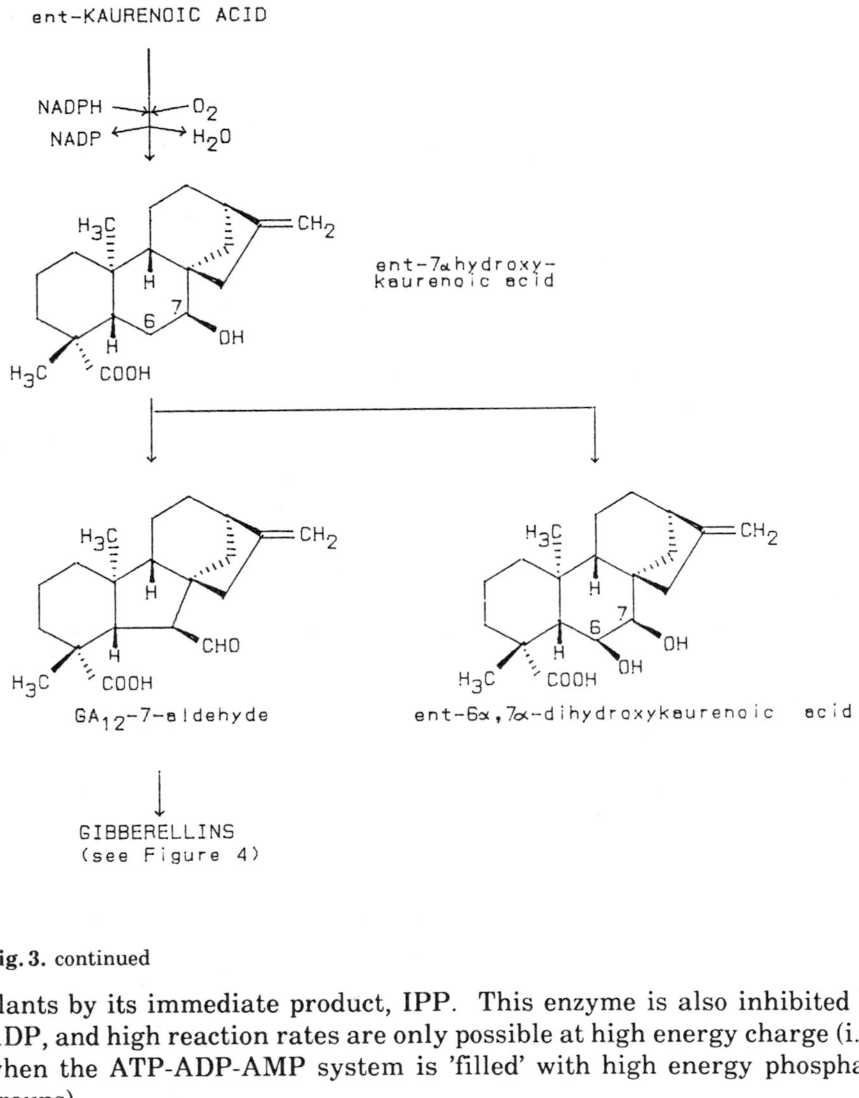

V. M. Sponsel

ent-KAURENOIC ACID

Fig. 3. continued

plants by its immediate product, IPP. This enzyme is also inhibited by ADP, and high reaction rates are only possible at high energy charge (i.e., when the ATP-ADP-AMP system is 'filled' with high energy phosphate groups).

Application of general enzyme inhibitors affects terpenoid synthesis. For instance sulphydryl reagents inhibit MVA kinase. Other compounds are more specific. Compactin (structure 4), a metabolite of *Penicillium brevicompactum*, is a competitive inhibitor of HMGCo-A reductase, but by affecting a reaction early in the pathway its effects on terpenoid biosynthesis are wide-ranging. In contrast, inhibitors which affect an enzyme which catalyses a reaction later in the pathway are more selective (10). For example compounds which inhibit the activity of *ent*-kaurene synthetase will, in theory, only inhibit the formation of cyclic diterpenes (see Fig. 2). The quaternary ammonium compounds AMO-1618 [2'-

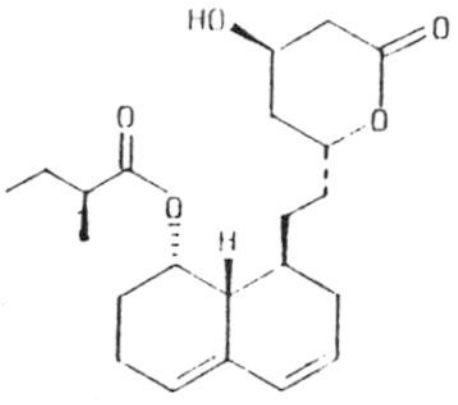

(4) Compactin

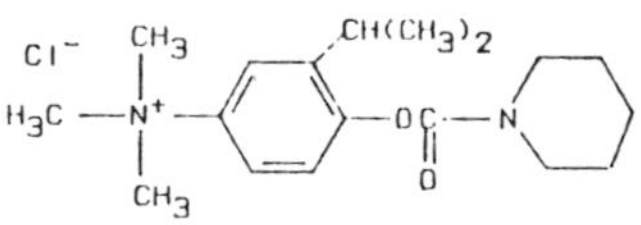

(5) AMO-1618

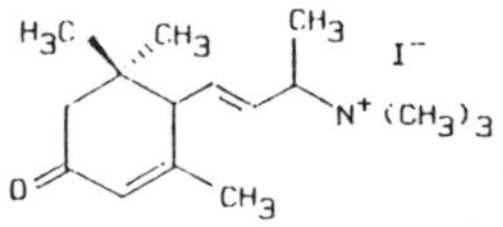

(6)

isopropyl-4'-(trimethylammonium chloride)-5'-methylphenylpiperidine-1-carboxylate] (structure 5) and N,N,N-trimethyl-1-methyl-(2',6',6'- trimethylcyclohex-2'-en-1'-yl)prop-2-enylammonium iodide (structure 6) are potent inhibitors of ent-kaurene synthetase A. Cycocel, also known as chlormequat chloride or CCC (2-chloroethyltrimethylammoniumchloride) (structure 7), which is used commercially to shorten stem length in a variety of crops also affects ent-kaurene synthetase A but is less effective and less specific than AMO-1618. The quaternary phosphorus-containing

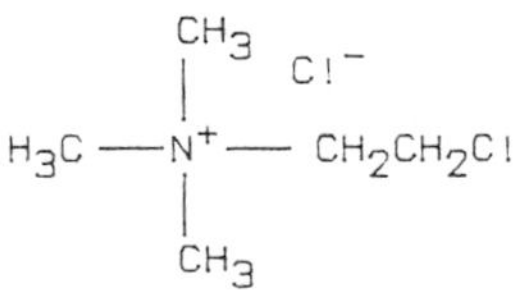

(7) Chlormequat

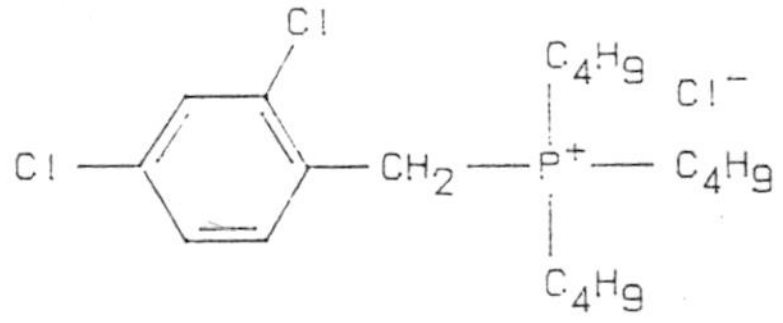

(8) Chlorphonium

compound Phosphon D, also known as chlorphonium chloride (tributyl-(2,4-dichlorobenzyl)phosphonium chloride) (structure 8) inhibits both the A and B activities of ent-kaurene synthetase. However the selectiveness and relative efficacy of all these growth retardants does appear to depend on the tissue to which they are applied (7). Naturally-occurring compounds which inhibit ent-kaurene synthetase A have been reported from castor bean (*Ricinus communis*) cell suspension cultures and sunflower (*Helianthus annuus*) seedlings (15,50).

Ancymidol [α-cyclopropyl - 4-methoxy - α- (pyrimidin - 5-yl) benzyl alcohol) (structure 9), together with the triazole group of growth retardants, e.g., Paclobutrazol [(2RS, 3RS)-1-(4-chlorophenyl)-4,4-dimethyl-2-(1 H-1,2,3-triazol-1-yl)penta-3-ol] (structure 10) inhibit the oxidation of *ent*-

(9) Ancymidol   (10) Paclobutrazol

kaurene to *ent*-kaurenoic acid. These compounds appear to interact with cytochrome P-450, thus inhibiting cytochrome P-450-dependent monooxygenases (41). Triazole fungicides of closely-related chemical structure inhibit the sequential oxidation of carbon-32 in ergosterol biosynthesis. This reaction is also known to be catalysed by similar monooxygenases.

Genetic and environmental control of gibberellin biosynthesis is covered elsewhere in this volume and is mentioned only briefly here. Cell-free systems from etiolated shoots of the dwarf-5 mutant of maize can convert MVA into *ent*- kaur-15-ene (structure 11) (29). Conversely, the major diterpene formed from MVA in cell-free systems derived from normal genotypes of maize is the usual isomer, *ent*-kaur-16-ene (structure 12). The mutant d-5 gene blocks the activity of *ent*-kaurene synthetase B so that the cyclization of CCP to *ent*-kaur-16-ene is prevented.

In *Pisum sativum ent*-kaurene-synthesising capacity is greater in tall cultivars than dwarf, but the biochemical basis for this difference has not been defined (7). Futhermore, *ent*-kaurene synthesis in pea seedlings is greatly enhanced by light, but the relationship between *ent*-kaurene synthesising capacity and stem growth in pea is far from clear (7).

(11) ent−kaur−15−ene
(ent−isokaurene)

(12) ent−kaur−16−ene
(ent−kaurene)

PATHWAYS FROM GA_{12}-ALDEHYDE

The biosynthetic pathway up to GA_{12}-aldehyde appears to be the same in all plants. In contrast, the conversion for GA_{12}-aldehyde to other GAs can vary from genus to genus. However, there is a basic sequence of reactions which is common to all pathways, namely the successive oxidation of carbon-20 leading to its elimination from the molecule and the formation of C_{19}-GAs.

The pathway, at its simplest, is as follows (see Fig. 4). Gibberellin A_{12}- aldehyde is first oxidised at C-7 to give the dicarboxylic acid, GA_{12}. A carboxyl group at carbon-7 is a feature of all GAs, and it is essential for biological activity. Further modifications involve the sequential oxidation of carbon-20. First the C-20 methyl group of GA_{12} is oxidised to a hydroxymethyl group (CH_2OH), which lactonises upon extraction and work-up to yield GA_{15}. The true intermediate is, however, the open-lactone form, and in the next step of the pathway it is oxidised to the C-20 aldehyde, GA_{24}. At GA_{24} there is a branch- point in the pathway. Carbon-20 can either be oxidised to the acid, giving GA_{25}, or more importantly it can be eliminated giving the C_{19}-GA, GA_9. The C_{19}-GAs have highest biological activity, and therefore can be considered as the objective to which the biosynthetic and metabolic pathways have been leading. The mechanism whereby C-20 is eliminated from the molecule has been discussed at length, but it is now accepted that C-20 is lost as CO_2 both in the fungus (11) and in cell-free systems from pea embryos (34).

The introduction of additional functional groups into the GA molecule can occur at any stage in this basic sequence of reactions. It is the position and order of insertion of all these additional substituents which distinguishes the metabolic pathways in different genera. Thus if GA_{12} becomes hydroxylated at carbon-3 and/or at carbon-13 then the GAs shown in Fig. 5 are formed on pathways comparable to the route from $GA_{12} \rightarrow GA_9$ discussed above. In actual fact the observed pathways are rarely discrete. They may converge or diverge to form a metabolic grid depending on the timing of hydroxylation (see Fig. 8).

The following discussion centres on two genera for which most information is available, namely pea and pumpkin. The data for pea are more straightforward and are considered first, although the study of GA metabolism in cell-free preparations was pioneered using the pumpkin system.

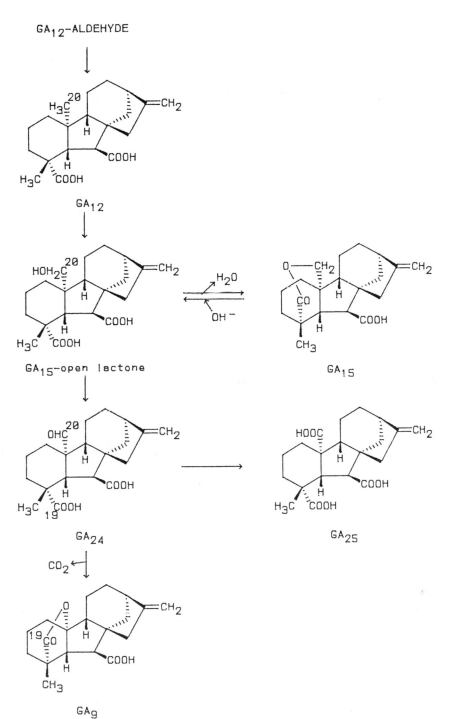

Fig. 4. The pathway from GA_{12}-aldehyde to the C_{19}-GA, GA_9.

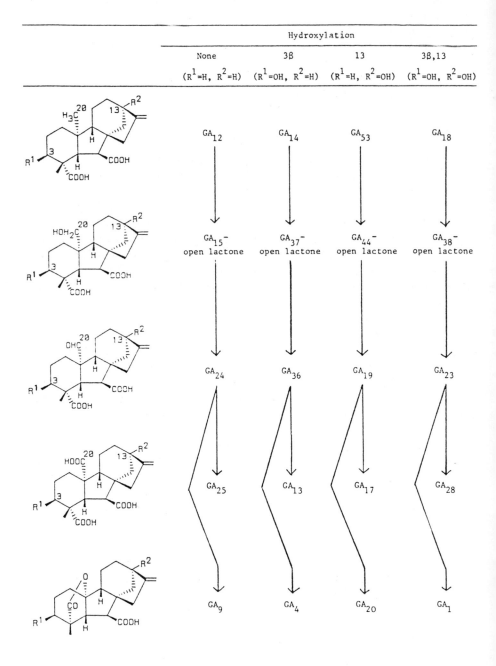

Gibberellin biosynthesis and metabolism

	Hydroxylation			
	None	3β	13	$3\beta,13$
	(R^1=H, R^2=H)	(R^1=OH, R^2=H)	(R^1=H, R^2=OH)	(R^1=OH, R^2=OH)
	GA_{12}	GA_{14}	GA_{53}	GA_{18}
	GA_{15}^- open lactone	GA_{37}^- open lactone	GA_{44}^- open lactone	GA_{38}^- open lactone
	GA_{24}	GA_{36}	GA_{19}	GA_{23}
	GA_{25}	GA_{13}	GA_{17}	GA_{28}
	GA_9	GA_4	GA_{20}	GA_1

Fig. 5. Metabolic pathways to C_{19}-GAs.

Pisum sativum.

There are two parallel metabolic pathways in pea seeds. One route, identical to the pathway shown in Fig. 4, leads to GA_9 whilst the second pathway leads to the 13-hydroxylated GA, GA_{20} (see Fig. 5). The pathways were established by comprehensive feeding and refeeding studies using cell-free preparations, coupled with investigations into the identity of native GAs in pea seeds. The work is described in some detail in the following paragraphs.

A high speed supernatant preparation from 18-day old pea embryos, together with Fe^{2+} and either NADPH or ascorbate, converted radio-labelled GA_{12} to GA_{15}, GA_{24}, GA_9 and GA_{51} (Fig. 4) (33). To establish individual steps within the pathway potential intermediates were re-fed. Gibberellin A_{15} was not itself metabolized but the open-lactone form, prepared by alkaline treatment of GA_{15}, was readily converted to GA_{24}, GA_9 and traces of GA_{51}. In turn GA_{24} was converted to GA_9 and some GA_{51}. Subsequently when GA_9 was incubated with a high speed supernatant from 29 day old embryos it was 2β-hydroxylated in good yield to give GA_{51} (structure 13). Thus the pathway $GA_{12} \rightarrow GA_{15}$ open lactone $\rightarrow GA_{24} \rightarrow GA_9 \rightarrow GA_{51}$ was established (see Figs. 5 and 6). GA_{24}, and not GA_{25}, was shown to be the immediate precursor of GA_9 in separate experiments.

Alternatively GA_{12}-aldehyde or GA_{12} was 13-hydroxylated by a low speed supernatant or by the washed microsomal pellet to give GA_{53} (Figs. 5 and 6) (33). Of the two substrates GA_{12} was the better. Both NADPH and molecular oxygen are essential for activity. Gibberellin A_{15}-open lactone and GA_{24} were 13-hydroxylated less efficiently. Moreover, GA_9 was not an acceptable substrate, although limited 13-hydroxylation of GA_9-open lactone was observed. Thus 13-hydroxylation probably occurs early in the pathway.

When GA_{53} was incubated with high speed supernatant in the presence of Fe^{2+} and ascorbate it was converted to GA_{44}, GA_{19} and GA_{20} (Figs. 5 and 6). Gibberellin A_{20} was 2β-hydroxylated by preparations from older embryos to give GA_{29} (structure 14) in *ca.* 75% yield. Individual steps were confirmed by refeeding experiments and the major pathway

(13) GA_{51} (14) GA_{29}

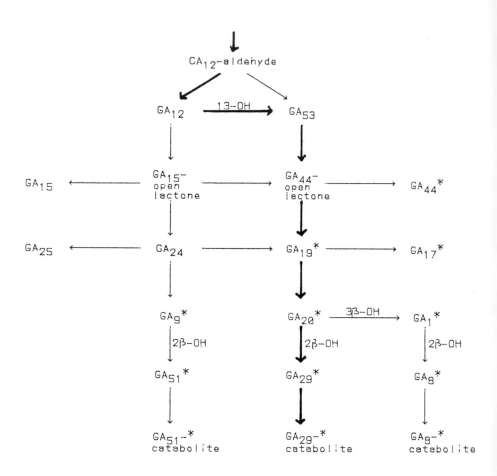

Fig. 6. GA metabolism in Pisum sativum. → = major pathway; * = endogenous GA; (N.B. 3β-hydroxylation occurs only in shoots).

$GA_{53} \rightarrow GA_{44}$-open lactone $\rightarrow GA_{19} \rightarrow GA_{20} \rightarrow GA_{29}$ was confirmed (Figs. 5 and 6) (33). The soluble enzymes involved in C_{20}-oxidation have the properties of dioxygenases. They require α-ketoglutarate and Fe^{2+} (see later).

Seven GAs are known to be native to developing pea seeds (47). These are the C_{20}-GAs, GA_{17}, GA_{19} and GA_{44}, and the C_{19}-GAs, GA_9, GA_{20}, GA_{29} and GA_{51}. All these GAs, with the exception of GA_{17}, were observed as metabolites in cell-free preparations (33) confirming that the pathways demonstrated *in vitro* do indeed operate *in vivo*. Quantitation of native GAs confirms that the 13- hydroxylation pathway is the major route to C_{19}-GAs in pea seeds.

The 2β-hydroxylation of GA_{20} to GA_{29} and of GA_9 to GA_{51} can be observed *in vivo* by injecting labelled substrate into intact seeds. Furthermore applied GA_{29} is converted almost quantitatively to an α,β-unsaturated ketone named GA_{29}-catabolite (structure 15) (46). Although GA_{20} and GA_{29} are located in the cotyledons of maturing seeds, the conversion of GA_{29} to GA_{29}- catabolite occurs only in the testa (45). Thus GA catabolism cannot be observed in cell-free preparations derived from pea embryos (33). Gibberellins which are 2β-hydroxylated (e.g. GA_{29} and GA_{51}) have low biological activity and their catabolism may constitute a means for their removal and/or disposal. An alternative method for the removal of 2β-hydroxylated GAs is by conjugation - for example GA_{29} may be converted to GA_{29}-glucosyl ether (structure 16) (see 46). However, conjugation is of minor importance in pea seeds.

(15) GA_{29}-catabolite   (16) GA_{29}-glucosyl ether

The 2β-hydroxylase from pea seeds has been partially purified and characterized. It is a soluble enzyme requiring α-ketoglutarate and Fe^{2+} for maximal activity (31). These properties are characteristic of a dioxygenase (26) (see later).

The early 13-hydroxylation pathway observed in pea seeds probably operates in pea vegetative tissues too. 13-Hydroxylated C_{19}-GAs are present, albeit in low levels, in seedlings, shoots and leaves of pea (9). However, an additional metabolic conversion namely the 3β-hydroxylation of GA_{20} (structure 17) to GA_1 (structure 18) is observed in shoots but not seeds of pea (Fig. 6). Gibberellin A_1 has high biological activity, and it has been implicated in the control of internode extension (32). A full discussion of this work can be found in chapter E4. 3β-Hydroxylation is therefore a very important metabolic conversion which has profound effects on plant growth. Like GA_{20}, GA_1 can also be 2β-hydroxylated giving GA_8, which in turn is oxidised to GA_8-catabolite (Fig. 6). Although the 2β-hydroxylated GAs and their catabolites have little biological activity they are important because the rate of their formation will regulate the amount of biologically active GA_1 available for growth.

(17) GA$_{20}$ (18) GA$_1$

Cucurbita maxima.

Endosperm preparations from developing pumpkin seeds yield very active GA-metabolising systems. Gibberellin A$_{12}$-aldehyde can be oxidised to GA$_{12}$ by either a microsomal mono-oxygenase of the type which catalyses conversions up to GA$_{12}$-aldehyde, or alternatively by a soluble enzyme akin to those involved in the pathway beyond GA$_{12}$. Conversions beyond GA$_{12}$-aldehyde and GA$_{12}$ are complex, and a metabolic grid can be constructed from the metabolic conversions which have been observed *in vitro* (Fig. 7).

Incubation of GA$_{12}$-aldehyde or GA$_{12}$ with a high speed supernatant in the presence of Fe^{2+} and a variety of cofactors led to the formation of GA$_{15}$, GA$_{24}$ and to their 3β-hydroxylated counterparts GA$_{37}$ and GA$_{36}$ (Figs. 5 and 7) (20). Manganese, which enhanced activity of microsomal enzymes earlier in the pathway, inhibited these soluble enzymes. An alternative but minor route to GA$_{37}$ and GA$_{36}$ via GA$_{14}$-aldehyde and GA$_{14}$ has also been observed. The 3β- hydroxylated C$_{20}$-intermediates appear to be the favoured substrates for further conversion. Thus GA$_{36}$ was converted to the C$_{19}$-GA, GA$_4$, and to the C-20 acid, GA$_{13}$ (Fig. 7). Both GA$_4$ and the 2β-hydroxylated derivative of GA$_{13}$, namely GA$_{43}$, accumulate as end products in incubations (Fig. 7). On the other hand the conversion of GA$_{24}$ to GA$_9$ has not been observed in pumpkin although the 3β-hydroxylation of GA$_9$ to GA$_4$ occurs readily.

The properties of the soluble enzymes catalysing the oxidation C-20 and those catalysing 2β- and 3β-hydroxylation have been studied in the *Cucurbita* system (26). These enzymes have the properties of dioxygenases. They have an absolute requirement for α-ketoglutarate, which acts as a co-substrate, and for Fe^{2+} and molecular oxygen. Reducing agents such as ascorbic acid are stimulatory, probably by maintaining iron in the reduced divalent form. Dioxygenases utilise molecular oxygen to oxidise both the substrate and the co-substrate, α-ketoglutarate. A brief summary of the properties of these and other enzymes involved in GA biosynthesis and metabolism is given in Table 1.

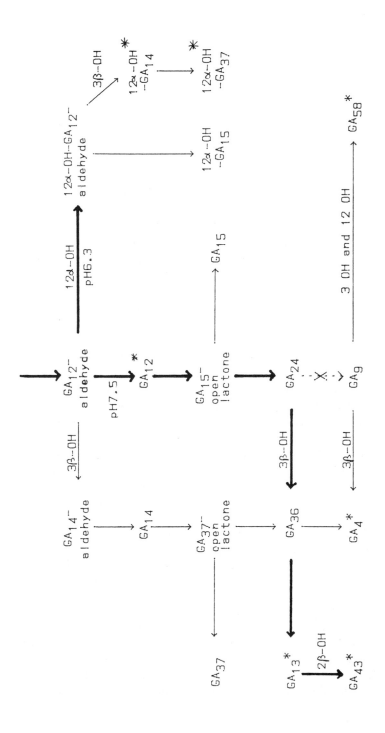

Fig. 7. GA metabolism in *Curcurbita maxima*. → = major pathway; * = endogenous GA.

Table 1. Enzymes catalysing reactions in the GA biosynthetic pathway—a summary of their properties.

Reaction or sequence of reactions	Enzyme properties		
	Location	Cofactors etc.	Ref.
MVA→ →ent-kaurene	soluble	ATP, Mn^{2+}/Mg^{2+}	see 4, also 39, 40
ent-kaurene→ → GA$_{12}$-aldehyde	microsomal	NADPH, O_2	23, 24
GA$_{12}$-aldehyde→	(1) microsomal	pH 7.5, NADPH, O_2	27
GA$_{12}$	(2) soluble	α-ketoglutarate, Fe^{2+}, O_2	20
sequential oxidation of carbon-20	soluble	α-ketoglutarate, Fe^{2+}, O_2	26
2β-hydroxylation	soluble	α-ketoglutarate, Fe^{2+}, O_2	31, 44
3β-hydroxylation	soluble	α-ketoglutarate, Fe^{2+}, O_2	26
12α-hydroxylation	microsomal	pH 6.3, NADPH, O_2	27
13-hydroxylation	microsomal	NADPH, O_2	33

If GA$_{12}$-aldehyde is incubated with a microsomal pellet the yield of products is dependent on the pH of the incubation mixture. At pH7.5, the one most commonly employed, GA$_{12}$-aldehyde is converted only to GA$_{12}$. However at pH6.2, the major product is 12α-hydroxy-GA$_{12}$-aldehyde (structure 19) together with GA$_{12}$ and GA$_{53}$ (27). The 12α-hydroxylation of GA$_{12}$-aldehyde appears to be catalysed by a microsomal mono-oxygenase requiring NADPH and O_2. 12α-Hydroxy-GA$_{12}$-aldehyde is not further metabolized by microsomal enzymes, but incubation with a high speed supernatant together with α-ketoglutarate and Fe^{2+} yields 12α-hydroxy-GA$_{15}$ (Fig. 7). Also formed are the 3β-hydroxylated products 12α-hydroxy-GA$_{14}$ and 12α-hydroxy-GA$_{37}$. In separate experiments deuterated GA$_9$ which was incubated with a low speed supernatant yielded GA$_4$ and GA$_{58}$ (structure 20) as products (27).

The products of *in vitro* metabolic studies using *C. maxima* are known to be native to developing pumpkin seeds (6), confirming that metabolism *in vitro* is a true representation of metabolism *in vivo*. Thus many of the GAs in developing seeds of pumpkin are 3β- and/or 12α-hydroxylated (6). GA$_{13}$ and GA$_{43}$ (see Fig. 7) are the major GAs in embryos and endosperm respectively, and several 12α-hydroxylated GAs, especially GA$_{58}$, also accumulate in both tissues. Whether the metabolic capacity of pumpkin embryos is similar to that of endosperm has not been established. Alternatively, GAs produced within the endosperm may be transported to the growing embryo. Pumpkin seeds, in marked contrast to those of pea,

(19) 12α-Hydroxy-
GA$_{12}$-aldehyde

(20) GA$_{58}$

do not accumualte 13-hydroxylated GAs, and 13-hydroxylating activity in cell-free systems from pumpkin is minimal.

Other seeds

GAs have been extracted from immature seeds/grain of a large number of plants of agronomic or horticultural importance together with related genera. Most frequently studied have been genera in the Leguminoseae (namely *Phaseolus, Pisum, Vicia, Vigna, Canavalia, Cytisus, Lupinus* and *Leucaena*), in the Cucurbitaceae (*Cucurbita, Cucumis, Lagenaria, Marah* and *Sechium*) in the Graminae (*Secale, Triticum, Hordeum, Oryza* and *Zea*) and in the Convolvulaceae (*Ipomoea, Pharbitis, Calonyction* and *Calystegia*). The nature of the GAs present in seeds of a particular genus allows predictions to be made as to the metabolic pathways which may operate in that genus (46). However detailed metabolic studies of the type conducted with *Pisum* and *Cucurbita* are lacking for most other genera. Nevertheless some generalizations can be made.

The GA metabolic pathways in maturing seeds are complex, and they lead to the accumulation of large amounts of several hydroxylated and poly-hydroxylated GAs in seeds of each genus. Two positions for hydroxylation in the ring system are particularly common, namely 3β- and 13-hydroxylation. To date 3β-hydroxylation has not been observed in pea seeds, and 13-hydroxylation is rarely observed in pumpkin and some other cucurbits, but apart from these exceptions 3β- and 13-hydroxylated GAs are observed in almost every other genus (1,2,13,14,16,30,35). 13-Hydroxylation seems to occur early in the pathway, whereas 3β-hydroxylation can apparently occur at many different stages in the pathway. 2β-Hydroxylation is also common. It occurs late in each pathway and leads to inactivation of a GA molecule. Hydroxylations at other positions in the GA ring system appear to be more specific. For example 12α-hydroxylation occurs in seeds of *Cucurbita maxima* in addition to 2β- and 3β-hydroxylation. In grain of barley (*Hordeum vulgare*) 12β- and 18-hydroxylation have been observed in addition to

$2\beta,3\beta$- and 13-hydroxylation (17). Indeed the major C_{19}-GAs which accumulate in barley grain are all hydroxylated at C-18. In grain of wheat (*Triticum aestivum*) the major C_{19}-GAs are 1β-hydroxylated, in conjunction with 3β- and/or 13-hydroxylation (16). 15β-Hydroxylation apparently occurs early in the GA pathway in seeds of sunflower (*Helianthus annuus*), for 15β-hydroxylated C_{20}- and C_{19}-GAs have been identified therein (36, and unpublished work). 15β-Hydroxylated GAs have also been identified in seeds of apple (*Malus sylvestris*) and pear (*Pyrus communis*) (see 3). It is interesting to note that in several instances the novel poly-hydroxylated GAs observed in maturing seeds are not present in vegetative tissues of the same genus.

In general, GA metabolism proceeds in parallel with seed development so that the more polar, highly oxidised GAs which occur late in the GA metabolic sequence accumulate during the later stages of seed development. The significance of the accumulation of large amounts of GAs during seed maturation, and of the diverse hydroxylation patterns observed have yet to be established. There is however some evidence that a large proportion of the GAs present in maturing seeds may be redundant. Thus pea seeds can undergo apparently normal maturation even though the levels of native GAs are very substantially reduced by the application of inhibitors of GA biosynthesis.

Vegetative tissues

The large accumulation of GAs in maturing seeds has been exploited by biochemists and physiologists so that GA metabolic pathways can be clearly defined. Furthermore seeds are such rich sources of enzyme activity that it has been possible to isolate some of the enzymes involved in GA metabolism, and to study their catalytic and regulatory properties. However, GA metabolism in vegetative tissues, and its control therein, is of more fundamental importance. Extraction of shoot tissues from a number of diverse genera e.g. *Pisum sativum, Zea mays, Agrostemma githago, Spinacia oleracea, Oryza sativa*, has revealed a similar sequence of 13-hydroxylated GAs in all of them (see 43). These results suggest that the pathway $GA_{53} \rightarrow GA_{44}$-open lactone $\rightarrow GA_{19} \rightarrow GA_{20}$, which was first observed in pea seeds and which was described earlier, may be common to shoot tissue of many, if not all, genera.

In most of these genera, GA_{20} appears to be a pivotal GA, and its fate has a profound effect on plant growth. Detailed metabolic studies using pea and maize shoots have shown that GA_{20} is 3β-hydroxylated to give GA_1 in tall genotypes (32,48) (Fig. 6). In the dwarf *le* mutant of pea and the *dwarf-1* mutant of maize 3β-hydroxylation of GA_{20} is blocked. Moreover GA_{20} applied to these dwarf mutants does not induce internode elongation indicating that GA_{20} is inactive *per se*, and that GA_1 controls internode elongation. Indeed Phinney (43) has suggested that GA_1 may be of widespread and fundamental importance in stem growth, and that it may fulfil the role of a "primary hormone".

The 3β-hydroxylation of GA_{20} appears to be closely regulated within the plant. This observation is consistent with the probable role for GA_1 as a primary hormone. Thus 3β-hydroxylation of GA_{20} occurs in pea shoots but not in pea seeds, suggesting that the *Le* gene is expressed only in shoots. In addition there is evidence that within pea shoots 3β-hydroxylation is also under ontogenetic and environmental control.

Alternatively in shoots of spinach (*Spinacia oleracea*) it is the formation of GA_{20}, rather than its metabolism, which appears to be under ontogenetic and environmental control (38). Thus the metabolism of GA_{19} to GA_{20} (see Fig. 6) is enhanced in long photoperiods, together with rapid stem elongation (bolting). There is, however, no evidence to date that GA_1 is present in spinach shoots, and the possibility exists that GA_{20} may be active *per se* in controlling shoot elongation in this genus.

CONCLUSIONS

A large proportion of the 72 known GAs have little or no biological activity. This is because there are structural requirements for GA activity (see 22), which probably represent the GA's ability to "fit" a receptor molecule. Some GAs such as the 3β-hydroxylated C_{19}-GAs, i.e., GA_1, GA_3, GA_4, and GA_7 appear to satisfy these requirements, and any of them might fulfil the role of "primary hormone". Other GAs, for instance the C_{20}-type, may achieve biological activity by virtue of their metabolism to C_{19}-GAs. Yet other GAs, such as GA_8, GA_{29}, GA_{34} and GA_{51}, are inactive due to the presence of a 2β-hydroxyl group, which is a deactivating substituent. Thus the numerous GAs present in any one plant or plant organ may represent a single bioactive GA, together with a number of its precursors and its metabolic products. The interrelationships of several higher plant GAs are summarized in Fig. 8.

Shoot tissues of the several genera so far examined contain a similar complement of 13-hydroxylated GAs including the 3, 13-dihydroxylated C_{19}-GA, GA_1. In these genera GA_1 is likely to be the primary bioactive GA. In other genera such as cucumber (*Cucumis sativus*) applied GA_1 has little biological activity (see 22), and its 13-desoxy counterpart, GA_4, may fulfil the primary role.

In seeds the situation is more complex. During early stages of seed development low levels of GAs, akin to those in vegetative tissue, may control fruit growth. However, later in development the large accumulation of novel polyhydroxylated GAs which are not known to occur in vetetative tissues may merely represent the elaboration of secondary metabolites which are without biological effect.

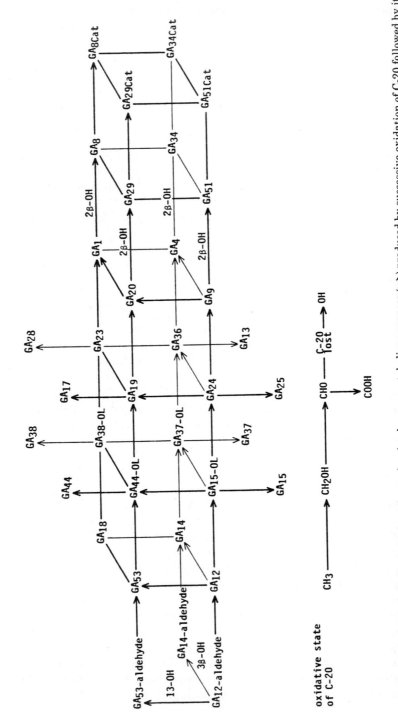

Fig. 8. Metabolic grid of GAs in higher plants showing known metabolic sequences (→) produced by successive oxidation of C-20 followed by its removal, combined with hydroxylation at C-13, C-3, and/or C-2. Not all reactions operate in all plants (see Figs. 6 and 7 for pathways in pea and pumpkin, respectively). Some other reactions are known but not shown. Potential sequences which have not been confirmed are depicted (—). Adapted from Kamiya et al. 1985 Abstracts, 12th Int. Conf. on Plant Growth Substances, Heidelberg. p. 9. (O.L.=open lactone; Cat=catabolite (αβ-unsaturated ketone).

Acknowledgements

I thank Dr. M.H. Beale for computer-graphics, and Professor J. MacMillan for useful comments on the manuscript.

References

1. Albone, K.S., Gaskin, P., MacMillan, J., Sponsel, V.M. (1984) Identification and localization of gibberellins in maturing seeds of the cucurbit *Sechium edule*, and a comparison between this cucurbit and the legume *Phaseolus coccineus*. Planta *162*, 560-65.
2. Arigayo, S., Sakata, K., Fujisawa, S., Sakurai, A., Adisewoju, S.S., Takahashi, N. (1983) Characterization of gibberellins in immature seeds of *Leucaena leucocephala* (Lmk) De. Wit. Agric. Biol. Chem. *47*, 2939-40.
3. Bearder, J.R. (1980) Plant hormones and other growth substances - their background, structures and occurrence. *In* Hormonal regulation of development. I, Molecular aspects of plant hormones. Encyclopaedia of Plant Physiology New Series, Vol. 9, pp. 9-112, MacMillan, J. ed., Springer-Verlag, Berlin, Heidelberg, New York.
4. Beytia, E.D., Porter, J.W. (1976) Biochemistry of polyisoprenoid biosynthesis. Ann. Rev. Biochem. *45*, 113-42.
5. Birch, A.J., Richards, R.W., Smith, H., Harris, A., Whalley, W.B. (1959) Studies in relation to biosynthesis-XXI. Rosenonolactone and gibberellic acid. Tetrahedron 7, 241-51.
6. Blechschmidt, S., Castel, U., Gaskin, P., Hedden, P., Graebe, J.E., MacMillan J. (1984) GC/MS analysis of plant hormones in seeds of *Cucurbita maxima*. Phytochem. *23*, 553-58.
7. Coolbaugh, R.C. (1983) Early stages of gibberellin biosynthesis. *In* The biochemistry and physiology of gibberellins. Vol. 1, pp. 53-98. Crozier, A. ed., Praeger, New York.
8. Cross, B.E., Grove, J.F., MacMillan, J., Mulholland, T.P.C. (1956) Gibberellic acid, Part IV. The structures of gibberic and *allo* gibberic acids and possible structures for gibberellic acid. Chem. and Ind. 1956, 954-55.
9. Davies, P.J., Emshwiller, E., Gianfagna, T.J., Proebsting, W.M., Noma, M., Pharis, R.P. (1982) The endogenous gibberellins of vegetative and reproductive tissue of G2 peas. Planta *154*, 266-72.
10. Dicks, J.W. (1979) Modes of action of growth retardants. *In* Recent developments in the use of plant growth retardants. Monograph 4, pp. 1-14. Clifford, D.R., Lenton, J.R. eds. British Plant Growth Regulator Group, Wantage.
11. Dockerill, R.C., Hanson, J.R. (1978) Fate of C-20 in C_{19}-gibberellin biosynthesis. Phytochem. *11*, 317-26.
12. Duncan, J.D., West, C.A. (1981) Properties of kaurene synthetase from *Marah macrocarpus* endosperm. Evidence for the participation of separate but interacting enzymes. Plant Physiol. *68*, 1128-34.
13. Durley, R.C., MacMillan, J., Pryce, R.P. (1971) Investigation of gibberellins and other growth substances in the seed of *Phaseolus multiflorus* and of *Phaseolus vulgaris* by gas chromatography-mass spectrometry. Phytochem. *10*, 1891-1908.
14. Fujisawa, S., Yamaguchi, I., Park, K.-H., Kobayashi, M., Takahashi, N. (1985) Qualitative and semi-quantitative analyses of gibberellins in immature seeds of *Pharbitis purpurea*. Agric. Biol. Chem. *49*, 27-33.
15. Gafni, Y., Schechter, I. (1981) Isolation of a kaurene synthetase inhibitor from castor bean seedlings and cell suspension cultures. Plant Physiol. *67*, 1169-73.
16. Gaskin, P., Kirkwood, P.S., Lenton, J.R., MacMillan, J., Radley, M.E. (1980) Identification of gibberellins in developing wheat grain. Agric. Biol. Chem. *44*, 158-93.
17. Gilmour, S.J., Gaskin, P., Sponsel, V.M., MacMillan, J. (1984) Metabolism of gibberellins by immature barley grain. Planta *161*, 186-92.

18. Graebe, J.E. (1980) GA-biosynthesis: the development and application of cell-free systems for biosynthetic studies. *In* Plant Growth Substances 1979, pp. 180-187, Skoog, F. ed. Springer Verlag, Heidelberg.

19. Graebe, J.E. (1982) Gibberellin biosynthesis in cell-free systems from higher plants. *In* Plant Growth Substances 1982, pp. 72-80, Wareing, P.F. ed. Academic Press, London.

20. Graebe, J.E., Hedden, P., Gaskin, P., MacMillan, J. (1974) Biosynthesis of gibberellin A_{12}, A_{15}, A_{24}, A_{36} and A_{37} by a cell-free system from *Cucurbita maxima*. Phytochem. *13*, 1433-40.

21. Graebe, J.E., Hedden, P., Gaskin, P., MacMillan, J. (1974). The biosynthesis of a C_{19}-gibberellin from mevalonic acid in a cell-free system from a higher plant. Planta *120*, 307-309.

22. Graebe, J.E., Ropers, H.J. (1978) Gibberellins. *In* Phytohormones and related compounds – a comprehensive treatise. Vol. 1, pp. 107-204, Letham, D.S., Goodwin, P.B., Higgins, T.J.V. eds. Elsevier/North Holland Biomedical Press, Amsterdam.

23. Hasson, E.P., West, C.A. (1976). Properties of the system for the mixed function oxidation of kaurene and kaurene derivatives in microsomes of immature seeds of *Marah macrocarpus*. Cofactor requirements. Plant Physiol. *58*, 473-78.

24. Hasson, E.P., West, C.A. (1976) Properties of the system for the mixed function oxidation of kaurene and kaurene derivatives in microsomes of the immature seed of *Marah macrocarpus*. Electron transfer components. Plant Physiol. *58*, 479-484.

25. Hedden, P. (1983) *In vitro* metabolism of gibberellins. *In* The biochemistry and physiology of gibberellins. Vol. *1*, pp. 99-150, Crozier, A. ed. Praeger, New York.

26. Hedden, P., Graebe, J.E. (1982) The cofactor requirements for the soluble oxidases in the metabolism of the C_{20}-gibberellins. J. Plant Growth Regulation 1, 105-116.

27. Hedden, P., Graebe, J.E., Beale, M.H., Gaskin, P., MacMillan, J. (1984) The biosynthesis of 12α-hydroxylated gibberellins in a cell-free system from *Cucurbita maxima* endosperm. Phytochem. *23*, 569-74.

28. Hedden, P., MacMillan, J., Phinney, B.O. (1978) The metabolism of the gibberellins. Ann. Rev. Plant Physiol. *29*, 149-192.

29. Hedden, P., Phinney, B.O. (1979) Comparison of *ent*-kaurene and *ent*-isokaurene synthesis in cell-free systems from etiolated shoots of normal and dwarf-5 maize seedlings. Phytochem. *18*, 1475-79.

30. Hiraga, K., Kawabe, S., Yokota, T., Murofushi, N., Takahashi, N. (1974) Isolation and characterization of plant growth substances in immature seeds and etiolated seedlings of *Phaseolus vulgaris*. Agr. Biol. Chem. *38*, 2521-27.

31. Hoad, G.V., MacMillan, J., Smith, V.A., Sponsel, V.M., Taylor, D.A. (1982) Gibberellin 2β-hydroxylases and biological activity of 2β-alkyl gibberellins. *In* Plant Growth Substances 1982, pp. 91-100, Wareing, P.F. ed. Academic Press, London.

32. Ingram, T.J., Reid, J.B., Murfet, I.C., Gaskin, P., Willis, C.L., MacMillan, J. (1984) Internode length in *Pisum*. The *Le* gene controls the 3β-hydroxylation of gibberellin A_{20} to gibberellin A_1. Planta *160*, 455-63.

33. Kamiya, Y., Graebe, J.E. (1983) The biosynthesis of all major pea gibberellins in a cell-free system from *Pisum sativum*. Phytochem. *22*, 681-90.

34. Kamiya, Y., Takahshi, N., Graebe, J.E. (1986) The loss of the C-20 carbon atom in C_{19}-gibberellin biosynthesis in a cell-free system from *Pisum sativum*. Planta, in press.

35. Kobayashi, M., Yamaguchi, I., Murofushi, N., Ota, Y., Takahashi, N. (1984) Endogenous gibberellins in immature seeds and flowering ears of rice. Agric. Biol. Chem. *48*, 2725-29.

36. MacMillan, J. (1984) Analysis of plant hormones and metabolism of gibberellins. *In* The biosynthesis and metabolism of plant hormones, pp. 1-16, Crozier, A., Hillman, J.R. eds. Soc. Exp. Biol. Seminar *23*, Cambridge U.P., Cambridge.

37. MacMillan, J., Takahashi, N. (1968) Proposed procedure for the allocation of trivial names to the gibberellins. Nature *217*, 170-71.

38. Metzger, J.D., Zeevaart, J.A.D. (1980) Effect of photoperiod on the levels of endogenous gibberellins in spinach as measured by combined gas chromatography – selected ion current monitoring. Plant Physiol. *66*, 844-46.

39. Ogura, K., Nishino, T., Seto, S. (1968) The purification of prenyl transferase and isopentenyl pyrophosphate isomerase of pumpkin fruit and some of their properies. J. Biochem. *64*, 197-203.

40. Ogura, K., Shinka, T., Seto, S. (1972) The purification and properties of geranyl pyrophosphate synthetase from pumpkin fruits. J. Biochem. *72*, 1101-1110.

41. Rademacher, W., Jung, J., Graebe, J.E., Schwenen, L. (1984) On the mode of action of tetcyclasis and triazole growth retardants. *In* Biochemical aspects of synthetic and naturally occurring plant growth regulators. Monograph *11*, pp. 1-11, Menhennet, R., Lawrence, D.K. eds. British Plant Growth Regulator Group, Wantage.

42. Railton, I.D., Fellows, B., West, C.A. (1984) *ent*-Kaurene synthesis in chloroplasts from higher plants. Phytochem. *23*, 1261-67.

43. Phinney, B.O. (1984) Gibberellin A_1, dwarfism and the control of shoot elongation in higher plants. *In* The biosynthesis and metabolism of plant hormones. pp. 17-41, Crozier, A., Hillman, J.R. eds. Soc. Exp. Biol. Seminar *23*. Cambridge U.P., Cambridge.

44. Smith, V.A., MacMillan, J. (1984) Purification and partial characterization of a gibberellin 2β-hydroxylase from *Phaseolus vulgaris*. J. Plant. Growth Regul. *2*, 251-64.

45. Sponsel, V.M. (1983) The localization, metabolism and biological activity of gibberellins in maturing and germinating seeds of *Pisum sativum* cv. Progress No. 9. Planta *159*, 454-68.

46. Sponsel, V.M. (1983) *In vivo* gibberellin metabolism in higher plants. *In* The biochemistry and physiology of gibberellins. Vol. *1*, pp. 151-250, Crozier, A. ed. Praeger, New York.

47. Sponsel, V.M. (1985) Gibberellins in *Pisum sativum*—their nature, distribution and involvement in growth and development of the plant. Plant Physiol. *65*, 533-38.

48. Spray, C.R., Phinney, B.O., Gaskin, P., Gilmour, S.J., MacMillan, J. (1984) Internode length in *Zea mays* L. The dwarf-1 mutation controls the 3β-hydroxylation of gibberellin A_{20} to gibberellin A_1. Planta *160*, 464-68.

49. West, C.A. (1973) Biosynthesis of gibberellins. *In* Biosynthesis and its control in plants, pp. 473-82. Milborrow, B.V. ed., Academic Press, London.

50. West, C.A., Shen-Miller, J., Railton, I.D. (1982) Regulation of kaurene synthetase. In: Plant growth substances 1982, pp. 81-90. Wareing, P.F. ed., Academic Press, London.

Appendix—Gibberellin Structures

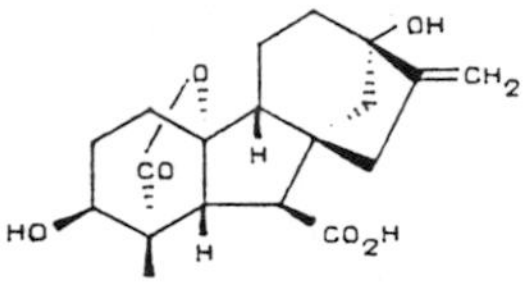

GA$_1$

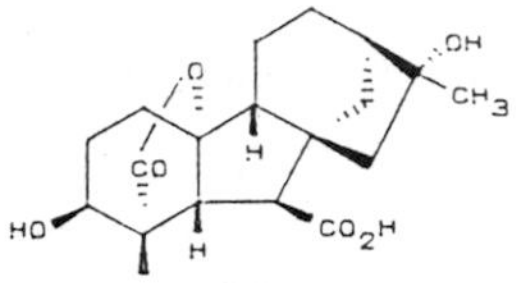

GA$_2$

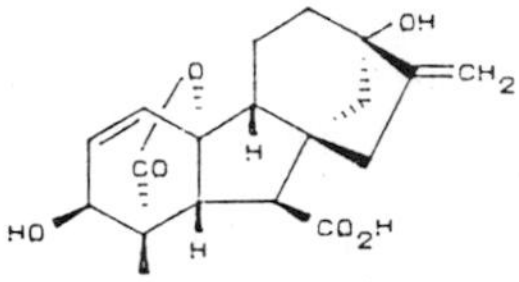

GA$_3$

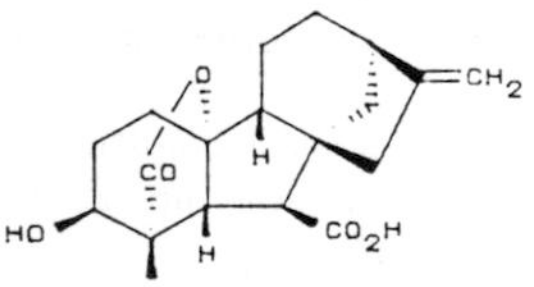

GA$_4$

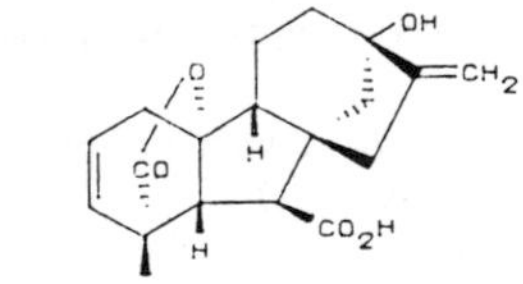

GA$_5$

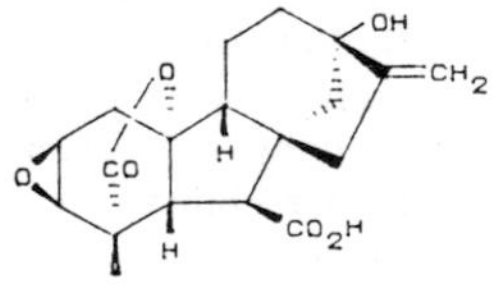

GA$_6$

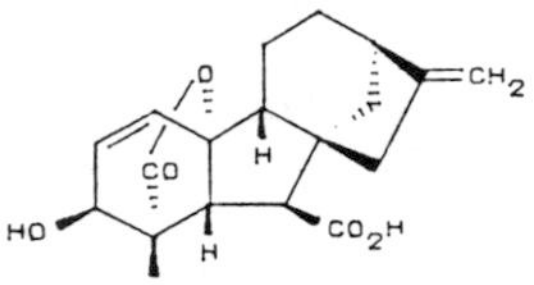

GA$_7$

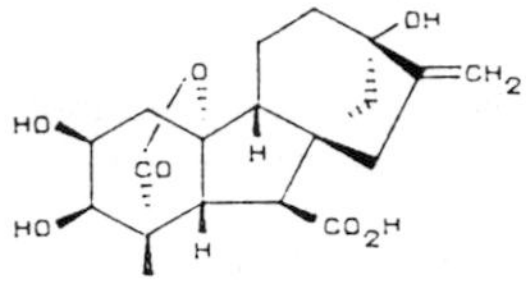

GA$_8$

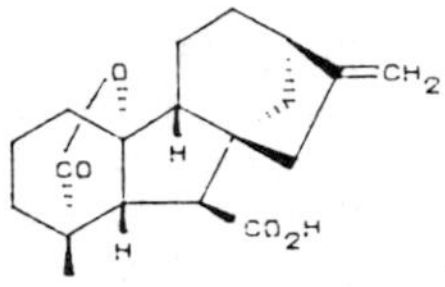

GA$_9$

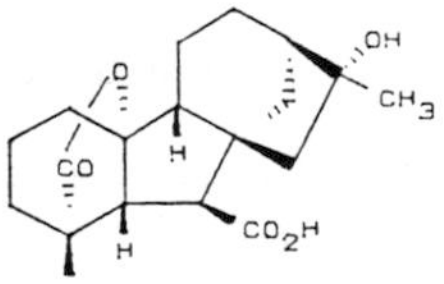

GA$_{10}$

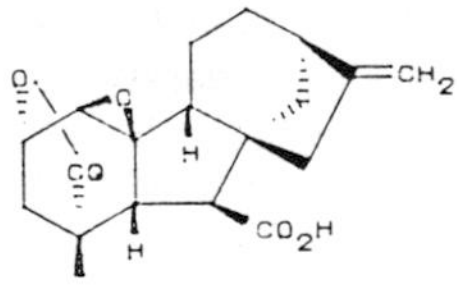

GA$_{11}$

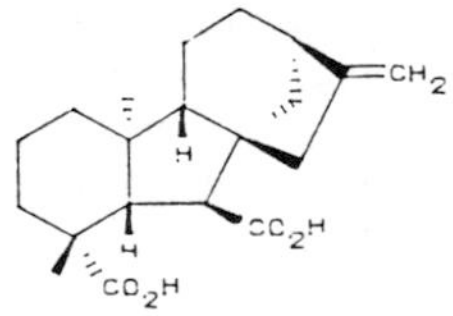

GA$_{12}$

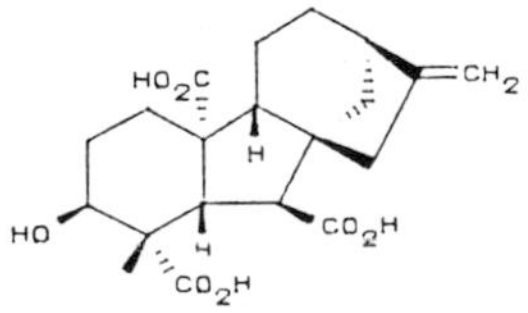

GA$_{13}$

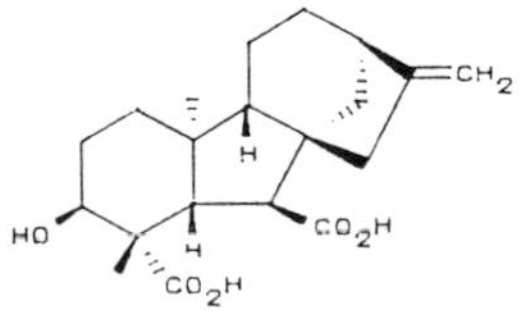

GA$_{14}$

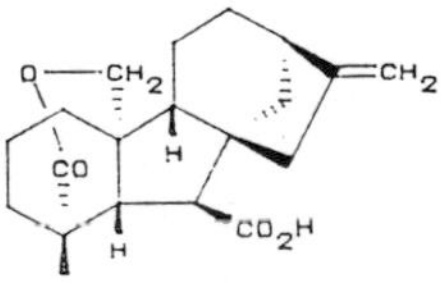

GA$_{15}$

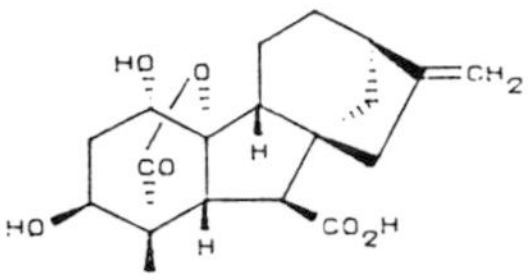

GA$_{16}$

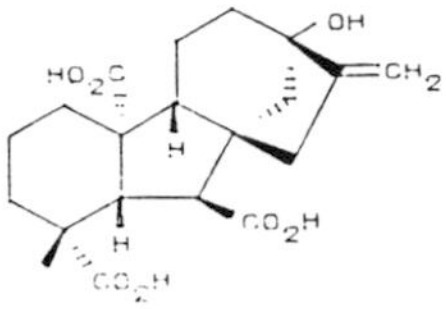

GA$_{17}$

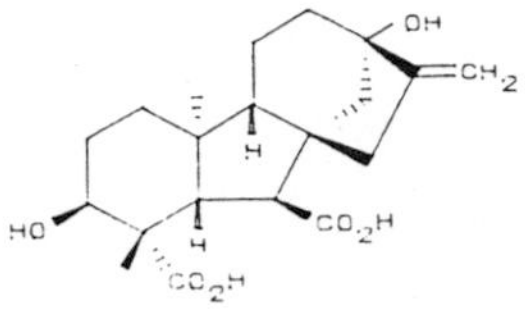

GA$_{18}$

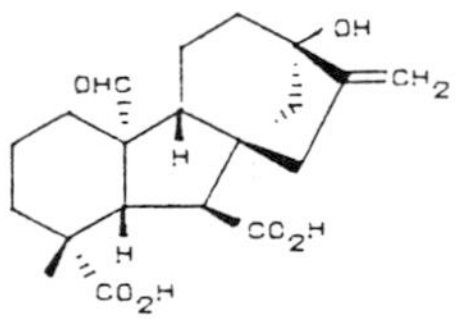

GA$_{19}$

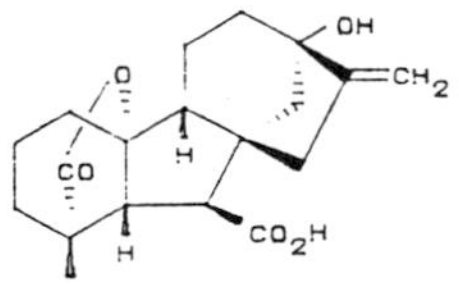

GA$_{20}$

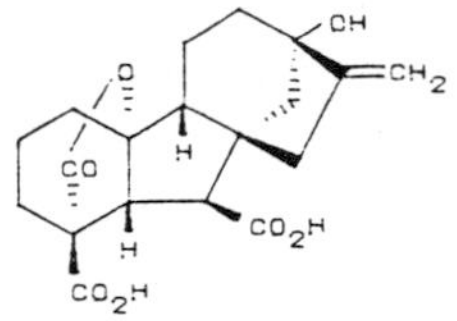

GA$_{21}$

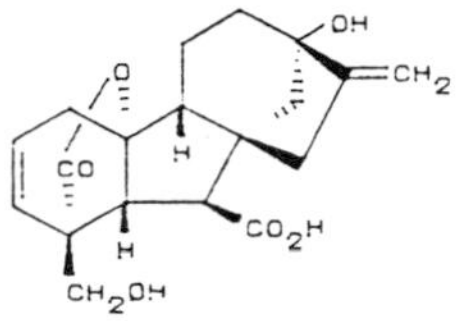

GA$_{22}$

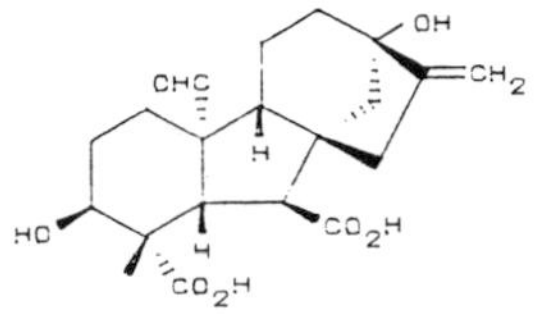

GA$_{23}$

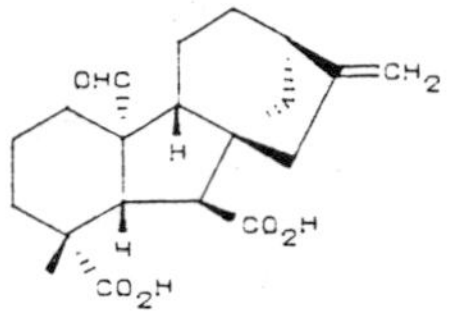

GA$_{24}$

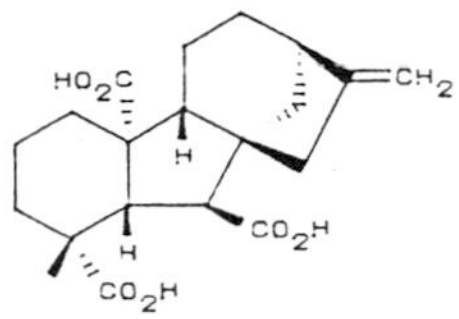

GA$_{25}$

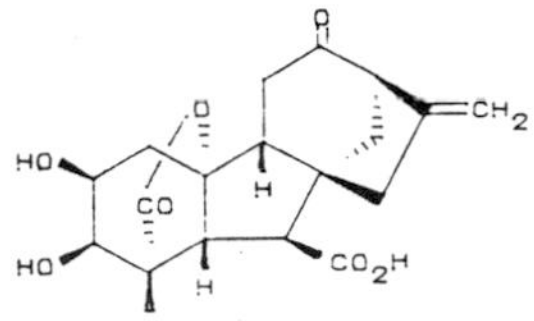

GA$_{26}$

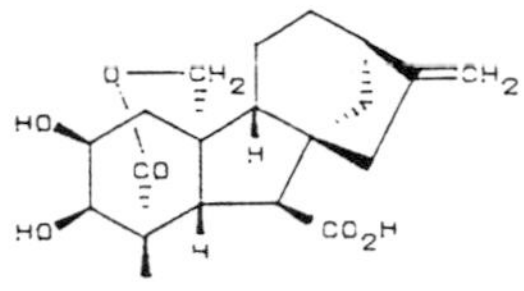

GA$_{27}$

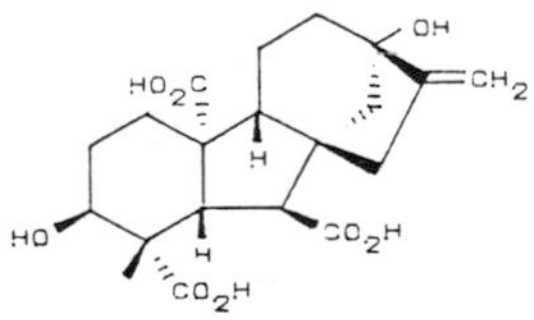

GA$_{28}$

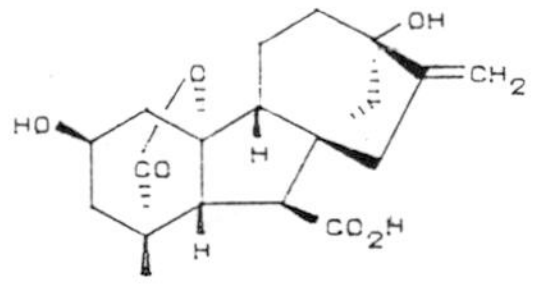

GA$_{29}$

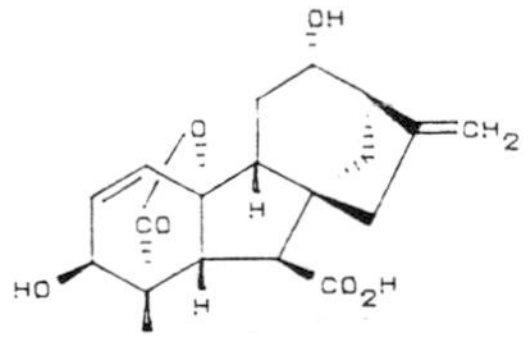

GA$_{30}$

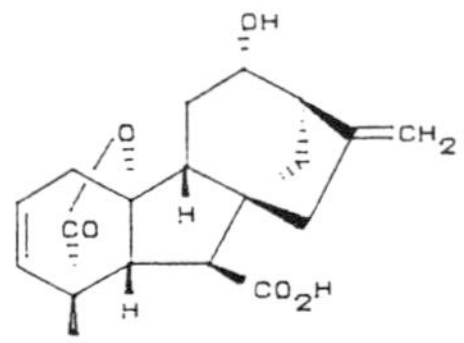

GA$_{31}$

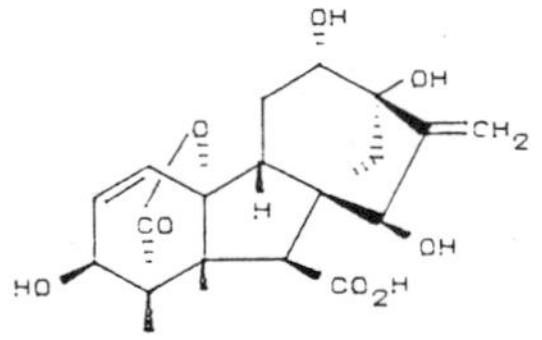

GA$_{32}$

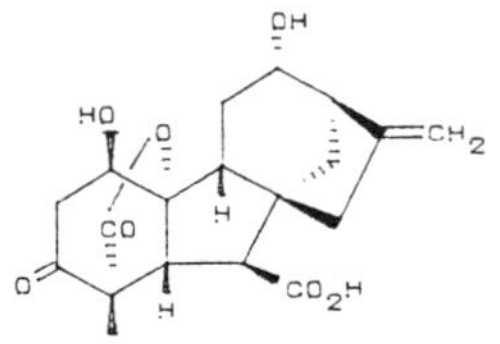

GA$_{33}$

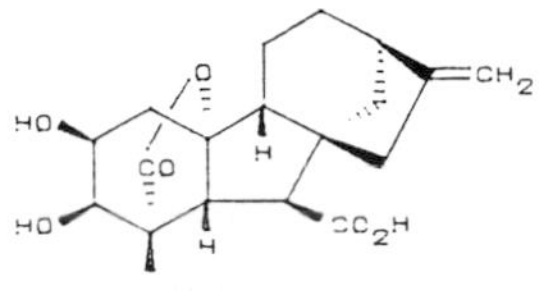

GA$_{34}$

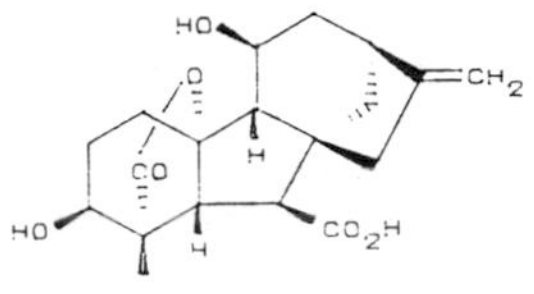

GA$_{35}$

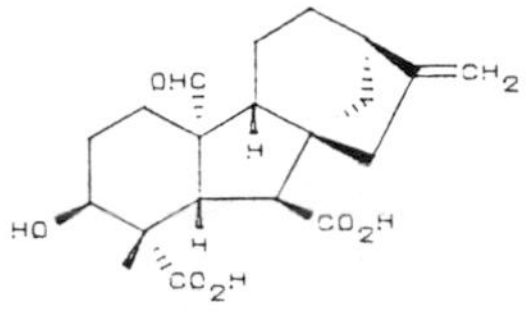

GA$_{36}$

73

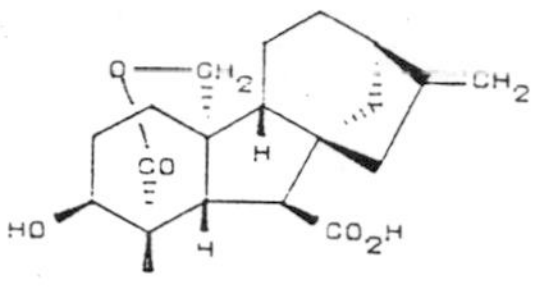

GA$_{37}$

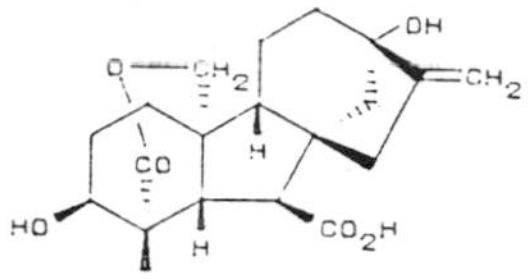

GA$_{38}$

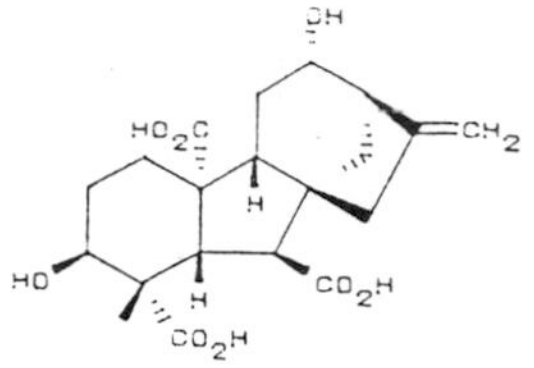

GA$_{39}$

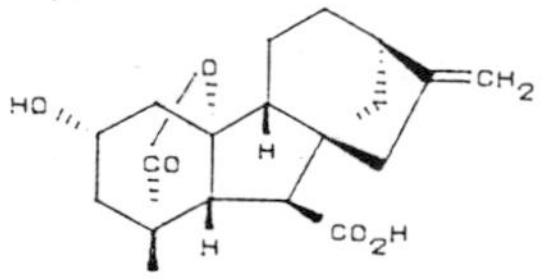

GA$_{40}$

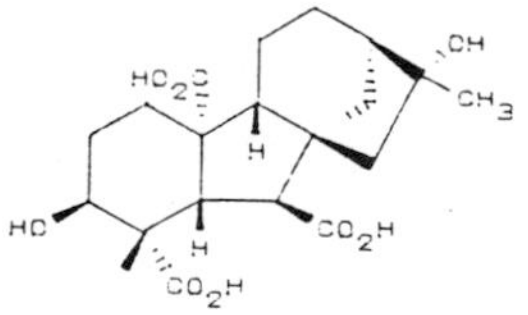

GA$_{41}$

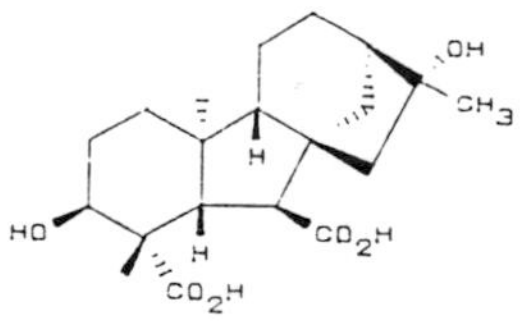

GA$_{42}$

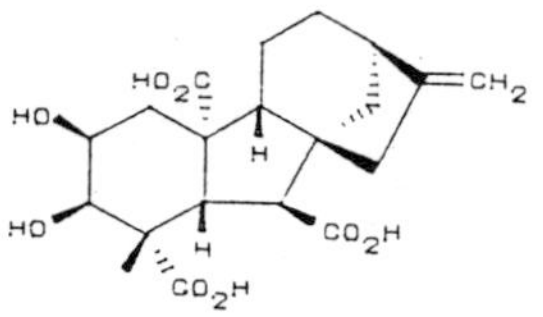

GA$_{43}$

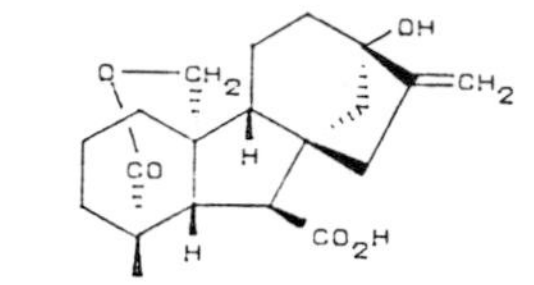

GA$_{44}$

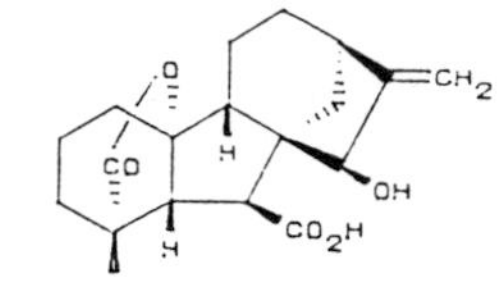

GA$_{45}$

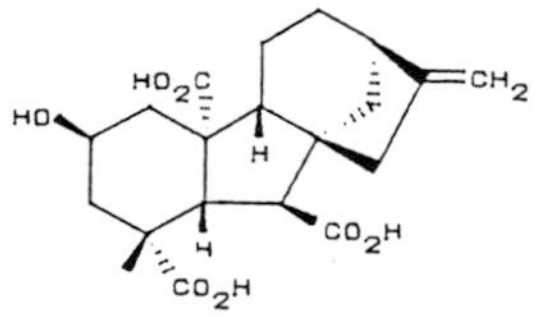

GA$_{46}$

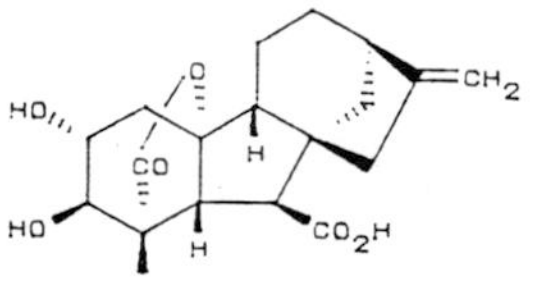

GA$_{47}$

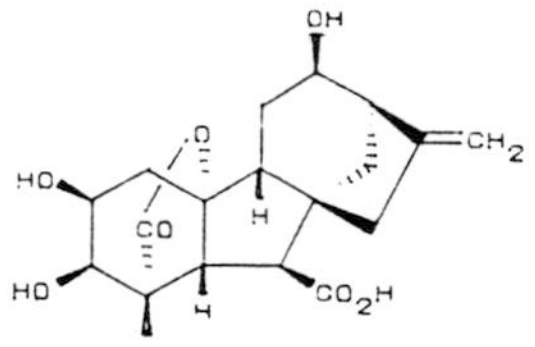

GA$_{48}$

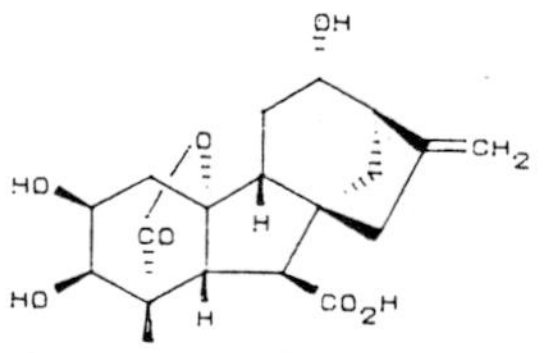

GA$_{49}$

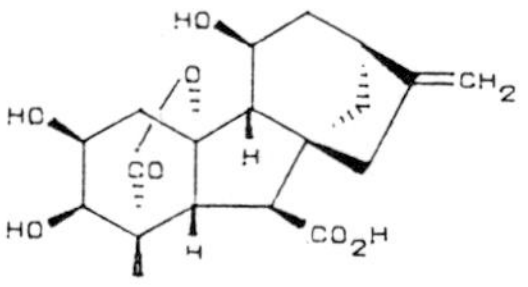

GA$_{50}$

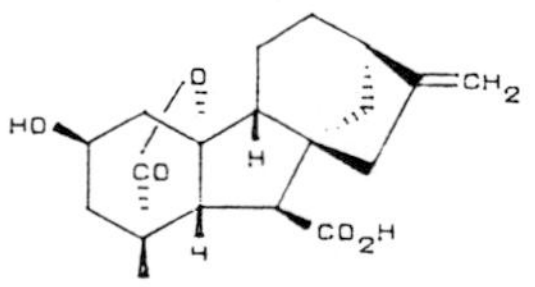

GA$_{51}$

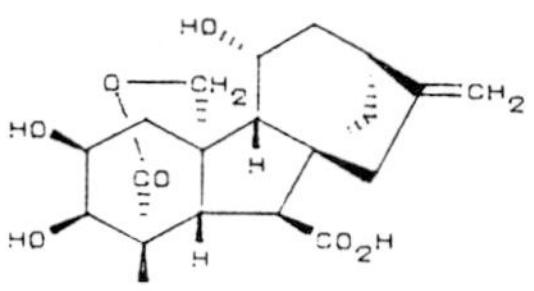

GA$_{52}$

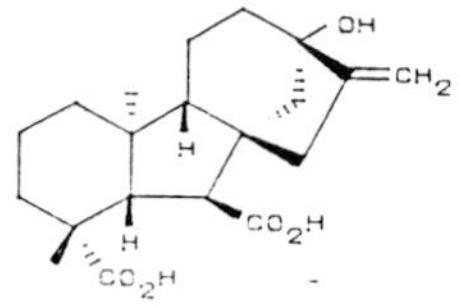

GA$_{53}$

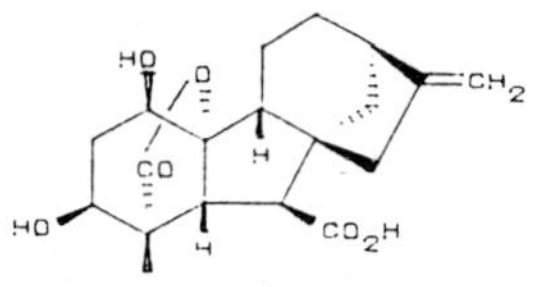

GA$_{54}$

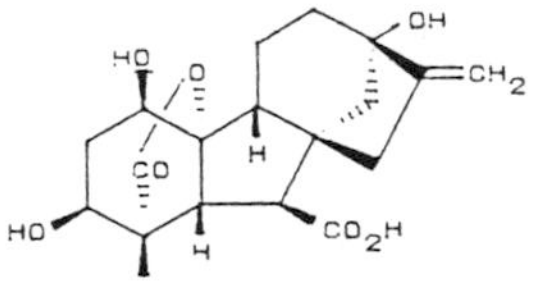

GA_{55}

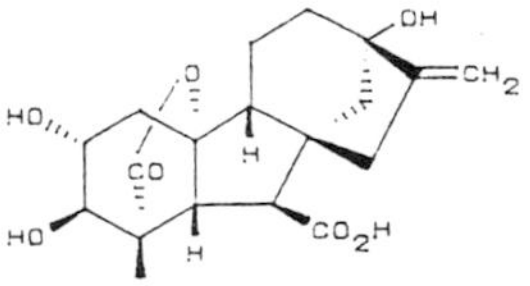

GA_{56}

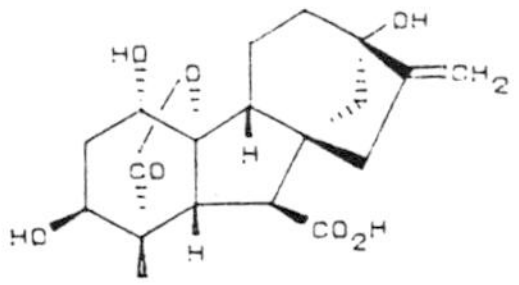

GA_{57}

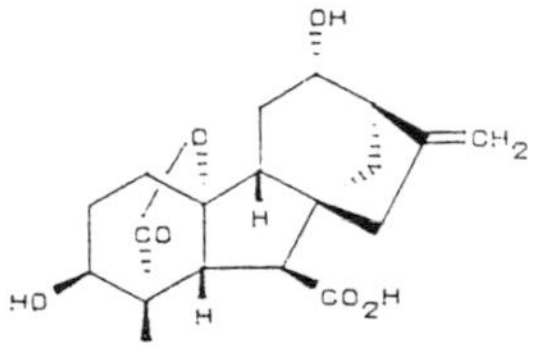

GA_{58}

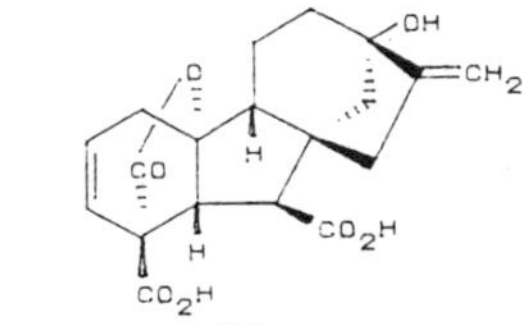

GA_{59}

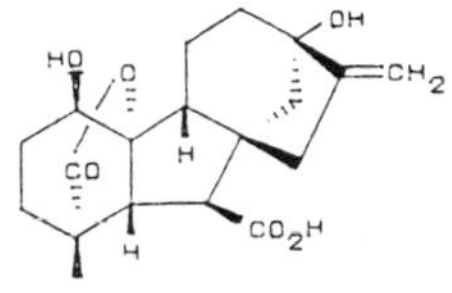

GA_{60}

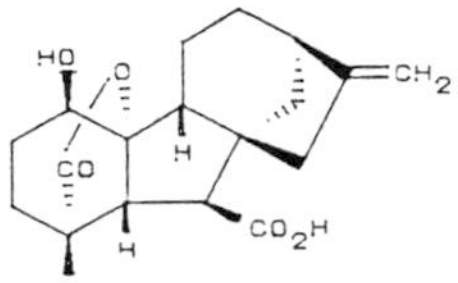

GA_{61}

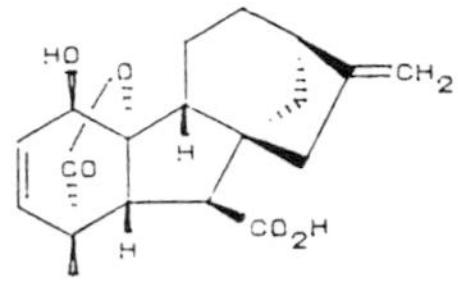

GA_{62}

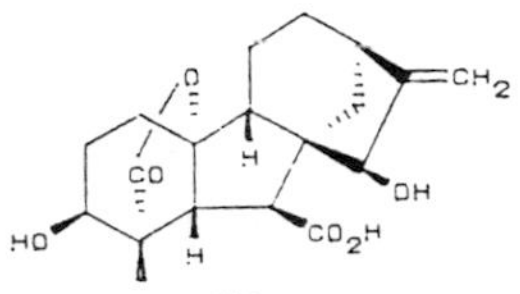

GA_{63}

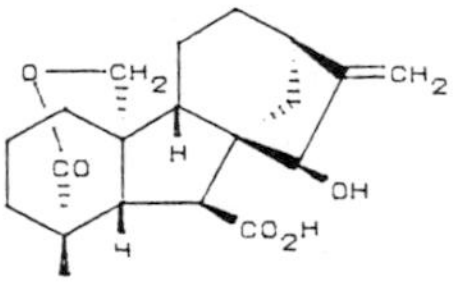

GA_{64}

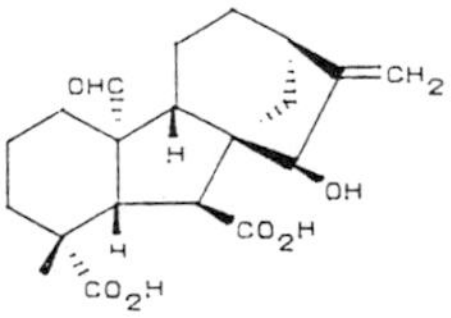

GA_{65}

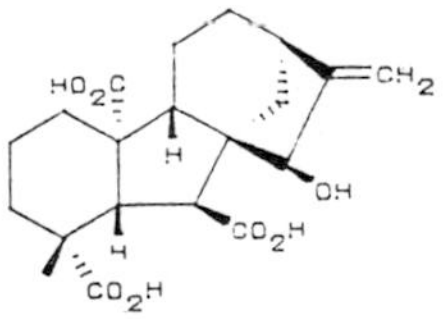

GA_{66}

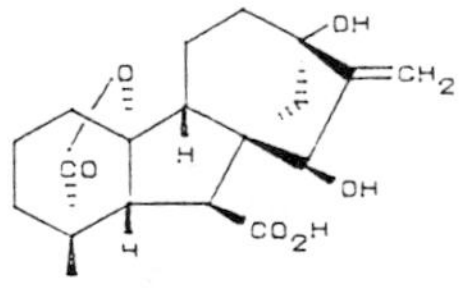

GA_{67}

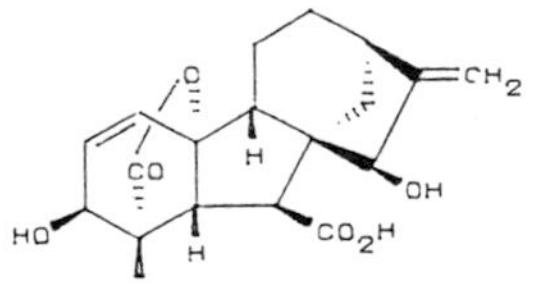

GA_{68}

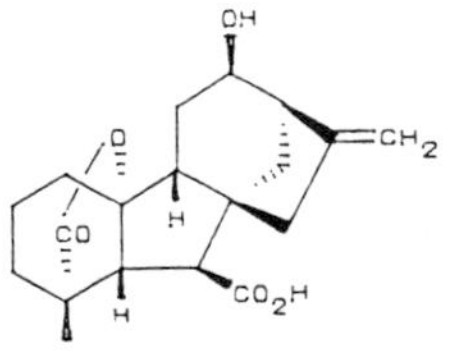

GA_{69}

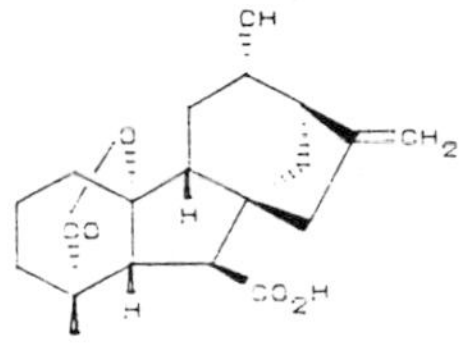

GA_{70}

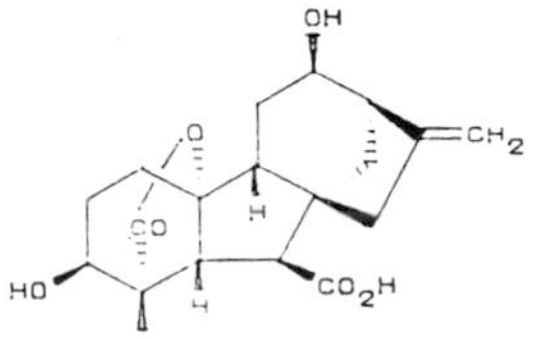

GA_{71}

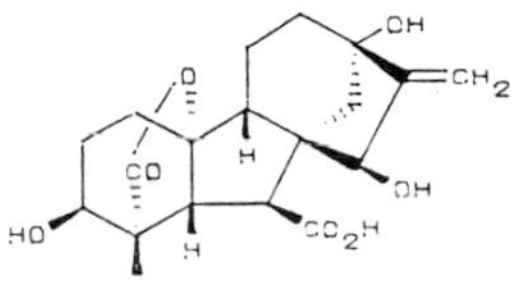

GA_{72}

B3. Cytokinin Biosynthesis and Metabolism

Brian A. McGaw*

Department of Botany and Microbiology, The University College of
Wales, Aberystwyth, Dyfed SY23 3DA, UK.

INTRODUCTION

The view that plant cell division is chemically controlled is not new,
indeed it can be traced back to the last century, but it was Haberlandt in
1913 (20) who provided the first experimental evidence for this concept.
He showed that phloem diffusates could stimulate parenchymatous potato
tuber cells to revert to a meristematic state. The identity of the active
substance/s in these diffusates was not established.

In the late 1940s and 1950s Folke Skoog and co-workers began a series
of investigations into the nutritional requirements of tissue cultures
derived from tobacco stem pith. On defined media in the presence of auxin
the pith tissue cells enlarged, but failed to divide. 'Normal' division was
restored, however, on the addition of several complex and undefined materials, most notably: coconut milk, vascular tissue extracts, autoclaved
DNA and yeast extracts (39). The conclusive identification of the first
active cell division promotor (or cytokinin as they are now known) was
acheived in 1959 (38) when 6-(furfurylamino) purine (Fig.1) was purified
from autoclaved herring sperm DNA. This compound, though one of the
most biologically active cytokinins (with activity recorded at levels of
10^{-9}M in tobacco callus growth bioassays (50)) is an artefactual
rearrangement product of heated DNA and is not found in plant tissues.
The first naturally occuring cytokinin was purified in 1963 by Letham
(26) from immature kernels of *Zea mays* and identified as 6-(4-hydroxy-3-
methylbut-*trans*-2-enylamino) purine more commonly known as zeatin.
Virtually all the naturally occuring cytokinins appear to be purine
derivatives (Fig.2) with a branched 5 carbon N^6 substituent (though this

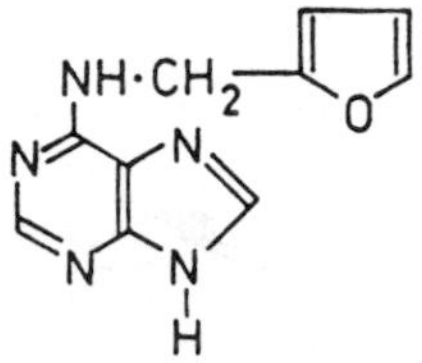

Fig. 1. 6-(furfurylamino) purine or kinetin

---

*Present address: The Rowett Research Institute, Greenburn Road, Bucksburn, Aberdeen
AB2 9SB, Scotland, UK.

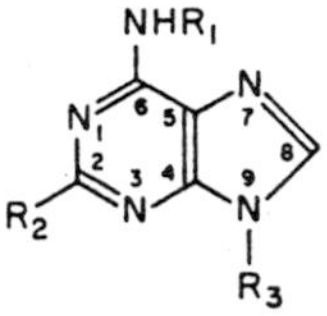

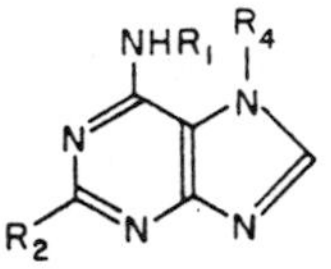

R$_1$	R$_2$	R$_3$	R$_4$	Trivial name	Abbreviation
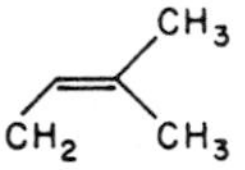 (isopentenyl: CH$_3$, CH$_2$, CH$_3$)	H	H	-	N^6(Δ^2-isopentenyl) adenine	iP
	H	ribosyl	-	N^6(Δ^2-isopentenyl) adenosine	[9R] iP
	CH$_3$S	ribosyl	-	2 methylthio N^6 (Δ^2-iso-pentenyl) adenosine	[2MeS9R] iP
	H	ribotide	-	N^6(Δ^2-isopentenyl) adeno-sine-5'- monophosphate	[9R-5'P] iP
			glucosyl	N^6 (Δ^2-isopentenyl) adenine-7-glucoside	[7G]iP
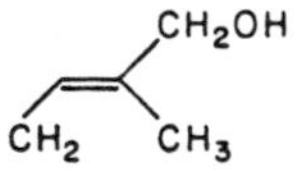 (trans-zeatin: CH$_2$OH, CH$_2$, CH$_3$)	H	H	-	*trans*-zeatin	Z
	H	ribosyl	-	*t*-zeatin riboside	[9R] Z
	H	glucosyl	-	*t*-zeatin-9-glucoside	[9G] Z
	H	-	glucosyl	*t*-zeatin-7-glucoside	[7G]Z
	H	alanyl	-	lupinic acid	[9Ala] Z
	H	ribotide	-	*t*-zeatin riboside-5'-monophosphate	[9R-5'P] Z
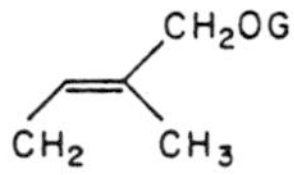 (CH$_2$OG, CH$_2$, CH$_3$)	H	H	-	zeatin-O-glucoside	(OG) Z
	H	ribosyl	-	zeatin riboside-O-glucoside	(OG)[9R] Z
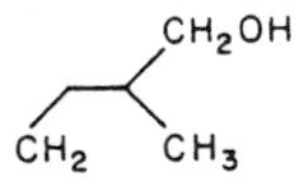 (dihydrozeatin: CH$_2$OH, CH$_2$, CH$_3$)	H	H	-	dihydrozeatin	(diH) Z
	H	ribosyl	-	dihydrozeatin riboside	(diH)[9R] Z
	H	glucosyl	-	dihydrozeatin-9-glucoside	(diH)[9G] Z
	H	-	glucosyl	dihydrozeatin-7-glucoside	(diH)[7G]Z
	H	alanyl	-	dihydrolupinic acid	(diH)[9Ala] Z
	H	ribotide	-	dihydrozeatin riboside-5'-monophosphate	(diH)[9R-5'P]Z
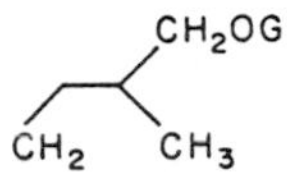 (CH$_2$OG, CH$_2$, CH$_3$)	H	H	-	dihydrozeatin-O-glucoside	(diHOG) Z
	H	ribosyl	-	dihydrozeatin riboside-O-glucoside	(diHOG)[9R]Z
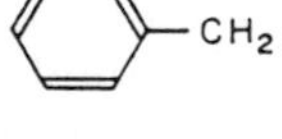 (benzyl: C$_6$H$_5$-CH$_2$)	H	H	-	N^6 (benzyl) adenine	BAP
	H	ribosyl	-	N^6 (benzyl) adenosine	[9R] BAP

Fig. 2. Free and tRNA cytokinin structures, nomenclature and abbreviations. *cis*--[9R]Z is a common cytokinin riboside in tRNA in which the -CH$_2$OH and -CH$_3$ in R$_1$ are reversed. Glucosyl (G) and ribosyl (R) refer to the β-D-glucopyranosyl and β-D-ribofuranosyl group.

may be lengthened by conjugation).

Cytokinins have been defined (56) "as substances which, in combination with auxin, stimulate cell division in plants and which interact with auxin in determining the direction which differentiation of cells takes". The term is now loosely used to cover all purine compounds that possess the necessary 5 carbon N^6 substituent; regardless of whether they exhibit cytokinin activity. Thus the biologically inactive (28) conjugates (i.e., the 7- and 9-β-D-glucosyl and 9-alanyl conjugates of zeatin and dihydrozeatin) are referred to as 'cytokinins' though they are probably the products of deactivation or control mechanisms designed to regulate levels of cytokinin activity in the plant (37).

There are now many bioassays available for the estimation of cytokinin activity. Some of these (i.e., tobacco pith callus and soybean callus growth, radish cotyledon expansion etc.) are directly related to the role of cytokinins in cell division, while other bioassay systems exploit their involvement in more specific metabolic processes (i.e., β-cyanin synthesis in *Amaranthus* seedlings, chlorophyll retention in oat leaves etc.). The structure/activity relationships of the various synthetic and naturally occuring cytokinins has been extensively studied in a number of different bioassay systems (28,31). There is broad agreement, in these systems, on the activity of the cytokinin bases and their various conjugates. Whilst iP, Z, (diH)Z, (see Fig. 2 for abbreviations), benzyladenine (BAP) and their 9-ribosyl (and in the case of Z and (diH)Z their O-glucosyl derivatives), are generally very active, cytokinin activity is markedly reduced in the 7- and 9-glucosyl and 9-alanyl conjugates.

OCCURRENCE, NOMENCLATURE AND MODES OF CYTOKININ ACTION

Figure 2 shows the structures of the major naturally occuring cytokinins, giving the trivial names for these compounds and a list of abbreviations. There are at least three currently used abbreviation systems, but for simplicity that proposed by Letham and Palni (27) (see Fig.2) should perhaps be generally adopted.

Several of the cytokinins in Fig.2 occur as components of tRNA, namely: [9R]iP, *cis*- and *trans*-[9R]Z, [2MeS 9R]iP, *cis*- and *trans*-[2MeS 9R]Z. The biologically active 2-methylthio compounds are exclusive to tRNA in higher plants, but [9R]iP, *trans*-[9R]Z and, occasionally the biologically inactive *cis*-[9R]Z have also been identified as free compounds (27). These cytokinins constitute only a small proportion of the 30 or so unusual bases that are known to occur in tRNA (30), but unlike the simple methylated bases (i.e., 6-methyl adenosine, 2-methyl adenosine etc.) which are found in several regions of the tRNA, they appear to be exclusively located adjacent to the 3' end of the anti-codon (30). Recent work has shown that these cytokinins always occur as the middle residue of a triple A sequence and in tRNA species corresponding to codons with the initial letter U (i.e., isoacceptors for Cys, Leu, Phe, Ser, Trp and Tyr) (30).

Since cytokinins (and indeed other classes of plant growth regulator) appear to be important in so many aspects of plant growth and development (23,51) a role in plant primary metabolic processes would go a long way towards explaining their modes of action. The occurrence of cytokinins in tRNA has inevitably led to the hypothysis that these compounds are involved in the control of protein biosynthesis. That these cytokinins increase the binding affinity of the aminoacyl tRNA to the ribosomes and facilitate codon recognition is a commonly cited mechanistic explanation of their mode of action.

Unfortunately this hypothysis fails to explain certain points. First, most of the biologically active naturally occuring cytokinins do not occur as constituents of tRNA. Second, when the highly active cytokinin benzyladenine (BAP) was externally applied to tobacco callus tissue its incorporation into tRNA was extremely low and non-specific in nature (i.e., not confined to the 3′ end of the anti-codon)(2). Third, the tRNA of cytokinin-requiring tobacco callus grown on BAP contains the usual complement of naturally occuring cytokinins (7).

The tRNA cytokinins may be of great importance in protein synthesis at the translational level, but it is clear that free cytokinin activity is not mediated *via* tRNA. There is strong evidence that externally applied cytokinins *do* stimulate protein synthesis in plants. The mechanism by which this occurs is not fully understood, but there are several possibilities. Firstly, cytokinins appear to cause an increased rate of RNA synthesis (including tRNA, rRNA and mRNA) perhaps by activation of chromatin-bound RNA polymerase (24). Secondly, cytokinins may act at the post-transcriptional level by the stimulation of polysome formation and/or the activation of polysomes in such a manner that increased recruitment of untranslated mRNA occurs. However, despite much work in these areas we remain a long way from understanding the exact mode of cytokinin action.

CYTOKININ ANALYSIS

The study of cytokinin biochemistry has benefited immensely from the development of modern analytical techniques (especially high-performance liquid chromatography and gas chromatography/mass spectrometry). For a detailed discussion of these aspects the reader is referred to Chapter D1 on instrumental methods in this volume.

Unfortunately many workers continue to rely on bioassay and co-chromatography for the quantitation and identification of cytokinins. It is

Table 1. Cytokinin levels in selected species and organs (expressed in n moles 100 g. fwt^{-1}).

Cytokinin	V. rosea Crown gall	Tobacco Crown gall	Radish Seed	L. Luteus Develop- ing Seed	L. Luteus Mature Seed	L. angusti- folius Seed	L. angusti- folius pod wall
Z	6.0	3.4	0.0	8.4	6.2		
[9R]Z	135.0	13.0	0.0	112.0	10.8	68.4	0.0
(diH)Z		0.7		188.7	0.01		
(diH)[9R]Z		2.2			42.5	56.7	45.3
[7G]Z		20.3	178.5				
[9G]Z	37.0	0.0	0.0				
(OG)Z	13.0	1.6		0.5	0.5	1.8	1.6
(OG)[9R]Z	64.0	2.4		10.5	5.9	7.2	7.6
(diHOG)Z				4.6	1.1	4.4	21.2
(diHOG)[9R]Z				76.1	39.1	71.8	213.6
[9R-5'P]Z		78.9					
(diH)[9R-5'P]Z		3.7					

over twenty years since the first isolation of zeatin from corn kernels (26) yet this compound has only been conclusively identified in a few plant species. There are less than ten plant tissues for which the application of rigorous analytical techniques have been applied to the construction of a total quantitative and qualitative picture (37). The data for some of these tissues are given in table 1. These data are essential to our understanding of the roles these compounds play in plant development. Coupled with an understanding of the sites of cytokinin action and the systems designed to control the levels or expression of cytokinin activity (i.e., biosynthesis, metabolism, transport and compartmentation) we can go some way towards ascribing functions to the different cytokinin structures. Unfortunately we know very little about the sites of cytokinin action and their cellular compartmentation. There is, however, a considerable amount of data relating to the biosynthesis and metabolism of these compounds in plant tissues and an attempt is made, in the remainder of this chapter, to relate these data to our understanding of the control of cytokinin activity and the possible roles of the various cytokinin structures.

CYTOKININ BIOSYNTHESIS

The Biosynthesis of tRNA Cytokinins

The biosynthesis of tRNA cytokinins and indeed other hypermodified tRNA bases (i.e., the methylated purines) are known to occur at the polymer level during post-transcriptional processing (21). The branched 5 carbon N^6 substituent of these cytokinins is derived from mevalonic acid pyrophosphate (10) which undergoes decarboxylation, dehydration and isomerisation to give Δ^2-isopentenyl pyrophosphate (iPP)(Fig.3). The latter then condenses with the relevant adenosine residue in the tRNA to give the [9R]iP moiety. It is not known whether the hydroxylation of the terminal methyl group (usually the *cis*-, but sometimes the *trans*-methyl) occurs in the Δ^2-isopentenyl pyrophosphate or in the [9R]iP residues of the tRNA polymer.

Consistent with the above model a Δ^2-iPP:tRNA-Δ^2-isopentenyl transferase has been partially purified from *Escherichia coli* which can utilise tRNA, but not oligoadenylic acids, adenosine-5'-monophosphate (AMP) or adenosine as substrates (4). On the other hand Holtz and Klämbt (22) have purified a more catholic enzyme from *Zea mays* which was able to isopentenylate tRNA, oligo and polyadenylic acids and adenosine. This work opens the interesting possibility that one enzyme may be responsible for the formation of both free and tRNA [9R]iP (subsequent modification of this cytokinin being dependent on enzymes which can discriminate between free and tRNA cytokinins).

The biosynthesis of the 2-methylthio derivatives has been extensively studied in *E. coli* and occurs at the polymer level after isopentenylation. Thiolation (cysteine being the source of the sulphur atom (18)) is followed by methylation (1). S-Adenosyl methionine is the donor of the methyl group in these compounds and also the hypermodified methylated purines and pyrimidines found in tRNA (18).

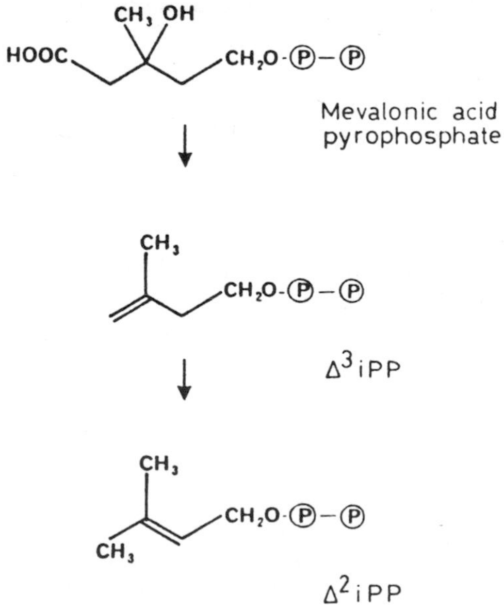

Fig. 3. The biosynthesis of Δ^2-isopentenyl pryophosphate.

The Biosynthesis of the Free Cytokinins

Via tRNA

Since tRNA contains cytokinins, biosynthesis *via* the hydrolysis of tRNA to its constituent mononucleotides is a possibility. In certain circumstances (i.e., *Agrobacterium tumefaciens* and *Corynebacterium fascians*, the causative organisms of crown gall and fasciation diseases of plants) the cytokinins present as constituent bases within tRNA (2-methylthio cytokinins, *cis*-[9R]Z, [9R]iP and indeed the hypermodified methylated purines) are found as free compounds (40,41). It is possible that tRNA hydrolysis may account for all the free cytokinins in these bacteria. In plant tissues, however, with the exception of [9R]iP and to a lesser extent *cis*- and *trans*-[9R]Z, which are common to both, the free and tRNA cytokinins are structurally distinct (e.g. free Z is mainly the *trans* isomer while Z present in tRNA is mainly the *cis* isomer). This remains one of the principal objections to the view that tRNA hydrolysis contributes to the pool of free cytokinins in plants. There is also further evidence against this view: first, the tRNA of cytokinin-requiring plant tissue cultures contains cytokinins (10), and secondly, tRNA 'turnover' rate is apparently not rapid enough in certain non-cytokinin-requiring tissues to account for the levels of endogenous cytokinins (21). In addition, the rate of incorporation of externally applied ^{14}C-adenine into free cytokinins is too high to be accounted for by tRNA 'turnover' (52).

Counter to these arguments are the possibilities that there is selective 'turnover' of tRNA subpopulations rich in [9R]iP or *trans*-[9R]Z (there is evidence of this in animal tumour tissues (5)) or that a *cis-trans* isomerase system exists which can convert tRNA derived *cis*-[9R]Z into its *trans* isomer. In studies on the 'turnover' of labelled RNA, and the kinetics of its conversion to labelled cytokinins, it has been claimed (32) that cytokinin biosynthesis can *only* be accounted for in terms of an indirect pathway. This view is based, however, on gross tRNA and oligonucleotide turnover. Since the levels of cytokinins made available by this turnover is not known it is not possible to assess its contribution to the free cytokinin pool (another difficult parameter to measure when metabolism and other biosynthetic routes are considered). Though most workers in this field would accept the possibility that tRNA hydrolysis may contribute to the free cytokinin pool there are few who would support the view that this is the principal or sole route by which free cytokinins in plants are derived.

The situation is further confused by recent studies on cytokinin biosynthesis in crown gall tumour tissues. Crown gall is a neoplastic disease of dicotyledonous plants caused by the incorporation of a section of DNA (known as T-DNA) from the tumour inducing or Ti plasmid of the causative organism *Agrobacterium tumefaciens* into the DNA of the host cell. These tumour tissues have greatly enhanced cytokinin levels (49) and have been used by several laboratories as model systems for the study of cytokinin biosynthesis and metabolism. The free cytokinins in these tissues (being mainly *trans*-Z derivatives) have little in common with those in the bacterial tRNA or culture media where the Z compounds are all *cis* isomers. This indicated that the T-DNA genes are involved in the *de novo* (direct) synthesis of cytokinins (probably of [9R-5′P]iP—see next section below—which is then stereospecifically hydroxylated by the plant to give *trans* zeatin compounds). The bacterial enzymes responsible for the synthesis of *cis* zeatin derivatives (either by tRNA hydrolysis or stereospecific *cis* hydroxylation of free [9R-5′P]iP) are presumably not coded for by any of the genes transferred on the T-DNA. However, the most abundant cytokinin in the tRNA of *Vinca rosea* crown gall tissue is *trans*-[9R]Z (44), while *cis*-[9R]Z and its 2-methylthio derivatives are the predominant cytokinins in the tRNA of untransformed *Vinca* callus and the cultured bacteria, respectively. This is circumstantial evidence for an indirect pathway via tRNA, but new and important evidence on the function of the genes in the T-DNA has conclusively demonstrated that one of the genes (gene 4) codes for a Δ^2-isopentenyl pyrophosphate : AMP-Δ^2-isopentenyl transferase (3) (for the biosynthesis of [9R-5′P]iP see next section below). Therefore, even in these tumour tissues, which provided perhaps the best evidence for a tRNA turnover pathway there seems little doubt that their enhanced cytokinin levels are due to the insertion of a *de novo* synthesising capability on the T-DNA.

De Novo (Direct) Cytokinin Biosynthesis

The above transferase enzyme from the T-DNA of *A. tumefaciens* is not the first enzyme to be investigated that could be responsible for *de novo* biosynthesis. Indeed an enzyme has been characterised in cell-free preparations of the slime mould *Dictyostelium discoideum* that catalyses the synthesis [9R-5′P]iP from AMP and Δ^2-iPP (54). More recently an AMP-Δ^2-isopentenyl transferase has been partially purified from cytokinin-autonomous tobacco callus tissue (8) and its substrate requirements have been studied in detail. The reaction does not occur at the base or nucleoside level (i.e., adenine or adenosine are not substrates) and the side chain must be the correct isopentenyl isomer and contain a pyrophosphate group (8)(Fig.4). The enzyme has not been studied with tRNA, oligoadenylic acids, polyadenylic acids or other potential polymeric substrates.

If this enzyme is responsible for the first step in cytokinin biosynthesis then the product of this reaction, [9R-5′P]iP, must be stereospecifically hydroxylated to form the *trans*-zeatin derivatives which predominate in plant tissues. This hydroxylation must occur with some rapidity since [9R-5′P]iP, [9R]iP and iP are rarely found as free compounds in plants. Similarily, when ^{14}C-adenine was fed to *Vinca rosea* crown gall tissue labelled zeatin derivatives were readily recovered, but no radioactivity was detected in HPLC fractions corresponding to the elution volume of the

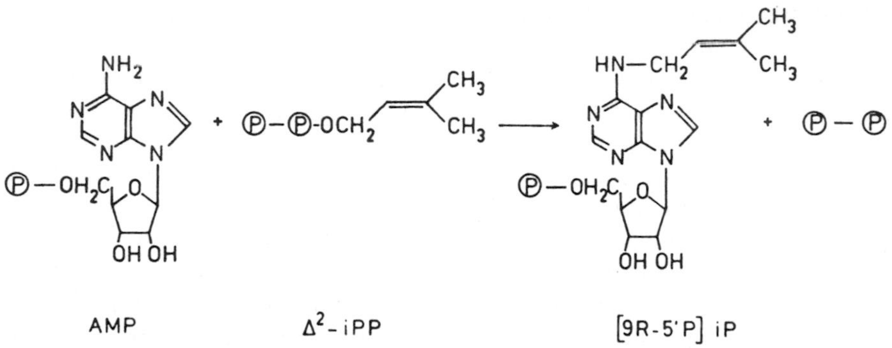

$$AMP \qquad \qquad \Delta^2\text{-iPP} \qquad \qquad [9R\text{-}5'P]\ iP$$

Fig. 4. *De novo* cytokinin biosynthesis.

iP cytokinins (46,52). There is good evidence that the hydroxylation reaction occurs at the nucleotide level. Firstly, following ^{14}C-adenine application to *V. rosea* crown gall tissue the peak of radioactivity in AMP was found to preceed that of [9R-5′P]Z and the amount of radioactivity in the [9R-5′P]Z always exceeded that of the [9R]Z or Z (52). Secondly, ^{14}C-iP has been supplied to several tissues and is stereospecifically *trans* hydroxylated (45). Most of the radioactivity in the cytokinin fractions was associated with the nucleotide [9R-5′P]Z. Very little iP or [9R-5′P]iP was recovered.

The results of these feeding experiments and the enzyme/cell free studies indicate that a *de novo* biosynthetic route is a strong possibliity in plant tissues. However, there are great difficulties in studying the biosynthesis of compounds that occur at extremely low levels whose putative precursors (i.e., AMP and Δ^2-iPP) are major products of plant metabolism. Very careful consideration should be exercised in interpreting these results. For instance, the 'high' incorporations of adenine into the cytokinins is taken as circumstantial evidence for a *de novo* route, but it is possible that aberrant synthesis is occuring as a result of exogenous application. Second, it is not possible to accurately measure specific activities of cytokinin metabolites that are probably not pure (i.e., co-chromatography of radioactivity with a cytokinin metabolite on HPLC is no guarantee that we are dealing with a radioactive cytokinin). In our experience quite high levels of 'incorporation' of ^{14}C-adenine into cytokinins can be observed in crude preparative HPLC runs. On further purification with different HPLC systems most of the radioactivity could be separated from the cytokinin fractions. We were unable to purify to constant specific activity. Future work in this area must attempt to adopt those standards set by biosynthetic studies in other areas of natural product chemistry. Convincing evidence for either biosynthetic route will not emerge until this approach is adopted.

CYTOKININ METABOLISM

The formation of the free cytokinins is presumed to begin with the biosynthesis of [9R-5'P]iP. This compound can be considered the precursor of the remaining twenty or so free cytokinins illustrated in Fig.2 and quantified, for certain tissues, in Table 1. As already discussed it appears that the [9R-5'P]iP is rapidly and stereospecifically hydroxylated to give the zeatin derivatives. This hydroxylation, therefore, is the first important metabolic event. From this point on many metabolic reactions occur with the consequent generation of aglycones, glucosides, ribosides, ribotides, amino acid conjugates, reduction and oxidation products (Fig.5). These metabolic events can be categorised under four broad headings, namely: conjugation, hydrolysis, reduction and oxidation. The occurrence of these events is discussed in relation to our knowledge of their enzymology and an attempt is made to assign a role to the various metabolites in the regulation and expression of cytokinin activity.

Conjugation

The various cytokinin conjugates are considered separately.

Ribosides and Ribotides

The cytokinin ribosides and their 5' mono-, di- and tri-phosphates are probably the most abundant naturally occuring cytokinins (28,37,49). Ribosyl conjugation is always confined to the 9 position of the purine ring. When ^{14}C labelled zeatin is externally applied [9R]Z and [9R-5'P]Z are usually important metabolic products (37). Where time course studies have been performed it has been found that the nucleotides are the dominant metabolites in the early stages following the feed and that rapid hydrolysis of these compounds then occurs to give more 'stable' or less biologically active products (i.e., N-glucosides or products of oxidative side chain cleavage). The metabolism of externally applied [9R]Z and [9R]iP have also been studied and again phosphorylation was followed by hydrolysis and further metabolism (25,29).

The enzymes responsible for the interconversions of the cytokinin bases, ribosides and ribotides have been studied in detail by Chen and co-workers using preparations from wheat germ. The following enzymes have been characterised: an adenosine phosphorylase which converts iP to [9R]iP (13), an adenosine kinase which converts [9R]iP to [9R-5'P]iP (9) and an adenosine phosphoribosyltransferase (9) which converts iP to its nucleotide directly. In all these cases the enzymes exhibited lower affinities for the N^6 substituted substrates ([9R]iP and iP) than for adenine or adenosine.

The ribosides and their aglycones (i.e., Z and iP) are extremely active in bioassay (28). Whether or not one or all of these compounds are the

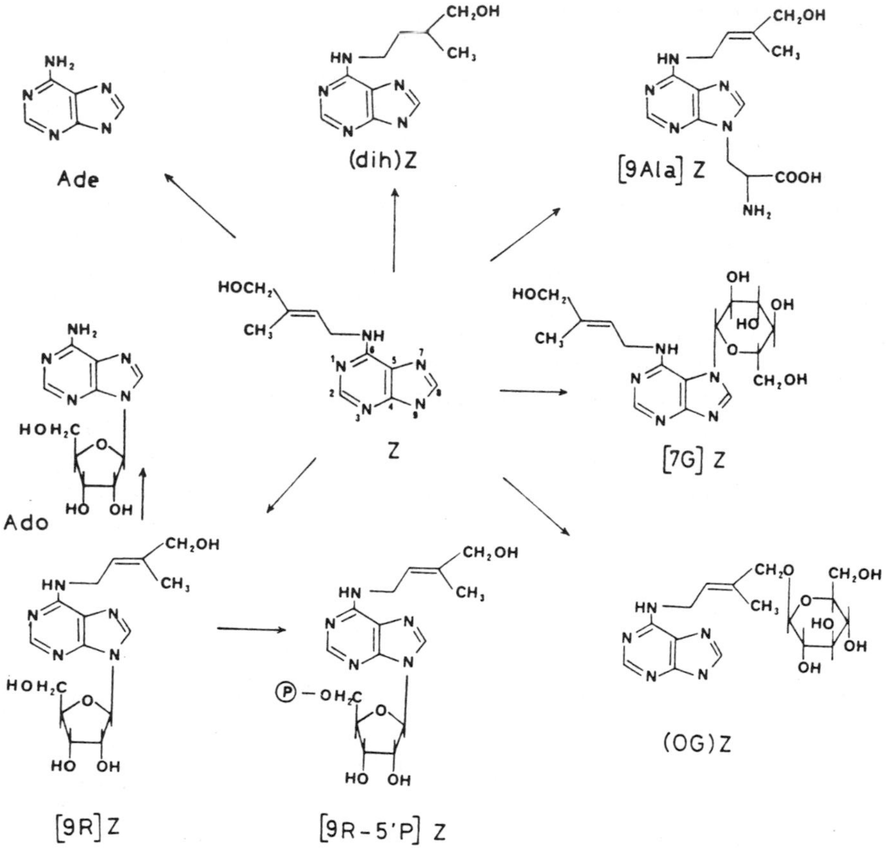

Fig. 5. Various metabolites of zeatin (Z). Ade and Ado refer to adenine and adenosine, respectively.

'active' molecules is not known because of the extensive interconversions that they undergo when externally supplied. The same applies to the readily hydrolysable nucleotides. The rapid formation of nucleotides following cytokinin application to plant tissues and their equally rapid hydrolysis indicates that they may be associated with uptake.

Glucosides

Unlike the ribosyl conjugates glucosylation is not confined to the 9 position of the purine ring (Fig.2). The 7- and 9-glucosides and side chain O-glucosides are major cytokinins in certain tissues (Table 1) and 3-glucosides (29,33) have also been detected as metabolites of externally applied cytokinins. Nothing is known about the enzymology of O-glucoside synthesis, but N-glucosylation has been studied in great detail in radish tissues (where [7G]Z is the predominant free cytokinin). Two glucosyltransferases, which separated on DEAE columns, have been partially purified from radish cotyledons (14). Both enzymes catalyse the formation of 7- and 9-glucosides of benzyladenine using UDPG or TDPG as the glucose source. However, the ratio of there products ([7G]BA/[9G]BA) was very different, being approximately 10 and 1.5. The latter enzyme has now been studied in relation to a large number of different naturally occuring and synthetic cytokinins (15). In all cases the rate of glucosylation and the [7G]/[9G] prouct ratio were determined. Interestingly, some cytokinins (e.g., cis-Z and trans-Z and BA) gave appreciable quantities of 9-glucosides, but others (e.g., (diH)Z and (OG)Z) gave only traces of 9-glucosyl products. There was little difference in the overall rates of N-glucosylation for these cytokinins.

In radish tissues N-glucosylation is the predominant fate of externally applied cytokinin. However, in contrast to the cell free work, [9G]Z was not identified as a metabolite when labelled zeatin was externally applied (36) and (diH)Z gave considerable quantities of its 3- and 9-glucosides. It is possible that other, as yet unidentified, enzyme systems are also involved in the formation of cytokinin glucosides.

N-Glucosyl conjugation is considered to be important in the regulation of cytokinin activity levels. The 7- and 9-glucosides are biologically inactive (28) and are extremely stable (i.e., are not hydrolysed to their active aglycones) in the tissues in which they are formed (47). Some plant tissues do not appear to form N-glucosides (either as endogenous compounds or on exogenous application of other cytokinins) and in these cases other methods of inactivation are adopted (i.e., oxidative side chain cleavage and/or the formation of amino acid conjugates).

In contrast the O-glucosides are biologically very active compounds. This may be due to the action of β-glucosidase enzymes and indeed several metabolic studies (36,37) with labelled (OG)Z indicate that it is readily hydrolysed to its aglycone. However, in several other studies it has been noted that exogenous application of labelled zeatin can lead to the formation of large amounts of O-glucosyl derivatives which remain unmetabolised over long periods (48). These findings have lead to the proposal that O-glucosides may be cytokinin storage forms, being stable and yet readily metabolised under certain conditions to yield biologically active cytokinins when required. Further evidence for this view has been found in work on the endogenous cytokinins of Phaseolus vulgaris during different stages of development (42). Decapitation of these plants leads to a rapid rise in the amount of (diHOG)Z in their leaves. On lateral bud development the levels of this compound fall dramatically.

Amino Acid Conjugates

[9Ala]Z (lupinic acid) and (diH)[9Ala]Z (dihydrolupinic acid) are minor endogenous cytokinins in the immature pod walls and root nodules of *Lupinus luteus* (53). The 9-alanyl conjugates of zeatin and benzyladenine have beem identified when these cytokinins were exogenously supplied to *Lupinus* spp. (48), immature apple seeds and derooted *Phaseolus* seedlings (27).

An enzyme, a β-(6-allylaminopurine-9-yl) adenine synthase, has been characterised (16) from developing seeds of *L. luteus* This enzyme utilises O-acetyl serine as the donor of the alanine residue, but is capable of conjugating a large number of different purine substrates (though the presence of an N^6 substituent greatly increases the rate of conjugation).

The role of these conjugates is probably similar to that of the N-glucosyl cytokinins. They are biologically inactive (28) and extremely stable compounds (48). As with glucosylation, the formation of amino acid conjugates is a common response of plant tissues to xenobiotic material and presumably, by rendering them more water soluble, facilitates their deposition in the vacuolé.

Hydrolysis

The hydrolysis of cytokinin ribosides and ribotides is a major component in the metabolism of externally applied cytokinins, especally in the early stages of experiments associated with uptake. Two 5'-ribonucleotidase enzymes (differing in molecular weight) which catalyse the hydrolysis of [9R-5'P]iP to [9R]iP (11) and an adenine nucleosidase (12) which converts [9R]iP to iP have been characterised in wheat germ. Unlike the conjugating enzymes from wheat germ, N^6 substitution made little difference to the enzyme affinity (i.e., AMP or adenosine were equally good substrates as [9R-5'P]iP or [9R]iP).

The N-glucoside and N-alanyl conjugates are extremely stable (47,48) in the tissues in which they are synthesised and therefore no enzymes capable of hydrolysing these compounds have been investigated. In certain circumstances, however, the O-glucosides are readily hydrolysed. Almond β-glucosidase (emulsin) preparations can cleave the O-glucosyl group in these compounds (though it is not capable of hydrolysing N-glucosyl conjugates), but this important enzyme has not been studied in any of the tissues where the endogenous cytokinins have been accurately quantified or their metabolism studied. Since the O-glucosides may occupy a key role as cytokinin storage forms this is an important area where future research could focus in an attempt to manipulate plant growth/development by controlling the size of the active cytokinin pool.

Reduction

Dihydrozeatin derivatives are commonly found in plant tissues and are frequent metabolites of applied zeatin (usually conjugated as their 9-ribosides, 9-ribotides or O-glucosides) (27,37). In bioassay (diH)Z and its conjugates are equally as active as their zeatin analogues (28). In studies where (diH)Z has been externally supplied to plants it appears to be more 'stable' than zeatin (43). This may be because it is not a substrate for cytokinin oxidase (an enzyme which cleaves the N^6 side chain—see next section). As a more 'protected' species (diH)Z may be important in the maintainance of cytokinin activity levels in an oxidative environment. There are no reports of (diH)Z derivatives being metabolised to zeatin compounds and the enzymology of side chain reduction has not been studied.

Oxidation

Oxidative side chain cleavage of externally applied Z, iP, [9R]Z and [9R]iP to give adenine, adenosine and adenine nucleotides is the major fate of these cytokinins in many tissues (37) (Fig. 6). Like the formation of N-glucosyl or N-alanyl conjugates, side chain cleavage leads to the irreversible loss of cytokinin activity and may be important in the regulation of cytokinin activity levels. An enzyme, cytokinin oxidase, has been partially purified from tobacco tissue (57), corn kernels and *Vinca rosea* crown gall tissue (35). In the latter tissues the specificity of the enzymes has been investigated with a large number of naturally occuring and synthetic cytokinins. Z, [9R]Z, iP, [9R]iP, [7G]Z, [9G]Z and [9Ala]Z all served as substrates for the enzymes, but side chain reduction (i.e., (diH)Z derivatives), the relocation of the Δ^2 double bond to Δ^3, the substitution of other functionalities (i.e., benzyladenine, kinetin) and the presence of an O-glucosyl group rendered the cytokinin resistant to oxidation. Some tissues (notably radish) seem to be unable to cleave the N^6 side chain. In these cases N-glucosylation is the mechanism by which cytokinin activity is controlled. In other tissues, for instance the *Lupinus* spp., both methods of inactivation are utilised (37).

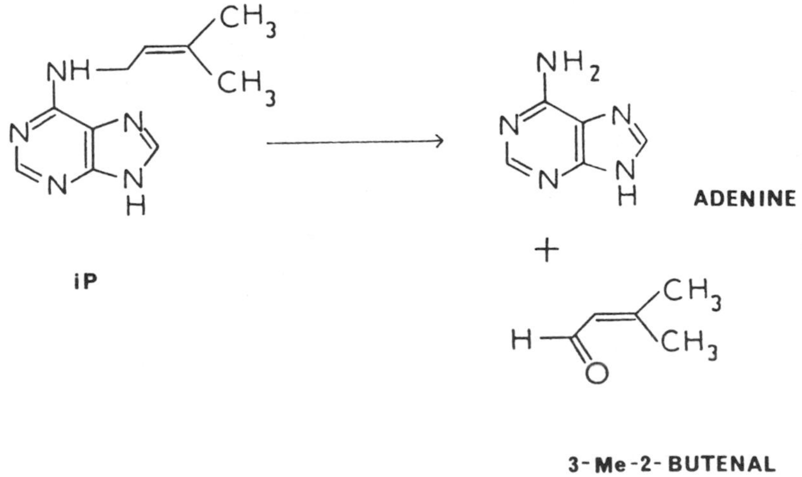

Fig. 6. The products of cytokinin oxidase action on iP.

The mechanism of the oxidative cleavage is not fully understood, but 3-methyl-2-butenal (6) and adenine (34) have been unambiguously identified by GC/MS as the products of cytokinin oxidase action on iP (Fig. 6). It has been postulated (57) that an unstable imine intermediate is formed and this appears to have been confirmed by physico-chemical studies (24a).

CONCLUSION

The study of cytokinin biochemistry, thanks to the advancement of modern analytical techniques, is poised for a period of rapid development. The purification and identification of cytokinin bases, ribosides and nucleotides is now a relatively straightforward matter. With the development of selected ion monitoring/isotope dilution mass spectrometric quantitation and radio-immunoassay we are now able to accurately quantify extremely low levels of cytokinins. For the first time, therefore, it is possible to study the endogenous metabolism of cytokinins without recourse to the artificial application of labelled material. It is hoped that these developments will eventually lead to an understanding of the mechanism and site/s of cytokinin action. If the quantitation of cytokinins at the cellular level can also be achieved we should then be in a position to explain how endogenous cytokinin activity levels are controlled.

References

1. Agris, P.F., Armstrong, D.J., Schäfer, K.P., Söll, D.(1975) Maturation of a hypermodified nucleoside in transfer RNA. Nucleic Acid Res. 2, 691-698.
2. Armstrong, D.J., Murai, N., Taller, B.J., Skoog, F.(1976) Incorporation of cytokinin N^6-benzyladenine into tobacco callus transfer ribonucleic acid and ribosomal ribonucleic acid preparations. Plant Physiol. 57, 15-22.
3. Barry, G.F., Rogen, S.G., Fraley, R.T., Brand, L.(1984) Identification of a cloned cytokinin biosynthetic gene. Proc. Natl. Acad. Sci. USA 81, 4776-4780.
4. Bartz, J.K., Söll, D.(1972) N^6-Δ^2-(isopentenyl) adenosine: biosynthesis in vitro in transfer RNA by an enzyme purified from Escherichia coli. Biochemie 54, 31-39.
5. Borek, E., Baliga, B.S., Gehrke, C.W., Kuo, G.W., Belman, S., Troll, W., Waalkes, T.P.(1977) High turnover rate of transfer RNA in tumor tissue. Cancer Res. 37, 3362-3366.
6. Brownlee, B.G., Hall, R.H., Whitty, C.D.(1975) 3-Methyl-2-butenal: an enzymatic product of the cytokinin, N^6-(Δ^2-isopentenyl) adenine. Can. J. Biochem. 53, 37-41.
7. Burrows, W.J., Skoog, F., Leonard, N.J.(1971) Isolation and identification of cytokinins located in the transfer ribonucleic acid of tobacco callus grown in the presence of 6-benzylaminopurine. Biochemistry 10, 2189-2194.
8. Chen, C-M.(1982) Cytokinin biosynthesis in cell-free systems. In: Plant growth substances 1982, pp. 155-164, Wareing, P.F., ed. Academic Press, London New York.
9. Chen, C-M., Eckert, R.L.(1977) Phosphorylation of cytokinin by adenosine kinase from wheat germ. Plant Physiol. 59, 443-447.

10. Chen, C-M., Hall, R.H.(1969) Biosynthesis of N^6-Δ^2-(isopentenyl) adenosine in the transfer ribonucleic acid of cultured tobacco pith tissue. Phytochemistry 8, 1687-1695.

11. Chen, C-M., Kristopeit, S.M.(1981) Metabolism of cytokinin: dephosphorylation of cytokinin ribonucleotide by 5'-nucleotidase from wheat germ cytosol. Plant Physiol. 67, 494-498.

12. Chen, C-M., Kristopeit, S.M.(1981) Metabolism of cytokinin: deribosylation of cytokinin ribonucleoside by adenine nucleosidase from wheat germ cells. Plant Physiol. 68, 1020-1023.

13. Chen, C-M., Petschow, B.(1978) Metabolism of cytokinin: ribosylation of cytokinin bases by adenine phosphorylase from wheat germ. Plant Physiol. 62, 871-874.

14. Entsch, B., Letham, D.S.(1979) Enzymatic glucosylation of the cytokinin, 6-benzylaminopurine. Plant Sci. Lett. 14, 205-212.

15. Entsch, B., Letham, D.S., Parker, C.W., Summons, R.E.(1979) Preparation and characterisation, using high performance liquid chromatography, of an enzyme forming glucosides of cytokinins. Biochim. Biophys. Acta 570, 124-139.

16. Entsch, B., Letham, D.S., Parker, C.W., Summons, R.E., Gollnow, B.E.(1979) Metabolism of cytokinins. In: Plant Growth Regulation 1979, pp. 109-115, Skoog, F., ed. Springer, Berlin Heidelberg New York.

17. Fittler, F., Hall, R.H.(1966) Selective modification of yeast seryl-tRNA and its effects on acceptance and binding functions. Biochem. Biophys. Res. Comm. 25, 441-446.

18. Geftner, M.L.(1969) The in vitro synthesis of 2'-O-methylguanosine and 2-methylthio N^6-(γ, γ-dimetylallyl) adenosine in transfer RNA of Escherichia coli. Biochem. Biophys. Res. Comm. 36, 435-441.

19. Geftner, M.L., Russell, R.L.(1969) Role of modifications in tyrosine transfer RNA: a modified base affecting ribosome binding. J. Mol. Biol. 39, 145-157.

20. Haberlandt, G.(1913) Zur physiologie der zellteilungen. Sitzungsber. K. Preuss. Akad. Wiss., 318-345.

21. Hall, R.H.(1973) Cytokinin as a probe of developmental processes. Ann. Rev. Plant Physiol. 24, 425-444.

22. Holtz, J., Klämbt, D.(1978) tRNA isopentenyltransferase from Zea mays L. Characterisation of the isopentenylation reaction of tRNA, oligo(A) and other nucleic acids. Hoppe-Seylers Z. Physiol. Chem. 359, 89-101.

23. Horgan, R.(1984) Cytokinins. In: Advanced Plant Physiology, pp. 90-116, Wilkins, M.B., ed. Pitman, London.

24. Kulaeva, O.N.(1981) Cytokinin action on transcription and translation in plants. In: Metabolism and Molecular Activities of Cytokinins, pp. 218-227, Guern, J., Péaud-Lenöel, C. eds. Springer, Berlin Heidelberg New York.

24a Laloue, M., Fox J.E. (1985) Characterisation of imine intermediate in the degradation of isopentenyl cytokinins by cytokinin oxidase from wheat. 12th Int. Conf. on Plant Growth Substances, Heidelberg. Abstracts p. 23

25. Laloue, M., Pethe-Terrine, C., Guern, J.(1981) Uptake and metabolism of cytokinins in tobacco cells: studies in relation to the expression of their biological activities. In: Metabolism and Molecular Activities of Cytokinins, pp. 80-96, Guern, J., Péaud-Lenöel, C. eds. Springer, Berlin Heidelberg New York.

26. Letham, D.S.(1963) Zeatin, a factor inducing cell division from Zea mays. Life Sci. 8, 569-573.

27. Letham, D.S., Palni, L.M.S.(1983) The biosynthesis and metabolism of cytokinins. Ann. Rev. Plant Physiol. 34, 163-197.

28. Letham, D.S., Palni, L.M.S., Tao, G.Q., Gollnow, B.I., Bates, C.M.(1983) Regulators of cell division in plant tissues. XXIX. The activities of cytokinin glucosides and alanine conjugates in cytokinin bioassay. J. Plant Growth Regulation 2, 103-115.

29. Letham, D.S., Tao, G.Q., Parker, C.W.(1982) An overview of cytokinin metabolism. In: Plant growth substances 1982, pp. 143-153, Wareing, P.F., ed. Academic Press, London.

30. Letham, D.S., Wettenhall, R.E.H.(1977) Transfer RNA and cytokinins. In: The Ribonucleic Acids, pp. 129-193, Stewart, P.R., Letham, D.S., eds. Springer, Berlin Heidelberg New York.

31. Matsubara, S.(1980) Structure-activity relationships of cytokinins. Phytochemistry 19, 2239-2253.

32. MaaBe, H., Klämbt, D.(1981) On the biogenesis of cytokinins in roots of Phaseolus vulgaris. Planta 151, 353-358.

33. McGaw, B.A., Heald, J.K., Horgan, R.(1984) Dihydrozeatin metabolism in radish seedlings. Phytochemistry 23, 1373-1377.

34. McGaw, B.A., Horgan, R.(1983) Cytokinin catabolism and cytokinin oxidase. Phytochemistry 22, 1103-1105.

35. McGaw, B.A., Horgan, R.(1983) Cytokinin oxidase from Zea mays kernels and Vinca rosea crown gall tissue. Planta 159, 30-37.

36. McGaw, B.A., Horgan, R., Heald, J.K.(1985) Cytokinin metabolism and the modification of cytokinin activity in radish. Phytochemistry 24, 9-13.

37. McGaw, B.A., Scott, I.M., Horgan, R.(1984) Cytokinin biosynthesis and metabolism. In: The Biosynthesis and Metabolism of Plant Hormones, pp. 105-133, Crozier, A., Hillman, J.R., eds. Cambridge University Press, Cambridge.

38. Miller, C.O., Skoog, F., Okomura, F.S., Saltza, M.H.von, Strong, F.M.(1956) Isolation, structure and synthesis of kinetin, a substance promoting cell division. J. Amer. Chem. Soc. 78, 1345-1350.

39. Miller, C.O., Skoog, F., Saltza, M.H.von, Strong, F.M.(1955) Kinetin, a cell division factor from deoxyribonucleic acid. J. Amer. Chem. Soc. 77, 1329-1334.

40. Morris, R.O., Regier, D.A., MacDonald, E.M.S. (1981) Analytical procedures for cytokinins: application to Agrobacterium tumefaciens. In:Metabolism and Molecular Activities of Cytokinins, pp. 3-16, Guern, J., Péaud-Lenöel, C., eds. Springer, Berlin Heidelberg New York.

41. Murai, N.(1981) Cytokinin biosynthesis and its relationship to the presence of plasmids in a strain of Corynebacterium fascians. In: Metabolism and Molecular Activities of Cytokinins, pp. 17-26, Guern, J., Péaud-Lenöel, C., eds. Springer, Berlin Heidelberg New York.

42. Palmer, M.V., Horgan, R., Wareing, P.F.(1981) Cytokinin metabolism in Phaseolus vulgaris L. I. Variation in cytokinin levels in leaves of decapitated plants in relation to bud outgrowth. J. Exp. Bot. 32, 1231-1241.

43. Palmer, M.V., Scott, I.M., Horgan, R.(1981) Cytokinin metabolism in Phaseolus vulgaris L. II. Comparative metabolism of exogenous cytokinins by detached leaves. Plant Sci. Lett. 22, 187-195.

44. Palni, L.M.S., Horgan, R.(1983) Cytokinins in transfer RNA of normal and crown-gall tissue of Vinca rosea. Planta 159, 178-181.

45. Palni, L.M.S., Horgan, R.(1983) Cytokinin biosynthesis in crown gall tissue of Vinca rosea: metabolism of isopentenyladenine. Phytochemistry 22, 1597-1601.

46. Palni, L.M.S., Horgan, R., Darrall, N.M., Stuchbury, T., Wareing, P.F.(1983) Cytokinin biosynthesis in crown gall tissue of Vinca rosea. Planta 159, 50-59.

47. Parker, C.W., Letham, D.S.(1973) Regulators of cell division in plant tissues. XVI. Metabolism of zeatin by radish cotyledons and hypocotyls. Planta 114, 199-218.

48. Parker, C.W., Letham, D.S., Gollnow, B.I., Summons, R.E., Duke, C.C., MacLeod, J.K.(1978) Regulators of cell division in plant tissues. XXV. Metabolism of zeatin in lupin seedlings. Planta 142, 239-251.

49. Scott, I.M., Horgan, R.(1984) Mass spectrometric quantification of cytokinin nucleotides and glycosides in tobacco crown gall tissue. Planta 161, 345-354.

50. Skoog, F., Hamzi, H.Q., Szweykowska, M., Leonard, N.J., Carraway, K.L., Fugii, T., Hegelson, J.P., Leoppky, R.W., (1967) Cytokinins: structure/activity relationships. Phytochemistry 6, 1169-1192.

51. Skoog, F., Schmitz, R.Y.(1979) Biochemistry and physiology of cytokinins. In:Biochemical Actions of Hormones, Vol. VI pp. 335-413, Litwack, G. ed. Academic Press, London.
52. Stuchbury, T., Palni, L.M.S., Horgan, R., Wareing, P.F.(1979) The biosynthesis of cytokinins in crown-gall tissue of *Vinca rosea*. Planta *147*, 97-102.
53. Summons, R.E., Letham, D.S., Gollnow, B.I., Parker, C.W., Entsch, B., Johnson, L.P., MacLeod, J.K., Rolfe, B.G.(1981) Cytokinin translocation and metabolism in species of the Leguminoseae: studies in relation to shoot and nodule development. In: Metabolism and Molecular Activities of Cytokinins, pp. 69-80, Guern, J., Péaud-Lenöel, C. eds. Springer, Berlin Heidelberg New York.
54. Taya, Y., Tanaka, Y., Nishimura, S.(1978) 5'-AMP is a direct precursor of cytokinin in *Dictyostelium discoideum*. Nature (London) *271*, 545-547.
55. Tepfer, D.A., Fosket, D.E.(1978) Hormone mediated translational control of protein synthesis in cultured cells of *Glycine max*. Develop. Biol. *62*, 486-497.
56. Wareing, P.F., Phillips, I.D.J.(1970) The Control of Growth and Differentiation in Plants, p.76. Pergamon Press, Oxford.
57. Whitty, C.D., Hall, R.H.(1974) A cytokinin oxidase in *Zea mays*. Can. J. Biochem. 52, 781-799.

B4. Biosynthesis and Metabolism of Ethylene

Thomas A. McKeon[1] and Shang-Fa Yang[2]

[1] Western Regional Research Center, Agricultural Research Service, Berkeley, California 94710, USA.
[2] Department of Vegetable Crops, University of California, Davis, California 95616, USA.

INTRODUCTION

Ethylene is a plant hormone that is involved in the regulation of many physiological responses (2). In addition to its recognition as a "ripening hormone", ethylene is involved in other developmental processes from germination of seeds to senescence of various organs and in many responses to environmental stresses.

In many ways, ethylene is the ideal plant hormone to investigate. As a simple gaseous hydrocarbon, it is readily isolated from plant material and it can be easily quantified down to 0.01 μ l/l using a gas chromatograph equipped with a flame ionization detector. Moreover, the levels of ethylene to which a plant is exposed can be controlled with a flow system. Thus, ethylene is far easier to work with than other plant hormones.

Ethylene was recognized as a plant-produced hormone over 50 years ago, yet the biosynthetic pathway of ethylene in plants remained elusive until the key intermediate, ACC, was shown to be the immediate precursor of ethylene. Although in the past decade much progress has been made in understanding ethylene biosynthesis and action, there are many challenges remaining. The purpose of this chapter is to describe both the progress in ethylene biochemistry and avenues for future research.

Abbreviations: ABA = abscisic acid;
ACC = 1-aminocyclopropane-1-carboxylic acid;
AEC = 1-amino-2-ethylcyclopropane-1-carboxylic acid;
AOA = aminooxyacetic acid;
AVG = aminoethoxyvinylglycine [L-2-amino-4-(2-aminoethoxy)-trans-3-butenoic acid];
IAA = indole-3-acetic acid;
KMB = 2-keto-4-methylthiobutyrate;
MACC = 1-malonylaminocyclopropane-1-carboxylic acid;
MTA = 5'-methylthio-adenosine;
MTR = 5-methylthioribose;
MTR-1-P = 5-methylthioribose-1-phosphate;
SAM = S-adenosylmethionine.

ELUCIDATION OF THE ETHYLENE BIOSYNTHETIC PATHWAY

The pathway for ethylene biosynthesis is shown below. Although a number of ethylene precursors were proposed after testing in plant tissue or in model systems, it was eventually shown that methionine was rapidly converted to ethylene in a chemical model system consisting of Cu^{2+} and ascorbic acid (29). Following up this work on the model system, Lieberman and coworkers showed that L-methionine labelled at the C-3,4 positions was readily converted by apple fruit tissue to labelled ethylene (29). Later, SAM was inferred as an ethylene precursor because the conversion of methionine to ethylene was inhibited by oxidative phosphorylation inhibitors, thus implying an energy(ATP)-dependent step in the biosynthesis of ethylene from methionine. Adams and Yang (5) confirmed this proposal by demonstrating that the labelled [^{35}S] methionine and [^{3}H-methyl] methionine released labelled MTA and its hydrolysis product MTR upon its conversion to ethylene in apple tissue. Thus, methionine must be converted into SAM before ethylene is released.

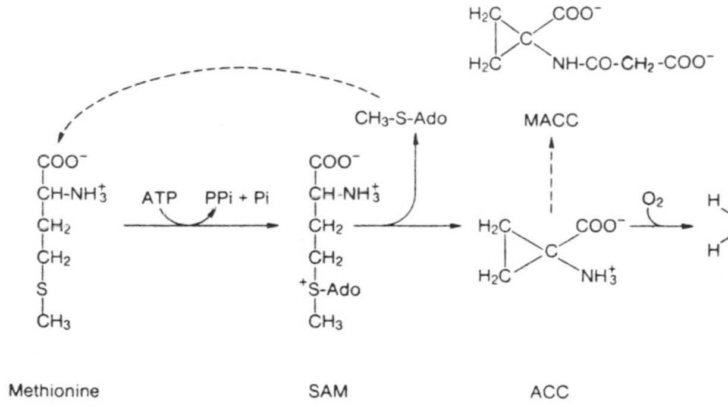

Methionine   SAM   ACC   Ethylene

The next step in the pathway is the conversion of SAM to ACC. Adams and Yang (6) identified ACC, MTA and MTR as the labelled products which accumulated when L-[U-^{14}C]methionine was incubated with apple tissue under anaerobic conditions which block ethylene production. Subsequent incubation of the tissue in air resulted in the production of labelled ethylene from the accumulated labelled ACC. Coinciding with these findings, Lürssen et al., (36), while screening a number of compounds as possible plant growth regulators, demonstrated that ACC dramatically stimulated ethylene production in plant tissues. Based on analogy to the chemical synthesis of ACC, they deduced that ACC would be derived from SAM and were thus able to propose the correct biosynthetic pathway for ethylene.

ACC had been isolated in 1957 from ripe cider apples and perry pears (16) and was postulated to be involved in ripening. However, interest in this unusual, nonprotein amino acid was not sparked until its recognition as an ethylene precursor.

In addition to its conversion to ethylene, ACC can be metabolized to N- malonyl-ACC (MACC). This conjugate was identified independently by two separate groups. Amrhein et al. (9, 10) sought alternate pathways for ACC metabolism based on the fact that some bacteria and yeast metabolize ACC to products other than ethylene. They found that ACC was metabolized to a conjugate which they identified as MACC. Hoffman et al. (22) sought an alternate pathway for ACC metabolism when they found that wilted wheat leaves lost more ACC than could be accounted for through conversion to ethylene, and subsequently demonstrated that this tissue could metabolize ACC to MACC.

An important consideration in ethylene biosynthesis is the limited amount of methionine present in plants. It was recognized that in order to maintain a high rate of ethylene production in apple fruit, the sulfur of methionine must be recycled back to methionine. It was first demonstrated that the 5'-methyl group of MTA is readily recycled to methionine (40), and then, using dual-labelled [^{35}S, ^{14}C-methyl]MTA, was shown that the CH_3S- group of MTA is converted as a unit to re-form methionine (5). Later work showed that the ribose moiety of MTA provides the carbon-skeleton for the 2-aminobutyrate portion of methionine (55, 62).

Enzymes Involved in the Biosynthesis and Regulation of Ethylene

Knowing how ethylene is synthesized is essential to understanding its regulation. The following section is subdivided under headings of the individual enzymes or enzymatic reactions associated with ethylene biosynthesis and regulation. Although the studies described in this section have been carried out with relatively crude enzyme preparations or *in vivo*, the existence of specific inhibitors for individual reactions has allowed considerable progress toward understanding of the biochemical regulation of ethylene biosynthesis.

ACC Synthase.

The ACC synthase, which catalyzes the conversion of SAM to ACC and MTA plays a key role in regulating ethylene production. Levels of ACC synthase are affected by changes in the growth environment, by changes in hormone levels and by physiological and developmental events (57).

The ACC synthase prepared from tomato fruit slices has been partially purified by ammonium sulfate fractionation and hydrophobic chromatography, and has been characterized (14, 60). The enzyme requires pyridoxal phosphate for activity and is sensitive to pyridoxal phosphate inhibitors, especially AVG ($K_i = 0.2 \mu M$) and AOA ($K_i = 0.8 \mu M$). These inhibitors have proven invaluable in studying the regulation of ethylene production by distinguishing effects on ACC synthase from effects on ACC conversion to ethylene (6). The preferred substrate for the ACC synthase is (-)-S-adenosyl-L-methionine (which has the S-configuration at the sulfonium position), the naturally occurring isomer of SAM, with a K_m of 20 μM, whereas (+)-SAM is an effective inhibitor ($I_{50} = 15$μM) (24). Recently, it has been shown that the ACC synthase is inactivated upon incubation with SAM and this process may be responsible for the short half-life of the ACC synthase *in vivo* (46a). The authors propose that the SAM, when activated by the ACC synthase, can irreversibly modify the enzyme in a "suicide-inactivation."

Because of the very low levels of the enzyme present with respect to other proteins and the instability of the enzyme, progress in the purification of ACC synthase has been slow. It is not clear whether the enzyme is inherently unstable or if the wounding process, which is routinely used to induce higher levels of the enzyme, induces hydrolytic enzymes resulting in a less stable synthase due to proteolytic nicking.

The ACC synthase plays a major role in regulating ethylene biosynthesis. Increased ethylene production is involved in developmental processes including germination, ripening, and senescence, and in stress responses to wounding, drought, waterlogging, chilling, toxic agents, infection or insect infestation (29, 57). In all these cases, it has been shown that the higher levels of ethylene are accompanied by increased ACC production, due to induction or activation of ACC synthase (57). This stimulation is shown by direct measurement of increased ACC levels and by the ability of AVG to block or diminish the increase in ACC synthesis and the accompanying increase in ethylene production. A true induction of ACC synthase is inferred since cycloheximide, an inhibitor of protein synthesis, effectively blocks the increased ethylene production (57) as do inhibitors of RNA synthesis (59). Moreover, tomato pericarp slices incubated with deuterium oxide (^2H_2O) develop ACC synthase with a greater buoyant density than that from control slices incubated with H_2O, showing that ACC synthase is synthesized *de novo* after wounding (4).

Ethylene production rates are influenced by ethylene and other plant hormones. Auxin, cytokinins, abscisic acid and ethylene all regulate ethylene production at the level of ACC synthesis, although they may exert their effects by different biochemical mechanisms.

Auxin (IAA) promotes ethylene production by inducing the synthesis of ACC synthase, resulting in higher levels of ACC; the increase in ethylene production parallels the increase in ACC, and treatment with AVG blocks the IAA-induced ethylene increase (58, 61). This auxin induction of ACC synthase is inhibited not only by two different protein synthesis inhibitors, cycloheximide and 2-(4-methyl-2,6-dinitroanilino)-N-methylpropionamide, but also by actinomycin D and α-amanitin, inhibitors of RNA synthesis. These data suggest that the induction of

ACC synthase occurs at the transcriptional level (59). It is, however, not clear whether IAA directly induces the synthase or if some other cellular process mediates the effect of IAA.

ABA effectively reduces wilting-induced ethylene production in wheat leaves (56). Pretreatment of wheat leaves with ABA inhibits ACC accumulation during the subsequent wilting treatment but does not significantly affect ACC levels in unwilted turgid leaves (37). In addition, when IAA-stimulated mung-bean hypocotyls are co-treated with ABA, ACC accumulation is blocked as a result of reduced ACC synthase activity (58). It appears that ABA does not repress ACC synthase in uninduced tissue but does inhibit the induction of the enzyme caused by wilting or by IAA treatment.

Conversely, cytokinins stimulate ethylene production in conjunction with other treatments that increase ethylene synthesis. Thus, ethylene production from IAA-treated (58) or water-stressed tissues (37) rises in response to application of benzyladenine as a result of greater ACC synthase activity and the consequently increased ACC accumulation. Because cytokinins alone do not markedly affect ACC levels, their effect on ACC synthase must be through some other factor which affects the level of induction. Since cytokinins and ABA affect ACC synthase levels only under induced conditions, the induction of ACC synthase will represent an ideal system for understanding the effect of these plant hormones on gene expression when probes for the genetic material of ACC synthase become available.

Depending on the tissue, ethylene can either promote ethylene production (autocatalysis) or inhibit ethylene production (autoinhibition). During autocatalysis in ripening fruits, ethylene affects ethylene production initially at the conversion of ACC to ethylene, although a massive increase in ACC synthesis occurs later (31). ACC synthase is the principal target of ethylene during autoinhibition. When excised grapefruit flavedo tissue was treated with ethylene for 10 hours, it evolved ethylene at 6% of that treated with air (control), and this autoinhibitory effect of ethylene resulted from inhibited ACC synthesis (43). In mungbean hypocotyls, AVG treatment greatly inhibited IAA-induced ethylene production, but caused a three- to four-fold increase in extractable ACC synthase activity (59), while incubation of the AVG-treated tissue with ethylene or with ACC effectively reduced the IAA- induced development of ACC synthase. These results suggest that ethylene treatment reduced the level of ACC synthase by inhibiting the synthesis of the enzyme or by enhancing the degradation of the enzyme (59).

When a tissue is treated with a known amount of a plant hormone, it is rarely determined how much of the hormone was taken up, translocated and metabolized by the tissue and this is the case with the studies described. Nevertheless, considerable progress has been made in understanding the role of hormones in ethylene biosynthesis. As analysis of plant hormones and their metabolism becomes more generally available, a clearer view of plant hormone interactions and their influence on ACC and ethylene production will result.

Ethylene-Forming Enzyme.
The conversion of ACC to ethylene is carried out by an oxidative enzyme or enzyme system which is generally referred to as the ethylene-forming enzyme (EFE). The EFE activity was first described by Adams and Yang (6), who trapped ACC in apple tissue incubated in a nitrogen atmosphere and demonstrated its conversion to ethylene under aerobic conditions. Because the EFE has not yet been isolated independent of intact cellular material (vacuoles, protoplasts or tissue), all characterization of the EFE has been done *in vivo*. This situation necessarily complicates interpretation of many experiments because observed effects on EFE can be attributed to direct or indirect action. Despite this problem, important information about EFE has been generated.

The EFE has a high affinity for ACC as its substrate. McKeon and Yang (38) demonstrated that EFE had a K_m of 66 µM for ACC, based on the dependence of ethylene production rate on the internal ACC concentration in pea hypocotyl segments which were fed exogenously with varying ACC levels, while Guy and Kende (19) obtained a similar K_m (61 µM) for EFE in pea-leaf vacuoles. When one of the ring hydrogens of ACC is substituted with an ethyl group, four stereoisomers of 1-amino-2-ethyl-cyclopropane-1-carboxylic acid (AEC) are generated:

(1R,2S)	(1S,2R)	(1S,2S)	(1R,2R)
100	1.2	0.5	0.5

Hoffman *et al.* (21) showed that in apple and cantaloupe fruit and etiolated mungbean hypocotyl the EFE preferentially utilized one of the stereoisomers, (1R, 2S)-AEC, for the synthesis of the ethylene analogue 1-butene. Both ACC and AEC appear to be degraded by the same enzyme, EFE, since both reactions are inhibited to the same extent under varying levels of oxygen depletion and Co^{2+} concentrations, and since, when both substrates are present simultaneously, they are mutually inhibitory. This stereoselectivity of EFE for AEC isomers has been employed as one of the criteria for demonstrating whether isolated enzymic systems correspond to the *in vivo* system (38). The other substrate of EFE is oxygen. The gas phase concentration of oxygen required for half-maximal activity has been reported to be 1.0% in flower petals (26), but the precise funciton of oxygen in the reaction is presently unknown.

Numerous inhibitors of EFE have been recognized. Cobaltous ion has no effect on ACC synthesis, yet is very effective in inhibiting EFE when applied in the range of 10 to 100 μM (36, 61). Alpha-aminoisobutyric acid, a structural analog of ACC, competitively inhibits the conversion of ACC to ethylene, although relatively high levels are required (46). Presumably, its structural resemblance to ACC is responsible for its inhibitory properties, and evidence indicates that α-aminoisobutyrate is also oxidized by EFE (33). Interestingly, cyclopropylamine and 1-aminocyclopropane-1-phosphonic acid, which are readily converted to ethylene chemically by the hypochlorite reagent used to assay ACC (35), are neither effective as inhibitors nor converted to ethylene *in vivo*. Free radical scavengers represent another class of inhibitors. Although many of these are effective in blocking the conversion of ACC to ethylene *in vivo* (57), they lack the specificity for targeted action desired in enzyme inhibitors.

While the conversion of ACC to ethylene by plants was initially considered to be a highly unusual reaction, it has turned out that there are a number of plant enzyme preparations (IAA oxidase, peroxidase, lipoxygenase, pea stem homogenates, microsomal preparations) that readily convert ACC to ethylene in the presence of various cofactors. While these systems are generally dependent on O_2, they lack the specificity and affinity for ACC characteristic of native EFE (19, 38, 53). These enzymic systems display K_m's for ACC ranging from 3 mM to 389 mM, indicating a much lower affinity for ACC relative to that found for EFE. Furthermore, these systems do not discriminate among the 2-ethyl-ACC stereoisomers as EFE does (19, 38, 53). It is likely that such systems generate "active-oxygen" species such as hydrogen peroxide or superoxide which in turn can react nonenzymatically with ACC to produce ethylene (39, 57).

Study of the reaction products of EFE have been of great importance in understanding the enzyme's mechanism. Noting that 1-phenylcyclopropylamine is oxidized chemically by various oxidants to ethylene and benzonitrile via the intermediacy of the nitrenium ion, it was proposed (7) that ACC is oxidized by EFE to form the corresponding nitrenium ion intermediate, which is then degraded into ethylene (derived from C-2,3 of ACC) and cyanoformic acid (derived from carboxyl and C-1 of ACC), the latter being further degraded spontaneously into HCN (derived from C-1 of ACC) and CO_2 (derived from the carboxyl group of ACC). Support for the proposed reaction products was provided by Peiser et al. (41), who showed that the carboxyl group of ACC is liberated as CO_2, whereas C-1 of

ACC yields HCN, which is then rapidly metabolized to yield β-cyano-alanine and asparagine:

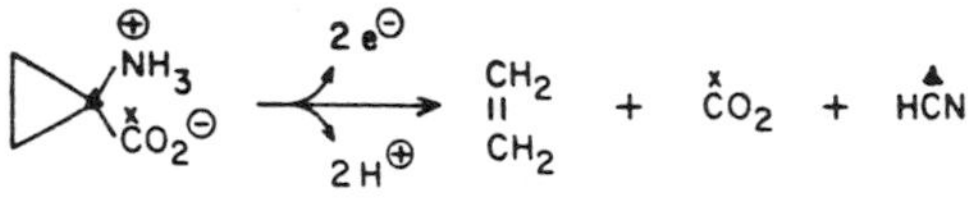

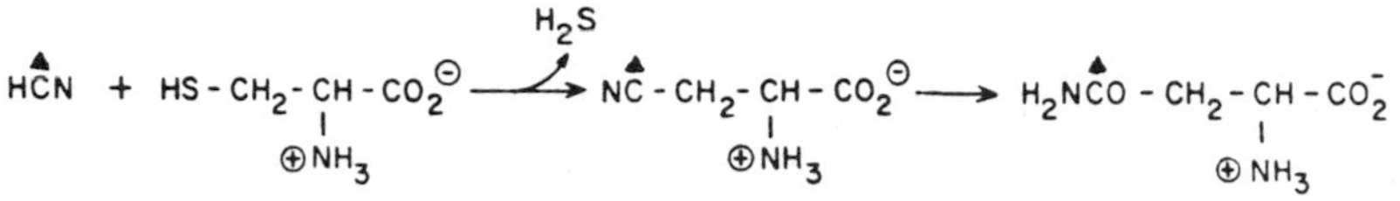

$$\text{ACC} \xrightarrow[2H^{\oplus}]{2e^{\ominus}} \begin{array}{c} CH_2 \\ \| \\ CH_2 \end{array} + CO_2 + HCN$$

$$HCN + HS-CH_2-CH(^{\oplus}NH_3)-CO_2^{\ominus} \xrightarrow{H_2S} NC-CH_2-CH(^{\oplus}NH_3)-CO_2^{\ominus} \longrightarrow H_2NCO-CH_2-CH(^{\oplus}NH_3)-CO_2^{-}$$

It should be noted that the conversion of ACC to CO_2, HCN and ethylene involves a two-electron oxidation. When cis- and trans-2,3-dideutero-ACC are oxidized with hypochlorite, cis- and trans-dideutero-ethylene, respectively, are released, confirming that this reaction occurs by a concerted mechanism, probably via nitrenium ion (8). However, when each of these dideutero-ACC's are incubated with apples slices, the label is scrambled and equal amounts of cis- and trans-dideutero-ethylene are produced:

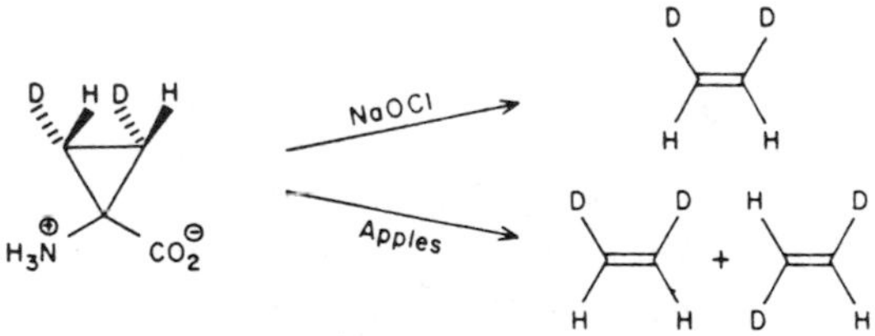

These observations indicate that the biological reaction does not proceed in a concerted manner, but occurs by a stepwise mechanism involving an intermediate that allows scrambling of the ring hydrogens. Since electrolytic oxidation of cis-2,3-dideutero-ACC also yields HCN and results in a scrambled label in the 1,2-dideutero-ethylene, as is observed in plant tissues, Pirrung (42) suggests that both the electrolytic and the EFE reactions occur by means of two one-electron oxidations of ACC through a cation radical, rather than through a nitrenium ion intermediate which is the product of a two-electron oxidation:

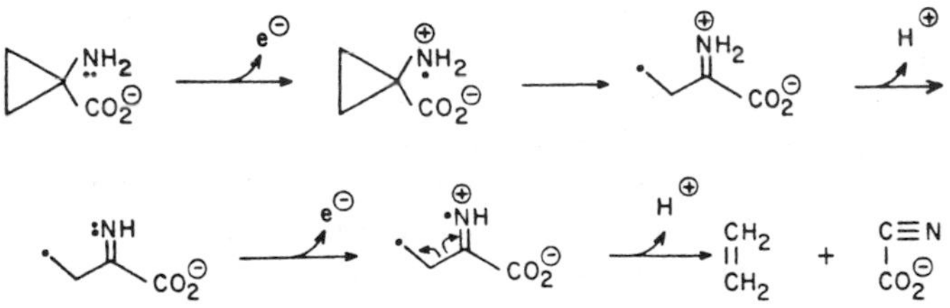

One problem with such an intermediate (and for the nitrenium ion as well) is that it is very reactive and would require stabilization *in vivo*. Baldwin et al. (11) found that several transition metal oxidants (Cu^{2+}, MnO_4^-, FeO_4^{2-}) produce ethylene from 2,3-dideutero-ACC with loss of stereochemistry as observed in plant tissues. They have therefore postulated that a metal ion is involved in the generation and stabilization of the cation radical.

Characterization of the isolated EFE is essential to understanding the regulation of ethylene biosynthesis, yet the isolation of EFE remains one of the key challenges of ethylene research. Intact protoplasts and vacuoles possess functional EFE properties (19), but upon rupture of the plasma membrane or vacuolar membrane, EFE activity is lost. Obviously, the secret to isolating EFE lies in knowing why EFE loses all detectable activity upon cell disruption. In addition to the failure to isolate EFE except in intact vacuoles, protoplasts or cells, the sensitivity of EFE to detergents and to chilling-treatment suggests a requirement for an intact membrane. There are several possible functions for this membrane: it has been proposed that EFE activity is coupled to a transmembrane proton flow (23) which could only occur in intact vesicles. This notion is supported by the inhibition of EFE by 2,4- dinitrophenol, an uncoupler of proton transport. Alternatively, EFE may require a membrane-bound electron transport system. Since the oxidation of ACC by EFE involves sequential electron transfers, disruption of the membrane could lead to loss of activity by interfering with the integrity of the electron transport chain. Finally, the membrane may protect the EFE by maintaining it in a high concentration of an as yet uncharacterized cofactor, or by otherwise protecting the EFE from inactivation.

The EFE, as measured by ethylene production in the presence of a saturating concentration of ACC, is present in most tissues of higher plants with the exception of unripe fruits (57). However, under some stress conditions, in response to ethylene, or in certain development stages (such as fruit ripening), the level of EFE increases markedly and effectively regulates ethylene production (57). Intact preclimateric (unripe) cantaloupe and tomato fruits have low levels of EFE that increase markedly following treatment with ethylene (31). The EFE also responds to environmental stress, increasing up to eight-fold within one hour of wilting treatment and decreasing upon rehydration (37). In response to high temperature ($>35°C$) EFE activity drops and at $40°C$ is lost (57).

ACC N-Malonyltransferase.

Because endogenous levels of ACC can increase during development or in response to stress, it seems logical that the plant would require some means to sequester ACC to prevent overproduction of ethylene. The N-malonylation of ACC serves this purpose, as ethylene production *in vivo* is

102

reduced by malonylation of ACC and promoted by blocking malonylation (30).

The ACC N-malonyltransferase has been isolated and partially purified from mungbean hypocotyls (25). The malonyl donor is malonyl-CoA, with a K_m of 0.25 mM; at concentrations greater than 0.75 mM, malonyl CoA inhibits the transferase (25). The K_m for ACC is 0.15 mM; AEC (32), non-polar D-amino acids (D-methionine, D-phenylalanine, and D-alanine) (25, 52) and α-aminoisobutyric acid (33) can also be malonylated. Based upon several observations it is thought that D-amino acid malonyltransferase and ACC malonyltransferase are the same enzyme (32, 52). These observations include (a) enzyme preparations malonylate both D-amino acids and ACC, (b) the K_m values of those amino acids serving as substrates of malonyltransferase agree with their corresponding K_i values when the same amino acids act as competitive inhibitors of ACC malonyltransferase, and (c) the (1R,2S)- and (1R,2R)-AEC isomers, which have a D-amino acid configuration, are more effective substrates and inhibitors of malonyltransferase than the (1S,2R)- and (1S,2S)-AEC isomers, which have an L-configuration.

The question arises as to the physiological role of MACC. It was initially hypothesized that MACC serves as a means for storing ACC in an unreactive form that could be hydrolyzed to ACC when needed for ethylene production. Such a system might, for example, be suitable for ethylene production in a germinating seed. Yet in germinating peanut seed, which contains a high level of MACC (50 to 100 nmole/g tissue), ethylene is derived almost exclusively from ACC produced *de novo* and MACC is converted to ethylene at less than 2% of the rate that ACC is (20). Previously, Satoh and Esashi (45) have reported that application of D-amino acids increased ACC content and promoted ethylene production in cocklebur cotyledons. These results can now be explained on the basis of D-amino acids inhibiting malonylation of ACC, resulting in a higher ACC level and thereby a higher ethylene production rate. When mungbean cotyledons are fed ACC and D-amino acids, MACC formation is inhibited, leading to higher endogenous levels of ACC, a concomitant increase in ethylene production, and malonylation of the D-amino acid (30). Thus, it is unlikely that MACC serves as a source of ACC; rather, it is a sink that allows depletion of ACC levels and hence reduces ethylene production.

The transferase is present in a wide range of plant tissues. Furthermore, the expression of malonyltransferase activity in pre-climacteric tomato fruit is markedly promoted by ethylene treatment (34), thereby providing an autoregulatory mechanism for limiting ethylene production.

Recycling of 5'-Methylthioadenosine to S-Adenosylmethionine.
As described earlier, the recycling of the methylthio- group from SAM is important to the maintenance of ethylene production due to the limited amounts of sulfur present in plants. While the pathway for recycling of MTA has not been fully elucidated in either plants or animals, the known products and intermediates of the pathway in plants (Fig. 1) and animals are similar with one exception: while animals use MTA phosphorylase to metabolize MTA to MTR-1-P in one step, plants utilize MTA nucleosidase, which converts MTA to MTR, and in turn, MTR kinase, which converts MTR to MTR-1-P (28). In both systems the MTR-1-P is then converted to methionine through several enzymatic steps which have been partially characterized (Fig. 1). The MTR-1-P is ultimately converted to KMB, in a series of enzyme reactions in which molecular oxygen is utilized and the C-1 carbon of MTR-1-P is lost as formic acid (28). Although at least two intermediates are produced in this step in animal tissue extracts (and presumably also plant extracts), the intermediates have not been identified. To complete the cycle, KMB is transaminated to methionine which is adenosylated to form SAM. The overall result of this cycle is that the ribose moiety of ATP furnishes the 4-carbon moiety of methionine from which ACC is derived; the CH_3S group of methionine is, however, conserved for continued regeneration of methionine (57). At this time, little is known about the regulation of this recycling pathway.

Since SAM is a key element in the biosynthesis of polyamines and in methylation reactions, its availability for ACC synthesis could also be a limiting factor in ethylene production. The enzyme ATP:L-methionine S-adenosyltransferase (methionine adenosyltranferase) is responsible for the conversion of methionine to SAM. The enzyme from plants has not been extensively studied, but partially purified methionine adenosyl-tranferase from pea seedling (1) appears to be similar to the enzyme from non-plant sources: in addition to similar K_m's for methionine (0.4 mM) and ATP (0.3 mM), the enzyme is also inhibited by high levels of SAM. Moreover, the transferase is inhibited by AMP and stimulated by ADP, suggesting possible regulation of the enzyme by adenylate energy charge (1). The enzyme in barley leaf appears to be localized in the cytosol (54) with no detectable activity found in isolated organelles.

There is some experimental evidence suggesting that SAM levels can regulate ethylene production. Senescing morning-glory flowers fed with selenomethionine produce ethylene at a greater rate than those fed with methionine (27). The proposed basis for this stimulation of ethylene production is the greater reactivity of selenomethionine than methionine with the methionine adenosyltransferase.

A further indication that the concentration of SAM may affect ethylene production can be inferred from studies on the inhibition of ethylene biosynthesis by polyamines. Inhibition of ACC synthesis by

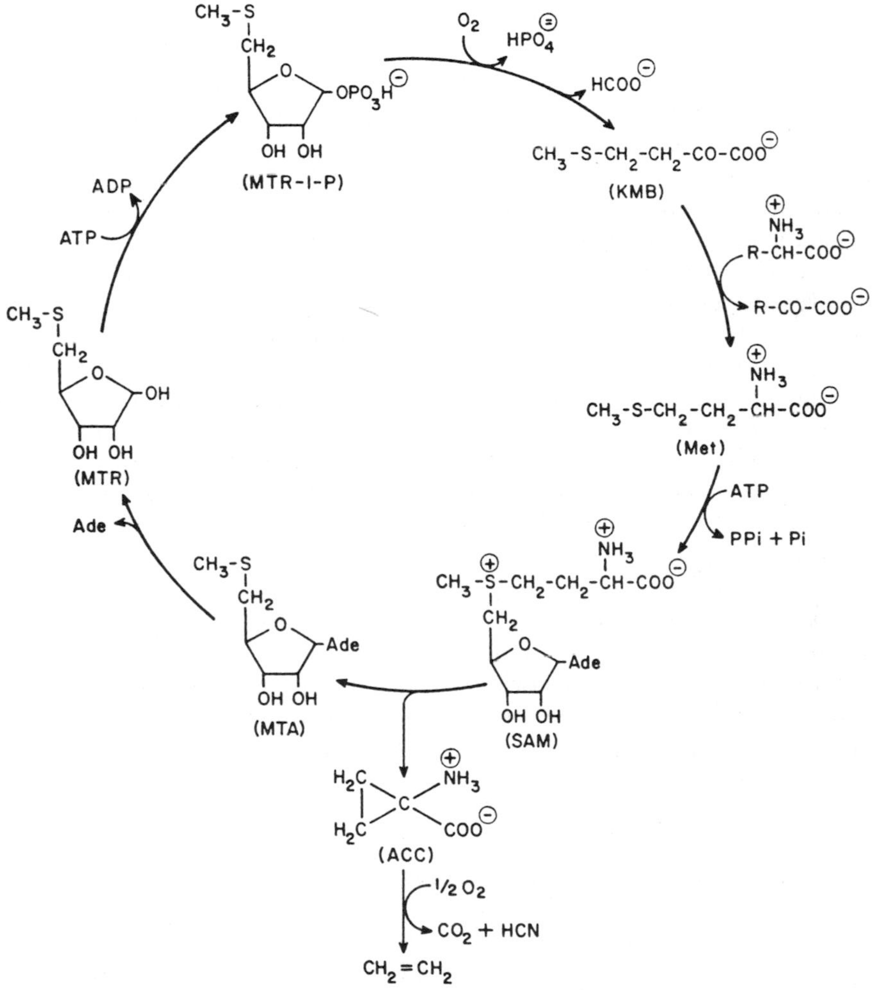

Fig. 1. Methionine cycle in relation to ethylene biosynthesis

AOA results in increased polyamine production while inhibition of polyamine biosynthesis results in increased ACC and ethylene levels (44). Thus, ACC and polyamine production may be mutually inhibitory, because ACC and polyamines are derived from the aminopropyl group of SAM and so compete for available levels of SAM. The function of adenosyltranferase activity in relation to regulation of ethylene biosynthesis remains to be clarified.

ETHYLENE METABOLISM

Although ethylene metabolism by plants was considered for some time to be an artifact arising from impure radiolabeled ethylene (2), it is now clear that in some plant tissues ethylene is oxidized (OX) to CO_2, in others it is incorporated in tissue (TI) by conversion to ethylene oxide and ethylene glycol, and in some plants both processes occur (51). The diversity of plants involved in ethylene metabolism suggests that it is a general phenomenon. In most instances, the rate of metabolism of ethylene is nearly first order with respect to ethylene even at fairly high levels of ethylene (> 40 µl/l), indicating a very high K_m; in the pea, the concentration of ethylene giving a half-maximal ethylene-metabolizing rate is approximately 1000 times the concentration necessary for half-maximal response in the pea growth test (12). However, the K_m for ethylene in TI by *Vicia faba* corresponds closely to the levels evoking physiological response (51). Although the TI and OX systems in pea are half-saturable only at very high levels of ethylene, Beyer (13, 51) has demonstrated that both oxygen deprivation and Ag^+ treatment, which are known to retard ethylene action, similarly inhibit ethylene metabolism, suggesting a connection between ethylene metabolism and the ethylene response. Thus, in several plants the capability to metabolize ethylene correlates with their responsiveness to ethylene during their development. However, the possibility that ethylene metabolism may be a nonessential consequence of ethylene action rather than an essential mediator of ethylene action cannot be ruled out. Thus, additional work is needed to ascertain the physiological role of ethylene metabolism.

ETHYLENE ANALOGUES AND ETHYLENE ANTAGONISTS

Burg and Burg (15) tested the ability of a number of ethylene analogues for ethylene-like action in the pea straight-growth test and found that the effectiveness of olefins that exert ethylene-like biological activity correlated with their ability to form a complex with Ag^+. They have, therefore, proposed that the ethylene receptor site contains a metal ion. Among ethylene analogues, propylene and acetylene were found to require 100 and 2800 times, respectively, the concentration of ethylene to give half-maximal response (Table 1); alkanes are, however, inactive with no detectable response to ethane at 300,000 µl/l (15). Sisler (47) extended the Burgs' idea by postulating that ethylene acts through a trans effect on ligand-metal ion coordination. He demonstrated that a number of such n-acceptors (e.g., carbon monoxide, isocyanide, phosphorus trifluoride), with chemical structures quite different from ethylene, were effective in

Table 1. A Comparison of Ethylene Analogues for Inducing Ethylene-like Response and for Inhibiting Ethylene Binding (13, 15).

Compounds	Relative Concentration for ½ Maximal Response	Relative Concentration for Inhibiting Ethylene Binding
Ethylene	1	1
Propylene	100	128
Carbon Monoxide	2,700	1,068
Acetylene	2,800	1,013
1-Butene	270,000	601,227

eliciting ethylene-like responses without any induction of ethylene synthesis.

Other than the natural inhibitors of ethylene action that have been postulated to be present in plants, there are three known types of antagonists that may be applied exogenously to inhibit ethylene action. These inhibitors have been used as diagnostic tests for ethylene action. The first, CO_2, prevents or delays many ethylene responses when ethylene concentration is low (1 µl/l or below). The mechanism of action is not known, but CO_2 has been suggested to be a competitive inhibitor of ethylene action presumably by competing with ethylene for the binding site, with a K_i of 1.5% (gas phase concentration) (15). CO_2 has been used commercially in controlled atmosphere storage of fruits where high CO_2 levels help to delay the ripening action of ethylene. The second, silver ion, inhibits ethylene action in a wide variety of ethylene-induced responses (13,51) and is used commercially to extend the shelf-life of cut carnations. It is thought that Ag^+ blocks ethylene action by interfering with ethylene binding. Finally, 2,5-norbornadiene, other cyclic olefins, and cis-butene inhibit ethylene action competitively. It is assumed that they compete with ethylene for the binding site to form olefin-receptor complexes that do not induce an ethylene-like response (50). Since 2,5-norbornadiene and other cyclic olefins are volatile and can readily diffuse away, they may be useful for the elucidation of ethylene action and for characterization of the ethylene binding site.

MODE OF ETHYLENE ACTION

Ethylene has been shown to induce the synthesis of new mRNA's and proteins (17). By analogy to other plant and animal hormones, it is thought that ethylene action is mediated by a receptor. However, debate ensues whether ethylene acts by binding directly to the receptor (49, 51) or through a metabolite of ethylene, e.g., ethylene oxide or ethylene glycol (12, 51).

Ethylene-binding sites have been isolated and partially purified from several plant sources including tobacco, beans and tomatoes (49, 51). The binding of ethylene to these sites is saturable at physiological levels of ethylene and the K_D for ethylene in the gas phase is 0.1 to 0.3 µl/l (or approximately 10^{-10} M in the liquid phase), in agreement with the value of 0.1 to 1 µl/l ethylene giving a half-maximal response in a number of plant systems (2, 15). In all cases, the ethylene-binding component is membrane bound and detergent soluble. It appears to be a protein based on heat- sensitivity, protease sensitivity, solubility and chromatographic behavior, and sensitivity to sulfhydryl agents (48, 51). The isolated receptor displays properties similar to those expected from studies of ethylene action *in vivo*. Ethylene binding to the isolated receptor is competitively inhibited by propylene at 128 times and by acetylene at 1013 times the concentration of ethylene (gas phase), corresponding to the *in vivo* activity of these analogues (Table 1). Moreover, 2,5-norbornadiene, which was found to be the most active competitive inhibitor of ethylene action (50), is similarly effective in blocking binding of ethylene to the receptor (48).

Currently, experimental evidence favors ethylene binding, as opposed to ethylene metabolism, as the general mediator of ethylene action. First, the products of ethylene metabolism do not evoke ethylene-like responses; ethylene oxide has no ethylene-like effect (51). Second, ethylene effects can be mimicked by some hydrocarbons (15) and counteracted by certain olefin antagonists (50) at levels which correlate to their ability to compete with ethylene binding to silver ion. Third, there are several inorganic compounds which cannot be metabolized to products related to any of the metabolic products of ethylene, yet evoke ethylene effects (47). Finally, CS_2, a potent inhibitor of ethylene oxidation, has no effect on ethylene action (3).

There is an important question remaining, however: What does the ethylene-binding protein do after ethylene binds? It is known that ethylene elicits many physiological responses (2). Furthermore, it has been shown that ethylene induces specific changes in genetic expression (17). Presumably, these changes would be mediated by the ethylene-binding protein. Although purification of ethylene-binding proteins is proceeding (48, 51), it remains to be shown that the isolated ethylene-binding protein functions *in vivo* and, if this is so, how this ethylene-receptor complex induces the known biochemical changes.

CONCLUSION

Considerable progress has been made in understanding the biosynthesis of ethylene and its regulation. All of this progress has been achieved by the use of intact tissue or relatively crude enzyme systems. At this time, the primary objectives in further elucidating the biochemical regulation of ethylene biosynthesis are the isolation and purification of the individual enzymes. The purified enzymes will greatly facilitate the isolation of the corresponding genetic material, allowing knowledge of the regulation of these enzymes and, therefore, ethylene biosynthesis to be extended to the genetic level. This knowledge is essential to the eventual use of molecular biology to predictably alter the developmental and environmental behavior of plants to improve their production and quality characteristics.

The mode of action of ethylene is not well understood. It is known that ethylene can be metabolized, but there is no strong evidence that ethylene metabolism is essential to ethylene action. Ethylene probably mediates its action through a binding protein. It is known that ethylene elicits many physiological responses and it has been shown that ethylene induces specific changes in genetic expression. All of these changes must somehow be mediated by the ethylene-binding protein. It remains to be shown how the simple hydrocarbon ethylene has such dramatic effects on plant development.

References

1. Aarnes, H. (1977) Partial purification and characterization of methionine adenosyltransferase from pea seedlings. Pl. Sci. Lett. *10*,, 381-390.
2. Abeles, F.B. (1973) Ethylene in Plant Biology, Academic Press, New York. 302 pp.
3. Abeles, F.B. (1984) A comparative study of ethylene oxidation in *Vicia faba and Mycobacterium paraffinicum*. J. Plant Growth Regul. *3*, 85-95.
4. Acaster, M.A., Kende, H. (1983) Properties and partial purification of 1-aminocyclopropane-1-carboxylate synthase. Plant Physiol. *72*, 139-145.
5. Adams, D.O., Yang, S.F. (1977) Methionine metabolism in apple tissue. Implication of S-adenosylmethionine as an intermediate in the conversion of methionine to ethylene. Plant Physiol. *60*, 892-896.
6. Adams, D.O., Yang, S.F. (1979) Ethylene biosynthesis: identification of 1-aminocyclopropane-1-carboxylic acid as an intermediate in the conversion of methionine to ethylene. Proc. Natl. Acad. Sci. USA *76*, 170-174.
7. Adams, D.O., Yang, S.F. (1981) Ethylene the gaseous hormone: mechanism and regulation of biosynthesis. Trends Biochem. Sci. *6*, 161-164.
8. Adlington, R.M., Baldwin, J.E., Rawlings, B.J. (1983) On the stereochemistry of ethylene biosynthesis. J. Chem. Soc., Chem. Commun. 290-292.
9. Amrhein, N., Breuing, F., Eberle, J., Skorupka, H., Tophof, S. (1982) The metabolism of 1-aminocyclopropane-1-carboxylic acids. *In* Plant Growth Regulators 1982 pp 249-58, ed. P.F. Wareing, Academic Press.

10. Amrhein, N., Schneebeck, D., Skorupka, H., Tophof, S. (1981) Identification of a major metabolite of the ethylene precursor 1- aminocyclopropane-1-carboxylic acid in higher plants. Naturwissenschaften 68, 619-620.
11. Baldwin, J.E., Jackson, D.A., Adlington, R.M., Rawlings, B.J. (1985) The stereochemistry of oxidation of 1-amino-cyclopropanecarboxylic acid. J. Chem. Soc., Chem. Commun. 206-207.
12. Beyer, E.M. Jr. (1975) $^{14}C_2H_4$: Its incorporation and metabolism by pea seedlings under aseptic conditions. Plant Physiol. 56, 273-278.
13. Beyer, E.M. Jr., Morgan, P.W., Yang, S.F. (1984) Ethylene. In Advanced Plant Physiology pp 111-26, ed M.B. Wilkins, Pitman Publishing Limited, London.
14. Boller, T., Herner, R.C., Kende, H. (1979) Assay for and enzymatic formation of an ethylene precursor, 1-amino-cyclopropane-1-carboxylic acid. Planta 145, 293-303.
15. Burg, S.P., Burg, E.A. (1967) Molecular requirements for the biological activity of ethylene. Plant Physiol. 42, 144-152.
16. Burroughs, L.F. (1957) 1-Aminocyclopropane-1-carboxylic acid: A new amino acid in perry pears and cider apples. Nature 179, 360-361.
17. Christoffersen, R.E., Laties, G.G. (1982) Ethylene regulation of gene expression in carrots. Proc. Natl. Acad. Sci. USA 79, 4060-4063.
18. Dodds, J.H., Heslop-Harrison, J.H., Hall, M.A. (1980) Metabolism of ethylene to ethylene oxide by cell-free preparations from Vicia faba L. cotyledons: effects of structural analogues and of inhibitors. Pl. Sci. Lett. 19, 175-180.
19. Guy, M., Kende, H. (1984) Conversion of 1-aminocyclopropane-1-carboxylic acid to ethylene by isolated vacuoles of Pisum sativum L. Planta 160, 281-287.
20. Hoffman, N.E., Fu, J.-R., Yang, S.F. (1983) Identification and metabolism of 1-(malonylamino)cyclopropane-1-carboxylic acid in germinating peanut seeds. Plant Physiol. 71, 197-199.
21. Hoffman, N.E., Yang, S.F., Ichihara, A., Sakamura, S. (1982) Stereospecific conversion of 1-amino-cyclopropanecarboxylic acid to ethylene by plant tissues. Conversion of stereoisomers of 1-amino-2-ethylcyclopropanecarboxylic acid to 1-butene. Plant Physiol. 70, 195-199.
22. Hoffman, N.E., Yang, S.F., McKeon, T. (1982) Identification of 1-(malonylamino) cyclopropane-1-carboxylic acid as a major conjugate of 1-aminocyclopropane-1-carboxylic acid, an ethylene precursor in higher plants. Biochem. Biophys. Res. Commun. 104, 765-770.
23. John, P. (1983) The coupling of ethylene biosynthesis to a transmembrane, electrogenic proton flux. FEBS Lett. 152, 141-143.
24. Khani-Oskouee, S., Jones, J.P., Woodard, R.W. (1984) Stereochemical course of the biosynthesis of 1-aminocyclopropane-1-carboxylic acid. I. Role of the asymmetric sulfonium pole and the α-amino acid center. Biochem. Biophys. Res. Comm. 121, 181-187.
25. Kionka, C., Amrhein, N. (1984) The enzymatic malonylation of 1-aminocyclopropane-1-carboxylic acid in homogenates of mung-bean hypocotyls. Planta 162, 226-235.
26. Konze, J.R., Jones, J.F., Boller, T., Kende, H. (1980) Effect of 1-aminocyclopropane-1-carboxylic acid on the production of ethylene in senescing flowers of Ipomoea tricolor Cav. Plant Physiol. 66, 566-571.
27. Konze, J.R., Kende, H. (1979) Interactions of methionine and selenomethionine with methionine adenosyltransferase and ethylene-generating systems. Plant Physiol. 63, 507-510.
28. Kushad, M.M., Richardson, D.G., Ferro, A.J. (1983) Intermediates in the recycling of 5-methylthioribose to methionine in fruits. Plant Physiol. 73, 257-261.
29. Lieberman, M. (1979) Biosynthesis and action of ethylene. Ann. Rev. Plant Physiol. 30, 533-591.

30. Liu, Y., Hoffman, N. E., Yang, S. F. (1983) Relationship between the malonylation of 1-aminocyclopropane-1-carboxylic acid and D-amino acids in mung-bean hypocotyls. Planta *158*, 437-441.
31. Liu, Y., Hoffman, N.E., Yang, S.F. (1985) Promotion by ethylene of the capability to convert 1-aminocyclopropane-1-carboxylic acid to ethylene in preclimacteric tomato and cantaloupe fruits. Plant Physiol. *77*, 407- 411.
32. Liu, Y., Su L.-Y., Yang, S.F. (1984) Stereoselectivity of 1-aminocyclopropane-carboxylate malonyltransferase toward stereoisomers of 1-amino-2-ethylcyclopropane-carboxylic acid. Arch. Biochem. Biophys. *235*, 319-325.
33. Liu, Y., Su, L.-Y., Yang, S.F. (1984) Metabolism of α-aminoisobutyric acid in mungbean hypocotyls in relation to metabolism of 1-aminocyclopropane-1-carboxylic acid. Planta *161*, 439-443.
34. Liu, Y., Su, L.-Y., Yang, S.F. (1985) Ethylene promotes the capability to malonylate 1-aminocyclopropane-1-carboxylic acid and D-amino acids in preclimacteric tomato fruits. Plant Physiol. *77*, 891-895.
35. Lizada, M.C.C., Yang, S.F. (1979) A simple and sensitive assay for 1-aminocyclopropane-1-carboxylic acid. Anal. Biochem. *100*, 140-45.
36. Lürssen, K., Naumann, K., Schröder, R. (1979) 1-Aminocyclopropane-1-carboxylic acid—an intermediate of the ethylene biosynthesis in higher plants. Z. Pflanzenphysiol. *92*, 285-294.
37. McKeon, T.A., Hoffman, N.E., Yang, S.F. (1982) The effect of plant-hormone pretreatments on ethylene production and synthesis of 1-aminocyclopropane-1-carboxylic acid in water-stressed wheat leaves. Planta *155*, 437-443.
38. McKeon, T.A., Yang, S.F. (1984) A comparison of the conversion of 1-amino-2-ethylcyclopropane-1-carboxylic acid stereoisomers to 1-butene by pea epicotyls and by a cell-free system. Planta *160*, 84-87.
39. McRae, D.G., Baker, J.E., Thompson, J.E. (1982) Evidence for involvement of the superoxide radical in the conversion of 1-aminocyclopropane-1-carboxylic acid to ethylene by pea microsomal membranes. Plant & Cell Physiol. *23*, 375-383.
40. Murr, D.P., Yang, S.F. (1975) Conversion of 5'-methylthioadenosine to methionine by apple tissue. Phytochemistry *14*, 1291-1292.
41. Peiser, G.D., Wang, T.-T., Hoffman, N.E., Yang, S.F., Liu, H.-W., Walsh, C.T. (1984) Formation of cyanide from carbon 1 of 1-aminocyclopropane- 1-carboxylic acid during its conversion to ethylene. Proc. Natl. Acad. Sci. USA *8,1* 3059-3063.
42. Pirrung, M.C. (1983) Ethylene biosynthesis. 2. Stereochemistry of ripening, stress, and model reactions. J. Am. Chem. Soc. *105*, 7207-7209.
43. Riov, J., Yang, S.F. (1982) Autoinhibition of ethylene production in citrus peel discs. Suppression of 1-aminocyclopropane-1-carboxylic acid synthesis. Plant Physiol. *69*, 687-690.
44. Roberts, D.R., Walker, M.A., Thompson, J.E., Dumbroff, E.B. (1984) The effects of inhibitors of polyamine and ethylene biosynthesis on senescence, ethylene production and polyamine levels in cut carnation flowers. Plant & Cell Physiol. *25*, 315-322.
45. Satoh, S., Esashi, Y. (1981) D-Amino-acid-stimulated ethylene production: molecular requirements for the stimulation and a possible receptor site. Phytochemistry *20*, 947-949.
46. Satoh, S., Esashi, Y. (1980) α-Aminoisobutyric acid: a probable competitive inhibitor of conversion of 1-aminocyclopropane-1-carboxylic acid to ethylene. Plant. Cell Physiol. *21*, 939-949.
46a Satoh, S., Esashi, Y. (1986) Inactivation of 1-aminocyclopropane-1-carboxylic acid synthase of etiolated mung bean hypocotyl segments by its substrate, S-adenosyl-L-methionine. Plant Cell Physiol. *27*, 285-291.
47. Sisler, E.C. (1977) Ethylene activity of some π-acceptor compounds. Tobacco Science *21*, 43-45.

48. Sisler, E.C. (1982) Ethylene-binding properties of a Triton X-100 extract of mung bean sprouts. J. Plant Growth Regul. *1*, 211-218.
49. Sisler, E.C., Goren, R. (1981) Ethylene binding – the basis for hormone action in plants? What's New in Plant Physiology *12*, 37-40.
50. Sisler, E.C., Yang, S.F. (1984) Anti-ethylene effects of *cis*-2-butene and cyclic olefins. Phytochemistry *23*, 2765-2768.
51. Smith, A.R., Hall, M.A. (1984) Mechanism of ethylene action. Pl. Growth Regulation *2*, 151-165.
52. Su, L.-Y., Liu, Y., Yang, S.F. (1985) Relationship between 1-aminocyclopropanecarboxylate malonyltransferase and D-amino acid malonyltransferase. Phytochemistry *24*, 1141-1145.
53. Venis, M.A. (1984) Cell-free ethylene-forming systems lack stereochemical fidelity. Planta *162*, 85-88.
54. Wallsgrove, R.M., Lea, P.J., Miflin, B.J. (1983) Intracellular localization of aspartate kinase and the enzymes of threonine and methionine biosynthesis in green leaves. Plant Physiol. *71*, 780-784.
55. Wang, S.Y., Adams, D.O., Lieberman, M. (1982) Recycling of 5'- methylthioadenosine-ribose carbon atoms into methionine in tomato tissue in relation to ethylene production. Plant Physiol. *70*, 117-121.
56. Wright, S.T.C. (1980) The effect of plant growth regulator treatments on the levels of ethylene emanating from excised turgid and wilted wheat leaves. Planta *148*, 381-388.
57. Yang, S.F., Hoffman, N.E. (1984) Ethylene biosynthesis and its regulation in higher plants. Ann. Rev. Plant Physiol. *35*, 155-189.
58. Yoshii, H. Imaseki, H. (1981) Biosynthesis of auxin-induced ethylene. Effects of indole-3-acetic acid, benzyladenine and abscisic acid on endogenous levels of 1-aminocylopropane-1-carboxylic acid (ACC) and ACC synthase. Plant & Cell Physiol. *22*, 369-379.
59. Yoshii, H., Imaseki, H. (1982) Regulation of auxin-induced ethylene biosynthesis. Repression of inductive formation of 1-aminocyclopropane-1-carboxylate synthase by ethylene. Plant & Cell Physiol. *23*, 639-649.
60. Yu, Y.-B., Adams, D.O., Yang, S.F. (1979) 1-Aminocyclopropane- carboxylate synthase, a key enzyme in ethylene biosynthesis. Arch Biochem. Biophys. *198*, 280-286.
61. Yu, Y.-B., Yang, S.F. (1979) Auxin-induced ethylene production and its inhibition by aminoethyoxyvinylglycine and cobalt ion. Plant Physiol. *64* 1074-1077.
62. Yung, K.H., Yang, S.F., Schlenk, F. (1982) Methionine synthesis from 5-methylthioribose in apple tissue. Biochem. Biophys., Res. Comm. *104* 771-777.

B5. Abscisic Acid Biosynthesis and Metabolism

Daniel C. Walton

Faculty of Environmental and Forest Biology, SUNY College of
Environmental Science and Forestry, Syracuse, New York 13210, USA.

INTRODUCTION

One of the questions which plant physiologists ask about a hormone is how its cellular levels are regulated. The concentration of a hormone, or of any other cellular constituent, will depend on its rate of synthesis and metabolism and on its rate of import into and export from the cell. Abscisic acid (ABA, Fig. 1) is a particularly interesting hormone with regard to the regulation of its levels, since they rise and fall dramatically in several kinds of tissues in response to environmental and developmental changes. When leaves of mesophytic plants are water stressed, ABA levels can rise from 10- to 50-fold within 4 to 8 hours, apparently due to a greatly increased rate of biosynthesis. When the plants are rewatered, the ABA levels drop to pre-stress levels within 4 to 8 hours. The drop in concentration is due to a reduced biosynthetic rate, a vigorous metabolism and possibly export from the leaves. In developing

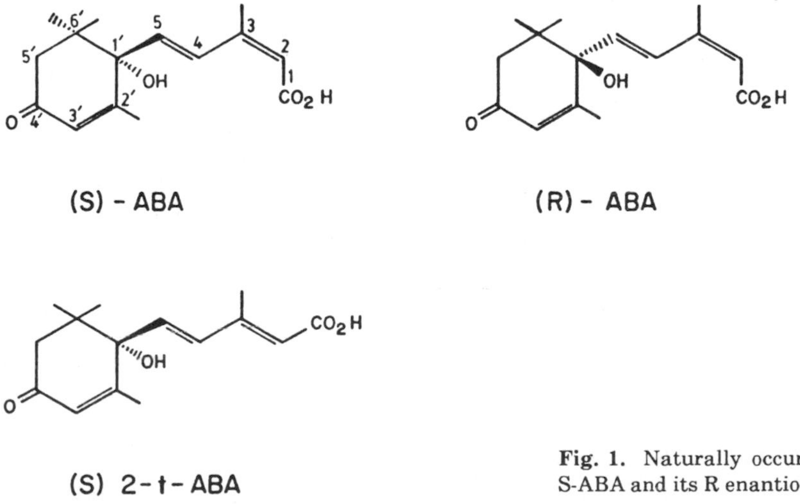

(S) - ABA

(R) - ABA

(S) 2-t-ABA

Fig. 1. Naturally occurring S-ABA and its R enantiomer.

113

seeds of various plants, ABA levels can rise a hundred-fold within a few days and then decline to low levels as the seeds mature and dessicate. Synthesis and metabolism, as well as import, are involved in changing the ABA levels. Dormant buds and seeds of woody plants accumulate high levels of ABA which then decrease when the tissues are exposed to low temperatures. A combination of synthesis, metabolism, import and export are probably involved in determining the ABA levels in these issues.

In order to understand how ABA levels are regulated by biosynthesis and metabolism, it is first necessary to know the identity of the compounds which are involved in the pathways. Enzymes which interconvert these compounds must be characterized and the enzyme co-factors identified. Only then is it possible to begin to determine how the pathways are regulated. In the case of ABA, we are still trying to determine the constituents of the pathways, particularly for biosynthesis. We have almost no knowledge of the enzymes involved. Consequently, our understanding of how these pathways are regulated will require considerably more knowledge than we currently possess.

A number of more detailed reviews of the subject matter included in this chapter have been written during the past decade. They include (17, 20, 21, 22, 27, 35, 36).

BIOSYNTHESIS

ABA is a sesquiterpene which like other sesquiterpenes has been shown to be derived from mevalonic acid (MVA) (31). Although ABA is a relatively simple molecule, the details of its biosynthesis have remained obscure. We do not even know at this time whether ABA is derived directly from a C-15 precursor such as farnesyl pyrophosphate (FPP), or whether it is derived from the cleavage product of a C-40 compound such as the carotenoid violaxanthin (Fig. 2). Why has there been so little progress in working out the ABA biosynthetic pathway since ABA was first described in 1965? As is the case for other hormones, ABA is usually present in the plant in very low concentrations. In most tissues the levels are from 10 to 50 ng/g fresh wt (4×10^{-8}M to 2×10^{-7}M). Only in water-stressed leaves, developing seeds and dormant buds and seeds are levels higher than 10^{-6}M. A second problem in studying ABA biosynthesis has been the poor incorporation of presumed precursors into ABA when they are applied in radioactive form. Even in water-stressed leaves in which ABA levels rise dramatically due to increased rates of synthesis, ^{14}C-MVA and $^{14}CO_2$ are poorly inorporated into ABA. Thus, one of the important tools used to study metabolic pathways in tissues has been of little use in studying ABA biosynthesis.

Although there have been suggestions that several C-15 compounds may be intermediates on a direct ABA biosynthetic pathway, no

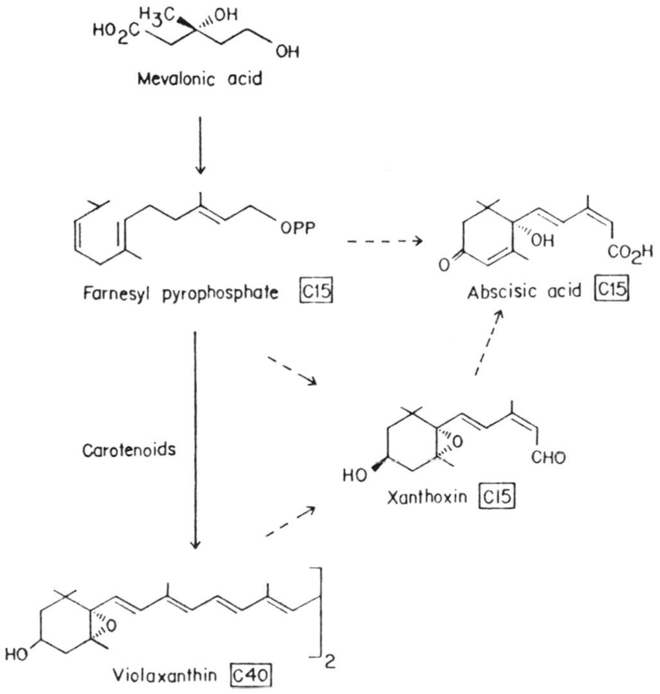

Fig. 2. Biosynthesis of ABA; C-15 vs C-40 pathways

convincing evidence has been produced that they are involved. The suggestion that ABA may be derived from a xanthophyll came from several observations. Violaxanthin can be photooxidized to xanthoxin, a C-15 compound with similarities to ABA in its carbon skeleton and in its ring oxygen substitution (Fig. 2). Xanthoxin has been shown to be a naturally-occurring compound in a variety of plants, and it was shown to be converted to ABA when it was fed in radioactive form to bean and tomato plants (6, 32, 33). Despite these observations we still lack evidence that ABA is derived fom xanthoxin, and even that the xanthoxin found in plants originated from a carotenoid. As indicated in Figure 2, xanthoxin could be derived direcly from FPP. The finding that avocado slices convert ^{3}H-MVA to both ABA and carotenoids, but apparently convert the carotenoid precursor ^{14}C-phytoene only to carotenoids, has been used as evidence against an indirect pathway from the carotenoids and their oxygenated derivatives, the xanthophylls (20). The inability to unequivocally demonstrate ABA intermediates, coupled with the poor conversions of presumtive precursors to ABA, brought work on biosynthesis in plants to a virtual halt by the late 1970s.

The discovery in 1977 that a rose pathogen, the fungus *Cercospora rosicola*, produces and excretes relatively large quantities of the naturally occurring S-enantiomer of ABA into its growth medium initiated work on the biosynthetic pathway in that organism (1). The hope was that the pathway in *C. rosicola* would be similar to or identical with the pathway in higher plants, and that the relatively large production of ABA by the fungus would allow its pathway to be determined more readily than can be done in higher plants. The discovery has also stimulated investigators to look for ABA production in other fungi with the result that at least 7 fungi have been reported to produce ABA (5, 29). Since only a relatively few fungi have been investigated for ABA production, it seems likely that the list will grow longer. The intriguing possibility that ABA may play a role in pathogenesis is still to be investigated.

While studies have been proceeding with *C. rosicola*, investigators have been restimulated to look at ABA biosynthesis in plants, using different techniques than had been used previously. The results of these studies suggest that ABA is produced from the cleavage of a carotenoid, although the identity of the carotenoids involved and definitive evidence for their participation is still lacking. The next 2 sections of this chapter will describe recent work done with *Cercospora* and with higher plants.

ABA Biosynthesis in *Cercospora*

The discovery that *C. rosicola* excretes a considerable amount of S-ABA into its growth medium led to the rapid development of a simple defined medium in which cell suspension cultures grow well, and into which they begin to excrete S-ABA after about 5 days growth (Fig. 3) (9, 28). The fungus converts both ^{14}C-acetate and ^{14}C-MVA into ABA with a reasonable yield unlike the results obtained with plants (2, 26).

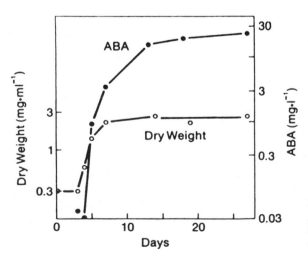

Fig. 3. Dry weight and ABA accumulation in cell suspension cultures of *C. rosicola*

When either of these compounds was fed in radioactive form to the fungus, another radioactive compound was isolated from the growth medium and identified as 1'-deoxy ABA (3, Fig. 4a) (26). When this compound was refed to the fungus it was converted to ABA in good yield. 1'-deoxy ABA was isolated from the fungal medium when neither 14C-acetate nor 14C-MVA had been added to the growth medium, so that it appears to be a naturally occurring fungal metabolite. Since ABA and 1'-deoxy ABA

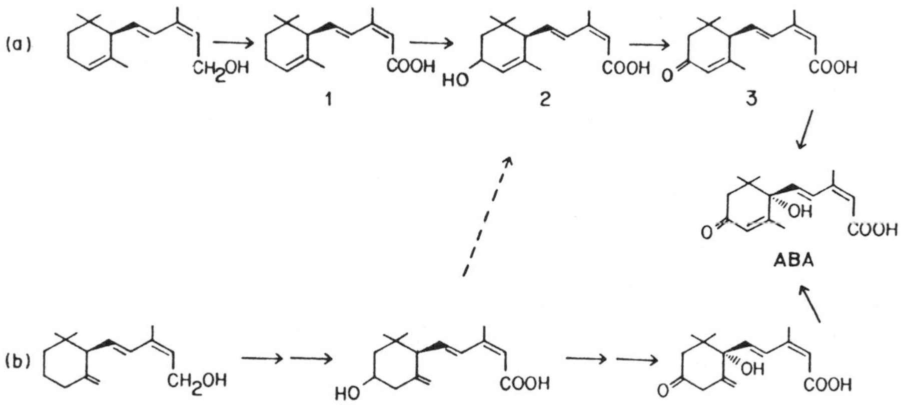

Fig. 4. Later stages of possible ABA biosynthetic pathways in (a) *C. rosicola* and (b) *C. cruenta*.

differ by only a hydroxyl group at the 1' position, it is assumed that 1'-deoxy ABA is an immediate precursor to ABA. Other compounds related to 1'-deoxy ABA, such as α-ionylidene acetic acid (1, Fig. 4a) and 4'-hydroxy-α-ionylidene acetic acid (2, Fig. 4a) are converted by the fungus to both 1'deoxy ABA and to ABA (27). Figure 4a shows one possble route for the later stages of ABA biosynthesis in *C. rosicola* based on these studies. The sequence of side-chain and ring oxidations shown in Figure 4 is not meant to imply a unique pathway, since it is possible that they could occur in one or more different sequences. Work with the related fungus, *C. cruenta*, has led to the suggestion that a variation of the apparent *C. rosicola* pathway may operate in this organism (29) (Fig. 4b). The primary difference between the 2 suggested pathways is in the initial location of the ring double bond after the cyclization of FPP. The figure also shows how the 2 pathways may be interrelated. Whether the 2 organisms really have the same pathways, or whether they vary slightly remains to be determined. Inhibitors of carotenoid biosynthesis did not affect the accumulation of ABA by *C. cruenta* so it is assumed that a direct pathway from FPP is involved (29). In neither fungus, however, has the presumptive first cyclic intermediate been isolated, so that the steps immediately after the formation of FPP are unknown.

117

After the discovery that α-ionylidene acetic acid and 1'-deoxy ABA may be intermediates on the ABA biosynthetic pathway in *C. rosicola*, these compounds were fed in radioactive form to several plant tissues (27). The results differed depending on the tissues fed. In the case of bean leaves, immature bean seeds, and avocado fruit, these compounds were not converted to ABA. α-ionylidene acetic acid was converted to 1'-deoxy ABA and various polar compounds were formed from both compounds. These results were similar to those previously obtained with barley plants (15). In the case of *Vicia faba* leaves, however, convincing evidence was obtained for the conversion of both α-ionylidene acetic acid and 1'-deoxy ABA to ABA. These results were obtained with both deuterium-labelled and tritium-labelled compounds. In the case of the former, the incorporation was demonstrated by combined gas chromatography-mass spectrometry which is usually considered to be definitive. It is not clear which of these results gives an accurate picture of ABA biosynthesis in plants. The lack of conversion of α-ionylidene acetic acid and 1'-deoxy ABA to ABA by several tissues may have resulted from a failure of these compound to reach the necessary cell compartment before they were metabolized to other compounds. This is the danger of drawing conclusions about biosynthetic intermediates solely on the basis of feeding experiments. On the other hand, it is possible that the results obtained with *V. faba* were anomalous and due to the presence of an enzyme which is not universally distributed in higher plants. These conflicting results suggest that other techniques must be used to determine whether these compounds are truly ABA intermediates in higher plants.

ABA Biosynthesis in Plants

Although the work with *C. rosicola* and *C. cruenta* suggest a direct C-15 pathway to ABA, recent evidence suggests that an indirect pathway from the carotenoids may be involved in higher plants. For example, several corn mutants have been described which lack the ability to synthesize carotenoids due to specific defects in their biosynthetic pathway. These mutants also have a reduced ability to accumulate ABA in their leaves and roots (25). In addition, inhibitors of carotenoid synthesis, such as fluridone and norflurazon, also inhibit the accumulaion of ABA under some conditions (12, 24, 30). The results of these experiments are suggestive of a precursor role for carotenoids, but they must be viewed with some caution. Plants which lack carotenoids and are grown in the light have bleached chlorophyll and show other changes in their chloroplasts. The inability of such plants to produce ABA may be an indirect effect of damage to their chloroplasts, rather than a lack of precursor carotenoid.

Experiments of a different type by Creelman and Zeevaart (4) also point to the possible role for xanthophylls as ABA precursors. These investigators water-stressed bean and *Xanthium* leaves in the presence of $^{18}O_2$, a heavy isotope of oxygen. Their rationale was that if 1'deoxy ABA is the immediate precursor to ABA in plants, as it apparently is in *C. rosicola*, then one ^{18}O atom should be incorporated into the ABA ring at the 1' position. This assumes that the hydroxyl oxygen is derived from O_2 as normally would be expected and as is the case for ABA made in *C. rosicola*. Their results showed, however, that only one atom of ^{18}O was incorporated into the ABA, but it was in the carboxyl group and not in the ring. The results suggest that (1) 1'-deoxy ABA is not the immediate precursor to ABA, at least in water-stressed leaves and (2) ABA formed in the water stressed leaves was derived from a prefomed precursor containing the oxygens which would become the 1' and 4' oxygens of ABA. One possible explanation for these results is that a xanthophyll, such as violaxanthin, was cleaved by oxygen to form an aldehyde containing ^{18}O. The aldehyde in the case of violaxanthin cleavage would be xanthoxin. If the xanthoxin were oxidized by dehydrogenases when converted to ABA, there would be one atom of ^{18}O in the carboxyl group. The second oxygen atom would have been obtained from water which would not contain ^{18}O. These conversions are summarized in Figure 5. Additional evidence that this hypothesis is feasible requires that xanthoxin as well as ABA be analyzed for ^{18}O under the conditions of the experiment. If xanthoxin is an ABA precursor derived from the cleavage of a xanthophyll during water stress, then its aldehyde oxygen should contain at least as much ^{18}O as the carboxyl group of ABA. Unfortunately this experiment has not yet

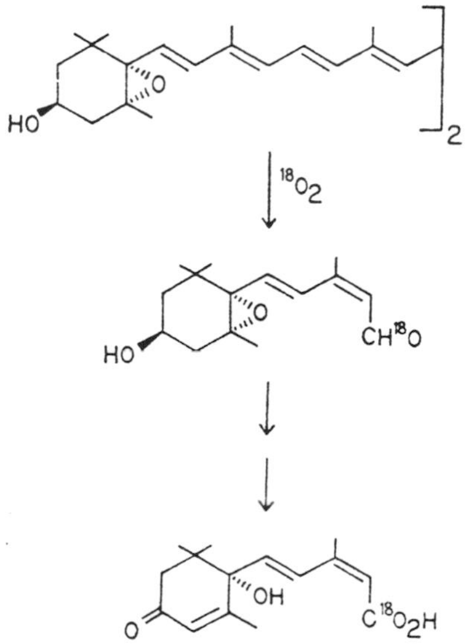

Fig. 5. Hypothetical cleavage of violaxanthin to xanthoxin by $^{18}O_2$ with subsequent conversion to ABA by dehydrogenases

been done because of the low levels of xanthoxin, even in water-stressed leaves.

As indicated earlier in this chapter, a major problem in studying the ABA biosynthetic pathway in higher plants has been the poor incorporation of radioactive precursors such as MVA and CO_2 into ABA. The poor incorporation of MVA led Milborrow (19) to propose that the chloroplast is the site of ABA synthesis on the assumption that transport of ^{14}C-MVA across the chloroplastic membrane was the rate-limiting step in incorporation. He reported that lysed chloroplasts incorporated ^{14}C-MVA into ABA, although the yields were very low (20). More recently, other investigators have reported that chloroplasts do not incorporate ^{14}C-MVA into ABA (10). One explanation for both of these results is that ABA is derived from preformed precursors present at high levels relative to ABA, and that the precursors are synthesized at low rates in mature leaves. Such precursors could be located within the chloroplasts, although this would not be necessary. The consequences of preformed precursors with such attributes would be that radioactive compounds will be incorporated into them very slowly and the radioactivity that is incorporated will be diluted by the high concentrations of precursor already present. Little radioactivity would appear in the ABA subsequently produced from the precursor. The major leaf xanthophylls such as lutein, violaxanthin and neoxanthin seem to fit such a desciption (Fig. 6). They are present at levels of more than 10^3 times greater than ABA in unstressed leaves and their rates of synthesis in mature leaves are low (8).

My colleagues and I have undertaken several experiments to test the hypothesis that xanthophylls are ABA precursors in water-stressed leaves (38). In the first type of experiment we attempted to determine whether the low incorporation of [14]CO_2 into ABA is at least consistent with its prior incorporation into a xanthophyll precursor. Bean plants were exposed to [14]CO_2-containing air for 24 hours, after which the plants were transferred to normal air for varying periods of up to 12 days before the leaves were stressed. Violaxanthin, lutein and ABA were isolated from the stressed leaves and their specific activities determined. The specific activities of all of the compounds were low, but the specific activity of ABA on a carbon atom basis was only 60 to 90% of that of the 2 xanthophylls regardless of the chase period. Reductions in the xanthophyll specific activities which occurred during the [12]CO_2 chase period were reflected in similar reductions in the ABA specific activities. The fact that the ABA specific activities were always less than those of lutein and violaxanthin is consistent with a precursor role for these compounds, and also consistent with the low incorporation of [14]CO_2 or ^{14}C-MVA into ABA.

A second type of experiment involved fluridone, which inhibits carotenoid biosynthesis by blocking dehydrogenation steps. Fluridone, and similar inhibitors have been used with mixed results in reducing ABA accumulation in plants. When seedlings are grown in the

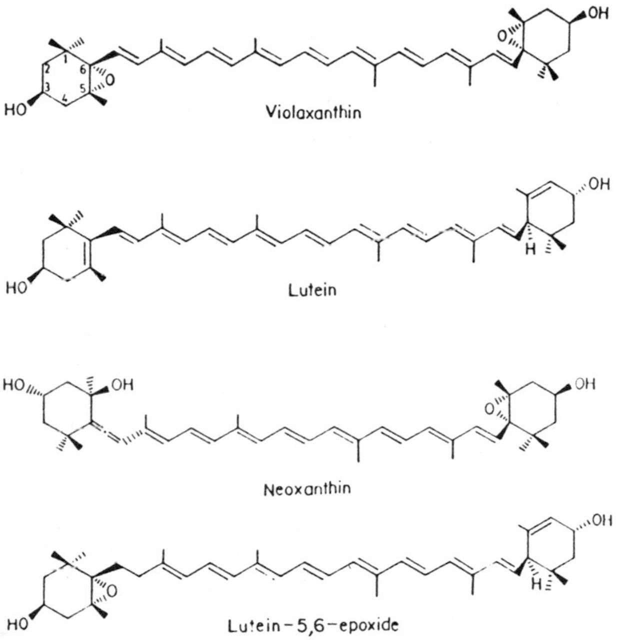

Fig. 6. Structures of several leaf xanthophylls

continuous presence of these compounds in the light ABA accumulation is strongly inhibited (12, 24). When the compounds are added to leaves which have greened, ABA accumulation does not appear to be affected (12, 30). There seem to be 2 possible explanations for these different results. One is that the total inhibition of carotenoid synthesis, with the resultant effects on the chloroplast, may inhibit ABA synthesis indirectly. On the other hand, the lack of an effect in green leaves may be due to a sufficient pool of carotenoids to support ABA synthesis without the necessity of ongoing carotenoid biosynthesis. We tried to distinguish between these 2 possibilities with our experiments. Bean leaves which were green and almost fully expanded were pretreated with fluridone and then exposed to $^{14}CO_2$. After a 24-hour labelling period, the leaves were water-stressed for 14 hours. ABA and several xanthophylls were isolated and both their total levels and specific activities measured. The results indicated that fluridone did not inhibit the accumulation of ABA in water-stressed leaves, but that the ABA specific activity was greatly reduced.

The reduction in ABA specific activity was similar to that of lutein, violaxanthin and neoxanthin. The reduction in specific activity was not due to a general inhibition of $^{14}CO_2$ incorporation, since this occurred to only about 5 to 10%. These results show that the formation of ABA from a preformed precursor was not inhibited by fluridone, but that the formation of ABA from $^{14}CO_2$ was inhibited and to about the same extent as was the synthesis of the xanthophylls. These results are also consistent with a xanthophyll precursor role. They suggest that the inability of fluridone and similar compounds to inhibit ABA synthesis in green leaves may be due to a sufficient pool of preexisting carotenoids. They also suggest that the lack of ABA synthesis in carotenoidless mutants and in bleached leaves could be due to a lack of carotenoids, rather than to an indirect effect on chloroplast development.

Xanthophylls with 2 different types of ring substitution are candidates for ABA precursors. Violaxanthin and lutein are representative of the 2 types. Violaxanthin contains 2 identical rings with a 3-hydroxyl group and a 5,6-epoxide. Neoxanthin and lutein-5,6-epoxide are other leaf xanthophylls which contain one ring with this substitution. Since these compounds contain 2 oxygens in their rings, no further ring oxygens need be added if they were cleaved to xanthoxin and then converted to ABA. Cleavage of lutein or lutein-5,6-epoxide by oxygen could produce one molecule of 4'-hydroxy-α-ionylidene acetaldehyde (Fig. 7). This compound, which apparently has not been isolated from plants, is one of the possible intermediates on the *C. rosicola* pathway. The results of the Creelman-Zeevaart experiment, however, indicated that ^{18}O was not incorporated into the ABA ring, which would be expected if 4'-hydroxy-α-ionylidene acetaldehyde is an ABA precursor.

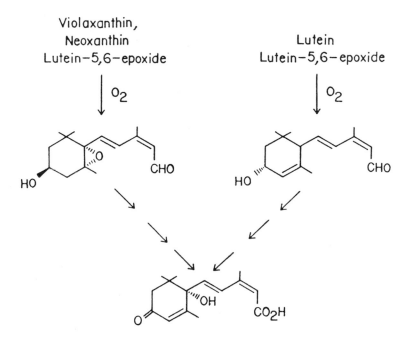

Fig. 7. Hypothetical cleavage of various xanthophylls by O_2 and subsequent conversion of the products to ABA

Although the experiments described in the previous paragraphs suggest that xanthophylls may be ABA precursors, they did not give any information about the identities of the xanthophylls. In order to test whether violaxanthin is an ABA precursor, intact leaves were treated so that the epoxide oxygens were partially replaced by ^{18}O (38). The leaves were then water-stressed and both violaxanthin and ABA were isolated and analyzed by mass spectrometry. If the ABA produced during the water stress period had been derived from violaxanthin containing ^{18}O in its epoxide group, we expected to observe ^{18}O at its 1'-hydroxyl group (Fig. 8). The results suggested that a portion of the ABA was derived from the violaxanthin which had been labelled wth ^{18}O, but that this violaxanthin accounted for only about 25% of the ABA produced. Whether the remainder of the ABA arose from other xanthophylls, from non-xanthophylls, or from violaxanthin which had not been labelled with ^{18}O remains to be determined. In addition to the experiment just described, we also repeated the Creelman-Zeevaart experiment. Our results were similar to theirs in that ^{18}O was incorporated in high yields into one of the carboxyl oxygens. In addition, however, there was evidence that 15 to 20% of the ring oxygens contained ^{18}O. Such incorporation suggests that ABA is derived, at least in part, from a precursor which requires that a ring oxygen be inserted when it is converted to ABA. As suggested, such a precursor could be lutein or lutein-5,6-epoxide.

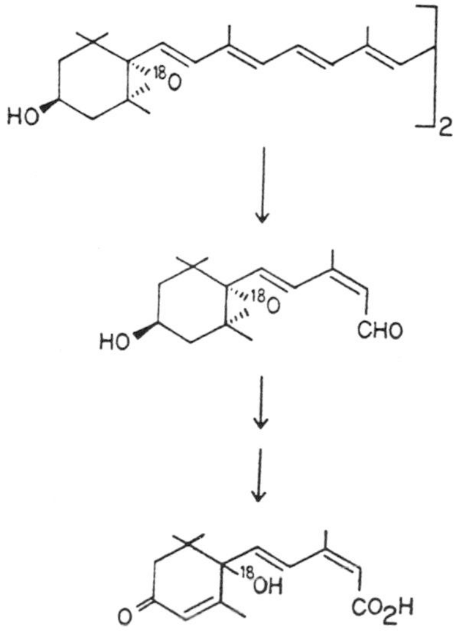

Fig. 8. Conversion of violaxanthin containing ^{18}O in the epoxide to ABA containing ^{18}O at the 1'-hydroxyl position

Clearly, there are now several lines of evidence which suggest that ABA is derived from xanthophylls in plants, although definitve evidence is still lacking. If xanthophylls are ABA precursors, there are still a formidable number of questions to be answered. Among these are: Is ABA derived only from xanthophylls in all tissues and under all conditions; which xanthophylls are precursors; which compounds are intermediates between the xanthophylls and ABA; where in the cell is ABA synthesized; how is synthesis controlled in stressed and non-stressed leaves as well as in other tissues. If definitive evidence can be obtained for the participation of xanthophylls as ABA precursors, it may be possible to obtain answers to the other questions in a shorter time than it has taken us to reach our present level of understanding.

ABA METABOLISM

Metabolites and Pathways

Research into ABA metabolism, which began shortly after its discovery, has been successful in identifying ABA metabolites. This success has been due in part to the early availability of ^{14}C-ABA, and also to the rapid and extensive metabolism of ^{14}C-ABA when it is applied to plant tissues.

In discussing the metabolism of a compound which is administered to a plant, we must be careful to distinguish between those metabolites which are later shown to occur naturally in the plant and those which are not. This distinction is particularly important in the case of ABA in which most feeding experiments have been done with a racemic mixture of the naturally occurring S-enantiomer and the unnatural R-enantiomer (Fig. 1). There is evidence that the 2 enantiomers are metabolized not only at different rates, but in some instances to different compounds. Care must be taken to determine that a metabolite has in fact been derived from the S-enantiomer. Ultimately, the isolation of the metabolite as a naturally occurring plant constituent is required in order to demonstrate that the compound is not an artifact of feeding. A number of ABA metabolites have been shown to occur naturally, while others have not been isolated from plants which have not been fed ^{14}C-RS ABA.

The initial ABA metabolites described were ABA glucose ester (ABA GE, 6, Fig. 10), phaseic acid (PA, 3, Fig. 9) and 6'-hydroxymethyl ABA (HM ABA, 2, Fig. 9). ABA glucose ester was first isolated as a naturally occurring compound from *Lupinus luteus* fruit and then shown to be produced when ^{14}C-ABA was fed to several plants (19). Although this compound is a naturally occurring ABA metabolite, feeding experiments with ^{14}C-RS ABA can exaggerate its importance since the R-enantiomer is often converted preferentially to the glucose ester. HM ABA was

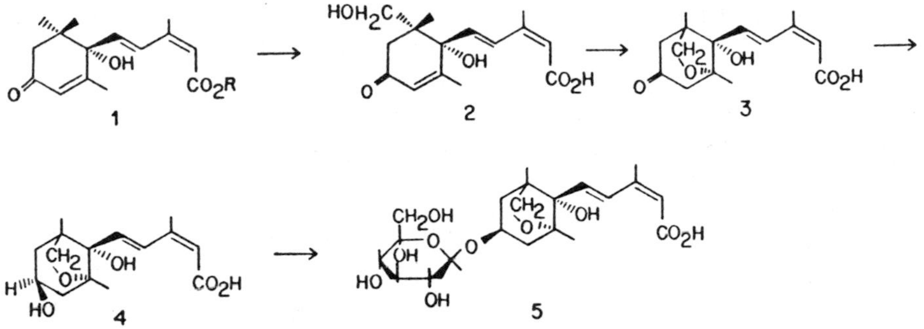

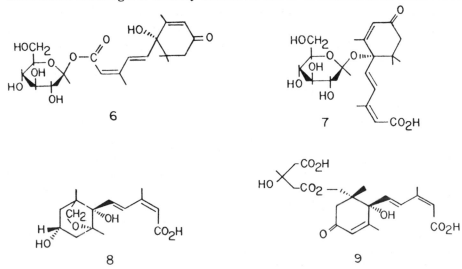

Fig. 9. Metabolic pathways for the conversion of ABA to DPA GS

isolated from tomato shoots which had been fed ^{14}C-ABA. This compound has been isolated only once, apparently because of the ease with which it rearranges to PA. PA had actually been isolated prior to the discovery of ABA, but its correct structure was only determined as the rearrangement product of HM ABA (18). The observation that HM ABA so readily rearranged to PA in the absence of enzymes, made it unclear whether PA was a natural ABA metabolite or an artifact of isolation. The discovery that an apparent PA reduction product dihydrophaseic acid (DPA, 4, Fig. 9) was present in very high concentrations in bean seeds suggested that PA is a naturally occurring ABA metabolite subject to further metabolism (37). Since the initial descriptions of PA and DPA, these compounds have been found in a wide variety of plants, and appear to be on the major pathway of ABA metabolism. An epimer of DPA, epi-DPA (8, Fig. 10) has also been shown to be a naturally occurring ABA metabolite although it usually occurs in lower concentrations than DPA

Fig. 10. Naturally occurring ABA metabolites not shown in Fig. 9

(39). Although DPA exists in high concentrations in some tissues, it is clear that it is not necessarily the endproduct of metabolism. Recently the 4'-glucoside of DPA (DPA GS, 5, Fig. 9) has been isolated from several tissues (13, 23) and there are indications from 14C-ABA feeding experiments that even it may be further metabolized.

Figure 10 shows 2 other naturally occurring ABA metabolites. The 1'-glycoside of ABA (ABA GS, 6) may be widespread in plants, although probably in low levels, and 3-hydroxy-3-methyl glutaryl HM ABA (9, Fig. 10) which has so far been isolated only from the seeds of *Robinia pseudoacacia*(17). Feeding experiments with ^{14}C-RS ABA show that other metabolites can be formed, such as base-labile conjugates of PA, DPA and epi-DPA. These compounds have not yet been characterized, nor have they been shown to occur naturally (17).

Effects of Metabolism on Physiological Activities

The identification of metabolites and the elucidation of metabolic pathways are intellectually challenging and interesting in their own right. In the case of a hormone, however, the plant physiologist is more interested in the role that metabolism plays in controlling the levels of active hormone. Metabolism can play a number of roles, including activation and inactivation of hormonal activities and conversion of a hormone to storage and/or transport forms. Unfortunately, it is often difficult to assess the effects of metabolism on hormone activity. The effectiveness of any applied substance will depend on factors besides its intrinsic activity. These will include its rate of uptake into the tissue, its metabolism to more or less active compounds after uptake, and its rate of entry into the proper cellular compartment.

We often do not know whether the results observed when a hormone is applied to plant tissue are indicative of its actual role in the plant, so it is hard to assess the significance of the apparent changes in the activities of the metabolites. We may not even be aware of the proper activity to measure. Regardless of these *caveats*, I think it is still useful to discuss how metabolism may affect physiological activities.

Inactivation

When analogues of ABA are compared with ABA in their ability to affect various physiological processes, it has been found that almost any change in the ABA molecule reduces the apparent activity. The activity of ABA in the full range of bioassays tested appears to depend on the presence of a free carboxyl group, a 2-cis,4-trans-pentadienoic side chain, a 4'-ketone and a double bond in the cyclohexane ring (36). Many of the metabolites described so far lack one or more of these functional groups. ABA GE lacks the free carboxyl, PA the ring double bond, and DPA and its further metabolites lack both the ring double bond and the 4'-keto group. HM-ABA, its 3-hydroxy-3-methyl glutaryl derivative and ABA-1'-glycoside all appear to have the necessary functional groups of ABA intact. Neither PA nor DPA appear to inhibit cell elongation in various tissues which indicates that the presence of the 4'-ketone and the ring double bond are both necessary for this activity. ABA glucose ester has been reported to inhibit cell elongation, but it seems likely that the activity depends on its hydrolysis to ABA as has been shown to be the case for other ABA esters. The 3-hydroxy-3-methyl conjugate of HM ABA has been reported to inhibt cell elongation. Whether the activity requires the intact molecule was not determined. PA has been reported to have only about 10% of ABA in an abscission bioassay, and has reduced activity in stomatal closure bioassays in several plants, although apparently no activity in closing the stomata of *V. faba*. DPA has no activity in any of the bioassays in which it has been tested. It seems reasonable to conclude that DPA synthesis, and that of its further metabolites, is a primary mode of ABA inactivation.

Activation

Although PA is inactive, or has a greatly reduced activity, in several bioassays compared with ABA, it does appear to be as active as ABA in reducing the GA_3-stimulated synthesis of α-amylase in barley aleurone layers. DPA, however, has no activity in this assay, so that the 4'-keto group, but not the ring double bond, is necessary for the results observed with PA and ABA. One explanation for PA activity in this assay is that both ABA and PA are able to bind to and activate the necessary receptor equally well. A second explanation is that PA is the active molecule and that the apparent ABA activity is derived from its conversion to PA, which does occur readily in barley aleurone layers. In order to distinguish between these 2 possibilities it will be necessary to block the conversion of ABA to PA. Either a specific inhibitor of ABA hydroxylation or a mutant which lacks the ability to metabolize ABA to PA will be required.

HM-ABA contains all of the ABA functional groups and has in addition a hydroxyl group at the 6'-methyl position. One intriguing possibility is that it is HM-ABA, rather than ABA, which is the active molecule in many of the ABA bioassays. Since all of the tissues which have been fed ^{14}C-ABA produced PA, and therefore HM-ABA, it is not possible to rule out an active role for HM-ABA. It is also not possible to test HM-ABA's activity directly. Even if it could again be isolated, it would be difficult to keep it from isomerizing to PA during any bioassay. As in the case of the barley aleurone bioassay, it will be necessary to find a specific inhibitor of ABA hydroxylation or a mutant that is unable to hydroxylate ABA.

Abscisic acid biosynthesis and metabolism

Storage and Transport Forms
One requirement that a metabolite must meet in order to be a storage or transport form of ABA is that it must be reconvertable to ABA. Of the various metabolites that have been described, only ABA GE and ABA GS seem likely to be able to meet this requirement, since esterase and glycosidase activity could release ABA from ABA GE and ABA GS, respectively. There is no evidence that either PA or DPA can be reconverted to ABA. ABA was shown to be released when ABA GE was fed to tomato plants. It is difficult, however, to determine whether hydrolysis of endogenous ABA GE occurs. The evidence obtained so far suggests that ABA GE is not a major source of ABA. When plants were subjected to a series of stress and rewatering cycles, ABA GE continued to increase in concentration (3). ABA GE has been shown to occur in the phloem and xylem. If it were hydrolyzed when it reached the sink tissue, it could be considered a transport form of ABA. ABA has also been shown to occur in the xylem and phloem, so it seems unlikely that ABA GE is necessary as a long distance transport form of ABA. There is no evidence about the fate of ABA GS. Since it appears to be a minor metabolite, it seems unlikely that it plays a major role as an ABA storage and/or transport form (17).

LOCALIZATION AND REGULATION OF METABOLISM

The capacity to metabolize applied ABA is widespread in plant tissues. ^{14}C-ABA has been shown to be metabolized by leaves, stems, roots, seeds and fruit and ABA metabolites have been isolated from all of these tissues (17, 35). Since ABA metabolites have also been found in the xylem and phloem, however, the occurrence of metabolites in a particular tissue does not necessarily indicate that it was formed there. There is less evidence about the subcellular localization of ABA metabolism. The conversion of ABA to HM ABA by a cell-free system obtained from the liquid endosperm of *Echinocystis lobata* was suggested to involve a mixed-function oxygenase (7). The report that the oxygen inserted into ABA to form HM ABA is derived from O_2 is further evidence for the participation of an oxygenase (4). Since mixed function oxygenase activity is generally associated with the endoplasmic reticulum, these results suggest a cytoplasmic location for the initial step of ABA metabolism. The conversion of PA to DPA in the *E. lobata* system appeared to involve a soluble enzyme, presumably also of cytoplasmic origin. The conversion of ABA by spinach mesophyll protoplasts to PA and DPA was shown to occur in the extraplastidic fraction, indicating that chloroplasts do not metabolize ABA (11).

An enzyme which catalyzes the transfer of glucose from UDP-glucose to form ABA GE has been isolated from cell suspension cultures of

Macleaya microcarpa (16). Presumably the enzyme is of cytoplasmic origin.

Because our knowledge of the ABA metabolic enzymes is limited, our knowledge of the regulation of metabolism is also limited. There are indications, however, of possible regulation in 2 systems. If barley aleurone layers are pretreated with 10^{-5}M ABA for several hours, the subsequent metabolism of exogenous ABA is from 2 to 5 times greater than it would have been without the pretreatment (34). The increased metabolic rate can be eliminated by inhibitors of protein and nucleic acid synthesis. These inhibitors also eliminate several new proteins which appear after ABA pretreatment. The suggestion has been made that ABA induces the synthesis of the ABA hydroxylating enzyme in this tissue since PA does not cause the new proteins to appear, nor does it increase ABA metabolism. The isolation of the hydroxylating enzyme from the aleurone layers has not been successful, so it has not been possible to demonstrate an increase of the enzyme directly. In water stressed leaves, ABA metabolism as well as ABA biosynthesis appears to be enhanced. There is no evidence, however, about the origin of this increased activity.

Acknowledgements

This work supported in part by NSF Grant PCM 8219122.

References

1. Asante,G., Merlini, Nasini,G. (1977) (+)-Abscisic acid, a metabolite of the fungus *Cercospora rosicola*. Experientia *33*, 1556.
2. Bennett,R.D., Norman, S.M., Maier, V.P. (1981) The biosynthesis of abscisic acid from (1,2-$^{13}C_2$) acetate in *Cercospora rosicola*. Phytochemistry *20*, 2343-2444.
3. Boyer,G.L., Zeevaart,J.A.D. (1982) Isolation and quantitation of -D-βglucopyranosyl abscisate from leaves of Xanthium and spinach. Plant Physiol. *70*, 227-231.
4. Creelman,R.A., Zeevaart,J.A.D. (1984) Incorporation of oxygen into abscisic acid and phaseic acid from molecular oxygen. Plant Physiol. *75*, 166-169.
5. Dorffling,K., Peterson,W. (1984) Abscisic acid in phytopathogenic fungi of the genera *Botrytis, Ceratocystis, Fusarium* and *Rhizoctonia*. Z. Naturforsch. *39*c, 683-684.
6. Firn, R.D., Burden,R.S., Taylor,H.F. (1972) The detection and the estimation of the plant growth inhibitor xanthoxin in plants. Planta *103*, 263-266.
7. Gillard,D.F., Walton,D.C. (1976) Abscisic acid metabolism by a cell-free preparation from *Echinocystis lobata* liquid endosperm. Plant Physiol. *58*, 790-795.
8. Goodwin,T.W. (1980) The biochemistry of carotenoids, Vol. 1, 2nd edition. Chapman and Hall: London and New York.
9. Griffin,D.H., Walton, D.C. (1982) Regulation of abscisic acid formation in *Mycosphaerella (Cercospora) rosicola* by phosphate. Mycologia *74*, 614-618.
10. Hartung,W., Heilmann,B. and Gimmler,H. (1981) Do chloroplasts play a role in abscisic acid synthesis? Plant Sci. Letters, *2*, 235-242.
11. Hartung,W., Gimmler,H. (1982) The compartmentation of abscisic acid (ABA), of ABA-biosynthesis, ABA-metabolism and ABA-conjugation. *In* Plant Growth Substances 1982, ed. P.F. Wareing, pp. 324-333. London: Academic Press.

12. Henson,I.E. (1984) Inhibition of abscisic acid accumulation in seedling shoots of pearl millet (Pennisetum americanum L.) following induction of chlorosis by norflurazon. Z. Pflanzenphysiol. *114*, 35-43.
13. Hirai,N. and Koshimizu,K. (1983) A new conjugate of dihydrophaseic acid from avocado fruit. Agric. Biol. Chem. *47*, 365-371.
14. Koshimizu,K., Inui,M., Fukui,H., Mitsui,T. (1968) Isolation of (+)-abscisyl-β-D-glucopyranoside from immature fruit of *Lupinus luteus*. Agric. Biol. Chem. 32, 789-791.
15. Lehmann,H., Schutte,H.R. (1976) Biochemistry of phytoeffectors. Biochem. Physiol. Pflanzen *169*, 55-61.
16. Lehmann,H., Schutte,H.R. (1980) Purification and characterization of an abscisic acid glucosylating enzyme from cell suspension cultures of *Macleaya microcarpa*. Z. Pflanzenphysiol., *96*, 277-280.
17. Loveys,B.R., Milborrow,B.V. (1984) Metabolism of abscisic acid. In *The Biosynthesis and Metabolism of Hormones*. ed. Alan Crozier and J.R. Hillman, pp. 71-104. Cambridge Univ. Press, Cambridge.
18. Milborrow,B.V. (1969) Identification of 'metabolite C' from abscisic acid and a new structure for phaseic acid. Chem. Commun. pp. 966-967.
19. Milborrow,B.V. (1974) Biosynthesis of abscisic acid by a cell-free system. Phytochemistry *13*,131-136.
20. Milborrow,B.V. (1974) The chemistry and physiology of abscisic acid. Ann. Rev. Plant Physiol. 25, 259-307.
21. Milborrow,B.V. (1978) Abscisic acid. In Phytohormones and Related Compounds: A Comprehensive Treatise, eds. D.S.Letham, P.B. Goodwin, and T.V.J. Higgins, vol. 1, pp. 295-347. Amsterdam, Oxford, New York: Elsevier/North Holland Biomedical Press.
22. Milborrow,B.V. (1983) Pathways to and from abscisic acid. In *Abscisic Acid* Ed. F.T. Addicott, pp. 79-111, Praeger: New York.
23. Milborrow,B.V. and Vaughan,G.T. (1982) Characterization of dihydrophaseic acid 4'-O-β-D-glucopyranoside as a major metabolite of abscisic acid. Australian Journal of Plant Physiology, *9*, 361-372.
24. Moore,R., Smith,J.D. (1984) Growth, graviresponsiveness and abscisic acid content of *Zea mays* seedlings treated with fluridone. Planta *162*, 342-344.
25. Moore,R., Smith, J.D. (1985) Graviresponsiveness and abscisic acid content of roots of carotenoid-deficient mutants of\Zea mays. Planta\164, 126-128.
26. Neill,S.J., Horgan,R., Lee,T.S., Walton,D.C. (1981) 3-methyl-5-(4'-oxo-2',6',6'-trimethyl-cyclohex-2'-en-yl)-2,4-pentadienoic acid, a putative precursor of abscisic acid from *Cercospora rosicola*. FEBS Lett. *128*, 30-32.
27. Neill,S.J., Horgan,R., Walton,D.C. (1984) Biosynthesis of abscisic acid. In *The Biosynthesis and Metabolism of Hormones*. ed. by Alan Crozier and J.R. Hillman. Cambridge Univ. Press, Cambridge, U.K..
28. Norman,S.M., Maier,V.P.,Echols,L.C. (1981) Development of a defined medium for growth of *Cercospora rosicola* Passerini. Appl. Environ. Microbiol. 41:981-85.
29. Oritani,T., Yamashita,K. (1985) Biosynthesis of (+)abscisic acid in *Cercospora cruenta*. Ag. Biol. Chem. *49*, 245-249.
30. Quarrie,S.A., Lister,P.G. (1984) Evidence of plastid control of abscisic acid accumulation in barley (*Hordeum vulgare* L.) Z. Pflanzenphysiol. *114*, 295-308.
31. Robinson,D.R., Ryback,G. (1969) Incorporation of tritium from $(4R)$-4-^{3}H-mevalonate into abscisic acid. Biochem. J. *113*, 895-897.
32. Taylor,H.F., Smith,T.A. (1967) Production of plant growth inhibitors from xanthophylls: a possible source of dormin. Nature *215*, 1513-1514.
33. Taylor, H.F., Burden,R.S. (1973) Preparation and metabolism of 2-^{14}C-cis, trans xanthoxin. J. Exptl. Bot. *24*, 873-880.
34. Tuan-hua,D.H., Uknes,S.J. (1982) Regulation of abscisic acid metabolism in the aleurone layers of barley seeds. Plant Cell Reports, *1*, 270-273.

35. Walton,D.C. (1980) Biochemistry and physiology of abscisic acid. Annual Review of Plant Physiology *31*, 453-489.
36. Walton,D.C. (1983) Structure-activity relationships of abscisic acid analogs and metabolites. *In Abscisic Acid* Ed. F.T. Addicott pp. 113-146, Praeger, New York.
37. Walton,D.C., Dorn,B. and Fey, J. (1973) The isolation of an abscisic acid metabolite, 4'-dihydrophaseic acid, from non-imbibed *Phaseolus vulgaris* seed. Planta, *112*, 87-90.
38. Walton,D.C., Li, Yi, Neill,S., Horgan,R. (1985) ABA biosynthesis: a progress report. *In* Current Topics in Plant Biochemistry and Physiology, Vol. 4 ed. D.D. Randall, D.G. Blevins, R.L. Larson, B.J. Rapp. Interdisciplinary Plant Biochemistry and Plant Physiology Program: Columbia, MO.
39. Zeevaart, J.A.D. and Milborrow,B.V. (1976) Metabolism of abscisic acid and the occurrence of *epi*-dihydrophaseic acid in *Phaseolus vulgaris*. Phytochemistry, *15*, 493-500

C. HOW HORMONES WORK

C1. Auxin and Cell Elongation

Robert E. Cleland

Department of Botany, University of Washington, Seattle, Washington
98195, USA.

INTRODUCTION

One of the most dramatic and rapid hormone responses in plants is the induction by auxins of rapid cell elongation in isolated stem and coleoptile sections. The response begins within 10 minutes, results in a 5-10 fold increase in the growth rate, and persists for hours or even days (36). It is hardly surprising that this may be the most studied hormonal response in plants.

How do auxins produce this response? To answer this, the process of cell enlargement must first be considered. Cell enlargement consists of two interrelated processes; osmotic uptake of water, driven by a water potential gradient across the plasma membrane, and extension of the existing wall area, driven by the turgor-generated stress within the wall. The process of cell enlargement can be described (31) by two equivalent equations:

$$dV/dt = Lp.\Delta\Psi \qquad\qquad \text{Equation 1}$$

$$dV/dt = m(P-Y) \qquad\qquad \text{Equation 2}$$

where dV/dt is the rate of increase in cell volume, Lp is the hydraulic conductivity, $\Delta\Psi$ is the water potential gradient across the plasma membrane, m is the wall extensibility, P is the turgor pressure and Y is the wall yield threshold (the turgor that must be exceeded for wall extension to occur). They can be combined into a third equation:

$$dV/dt = \frac{m.Lp}{m+Lp}\left(\Psi_a - \Pi - Y\right) \qquad\qquad \text{Equation 3}$$

where Ψ_a is the apoplastic water potential and Π is the osmotic potential of the cell. In the absence of auxin, the growth rate is low because of a low m, low Lp, low P or high Y (or a combination of these). When auxin *initiates* rapid cell enlargement, it must do so by increasing m, Lp, Ψ_a or P, or by decreasing π or Y. No matter what the initial effect of auxin is (e.g., gene activation, ATPase activation, or change in membrane permeability) increased cell enlargement can be initiated *only* if the ultimate effect is a change in one of these cellular growth parameters.

Once rapid cell enlargement is initiated, other changes are required if the rate is to be maintained. For example, as water is taken up in the process of cell enlargement, will rise, with the result that the growth rate will decline. This can be prevented if osmoregulation occurs; i.e., if osmotic solutes are generated (by hydrolysis of starch or by photo-synthesis) or taken into the cell in amounts sufficient to maintain. Likewise, as the walls extend, their mechanical properties will change. This can be counteracted by synthetic processes which maintain the ability of the walls to undergo biochemical wall loosening (and thus an increase in m). Auxin may influence these maintenance processes, as well as initiating rapid cell enlargement. The overall process of cell enlargement, then, is a complex series of events, several of which may be influenced by auxin. If the role of auxin is to be understood, it is essential to separate the effects of auxin on the initial growth response (IGR) from its effects on the prolonged growth response (PGR) and on the processes involved in growth maintenance. This can be done by examining the growth response in the first 30-90 minutes (IGR) or after steady-state growth has been achieved, e.g., after 3-6 hours (PGR).

Another factor that must be considered is the heterogeniety of auxin-responsive tissues. Consider the anatomy of a dicot seedling stem. On the outside there is a layer of epidermis, under which are several subepi-dermal layers of collenchyma-like cells. Inside of this are the large, parenchyma cells of the cortex in which are embedded the vascular strands. Studies utilizing slit stems or bored-out stem sections have shown that the cells differ in their involvement in auxin-induced growth (35). The subepidermal layers are the major auxin-receptive cells, whose cellular growth parameters must be altered by auxin. The remaining cells simply enlarge in concert with these subepidermal cells, even though their cellular growth parameters may not be directly affected by auxin. (This fact is the basis of the "slit-pea test", which used to be one of the standard bioassays for auxin). In *Avena* coleoptiles, on the other hand, all cell layers appear to be more nearly equally important in regulating the rate of cell enlargement (13). In studying auxin-action, it is important to focus on the "driver" cell layers, and not on the remainder of the cells. Many of the apparent contradictions in the literature may be due to the use of whole tissues rather than just the auxin-response tissues in auxin studies.

Characteristics of Auxin-induced Cell Enlargement

Let us now consider some of the characteristics of auxin-induced cell enlargement (Table 1). First consider the time-course (Fig. 1). Upon addition of auxin there is a lag of at least 8 minutes before the growth rate begins to increase. Then the rate rises until a maximum is reached after 30-60 minutes. The length of the lag can be increased from 8-10 minutes by lowering the temperature or by using suboptimal auxin concentrations, but it cannot be decreased below 8-10 minutes by raising the

Table 1. Characteristics of the Initial Growth Response to Auxin

1) Minimum lag 8–10 minutes

2) Growth rate $= \log$ (IAA)

3) Requires:
    a) ATP synthesis
    b) Active ATPase
    c) Continual protein synthesis
    d) Turgor in excess of yield threshold

4) Not required:
    a) Exogenous sugars
    b) Exogenous K^+ or Ca^{++}

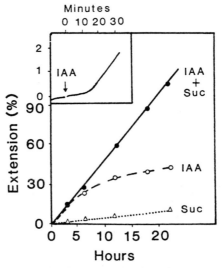

Fig. 1. Time course of auxin-induced growth of *Avena* coleoptile sections. Section incubated in 10 mM phosphate buffer, pH 6.0, $\pm$ 2% sucrose and with 10 uM IAA added at time zero. Rapid elongation is initiated after an 8-10 minute lag. The initial rate is independent of the presence or absence of absorbable solutes (sucrose), but absorbable solutes are required for continual rapid elongation.

temperature, using superoptimal auxin or by removal of the cuticle surrounding the tissue (19). Thus we must conclude that the lag is not the time needed for auxin simply to penetrate to its site of action. This is further indicated by the fact that in some, but not all cases, the growth rate actually decreases slightly immediately after addition of auxin (19). The lag is independent of the rate of protein synthesis; thus it is unlikely to reflect the time required to synthesize some new protein (20). The rapidity with which the maximum growth rate is achieved, and the actual rate, are dependent on the tissue and can be influenced by the past history of the tissue. For example, freshly cut maize coleoptile sections respond slowly and rather poorly to auxin, while sections "aged" in water for 3 hours are far more responsive (52). The maximum growth rate in this tissue is 8-10 %/hour. In contrast, oat coleoptiles show no such change in sensitivity to auxin, but have a maximum growth rate of only 4-5 %/hour (7). Once a maximum rate is achieved, it can be maintained for up to 18 hours in oat coleoptiles, as long as auxin and absorbable solutes are present (steady-state kinetics) (7). In most dicot stem sections, however, there is a decrease in the growth rate after the first maximum is reached, followed by a rise to a second, lower steady-state rate (peak-and-valley kinetics) (51) (Fig. 2). The initial peak in the growth response is the IGR, while the subsequent enhanced growth rate is the PGR. Unlike coleoptile

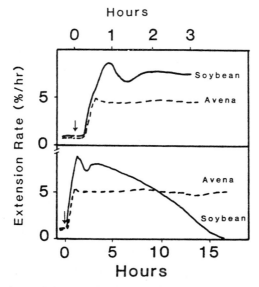

Fig. 2. Comparison of growth kinetics for *Avena* coleoptile and soybean hypocotyl sections, incubated with IAA (10 uM) and sucrose (2%). The growth rate is enhanced after a lag of 10-12 minutes for both. The rate for *Avena* sections reaches a maximum after 30-45 minutes and then remains nearly constant for 18 hours (steady-state kinetics). The rate for soybean hypocotyls increases to a first maximum after 45-60 minutes, then declines before climbing to a second maximum (peak-and-valley kinetics), after which it steadily falls over the next 16 hours as the turgor falls due to dilution of the osmotic solutes during growth.

tissues, dicot stems sections rarely maintain a constant growth rate for more than 2-3 hours; thereafter the growth rate falls continuously so that by 16-20 hours growth has ceased (23).

In most tissues both the first maximum growth rate and the average growth rate over the first 2-4 hours are proportional to the log of the external auxin concentration over a range of about $2\frac{1}{2}$ decades, but the concentration range varies from tissue to tissue (Fig. 3). Thus for oat

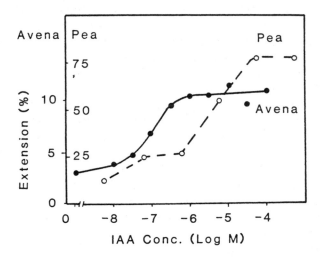

Fig. 3. A comparison of the auxin concentration curves for *Avena* coleoptile (7) and green pea stem sections (23). For both, the growth rate is proportional to the log(IAA) over a 300-fold range, but the range differs between the two tissues.

coleoptiles the range is from 3 nM to about 1 uM (7), while for light-grown pea stems it is 300 nM to 50 uM (23). This initial rate of auxin-induced growth is not increased by the addition of any particular ion or sugar; auxin, alone, is sufficient to initiate this response (10). On the other hand, the maintenance of rapid cell enlargement in coleoptile tissues requires the presence of some solute that can be absorbed. KCl, NaCl, glucose or sucrose are all equally effective at optimal concentrations (45). For most dicot stem sections, however, addition of either ions or sugars is without effect on the growth rate. This correlates with the fact that only low levels of sugars or ions are taken into dicot stem sections, with the result that is not maintined, and rapid cell elongation cannot occur over prolonged periods of time (13).

Auxin-induced cell enlargement is an energy-requiring process. All inhibitors of ATP synthesis (e.g., KCN, DNP, azide) or ATPase activity (vanadate, DCCD, DES) block auxin-induced growth within minutes (19, 36). These data indicate that the energy of ATP drives some critical step in cell enlargement, and that an ATPase must be involved in some step. All inhibitors of protein synthesis so far tested also inhibit auxin-induced growth within minutes after they inhibit protein synthesis (2). It has been suggested that some short-lived protein is required for auxin action (6), but the possibility that it is protein synthesis, in general, rather than any specific protein that is required has not been eliminated. RNA synthesis inhibitors are not nearly as effective (13).

Another requirement for auxin action, at least in coleoptile tissues, is a sufficient cell turgor. Addition of osmoticum sufficient to reduce (Ψ_a-π-Y) to zero inhibits elongation by eliminating the driving force for wall extension, but it also largely blocks the ability of auxin to increase **m**. This is shown (Fig. 4) by the fact that when turgor is restored after a period of time with auxin at reduced turgor, there is no burst of growth, as would be expected if auxin had been acting normally during the period of low turgor (39).

In conclusion, then, we can say that the initiation of auxin-induced growth requires the continued presence of auxin, a continued supply of ATP and active ATPases, protein synthesis and turgor in excess of Y. Any mechanism to explain auxin-induced elongation must take into account these requirements, as well as the 8-10 minute lag which always occurs.

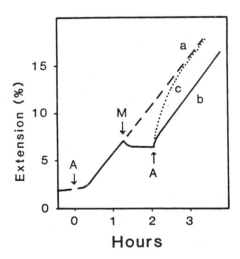

Fig. 4. Demonstration that turgor is required for auxin-induced wall loosening in *Avena* coleoptile sections (lack of stored growth). Sections were treated in water with IAA (10 uM) at A, then transfered to IAA + 0.2 M mannitol at M. Curve a is a projection of the growth if the section had remained in A. Upon return to A, rapid growth resumes (b). If auxin-induced wall loosening had occurred during the period of low turgor the extension would have followed curve c, since the growth potential would have been stored up.

THE INITIATION OF CELL ENLARGEMENT BY AUXIN

The Cellular Parameters

Whenever auxin alters the rate of cell enlargement, it can only do so by changing one or more of the cellular parameters in equations 1-3. For example, an increase in the growth rate requires an increase in wall extensibility (m), an increase in turgor (P) or a decrease in the wall yield threshold (Y) in order to satisfy equation 2. Which of these is affected by auxin? Only P can be measured directly, using the micropressure probe. In pea stem sections, the only tissue to be tested with this technique so far, addition of auxin causes a large increase in the growth rate without any change in P (16). Two indirect tests for the effect of auxin on Y agree (10, 15), in that there is no change in Y in response to auxin over at least short time periods. A number of tests have been devised to assess the wall extensibility, including the Instron technique, stress-relaxation, creep tests, bending assays and most recently, turgor relaxation. None of these tests is a direct measure of m, but all agree that auxin causes a large increase in m whenever it promotes cell elongation (5, 11). We can say with some confidence, then, that auxin induces rapid cell enlargement by increasing wall extensibility.

What does an increase in m mean? Cell walls consist of a tangle of polysaccharide chains. When under tension, they behave as a viscoelastic material, and extend to a limited extent before crosslinks and chain entanglements stop further extension (5, 46). For extension to occur, a crosslink or entanglement must be cleaved biochemically; each of these "wall loosening events" permits a small amount of additional extension to take place. Wall extensibility is primarily the rate at which these wall loosening events occur. Auxin must increase m by increasing the rate of wall loosening events. How?

One problem is that we do not know what bonds are involved. They might be hydrogen bonds, ionic crosslinks or covalent bonds. Hydrogen bonds can probably be eliminated, since agents such as urea that break hydrogen bonds do not loosen cell walls (48). The possibility that the critical bonds consist of calcium crosslinks between pectic chains has been widely considered, and these bonds may be important in the extensibility of dicot stem cell walls (32). However, in at least *Avena* coleoptiles calcium crosslinks have been shown to be of little importance in wall strength (14). It is more likely that the critical bonds involve one or more of the wall polysaccharides. Bonds within the cellulose microfibrils seem unlikely, since cellulases do not promote cell enlargement (42). On the other hand, xyloglucans seem to undergo some bond breakage in response to auxin, since the molecular weight has been shown to decrease in Azuki bean walls (33), and more xyloglucan is soluble after auxin treatment of pea stem sections (30). Bonds within other hemicelluloses such as the arabinogalactans are also possibilities. It has also been suggested that wall loosening might involve the intercalation of new polysaccharide into the existing wall so as to add new wall area, in a manner analogous to the expansion of bacterial cell walls (5). However, the lack of any obvious correlation between wall synthesis and cell enlargement (1) suggests that this is not correct. Thus our present data, while far from complete, supports the concept that wall loosening involves the cleavage of some bond within one of the hemicellulose components of the wall.

Wall Loosening Factors: Concept and Evidence

It is generally agreed that in order for auxin to induce cell elongation it must first attach to some receptor. The location of this receptor is still uncertain, since auxin receptors have been found in many places in plant cells (41). What is certain is that the receptors are not located in the cell wall. And yet, it is the wall which undergoes the biochemical change which leads to cell enlargement. This can only mean that there is some communication between the cell, when stimulated by auxin, and the cell wall, which leads to wall loosening. This communication must be by means of one or more chemicals, which we can call wall loosening factors (WLF) (Fig. 5).

What are the WLFs which are responsible for auxin-induced growth? This is a subject which has caused great controversy in recent years. To date, only only one WLF has been positively identified; protons. In 1970–71 Hager et al. (26) and Rayle and Cleland (38) independently suggested that when coleoptile or stem cells are stimulated by auxin, they excrete protons into the apoplastic solution, where the lowered pH activates polysaccharide hydrolases which loosen the walls (Fig. 5). This

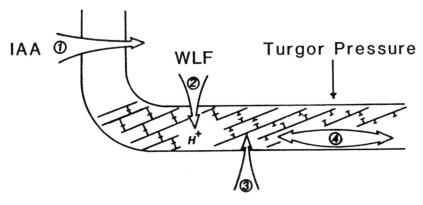

Fig. 5. The wall-loosening factor (WLF) concept and the acid-growth theory. Auxin enters the cell (1) and interacts with a receptor. A WLF is then exported to the wall (2), where it induces wall loosening (3). In acid-growth, the WLF is H^+, and the resulting lowered apoplastic pH activates wall polysaccharidases which cleave load-bearing bonds in the wall (3), permiting turgor-driven wall expansion (4).

"acid-growth theory" can be tested by means of four predictions. If it is correct, it should be possible to show that: 1) auxin causes growing cells to excrete protons, 2) addition of acid to tissues should substitute for auxin and induce rapid cell enlargement, 3) neutral buffers infiltrated into the walls should prevent auxin-induced growth by preventing the decline in wall pH, and 4) any other agent which induces proton excretion should also induce rapid cell elongation. The theory has been tested for only a limited number of tissues to date. With *Avena* coleoptiles, the most thoroughly tested material, the theory is completely confirmed. The apoplastic solution has a pH of about 5.7 in the absence of auxin. Upon addition of auxin, the pH begins to decrease after a lag of about 8-10 minutes, and comes to a new equilibrium at about 4.7 (8). The rate of proton excretion is closely correlated with the growth rate (8). When isolated cell walls are placed under tension, they undergo little irreversible extension at pH 5.7, but at pH 4.7 they elongate rapidly (39). Neutral buffers almost completely inhibit auxin-induced growth, as long as they are able to penetrate into the wall space (18). Finally, the phytotoxin fusicoccin (FC), which causes plant cells to excrete acid at a great rate, causes *Avena* coleoptiles to elongate at rates greater than that induced by auxin (9). Almost identical results have been obtained with soybean hypocotyl sections (40).

With maize coleoptile sections, on the other hand, the evidence suggests that protons are not the only WLF. While auxin does cause proton excretion, added acid does cause some growth and neutral buffers partly inhibit auxin-induced growth, the predicted quantitative relationship between apoplastic pH and growth is not found (29). It would appear that in maize coleoptiles there must be an additional WLF.

As yet, there is no direct evidence for any other WLF. Perhaps the most likely is calcium ions. Calcium ions are known to inhibit the enzymatic wall loosening process (14); removal, by uptake into the cell, would relieve this inhibition and thus promote growth. In dicot walls, where calcium crosslinks may be important in the strength of the wall (32), uptake of calcium would break such links and lead to a second type of wall loosening. Unfortunately, attractive as this idea is, there is really no direct evidence to support or reject it. Other possible WLFs include the secretion of polysaccharidases or the removal of small organic molecules that might inhibit such enzymes.

The Mechanism of Auxin-Induced Proton Excretion

Although protons may not be the only WLF, there can be little doubt that they are an important WLF in some tissues, and that auxin does enhance their excretion from plant cells. The question is, how? Hager et al (26) originally suggested that auxin might activate a proton ATPase, localized in the plasma membrane (Fig. 6). This pump could be

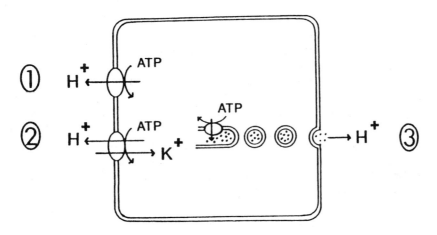

Fig. 6. Three possible mechanisms for auxin-induced proton excretion from plant cells. A) Auxin activates an electrogenic ATPase in the plasma membrane, which exports H^+, and hyperpolarizes the membrane potential. B) Auxin activates an electroneutral ATPase, which exchanges H^+ for K^+. C) Auxin activates an ATPase in the ER, acidifying the lumen of the ER. Vesicles produced from the ER then carry the acid to the plasma membrane, where the contents are released to the apoplastic solution (bucket-brigade theory).

electroneutral, exchanging H^+ for K^+, or it could be electrogenic, only exporting protons from the cell. This brilliant idea is supported by considerable data. As already indicated, inhibitors of ATP synthesis or ATPase action always block auxin-induced proton excretion. It is now well established that the plasma membrane of plant cells does possess a proton ATPase, which is responsible for the electrogenic export of protons (44). If the activity of this ATPase was increased, one would expect the membrane potential of the cells to be hyperpolarized by auxin, with a time course identical to that for proton excretion. In fact, this is just what is observed (3). There is first a slight depolarization, followed by hyperpolarization starting after about 8-10 minutes and ultimately reaching 25 mV. Let us assume, at this point, that the proton excretion is mediated by an ATPase.

Does auxin directly activate this ATPase? Probably not. Even though auxin can alter the efficiency of ATP-driven proton transport into vesicles from a plasma membrane-enriched membrane fraction from pea roots (22), it is doubtful whether auxin is acting directly on the ATPase. For example, the plasma membrane ATPase itself is not an auxin receptor (17), although the auxin receptor in the plasma membrane may be associated with the ATPase so closely that it regulates the activity of this enzyme. The 10 minute lag before the onset of auxin-induced proton excretion, and the requirement for continued protein synthesis cannot easily be fitted into such a scheme. It is more likely that the activation of the ATPase is indirect; the consequence of some other action of auxin. Several possibilities have been suggested. The first is that auxin might act to lower the cytoplasmic pH; it is known that a decrease in the cytoplasmic pH does increase the activity of the electrogenic ATPase (43). This might occur via a release of acid from the vacuole, for example. The evidence for this mode of action is conflicting at present, as measurements of the cytoplasmic pH by the ^{31}P-NMR technique have shown no effect of auxin on the cytoplasmic pH (47), while measurements by fluorescent dyes or by pH-microelectrodes indicate that the cytoplasmic pH does decrease by up to 0.2 pH units after auxin treatment (4). A second possibility is that auxin might alter the cytoplasmic calcium concentration, perhaps by facilitating release of calcium ions from the vacuole (27). The increased calcium, in turn, might activate a calmodulin-requiring protein kinase to phosphorylate some essential membrane proteins such as the ATPase. Attractive as this hypothesis is, there is as yet no direct evidence to support it. These two theories have the attraction that they would explain the lag, but they do not explain the requirement for protein synthesis.

Another possibility is the "bucket-brigade" theory (37). It has been known that vesicles from the ER and golgi transport wall precursors and extracellular proteins to the plasma membrane and then release their contents to the apoplastic solution. It is possible that the ER contains an ATPase which is only active in the presence of auxin (auxin binding sites have been detected on the ER in some tissues(37)). In the absence of auxin, vesicles produced from the ER would be neutral in pH, but in the presence of auxin, the lumen of the ER would become acidic, and any new vesicles produced from the ER would carry acid to the cell wall (Fig. 6). The lag would be a measure of the time for the first of these acid-containing vesicles to be formed and transported to the plasma membrane, while the requirement for protein synthesis would reflect the necessity for new proteins to form the new vesicles. Again, there is a noticable lack of direct evidence to support such an idea, and doubts have been raised as to whether auxin-sensitive cells contain enough vesicles to provide for the known rate of efflux of protons. We must conclude that the mechanism of auxin-induced proton excretion is yet to be established.

THE MAINTENANCE OF AUXIN-INDUCED GROWTH

Any analysis of long-term auxin-induced growth shows that it is more complex than the initial growth response. It has been suggested (50) that the prolonged growth response (PGR) involves a completely different mechanism than the initial growth response (IGR). The fact that exogenous acid induces only a short-lived growth response has been taken as evidence that the acid-growth response is relevant to the IGR (if at all) and not to the PGR. This suggestion seems unlikely, however, as all the characteristics of the IGR are present during the PGR. For example, in both *Avena* (18) and soybean hypocotyl sections (40) neutral buffers are as effective in inhibiting the PGR as the IGR. Likewise, auxin-induced proton excretion is just as great during the PGR as the IGR (13).

It is more likely that the PGR simply requires additional steps and processes, some of which are affected by auxin. As the cells start to enlarge and take up water, the dilution of the osmotic solutes will reduce the effective turgor unless additional solutes are taken up or are manufactured in the cells. Thus when *Avena* coleoptile sections are incubated with auxin but without any absorbable solutes, the growth rate declines after the first hour in parallel with the decline in osmotic concentration, while in the presence of absorbable solutes such as sucrose or KCl, both the growth rate and the osmotic concentration remain nearly constant for hours (45). The rate of solute uptake into *Avena* coleoptile cells is definitely greater in the presence of auxin, but this is not, in fact, a response to auxin but only a response to the growth itself (45). In intact stems, on the other hand, auxin may play a role in facilitating movement of solutes through the phloem from source to sink, in this case the growing zone (34). The nature of this auxin effect, or even its location in the tissue, is still very much in doubt.

A second important parameter is the capacity of cells to undergo auxin-induced growth (12). For example, the capacity of *Avena* coleoptile sections to undergo the IGR (i.e., the acid-growth) can vary considerably. This capacity for acid-induced growth (CAWL) is measured by determining the extension rate of isolated frozen-thawed sections when placed under constant tension and at a constant acidic pH. The CAWL declines with time as live sections are incubated without auxin, while auxin causes a slow increase in CAWL for at least 6-8 hours (Fig. 7). The result is that *Avena* coleoptile cells that have been growing for 8 hours in auxin have almost twice the capacity to extend in response to a particular turgor and acidic pH as do cells that have been lacking auxin for the same period of time. It should be noted that this CAWL is a reversible parameter; it can increase or decrease in response to the presence or absence of auxin.

Cells can also undergo a progressive, irreversible loss of the ability to grow in response to auxin. In mung bean hypocotyls, for example, sections removed at varying distances from the apex (and thus of increasing age), have progressively decreased capacity to undergo auxin-induced growth (24). It would appear that this loss in capacity is irreversible, as the sections did not develop any increase in growth rate with prolonged incubation with auxin.

What is the cause of the change in capacity for auxin-induced growth? In some cases, such as the maize mesocotyl, the decrease in growth

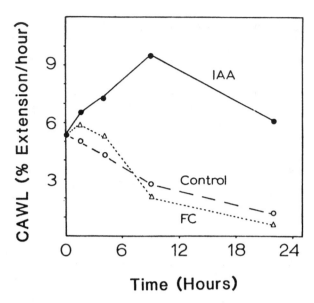

Fig. 7. The effect of auxin on the capacity for acid-induced wall loosening (CAWL) of *Avena* coleoptile sections (12). Sections were incubated 0-20 hours in phosphate buffer without (control) or with 10 µM IAA or fusicoccin (FC). The sections were then frozen-thawed, and the rate of extension in response to 20 g tension at pH 3.0 was determined.

potential is paralleled by a decrease in auxin-binding sites (53). In other cases, there is no lack of auxin-induced proton excretion, but rather a lack of ability of the cell walls to undergo loosening (24). One possibility is that the walls become irreversibly crosslinked, and thus inextensible, due to the deposition of lignin or hydroxyproline-proteins, or due to diferulate or isodityrosine links between existing proteins (21). A careful study correlating the capacity of stem cells to undergo auxin-induced growth with the appearance of these crosslinks has yet to be made.

A second possibility is that at least the reversible capacity is related to a need to regenerate the wall structure lost during the extension process. As the wall extends, it will thin and become more entangled. The intercalation of new wall polysaccharides, while not causing extension in themselves, may restore to the wall the capacity for extension. Likewise, during extension there may be a loss of critical wall loosening enzymes. One role of auxin may be to promote their synthesis. One of the most exciting discoveries in the auxin area is the finding that a small number of gene transcripts are promoted strongly by auxin (25, 49). With several of these mRNAs, the promotion begins coincident with the start of proton excretion (i.e., about 10 minutes after addition of auxin) and reaches a maximum at about the time that the growth rate also reaches its maximum. Other transcripts are induced more slowly, and reach maxima only after several hours (Fig. 8). Could any of these be required to maintain the capacity for auxin-induced growth? As yet we do not know the identity of any of these transcripts. Since we know of a number of enzymes (28) which are induced by auxin (e.g., certain peroxidase, cellulases and ACC synthase), which are not directly involved in cell enlargement, it would be premature to assign to any of these transcripts a role in cell enlargement. Nevertheless, it is tempting to speculate that some of these transcripts will prove to be proteins which are required for the maintenance of the wall loosening capacity, and which are induced by auxin in parallel with the induction of proton excretion. Induction of just proton excretion by agents such as fusicoccin will induce the IGR, but not the PGR, and thus will not duplicate the ability of auxin to promote long-term rapid cell enlargement.

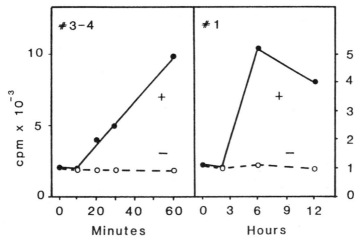

Fig. 8. The effect of auxin on specific mRNAs from pea stem sections (49). Sections were incubated 0-12 hours $\pm$ IAA, then mRNA was isolated, translated with ^{32}(S)-methionine, and proteins were separated on 2-D gels. The amount of transcript produced for two specific transcripts is shown; one of which is induced with a lag of only 10 minutes (#3-4) and a second which has a 2-hour lag (#1).

CONCLUSIONS AND DIRECTIONS FOR FUTURE RESEARCH

While exceptions may exist, the general pattern of auxin-induced cell enlargement is as follows. When auxin is given to a receptive tissue, no consistent response can be detected for 8-10 minutes. Thereupon two events occur, more or less simultaneously. One is the increased activity of an electrogenic proton pump, resulting in acidification of the apoplastic solution and hyperpolarization of the membrane potential. A plasma membrane-associated ATPase is probably involved, although the mechanism by which auxin activates this enzyme is uncertain. The resulting decreased apoplastic pH activates an as yet unidentified polysaccharidase, which cleaves load-bearing bonds in the cell wall and permits turgor-driven wall extension to occur. The second process is the appearance of a few new mRNAs. The enzymes that they code are unknown, and their role in auxin- induced cell elongation is uncertain, but it seems likely that whatever mechanism leads to proton excretion after 8-10 minutes also leads to gene activation. After rapid cell elongation is initiated, its continuance requires several additional processes, some of which are also auxin sensitive. These include osmoregulation to maintain turgor pressure, and maintenance of the capacity of the walls to undergo wall loosening.

Future studies must focus on the holes in this scheme. What is auxin doing during that first 8-10 minutes? Is it altering cytoplasmic pH or cytoplasmic Ca^{++}? Is it altering membrane protein phosphorylation or phosphoinositol turnover? What wall polysaccharidase is being activated by the apoplastic acidity, and what bonds are being cleaved? Are there other wall loosening factors than protons? What are the exact requirements for continued auxin-induced growth? Why does it cease eventually? Experiments, not arguments, will answer these questions.

References

1. Baker, D.B., Ray, P.M. (1965) Direct and indirect effects of auxin on cell wall synthesis in oat coleoptile tissue. Plant Physiol. 40, 345-352.
2. Bates, G.W., Cleland, R.E. (1979) Protein synthesis and auxin-induced growth; inhibitor studies. Planta 145, 437-442.
3. Bates, G.W., Goldsmith, M.H.M. (1983) Rapid response of the plasma- membrane potential in oat coleoptiles to auxin and other weak acids. Planta 159, 231-237.
4. Brummer, B., Felle, H., Parish, R.W. (1984) Evidence that acid solutions induce plant cell elongation by acidifying the cytosol and stimulating the proton pump. FEBS Lett. 174, 223-227.
5. Cleland, R.E. (1971) Cell wall extension. Annu. Rev. Plant Physiol. 22, 197-222.
6. Cleland, R.E. (1971) Instability of the growth-limiting proteins of the Avena coleoptile and their pool size in relation to auxin. Planta 99, 1-11.
7. Cleland, R.E. (1972) The dosage response curve for auxin-induced cell elongation: a reevaluation. Planta 104, 1-9.
8. Cleland, R.E. (1975) Auxin-induced hydrogen ion excretion: correlation with growth, and control by external pH and water stress. Planta 127, 233-42.
9. Cleland, R.E. (1976) Fusicoccin-induced growth and hydrogen ion excretion of Avena coleoptiles: relation to auxin responses. Planta 128, 201-206.
10. Cleland, R.E. (1977) The control of cell enlargement. In Integration of activity in the higher plant, pp. 101-115, Jenning, D.H., ed. Cambridge Press, Cambridge.
11. Cleland, R.E. (1981) Wall extensibility: hormones and wall extension. In: Plant Carbohydrates, Vol. II., Extracellular carbohydrates. Encyl. Plant Physiol., N.S. 13B, pp. 255-273, Tanner, W., Loewus, F.A. ed. Springer, Heidelberg.
12. Cleland, R.E. (1983) The capacity for acid-induced wall loosening as a factor in the control of Avena coleoptile cell elongation. J. Exp. Bot. 34, 676-680.
13. Cleland, R.E. Unpublished data.
14. Cleland, R.E., Rayle, D.L. (1977) Reevaluation of the effect of calcium ions on auxin-induced elongation. Plant Physiol. 60, 709-712.

15. Cosgrove, D.J. (1985) Cell wall yield properties of growing tissues: evaluation by in vivo stress relaxation. Plant Physiol. 78, 347-356.
16. Cosgrove, D.J., Cleland, R.E. (1983) Osmotic properties of pea internodes in relation to growth and auxin action. Plant Physiol. 72, 332-338.
17. Cross, J.W., Briggs, W.R., Dohrmann, U.C., Ray, P.M. (1978) Auxin receptors of maize coleoptile membranes do not have ATPase activity. Plant Physiol. 61, 581-584.
18. Durand, H., Rayle, D.L. (1973) Physiological evidence for auxin- induced hydrogen-ion secretion and the epidermal paradox. Planta 114, 185-193.
19. Evans, M.L. (1985) The action of auxin on plant cell elongation. Critical Rev. Plant Sci. 2, 317-365.
20. Evans, M.L., Ray, P.M. (1969) Timing of the auxin response in coleoptiles and its implications regarding auxin action. J. Gen. Physiol. 53, 1-20.
21. Fry, S.C. (1986) Cross-linking of matrix polymers in the growing cell walls of Angiosperms. Annu. Rev. Plant Physiol. 37, in press.
22. Gabathuler, R., Cleland, R.E. (1985) Auxin regulation of a proton translocating ATPase in pea root plasma membrane vesicles. Plant Physiol. 79, 1080-1085.
23. Galston, A.W., Baker, R.S. (1951) Studies on the physiology of light action. IV. Light enhancement of auxin-induced growth in green peas. Plant Physiol. 26, 311-317.
24. Goldberg, R., Prat, R. (1981) Development of the response to growth regulators during the maturation of mung bean hypocotyls. Physiol. Veg. 19, 523-532.
25. Hagen, G., Kleinschmidt, A., Guilfoyle, T. (1984) Auxin-regulated gene expression in intact soybean hypocotyl and excised hypocotyl sections. Planta 162, 147-153.
26. Hager, A., Menzel, H., Krauss, A. (1971) Versuche und Hypothese zur Primarwirkung des Auxins beim Streckungswachstum. Planta 100, 47-75.
27. Hertel, R. (1983) The mechanism of auxin transport as a model for auxin action. Z. Pflanzenphysiol. 112, 53-67.
28. Higgins, T.J.V., Jacobsen, J.V. (1978) The influence of plant hormones on selected aspects of cellular metabolism. In: Phytohormones and related compounds, Vol. I, pp. 467-514, Letham, D.W., Goodwin, P.B., Higgins, T.J.V., eds. Elsevier, Amsterdam.
29. Kutschera, U., Schopfer, P. (1985) Evidence against the acid growth theory of auxin action. Planta 163, 483-493.
30. Labovitch, J.M., Ray, P.M. (1974) Relationship between promotion of xyloglucan metabolism and induction of elongation by IAA. Plant Physiol. 54, 499-502.
31. Lockhart, J.A. (1965) An analysis of irreversible plant cell elongation. J. Theor. Biol. 8, 264-275.
32. Nakajima, N., Morikawa, H., Igasashi, S., Senda, M. (1981) Differential effect of calcium and magnesium on mechanical properties of pea stem cell walls. Plant Cell Physiol. 22, 1305-1315.
33. Nishitani, K., Masuda, Y. (1983) Auxin-induced changes in cell wall xyloglucans. Plant Cell Physiol. 24, 345-355.
34. Patrick, J.W. (1982) Hormone control of assimilate transport. In: Plant growth substances 1982, pp. 669-678, Wareing, P.F., ed. Academic Press, New York.
35. Pearce, D., Penny, D. (1983) Tissue interactions in IAA-induced rapid elongation of lupin hypocotyls. Plant Sci. Lett. 30, 347-353.
36. Penny, P., Penny, D. (1978) Rapid responses to phytohormones. In: Phytohormones and related compounds, Vol. II, pp. 537-97, Letham, D.S., Goodwin, P.B., Higgins, T.J.V., eds. Elsevier, Amsterdam.
37. Ray, P.M. (1977) Auxin-binding sites of maize coleoptiles are localized on membranes of the endoplasmic reticulum. Plant Physiol. 59, 594-599
38. Rayle, D.L., Cleland, R.E. (1970) Enhancement of wall loosening and elongation by acid solutions. Plant Physiol. 46, 250-253.
39. Rayle, D.L., Cleland, R.E. (1972) The in-vitro acid-growth response: relation to in-vivo growth responses and auxin action. Planta 104, 282-296.

147

40. Rayle, D.L., Cleland, R.E. (1980) Evidence that auxin-induced growth of soybean hypocotyls involves proton excretion. Plant Physiol. *66*, 433-37.
41. Rubery, P.H. (1981) Auxin receptors. Annu. Rev. Plant Physiol. *32*, 569-596.
42. Ruesink, A.W. (1969) Polysaccharidases and the control of cell wall elongation. Planta *89*, 95-107.
43. Sanders, D., Hansen, U-P., Slayman, C.L. (1981) Role of the plasma membrane proton pump in pH regulation in non-animal cells. Proc. Natl. Acad. Sci. USA *78*, 5903-5907.
44. Spanswick, R.M. (1981) Electrogenic ion pumps. Annu. Rev. Plant Physiol. *32*, 267-289.
45. Stevenson, T.T., Cleland, R.E. (1981) Osmoregulation in the *Avena* coleoptile in relation to growth. Plant Physiol. *67*, 749-753.
46. Taiz, L. (1984) Plant cell expansion: regulation of cell wall mechanical properties. Annu. Rev. Plant Physiol. *35*, 585-657.
47. Talbolt, L.D., Robert, J.K.M., Ray, P.M. (1984) Effect of IAA- and fusicoccin-stimulated proton extrusion on internal pH of pea cells. Plant Physiol. *75*, S41.
48. Tepfer, M., Cleland, R.E. (1979) A comparison of acid-induced cell wall loosening in *Valonia ventricosa* and in oat coleoptiles. Plant Physiol. *63*, 898-902.
49. Theologis, A., Ray, P.M. (1982) Changes in messenger RNAs under the influence of auxin. In: Plant growth substances 1982, pp. 43-57, Wareing, P.F., ed. Academic Press, New York.
50. Vanderhoef, L.N., Dute, R.R. (1981) Auxin-regulated wall loosening and sustained growth in elongation. Plant Physiol. *67*, 146-149.
51. Vanderhoef, L.N., Stahl, C.A. (1975) Separation of two responses to auxin by means of cytokinin inhibition. Proc. Natl. Acad. Sci. USA *72*, 1822-1825.
52. Vesper, M.J., Evans, M.L. (1978) Time-dependent changes in the auxin sensitivity of coleoptile segments. Apparent sensory adaptation. Plant Physiol. *61*, 204-208.
53. Walton, J.D., Ray, P.M. (1981) Evidence for receptor function of auxin binding sites in maize. Red light inhibition of mesocotyl elongation and auxin binding. Plant Physiol. *68*, 1334-1338.

C2. The Control of Gene Expression by Auxin

Gretchen Hagen

Botany Department, University of Minnesota, St. Paul, Minnesota 55108, USA.

INTRODUCTION

Auxin has been shown to be involved in a variety of diverse plant growth and developmental responses including cell elongation, cell division and cell differentiation (24). The mechanism(s) by which auxin affects such diverse processes is unknown, but, over the past 30 years, several major hypotheses have been proposed and have directed the efforts of auxin biologists. These hypotheses will be discussed briefly as an introduction to a discussion of a current molecular approach being taken to elucidate the mechanism of auxin action.

In the 1950s, the observation that treatment of tobacco pith tissue with auxin caused marked increases in RNA and DNA content within 48 hours led to the suggestion that auxin affects nucleic acid synthesis (21) (the gene activation or gene expression hypothesis). This was an attractive hypothesis, as it suggested a general mechanism of action for diverse auxin responses. As a result, research efforts were directed to define the molecular sequence of events from auxin application to changes in nucleic acid metabolism, and to determine what role the newly synthesized macromolecules play in growth and development (for review, see refs. 11,12,16,17).

A wide variety of plant species have been used in the investigation into the validity of the gene activation hypothesis and, although many species respond to auxin in a similar manner at the molecular level, there are exceptions (17). To simplify a considerable amount of data and to provide continuity with results that will be reported later in this chapter, the only early work concerning the gene activation hypothesis that will be summarized will be that on the auxin-treated soybean hypocotyl and excised soybean hypocotyl sections.

When a three day old etiolated soybean seedling is sprayed with high concentrations of auxin (most frequently, the synthetic auxin 2,4-dichlorophenoxyacetic acid or 2,4-D at 1–2mM), cell division in the apical meristem and cell elongation in the elongation zone directly below the apical meristem cease. In general, at these levels of applied auxin, normal growth of intact plants is inhibited(17). Within 24 hours after auxin application, the basal or mature region of the hypocotyl (a normally quiescent region below the elongation zone) begins to enlarge radially with massive cell proliferation. Concomitant with these morphological changes in the mature region of the hypocotyl are dramatic increases in DNA, RNA, and protein content (17,30). The increase in RNA content is mainly the result of an increase in ribosomal RNA (rRNA; 17) which contributes to the increase in the pool of ribosomes. Based on studies of RNA polymerase I (the enzyme involved in the transcription of rRNA genes) activity (10) and enzyme amounts, it has been concluded that both the rate of RNA chain initiation and chain propagation by RNA polymerase I are enhanced in the soybean hypocotyl within 24-48 hours of auxin application (30). Further, it has been shown that the auxin-induced increase in RNA polymerase I enzyme amount arises from *de novo* synthesis resulting in an enzyme with a subunit structure that is indistinguishable from untreated hypocotyl enzyme (12). Recently, the increase in rRNA and ribosome accumulation have been shown to be correlated with a 3- to 8-fold increase in the amount of translatable ribosomal protein messenger RNA (mRNA) following the treatment of soybean seedlings with 2,4-D for 24 hours (8,9). The activity of RNA polymerase II (the enzyme involved in the transcription of mRNA genes) remains constant (10,11) and the amount of enzyme increases slightly (12) following auxin treatment of soybean hypocotyl. This increase also results from *de novo* synthesis and no subunit modification is observed in the newly synthesized enzyme (12).

Due to the ease in handling and experimental manipulation, many studies have been performed on excised plant organs. The molecular effects on soybean hypocotyl basal sections that have been incubated in the presence of auxin for 12 hours include an increase in RNA (predominantly rRNA) synthesis, ribosomes, protein synthesis and chromatin-bound RNA polymerase I activity (30), and these effects are similar to those observed in the intact hypocotyl. One distinct difference with the incubated, basal sections is that there is no auxin-induced radial enlargement and cell proliferation during incubation, as is seen in the intact seedling which has been sprayed with auxin. Excised, basal sections are useful, nonetheless, for an investigation into the events occurring prior to activation of cell division as a result of exposure to auxin.

When sections from the elongating zone of the soybean hypocotyl are incubated in the presence of auxin, rapid cell elongation is observed and can easily be measured. Because of the rapid response and easy assay for growth, a majority of studies have focused on the effects of auxin on cell elongation. In contrast to the basal sections, the levels of RNA, ribosomes and protein synthesis measured during incubation of soybean hypocotyl elongating sections in auxin are maintained at those levels found in the intact hypocotyl elongation zone. Since RNA and protein synthesis were proposed to be an integral part of growth by cell elongation, a number of inhibitor studies were performed using mainly the RNA synthesis inhibitor actinomycin D and the protein synthesis inhibitor cycloheximide. These studies demonstrated that continued RNA and protein synthesis are required for continued cell elongation, and the inhibitor data using auxin-treated elongating sections suggested that the ability of auxin to increase the rate of cell elongation is dependent upon newly synthesized RNA and protein (17).

The gene activation hypothesis for auxin action as it relates to the process of cell elongation was reexamined as a result of the development of a sensitive optical method for measuring rapid changes in elongation rates in plant organs (6). The results showed that in many plant species, the response to added auxin has a latent period of about 10 minutes before an increase in elongation rate is observed (5). It was argued that if the initial response to added auxin was the *de novo* synthesis of RNA and proteins necessary for cell enlargement, these macromolecules would have to be synthesized during the latent period before cell enlargement was observed (6). Although there are some reports of incorporation of RNA precursors into RNA within minutes of auxin addition to incubated sections, many organs do not show increased incorporation for 10-60 minutes after auxin addition and some organs do not show the increase at all (17). From elongation rate kinetic data, RNA synthesis inhibitor studies and a consideration of half-lives of RNA and proteins, it was concluded that auxin probably does not act primarily at the level of gene activation, although control by auxin at the translational level was not ruled out (6). This conclusion caused a shift in the focus of auxin research towards the involvement of auxin in cell elongation at the level of the cell wall (4,5,20) and away from auxin involvement in gene activation during cell elongation, division and differentiation. It was suggested that the initial action of auxin was to cause cell walls to loosen (4), and when it was proposed that the factor responsible for this wall loosening was hydrogen ions, the acid-growth or cell wall acidification hypothesis was formulated (15,19). This hypothesis states that auxin acts to regulate the pH of the cell wall (by proton extrusion) such that wall loosening enzymes are activated and cell enlargement can occur.

Evidence in support of the wall acidification hypothesis has come primarily from studies on acid-induced cell elongation (20), but the conclusion that acid-induced elongation is identical to auxin-induced elongation has been challenged (25). A detailed analysis of the early kinetics of auxin-induced elongation revealed that there are two elongation responses, one which is rapid and transient (does not attain steady state) and one which has a latent period longer than the first response but does attain a steady state rate following an initial rise (25). The evidence for this conclusion came from a.) a reexamination of kinetic data on cell elongation and preparation of summation constructs from that data, b.) from experiments using cytokinin (an inhibitor of cell elongation) and auxin analogs and c.) studies on acid-induced growth. Consideration of all the data led to the "dual sites" hypothesis, which states that in its regulation of cell elongation, auxin maintains the cell wall in a chemical and/or physical state to allow continued cell elongation, while also affecting gene expression for the production of specific macromolecules required for permanent cell elongation (25).

CURRENT APPROACHES

In the last few years, there has been renewed interest in the relationship between auxin action and gene expression, and major advances have been made in defining the level(s) of control of gene expression by auxin. The application'of numerous highly sensitive experimental techniques that are available to molecular biologists has facilitated the analysis of macromolecular changes that occur within minutes after auxin application. Some of the techniques used in these studies will be described briefly and results will be summarized.

PROTEIN SYNTHESIS STUDIES

Changes in protein profiles have been observed in the soybean hypocotyl following brief exposure to auxin (29). For these studies, incubation of soybean hypocotyl sections in media containing a radioactive amino acid precursor was carried out in the presence or absence of added auxin. The *in vivo* radiolabeled polypeptides were isolated and separated by one- and two-dimensional polyacrylamide gel electrophoresis. In the two-dimensional (2D) gel system, polypeptides are separated by isoelectric focusing in the first dimension and by molecular weight in the second dimension, allowing heterogeneous populations of proteins to be analyzed with high resolution. The radioactivity associated with the polypeptides is visualized by autoradiographic exposure to Xray film. The spectrum of labeled polypeptides isolated from soybean sections that were incubated in the presence or absence of auxin is broad, and auxin treatment appears to induce the synthesis of certain polypeptides while repressing that of others (29). A major disadvantage of *in vivo* labeling is that the analysis is dependent upon the time that is required to achieve sufficient radiospecific activity of the polypeptides. In the case of soybean, this was 1–3 hr, which precludes an analysis of rapid changes in polypeptide patterns resulting from auxin treatment. A second disadvantage is the extensive incorporation of radioactive label into wounded ends of the hypocotyl sections, which may mask the auxin responses within the internal regions of the sections.

An alternative method that has been used to study changes in protein profiles is *in vitro* translation in cell-free extracts. For this, purified polyadenylated RNA (poly(A)+RNA, presumptive mRNA) is translated *in vitro* using either a wheat germ extract or rabbit reticulocyte lysate in the presence of a labeled amino acid precursor. The labeled translation products are resolved on the 2D gel system described above (Fig. 1). It is possible, then, to isolate RNA from organs that have had brief exposure to auxin, translate the RNA *in vitro* and examine the spectrum of

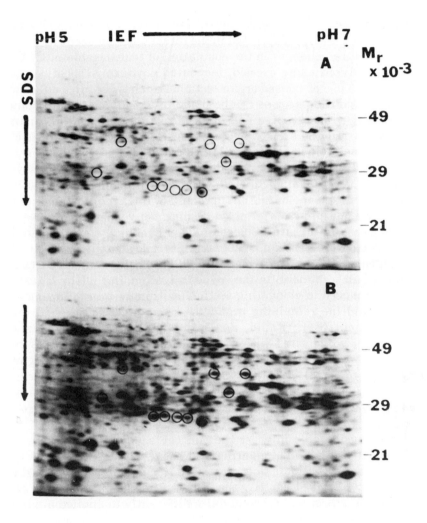

Fig. 1. Two-dimensional polyacrylamide gel analysis of *in vitro* translation products from poly (A)+RNA isolated from intact soybean hypocotyl of seedlings that were untreated (A) or (B) treated with 2,4-D for 2 hr. Several of the *in vitro* translation products that are induced with auxin treatment are indicated.

polypeptides that are potentially found in the cells at any point after auxin application.

In vitro translation of mRNAs isolated from auxin-treated hypocotyls of intact soybean seedlings and a 2D gel analysis of the translation products showed that up to 40 mRNAs may undergo either an upward or downward shift in relative concentration in response to auxin treatment, and changes can be observed within 5 hr after auxin treatment (2). The vast majority of the in vitro translation products from auxin-treated and untreated soybean seedlings did not differ. These results are compatible with nucleic acid hybridization studies which indicated that auxin treatment did not affect the concentration of the large majority of the 40,000 different mRNAs expressed in the intact soybean seedling (2).

Auxin-induced changes in the population of translatable mRNA have also been analyzed using excised, incubated soybean hypocotyl sections (30,31,32). The use of excised organ sections for these studies offers many advantages, and comparisons can be made to the intact seedling since it has been shown that early auxin-induced events are similar in intact soybean hypocotyl and excised hypocotyl sections (30). An analysis of the pattern of translation products on 2D gels revealed that a similar, specific set of 10-12 in vitro translation products is induced in the intact soybean hypocotyl as well as in incubated hypocotyl sections (from both the elongating zone and the basal, mature zone of the hypocotyl), while most of the translation products show no change between auxin-treated and untreated samples. Figure 1 shows a 2-D gel of the translation products from untreated (A) and auxin-treated (B) intact soybean hypocotyl.

Changes in the in vitro translation products can occur rapidly following auxin treatment. In the soybean system, one of the translation products is induced in elongating sections within 15 min following auxin treatment, and the remaining products are induced within 30 min of auxin treatment (31). Rapid, auxin-induced changes in translation products are also detected within 20 min of treatment of excised pea epicotyl sections (22) and within 10 min of treatment of elongating maize coleoptile sections(33). While some of these changes appear rapidly enough to be implicated in the process of cell enlargement (the "classical" rapid response to auxin; 5), the function of these rapidly-induced products is unknown.

It is of considerable interest that, in the soybean system, the early response to auxin in both elongating and basal sections as well as in the intact seedling is the induction of a similar, specific set of translation products. As described earlier, the elongating and basal zones of the intact seedling ultimately respond quite differently to applied auxin (cell elongation verses cell division), and the observation of a similar, early response to auxin has led to the suggestion that there is a common, primary site of auxin action (32).

RNA STUDIES

Modulation of *in vitro* translation products following auxin treatment could result from a variety of factors, including an alteration in mRNA levels for specific polypeptides (reflecting changes in transcription, processing and/or stability) or alteration in the efficiency of *in vitro* translation of specific mRNAs. Research has been initiated to select those mRNAs that are specifically modulated by auxin, for use as probes in defining the level of control of the auxin-induced changes. Because the total mRNA population isolated from an organ is usually heterogeneous, it is often not possible to fractionate mRNAs and recover a single mRNA species. Therefore, to identify specific auxin-regulated mRNAs, recombinant DNA technology has been used (for specific details on recombinant DNA methods, see ref. 18). Briefly, total poly(A) + RNA from auxin-treated organs is used as a template to synthesize a complementary DNA copy (complementary DNA or cDNA). The cDNA is made double stranded, inserted into a plasmid (or phage) DNA vector and transformed into a bacterial host, where the recombinant DNA molecules are replicated. Those bacteria containing a single recombinant DNA vector represent individual members of a library of cDNA clones. To distinguish which of the bacterial colonies contains auxin-inducible cDNA sequences, a differential hybridization screening procedure has been used. For this, the nucleic acids within individual, transformed bacterial colonies are isolated and hybridized with radiolabeled probes derived from mRNA isolated from either auxin-treated or untreated organs. The separate hybridization of nucleic acids from each cDNA clone in the library to the labeled, auxin-treated and untreated mRNA probes should resolve differences in levels of hybridization, reflecting differences in abundance of specific mRNAs within the organs. Since each of the cDNAs that was originally produced is in a separate bacterial cell, those that show a differential hybridization response between auxin-treated and untreated organ probes can be isolated and purified. These cDNA clones represent single mRNA species and are sensitive probes for use in a detailed analysis of the changes in gene expression by auxin.

Using the basic strategy described above, 12 cDNA clones were selected by differential hybridization screening as sequences representing poly(A)+RNAs that are reduced in concentration (down-regulated) following auxin treatment of soybean hypocotyl (1,3). Six, independent (non-related) cDNA clones representing poly(A) + RNAs that are increased in concentration (up-regulated) following auxin treatment of soybean hypocotyl have been described (14,26). Further, 2 cDNA clones representing auxin-inducible sequences from pea have been characterized (23).

The 12 auxin down-regulated cDNA clones from soybean were shown to fall into three groups, based on RNA blot hybridization analysis (1,3). For these studies, poly(A)+RNA was isolated from treated or untreated organs, fractionated according to size by agarose gel electrophoresis and transferred and fixed to a nitrocellulose filter. The cDNA clones were radiolabeled and hybridized to the RNA on the nitrocellulose filter, and the radioactivity associated with specific hybrids was visualized by autoradiographic exposure to Xray film. Using molecular weight markers, the size of the RNA species to which each cDNA clone hybridizes was determined. Seven of the auxin down-regulated clones hybridize to a RNA species of 1000 bases (group A), 4 clones hybridize to a RNA of 350 bases (group B) and 1 clone hybridizes to a RNA of 750 bases (group C). Using solution hybridization to quantitate, it was shown that sequences in group A undergo a 100-fold reduction, group B, a 10-fold reduction and group C, a 15-fold reduction in relative concentration following auxin treatment. By RNA blot hybridization, it was shown that RNA levels for all three groups of sequences are reduced within 4 hr after auxin application. Using hybridization/translation techniques, in which the cDNA clones are used to select homologous RNAs that are subsequently eluted and translated in an *in vitro*, cell-free translation system, it was shown that group A sequences encode a 33 kilodalton peptide, group B, a 10 kilodalton peptide and group C, a 25 kilodalton peptide.

The 6 cDNA sequences representing auxin, up-regulated mRNAs from soybean (14,26) have been characterized by RNA blot hybridization analyses (14,26,28; Fig. 2), as well as RNA slot blot analyses, in which the poly(A)+RNA is directly fixed to a nitrocellulose filter without prior fractionation and hybridized to the cDNA clones (14). The radioactivity associated with the slot blots was counted by liquid scintillation and used

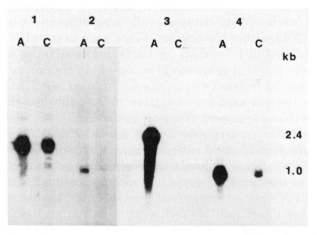

Fig. 2. RNA blot hybridization analysis of poly (A)+RNA from intact soybean hypocotyl. Poly (A)+RNA from untreated (C) or 2,4-D treated (A) hypocotyl was fractionated by agarose gel electrophoresis, transferred to nitrocellulose and hybridized to radiolabeled soybean cDNA clones 1) pGH1, 2) pGH2, 3) pGH3 and 4)pGH4 (14). RNA size in kilobases (kb) is indicated.

to quantitate the induction process. The results of studies on the 6 soybean cDNA clones are summarized below.

The levels of the auxin-inducible RNA sequences in soybean organs that have not been removed from their normal auxin source range from undetectable (14) to relatively high (14,26). When organ sections are depleted of endogenous auxin by incubation in media lacking auxin, the RNA levels for these up-regulated sequences generally decline (14,26) or remain low (14). Upon addition of auxin, RNA levels increase rapidly (within 15-30 min) and continue to increase for several hours. The relative concentrations of the RNAs increase from several fold (up to 10-fold; 14,26) to, in one case, several hundred fold (14) in response to auxin, although even at maximum auxin stimulation, these levels represent only 0.01% or less of the total poly(A)+RNA. RNA levels for 2 of the cDNA clones are highest in the elongating zone of the soybean hypocotyl relative to the apical and basal zones, suggesting that these RNAs are developmentally regulated (26). Relative RNA levels for the other 4 soybean cDNA clones appear approximately equivalent in both the elongating and basal sections of the untreated hypocotyl, and, following auxin-depletion of the sections, the levels are increased in both organ sections with auxin addition (14). The concentration of 4 of the auxin-responsive RNAs is also rapidly increased in the hypocotyl of intact soybean seedlings that have been treated with auxin (14). Figure 2 shows the relative RNA levels from soybean hypocotyl of intact seedlings that were either untreated (C) or sprayed with 2,4-D (A) for 4 hr.

RNA levels for the soybean sequences are increased by treatment with the naturally occurring auxin indole-3-acetic acid (IAA; 14,28) and with the synthetic auxins α-naphthalene acetic acid (NAA; 28), 2,4-D (14,26,28), and 2,4,5-trichlorophenoxyacetic acid (2,4,5-T; 14), but not increased by ethylene (14,28). The RNA levels for some of these sequences also show an increase following 2,4-D treatment of plumules from soybean, mung bean (Fig. 3) and pea (Hagen and Guilfoyle, unpublished). If cell elongation is inhibited by the addition of cytokinin to auxin-treated, soybean hypocotyl sections, RNA levels for the 2 auxin-regulated sequences examined are not altered (28). Additionally, if cell elongation is stimulated with the fungal toxin fusicoccin in the absence of added auxin, auxin-regulated mRNAs are unaffected (28), suggesting that the auxin-induced increase in these mRNAs is not simply a result of enhanced rates of cell elongation and is strictly dependent on auxin.

RNA blot hybridization analyses have also been used to characterize the auxin-responsive pea sequences (23). The induction of the specific sequences occurs rapidly (within 15 min) following auxin treatment. This response is specific to auxins (NAA, 2,4-D, IAA), as no induction is observed with a.) several nonauxin analogs, b.) other plant growth regulators, including gibberellic acid, abscisic acid and ethylene, or c.) environmental stresses such as anaerobiosis, heat shock or cold shock.

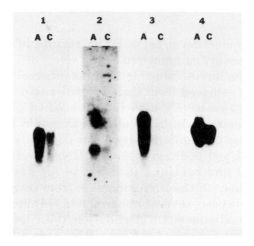

Fig. 3. RNA blot hybridization analysis of poly (A)⁺RNA from mung bean plumules. Poly (A)⁺RNA from untreated (C) or 2,4-D treated (A) mung bean plumules was fractionated by agarose gel electrophoresis, transferred to nitrocellulose and hybridized to radiolabeled soybean cDNA clones 1) pGH1, 2) pGH2, 3) pGH3 and 4) pGH4 (14).

Three protein synthesis inhibitors (cycloheximide, emetine and anisomycin) were, however, shown to be effective in the induction of the mRNA levels, suggesting that the auxin-responsive pea sequences are regulated by a protein with a short half-life. The induction of these sequences by auxin is unaffected by the simultaneous addition of various inhibitors of auxin-mediated proton secretion (mannitol, vanadate, cerulein and cycloheximide). This result strongly suggests that the induction of these sequences is not dependent on hydrogen ion secretion and cytoplasmic pH changes.

Taken together, these data demonstrate that auxin can rapidly and specifically modulate mRNA levels in a variety of plant organs from several plant species.

TRANSCRIPTION STUDIES

The increase in mRNA levels of the auxin-inducible sequences could be the result of increased transcription rates or the result of post-transcriptional events such as an increase in the processing of a precursor transcript, increased transport of the mRNA from the nucleus to the cytoplasm or stabilization of the mature mRNA species. Preliminary data from studies on the mechanism of control of the auxin down-regulated soybean hypocotyl sequences has suggested that those sequences are modulated by a post-transcriptional process (3). Four of the auxin up-regulated soybean hypocotyl sequences were investigated to determine whether transcriptional control was involved in their regulation by auxin

(13,14). For these studies, nuclear transcripts synthesized and radiolabeled *in vitro* in nuclei isolated from auxin-treated or untreated soybean plumules were extracted and hybridized to auxin-regulated cDNA sequences which were slot blotted onto nitrocellulose. The blots were exposed to Xray film and the radioactivity associated with the hybrids was quantitated by densitometric scanning of the slots on the film. In isolated nuclei, RNA polymerase II transcripts (presumptive mRNAs) result primarily from elongation of RNA chains that were presumably initiated *in vivo* (7). Introduction of a radioactive ribonucleoside triphosphate in the *in vitro* nuclei incubation mix should allow the labeling of RNA transcripts that continue to elongate *in vitro* (nuclear "run-off" transcription). The labeled products should, therefore, reflect the spectrum of RNAs being transcribed *in vivo* at the time of nuclei isolation. Results of the *in vitro* transcription studies with the auxin up-regulated clones are summarized below.

Transcription rates for all four soybean auxin-regulated clones increase rapidly 10-to 100-fold (depending on the specific clone analyzed) following auxin treatment. Increased transcription rates are at least half-maximal by 15 min after auxin application, and, for one sequence, an increase in transcription rate can be detected within 5 min after auxin application. Based on α-amanitin inhibition (at 1 μg/ml) of the hybridization signal, this transcription is carried out by RNA polymerase II (14).

In 2,4-D dose response studies, a linear or near-linear increase in transcription was observed over several logs of increasing 2,4-D concentration, and, for 3 clones, the transcriptional response was not saturated at 10^{-3} M 2,4-D. As little as 10^{-7} to 10^{-8} M 2,4-D increased the transcription rates for 3 of the sequences. In other studies using 2,4-D, auxin-regulated transcriptional responses were shown to occur in nuclei isolated from plumules of mung bean (*Vigna radiata*) and green bean (*Phaseolus vulgaris*), as well as from hypocotyl (14) and roots (Hagen and Guilfoyle, unpublished) of soybeans.

Other auxins, including IAA, NAA and 2,4,5-T were effective in increasing transcription rates, whereas 2 nonauxins related to IAA (i.e., tryptophan and indolealdehyde) produced no transcriptional response with these 4 auxin-regulated clones. Other parameters have been tested and shown not to induce transcription of the 4 soybean sequences[1]. Results from studies using the protein synthesis inhibitor cycloheximide indicate that protein synthesis is not required for the activation of transcription of the four auxin-regulated soybean sequences.

---

[1] Treatments ineffective in the transcriptional induction of auxin-regulated mRNA sequences in soybean include: a) protein synthesis inhibitors chloramphenicol and cyclo-heximide; b) hormones GA_3, ethylene, and isopentenyladenosine, and the plasma-membrane proton-pump promoter fusicoccin; and c) stress by heat or cold shock. (Data from (11) or unpublished results of Hagen and Guilfoyle or Kaye and Guilfoyle).

These results demonstrate that the auxin-induced increase in mRNA concentration of these sequences is controlled, at least in part, at the level of transcription; other possible levels of control that may be involved have yet to be investigated. The observations that the induction of transcription is rapid (occurring within 5-15 min), selective (only a few cDNA clones were shown to be affected by auxin) and direct (not dependent on protein synthesis) suggest that these transcription events may be close to the primary site of action for auxin.

GENE ISOLATION AND FUTURE DIRECTIONS

A more complete understanding of the control of gene expression by auxin requires the characterization of the genes that are transcribed in response to auxin. A preliminary examination of the genomic organization of 4 of the soybean auxin-responsive sequences reveals a certain level of complexity (Fig. 4). For these studies, total soybean nuclear DNA was cleaved into fragments using a restriction endonuclease, the fragments separated according to size by agarose gel electrophoresis and transferred and fixed to a nitrocellulose filter. The cDNA clones were radiolabeled, hybridized to the genomic fragments on the filter and the radioactivity associated with the hybrids visualized by autoradiographic exposure to Xray film. As shown in Fig. 4, multiple genomic fragments hybridize to each of the cDNA clones and some of these fragments are at least twice the size of the mature RNA species to which the cDNA clones hybridize (14). These results suggest that there may be multiple gene sequences that share homology with each of the cDNA

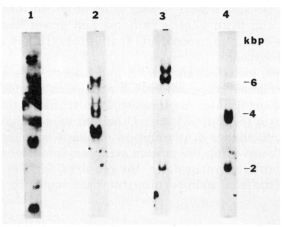

Fig. 4. Soybean genomic DNA blot analysis. Soybean genomic DNA was cleaved with the restriction endonuclease Eco RI, separated by agarose gel electrophoresis, transferred to nitrocellulose and hybridized to radiolabeled soybean cDNA clones 1) pGH1, 2) pGH2, 3) pGH3 and 4) pGH4 (14). DNA fragment size in kilobase pairs (kbp) is indicated.

clones (a multigene family) and that the genes may be very large (i.e., may contain noncoding, intervening sequences). Evidence for a small, multigene family for 2 of the soybean auxin-responsive clones comes from hybridization/translation experiments, where multiple *in vitro* translation products are produced from the RNA(s) selected by a single cDNA clone (27).

Soybean genomic fragments for all 6 auxin, up-regulated cDNA clones have been isolated, purified and are being sequenced (27; Hagen and Guilfoyle, unpublished). Nucleotide sequence data of the genes and RNAs (as determined from the cDNA sequences) will provide information on gene structure and organization, and, where multigene families are shown to exist, should reveal whether all members of the family are expressed. Sequence data will also be crucial for the analysis of transcription and RNA processing. Sequence comparisons of many auxin-regulated genes may help to locate regions that are potentially important for transcriptional regulation by auxin. It is possible, for example, that all auxin-regulated genes contain common sets of specific nucleotides. Identification of important sequences could be approached by inserting putative auxin regulatory sequences (and fragments of these sequences containing defined deletions) in promoter-requiring expression vectors and assessing which nucleotides are essential for transcriptional induction by auxin.

From the nucleotide sequence of the polypeptide coding regions of the auxin-regulated genes, deduced amino acid sequences can be derived; these may give some clues as to the types of polypeptides that are rapidly synthesized in response to auxin. The coding sequences can also be inserted into expression vectors for the production of polypeptides, and the purified polypeptides used in the production of antibodies. These antibodies will be sensitive tools in cellular localization studies, and for determining the functional role of these auxin-regulated sequences.

The results from an analysis of auxin-responsive soybean sequences clearly demonstrate that auxin affects gene expression at the level of transcription. Using the soybean sequences, the role of auxin in other possible levels of control (i.e., post-transcription or translation) can now be addressed. One goal in the molecular approach to auxin action will be to isolate and characterize a large number of auxin-responsive mRNAs from other dicots as well as from monocots, and assess their regulation by auxin. These sequences will be valuable probes for a high resolution analysis of the mechanism(s) of auxin action, which, undoubtedly, will include studies on auxin receptors, on factors which may specifically interact with DNA to allow rapid transcription of certain genes and on the function of the RNAs and/or polypeptides that are regulated by auxin.

Acknowledgments

I wish to thank Tom Guilfoyle and Bruce McClure for critical reading of the manuscript, and Linda Zurfluh for providing the data presented in Figure 1. Results presented in Figures 2-4 and footnote 1 were generated with research grant support from the National Science Foundation (PCM-8208496 to Tom Guilfoyle) and Agrigenetics.

References

1. Baulcombe, D.C., Key, J.L. (1980) Polyadenylated RNA sequences which are reduced in concentration following auxin treatment of soybean hypocotyls. J. Biol. Chem. *255*, 8907-8913.
2. Baulcombe, D., Giorgini, J., Key, J.L. (1980) The effect of auxin on the polyadenylated RNA of soybean hypocotyls. *In* Genome organization and expression in plants, pp. 175-85, Leaver, C.J., ed. Plenum, New York.
3. Baulcombe, D.C., Kroner, P.A., Key, J.L. (1981) Auxin and gene regulation. *In* Levels of genetic control in development, pp. 83-97, Subtelny, S., Abbott, U.K., eds. Liss, New York.
4. Cleland, R. (1971) Cell wall extension. Ann. Rev. Plant Physiol. *22*, 197-222.
5. Evans, M.L. (1974) Rapid responses to plant hormones. Ann. Rev. Plant Physiol. *25*, 195-223.
6. Evans, M.L., Ray, P.M. (1969) Timing of the auxin response in coleoptiles and its implications regarding auxin action. J. Gen. Physiol. *53*, 1-20.
7. Evans, R.M., Fraser, N., Ziff, E., Weber, J., Wilson, M., Darnell, J.E. (1977) The initiation sites for RNA transcription in Ad2 DNA. Cell *12*, 733-739.
8. Gantt, J.S., Key, J.L. (1983) Auxin-induced changes in the level of translatable ribosomal protein messenger ribonucleic acids in soybean hypocotyl. Biochem. *22*, 4131-4139.
9. Gantt, J.S., Key, J.L. (1985) Coordinate expression of ribosomal protein mRNAs following auxin treatment of soybean hypocotyls. J. Biol. Chem. *260*, 6175-6181.
10. Guilfoyle, T.J., Lin, C.Y., Chen, Y.M., Nagao, R.T., Key, J.L. (1975) Enhancement of soybean RNA polymerase I by auxin. Proc. Natl. Acad. Sci. USA *72*, 69-72.
11. Guilfoyle, T J., Key, J.L. (1977) Purification and characterization of soybean DNA-dependent RNA polymerases and the modulation of their activities during development. In: Nucleic acids and protein synthesis in plants, pp. 37-63, Bogorad, L., Weil, J., eds. Plenum, New York.
12. Guilfoyle, T., Olszewski, N., Zurfluh, L. (1980) RNA polymerases and transcription during developmental transitions in soybean. *In* Genome organization and expression in plants, pp. 93-104, Leaver, C.J., ed. Plenum, New York.
13 Hagen, G., Guilfoyle, T.J. (1985) Rapid induction of selective transcription by auxins. Mol. Cell. Biol. *5*, 1197-1203.
14. Hagen, G., Kleinschmidt, A., Guilfoyle, T. (1984) Auxin-regulated gene expression in intact soybean hypocotyl and excised hypocotyl sections. Planta *162*, 147-153.
15. Hager, A., Menzel, H., Krauss, A. (1971) Experiments and hypothesis concerning the primary action of auxin in elongation growth. Planta *100*, 47-75.
16. Jacobsen, J.V. (1977) Regulation of ribonucleic acid metabolism by plant hormones. Ann. Rev. Plant Physiol. *28*, 537-564.
17. Key, J.L. (1969) Hormones and nucleic acid metabolism. Ann. Rev. Plant Physiol. *20*, 449-474.
18. Maniatis, T., Fritsch, E.F., Sambrook, J. (1982) Molecular cloning. Cold Spring Harbor Laboratory, Cold Spring Harbor, New York.
19. Rayle, D.L., Cleland, R. (1970) Enhancement of wall loosening and elongation by acid solutions. Plant Physiol. *46*, 250-253.

20. Rayle, D.L., Cleland, R. (1977) Control of plant cell enlargement by hydrogen ions. Curr. Topics Dev. Biol. *11*, 187-214.
21. Silberger, J., Skoog, F. (1953) Changes induced by indoleacetic acid in nucleic acid content and growth of tobacco pith tissue. Science *118*, 443-444.
22. Theologis, A., Ray, P.M. (1982) Early auxin-regulated polyadenylated mRNA sequences in pea stem tissue. Proc. Natl. Acad. Sci. USA *79*, 418-421.
23. Theologis, A., Huynh, T.V., Davis, R.W. (1985) Rapid induction of specific mRNAs by auxin in pea epicotyl tissue. J. Mol. Biol. *183*, 53-68.
24. Thimann, K. (1977) Hormone action in the whole life of plants. University of Massachusetts Press, Amherst, Massachusetts.
25. Vanderhoef, L.N. (1980) Auxin-regulated cell enlargement: is there action at the level of gene expression? *In* Genome organization and expression in plants, pp. 159-73, Leaver, C.J., ed. Plenum, New York.
26. Walker, J.C., Key, J.L., (1982) Isolation of cloned cDNAs to auxin-responsive poly (A)+RNAs of elongating soybean hypocotyl. Proc. Natl. Acad. Sci. USA *79*, 7185-7189.
27. Walker, J C., Key, J.L. (1983) Organization and expression of two small multigene families encoding auxin-regulated mRNAs in soybean (*Glycine max*). (Abstract) J. Cell Biol. *97*, 415a.
28. Walker, J C., Legocka, J., Edelman, L., Key, J.L. (1985) An analysis of growth regulator interactions and gene expression during auxin-induced cell elongation using cloned complementary DNAs to auxin-responsive messenger RNAs. Plant Physiol. 77, 847-850.
29. Zurfluh, L.L., Guilfoyle, T.J. (1980) Auxin-induced changes in the patterns of protein synthesis in soybean hypocotyl. Proc. Natl. Acad. Sci. USA 77, 357-361.
30. Zurfluh, L.L., Guilfoyle, T.J. (1981) Auxin-induced nucleic acid and protein synthesis in the soybean hypocotyl. *In* Levels of genetic control in development, pp. 99-118, Subtelny, S., Abbott, U.K., eds. Liss, New York.
31. Zurfluh, L.L., Guilfoyle, T.J. (1982) Auxin-induced changes in the population of translatable messenger RNA in elongating sections of soybean hypocotyl. Plant Physiol. *69*, 332-337.
32. Zurfluh, L.L., Guilfoyle, T.J. (1982) Auxin-and ethylene-induced changes in the population of translatable messenger RNA in basal sections and intact soybean hypocotyl. Plant Physiol. *69*, 338-340.
33. Zurfluh, L.L., Guilfoyle, T.J. (1982) Auxin-induced changes in the population of translatable messenger RNA in elongating maize coleoptile sections. Planta *156*, 525-527.

C3. Gibberellin and Abscisic Acid in Germinating Cereals

John V. Jacobsen and Peter M. Chandler

CSIRO Division of Plant Industry, PO Box 1600, Canberra, ACT 2601, Australia.

INTRODUCTION

Study of the response of cereal aleurone to gibberellic and abscisic acids (GA and ABA, respectively), particularly with reference to α-amylase synthesis, has made a significant contribution to our understanding of GA action in plant cells especially as it relates to the control of protein synthesis. While much of the work has been carried out using isolated aleurone from a single cultivar of barley ("Himalaya"), it seems so far that the principles which have emerged from this system can be applied to *in vivo* behaviour of other cereal grains.

This article is a synopsis of our understanding of hormone action in aleurone. It briefly discusses the role of GA and ABA in grain development, and the influence of environment and genotype of the developing grain on hormone responsiveness of mature aleurone. For background information and for more comprehensive documentation and referencing of earlier studies, the reader is referred to recent reviews (4,5,8,28,30,56).

The incubation of isolated aleurone layers in media containing specified concentrations of GA and/or ABA, has been extensively studied as a model for hormone action in plants. The main observation is that GA (coming from the developing embryo in the case of a germinating seed, or added to the media in the case of isolated aleurone layers) can stimulate aleurone cells to secrete a range of hydrolytic enzymes, in particular α-amylase (Fig. 1). These enzymes are responsible for the mobilization of stored endosperm reserves which provide the growing seedling with a supply of fixed carbon and reduced nitrogen. The interest in ABA lies in the observation that it can prevent the action of GA if present in excess, and induces its own set of proteins in isolated aleurone.

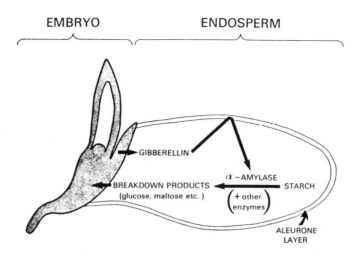

Fig. 1. Schematic diagram illustrating some of the principal features associated with reserve mobilization in a wheat or barley grain following germination. Gibberellin produced by the embryo stimulates cells of the aleurone layer to synthesize and secrete α-amylase (and other hydrolases) which degrades starch in the endosperm providing sugars for the young seedling.

HORMONES IN DEVELOPING GRAIN

While the responses of mature isolated aleurone to applied GA or ABA are fairly well understood, quantitative aspects of the response can vary between different harvests of the same cultivar. This observation suggests that environmental conditions during grain development can influence the eventual hormone responsiveness of aleurone. For example, it has been shown that aleurone from barley grown in relatively high temperatures produces high levels of α-amylase in the absence of added GA (47). In normal development, cereal seeds accumulate significant levels of both GA and ABA (see below); since the activity of these hormones during grain development, or their residual levels in the dry seed may influence subsequent aleurone behaviour during and following germination, we will begin by considering the role of GA and ABA in the developing grain before going on to discuss the response of aleurone from mature seeds.

Gibberellins in Developing Grain

The accumulation of GA activity in developing grains of wheat and barley (Fig. 2) occurs as fresh weight increases and peak levels (100->700 pg GA_3 equiv. per grain) are attained prior to maximum dry weight (43,50). Most of the GA present during development is in the endosperm (which includes the aleurone), testa and inner pericarp. As maturation

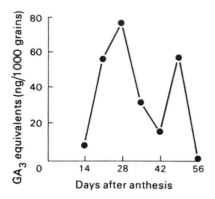

Fig. 2. Bioassay of gibberellin activity in developing grains of field-grown wheat (cv Kleiber). The second peak of GA activity (7 weeks after anthesis) is seen in some crops but not in others. (Data from 50).

proceeds, GA levels fall until in the near-mature grain the embryo contains the greatest amount of GA, although the total GA content of the seed at this stage is only a few percent of peak levels. Under normal conditions there is very little GA present in the dry wheat or barley seed.

Although there are similarities between wheat and barley in their pattern of GA accumulation during development, the actual GA species present are not the same as revealed by capillary GC-MS (17,18). Both the variety of GAs present and their species specificity add to the difficulties of ascribing any role to GA during grain development. It has been suggested that the α-amylase appearing in the endosperm and aleurone of developing barley grains is produced in response to locally-synthesized GA, since application of CCC (an inhibitor of GA synthesis) to the developing grains reduced production of α-amylase (14). Apart from this observation, however, there is little to connect GA with any developmental changes.

Although it is commonly believed that variation in behaviour of mature aleurone may be related to its hormonal status, there is little experimental support for this view. There is also little information available concerning the influence of environmental factors (such as the high temperature studies described above) on GA levels in the mature seed and the subsequent effect this may have on aleurone response. In maize there is an interesting situation which may be a useful model (23). Excised maize endosperm produces α-amylase and protease in the absence of added GA. There is a small stimulation in enzyme production following addition of GA, and production of both amylase and protease is markedly reduced by ABA. These data are consistent with a model in which it is proposed that the high level of constitutive hydrolase production by maize endosperm in the absence of added GA is due to relatively high endogenous levels of GA. To test this model a dwarf mutant of maize (d_5) deficient in endogenous GA was examined; it was found that 10 mM GA_3 caused increases of 500% in protease production and 300% in α-amylase production relative to increases of 30% and 40% respectively in a standard hybrid line. These results suggest that endogenous kernel GA levels can influence the apparent response to added GA, although to be certain of this we need to know that the response is truly an effect of the d_5 mutation (by comparing isogenic lines), and furthermore that the d_5 mutation is actually reducing grain GA levels.

Abscisic Acid in Developing Grain

Cereal grains also accumulate ABA during development with maximal levels being reached generally after peak GA levels and just before the beginning of dehydration (see Fig. 3 in Chapter E13). Levels of ABA then fall rapidly as the grain dries out. The absolute levels of ABA which are reached (50-200 µg/kg fresh weight) correspond to concentrations of 0.5 to 2 µM assuming uniform grain distribution (36). Such levels are at least an order of magnitude higher than maximum GA levels (when expressed as GA_3 equivalents/grain) and exceed the levels necessary for the induction of ABA-induced polypeptides in mature barley aleurone (Ariffin Z. and Chandler P.M., unpublished observations). Although these levels are high, they are only 1-10% of corresponding ABA levels in grains of legumes.

The increase in grain ABA levels during development could arise from *de novo* synthesis of ABA within the grain, or from transport of ABA synthesized elsewhere in the plant into the grain via phloem, or from a combination of these processes. There is evidence to support the operation of both sources. Evidence for *de novo* synthesis within the grain comes from two types of experiments. The first involves feeding radioactive precursors of ABA to developing grains and monitoring incorporation of radioactivity into ABA (40). In the second type of experiment immature ears of wheat were cultured and levels of ABA monitored during grain growth (35). The results of these two studies indicate that developing wheat grains have the enzymatic capacity to convert ABA precursors into ABA, and in addition can accumulate normal levels of ABA in culture when separated from the bulk of the parent plant.

As well as these observations, suggesting the occurrence of ABA synthesis in the developing grain, there is also evidence that ABA can be transported from leaves of wheat and barley to the maturing grain. In wheat a small proportion of radiolabelled ABA injected into leaves appeared in grain (13), and in barley it was observed that water-stressing leaves during grain development could increase grain ABA levels by up to 100% (21).

Which of these two sources of ABA in developing grain is of the greater importance is difficult to assess. In maize there exists a class of mutants (viviparous) which are characterized by premature sprouting of developing kernels through either a block in ABA synthesis or in apparent ABA sensitivity. Kernels homozygous for this class of mutation developing on cobs from heterozygous parent plants presumably receive the same amount of ABA via phloem as do other seeds (heterozygous or homozygous wild type) yet they still sprout prematurely. Although ABA levels have not been investigated in such situations, it seems likely that in this case the developing kernel must make a significant contribution to its own ABA levels, since grains homozygous for the defect in ABA synthesis fail to accumulate sufficient ABA to prevent premature sprouting.

It is well known that water stress leads to ABA accumulation in leaf tissues. However it is unlikely that water stress associated with dehydration of cereal grain is responsible for the normal developmental accumulation of ABA, since levels of ABA have peaked and are beginning to fall before desiccation commences. There is evidence that the fall in ABA levels occurs through enhanced metabolism of ABA, since at the time ABA levels are declining, there is a rise in levels of ABA metabolites such as "bound" ABA, phaseic acid and dihydrophaseic acid. Premature drying of the grain results in a premature decline in ABA levels (35).

The function of ABA in the developing grain appears to be twofold: the promotion of storage protein synthesis and the prevention of premature germination (58). These functions are discussed more fully in Chapter E13. A line of evidence supporting the proposal that ABA may be involved in inhibiting germination prior to grain dehydration comes from a consideration of the effect of ABA on gene expression. It has been shown (see "Control of transcription"—p. 177) that ABA reverses the stimulatory effect of GA on transcription of α-amylase genes in aleurone. Other work indicates that one of the mRNAs which accumulates in aleurone in response to ABA is an α-amylase inhibitor (44). We have discussed above evidence which indicates that during normal development of barley grains α-amylase is produced in response to synthesis of GA in the grain. Since GA is a potent stimulator of germination, even of dormant seed, it is intriguing that some of the early events following normal germination (stimulation of transcription of α-amylase genes and production of α-amylase) are reversed by treatment with ABA (prevention of GA-induction of α-amylase gene transcription, and induction of an inhibitor which may negate any α-amylase which was already made). These effects of ABA on gene expression suggest that the presence of the hormone in developmentally mature grains would reduce the tendency for premature germination to occur.

In vivo evidence for the effect of kernel ABA levels/sensitivity on expression of α-amylase genes is suggested by the profile of α-amylase activity in developing kernels of the viviparous (vp) mutants of maize referred to earlier. Levels of α-amylase activity are similar in control and vp kernels developing on the same cob until about 30 days after pollination. At later stages of development, however, α-amylase activity falls in control kernels, but continues to increase in vp kernels (61). These results are consistent with either the lack of production of ABA associated with some vp mutants (e.g., vp_5) or apparent insensitivity to ABA of other vp mutants (e.g., vp_1), and suggest that the decline in α-amylase expression during maturation of normal kernels is in response to increasing ABA levels.

Before we can reliably assess the biological significance of the profiles in GA and ABA which occur during grain development information is needed in at least three areas. First, we have little knowledge at present of the acquisition of aleurone and embryo sensitivity to GA and ABA during grain development, and therefore little is known about the effects of these hormones in controlling gene expression during development. Cloned DNA probes for both GA-induced and ABA-induced mRNAs (from mature aleurone) are now available for this to be studied. Second, most of the studies measuring GA and ABA accumulation during seed development have used commercial cultivars. It is likely that such varieties have been exposed to direct or indirect selection by breeders for specific germination characteristics. If the GA/ABA accumulation profiles we observe in these cultivars originated in wild grasses as a mechanism for controlling tendency to sprout or remain dormant, this mechanism may no longer be relevant to domesticated forms. Third, it would aid our understanding greatly if more mutants which showed altered patterns of accumulation of, or sensitivity to, GA and ABA in seeds, especially of wheat and barley, were available and studied. These may then allow specific aspects of grain development to be associated with changes in hormone levels or hormone sensitivity.

EMBRYONIC CONTROL OF ENDOSPERM FUNCTION

Gibberellin Production During Germination

The high levels of GA present in developing grain fall during maturation so that the dry grain usually contains very low levels. Following germination GA levels again rise with GA_1 being the predominant species in barley although low amounts of other GAs (GA_3, GA_{17}, GA_{19}, GA_{34} and GA_{48}) are also present (17,49). For the first 48h, the scutellum is the major site of GA production, and since this process in inhibited by the GA synthesis inhibitor CCC it is likely that GA arises by new synthesis. At later stages of germination the embryonic axis may be a more important source of GA. If the embryo is removed from the endosperm earlier than approx. 24h after germination, there is no increase in GA in the endosperm and very little subsequent production of α-amylase. This suggests that it is GA from the embryo which stimulates production of α-amylase in the aleurone (Fig. 1). The pathway by which GA from the scutellum reaches the aleurone is not well understood; it has usually been thought of as occurring by diffusion in free space, but the possibility of symplasmic movement via plasmodesmata between aleurone cells has also been suggested (5).

It has been proposed from results using immunoassay for GAs (6) that α-amylase production by aleurone of barley cv. Himalaya occurs in response to GA_4 synthesized in aleurone in response to some unidentified component from the embryo. This suggestion is novel in that for the first time it invokes a direct role for GA synthesis by the aleurone in the control of α-amylase synthesis. However, the "classical" view, which does not involve aleurone GA synthesis (Fig. 1) should not be discarded, since others have not been able to find GA_4 in germinating barley caryopses (17,63) or to confirm that GA_1-promoted production of α-amylase by aleurone is sensitive to inhibitors of GA synthesis (20).

Patterns of Endosperm Breakdown During Germination

In the past there has been controversy, which still persists to some extent, concerning the relative importance of the scutellum versus the aleurone layer in terms of hydrolase production, and the extent to which this is GA-dependent. Differences may have arisen in part because different cereals were being compared (see below), but in addition different techniques were being used to monitor the early stages of endosperm breakdown and/or hydrolase production.

Morphological evidence (based on scanning electron microscopy) indicated that initial starch degradation in barley occurred adjacent to the scutellar epithelium (39,19). From these studies it was concluded that the scutellum was an important source (perhaps the major source) of hydrolases involved in endosperm degradation, particularly at early stages of endosperm hydrolysis. The contribution made by the scutellum and aleurone of barley was also assessed using dissection experiments in which various tissues were cultured and assayed for α-amylase production. Tissues containing scutellum and aleurone cells contained 5 times as much α-amylase activity as tissues with scutellum cells alone (48,38). Applied GA was also more effective in α-amylase production in tissues containing aleurone cells. Another approach was to excise aleurone and scutellum tissues from germinating barley grains and measure their subsequent rate of α-amylase secretion over a short time in culture (51). Rates of secretion by excised aleurone layers were higher than by isolated scutella and peaked at different times (3-4d versus 1d after germination, respectively) (Fig. 3). It was calculated that the scutellum could account for a maximum of 5-10% of the total α-amylase activity found in the starchy endosperm.

In barley, therefore, there seems little doubt that the aleurone is the major source of α-amylase although a small amount may come from the scutellum especially at earlier stages in seedling development. The

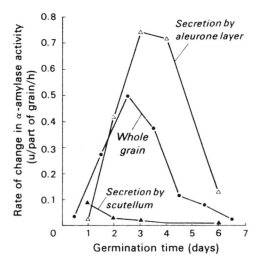

Fig. 3. Rates of α-amylase secretion by isolated scutella or aleurone layers from barley grains germinated for the indicated times. Also shown is the rate of change of α-amylase activity in starchy endosperm of the whole grain. (From 51).

situation in some of the tropical cereals, e.g. sorghum, maize and rice is somewhat different in that the scutellum is apparently the major source of α-amylase (1,4,15) and this is largely insensitive to applied GA. It would be of interest to determine whether this insensitivity reflects high endogenous levels of GA as suggested earlier in the case of maize.

In assessing the relative importance of scutellum versus aleurone in hydrolase production, particularly by microscopic, histochemical or immunohistochemical means, one potential difficulty of interpretation is that enzymes secreted by either tissue may not remain adjacent to their source but may be re-located within the endosperm, for instance by preferentially accumulating in cavities (48). In such a situation the sites of hydrolysis are not necessarily adjacent to the sites of hydrolase production. It may now be possible to assess relative mRNA levels for particular hydrolases in scutellum versus aleurone by a technique called tissue section hybridization. By hybridizing radioactive cloned DNA molecules for particular hydrolases (e.g. α-amylase) to lightly fixed longitudinal sections of germinating grains, it should be possible to assess the relative amounts of mRNAs for the particular hydrolase in different tissues. While the relative mRNA levels may not necessarily reflect levels of hydrolase production (and subsequent endosperm hydrolysis) they should provide a clear indication of the responsiveness of, say, α-amylase genes in different tissues to induction by GA.

ALEURONE RESPONSE

Reserve Degradation

During early seedling development, the stored reserves (protein, carbohydrate, lipid and mineral complexes) in the starchy endosperm and aleurone are degraded and mobilized via the scutellum to the developing seedling. The battery of enzymes required for eventual liquefaction of the stored reserves derive chiefly from the aleurone layer and many are under some form of GA control (see below).

Hydrolysis of walls of the aleurone cells and of endosperm cells is probably necessary for efficient release of hydrolytic enzymes from aleurone and for maximal access of the hydrolytic enzymes to starch grains and surrounding protein matrix in the endosperm. In addition, however, hydrolysis of cell walls represents a significant reserve mobilization in itself, and it has been calculated that 18.5% of the carbohydrate used by 6 day barley seedlings originated from aleurone and endosperm cell walls (42).

Hydrolysis of aleurone cell walls is first visible around the middle cells of the three-cell-thick barley aleurone (for review see 5). Hydrolysis continues but it leaves an innermost layer of wall material which is apparently resistant to degradation. This layer could be similar to cell wall connections which persist around plasmodesmata connecting aleurone cells, and it may contribute mechanical stability to the aleurone layer as a whole.

Arabinoxylans and $(1\rightarrow3)(1\rightarrow4)$-$\beta$-glucans are the major cell wall polysacharides of barley aleurone, and endosperm cell walls appear to be similar. There are at least three arabinoxylan degrading enzymes which are released from aleurone in response to GA_3, and of these endo-$(1\rightarrow4)$-β-xylanase is the one most likely to be involved in the initial cell wall hydrolysis. However there is a discrepancy (5) between the relatively late appearance of endoxylanase activity and the relatively early beginning of cell wall hydrolysis which at present is unresolved. The other major polysaccharide is believed to be degraded by two endo-$(1\rightarrow3)(1\rightarrow4)$-$\beta$-glucanases which have been recently characterized (62) and which also derive from aleurone.

The starch deposits of wheat and barley exist in two size classes of starch granules embedded in a protein matrix through the endosperm. The two glucose polymers forming the starch are amylose, a linear $(1\rightarrow4)$-α-glucan and amylopectin, a highly branched form of amylose containing $(1\rightarrow6)$-α-branches to other polymers. Following germination of barley the small starch granules are first degraded, primarily from the surface. Large starch granules frequently suffer initial surface pitting, but hydrolysis may then occur inside the granule and proceed through a hollow shell stage.

Starch hydrolysis proceeds through the action of several different enzymes. α-Amylase, an endohydrolase is probably important in the initial degradation of amylose and amylopectin, and there is evidence (37) that different isozyme groups of α-amylase have different *in vitro* activities on intact starch grains. β-amylase and α-glucosidase are both exohydrolases, releasing maltose and glucose residues respectively. Like α-amylase these enzymes are inactive at branch points in amylopectin, which are hydrolysed by limit dextrinase.

In contrast to α-amylase, α-glucosidase and limit dextrinase, which are all synthesized in the aleurone layer following germination, β-amylase exists in the endosperm of the dry grain in an inactive form, disulphide bonded to a component of protein bodies (59). The increase in β-amylase activity which follows germination is thought to result from the activity of a GA-induced proteinase which cleaves the β-amylase-protein body complex releasing active β-amylase.

Enzyme Response

The observation that isolated aleurone layers of barley respond to exogenously applied GA in a manner similar to intact aleurone layers following seed germination has made this a favourite system for studying the effect of GA on the activity of the hydrolases associated with mobilization of seed reserves. From a theoretical viewpoint we can envisage GAs having either a direct effect on the amount, cellular location, or activity of an enzyme, or an indirect effect as found for the appearance of β-amylase activity in the endosperm (see preceding section). Jacobsen (28) has summarized the response to GA of about 18 aleurone enzymes as falling into essentially four groups. The first group is characterized by a rapid change in activity of the enzyme, and presumably little change in amount. Enzymes involved in phospholipid metabolism fall into this group, and at present there is little information concerning the mechanism of their regulation. The second group includes enzymes whose synthesis and secretion are stimulated by GA such as α-amylase and protease (Fig. 4). Many aspects of hormonal control of gene expression in aleurone have been studied with particular reference to α-amylase, and this topic will dominate the remainder of this Chapter. In the third group, which includes β-glucanase, acid phosphatase and ribonuclease, fall those enzymes which increase in activity by new synthesis in the absence of added GA, but which may show an additional increase in activity if GA is present. Finally, there are some enzymes (xylopyranosidase and arabinosidase) whose level is constant in the absence of GA, but increases in the presence of GA and also show secreted activity. In enzymes of the last two groups the precise relationship between isozymes formed in the absence or presence of GA is frequently not known, so it is difficult to assess whether GA is stimulating levels of

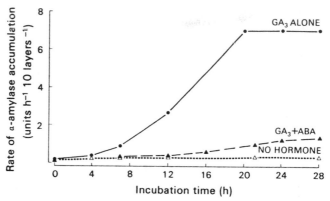

Fig. 4. Rates of α-amylase accumulation in isolated barley aleurone layers (plus surrounding medium) incubated for the indicated time without hormone, in the presence of 1 μM GA₃, or in the presence of 1 μM GA₃ plus 25 μM ABA. (From 24).

existing isozymic forms or causing *de novo* accumulation of new isozymes.

α-AMYLASE AND ITS GENETICS

Since the first demonstration that barley α-amylase could be resolved into a number of isozymes by electrophoresis, there have been many similar reports of multiple forms of the enzyme from a number of cereals. This has raised questions about the cause of this heterogeneity. Is it due to electrophoretic artifact, to post-translational modification of a single polypeptide chain, to multiple processing of a single α-amylase gene transcript, or does it result from the expression of a number of α-amylase genes? In the context of GA action, perhaps the principal question is whether the GA-promoted synthesis of multiple α-amylase isozymes involves only one or does it involve several sites and perhaps mechanisms of action? Such considerations have evoked studies of the nature of the α-amylase isozymes.

The α-amylases of barley have been studied extensively and many of the characteristics have been described previously (11,31 for review see 28). A result of major importance is that the isozymes separate into two isoelectric point (pI) groups and that these two groups of isozymes differ in many ways while isozymes within groups are more alike (Fig. 5). All of these results, but perhaps particularly the proteolytic fingerprinting studies, led to the belief that amino acid sequence differences existed between isozyme groups and perhaps even within groups. This was supported by cell free mRNA translation studies which indicated that synthesis of the two groups of isozymes was directed by two separate mRNA populations. Ultimate proof of genetic differences between

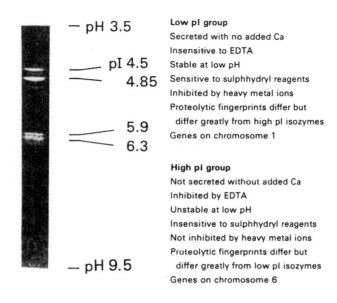

— pH 3.5

— pI 4.5

— 4.85

— 5.9

— 6.3

— pH 9.5

Low pI group
Secreted with no added Ca
Insensitive to EDTA
Stable at low pH
Sensitive to sulphhydryl reagents
Inhibited by heavy metal ions
Proteolytic fingerprints differ but
  differ greatly from high pI isozymes
Genes on chromosome 1

High pI group
Not secreted without added Ca
Inhibited by EDTA
Unstable at low pH
Insensitive to sulphhydryl reagents
Not inhibited by heavy metal ions
Proteolytic fingerprints differ but
  differ greatly from low pI isozymes
Genes on chromosome 6

Fig. 5. Characteristics of the low and high pI α-amylase isozyme groups from Himalaya barley. The diagram shows an isoelectric focussing gel pattern of purified α-amylase. (Partly from 31).

isozyme groups came from the demonstration that the genes for high and low pI groups of isozymes occur on different chromosomes, chromosomes 6 and 1 respectively (10,46). Thus, there is little doubt now that barley α-amylase is encoded by two (sets of) structural genes.

Recently, use of recombinant DNA techniques has provided additional support for the existence of multiple α-amylase genes in barley. cDNA clones complementary to α-amylase mRNA have been used to probe restricted and size-fractionated genomic DNA of Himalaya barley, and about seven DNA fragments have been found to contain α-amylase gene sequences. Also, base sequencing and restriction enzyme mapping studies of α-amylase cDNA clones isolated in various laboratories have demonstrated base sequence heterogeneity among clones. As for the α-amylase proteins, the clones appear to fall into 2 classes. Studies of amino acid sequence and composition, show that one class corresponds to the low pI group of α-amylases and the other to the high pI group (12,57). Base sequence homology between clones of the two groups is about 75%, while clones with homology of 90-95% have been found and appear to specify α-amylase isozymes from the same group. From cDNA analyses, it seems likely that some if not all of the α-amylase isozyme heterogeneity can be ascribed to the existence of two multigene families occurring on separate chromosomes.

The situation in wheat is very similar to that in barley. Wheat isozymes fall into two pI groups and although there are many more isozymes in wheat, due to the presence of the A, B and D genomes, most of them have been located on chromosomes 6 and 7 (16) which correspond to chromosomes 6 and 1 of barley respectively. Thus although the characteristics of the groups of α-amylase enzymes from wheat and barley have not been compared in any great detail, they do show genetic similarities.

The studies described above indicate that GA regulates the expression of α-amylase genes located on two different chromosomes. For other proteins, using wheat-barley addition lines, the gene for an unidentified GA-regulated aleurone protein has been located on chromosome 4 (Jacobsen, unpublished data). In addition, enzymes of the same activity as GA-inducible enzymes (although not necessarily the same isozymes) occur on various chromosomes of wheat and barley (9,22). Cloning of mRNAs from de-embryonated wheat grains treated with GA has revealed the presence of at least seven different mRNAs which increase in abundance with GA treatment (7). Therefore, it would appear that GA regulates the expression of a suite of genes, at least some of which are spread widely throughout the genome.

HORMONE PERCEPTION

It is axiomatic that for a cell to respond to a hormone, the cell must perceive the hormone. Although a considerable amount of information about transcriptional and post-transcriptional events in α-amylase induction has accrued in recent times (see later), little is known about hormone perception and the events which lead to regulation of gene transcription and other hormone-associated changes. As in the case of the steroid hormone receptor proteins in animal cells, GA perception in aleurone can be envisaged as involving a protein of some sort (see Chapter C4 and 56). One line of study of the induction of GA-sensitivity in wheat aleurone by environmental factors envisages plasma membrane proteins as primary perception sites for GA (8) but although such a model is consistent with a number of studies, there is no direct evidence that such proteins exist or that interaction between them and GA has any biological consequences. A variation of this concept envisages perception involving a cytoplasmic protein which may or may not also be a membrane protein, but again no such entity has been demonstrated. In line with a body of evidence showing that GA interacts with membrane lipid molecules, (see 56) a second concept envisages the membrane lipids themselves as taking part in GA perception. Such membrane perturbations could have a number of consequences, perhaps one of them involving changes in expression of the genome. Probably more is known about the biochemical events of GA control of α-amylase induction in aleurone than for any hormonal response in any plant tissue, yet much of the hormone binding literature concerned with plants is about other hormones and systems. Photoaffinity-labelled hormones probably offer most promise as probes for hormone binding sites. Photoaffinity studies have so far been done on auxin and ABA binding but photoaffinity labelled gibberellins are currently being assessed for use in aleurone studies. Since a good deal is known about the events of α-amylase induction in aleurone, there exists the possibility that the biological significance of any GA binding site may be assessable. The development of GA-responsive protoplasts from mature aleurone cells (see next section) will facilitate the use of photoaffinity labelled probes.

ACTION OF GIBBERELLIN AND ABSCISIC ACID

Control of Transcription

In keeping with studies on animal cells, it has been envisaged that plant hormones may act solely or in part by controlling transcription of genes, and thus levels of mRNA, which would, in turn, regulate rates of synthesis of specific hormone-induced proteins (for review see 30). For aleurone cells, there is evidence which shows that inhibitors of transcription prevent α-amylase induction and that changes in nucleic acid synthesis, particularly mRNA synthesis, can be detected (for review see 28). However, it has been difficult to relate the GA-induced changes in RNA synthesis to α-amylase mRNA. Recently, techniques have been developed which permit the isolation of transcriptionally active nuclei from cells and the detection of synthesis of specific gene transcripts. These techniques have been applied to the study of gibberellin action in barley (29) and oat (64) aleurone cells.

Because it is very difficult to isolate nuclei in good yield from intact aleurone cells, it was necessary to produce aleurone cell protoplasts from which nuclei could be easily isolated. Procedures for preparing protoplasts were devised and the oat and barley protoplasts prepared were responsive to gibberellin (Fig. 6) in essentially the normal way (26,32). Barley protoplasts synthesized and secreted the same α-amylase isozymes as intact aleurone cells, the efficiency of production was about the same as intact cells under similar osmotic conditions, and the cytoplasmic changes occurring in protoplasts were also similar to those of intact aleurone cells. Thus there was reason to believe that such protoplasts were suitable for the study of the normal events of GA-promoted α-amylase synthesis.

Nuclei were easily isolated from control and hormone-treated protoplasts and used for transcription studies. The nuclei were transcriptionally active and gene transcripts, particularly α-amylase transcripts were detected and quantified in the newly synthesized RNA. Nuclei from GA-treated barley cells make less total RNA, more α-amylase

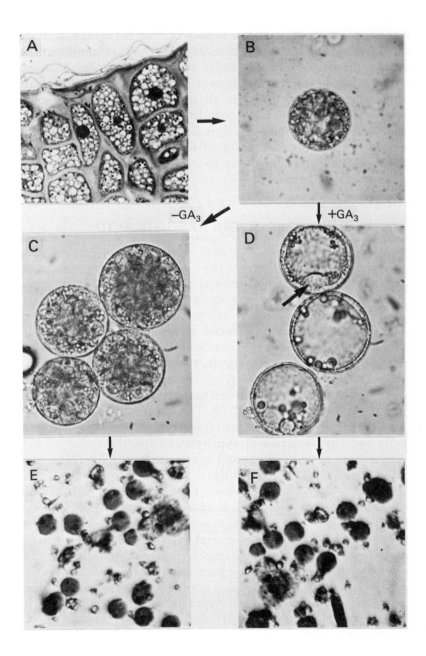

Fig. 6. The preparation of nuclei from aleurone cells. Protoplasts (B) are prepared from intact aleurone cells (A) and incubated in the presence and absence of GA_3 for 24 h. In the absence of the hormone (C) the cells swell and individual aleurone grains become visible while in its presence (D), the cells vacuolate and the nucleus (arrow) becomes visible in the parietal cytoplasm. The cells are physically lysed and the nuclei (E, F) are recovered by centrifugation. (A from 33, B-D from 32, E-F unpublished).

178

mRNA but less rRNA than nuclei from control cells (Fig. 7). If ABA was present during incubation of protoplasts with GA at levels which were inhibitory to α-amylase synthesis, all of the effects of GA on transcription were prevented. The studies on barley and oats have given essentially similar results and they indicate that both GA and ABA control α-amylase synthesis at least in part by controlling events within the nuclei, probably transcription of the α-amylase genes. The effects of GA and ABA on α-amylase gene transcription are entirely consistent with the effects of the hormones on levels of α-amylase mRNA (see Section 6.2).

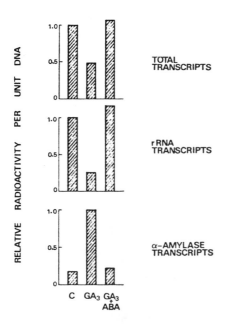

Fig. 7. Comparison of the synthesis of various transcripts by nuclei isolated from aleurone protoplasts which have been incubated in the presence (GA_3, GA_3 + ABA) and absence (C) of hormones for 24 h. (Data from 29).

These studies employed cDNA probes and hybridization conditions which probably favoured assay of transcripts from all α-amylase genes. By using appropriate hybridization conditions and more selective cDNA probes, it would be possible to measure the transcription of each major group of α-amylases genes and perhaps even individual genes. Selective hybridization methods have been used with great benefit in the study of α-amylase mRNA levels in intact cells (see below).

Change in mRNA Levels

It follows from the previous section that if the GA-enhanced transcription of α-amylase genes is a major factor in the control of α-amylase synthesis, then one would expect the level of α-amylase mRNA to accumulate in the presence of GA. There is an abundance of evidence indicating that this occurs.

The first assay which was specific for α-amylase mRNA was developed in 1976 (25) and involved indirect assay of the message. Isolated RNA was translated using cell-free techniques and the amount of newly synthesized α-amylase polypeptides was used as a measure of the level of α-amylase mRNA present. Results of similar experiments from a number of laboratories agree that GA causes translatable mRNA for α-amylase to accumulate (see 28) (Fig. 8). These studies did not discriminate between different α-amylase mRNAs because α-amylase migrated as a single component on the acrylamide gels used to separate the newly synthesized polypeptides. However, more recently, using two dimensional gel electrophoresis, it has been demonstrated that *in vitro* synthesized α-amylase has components corresponding to both the high and low pI α-

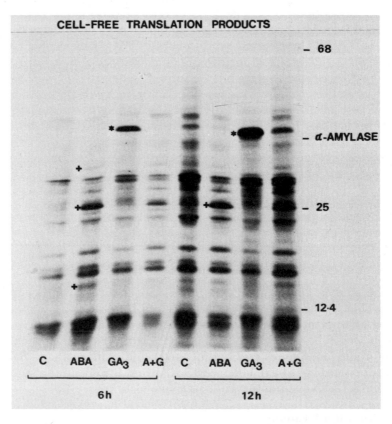

Fig. 8. Fluorograph of an SDS-gradient polyacrylamide gel of the cell-free translation products synthesized in the wheat germ system using total RNA isolated from barley aleurone layers incubated for 6 and 12 h with hormones or without (C). The asterisks show the α-amylase-related (precursor) polypeptide and the (+) signs locate ABA-specific polypeptides. (From 24).

amylase groups thus indicating that GA causes mRNA for both α-amylase groups to accumulate.

The cell-free mRNA translation studies presented the opportunity of observing effects of GA and ABA on mRNA species other than α-amylase mRNA. It is clear from Fig. 8 that there are changes in mRNAs other than α-amylase. The amounts of some polypeptides increase and those of others decrease with time after adding GA indicating that GA may cause many changes in the translatable mRNA population, some increasing, some decreasing and some remaining steady. Several studies examined the effects of ABA in conjunction with GA and although there was some discrepancy in results, the consensus result seems to be that ABA inhibits the accumulation of translatable α-amylase mRNA (Fig. 8). At the same time, ABA causes increased levels of some mRNAs and has no effect on others. It has also been shown that ABA-induced mRNAs are prevented from appearing by GA. These studies indicate that, in principle, GA and ABA act in similar but opposing ways at the level of mRNA. Both cause increased levels of some mRNAs and decreased levels or no change in others and each hormone appears to antagonize the other's action.

The availability of α-amylase cDNA clones presented the opportunity for direct analysis of mRNA and the possibility of assaying total α-amylase gene transcripts as opposed to translatable transcripts which perhaps did not reflect the total. By using the dot blot or Northern blot techniques, hybridization analysis of RNA from control and hormone treated aleurone has provided evidence which supports the cell-free translation studies. Results of several studies on barley indicate that total α-amylase mRNA accumulates in the presence of GA, but not in its absence, and that ABA inhibits α-amylase mRNA accumulation (Fig. 9). Hybridization analyses of the accumulation of α-amylase mRNA with time have employed cDNA probes which appear to discriminate between the high and low pI groups of α-amylase mRNAs. The results of these

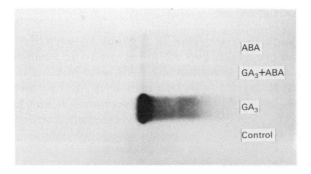

Fig. 9. Hybridization of an α-amylase cDNA (pHV19) clone to size fractionated RNA isolated from aleurone layers which were incubated with and without (control) hormones for 24 h. (From 12).

studies (27,54) can be generalized as follows. The mRNA for the low pI α-amylase is already present in hydrated aleurone, it increases gradually in response to GA increasing by about 10-20 fold over 24 h and responds maximally to a low level (about 10^{-8} M) of GA (Fig. 10). In contrast,

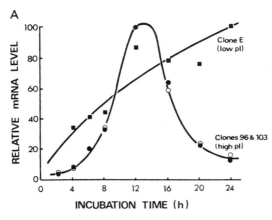

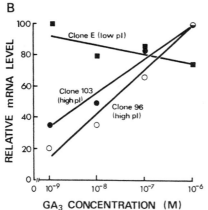

Fig. 10. Assay of mRNA for the high and low pI α-amylase groups in barley aleurone layers. A. RNA was isolated from aleurone treated with 1 μM GA for different times, dot blots were prepared and probed with radioactive group specific α-amylase cDNA clones. The dot-blot data was quantified by densitometry. B. RNA was isolated from aleurone treated with different GA concentrations for 12 h and processed as for A. (Data plotted from Table 1 of 27).

mRNA for the high pI α-amylase is present in very low amounts in hydrated tissue, increases by 50-100 fold over 12-16 h and then decreases, and responds maximally to a high (10^{-6} M) level of GA. In many ways, these results are in accord with the responses of the low and high pI groups of α-amylase isozymes to GA (see section 6.4) and they reinforce the notion that the two groups of α-amylase genes are differentially regulated

Changes in Protein Synthesis

Many of the effects of hormones on mRNA levels in aleurone are reflected in their effects on the pattern of protein synthesis which is extensively modified by GA and ABA. *In vivo* pulse labelling studies (Fig.

11) show that, in the presence of GA, a-amylase rapidly becomes the major protein synthesized and that the synthesis of many polypeptides is depressed relative to a-amylase. ABA has two effects: it suppresses the GA-promoted changes and also promotes the synthesis of several polypeptides. Other results show that excess GA over ABA suppresses the synthesis of ABA-specific polypeptides. In principle, these effects of GA and ABA on protein synthesis are similar in many respects to those on mRNA levels described in the previous section. Here, both GA and ABA promote the synthesis of specific polypeptides and each antagonizes the other's action. In general, these results indicate that many of the effects of both hormones on protein synthesis are likely to be mediated, at least in part, by regulation of mRNA levels, as in the case of a-amylase.

Changes in a-Amylase Isozymes

In barley aleurone, added GA causes the onset of accumulation of a-amylase within a few hours, and this continues for 30-40 hours. Most of

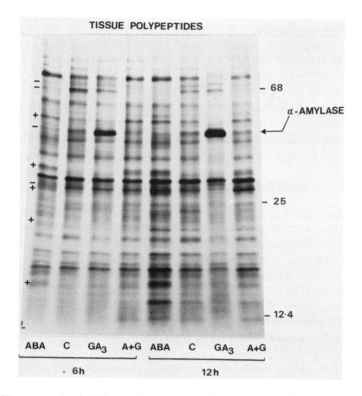

Fig. 11. Fluorograph of SDS-gradient polyacrylamide gel of *in vivo* pulse-labelled polypeptides synthesized by barley aleurone layers incubated with and without (C) hormones for 6 and 12 h. The (−) signs refer to polypeptides which were decreased by ABA and the (+) signs to those which were increased by ABA. (Data from 24).

the enzyme is secreted from the aleurone tissue. When α-amylase accumulation is examined by separation of isozymes, it becomes evident that different isozymes do not accumulate at the same rate. Isozymes within the low pI group and the high pI group behave similarly but the groups behaved differently. The low pI group appears first during incubation and the high pI group second (Fig. 12A), although the latter accumulates rapidly and becomes the dominant isozyme group. The low pI group appears at low GA concentration and shows relatively little response to increasing GA levels, while the high pI isozymes are not present in 10^{-8} M GA but show a dramatic response to increasing GA (Fig. 12B). These results complement those shown in Fig. 10. It is evident that the timing of the onset and the subsequent rates of accumulation of the mRNAs and isozymes for each group are substantially in accord. A similar parallel can be drawn between the sensitivities to GA concentration of the mRNA levels and the isozymes for each group. This comparison adds confidence to the identification of the cDNA clones, it indicates that the syntheses of the two isozyme groups are regulated primarily by the levels of their individual mRNAs and it again indicates that GA differentially regulates expression of the two groups of genes.

In wheat, analysis of α-amylase isozyme production has not yet progressed as far as for barley but it is known that production of the two α-amylase groups is also not synchronous. In germinating grain, the high pI isozymes appear first which is the reverse of enzyme production by isolated barley aleurone. However, the relative productions of the two groups of α-amylases requires clarification because there is evidence that it can change depending on whether enzymes are produced in germinating grain, or in GA treated de-embryonated grain or isolated aleurone layers.

Translation—a Role for Hormonal Control?

The studies relating mRNA levels to synthesis of α-amylase have shown that the accumulation of α-amylase mRNA is correlated with the rate of α-amylase synthesis under a number of experimental conditions (for review see 28). This indicates that α-amylase mRNA is probably translated without delay and that α-amylase synthesis is regulated in the main by the level of its mRNA. However, there is evidence which casts doubt on the apparent simplicity of this process.

It has been estimated using cell free translation techniques that α-amylase mRNA accumulates to a maximum of about 20% of the total translatable mRNA. However, at the same time, *in vivo* protein labelling studies indicate that α-amylase constitutes approx 50-60% of total protein synthesis. It would thus appear that α-amylase synthesis constitutes a greater proportion of protein synthesis than its mRNA does of total translatable mRNA. It would therefore seem that protein synthesis is being regulated by factors in addition to mRNA levels.

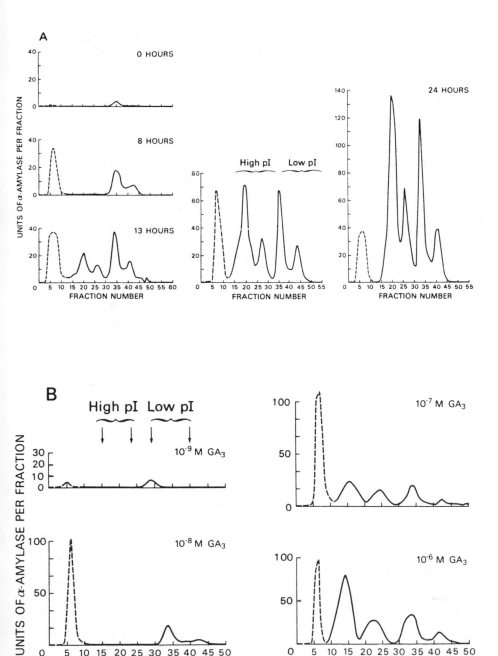

Fig. 12. DEAE-cellulose fractionation of total amylase produced by aleurone layers incubated (A) with 1 µM GA for various times and (B) with various concentrations of GA for 24 h. The dashed peaks are β-amylase and all others are α-amylase. (From 31).

185

One explanation contends that the initiation of translation of α-amylase mRNA is very efficient and that this explains the "gearing" up of α-amylase synthesis. Such efficiency may well be a function of the mRNA "translation-initiation" nucleotide sequences, but it may also be facilitated by hormone dependent initiation factors. Alternatively, the absolute rate of α-amylase synthesis may remain constant and it may become such a high proportion of protein synthesis because the synthesis of many polypeptides is suppressed. The latter is perhaps the preferable explanation because the rate of accumulation of α-amylase activity reaches a constant level at about the time mRNA level becomes maximal. Thus the most likely interpretation of these results at present is that they are due to suppression of synthesis of many polypeptides. Some of this suppression can be explained by the fact that the levels of some mRNAs are reduced in the presence of GA (Fig. 8), but this is not necessarily the only explanation. GA might suppress translation of some mRNAs as well.

The case for ABA acting on translation of α-amylase mRNA is also not strong (for discussion, see 28 and references therein). It is based partly on studies interpreted to mean that ABA can inhibit α-amylase synthesis at a time when α-amylase synthesis is not dependent on RNA synthesis. Presumably, ABA could not be acting by inhibiting RNA synthesis. However, different laboratories have obtained different results and the case for action of ABA at the level of translation is best described at present as disputed.

In summary, it appears that there is not yet a strong case for hormonal control of α-amylase mRNA translation in aleurone.

Multilevel Action by ABA

It can be seen from the foregoing that ABA antagonizes GA action and that there are two major facets to its action—one preventative and one promotive. It prevents GA action at the level of transcription and this probably explains the suppression of GA-induced mRNAs and, at least in part, the suppression of GA-induced protein synthesis. The promotive action involves the ABA-induced increase in mRNAs and proteins, many of which are probably related. One aspect of the promotive role of ABA has been related to an anti-α-amylase action. An ABA-induced endogenous α-amylase inhibitor has been identified (44,45,60). The inhibitor reacts only with high pI α-amylase. It is present in mature starchy endosperm and it presumably accumulates during grain development in response to endogenous ABA (see section 2.2). In developing grain, the inhibitor may serve to inhibit α-amylase during starch synthesis or to prevent starch degradation by α-amylase arising in the aleurone during premature sprouting of the grain. When ABA exists in mature aleurone, the inhibitor could play a similar role during germination. Whatever the situation, it is evident that ABA can deal with α-amylase synthesis at several levels, not only preventing synthesis but also giving rise to a "mopping up" protein to cope with already existing enzyme.

Enzyme Processing and Secretion

Following synthesis of α-amylase protein, it is processed, transported to the plasmalemma and released from the aleurone cell (exocytosis). Despite a number of studies over the last 10 years or so, many of the essential features of these processes have eluded detection. In animal cells, secretory proteins are synthesized on rough endoplasmic reticulum (ER), polypeptides are segregated into the lumen of the ER, the polypeptide passes through the Golgi apparatus and into secretion vesicles which then fuse with the plasmalemma. Processing of the proteins during this process includes co-translational excision of a signal polypeptide, glycosylation in the ER and subsequent modification of the glycosylation in the Golgi body (for reviews see 4,52).

There is some evidence which indicates that, at least in part, the early events of α-amylase synthesis and transport may be similar to those of animal secretory proteins. Comparison of *in vivo* and *in vitro* synthesized α-amylase polypeptides from wheat and barley provides evidence that the newly synthesized (unprocessed) polypeptide contains a signal sequence which is subsequently lost. The amino acid sequence has been deduced from base sequencing of α-amylase clones. The putative leader sequences from two α-amylase clones, one apparently a high pI isozyme clone and the other a low one, are very different (matching in only 10 out of 23 amino acids) although both are hydrophobic in nature. In addition α-amylase from barley has been shown to be a glycoprotein, containing N-acetylglucosamine, glactose, mannose, glucose and fucose (41,53). Although there is no direct evidence that α-amylase synthesis occurs in RER, stacks of RER are prominent in aleurone cells secreting α-amylase, the four major α-amylase isozymes have been located in the lumen of the RER, and it has been shown that these molecules are destined for secretion. In addition to this, GA-promoted acid phosphatase has been localized cytochemically in ER. These studies indicate that the early events of the secretion pathway in aleurone are substantially similar to those in animal cells.

However, the route followed after the ER lumen is not known. Possibilities are shown in Fig. 13. One possibility (see 1 on Fig. 13) is that the enzyme flows from the ER directly into some kind of ER-derived secretory vesicle which empties its contents through the plasmalemma. Although phenomena like the fragmentation of ER into (secretory?) vesicles and the apparent fusion of various vesicles with the plasmalemma have been described in electron microscope studies, there is no good evidence that such vesicles contain α-amylase and there is no

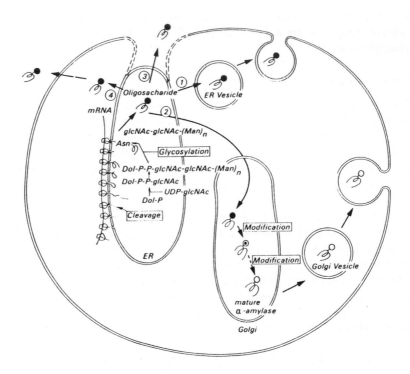

Fig. 13. Possible pathways for α-amylase secretion from aleurone cells. (Adapted from 2 and 4).

guarantee that such structures are not artifacts of tissue preparation. Several studies have described α-amylase-containing vesicles from homogenates of aleurone cells and the vesicle membranes had ER-like characteristics, but it is possible that they are artifacts of cell disruption rather than authentic secretory vesicles. In general, organelle fractionation studies have not been able to unequivocally identify a secretory vesicle.

Another possibility is that the transport pathway involves the Golgi apparatus (see 2 on Fig. 13). Glycosylation of secretory proteins in animal cells occurs in the ER and in the Golgi apparatus, and there is evidence that while glucosamine and mannose are added in the ER, galactose, glucose and fucose are added in the Golgi (see 52). If this is true for aleurone, it follows that Golgi may be involved in α-amylase transport and processing since α-amylase contains galactose, glucose and fucose. Electron microscope studies of chemically fixed tissue led to the conclusion that Golgi bodies are not as common as one might expect in cells secreting α-amylase and thus may not be involved, but more recent freeze fracture studies show that Golgi are abundant. Arguments against the involvent of Golgi include the contention that the low level of α-amylase glycosylation tends to make Golgi involvement unnecessary, and the fact that wheat α-amylase can be synthesized and secreted by yeast cells, which do not have Golgi bodies (see 52). ATPase is a GA-promoted enzyme in aleurone, and using cytochemistry and electron microscopy it has been shown to exist in Golgi bodies, as well as ER, in GA-treated cells. However, whether the Golgi enzyme is part of the GA-response, or whether it is part of the transport pathway is unknown. Monensin, an ionophore which interferes with Golgi function and secretion in animal cells, also blocks release of α-amylase from rice scutellum and barley aleurone cells and causes distention of the Golgi (see 3, R.L. Jones, pers. comm.). These results seem to add weight to the argument for Golgi involvement.

Two other possibilities do not involve vesicles of any sort. One envisages the enzyme flowing from ER directly to the periplasmic space through connections between ER and the plasmalemma (see 3 on Fig. 13). Cytochemical studies of acid phosphatase in GA-treated aleurone cells show the ER (containing the enzyme) and the plasmalemma in close proximity in places, but no definite continuity between the two membrane systems was evident. Indeed, reports of such membrane continuity are rare. The other possibility (see 4 on Fig. 13) involves release of the enzyme from the ER into the cytoplasm and diffusion through the cytoplasm to the plasmalemma. While such a mechanism is unattractive from the point of view of possible cytoplasmic damage caused by GA-induced hydrolases, it cannot be ruled out.

It is also possible that different α-amylase isozymes may follow different pathways. Studies of glycosylation of α-amylase have so far only dealt with total preparations and it has been determined that the sugars glucose, mannose and n-acetylglucosamine occur at the rate of only a half residue per mole of protein. Results of cDNA sequencing studies (reviewed in 2) indicate that whereas a low pI isozyme sequence has no glycosylation site in the mature polypeptide, a high pI isozyme sequence does. Such results lead to the thought that perhaps the low pI α-amylase isozymes are not glycosylated and the high pI isozymes are, and that this may reflect differences in the intracellular transport pathway. Could it be that the high pI isozymes travel via the Golgi apparatus and the low pI isozymes via some other route?

Whether GA exercises control over any of the transport processes is unknown but there is evidence that Ca^{++} ions may exercise some control, principally over secretion of the high pI isozymes. Recent studies have shown that Ca^{++} does not exercise any control over mRNA production, which seems to be the preserve of GA, and that it acts at or subsequent to translation. Whether Ca^{++} regulates synthesis and/or stability of the newly synthesized high pI polypeptide or whether it is involved with glycosylation, transport or exocytosis is yet to be determined.

Clearly there is still much to be learnt about α-amylase transport and secretion in aleurone cells. Organelle isolation studies will undoubtedly be useful although they are subject to the criticism that any α-amylase localization may be an artifact of tissue disruption and that organelles may be difficult to identify. A very useful adjunct to such studies would involve immunocytochemistry coupled with electron microscopy using fixation methods which minimize enzyme migration and give good structural preservation. GA-responsive protoplasts appear to offer an excellent opportunity for such studies. Any such study should ultimately employ antibodies which are isozyme specific. Studies of α-amylase secretion have been hindered by the presence of thick aleurone cell walls which trap large quantities of secreted enzyme. Consequently, it has been difficult to quantify secreted enzyme while, in aleurone layer homogenates, the intracellular enzyme is diluted by the secreted enzyme in the cell walls. Having no cell walls, protoplasts may also be of value in these studies. The kinetics of secretion can be studied much better than before and enzyme preparations rich in intracellular α-amylase can be obtained from washed protoplasts. Protoplasts can also be easily and rapidly lysed and they offer rapid access to cellular organelles.

References

1. Aisien, A.O., Palmer, G.H. (1983) The sorghum embryo in relation to the hydrolysis of the endosperm during germination and seedling growth. J. Sci. Food Agric. *34*, 113-121.
2. Akazawa, T., Hara-Nishimura, I. (1985) Topographic aspects of biosynthesis, extracellular secretion, and intracellular storage of proteins in plant cells. Ann. Rev. Plant Physiol. *36*, 441-472.
3. Akazawa, T., Mitsui, T. (1985) Biosynthesis, intracellular transport and secretion of α-amylase in rice seedlings. In: New Approaches to Research on Cereal Carbohydrates Hill, R.D., Munck L., ed. Elsvier, Amsterdam. pp. 129-137.
4. Akazawa, T., Miyata, S. (1982) Biosynthesis and secretion of α-amylase and other hydrolases in germinating cereal seeds. Essays in Biochem. *18*, 40-78.
5. Ashford, A.E., Gubler, F. (1984) Mobilization of polysaccharide reserves from endosperm. In: Seed Physiology Vol. 2, pp. 117-162, Murray, D., ed. Academic Press, New York.
6. Atzorn, R., Weiler, E.W. (1983) The role of endogenous gibberellins in the formation of α-amylase by aleurone layers of germinating barley caryopses. Planta *159*, 289-299.
7. Baulcombe, D.C., Buffard, D. (1983) Gibberellic-acid-regulated expression of α-amylase and six other genes in wheat aleurone layers. Planta *157* 493-501.
8. Black, M., Chapman, J., Norman, H. (1982) The ability of wheat aleurone tissue to participate in endosperm mobilization. *In*: Recent advances in phytochemistry Vol. 17 Mobilization of reserves in germination, pp. 193-211, Nozzolillo, C., Lea P.J., Loewus, F.A. ed. Plenum New York, London.
9. Brown A.H.D. (1983) Barley. *In*: Isozymes in plant genetics and breeding. Pt. B. pp. 57-77. Tanksley, S.D. Orton, T.J. eds. Elsevier, Amsterdam.
10. Brown, A.H.D., Jacobsen, J.V. (1982) Genetic basis and natural variation of α-amylase isozymes in barley. Genet Res. *40*, 315-324.

11. Callis, J., Ho, T-H.D. (1983) Multiple molecular forms of the gibberellin-induced α-amylase from the aleurone layers of barley seeds. Arch. Biochem. Biophys. 224, 224-234.

12. Chandler, P.M., Zwar, J.A., Jacoben, J.V., Higgins, T.J.V., Inglis, A.S. (1984) The effects of gibberellic acid and abscisic acid on α-amylase mRNA levels in barley aleurone layers: Studies using an α-amylase cDNA clone. Plant. Molec. Biol. 3, 407-418.

13. Dewdney, S.J., McWha, J.A. (1978) The metabolism and transport of abscisic acid during grain fill in wheat. J. Exptl. Bot. 29, 1299-1308.

14. Duffus, C.M. (1969) α-amylase activity in the developing barley grain and its dependence on gibberellic acid. Phytochemistry 8, 1205-1209.

15. Dure, L.S. (1960) Site of origin and extent of activity of amylases in maize germination. Plant. Physiol. 35, 925-934.

16. Gale, M.D., Spencer, D.M. (1977) The location of chromosomal control of GA induced enzyme release by distal half-grains of wheat. Biochem. Genet. 15, 47-57.

17. Gaskin, P., Gilmour, S.J., Lenton, J.R., MacMillan, J., Sponsel, V.M. (1984) Endogenous gibberellins and kaurenoids identified from developing and germinating barley grains. J. Plant Growth Regul. 2, 229-242.

18. Gaskin, P., Kirkwood, P.S., Lenton, J.R., McMillan, J., Radley, M.E. (1980) Identification of gibberellins in developing wheat grain. Agric. Biol. Chem. 44, 1589-1593.

19. Gibbons, G.C. (1981) On the relative role of the scutellum and aleurone in the production of hydrolases during germination of barley. Carlsberg Res. Commun. 46, 215-225.

20. Gilmour, S.J., MacMillan, J. (1984) Effect of inhibitors of gibberellin biosynthesis on the induction of α-amylase in embryoless caryopses of Hordeum vulgare cv Himalaya. Planta 162 89-90.

21. Goldbach, H., Goldbach, E. (1977) Abscisic acid translocation and influence of water stress on grain abscisic acid content. J. Exptl. Botany 28, 1342-1350.

22. Hart, G.E. (1979) Genetical and chromosomal relationships among the wheats and their relatives. Stadler Symp. 11, 9-29.

23. Harvey, B.M.R., Oaks, A. (1974) The role of gibberellic acid in the hydrolysis of endosperm reserves in Zea mays. Planta 121, 67-74.

24. Higgins, T.J.V., Jacobsen, J.V., Zwar, J.A. (1982) Gibberellic acid and abscisic acid modulate protein synthesis and mRNA levels in barley aleurone layers. Plant Molec. Biol. 1 191-215.

25. Higgins, T.J.V., Zwar, J.A., Jacobsen, J.V. (1976) Gibberellic acid enhances the level of translatable mRNA for α-amylase in barley aleurone layers. Nature 260, 166-169.

26. Hooley, R. (1982) Protoplasts isolated from aleurone layers of wild oat (Avena fatua L.) exhibit the classic response to gibberellic acid. Planta 154, 29-40.

27. Huang, J., Swegel, M., Dandekar, A.M., Muthukrishnan, S. (1984) Expression and regulation of α-amylase gene family in barley aleurones. J. Molec. Appl. Genet. 2, 579-588.

28. Jacobsen, J.V. (1983) Regulation of protein synthesis in aleurone cells by gibberellin and abscisic acid. In: The biochemistry and physiology of gibberellins, Vol 2 pp. 159-187. Crozier, A., ed. Praeger, New York.

29. Jacobsen, J.V., Beach, L.R. (1985) Evidence for control of transcription of α-amylase and ribosomal RNA genes in barley aleurone protoplasts by gibberellic acid and abscisic acid. Nature 316, 275-277.

30. Jacobsen, J.V., Higgins, T.J.V. (1978) The influence of phytohormones on replication and transcription. In: Phytohormones and related compounds: A Comprehensive Treatise. Vol. 1, The Biochemistry of Phytohormones and Related Compounds, pp. 515-582, Letham, D.S., Goodwin, P.B., and Higgins, T.J.V. eds. Elsevier/North Holland, Amsterdam.

31. Jacobsen, J.V., Higgins, T.J.V. (1982) Characterization of the α-amylases synthesized by alcurone layers of Himalaya barley in response to GA₃. Plant Physiol 70, 1647-1653.
32. Jacobsen, J.V., Zwar, J.A., Chandler, P.M. (1985) Gibberellic acid responsive protoplasts from mature aleurone of Himalaya barley. Planta 163, 430-428.
33. Jones, R.L. (1969) The fine structure of barley aleurone cells. Planta 85, 359-375.
34. King, R.W. (1976) Abscisic acid in developing wheat grains and its relationship to grain growth and maturation. Planta 132, 43-51.
35. King, R.W. (1979) Abscisic acid synthesis and metabolism in wheat ears. Aust. J. Plant. Physiol. 6, 99-108.
36. King, R.W. (1982) Abscisic acid in seed development In: The physiology and biochemistry of seed development, dormancy and germination, pp. 157-181, Khan, A.A., ed. Elsevier Biomedical Press, Amsterdam.
37. MacGregor, A.W. (1980) Action of malt α-amylases on barley starch granules. MBAA Technical Quarterly 17, 215-221.
38. MacGregor, A.W., MacDougall, F.H., Mayer, C., Daussant, J. (1984) Changes in level of α-amylase components in barley tissues during germination and early seedling growth. Plant Physiol. 75, 203-206.
39. MacGregor, A.W., Matsuo, R.R. (1982) Starch degradation in endosperms of barley and wheat kernels during initial stages of germination. Cereal Chem. 59, 210-216.
40. Milborrow, B.V., Robinson, D.R. (1973) Factors affecting the biosynthesis of abscisic acid. J. Exptl. Botany 24, 537-548.
41. Mitchell, E.D. (1972) Homogeneous α-amylase from malted barley. Phytochem. 11, 1673-1676.
42. Morrall, P., Briggs, D.E. (1978) Changes in cell wall polysaccharides of germinating barley grains. Phytochemistry 17, 1495-1502.
44. Mounla, M.A.K. (1978) Gibberellin-like substances in parts of developing barley grain. Physiol. Plantarum 44, 268-272.
44. Mundy, J. (1984) Hormonal regulation of α-amylase inhibitor synthesis in germinating barley. Carlsberg Res. Commun. 49, 439-444.
45. Mundy, J., Svendsen, I., Hejgaard, J. (1983) Barley α-amylase/subtilisin inhibitor. I Isolation and characterization. Carlsberg Res. Commun. 48, 81-90.
46. Muthukrishran, S., Gill, B.S., Swegle, M., Chandra, G.R. (1984) Structural genes for α-amylases are located on barley chromosomes 1 and 6. J. Biol. Chem. 259, 13637-13639.
47. Nicholls, P.B. (1982) Influence of temperature during grain growth and ripening of barley on the subsequent response to exogenous gibberellic acid. Aust. J. Plant. Physiol. 9, 373-383.
48. Palmer, G.H. (1982) A reassessment of the pattern of endosperm hydrolysis (modification) in germinated barley. J. Inst. Brew. 88, 145-153.
49. Radley, M. (1967) Site of production of gibberellin-like substances in germinating barley embryos. Planta 75, 164-171.
50. Radley, M. (1976) The development of wheat grain in relation to endogenous growth substances. J. Exptl. Botany 27, 1009-1021.
51. Ranki, H., Sopanen, T. (1984) Secretion of α-amylase by the aleurone layer and the scutellum of germinating barley grain. Plant Physiol. 75, 710-715.
52. Robinson, D.G. (1985) Plant Membranes. John Wiley and Sons, New York.
53. Rodaway, S.J. (1978) Composition of α-amylase secreted by aleurone layers of grains of Himalaya barley. Phytochem. 17, 385-389.
54. Rogers, J.C. (1985) Two barley alpha amylase gene families are regulated differently in aleurone cells. J. Biol. Chem. 260 3731-3738.
55. Slominski, B., Rejowski, A., Nowak, J. (1979) Abscisic acid and gibberellic acid contents in ripening barley seeds. Physiol. Plant 45, 167-169.
56. Stoddart, J.L., Venis, M.A. (1980) Molecular and subcellular aspects of hormone action. In: Hormonal regulation of development I Molecular aspects of plant hormones.

Encyclopedia of Plant Physiology, Vol. 9, pp 445-510, MacMillan, J., ed. Springer-Verlag, Berlin Heidelberg New York.

57. Svensson, B., Mundy, J., Gibson, R.M., Svendsen, I. (1985) Partial amino acid sequences of a-amylase isozymes from barley malt. Carlsberg Res. Commun. *50*, 15-22.
58. Triplett, B., Quatrano, R.S. (1982) Timing, localization and control of wheat germ agglutinin synthesis in developing wheat embryos. Developmental Biology *91*, 491-496.
59. Tronier, B., Ory, R.L. (1970) Association of bound beta-amylase with protein bodies in barley. Cereal Chem. *47*, 464-471.
60. Weselake, R.J., MacGregor, A.W., Hill, R.D. (1983) An endogenous a-amylase inhibitor in barley kernels. Plant Physiol. *72*, 809-812.
61. Wilson, G.F., Rhodes, A.M., Dickinson, D.B. (1973) Some physiological effects of viviparous genes vp_1 and vp_5 on developing maize kernels. Plant Physiol. *52*, 350-356.
62. Woodward, J.R., Fincher, G.B. (1981) Purification and chemical properties of two 1,3;1,4-β-glucan endohydrolases from germinating barley. Eur. J. Biochem. *121* 663-669.
63. Yamada, K. (1982) Determination of endogenous gibberellins in germinating barley by combined gas chromatography—mass spectrometry. J. Am. Soc. Brew. Chem. *40*, 18-25.
64. Zwar, J.A., Hooley, R. (1986) Hormonal regulation of a-amylase gene transcription in wild oat (*Avena fatua* L.) aleurone protoplasts. Plant Physiol. *80*, 459-463.

C4. Hormone Binding and Its Role in Hormone Action

Kees R. Libbenga and Albert M. Mennes

Department of Plant Molecular Biology, Botanical Laboratory, University of Leiden, 2311 VJ Leiden, Netherlands.

INTRODUCTION

Communication between cells in multicellular organisms is required to regulate their differentiation and organization into tissues, to control their growth and division and to regulate their diverse activities. When the organism becomes more complex during its development, communication between its different parts requires signaling systems which operate over relatively long distances. Hormonal systems belong to such long-range communication systems. Both plants and animals make use of hormonal systems, but in higher plants, which do not possess a nervous system, long-range communication is largely dependent upon a complex hormonal system.

In a hormonal system, cells of the different tissues and organs do not only transmit signals, but they are also capable of detecting signals which they receive from other parts and responding to those signals in their own characteristic way. In this chapter we will discuss the molecular mechanisms by which target cells for plant hormones translate the signals into a specific response. Assuming that those mechanisms are largely unknown, which is very close to reality, how should one proceed to unravel them? Of course, the first thing to do is to study what we know about such mechanisms from investigations with other organisms and to try to elucidate a few general principles. As a working hypothesis one might then assume that these principles are, *mutatis mutandis*, also valid for higher plants. It is along this line that we will discuss the mechanism of primary action of plant hormones.

We know from other organisms that target cells are equipped with a distinctive set of receptors for detecting a complementary set of chemical signals. Receptors are (glyco) proteins which specifically and reversibly bind chemical signals but, unlike enzymes, do not convert them chemically. Upon binding, the receptor molecules are, through a conformational change, transformed into an activated state. This causes the initiation of a molecular program that ultimately leads to the characteristic response. Thus, receptor proteins act both as signal detectors and transducers.

Hormones often have pleiotropic effects, i.e., different types of target cell all respond to the same set of signals, but in a different way. In many cases these types of target cell have similar perception-and-transduction mechanisms, but the molecular programs which are elicited by these mechanisms are different. This is illustrated in a very simplified way by Fig. 1. In this example, receptor activation triggers a molecular program which is simply a direct activation of a distinct set of enzymes. If the set of responsive enzymes is different in another type of target cell, then the same signal elicits a different response via a similar perception-and-transduction chain.

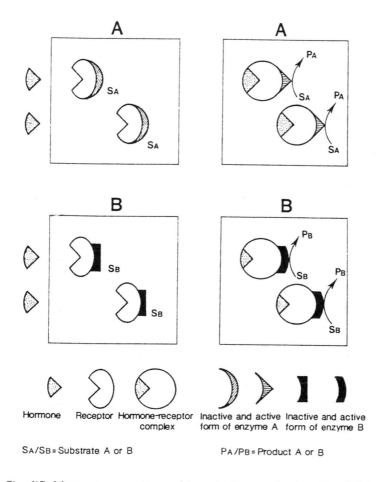

SA/SB = Substrate A or B   PA/PB = Product A or B

Fig. 1. Simplified hormone perception and transduction mechanism. Two different target cells, A and B respectively, contain the same receptor protein for a certain hormone. In cell A enzyme A, and in cell B enzyme B is transformed into an active state upon binding of the hormone. In this way the same signal triggers a different molecular programme in different target cells.

Of course, hormones do not only modulate enzyme activities in target cells. In general, most chemical signals ultimately influence target cells either by altering the properties (activities) or rates of synthesis of existing proteins or by altering the synthesis of new ones. Moreover, perception-and-transduction chains may be more complex than the ones shown in Fig. 1. This can be inferred from current models on perception and transduction of animal hormones. These models also show that, although each hormone is detected by a specific receptor, transduction follows only a limited number of pathways. Figs. 2 and 3 represent two major pathways of hormone transduction in animal systems. In one pathway (Fig. 2) the receptors, are localized at the plasmamembrane, with the hormone-binding moiety facing outside the cell. These receptors function as sensory systems for external hormone levels and transduce the signal into intracellular signals, either via activation of adenylate cyclase which converts ATP into cyclic AMP and/or activation of phospholipase C which converts the membrane lipid phosphatidylinositol-4,5-biphosphate into inositol triphosphate (IP3) and diacylglycerol (DG). These signals activate c-AMP-dependent protein kinase, liberate Ca^{2+} from internal sources thus increasing the levels of free Ca^{2+}, and activate protein kinase C, respectively. These processes somehow trigger the cell's response, which may include alterations in gene expression. Many peptide hormones, for example, follow such transduction routes.

Fig. 3 represents a transduction pathway for hormones, like steroids, which are readily taken up by the target cells by simple diffusion through the plasmamembrane. These target cells are equipped with internal receptors detecting intracellular hormone levels. These receptors are regulatory proteins which may directly interact with target-cell-specific non-histone proteins and DNA sequences of the chromatin. This interaction results in increased rates and/or altered patterns of gene transcription. Target cells for steroid hormones also have plasma-membrane-bound receptors which detect external hormone levels and might be responsible for steroid-induced rapid responses in Na^+/H^+ fluxes over the plasmamembrane and in c-AMP and Ca^{2+} levels.

Transduction chains as shown by Fig. 2 and 3 are multistep regulatory circuits, of which many details are still unknown. The above mentioned models do not explain the complex responses to hormones, such as cell division. However, they give us an insight into the first process of perception and transduction of the signals producing these complex responses. The aim of plant-hormone receptor research is to unravel these initial events in plant-hormone action. We do not pretend that perception and transduction of plant hormones are necessarily identical with those of animal hormones. However, we want to propose the following working hypothesis:

a. Plant hormones are detected by specific receptor proteins.

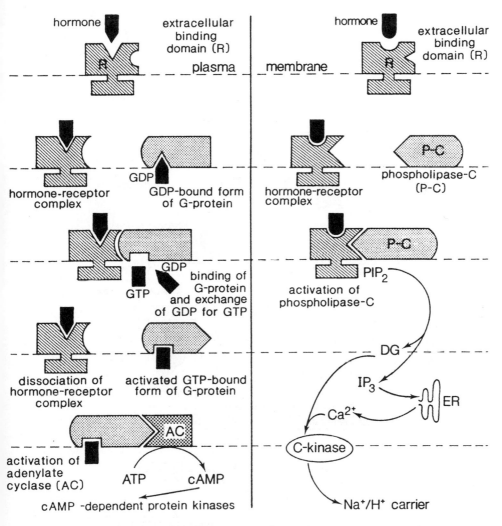

Fig. 2. Schematic diagram of the perception and transduction mechanisms for animal peptide hormone. In the left part of the picture a cell-surface receptor is shown. It consists of an extracellular specific binding domain coupled to a transmembrane domain. Upon binding of the hormone the receptor domain exposes a binding site for the inactive GDP-form of a G-protein. Binding results in a conformational change in the G-protein, thus exchanging GDP for GTP. The formed active GTP-G-protein complex dissociates from the receptor and binds to adenylate cyclase which is activated to produce the second messenger cAMP from ATP. This second messenger stimulates regulatory phophorylations.

In the right part of the picture the cell-surface receptor, upon activation by a hormone, stimulates a membrane phospholipase-C, which hydrolyzes inositol-containing phospholipids like phosphatidylinositol 4,5-biphosphate (PIP_2) present in the membrane, to inositol triphosphate (IP_3) and diacylglycerol (DG). IP_3 acts as a second messenger in the cytoplasm in mobilizing intracellular calcium from the endoplasmic reticulum. The second regulatory signal DG operates within the plane of the membrane to activate, in concert with the elevated concentration of calcium ions in the cytoplasm, a protein kinase C. One of its functions which has been proposed is the activation of the Na^+/H^+ exchange carrier to increase the cytoplasmic pH.

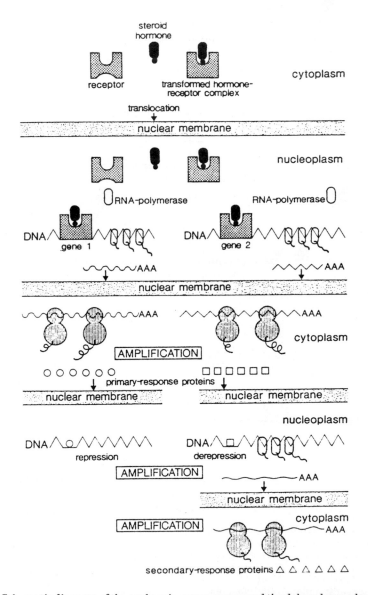

Fig. 3. Schematic diagram of the early primary response and the delayed secondary response to a steroid hormone. Steroid hormones bind to their receptor in the cytoplasm thus inducing a transformation resulting in a translocation of the complex into the nucleus. Some steroid receptors are always present in the nucleus and therefore do not show translocation. The hormone-receptor complex binds specifically and with high affinity to regulatory proteins and DNA sequences of the chromatin. The result is an increase in RNA-polymerase II activity producing gene-specific mRNAs which are translated in the cytoplasm into primary-response proteins, thus amplifying the signal. Each of these proteins may in turn activate or repress other genes thus producing an amplified secondary response.

b. The transduction of plant hormones follows only a limited number of major pathways.

c. The major transduction chains for plant hormones may have features in common with those represented by Fig. 2 and 3. Indeed, some regulatory systems appear to be highly conservative, the Ca^{2+}/calmodulin system being an example.

PERCEPTION AND TRANSDUCTION OF PLANT HORMONES: AUXINS

In the preceding section we have shown that in order to understand plant-hormone action at the molecular level we have to think in terms of signal perception, transduction and response. Therefore, we have to search for receptors, which by definition are the signal-percepting molecules, and then describe how they are involved in the transduction of the signal.

Auxins, often in combination with other hormones, influence many processes in higher plants. In this section we want to discuss auxin-receptor research on two examples of auxin action:

- Initiation and maintenance of tissue proliferation and formation of adventitious roots or shoots (regeneration).
- Stimulation of cell elongation.

For complete reviews on auxin receptors the reader is referred to (8, 24, 27, 42, 47, 55).

Proliferation and Regeneration

A classical model system for the study of hormonal control of proliferation and regeneration is cultured stem-pith tissue from *Nicotiana tabacum*. In 1957 it was demonstrated for the first time that proliferation and regeneration in this tissue can be controlled by exogenous auxins and cytokinins (46). These two classes of plant hormones are both required for proliferation; at relatively high auxin concentrations ($[IAA] \approx 10^{-5}$M) and low cytokinin concentrations ($[kinetin] \approx 10^{-7}$M) the proliferating tissues produce roots, whereas at relatively low auxin concentrations ($\approx 10^{-7}$M) and high cytokinin concentrations ($\approx 5 \times 10^{-6}$M) they produce shoots. At intermediate auxin: cytokinin ratios (*ca.* 10^{-5}M IAA and 10^{-6}M kinetin) only proliferation occurs. Although this discovery initiated a vast amount of fundamental and applied research on the hormonal control of proliferation and regeneration, we still know very little about how auxins and cytokinins work and interact at the molecular level.

Several years ago it was decided in the authors' laboratory to approach this problem by unraveling the perception and transduction chains of these hormones. We started with a search for high-affinity auxin-binding sites in 3-week old callus tissues, derived from freshly isolated tobacco stem pith. A flow scheme for preparing various cell fractions is shown in Fig. 4. The results of extensive binding experiments

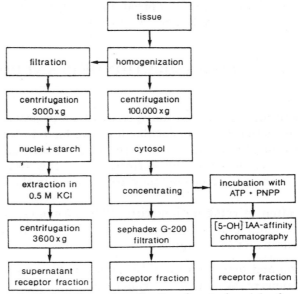

Fig. 4. Flow scheme for the isolation and purification of a cytoplasmic/nuclear auxin receptor protein.

have been summarized in Table 1. The callus tissue contains three classes of auxin-binding proteins, which can be distinguished by their binding behaviour and their location. Two of these classes of proteins are membrane-bound and probably localized on the plasmalemma. One has a high affinity for the auxin-transport inhibitor naphthylphthalamic acid (NPA) and binding takes place rapidly at 0°C at an optimal pH of 4; the natural auxin IAA shows low-affinity binding, which can only be demonstrated in competition experiments with radiolabeled NPA (32). Because of its high affinity for NPA this class is called NPA-binding proteins.

The other membrane-bound auxin-binding proteins can be distinguished from the NPA-binding proteins by their temperature- and time-dependent binding behaviour. With 10^{-7}M IAA maximal binding is reached after 60 min at 25°C at an optimal pH of 5, and it does not bind NPA. This binding protein exhibits complex binding and its concentration is much higher than that of the NPA-binding protein (56,57). In solubilized fractions of homogenized tissue and in salt extracts of isolated nuclei a third binding protein is present in apparently very low concentrations. The binding of active auxins to this protein is also temperature- and time-dependent. At 2.5 nM IAA, maximum binding is reached after 45 min at

Table 1. General characteristics of auxin receptors in tobacco tissues. From (52).

General characteristics	Receptors		
	Membrane bound		Soluble
	Auxin recep.	NPA* recep.	Auxin recep.
Presence in tissues:			
Stem pith	+	+	−
Stem-pith callus	+	+	+
Leaves	+	+	$nd^\dagger$
Protoplasts (2 days)	+	+	−
Shoot tips	$nd^\dagger$	$nd^\dagger$	+
Cell-suspension cul.	+/−	+	+
Location in the cell	Plasmalemma	Plasmalemma	Cytoplasm Nucleus
pH optimum of binding	5.0	4.0	7.5
Specificity:			
Ka for IAA (M^{-1})	6×10^4	5×10^3	1.6×10^8
1-NAA	1×10^7	2×10^4	1×10^8
2-NAA	1×10^6	1×10^5	1×10^7
2,4-D	3×10^5	5×10^5	3×10^7
TIBA*	1.5×10^4	8.9×10^4	1×10^8
NPA*	$< 1 \times 10^3$	$3-5 \times 10^8$	$< 1 \times 10^5$
Fusicoccin	$< 1 \times 10^3$	1.1×10^4	$nd^\dagger$
Tryptophan	$< 1 \times 10^3$	$< 1 \times 10^3$	$< 1 \times 10^5$
Concentration (pM/mg protein)	50	1.6	0–0.2

† nd = not determined.
* NPA = naphthylphthalamic acid } auxin transport inhibitors
  TIBA = triiodobenzoic acid

an optimal pH of 7.5. The affinity of this binding protein for IAA is much higher than that of the membrane-bound auxin-binding protein. This protein is called the cytoplasmic/nuclear auxin-binding protein (36,37).

Cytoplasmic/Nuclear Auxin-binding Proteins
   Auxin is readily taken up by the cells and one class of specific high-affinity auxin-binding proteins is present in both the cytoplasm and the nucleus. Therefore, as a working hypothesis it was assumed that a major pathway of auxin perception and transduction roughly follows the scheme as established for steroid hormones, i.e., auxin directly controls transcriptional activity in nuclei via coupling to a cytoplasmic/nuclear receptor (Fig. 3). In order to verify this working hypothesis the influence of Sephadex-G200 receptor(R)-fractions (Fig. 4) on the transcriptional activity of isolated nuclei (6, 34) was studied. It was found that addition of R-fractions resulted in a reproducible auxin-dependent stimulation of RNA-polymerase-II activity, provided that a minimum concentration of specific binding protein is present (Fig. 5, 51). At present it is not clear whether this is an overall stimulation or that it is a stimulation of the

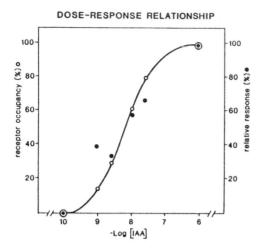

DOSE-RESPONSE RELATIONSHIP

Fig. 5. Dose-response relationship of the stimulation of transcription in isolated nuclei by a high-affinity binding protein fraction from the cytosol of tobacco callus. From (52).

transcription of specific sets of genes. Comparable results were obtained with a 2,4-D-dependent cell suspension line from *Nicotiana tabacum* (1). Moreover, it is found that in early-stationary phase cells most binding activity is present in the cytosol, whereas in rapidly dividing log-phase cells most specific auxin-binding is present in salt extracts of isolated nuclei. This observation is extremely interesting in that it resembles the apparent translocation of occupied steroid receptors to the nuclei in hormone-activated target tissues. However, one should be very careful in drawing conclusions on receptor concentrations from *in vitro* binding experiments. The experience with the cytoplasmic/nuclear binding proteins tells us that even with highly standardized procedures, the number of binding sites per mg of protein varies considerably and often is below the detection level of the binding assay. This indicates that unknown factors might influence the recovery of the high-affinity binding sites. For example, we have found that the amount of specific IAA-binding in crude protein extracts can be significantly increased by adding MgATP and/or excess artificial substrate for phosphatases (p-nitrophenyl-phosphate) to the binding-assay medium (52). The results of these experiments have been summarized in Fig. 6. These observations suggest that the auxin-binding protein might be liable to affinity modulation by ATP-dependent phosphorylation and by dephosphorylation, transforming the binding proteins into a high- or a low-affinity form respectively. The phosphatase activity in the preparations—which may vary from experiment to experiment—has probably been a strong disturbing factor in the recovery of high-affinity auxin-binding in the original experiments. Such an affinity modulation has also been proposed for steroid receptors (54). This putative auxin-receptor protein could also be detected in various kinds of proliferating tissue from *Nicotiana tabacum* (Table 1). An examination of

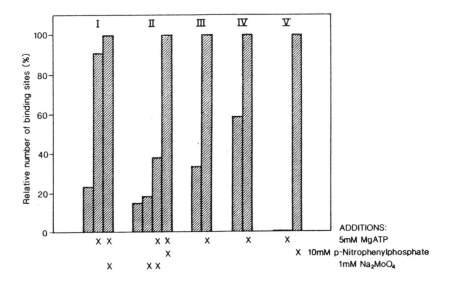

Fig. 6. The number of specific binding-sites computed from Scatchard analysis of IAA binding to soluble fractions isolated from several tissues. The crosses under the bars indicate which substances mentioned in the right column were added to the fractions just before performing the binding assay. The maximal number of binding sites was: 5.6×10^{-11} M in I; 27.8×10^{-11} M in II; 7.9×10^{-11} M in III; 12.1×10^{-11} M in IV; 8.8×10^{-1} M in V. From (52).

different tobacco shoot tissues revealed that this binding protein is present in shoot tips, but apparently not in mature tissue like stem pith (5). A reexamination of the latter tissue is required because we cannot exclude the possibility that the binding proteins are present after all, but predominantly in their low-affinity form.

We have tried to purify the cytoplasmic/nuclear binding sites by affinity chromatography after activation of the binding sites with MgATP in the presence of p-nitrophenylphosphate. Addition of a fraction, which was eluted from the affinity column by IAA, to isolated nuclei has resulted in a significant stimulation of transcription on several occasions (Fig. 7). In an attempt to label the proteins bound to the column, ^{32}P-ATP was added to crude preparations prior to affinity chromatography. Among the 12 to 15 proteins detected by silver staining upon denaturing gel electrophoresis of the fraction eluted by IAA in the presence of 0.3 M NaCl, 3 proteins were labeled. However, it has still to be proven that the phosphorylated protein fractions indeed contain the receptor.

Membrane-bound Auxin-binding Proteins

In order to establish a receptor function for membrane-bound hormone-binding proteins, one first needs to know on which membrane system they are localized. Indirect evidence indicates that the membrane-

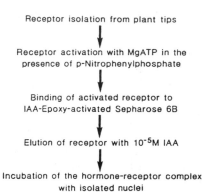

Receptor isolation from plant tips

↓

Receptor activation with MgATP in the
presence of p-Nitrophenylphosphate

↓

Binding of activated receptor to
IAA-Epoxy-activated Sepharose 6B

↓

Elution of receptor with 10⁻⁵M IAA

↓

Incubation of the hormone-receptor complex
with isolated nuclei

Fig. 7. Flow scheme for the purification of the activated receptor by means of affinity chromatography, and the stimulation of *in vitro* transcription by the eluted fraction.

	dpm	% stimulation
Nuclei + IAA	1439	
Nuclei + eluted hormone receptor complex	2448	70

bound auxin-binding proteins from tobacco tissues are localized on the plasmamembrane. This evidence stems from experiments with tobacco leaf-mesophyll protoplasts. The binding proteins can be detected in microsome fractions from leaf mesophyll tissue, but they are missing, or at least cannot be detected, in freshly isolated protoplasts from that tissue. Upon culturing these protoplasts the binding proteins can be detected again just after the onset of cell divisions (57). It was found that the binding proteins are not destroyed by osmotic stress, but most probably by proteolytic enzymes that are present in the cellulase and macerozyme preparations used for protoplast isolation (57). This is further supported by the observation that pure protease preparations have the same effect as the enzymatic preparation of protoplasts. If these enzymes do not penetrate the plasmalemma of the protoplasts—and up to now there are no reports describing endocytosis of proteins into protoplasts—then only the proteins in the cell wall and at the external face of the plasmalemma will be degraded. Furthermore, it is known from animal systems, that proteins taken up by endocytosis are usually transferred to lysosomes and destroyed. We can exclude a possible localization in the cell wall, because it was found that cell-wall fragments from tobacco callus are pelleted at low-speed centrifugation, while binding proteins are not. Moreover, the density, in linear sucrose gradients, of particles containing the binding proteins is much lower than that of cell-wall fragments (56). Hence, an obvious explanation for the disappearance of binding proteins by exogenous proteases is that they are located at the external face of the plasmalemma. Such a location of the binding sites is in agreement with the pH optimum of 5 for auxin binding.

As distinct from NAA+kinetin-requiring cell-suspension lines from tobacco, 2,4-D-dependent cell lines from the same source do not have detectable amounts of membrane-bound auxin-binding proteins. These cell lines seem to have lost their ability to regenerate roots. However, after subculture on medium with NAA + kinetin the binding proteins reappear and the ability to regenerate roots is restored (Fig. 8). The 2,4-D-

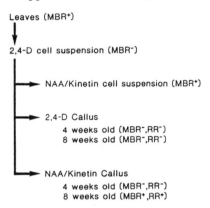

Leaves (MBR⁺)

2,4-D cell suspension (MBR⁻)

NAA/Kinetin cell suspension (MBR⁺)

2,4-D Callus
4 weeks old (MBR⁻,RR⁻)
8 weeks old (MBR⁻,RR⁻)

NAA/Kinetin Callus
4 weeks old (MBR⁻,RR⁻)
8 weeks old (MBR⁺,RR⁺)

Fig. 8. Sequence of cultures derived from tobacco leaves. MBR$^+$/MBR$^-$ = presence and absence of specific membrane-bound NAA binding; RR$^+$/RR$^-$ = appearance and non-appearance of roots in the culture within two weeks after transfer to solid medium containing 10^{-4} M IAA and 10^{-6} M kinetin, respectively.

dependent cell lines still possess the cytoplasmic/nuclear auxin-binding proteins and the NPA-binding proteins (31). On the basis of these results it is tempting to assume that the membrane-bound auxin-binding proteins might have some function in auxin-induced root regeneration. The high concentrations of IAA which are generally required for root regeneration are in agreement with the apparently low affinity of this binding protein for natural auxins.

A biochemical receptor function of the membrane-bound auxin-binding proteins has not yet been established. Preliminary experiments, aimed at the identification of some auxin-sensitive membrane-bound functions, show that *in vitro* phosphorylation of membrane components can be stimulated by auxin (Table 2). The phosphorylation of at least two membrane proteins is stimulated (31). At present, we do not exclude the possibility that the overall stimulation of membrane phosphorylation is partly due to phosphorylation of membrane lipids.

NPA-binding Proteins

It is generally believed that naphthylphthalamic acid (NPA) binds to plasmamembrane-bound auxin-efflux carriers. These carriers are assumed to bind auxin at the cytoplasmic site of the plasmamembrane, transport it through the membrane and release it into the cell-wall free space. Indeed, auxin efflux from preloaded cell suspension cells can be significantly decreased by low external NPA concentrations (31). It is

Table 2. Specificity of auxin-induced phosphorylation and comparison with receptor occupancy. From (31).

Compound (10^{-5}M)	% phosphorylation*		% occupied binding sites	$K_a (M^{-1})$
1-NAA	2.78	100	98	5×10^6
2-NAA	1.22	44	91	1×10^6
2,4-D	1.02	37	75	3×10^5
IAA	0.67	24	38	6×10^4
Tryptophan	0.24	9	< 1	$< 10^3$

*Left column: absolute phosphorylation expressed as % of total amount ^{35}S-ATP present in assay. A background of 3% was subtracted. Right column: relative phosphorylation. The maximum of 2.78% (1-NAA) was taken as 100%.

therefore probable that the NPA-binding proteins have an auxin-transport function. We could not establish a role for this auxin-export carrier in cultured cells. For example, addition of saturating concentrations of NPA to the medium of the cell-suspension cultures neither affected the doubling-rate nor the total cell yield (31). As has been found for the membrane-bound auxin-binding proteins, freshly isolated protoplasts from mesophyll tissue do not possess the NPA-binding proteins in detectable amounts. These binding proteins also reappear after the onset of cell divisions (28).

Concluding Remarks

The evidence obtained thus far indicates that tobacco cells are equipped with genetic information coding for at least two classes of auxin-binding proteins and a class of putative auxin-export carriers.

The cytoplasmic/nuclear binding proteins are probably receptors which are capable of detecting *relatively low intracellular* concentrations of auxin. The occupied receptors presumably stimulate the transcription of genes that are either directly and/or indirectly involved in the cell's response to auxin, i.e. cell division. How the auxin-receptor complexes interact with nuclear chromatin and which genes are involved is not known. Up to now the data obtained for the cytoplasmic nuclear auxin-receptor agree well with the model for steroid hormones (Fig. 3). The membrane-bound auxin-binding proteins may be receptors which are capable of detecting only *relatively high extracellular* auxin concentrations. The occupied receptors somehow transduce the auxin-signal into intracellular signals, which trigger the response, i.e., possibly programming the cells in such a way that the proliferating tissue acquires the potential to regenerate roots.

If these two classes of auxin-binding proteins constitute the receptor system by which auxin controls proliferation and root formation, then, since cytokinins do not interfere with the binding of auxin to these receptors, the cells must also be equipped with a distinct set of cytokinin receptors. The interaction between auxins and cytokinins occurs probably at the level of signal transduction. Unfortunately, at present nothing is known about cytokinin receptors in the material discussed in this section (see also the section on cytokinin receptors).

Cell Elongation

The experimental systems most widely used in the study of the mechanism of auxin action are those in which auxin seems to be a major limiting factor, i.e., the classical auxin bioassays. In auxin bioassays (coleoptile and stem segments from etiolated seedlings) exogenous auxin stimulates growth by cell elongation in a concentration-dependent way. In auxin-receptor research coleoptiles from *Zea mays* have been studied most extensively. Therefore the results obtained with this system may give an impression of the progress that has been made in the isolation and characterization of auxin receptors involved in cell elongation.

Up to now, auxin-binding proteins in maize coleoptiles have only been detected in microsome fractions. The properties of these binding proteins as described by different investigators have been summarized in Table 3. Most reports agree that *in vitro* specific binding of auxin occurs rapidly at 0°C at an optimal pH of 5.5. With the aid of gradient-centrifugation techniques and the use of membrane-marker enzymes it was found that most binding sites are located on the endoplasmic reticulum (ER) (41). However, besides the ER, the plasmamembrane and the tonoplast possibly contain high-affinity auxin-binding sites as well (2,11). The binding proteins can easily be solubilized by either acetone or triton X-100 treatment of the microsome preparations (9, 53). Recently, solubilized binding proteins from crude microsome fractions were purified to a high degree by a combination of ligand-affinity chromatography and immunological methods (29). This purified auxin-binding protein (ABP) has a sharp binding optimum at pH 5.5. It seems to be a 40 kilodalton dimer in its native form and it has a higher affinity for NAA ($K_d \approx 5.7 \times 10^{-8}$M) than is found for NAA binding to microsome fractions ($K_d \approx 4 \times 10^{-7}$M). Scatchard analysis of NAA binding in the purified binding-protein preparations reveals only one class of binding sites. With monospecific antibodies against the purified binding protein (IgGanti-ABP) and using indirect immunofluorescence labelling of microscopic preparations of fixed coleoptile segments, it could be shown that the binding proteins are localized within the outer epidermal cells. IgGanti-ABP at 10^{-8}M specifically inhibits auxin-induced growth in coleoptile segments and it strongly reduces the auxin response of split coleoptile sections. Two conclusions are drawn from these observations. 1) The binding protein is involved in auxin-induced elongation growth; 2) The binding protein has to be located at the external face of the

Table 3. Characteristics of membrane-bound binding sites. From (31)

Material	Frac-tion[a]	Ligand	$K_d(\mu M)$	Binding sites conc.[b]	pH	Temp. (°C)	Time[c]	Reference
Zea								
coleoptiles	CM	IAA	3–4	30	6	0	<30	Hertel *et al.* (1972), Planta 107,325
	CM	IAA	1.7	51	5.5	0	<15	Batt, *et al.* 1976, Planta 130, 7
	CM	IAA	5.8	96	5.5	0	<15	Batt, *et al.* 1976, Planta 130, 7
	CM	NAA	1–2	40	6	0	<30	Hertel *et al.* (1972)
	CM	NAA	0.5–0.7	30–50	5.5	0	<15	Ray *et al.* (1977) Plant Physiol. 59, 357
	CM	NAA	0.5	38[d]	5.5	20	10	Moloney and Pilet (1981), Planta 153, 447
	CM	NAA	0.15	38	5.5	0	<15	Batt, *et al. (1976)*
	CM	NAA	1.6	96	5.5	0	<15	Batt, *et al. (1976)*
	CM	NAA	0.15	100[e]	5.5	0	<15	Murphy (1980), Planta 149, 417
	H	NAA	1.16	32	5.5	0	<15	Batt and Venis (1976)
	H	NAA	0.5	–	5.5	0	<15	Murphy (1980)
	L	NAA	0.39	24	5.5	0	<15	Batt and Venis (1976)
	L	NAA	0.44	–	5.5	0	<15	Murphy (1980)
	ER	NAA	1.7	15	6	0	5	Normand *et al.* (1977)[g]
	ER	NAA	0.4	40	5.5	0	<30	Ray (1977)
	TP	NAA	1.3	20	5.5	0	<30	Dohrmann *et al.* (1978)
mesocotyls	CM	NAA	0.75	52	6	4	5	Normand *et al.* (1977)
	ER	NAA	0.2	1.9[f]	5.5	–	–	Walton and Ray. (1981)
roots	PM	NAA	0.8	16	5.5	20	10	Moloney and Pilet (1981)
	PM	NAA	1.2	19	5.5	20	10	Moloney and Pilet (1981)

a CM = crude membrane, PM = plasmamembrane, ER = endoplasmic reticulum
   TP = tonoplast, H = heavy band, L = light band
b Concentration expressed as pmol/g fresh weight unless otherwise indicated.
c Time expressed in minutes
d Not given by authors but determined by us from their Scatchard plots.
e Maximum value.
f pmol/mg protein.
g In: Regulation of Cell Membrane Activities in Plants. Marrè, E., Ciferri, O. eds. Elsevier Amsterdam, pp. 185-202.

plasmamembrane, because it seems highly unlikely that the IgGanti-ABP reaches the cytoplasm of living cells (30).

If these auxin-binding proteins are indeed receptors involved in elongation growth then their relatively high concentration within the outer epidermal cells raises the interesting question of whether this tissue is the main auxin-sensitive part of the coleoptile. The answer to this question might be affirmative, since removal of the outer epidermis seems to make the coleoptile sections substantially less sensitive to auxin (40). Further circumstantial evidence for a receptor function of membrane-bound auxin-binding sites in cell elongation is provided by results from maize mesocotyls. Microsome fractions from mesocotyls contain an abundant class of high-affinity auxin-binding proteins with almost similar properties to those in microsome fractions from coleoptiles.

Mesocotyls from etiolated and red-light-treated maize seedlings show a large difference in the amounts of auxin-binding proteins, either expressed per mg of membrane protein or per g fresh weight. Red light considerably reduces the amount of binding proteins (30–50%) and reduces the sensitivity to auxin-induced elongation. This treatment does not alter the affinity but only the number of binding proteins. Auxin-dose-response curves obtained from mesocotyl segments isolated from either irradiated or etiolated seedlings do not show a difference in the auxin concentration at which half-maximum response is obtained, but only in the maximum-response value. This can be expected if the response is proportional to the amount of occupied receptors (59).

Plasmamembrane-bound ATP-driven proton pumps are assumed to be involved in auxin-induced acidification of the apoplast in elongating cells (7). This suggests that such proton pumps might be targets for auxin receptors. A rather direct approach to verify this hypothesis is to isolate receptors and pumps and to reconstitute a functional system by incorporating these putative components of the auxin-transduction chain into artificial membranes. Only one report has appeared describing such an approach (50). The auxin-binding proteins from maize coleoptiles were solubilized and partially purified. The purified fractions exhibited both high-affinity auxin binding and ATPase activity. These proteins were incorporated into artificial membranes in which transmembrane currents were measured. ATP, binding proteins + ATPases and NAA were added to the membranes in different sequences. Only after all components had been added was an increase in transmembrane current observed. In all, four experiments were performed:

1. Addition in the sequence binding protein + ATPase–ATP–NAA. A significant, sharp increase in current occurs after addition of NAA, with no apparent lag time. This is in agreement with the rapid binding of auxin to the binding site.
2. Binding protein + ATPase–ATP–Benzoic acid–NAA. As in 1: the benzoic acid has no influence whatsoever, which is consistent with the fact that it does not bind to the binding protein.
3. ATP–NAA–binding protein + ATPase. A very small, but sharp, increase 2 min after addition of the binding protein + ATPase. The fact that a lag time of ca 2 min is found is not surprising, since the binding proteins + ATPases have to be incorporated into the membrane before an effect can be brought about.
4. Binding protein + ATPase plus ATP and NAA simultaneously. As in 1.

Only in the experiments in which NAA was added as the last component is a convincing, immediate increase in current observed. Unfortunately no experiment was done in which binding protein + ATPase and NAA were added before the addition of ATP. The optmimum pH for stimulation (pH 5.3) is in good agreement with the optimum pH of binding (pH 5.5). Finally, the K_d of the binding, ca $10^{-7}M$, is close to the lower detection limit of an NAA effect (ca $10^{-7}M$).

This very interesting approach requires further improvement: the active components in the binding protein + ATPase preparations should be further purified and characterized, and ATP-dependent transmembrane proton fluxes can perhaps be approached more directly by studying the kinetics of proton accumulation in sealed artificial-membrane vesicles.

Auxin Receptors and the Acid-Growth Theory

Over the past *ca.* 15 years the study of auxin action in cell elongation has been dominated by the so-called acid-growth theory. According to this theory auxin induces acidification of the free space in the cell wall, presumably by the activation of plasmamembrane-bound proton pumps (7). The increase in proton concentration brings about an increase in the plasticity of the cell wall, thus causing a rapid increase in elongation rate of the tissues. This theory does, however, not lack seemingly contradictory data (23) (see Chapter C1), so that the mechanism of action of auxin is not completely solved.

If the conclusions drawn from the experiments with IgGanti-ABP are correct, then the auxin-responsive cells are equipped with receptors at the external face of the plasmamembrane that can detect auxin-signals outside the plasmamembrane and translate them into a response. In our opinion, an attractive possibility is that these auxin receptors are the first elements in a transduction chain that might be comparable to the IP3 and DG generating system as recently discovered in animal cells (see Introduction). This would mean that auxin transduction is a subtle and complicated system which, among other things, regulates cytoplasmic Ca^{2+} levels and plasmamembrane-bound protein complexes, such as particular classes of proton pumps, probably by phophorylation processes. According to this view the vivotoxin fusicoccin apparently unsettles the system by overstimulation of proton pumps.

In both plants and animals calcium plays an important role as sensory transducer of conditions outside the plasmamembrane and as modulator of cytoplasmic activity. Although the role of Ca^{2+} in auxin-induced cell elongation is still a matter of dispute (7), we must realise that changes in cytoplasmic Ca^{2+} levels may be rapid and transient. Interestingly, auxin-regulated cell elongation can be inhibited by calmodulin-binding drugs (12).

There is some evidence showing that phophorylation of membrane proteins is a rapid effect in auxin action. For example, protein kinase activity in plasmamembrane-enriched microsome preparations from soybean hypocotyls can be stimulated by auxin (2,4-D) at physiological concentrations, whereas structurally related non-active auxin is inactive.

This stimulation results in the phophorylation of a few membrane proteins, with apparent molecular weights of 45 and 50 kilodaltons, within *ca.* 10.sec. (54).

Concluding Remarks

There is at least circumstantial evidence that maize coleoptiles and mesocotyls contain membrane-bound auxin-receptor proteins involved in elongation growth. It seems that functional receptors are located at the external face of the plasmamembrane and it is assumed that the ER-bound auxin-binding sites may represent precursors of functional receptors which undergo a maturation process to a final form in the plasmamembrane (54). The fact that monospecific antibodies are available makes it now possible to establish the intracellular localization of the auxin receptors more directly by means of electronmicroscopical immuno-labelling techniques. High affinity membrane-bound auxin-binding sites have also been described for other tissues both from monocotyledons and dicotyledons. In those cases where localization studies have been performed the plasmalemma is tentavely indicated as the membrane system which contains the binding sites. Most of the binding sites bind the ligand rapidly at 0°C at a pH optimum varying between pH 4 to 6.5. All these binding sites need further characterization and it remains to be proven that they have a receptor function. In unravelling the auxin-transduction chain in cell elongation, one must be alert not to be limited by the acid-growth theory. An open mind for alternative concepts and research strategies is required.

OTHER PLANT HORMONES

To an even greater extent than in auxin-receptor research, the study of transduction chains of the other plant hormones has stuck to the characterization of high-affinity binding sites. For this reason we shall discuss these hormones only briefly.

Cytokinins

Although particulate and soluble high-affinity binding sites have been characterized for a number of higher-plant systems (22), we will confine ourselves to the most studied cytokinin-binding proteins in wheat germ. This system contains a binding protein (CBF-1 = wheat germ cytokinin-binding protein) which is highly specific for 6-substituted purine cytokinins. This protein appears to be loosely bound to ribosomes (14). A similar, if not identical, protein and a 30 kilodalton high-affinity cytokinin-binding protein (CBF-2) are present in an apparently free form in the cytosol. The CBF-1 protein from the ribosomal fraction has been purified to a very high degree by ligand-affinity chromatography. The purified CBF-1 appears to consist of 4 subunits (13). A similar high-affinity cytokinin-binding protein (CBP) has been described by others (39). However, the molecular weight, subunit complexity and behaviour on ion-exchange chromatography of CBP are different from CBF-1. Re-examination of the subunit structure of CBF-1 has shown a variation in the structure of the cytokinin-binding moiety depending on the source of the wheat germ (15).

A protein kinase has been isolated and characterized that catalyses the phosphorylation of CBP (38). This reminds us of the affinity modulation proposed for the cytoplasmic/nuclear auxin receptors from tobacco cells, although in this case the consequences of the phophorylation on the properties of CBP are unknown.

The association of CBF-1 and CBP with ribosomes suggests that these binding proteins might have a receptor function in translation processes. However, at present no conclusive evidence for such a function exists.

Finally, we will briefly mention a report on an *in vitro* cytokinin-dependent stimulation of nuclear transcription by a cytokinin-binding protein isolated from excised barley leaves by means of affinity chromatography (25). These results are comparable to our own data (51).

Gibberellins

The gibberellin-induced initiation and control of hydrolytic enzymes in the cereal aleurone tissue is one of the better known hormonal responses in plants. However, GA perception and transduction in this system is largely unknown. Even the number of studies on the identification and characterization of specific high-affinity GA-binding sites is extremely limited and the published reports should all be considered as preliminary.

A relatively old report (21) shows that aleurone-grain-rich fractions from wheat aleurone tissue contain specific GA_1-binding sites in a concentration of 0.45 pmoles per mg protein and with a Kd of 1.5×10^{-6}M. The presence of Ca^{2+} ions is absolutely required for binding and abscisic acid strongly inhibits *in vitro* binding. These latter findings correlate with the requirement of Ca^{2+} in GA-controlled synthesis of α-amylase and the inhibition of GA-action by abscisic acid *in situ*. Of course, Ca^{2+} might also play an important role in later steps of the GA-transduction chain.

Incubation of isolated aleurone from barley with [3H]GA_1 at low temperatures (1–1.5°C), at which metabolism and accumulation of applied GA_1 cannot be detected, results in a saturable retention of the hormone. The retention is unaffected by GA_8, which is inactive in the aleurone system. After transfer of the aleurone layers to GA_1-free medium and raising the temperature to 25°C α-amylase production is induced, thus showing that the retained GA_1 is physiologically active (23).

The aleurone system is not only highly attractive for GA-receptor research because the molecular responses to the hormone are relatively well known, but also because sensitivity mutants are available. Recently it was shown that preincubation of deembryonated seeds or isolated aleurone layers from GA_3-insensitive Rht3 dwarf wheats for 20 h at 5°C, brings about a dramatic increase in GA_3-sensitivity of the aleurone layers. The aleurone layers from the mutant strains show a reduction in the amount of GA_3-induced α-amylase, but are not altered with respect to the time course of α-amylase production, the relative amounts of isozyme production by the α-amylase structural genes, the starch liquefaction capacity of the enzymes, uptake and metabolism of GA_3, levels of endogenous inhibitors and cellular metabolism (43). This indicates that GA-receptors and/or other components of the GA_3-transduction chains are affected in the mutants. In addition, the fact that GA-sensitivity can be restored by low-temperature treatment, makes these mutant strains potent tools for the unravelling of the GA perception and transduction chain in aleurone layers.

An interesting report has appeared providing evidence that GA (and ABA) controls α-amylase and ribosomal RNA synthesis by regulating gene transcription (20). These results were obtained in run-off transcription studies with nuclei isolated from aleurone cell protoplasts, incubated for 24 h with and without GA (and ABA). It, therefore, seems worthwhile to analyze these protoplasts for the presence of high-affinity GA-binding proteins, and to see if these results can be reproduced in reconstitution experiments with nuclei isolated from untreated aleurone cells.

Abscisic acid

Apart from two preliminary studies (10,18) there is only one report that provides strong evidence for the presence of ABA receptors in a particular class of ABA-target cells, i.e., guard cells (19). ABA affects guard cells by stimulating efflux of K^+, their major osmotically active constituent, thus inducing stomata closure. High-affinity binding of [³H]cis(+)ABA, the physiologically active enantiomer, to guard-cell protoplasts from *Vicia faba*, could be demonstrated by photoaffinity labelling. The binding proteins have an apparent Kd for ABA of $3-4\times10^{-9}$M, which corresponds well with the ABA concentration ($\approx5\times10^{-9}$M) at which half-maximum response is obtained in stomatal-closure bioassays. Upon sodium dodecyl sulphate polyacrylamide gel electrophoresis the binding proteins resolve into three species: A (Mw 20.2 kilodaltons), B (Mw 19.3 kilodaltons) and C (Mw 24.3 kilodaltons). At an alkaline pH in the medium ABA binds preferentially to A, whereas at an acidic pH most ABA binds to B and C. In such a system binding is largely independent of apoplasmic pH and this corresponds well with the observation that ABA induces stomatal closure at alkaline as well as at acidic pH. There is a close correlation between physiological activity and the ability to displace [³H]ABA from its high-affinity binding sites, for a range of ABA analogues. Mild tryptic treatment of guard-cell protoplasts before incubation with ABA completely eliminates the ABA-binding sites, but leaves the protoplasts intact. These results indicate that the binding sites are proteins that are located at the plasmalemma of guard cells, with the ABA-binding moiety facing the apoplasmic space. As the reader will recall, a similar location has been assumed for membrane-bound auxin-binding proteins in tobacco cells and maize coleoptiles.

The guard-cell system looks very promising for future receptor research.

Ethylene

Specific high-affinity membrane-bound ethylene-binding sites have been described for various tissues (3,4,44,45,49). The binding kinetics differ from those of binding sites for other plant hormones by low association and dissociation rates. It is not clear what receptor function these binding sites might have and future progress will depend on the availability of suitable target systems.

SUMMARY AND CONCLUSIONS

- Specific high-affinity binding proteins have been described for all major classes of plant hormones. There is good, albeit circumstantial evidence that at least the membrane-bound binding proteins in maize coleoptiles, tobacco cells and broad bean guard cells have a receptor function in auxin-controlled elongation, auxin-induced root regeneration and abscisic acid-controlled stomatal closure respectively. These putative receptors appear to be localized at the plasmamembrane with the binding moiety facing the apoplast, thus constituting sensory systems for hormone levels outside the cells.
- The fact that hormones can easily be taken up by plant cells suggests that these cells might also be equipped with intra-cellular receptors. It is not unlikely that the cytoplasmic/nuclear auxin-binding proteins in tobacco cells and perhaps the cytoplasmic cytokinin-binding sites in wheat germ and in barley leaves represent such receptors.
- At present no conclusive evidence as to a biochemical function of the plant hormone-binding sites exists, although all evidence obtained thus far indicates that the cytoplasmic/nuclear binding proteins in tobacco cells and the cytokinin-binding proteins in barley leaves are directly involved in transcriptional activity in the nuclei. In this respect it is important to note that over the past few years it has been found that

auxin is capable of stimulating the transcription of specific genes with response times measured in minutes (17,48,58) (see Chapter C2).

- Although it is not certain that auxin-induced acidification is a primary trigger of the rapid auxin response in cell elongation, regulation of apoplasmic pH by modulation of proton pumps may be part of the auxin-transduction chain.
- One may wonder how many pathways for plant-hormone perception and transduction do exist. Perhaps there are only a few major pathways, possibly based on very conservative principles of hormone perception and transduction in higher eukaryotes. As a working hypothesis we will propose at least two major pathways:

1. Intracellular receptor proteins which are directly involved in gene expression either on the transcriptionial and/or translational level.
2. Plasmamembrane-bound receptor proteins which function as sensory systems for external hormone levels and which transduce intracellular signals, possibly via the phosphatidylinositol pathway. The secondary signals control the cell's activity via modulation of cytoplasmic Ca^{2+} levels and protein kinase activity. Some evidence supporting this view has recently been presented (35, 60).

References

1. Bailey, H.M., Barker, E.J.D., Libbenga, K.R., Van der Linde, P.C.G., Mennes, A.M., Elliott, M.C. (1985). An auxin receptor in plant cells. Biol. Plant. 27, 105-109.
2. Batt, S., Venis, M.A. (1976). Separation and localization of two classes of auxin binding sites in corn coleoptile membranes. Planta 130, 15-21.
3. Bengochea, T., Acaster, M.A., Dodds, J.H., Evans, D.E., Jerie, P.H., Hall, M.A. (1980). Studies on ethylene binding by cell-free preparations from cotyledons of *Phaseolus vulgaris* L. II. Effect of structural analogues of ethylene and of inhibitors. Planta 148, 407-411.
4. Bengochea, T., Dodds, J.H., Evans, D.E., Jerie, P.H. Niepel, B., Shaari, A.R., Hall, M.A. (1980). Studies on ethylene binding by cell-free preparations from cotyledons of *Phaseolus vulgaris* L. I. Separation and characterisation. Planta 148, 397-406.
5. Bogers, R.J., Kulescha, Z., Quint, A., Van Vliet, Th. B., Libbenga, K.R. (1980). The presence of a soluble auxin receptor and the metabolism of 3-indoleacetic acid in tobacco-pith explants. Plant Sci. Lett., 19, 311-317.
6. Bouman, H., Mennes, A.M., Libbenga, K.R. (1979). Transcription in nuclei isolated from tobacco tissues. FEBS Letts. 101, 369-372.
7. Cleland, R.E. (1982). The mechanism of auxin-induced proton efflux. In: Plant Growth Substances 1982, pp. 23-31, Wareing, P.F. ed. Academic Press, London, New York..
8. Cross, J.W. (1985). Auxin action: the search for the receptor. Plant, Cell and Environment 8, 351-359.
9. Cross, J.W., Briggs, W.R. (1979). Solubilized auxin-binding protein. Planta 146, 263-270.
10. Curvetto, N., Delmastro, S., Brevedan, R. (1982). Guard cell protoplast as cell hormone target. Plant cell affinity chromatography. Plant Physiol. 69, Suppl. 40.
11. Dohrmann, U., Hertel, R., Kowalik, H. (1978). Properties of auxin binding sites in different subcellular fractions from maize coleoptiles. Planta 140, 97-106.
12. Elliott, D.C., Batchelor, S.M., Cassar, R.A., Marinos, N.G. (1983). Calmodulin-binding drugs affect responses to cytokinin, auxin, and gibberellic acid. Plant Physiol. 72, 219-224.
13. Erion, J.L., Fox, J.E. (1981). Purification and properties of a protein which binds cytokinin-active 6-substituted purines. Plant Physiol. 67, 156-162.

14. Fox, J.E., Erion, J. (1977). Cytokinin-binding proteins in higher plants. In: Plant Growth Regulation, proc. 9th Int. Conf. on Plant Growth Substances, Lausanne, pp. 139-146, Pilet, P.E., ed. Springer-Verlag, Berlin, Heidelberg, New York.

15. Fox, J.E., Gregerson, E. (1982). Variation in a cytokinin binding protein among several cereal crop plants. In: Plant Growth Substances 1982, pp. 207-214, Wareing, P.F., ed. Academic Press, London, New York.

16. Grody, W.W., Schrader, W.T., O'Malley, B.W. (1982). Activation, transformation, and subunit structure of steroid hormone receptors. Endocrine Reviews, 3, 141-159.

17. Hagen, G., Guilfoyle, T.J. (1985). Rapid induction of selective transcription by auxins. Mol. Cell. Biol. 5, 1197-1203.

18. Hocking, T.J., Clapham, J., Cattell, K.J. (1978). Abscisic acid binding to subcellular fractions from leaves of Vicia faba. Planta 138, 303-304.

19. Hornberg, C., Weiler, E.W. (1984). High-affinity binding sites for abscisic acid on the plasmalemma of Vicia faba guard cells. Nature 310, 321-324.

20. Jacobsen, J.V., Beach, L.R. (1985). Control of transcription of α-amylase and rRNA genes in barley aleurone protoplasts by gibberellin and abscisic acid. Nature 316, 275-277.

21. Jelsema, C.L., Ruddat, M., Marré, D.J., Williamson, F.A. (1977). Specific binding of gibberellin A₁ to aleurone grain fractions from wheat endosperm. Plant & Cell Physiol. 18, 1009-1019.

22. Keim, P.S., Erion, J. Fox, J.E. (1981). The current status of cytokinin-binding moieties. In: Metabolism and Molecular Activities of Cytokinins, pp. 179-190, Guern, J., Péaud-Lenoël, C.eds. Springer-Verlag, Berlin.

23. Keith, B. Boal, R., Srivastava, L.M. (1980). On the uptake, metabolism and retention of [³H] gibberellin A₁ by barley aleurone layers at low temperatures. Plant Physiol. 66, 956-961.

24. Kende, H., Gardner, G. (1976). Hormone binding in plants. Ann. Rev. Plant Physiol. 27, 267-290.

25. Kulaeva, O.N. (1985). Hormonal regulation of transcription and translation in plants. In: Proc. 16th FEBS meeting at Moscow, Part C, pp. 391-396. Ovchinnikov, Y.A., ed., VNU Science Press.

26. Kutschera, U., Schopfer, P. (1985). Evidence against the acid-growth theory of auxin action. Planta 163, 483-493.

27. Libbenga, K.R., Maan, A.C., Van der Linde, P.C.G., Mennes, A.M. (1985). Auxin receptors. In: Hormones, Receptors and Cellular Interactions in Plants. pp. 1-68,. Chadwick, C.M., Garrod, D.R. eds. Cambridge University Press.

28. Libbenga, K.R., Maan, A.C., Van der Linde, P.C.G., Mennes, A.M., Harkes, P.A.A. (1985). Auxin receptors in tobacco leaf protoplasts. In: The Physiological Properties of Plant Protoplasts. pp. 219-225, Pilet, P.E., ed. Springer-Verlag, Berlin, Heidelberg.

29. Löbler, M., Klämbt, D. (1985). Auxin-binding protein from coleoptile membranes of corn (Zea mays L.). I. Purification by immunological methods and characterization. J. Biol. Chem. 260, 9848-9853.

30. Löbler, M., Klämbt, D. (1985). Auxin-binding protein from coleoptile membranes of corn (Zea mays L.). II. Localization of a putative receptor. J. Biol. Chem. 260, 9854-9859.

31. Maan, A.C. (1985). Analysis of auxin binding to plant membranes. Ph.D. thesis, University of Leiden.

32. Maan, A.C., Kühnel, B., Beukers, J.J.B., Libbenga, K.R. (1985). Naphthylphthalamic acid-binding sites in cultured cells from Nicotiana tabacum. Planta 164, 69-74.

33. Maan, A.C., Van der Linde, P.C.G., Harkes, P.A.A., Libbenga, K.R. (1985). Correlation between the presence of membrane-bound auxin binding and root regeneration in cultured tobacco cells. Planta 164, 376-378.

34. Mennes, A.M., Bouman, H., Van der Burg, M.P.M., Libbenga, K.R. (1978). RNA synthesis in isolated tobacco callus nuclei, and the influence of phytohormones. Plant Sci. Lett. 13, 329-339.

35. Morré, D.J., Sandelius, A.S. (1985). Phosphatidylinositol turnover in isolated soybean membranes and response to auxin. Abstracts 12th Int. Conf. on Plant Growth Substances, Heidelberg, p. 62.

36. Oostrom, H., Kuleschà, Z., Van Vliet, Th. B., Libbenga, K.R. (1980). Characterization of a cytoplasmic auxin receptor from tobacco pith callus. Planta 149, 44-47.

37. Oostrom, H., Van Loopik-Detmers, M.A., Libbenga, K.R. (1975). A high affinity receptor for indoleacetic acid in cultured tobacco pith explants. FEBS Lett. 59, 194-197.
38. Polya, G.M., Davies, J.R. (1983). Resolution and properties of a protein kinase catalyzing the phosphorylation of a wheat germ cytokinin-binding protein. Plant Physiol. 71, 482-488.
39. Polya, G.M., Davis, A.W. (1978). Properties of a high-affinity cytokinin-binding protein from wheat germ. Planta 139, 139-147.
40. Pope, D.G. (1982). Effect of peeling on IAA-induced growth in Avena coleoptiles. Ann. Bot. 49, 495-501.
41. Ray, P.M. (1977). Auxin-binding sites of maize coleoptiles are localized on membranes of the endoplasmic reticulum. Plant Physiol. 59, 594-599.
42. Rubery, P.H. (1981). Auxin receptors. Ann. Rev. Plant Physiol. 32, 569-596.
43. Singh, S.P., Paleg, L.G. (1984). Low temperature induction of hormonal sensitivity in genotypically gibberellic acid-insensitive aleurone tissue. Plant Physiol. 74, 437-438.
44. Sisler, E.C. (1982). Ethylene-binding properties of a Triton X-100 extract of mung bean sprouts. J. Plant Growth Reg. 1, 211-218.
45. Sisler, E.C. (1982). Ethylene-binding in normal, rin, and nor mutant tomatoes. J. Plant Growth Reg. 1, 219-226.
46. Skoog, F., Miller, C.O. (1957). Chemical regulation of growth and organ formation in plant tissues cultured in vitro. Symp. Soc. Exp. Biol. 11, 118-131.
47. Stoddart, J.L., Venis, M.A. (1980). Molecular and subcellular aspects of hormone action. In: Encyclopedia of Plant Physiology, vol. 9, Hormonal Regulation of Development, pp. 445-510, MacMillan, J. ed. Springer-Verlag, Berlin.
48. Theologis, A., Huynh, Th.V., Davis, R.W. (1985). Rapid induction of specific mRNAs by auxin in pea epicotyl tissue. J. Mol. Biol. 183, 53-68.
49. Thomas, C.J.R., Smith, A.R., Hall, M.A. (1984). The effect of solubilisation on the character of an ethylene-binding site from Phaseolus vulgaris L. cotyledons. Planta 160, 474-479.
50. Thompson, M. Krull, U.J., Venis, M.A. (1983). A chemo-receptive bilayer lipid membrane based on an auxin-receptor ATPase electrogenic pump. Biochem. Biophys. Res. Comm. 110, 300-304.
51. Van der Linde, P.C.G., Bouman, H., Mennes, A.M., Libbenga, K.R. (1984). A soluble auxin-binding protein from cultured tobacco tissue stimulates RNA synthesis in vitro. Planta 160, 102-106.
52. Van der Linde, P.C.G., Maan, A.C., Mennes, A.M., Libbenga, K.R. (1985). Auxin receptors in tobacco. In: Proc. 16th FEBS meeting at Moscow, Part C, pp. 397-403, Ovchinnikov, Y.A. ed., VNU Science Press.
53. Venis, M.A. (1977). Solubilization and partial purification of auxin-binding sites of corn membranes. Nature 226, 268-269.
54. Venis, M. A. (1977). Receptors for plant hormones. Adv. Bot. Res. 5, 53-88.
55. Venis, M. A. (1985). Hormone binding in plants. Longman Inc., New York–London.
56. Vreugdenhil, D., Burgers, A., Libbenga, K.R. (1979). A particle-bound auxin receptor from tobacco pith callus. Plant Sci. Lett. 16, 115-121.
57. Vreugdenhil, D., Harkes, P.A.A., Libbenga, K.R. (1980). Auxin-binding by particulate fractions from tobacco leaf protoplasts. Planta 150, 9-12.
58. Walker, J.C., Legocka, J., Edelman, L., Key, J.L. (1985). An analysis of growth regulator interactions and gene expression during auxin-induced cell elongation using cloned complementary DNAs to auxin-responsive messenger RNAs. Plant Physiol. 77, 847-850.
59. Walton, J.D., Ray, P.M. (1981). Evidence for receptor function of auxin-binding sites in maize. Plant Physiol. 68, 1334-1338.
60. Zbell, B. (1985). An auxin-mediated control of an intracellular proton-pump via reversible protein phosphorylation and its consequence for the primary action of auxin. Abstracts 12th Int. Conf. on Plant Growth Substances, Heidelberg, p. 62.

APPENDIX—METHODS IN RECEPTOR RESEARCH

In this section we will describe some general methods in hormone-receptor research.

A search for receptors starts with the identification of binding sites in target cells and tissues. In many cases binding sites are being studied using crude or partly purified cell fractions (microsome fractions, cytosolic fractions, etc.). The probes to be used in such studies are radiolabelled ligands with a high specific activity. In order to identify binding sites as putative receptors a number of criteria have to be fulfilled: the binding of ligands to the sites has to be reversible and of high affinity, low capacity and high specificity. Up till now, all real receptors appear to be (glyco)-proteins, so one should also check for the protein nature of the binding sites (for example by using protein degrading enzymes).

The following scheme shows the simple binding of a ligand to its receptor:

$$H^* + R \quad \underset{k-1}{\overset{k+1}{\rightleftharpoons}} \quad H^*R \tag{1}$$

At equilibrium:

$$H^*R = R_t \frac{H^*}{H^* + K_d} \approx R_t \frac{H_t^*}{H_t^* + K_d} \tag{2}$$

H^* is the concentration of free radiolabelled ligand; R is the concentration of unoccupied receptors, H^*R is the concentration of occupied receptors; R_t is the total concentration of receptors; $K_d = \frac{k-1}{k+1}$ is the dissociation constant and its reciprocal value, i.e. $K_a = \frac{k+1}{k-1}$, is the affinity constant; H_t^* is the total concentration of radiolabelled ligand. In the case of receptors, R_t is, in general, many orders of magnitude lower than H_t, i.e., binding is of low capacity, so H^* approximately equals H_t^*.

From equation (2) it can easily be inferred that the concentration of ligand at which half of the number of receptors are occupied equals K_d. Since hormones are regulatory signals which are active at low concentrations (nanomolar to micromolar range), the K_d will be low, or conversely the K_a will be high, i.e. binding is of high affinity.

In order to verify the specificity of the binding one performs competition experiments with non-active ligands. These are ligands which neither evoke nor reversibly inhibit the response obtained by the ligand of interest. Consider the following binding scheme:

$$H^* + R \quad \underset{k-1}{\overset{k+1}{\rightleftharpoons}} \quad H^*R$$
$$I + R \quad \underset{k-2}{\overset{k+2}{\rightleftharpoons}} \quad IR \tag{3}$$

Symbols as in (1) and (2), except that I is the concentration of free (non-labelled) non-active ligand which competes with H^* for the same binding sites;

$$K_{d_1} = \frac{k+1}{k-1} \quad ; \quad K_{d_2} = \frac{k+2}{k-2}$$

At equilibrium:

$$H^*R = R_t \frac{H^*}{H^* + K_{d_1} + \dfrac{K_{a_2}}{K_{a_1}} \cdot I} \tag{4}$$

218

If $I = 0$, then:

$$H*R = R_\tau \frac{H*}{H* + K_{d_1}} \qquad (5)$$

Now we determine the concentration of I (I_{50}) at which the concentration of H*R is reduced to 50% of that without I, thus:

$$R_\tau \frac{H*}{H* + K_{d_1} + \frac{K_{a_2}}{K_{a_1}} \cdot I_{50}} = 0.5\ R_\tau \frac{H*}{H* + K_{d_1}} \qquad (6)$$

From equation (6) it follows that:

$$I_{50} = \frac{K_{a_1} \cdot H* + 1}{K_{a_2}} \qquad (7)$$

If $K_{a_1} H* \ll 1$, or $H* \ll K_{d_1}$, then (7) simplifies to:

$$I_{50} \approx \frac{1}{K_{a_2}} = K_{d_2} \qquad (8)$$

In practice, if $H* \leq 0.1 \cdot K_{a_1}$, then I_{50} is a fair approximation of K_{d_2}.

If R is a real receptor for H*, then we may expect that K_{a_2} will be relatively low, i.e. the binding sites have low affinity towards I.

Crude or partially purified preparations contain many components which retain the ligand with low affinity and high capacity. This is called non-specific binding. Now, consider the following binding scheme:

$$\left.\begin{aligned} H* + R_1 \xrightleftharpoons[k-1]{k+1} H*R_1 \\[2ex] H* + R_2 \xrightleftharpoons[k-2]{k+2} H*R_2 \end{aligned}\right\} \qquad (9)$$

In this scheme the non-specific binding sites have been pooled as one class with dissociation constant K_{d_2} and total number of binding sites $R_{2\tau}$.
The total concentration of bound radioactive ligand is:

$$B* = H*R_1 + H*R_2$$

At equilibrium:

$$B* = R_{1\tau} \frac{H*}{H* + K_{d_1}} + R_{2\tau} \frac{H*}{H* + K_{d_2}} \qquad (11)$$

In a parallel trial we also add unlabelled ligand (H) to the binding assay. In this case at equilibrium:

$$B* = R_{1\tau} \frac{H*}{H* + H + K_{d_1}} + R_{2\tau} \frac{H*}{H* + H + K_{d_2}} \qquad (12)$$

(In this case it is assumed that the affinity constants for the labelled and unlabelled ligand are equal).

Hormone binding

At increasing concentrations of H, the first term in the right-hand part of equation (12) will become negligibly small, whereas non-specific binding will almost not be affected, because K_{d_2} is high. At relatively high concentrations of H ($H \gg H^*$, $H \gg K_{d_1}$, but $H \ll K_{d_2}$):

$$B^* \approx R_{2t} \frac{H^*}{K_{d_2}} \tag{13}$$

Hence, one can determine the amount of unspecifically bound labelled ligand by adding excess unlabelled ligand to a parallel trial.

After correction for unspecific binding at each H^*, one can estimate the dissociation constant and the concentration of R_{1t} from equation (2).

To this purpose one usually transforms equation (2) into a linear relationship, for example following the method of Scatchard. If we denote H^*R as B^* and H^* as F^*, then:

$$B^* = R_t \frac{F^*}{F^* + K_d} \quad , \text{ or} \tag{14}$$

$$\frac{B^*}{F^*} = -K_a B^* + KaR_t \tag{15}$$

Plotting $\dfrac{B^*}{F^*}$ against B^*, yields a straight line with negative slope K_a and the intercept with the abscissa equals R_t. Binding parameters can thus be estimated by linear regression.

If more than one class of receptors is present (let us assume that there are two of such classes) then the Scatchard plot is a hyperbola with asymptotes:

$$\left. \begin{array}{l} \dfrac{B^*}{F^*} = -K_{a_1} B^* + K_{a_1} \cdot R_{1t} \\[2ex] \dfrac{B^*}{F^*} = -K_{a_2} B^* + K_{a_2} \cdot R_{2t} \end{array} \right\} \tag{16}$$

In this case non-linear regression methods are required to estimate the binding parameters for each class of receptors. Scatchard plots are frequently analysed incorrectly and, moreover, one should be very careful in interpreting non-linear Scatchard plots, because they might also indicate cooperative binding.

For the determination of the amount of bound ligand, quite a number of methods are being used: equilibrium dialysis, separation of bound from free ligand by high-speed centrifugation (in case of microsomes) or adsorption of free ligand to charcoal coated with a molecular sieve such as dextran (in case of soluble binding proteins). Also various ultrafiltration techniques are being used.

Most of what we know about plant hormone receptors is based on binding experiments. In other words, in most cases only binding sites have been identified. However, if all criteria for putative receptors have been fulfilled, then we still have to proof a receptor function for these binding sites. Although no standard procedure exists, one can indicate a few steps in this important part of receptor research:

- Purification of the binding proteins by convential isolation and separation techniques, including ligand-affinity chromatography and photo-affinity labelling. A major problem is that receptor proteins occur in extremely low concentrations (picomoles per mg of protein) and are liable to (enzymatic) degradation and inactivation. Moreover, membrane-bound binding proteins have first to be solubilized. This may raise serious problems if they are highly hydrophobic integral membrane proteins, as for example ethylene binding sites.
- Preparation of (monoclonal) antibodies against purified binding proteins for characterization, intracellular localization and analysis of receptor function. Although plant hormone-receptor immunochemistry has got a strong impulse from a few studies, it is still in its infancy.

- Correlation between hormone sensitivity and levels of binding sites in tissues during development of plants, including so-called hormone-sensitivity mutants. Antibodies are essential probes for such studies. The results might give strong indications as to the physiological function of the binding sites.
- Reconstitution of *in vitro* hormone-sensitive systems. This requires purified native binding proteins and their primary biochemical targets. The nature of putative targets can be derived from the growing body of evidence on rapid hormone-induced biochemical responses (response times in seconds and minutes) and from the intracellular localization of the binding sites. In case of membrane-bound binding sites reconstitution might require incorporation of binding sites and targets into artificial membranes. In general, reconstitution experiments are extremely difficult to perform. To mention only one problem, a rapid biochemical response might still involve a few (unknown) intermediary steps. A good example is the modulation of adenylate cyclase by animal peptide hormones, which requires interactions of hormone-receptor complexes with G-proteins as intermediary steps.

D. HORMONE ANALYSIS

D1. Instrumental Methods of Plant Hormone Analysis.

Roger Horgan

Department of Botany and Microbiology, University College of Wales, Aberystwyth, Dyfed, SY23 3DA, Wales, U.K.

INTRODUCTION

From information presented in previous chapters it will be clear to readers that plant hormones are, as a rule, present at very low levels in most plant tissues. Whilst relatively high levels of some hormones are found in immature seeds of certain species (e.g., GAs in developing pea seeds (10)) even these levels are low when compared with the levels of most plant secondary metabolites. Thus while many alkaloids, terpenoids and phenolics may be present at levels of mgs. per gm. dry weight of plant material, plant hormones are usually present at several hundred to several thousand fold lower levels. It is not suprising therefore that knowlege of the chemical identity of plant hormones has been limited by the techniques available for their isolation in a pure state and by the sensitivity of the spectroscopic techniques required to elucidate their chemical structure.

In the last ten years there have been spectacular improvements in the sensitivity of spectroscopic methods of structure determination and corresponding increases in performance in chromatographic techniques, principally via the development of high performance liquid chromatography (HPLC) and capillary column gas-liquid chromatography (GLC). This improvement in methodology can be clearly seen if one compares the isolation and identification of zeatin by Letham in 1963 (15), where 60 kg of plant material had to be extracted and purified by traditional chromatographic methods to yield the mg. of material needed for spectroscopic studies, with the identification of 1'-deoxy ABA as a precursor of abscisic acid (ABA) in the fungus *Cercospora rosicola* (25) where, after purification by HPLC, identification was possible at the µg level.

This chapter is concerned with the application of modern instrumental techniques to the isolation, identification and quantitation of plant hormones. Clearly the theoretical background to these techniques is beyond the scope of this work and readers are refered to suitable texbooks for this information (e.g., 9, 28, 19). However, it is very important for a critical understanding of the methods used in the identification of plant hormones to appreciate the inherent limitations of the various techniques and so these will be touched upon in the relevant sections. In particular it is necessary to appreciate the importance of sample purity to the interpretation of spectroscopic data. Even with the most sophisticated instrumentation, correct indentifications can only be made if the spectroscopic data obtained is relevant to the hormone under investigation. Two examples of mistaken identities provide informative reading on this point (33,5).

The chapter is organised in the chronological order in which a real analysis of a plant hormone would probably proceed. First, the compound would have to be isolated in a sufficiently pure form, second, its structure would have to be determined by appropriate methods, and finally a strategy would need to be devised for its quantitative mesurement.

ISOLATION AND PURIFICATION OF PLANT HORMONES

Extraction and Preliminary Purification

The methods of extraction and preliminary purification of plant hormones using traditional methods such as solvent partitioning, ion exchange chromatography and, paper and thin layer chromatography will not be discussed in this chapter as strictly speaking they fall outside the area of instrumental methods and many of these methods are being superseded by HPLC based methods. Nevertheless it is often necessary to revert to older methods particularly with plant extracts that are too large for the initial use of HPLC. The readers attention is drawn to the extremely comprehensive treatment of these methods by Yokota et al. (34).

Bioassays

Although the bulk of this chapter is concerned with the use of physical methods for the detection of plant hormones it should be noted at this point that the primary detection of any novel plant hormones is dependent on biosassay. Bioassays are also necessary when studying the hormone content of novel plant materials particularly with regard to gibberellins and cytokinins. Because of the trace nature of plant hormones in most extracts, direct physico-chemical detection is impossible during the early stages of purification. In the case of the gibberellins even detection at the latter stages of purification is difficult due to the low wavelength and low extinction coefficient of UV absorption by these compounds. In these situations bioassays have to be used to detect the compounds of interest and to monitor the purification process. The choice of suitable bioassays can be problematical. Ideally a bioassay should respond quantitatively to all the members of a certain group of plant hormones, be highly selective towards a that particular group of compounds, have high sensitivity and not be inhibited by other compounds in the plant extract. These conditions are never met in practice. In the case of the gibberellins the most suitable assays are probably the dwarf maize assay (26) and the Tanginbozu dwarf rice assay (22). Both respond well to a good range of GAs and the former assay is particularly suitable for relatively crude plant extracts where the presence of inhibitors may interfere with other assays. In the case of the cytokinins, which by definition are promoters of cell division in plant callus cultures, the Tobacco callus (23) or the Soybean callus (20) bioassay is required to unambiguously reveal the presence of cytokinins in a plant extract. However these assays require about 21 days for growth and so the more rapid but probably less specific *Amaranthus* betacyanin bioassay (3) is frequently used for routine monitoring of cytokinin activity during purification processes.

It should be pointed out that although bioassays are necessary for the detection of plant hormones they are now generally accepted to be unsuitable for quantitative work. The presence of inhibitors in most plant extracts and the logarithmic nature of the reponse of most assays makes for very inaccurate and imprecise measurements. To large extent realisation of the quantitative limitations of bioassays has provided the stimulus for the development of the physical methods of quantitation described in this chapter.

High Performance Liquid Chromatography

HPLC is distinguished from traditional chromatography by its high efficiency and resolution, and speed of separation. The first two improvements are achieved by the utilisation of small particle size stationary phase materials at the cost of relatively complex and expensive instrumentation. This results mainly from the high liquid pressures required to achieve fast flow rates through small particle size columns. The resolution of a system is largely dependent on the chemistry of the mobile and stationary phases. The relative importance of the chemical (separation factor a and capacity factor k') and physical (efficiency N) terms in governing the resolution (Rs) can be seen from the chromatographic equation:

$$R_s = \frac{1}{4}\left(\frac{a-1}{a}\right)\left(\frac{k'}{k'+1}\right)\sqrt{N}$$

Optimising a and k' can only be achieved through careful selection of mobile and stationary phases. The power of HPLC and high resolution capillary GLC stems from the hugh increases in N possible in these systems. From the point of view of plant hormone analysis any increase in resolution of a chromatographic system is valuable since it increases the chances of separating the compound(s) of interest from other interfering materials.

In assessing HPLC methods for plant hormone analysis it is important to appreciate the presence of two often conflicting factors. Firstly the need to separate the hormone from other compounds present in the extract and secondly the abilty of the system to separate closely related hormones. In general most HPLC systems for plant hormones have been developed using the second critera. Whilst this is a useful indicator of the resolution of the system it should be borne in mind that when used preparatively on a realtively crude plant extract the desired compound may not be obtained in a degree of purity sufficient for unambiguaous interpretation of subsequent spectroscopic data. The first attempt to develop a HPLC system for plant hormones was directed at improving the resolution of an open column partition chromatography system for gibberellins (27). Although this work illustrated the potential of HPLC it has not found widespread use due to the technical complexity of using aqueous stationary phases. The most important development in HPLC as far as plant hormones are concerned was the commercial production of so called 'bonded reverse phase materials'. In reverse phase chromatography the support material, usually silica, is coated with a liquid phase of very low polarity. In aqueous solution low polarity compounds will partition into the stationary phase and will only be eluted with solvents of lower polarity than water.

For plant hormone HPLC the most frequently used stationary phase is octadecylsilane (ODS) which is covalently bonded to microparticulate or microspherical silica. Solvents are usually binary mixtures or gradients of water (weak solvent) and the lower alcohols or acetonitrile (strong solvents). Since the selectivity of reverse phase HPLC is at its best when the compounds being analysed are un-ionised, acidic aqueous phases are used for the analysis of acid plant hormones such as GAs, ABA and IAA, whereas neutral buffers are prefered to suppress the ionisation of basic compounds such as cytokinins. The great value of reverse phase HPLC for the isolation of plant hormones lies in the enormous range of compound polarities that can be accomodated on a single stationary phase if gradient elution is used. Under reversed phase conditions polar compounds will elute before non-polar compounds. Thus polar plant hormone conjugates and free compounds may be fractionated in a single HPLC step. Gradient elution reverse phase HPLC therefore provides a very valuable method for the preliminary purification of plant extracts. Indeed reverse phase HPLC on reasonably large size columns (10 to 20 mm. ID) is gradually superseding the older methods of solvent partitioning, paper chromatography and open column chromatography as the first step in plant hormone purification. With columns of this size injection sizes of several mls. can be used and if the initial solvent is of low polarity the sample will concentrate at the end of the column. With subsequent gradient elution suprisingly good resolution can be achieved with very crude samples. A typical chromatogram resulting from the use of this technique is shown in Fig.1 . An 80% methanol extract of bean leaves was

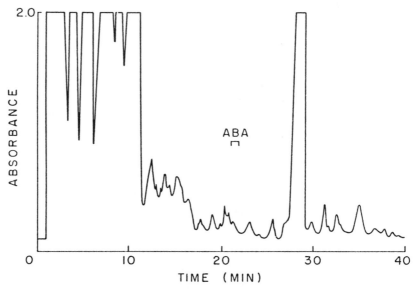

Fig.1. Reverse phase HPLC of an extract of *Phaseolus vulgaris* leaves on a 150×10mm column of Spherisorb ODS-2 using a linear gradient of 20% methanol in 0.1 M acetic acid to 100% methanol over 40 min. at a flow rate of 5 ml.min⁻¹. The fraction denoted by the bar, at the retention time of an ABA standard, was collected and analysed as described in the text.

chromatogaphed on a ODS reverse phase column (10×150mm) with a linear gradient of 20% methanol in 0.1 M acetic acid to 100% methanol over 40 mins. The chromatogam shown in Fig. 1 is the absorbance trace obtained by monitoring the column eluate at 265 nm. The fraction corresponding to the elution time of ABA was collected, methylated with diazomethane and examined by GC-MS. Methyl ABA was clearly identified by its mass spectrum.The superior resolution of HPLC over conventional open column chromatography is illustrated in Fig.2 which compares the reverse phase chromatography of a mixture of cytokinins on LH 20 Sephadex and on an octadecyl silica HPLC material. Details of reverse HPLC systems for the separation of gibberellins (14), cytokinins (12) and, ABA and IAA (7) have been published.

It should be pointed out that at the stage of initial purification the collection of the correct fraction(s) from a preparative HPLC column will be on the basis of the elution time of a standard compound if a known hormone is being purified or on the basis of biological activity, resulting from bioassaying successive fractions, if a new compound is being isolated. With a crude extract the ususal detectors UV, RI are of little value in

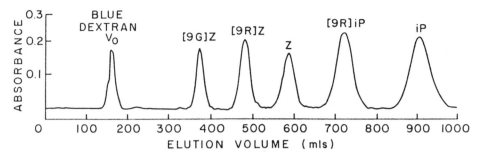

Fig.2a. Separation of a series of cytokinins on a 80×2.5 cm column of LH20 Sephadex eluted with 35% ethanol at a flow rate of 30 ml hr^{-1}.

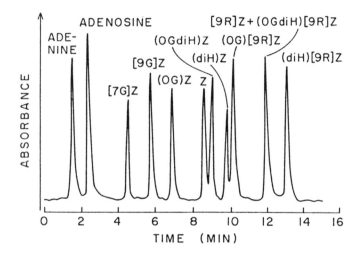

Fig.2b. Separation of a series of cytokinins by HPLC on a 150×4.5mm column of Spherisorb ODS-2 using a linear gradient of 5% acetonitrile in water (pH adjusted to 7 with triethyl-ammonium bicarbonate) to 20% acetonitrile over 30 mins at a flow rate of 2ml min^{-1}.

locating the compound(s) of interest. This point will be discussed later in the context of the quantitative measurement of plant hormones.

Gas-Liquid Chromatography

GLC is usually used as the last chromatographic step of a hormone analysis, either utilising a specific detector on the GLC instrument or in combination with mass spectrometry (GC-MS). In all cases the specificity of the analysis is affected by the resolution of the GLC system itself. It is instructive to compare the efficiency term N in the chromatographic expression for the traditional packed chromatographic column, where the liquid phase is depositied on a particulate support and a wall coated capillary column where the phase is coated as a very thin layer on the wall of the column. Although the value N, expressed as theoretical plates per meter is similar for the two types of column (i.e., about 1500), the practicality of using long capillary columns, typically 25 meters, means that these columns can exhibit efficiencies of say 50000 theoretical plates compared to about 5000 theoretical plates for 3 or 4 meter packed column. In seeking to identify and quantify plant hormones in complex extracts the increased resolution obtained with capillary columns makes a very valuable contribution to the analytical system. For most GC-MS analyses of plant hormones, capillary columns, particularly those in which the stationary phase is chemically bonded to the wall of the column, would be used. Fig. 3a shows the separation of a group of permethylated cytokinins on a 1.5 m packed GLC column. Fig. 3b shows the separation of the same compounds on a 12 m bonded phase capillary column.

IDENTIFICATION OF PLANT HORMONES

The problems of the chemical characterisation of plant hormones need to be considered at two levels of difficulty. At the easier level there is the problem of confirming the identity of known plant hormones in extracts of plant in which they have not previously been identified. At the much more difficult level there is the problem of identifying unknown compounds exhibiting biological activity corresponding to one of the known classes of plant hormone or possibly a previously undescribed form of biological activity. For confiming the identity of known plant hormones there is no doubt that mass spectrometry (MS) and in particular combined gas chromatography–mass spectrometry (GC–MS) is the technique 'par excelence'. Whilst GC–MS also plays an important role in the tentative identification of novel compounds complete identification at this level frequently requires the use of additional techniques and eventually requires confirmation of the proposed structure by unambiguous chemical synthesis.

GC–MS Identification of Plant Hormones

The coupling of the outlet of a gas chromatography column to the source of a mass spectrometer enables mass spectra to be obtained as individual components elute from the gas chromatograph. Since modern mass spectrometers can scan a decade of mass (e.g.,300-30 amu(atomic mass units)) in less than one second, several mass spectra may be obtained on a single GC peak. With the addition of computerised aquisition, storage and processing of the mass spectra enormous amounts of information may be obtained about the chemical nature of complex plant extracts. Ideally the GLC system should be able to resolve all the

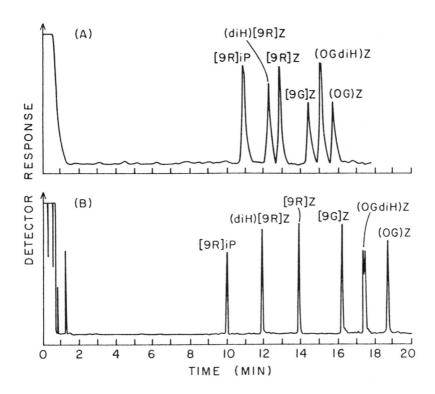

Fig. 3. GLC separation of a series of permethyl cytokinins on (A) a conventional packed column (1.5 m × 4 mm 3% OV1) and (B) a 12m × 0.3 mm bonded phase capillary column.

components of the mixture under analysis. However, with suitable data processing software individual components can be identified even if they coelute from the GC column. Thus provided the mass spectrum of the plant hormone under investigation is available as a reference, and additionally if possible its retention time (Kovats retention index) on the GLC system being used, it is a relatively easy matter to confirm the presence of a known compound in a partially purified plant extract. Details of the theory and operation of GC–MS systems and associated data systems may be found in Rose and Johnson (28). In this section some of the features of the mass spectra of members of the main classes of plant hormones will be described in relation to the use of MS and GC–MS for their identification.

With the exception of ethylene, GLC of all plant hormones requires that they first be converted to volatile derivatives. In general carboxylic acids are converted to methyl esters by diazomethane, alcohol groups to trimethylsilyl (TMS) derivatives with a reagent such as bis-trimethylsilyl acetamide (BSA) or to methyl ethers with the methyl suphonyl anion and methyl iodide (DMSO⁻/CH$_3$I) and amino groups TMSed or methylated as above. Thus in identifying known compounds mass spectra are rarely recorded for the free compound but usually via GC–MS of one or more derivatives.

Auxins

The mass spectra of a large number of IAA derivatives have been recorded (8, 32, 2) and thus provide information for identifying these compounds in plant extracts. The mass spectra of compounds related to IAA are usually very simple and are typified by the spectrum of IAA methyl ester and its TMS derivative shown in Fig.4. The ion corresponding to the

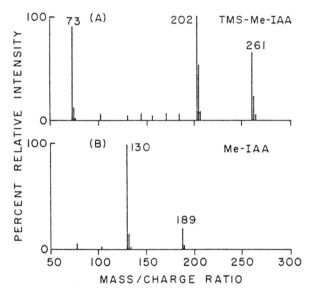

Fig. 4. Electron impact mass spectra of (A) trimethylsilyl-IAA methyl ester and (B) IAA methyl ester.

unfragmented molecule, the M^+ or molecular ion is apparent in both spectra. The great stability of the indole nucleus is reflected in the base peaks which have the structures shown in Fig. 5. The presence of one of these spectra in the GC–MS run of a suitably derivatised plant extract, at the correct retention time, may be taken as conclusive proof for the existence of IAA in the extract.

Gibberellins

GC-MS was first used for plant hormone analysis by MacMillan and his co-workers in identifying gibberellins (17). Although basically similar in structure, many of the gibberellins exhibit widely differing substitution patterns which are reflected in the mass spectra of their methyl-TMS derivatives (4). Whilst certain ions are very diagnostic of hydroxyl groups at

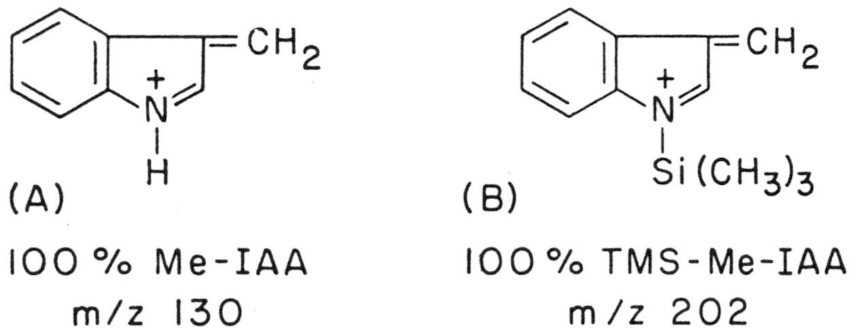

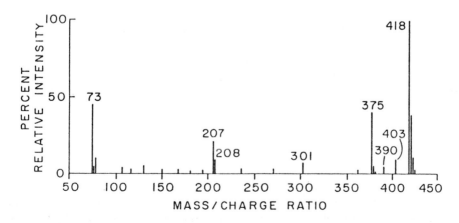

(A)
100 % Me-IAA
m/z 130

(B)
100 % TMS-Me-IAA
m/z 202

Fig. 5. Structures of the base peaks in the electron impact mass spectra of (A) IAA methyl ester and (B) trimethylsilyl-IAA methyl ester.

certain positions in the molecule eg 207/208 for 13-OH and 129 for 3-OH the unambiguous identification of a known GA requires careful comparison of the mass spectrum with that of a standard compound and often requires the addition information provided by the Kovats retention index (1). Because of their cyclic structures the molecular ions of many methyl/TMS GAs are very prominent and the spectra are very simple. This is illustrated by the spectrum of methyl/TMS GA$_{20}$ shown in Fig. 6. The presence of the 13-OH group is clearly indicated by the prominent ions at m/z 207 and 208.

Cytokinins

Both TMS and permethyl derivatives of cytokinins exhibit good GLC properties and have mass spectra which provide excellent fingerprints for

Fig. 6. Electron impact mass spectrum of trimethylsilyl GA$_{20}$ methyl ester.

the identification of known compounds in plant extracts by GC–MS (21, 16, 31). From the point of view of interpretation of fragmentation patterns the mass spectra of permethyl cytokinins are easier to understand than those of the TMS compounds. Maximum structural information is shown by the MS of underivatised cytokinins introduced directly into the source of the mass spectrometer using a heated probe. The mass spectrum of underivatised zeatin is shown in Fig. 7. The

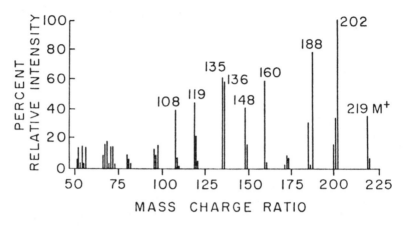

Fig. 7. Electron impact mass spectrum of zeatin.

molecular ion at m/z 219 is clearly visible and suggests the presence of an odd number of nitrogen atoms on the molecule. The ions at m/z 202 and 188 arise by losses of fragments of 17 and 31 amu from the molecular ion and indicate the presence of a primary -OH group. The presence of strong ions at m/z 136, 135, 119 and 108 provide strong evidence that the molecule contains an adenine moeity. The location of the side chain on the 6-amino group is indicated by the presence of an ion at m/z 148 and the presence of an additional alkyl group in the side chain is indicated by an ion at m/z 160.

Readers should note that detailed analysis of the fragmentation of zeatin is largely retrospective and that mass spectra of unknown compounds are often very difficult to interpret. In the case of zeatin, mass spectrometry alone would not be sufficient to completely characterise the side chain.

Abscisic Acid and Related Compounds

The mass spectral fragmentation of methyl ABA, shown in Fig.8, has been examined in considerable detail (11). Whilst the molecular ion is very weak there are a sufficient number of characteristic ions to enable the compound to be easily identified in plant extracts. It should be noted that methyl ABA and its 2-trans isomer have identical mass spectra. Thus identification of ABA in a plant extract should also involve the information provided by the GLC retention time.

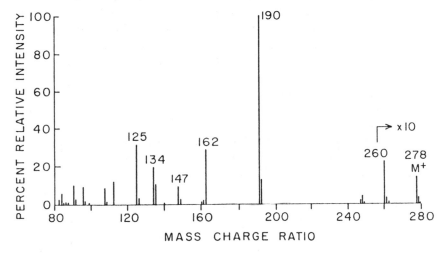

Fig. 8. Electron impact mass spectrum of methyl abscisate.

QUANTITATIVE MEASUREMENT OF PLANT HORMONES

The ultimate aim of most analyses of plant hormones is to measure precisely and accurately the level of the compound(s) of interest under certain known physiological conditions. Whilst theoretically any physico-chemical detector with a defined quantitative response to the particular compound(s) may appear suitable for this purpose the practical problems of hormone measurement are enormous. In general these stem from the fact that plant hormones are nearly always trace components in any crude plant extract and are present together with many structuraly similar compounds. Thus the selectivity and sensitivity of the measurement system is extremely important. All currently used physico-chemical methods for measuring plant hormones use a chromatographic system, HPLC or GLC coupled to detectors with varying degrees of sensitivity and selectivity. Detectors with a low degree of selectivity, e.g., flame ionisation with GLC or single wavelength UV with HPLC, necessitate the use of very pure extracts to ensure that the peak being measured contains only the compound of interest. For this reason these detectors are rarely used to measure hormones in plant extract and results obtained with these detectors should be judged very critically. The recent advent of diode array UV/VIS detectors, which can record whole spectra on HPLC peaks or even sections of HPLC peaks as they pass through a cell, may offer a way to selectively measure plant hormones directly in HPLC eluates. However these instruments have not yet been evaluateed in this context.

An example of the use of a highly sensitive and specific detector may be seen in the use of an electron capture detector (ECD) for the GLC determination of ABA (29). The unsaturated carbonyl group of ABA renders it electron capturing and, in a suitably partially purified plant extract, ABA may often be the only compound at the correct GLC retention time to exhibit this property and therefore can be accurately measured. In a similar way the use of a nitrogen specific detector has been proposed as a method of measuring cytokinins although the method had not yet found widespread use.

IAA exhibits a very strong and characerstic fluorescence with an excitation maximum at 280nm and an emission maximun at 350nm. A sensitive spectrofluorimeter with a narrow band width for excitation and emission therefore a makes a sensitive and selective HPLC detector for IAA. With the correct instumentation IAA can be measured at the pg. level in relatively crude plant extracts (6).

In general plant extracts need to be purified to varying degrees before measurements can be made. During the purification process losses of material occur and therefore these have to be accounted for in the final result. Thus methods for the reliable quantitative measurement of plant hormones require the use of a suitable internal standard. In principal, an internal standard may be any compound sufficiently similar in its chemical and physical properties to the compound of interest to pass through the extraction and purification procedures with the same degree of loss. However, it must be sufficiently different from the compound of interest to be measured independently in the final quantitative step. Thus by adding a known amount of internal standard at the initial extraction step and measuring it accurately at the final step its percentage recovery may be calculated. Provided the first criterion above has been met this will be identical to the recovery of the endogenous compound and thus the losses of this during workup may be corrected for.

The need to meet the two somewhat conflicting critera mentioned above place several restrictions on types of internal standards suitable for plant hormone quantitation. Thus isomers and homologues are not usually suitable internal standards. For example 2-*trans* abscisic acid (2T-ABA) has different solubility in several organic solvents to ABA and may not be recovered to the same degree on solvent partitioning of plant extracts. It separates from ABA in a number of chromatographic systems and although this makes it a suitable internal standard from the point of independent measurement at the last step, it introduces severe complications when purifying plant extracts by TLC or HPLC.

The most suitable internal standards for plant hormone determinations are isotopically labelled versions of the compounds themselves. Compounds may be labelled with radioactive and/or heavy isotopes. The techniques involved in using these different types of internal standard depend on the methods necessary to detect and measure the internal standard and the hormone itself. Radioactive internal standards may be measured by liquid scintillation counting. Thus the recovery of labelled hormone in the course of purification procedure is simply calculated as ratio of the DPM added at the intial extraction phase to the DPM present at the final measurement stage. If this is assumed to be the same as the recovery of the endogenous compound, then multiplication of the final measured amount of endogenous compound by this ratio will give the amount present in the original extract. Superficially this appears to be a very straightforward method limited only by the availability of suitable radioactive hormones. In practice, however, it is subject to certain limitations. Whilst the measurement of amount of internal standard is specific for the radioactive material the same is not true for the measurement of the endogenous compound. Measurement systems such as GLC or HPLC with UV or flourescence detection will respond to both labelled and unlabelled molecules of hormone. Thus the amount of hormone measured will be the sum of the endogenoous material plus the internal standard. Since the amount of internal standard, at this stage, will be know from the radioactivity present in the extract, it can be subtracted from the amount measured by the particular detection system in use to give the correct value for the endogenous hormone. If the specific activity of the internal standard is high (usually only possible with ^{3}H-labelled materials) the amount that has to be added to an extract to give a measurable number of counts in the final analysis may be so small that its contribution to the endogenous compound may be neglected. On the other hand, if the specific activity of the internal standard is low it may effectively overwhelm the endogenous compound to such an extent that the amount cannot be accurately measured. It is also important that the internal standard remains radiochemically pure throughout the procedure and any labelled compounds formed by chemical or biochemical transformations during the workup are separated from it before the final counting. The need to measure both radioactivity and amount of hormone present means that two independent measurements need to be made and this introduces a greater degree of error into the proceedings.

In general the best methods for the quantitation of plant hormones use heavy isotope-labelled internal standards. Selected examples of these are shown in Fig. 9. Since mass spectrometers can easily disinguish between isotopically labelled molecules GC–MS can be used to measure the relative proportions of an endogenous compound and its heavy isotope-labelled internal standard. Thus if the ratio of these in the final sample is know together with the amount of internal standard added at the initial extraction stage the simple expression:

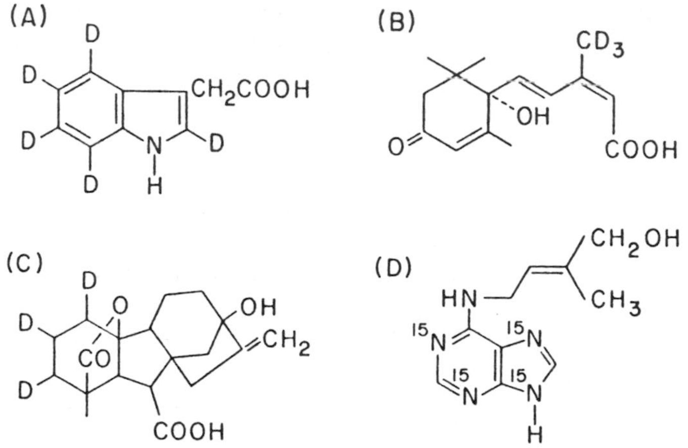

Fig. 9. Structures of some heavy isotope labelled plant hormones used as internal standards. (A) [^{2}H$_5$]-IAA. (B) [^{2}H$_3$]-ABA. (C) [^{2}H$_3$]-GA$_{20}$. (D) [^{15}N$_4$]-zeatin.

Endogenous amount =

$$\frac{\text{Endogenous ion intensity}}{\text{Internal standard ion intensity}} \times \text{Amount of internal standard added}$$

can be used to calculate the original amount of endogenous hormone in the extract. Heavy isotope internal standards have been used to measure ABA (24), IAA (18), cytokinins (30) and gibberellins (13) in plant extracts. In practice the ratio of endogenous to heavy isotope-labelled compound may be measured from suitable ions in a full mass spectrum obtained from a GC–MS analysis. Because there is often slight separation of the labelled and unlabelled compounds on the GC, a single mass spectrum taken at one point on a GC peak may give an erroneous value of the isotope ratio.

Measurements of this type are usually made with the mass spectrometer operating in the selected ion monitoring (SIM) mode. In this mode the mass spectrometer is adjusted to monitor one or two ions of the endogenous compound and the equivalent mass shifted ions of the internal standard. Because the ions are close together in mass they can be selectively monitored by electrical rather than magnetic scanning they can be cyclically selected very rapidly and measured with a much improved signal to noise ratio. Thus many measurements of the isotope ratios may be made over a single GLC peak with much greater sensitivity than can be achieved from the full mass spectra method. The ratio of the areas under the relevant peaks can be used as measure of the the relative amounts of endogenous compound to internal standard. Thus the amount of endogenous hormone originally present in the extract can be calcuated as described above. A SIM trace from the GC–MS of extract of ABA from

Phaseolus vulgaris leaves is shown in Fig.10. The ions monitored are the m/z 190 ion from the endogenous methyl-ABA and the coresponding ion at m/z 193 from the $^2H_3$ internal standard shown in Fig. 9.

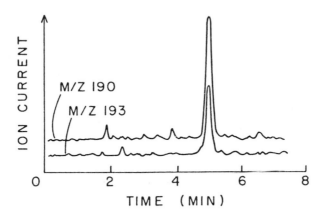

Fig. 10. SIM trace from the determination of ABA in an extract of *Phaseolus vulgaris* leaves utilising [$^2H_3$]-ABA as an internal standard. The peak for the ion at m/z 190 is due to endogenous methyl abscisate and the peak at m/z 193 is due to the internal standard.

An obvious limitation of this method is the presence of interfering ions from other compounds with the same GC retention time as the hormone. Before the SIM method is used full MS should be obtained from a sample of the hormone purified in an identical manner to that used for the quantitative determinations. If potentially interfering ions are observed the purification method has to be modified to eliminate the interfering compound(s).

It can be seen from the foregoing account that the identification and measurement of plant hormones is a technically difficult problem. In this chapter it has only been possible to discuss briefly the procedures involved. Whilst the methods described are in general use in many other areas of analytical biochemistry, their application to plant hormone analysis is not necessarily as straightforward as this account may suggest. Because of the complex chemical nature of plant extracts and the low levels of many plant hormones analytical methods often have to be operated at their limits of sensitivity and resolution. In addition different plant species often pose different analytical problems due to qualitative and quantitative differences in interfering compounds. These factors have to be taken into account when applying the techniques discussed in this chapter.

References

1. Albone, K.S., Gaskin, P., MacMillan, J., Sponsel, V.M. (1984) Identification and localisation of gibberellins in maturing seeds of the cucurbit *Sechium edule*, and a comparison between this cucurbit and the legume *Phaseolus coccineus*. Planta *162*, 560-565.
2. Badenoch-Jones, J., Summons, R.E., Entsch, B., Rolfe, B.G., Parker, C.W., Letham, D.S. (1982) Mass spectrometric identification of indole compounds produced by *Rhizobium* strains. Biomed. Mass Spectrometry *9*, 430-436.
3. Biddington, N.L. and Thomas, T.H. (1973) A modified *Amaranthus* betacyanin bioassay for the rapid determination of cytokinins in plant tissues. Planta *111*, 183-186.
4. Binks, R., MacMillan, J., Pryce, R.J. (1969) Plant hormones VIII. Combined gas chromatography-mass spectrometry of the methyl esters of gibberellins A_1 to A_{24} and their trimethylsilyl ethers. Phytochemistry *8*, 271-284.
5. Bowman, W.R., Linforth, R.S.T., Rossall, S., Taylor, I.B. (1984) Accumulation of an ABA analogue in the wilty tomato mutant, *flacca*. Biochemical Genetics *22*, 369-378.
6. Crozier, A., Loferski, J., Zaerr, J., Morris, R.O. (1980) Analysis of picogram quantities of indole-3-acetic acid by high performance liquid chromatography-fluorescence procedures. Planta *150*, 366-370.
7. Durley, R.C., Kannangara, T., Simpson, G.M. (1982) Leaf analysis for abscisic, phaseic and 3-indolylacetic acids by high performance liquid chromatography. J.Chromatography *236*, 181-188.
8. Ehmann, A. (1974) Identification of 2-O-(indole-3-acetyl)-D- glucopyranose, 4-O-(indole-3-acetyl)-D-glucopyranose and 6-O- (indole-3-acetyl)-D-glucopyranose from kernels of *Zea mays* by gas liquid chromatography-mass spectrometry. Carbohydrate Research *34*, 99-114.
9. Engelhart, H. (1978) High Perfomance Liquid Chromatography. Springer-Verlag, Berlin.
10. Frydman, V.M., Gaskin, P., MacMillam, J. (1974) Qualitative and quantitative analysis of gibberellins throughout seed maturation in *Pisum sativum* cv. Progress No.9. Planta *118*, 123-132.
11. Gray, R.T., Mallaby, R., Ryback, G., Williams, V.P. (1974) Mass spectra of methyl abscisate and isotopically labelled analogues. J. Chem. Soc Perkin Trans. *2*, 919-924.
12. Horgan, R., Kramers, M.R. (1979) High-performance liquid chromatography of cytokinins. J.Chromatography *173*, 263-270.
13. Ingram, T.J., Reid, J.B., Murfet, I.C., Gaskin, P., Willis, C.L., MacMillan, J. (1984) Internode length in *Pisum*. The *Le* gene controls the 3-hydroxylation of gibberellin A_{20} to gibberellin A_1. Planta *160*, 455-463.
14. Koshioka, M., Harada, J., Takeno, K., Noma, M., Sassa, T., Ogiyama, K., Taylor, J.S., Rood, S.B., Legge, R.L., Pharis, R.P. (1983) Reverse phase C18 high-performance liquid chromatography of acidic and conjugated gibberellins. J. Chromatography *256*, 101-115.
15. Letham, D.S., Shannon, J.C., MacDonald, I.R.C. (1964) The structure of zeatin, a (kinetin like) factor inducing cell division. Proc. Chem. Soc. London 230-231.
16. MacLeod, J.K., Summons, R.E., Letham, D.S. (1976) Mass spectrometry of cytokinin metabolites. Per(trimethylsilyl) and permethyl derivatives of glucosides of zeatin and 6-benzylaminopurine. J.Org.Chem. *41*, 3959-3967.
17. MacMillan, J., Pryce, R.J., Eglinton, G., McCormick, A. (1967) Identification of gibberellins in crude plant extracts by combined gas chromatography-mass spectrometry. Tetrahedron Lett. 2241-2243.
18. Magnus, V., Bandurski, R.S., Schulze, A. (1980) Synthesis of 4,5,6,7 and 2,4,5,6,7 deuterium labelled indole-3-acetic acid for use in mass spectrometric assays. Plant Physiol. *66*, 775-781.
19. Millard, B.J. (1979) Quantitative mass spectrometry. Heyden, London.

20. Miller, C. O. (1963) Kinetin and kinetin-like compounds. In: Modern methods of plant analysis. Linskins, H.F. and Tracey, M. V. eds. Vol. VI, pp 194-202. Springer: Berlin-Heidelberg- New York.

21. Morris, R.O. (1977) Mass spectrometric identification of cytokinins. Glucosyl zeatin and glucosyl ribosylzeatin from *Vinca rosea* crown gall. Plant Pysiol. *59*, 1029-1033.

22. Murakami, Y. (1968) The microdrop method, a new rice seedling test for gibberellins and its use for testing extracts of rice and morning glory. Bot. Mag. *79*, 33-43.

23. Murashige, T. and Skoog, F. (1962) A revised medium for rapid growth and bioassays with tobacco tissue cultures. Physiol. Plant. *15*, 473-497.

24. Neill, S.J., Horgan, R., Heald, J.K. (1983) Determination of the levels of abscisic acid-glucose ester in plants. Planta *157*, 371-375.

25. Neill, S.J., Horgan, R., Lee, T.S., Walton, D.C. (1981) 3- methyl-5(4 -oxo-2 ,6 ,6 - trimethylcyclohex-2 -en-1 -yl)-2,4 pentadienoic acid from *Cercospora rosicola*. FEBS Letters *128*, 30-32.

26. Phinney, B.O. (1956) Growth response of single-gene mutants of maize to gibberellic acid. Proc. Natl. Acad. Sci. USA. *42*, 185-186.

27. Reeve, D.R., Crozier, A. (1978) The analysis of gibberellins by high performance liquid chromatography, in Isolation of Plant Growth Substances, SEB Seminar Series, Vol.4 ed. J.R. Hillman, Cambridge University Press, pp.41-77.

28. Rose, M.E., Johnson, R.A.W. (1982) Mass spectrometry for chemists and biochemists. Cambridge University Press.

29. Saunders, P.F. (1978) The identification and quantitative analysis of abscisic acid in plant extracts, in Isolation of Plant Growth Substances, SEB Seminar Series, Vol.4 ed. J.R.Hillman, pp. 115-134.

30. Scott, I.M., Horgan, R. (1984) Mass spectrometric quantification of cytokinin nucleotides and glycosides in tobacco crown gall tissue. Planta *161*, 345-354.

31. Scott, I.M., Horgan, R., McGaw, B.A. (1980) Zeatin-9-glucoside a major endogenous cytokinin of *Vinca rosea* L. crown gall tissue. Planta *149*, 472-475.

32. Udea, M., Bandurski, R.S. (1974) Structure of indole-3-acetic acid myoinositol esters and pentamethyl-myoinositols. Phytochemistry *13*, 243-253.

33. Van Staden, J., Drewes, S.E. (1974) Identification of cell division inducing compounds from coconut milk. Physiol. Plant. *32*, 347-352.

34. Yokota, T., Murofushi, N., Takahashi, N. (1980) Extraction, purification and identification, in Encyclopedia of Plant Physiology, Vol.9, Molecular aspects of plant hormones. ed. J.MacMillan, Springer-Verlag, Berlin, pp.113-201.

D2. Immunoassay Methods of Plant Hormone Analysis

Valerie C. Pence and John L. Caruso

Department of Biological Sciences, University of Cincinnati, Cincinnati, Ohio 45221, USA.

INTRODUCTION

There has been much recent interest in the application of immunoassays to the quantitation of plant hormones. Immunoassays are based on the ability of animals to produce proteins (*antibodies*) which recognize and bind to specific compounds (*antigens*) foreign to the animal. Such antibodies can be raised against plant hormones and used for hormone quantitation. The technique was first applied to plant hormones in an assay measuring the inactivation of hormone-labeled bacteriophage by antibodies raised against the hormone (12), but it has been radio-immunoassays (RIAs) and enzyme immunoassays (EIAs) which have been most widely used in this field (44).

The specificity and sensitivity of antibodies make immunoassays attractive for the quantitation of hormones from small amounts of plant tissues. Because of this, and the fact that large numbers of samples can be processed simultaneously, the expectation has been that immunoassays should allow rapid quantitation of hormone levels in relatively crude extracts, in contrast to the more extensive clean-up needed by most physico- chemical methods. In reality, application of these techniques has not been accomplished as easily as first envisioned, since in most cases, tissue extracts require some purification before they can be analyzed by immunoassay. As the characteristics of immunoassays for plant hormones become more well-defined, however, their use in providing information on the role of hormones in plant development should be enhanced.

This chapter will review some of the major features of these immunoassays, the characteristics and problems of the assays for specific hormones, and their application to plant tissue extracts. In addition, other possible applications of immunochemicals to plant hormone research will be discussed. For more detailed treatment of various aspects of the assays as well as specific methods, the reader is directed to the reports cited in the following text and recent reviews (43, 4).

ANTIBODY PRODUCTION

Since plant hormones are *haptens*, or small molecules which on their own cannot elicit an antibody response, they must be covalently linked to a large protein, commonly an albumin (BSA, HSA, OA), before use. A solution of the hormone-protein antigen is mixed with an oily adjuvant which slows the release of the antigen after injection. A series of injections are then made into an appropriate animal to stimulate antibody production.

Rabbits have been the most commonly used animal for *polyclonal antibody* (PAb), or antiserum, production. The animals are injected at several sites on the back and boosted for a series of weeks. Other injection sites, such as footpads, have been used, but are not recommended because of the discomfort to the animal. Bleedings can then be made from the ear, and the antibody-containing serum isolated. A number of spleen cells are involved in producing antibodies to the antigen, and these antibodies will differ in the *epitope*, or portion of the antigen molecule, against which they are directed. As a result, some of the antibodies may be less specific than others, if they react against an epitope shared by related compounds. The PAb preparation is thus a mixture of more and less specific antibodies, as well as other, unrelated, antibodies.

The production of *monoclonal antibodies* (MAbs) allows the isolation of individual antibodies and those which are most specific for the hormone can be selected (see 33, for review). Mice are most often used, and undergo a series of injections similar to those used for rabbits. The mouse is then killed and the spleen removed and aseptically macerated in a sterile medium. The spleen cells, which produce the antibodies, cannot continue growth in culture on their own. Thus, they are fused with mouse tumor (myeloma) cells, which grow well in culture. The fusion products, or hybridoma cells, are capable both of growing in tissue culture and of producing antibodies. A series of dilution cultures isolate clones from single *hybridoma* cells, each of which synthesizes only one form of an antibody against the hormone.

Antibodies from a variety of clones can be tested and those which are most specific for the antigen selected for use in the immunoassay. The hybridoma lines can be stored frozen, and, as needed, grown for the production of antibodies, either in tissue culture or by injection into mice. In the latter case, the cells grow into large tumors, the mice are killed, and tumor fluid (ascites) containing the antibodies is removed.

Dilutions of the rabbit serum or mouse ascites may be used directly in the immunoassays, or they can be partially purified by ammonium sulfate precipitation. In some cases, affinity purification is used to improve activity, particularly with PAbs.

The decision to use MAbs or PAbs must be based on a number of factors, including the greater initial cost and facilities needed for producing MAbs, as well as any improvement in specificity the MAbs may provide for measurement of a particular hormone. It is likely that, as the use of immunoassays in plant hormone research increases, so will the sources of antibodies. Thus, the production of MAbs or PAbs by every lab desiring to run immunoassays will not be necessary. Several commercially produced MAbs are currently available, providing the researcher with the choice of purchasing antibodies, a decision likely to be influenced by a laboratory's time and economic constraints.

THE IMMUNOASSAYS

Both RIAs and EIAs have been used in the measurement of plant hormones. These differ primarily in the tracer or labeled hormone used-- radioactive isotopes in the case of the RIA and enzymes in the EIA. The EIA has the advantage of avoiding the use of radioisotopes, and thus is safer and is less expensive, requiring less space and equipment. By utilizing enzymes, however, the EIA is subject to factors which affect enzyme activity, both in the synthesis of the hormone-enzyme conjugate and in the running of the assay (20). The enzyme tracer may also introduce into the assay the same chemical linkage bridge between hormone and protein which was present in the original antigen. When PAbs are used, antibodies may be present which recognize this bridge in preference to the hormone itself, causing preferential binding of the tracer over the free hormone, thus decreasing the sensitivity of the assay (44). In choosing between the two systems, available equipment, expertise, and antibodies must all be considered. Whereas both EIAs and RIAs may be viable options for most laboratory use, the non-isotopic character of EIAs should allow the expansion of the technique into less sophisticated settings.

There are a number of possible protocols for immunoassays, but Figure 1 outlines that which has been most commonly used for plant hormone assays. Labeled and unlabeled hormones are mixed with the antibodies to compete for binding sites. The antibodies with their bound components are then separated from the unbound hormone and tracer. The antibody may be precipitated with ammonium sulfate, or it may be adsorbed onto a polystyrene substrate. Unbound materials are removed by washing. An EIA in which the antibody is immobilized by adsorption onto a solid substrate is called an enzyme-linked immunosorbent assay, or ELISA, and these have been the EIAs used in plant hormone analyses, to date.

The amount of tracer is measured, either as counts per minute (RIA) or as enzyme activity (EIA). Tritium is the most commonly used radiolabel, although ^{125}I has also been used. The most commonly used enzyme, alkaline phosphatase (AP), is measured as activity in converting

242

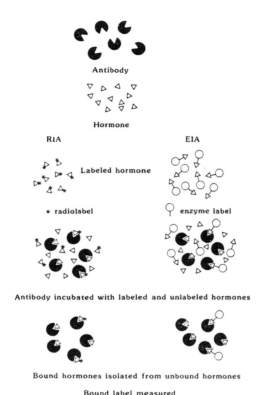

Fig. 1. Generalized scheme of immunoassays for plant hormones.

the substrate, *p*-nitrophenylphosphate, to *p*-nitrophenol, which is colorimetrically measured. The amount of bound tracer is inversely proportional to the amount of unlabeled hormone with which it competes in the incubation medium; thus, the color produced in the above reaction is also inversely proportional to the amount of unlabeled hormone present. Standard curves are often presented as the concentration of standard hormone vs. the logit B/B_0 ($=\ln {}^{(B/B_0)}/_{1-(B/B_0)}$), rather than absorbance or cpm, as this transformation increases the linearity of the curve.

IMMUNOASSAYS FOR PLANT HORMONE ANALYSIS

Both RIAs and EIAs have been developed for all four major classes of non-volatile plant hormones: indole-3-acetic acid (IAA), abscisic acid (ABA), cytokinins, and gibberellins (GAs). The sensitivities of these assays compare favorably with those of physico-chemical methods of detection, ranging from 0.3 pmol in the RIA for IAA to as little as 0.001 pmol in EIAs for GAs (Table 1).

Table 1. Some reported sensitivities of immunoassays for plant hormones.

Hormone	Assay	Sensitivity (pmol)	Reference
IAA	EIA	0.02	45
	RIA	0.3	19
ABA	EIA	0.02	21
	RIA	0.04	17
ZR	EIA	0.015	24
	RIA	0.03	40
GAs	EIA	0.001	2
	RIA	0.005	46

IAA

Two types of antigens have been used for generating antibodies to IAA. The first is an IAA protein conjugate synthesized using the Mannich formaldehyde reaction (28), which couples the ring of IAA to the carrier protein, presumably at the indolic nitrogen of IAA (IAA-N). The second is the coupling of IAA to the protein via the carboxyl group (IAA-C), with the mixed anhydride reaction (41). Because of the very different epitopes presented in the antigens produced by these two methods, the resulting antibodies differ in their cross-reactive properties. PAbs raised against the IAA-C antigen show a high degree of cross-reactivity with other carboxyl-linked IAA conjugates, including several naturally-occurring conjugates (Table 2). They are also poorly reactive with free IAA unless the charged carboxyl group is neutralized by methylation, and thus, samples are routinely methylated prior to the IAA-C assays. PAbs raised against the IAA-N antigen are much more specific, with major cross-reactivity seen only with the synthetic auxin naphthaleneacetic acid (Table 2). Such PAbs also recognize free IAA, and methylation, which commonly employs highly toxic and explosive diazomethane, can be avoided. Antibodies to IAA-N, however, have been reported to develop at much lower titers than IAA-C antibodies (28,41) and so have not been as widely used as the latter.

There have been several approaches to the problem of measuring free, as opposed to conjugated, IAA. The first has been the use of acidic partitioning of the extract with an organic solvent. At pH 3.0, much of the free IAA is separated from the conjugates. Secondly, MAbs against the IAA-C antigen can be selected which are much more specific than the IAA-C PAbs, although there may still be some cross-reactivity with selected conjugates, depending on the antibody (Table 2) (22, and this laboratory, unpublished). A third approach is the use of affinity purification to concentrate IAA-N PAbs, compensating for the low antibody titers often obtained. This is accomplished by washing the

Table 2. Comparison of cross-reactivities of antibodies raised against IAA.

Compound[1]	Cross-reactivity (%)			
	Polyclonal		Monoclonal	
	IAA-N[2]	IAA-C[2]	IAA-C[3]	IAA-C[4]
IAA	100	0.1	<0.01	0
IAA-CH$_3$	0.3	100	100	100
IAAsp	<0.1	0.4	–	–
IAAsp-CH$_3$	–	27	4.3	0.5
IAAla	<0.1	10	–	–
IAAla-CH$_3$	–	62	5.4	1.5
IAGly[5]	0.2	100	0.2	58
IAPhe	0.1	7	1.9	0.6
IAGlt	0.3	68	12.7	–
IAVal	<0.1	25	3.4	–
IALeu	<0.1	7	2.9	–
5-OH-IAA	0.3	<0.1	0.4	0.2
IAAmide	<0.1	17	0.3	1
IAN	<0.1	0.9	2.1	1
I-3-MetOH	<0.1	3	–	–
I-3-EtOH	<0.1	0.5	–	0.04
ILA	<0.1	<0.1	0.08	0.5
PAA	0.4	<0.03	0.02	0.01
IBA	<0.1	0.7	<0.01	1
NAA	22	0.2	1.4	0.1
2,4-D	0.09	0.06	0.7	0.01
TIBA	<0.1	<0.1	<0.01	–
Tryptamine	<0.1	<0.1	<0.01	–
Tryptophan	<0.1	<0.03	<0.01	0.04

[1] Abbreviations: IAAsp, indoleacetyl aspartate; IAAla, indoleacetylalanine; IAGly, indoleacetylglycine; IAPhe, indoleacetylphenylalanine; IAGlt, indoleacetylglutamic acid; IAVal, indoleacetylvaline; IALeu, indoleacetylleucine; 5-OH-IAA, 5-hydroxyindole-3-acetic acid; IAAmide, indoleacetamide; IAN, indoleacetylnitrile; I-3-MetOH, indole-3-methanol; I-3-EtOH, indole-3-ethanol; ILA, indolelactic acid; PAA, phenylacetic acid; IBA, indolebutyric acid; NAA, naphthaleneacetic acid; 2,4-D, 2,4-dichlorophenoxyacetic acid; TIBA, triiodo-benzoic acid.
[2] Reference 25.
[3] This laboratory, unpublished.
[4] Reference 22.
[5] The following compounds were methylated for the IAA-C assays.

antibody preparation through an agarose column to which IAA has been covalently linked with the same reaction used to produce the antigen (25). After elution from the column, the antibody shows marked improvement in the ELISA (Table 3).

Table 3. Comparison of crude and affinity purified IAA-N antibody preparations in the ELISA (25).

Antibody	Coating Concentration (μg/well)	Reaction Time (min)	Absorbance in ELISA[1]		% Inhibition by IAA
			Without IAA	With IAA[2]	
Crude	10	45	0.551	0.416	25
Affinity purified	0.8	38	1.871	0.249	87

[1] Absorbance is inversely proportional to free hormone concentration.
[2] Saturating concentration.

For EIAs, the enzyme tracer is conjugated to IAA either on the ring or via the carboxyl group, depending on which assay is used. The IAA-N-AP tracer is synthesized, as is the IAA-N antigen, using the Mannich reaction (25), but the IAA-C-AP has been produced using a carbodiimide (45) rather than the mixed anhydride reaction.

ABA

The quantitation of ABA by immunoassay has presented a problem similar to that seen with IAA. Conjugation of ABA to a protein carrier in the production of the antigen is most easily accomplished at the carboxyl group, using either the mixed anhydride reaction (42) or a carbodiimide (8, 36, 38). The resulting antibodies, however, cross react with naturally occurring ABA conjugates which are also linked at the carboxyl group. Antibodies to carboxyl-linked ABA have a higher affinity for ABA methyl ester than for free ABA, although the difference is generally less than 3-fold, much less than the difference seen between free and methylated IAA. In some reports, however, tissue samples are methylated in order to maximize the sensitivity of the assay (42). By linking ABA to the carrier protein at the C4' keto group, antibodies are obtained which recognize free but not conjugated ABA (39). The synthesis of the ABA-C_4' antigen is a multistep procedure, however, and has not been as widely used as carboxyl-linked ABA. High specificity for free ABA has also been obtained with MAbs raised against carboxyl-linked ABA (21).

Another consideration in the synthesis of antibodies to ABA has been their differing reactivities with (+)-ABA and (–)-ABA ((S)-ABA and (R)-ABA). In the first assays reported, PAbs were raised against antigen synthesized from the racemic mixture, (±)-ABA. When tested against the individual enantiomers, these antibodies showed a preference for

(–)-ABA, rather than the naturally-occurring (+)-ABA, thus greatly reducing the sensitivity of the assay (36). Synthesis of PAbs raised against small amounts of carboxyl-linked (+)-ABA, a much more expensive compound, showed very little reactivity with (–)-ABA (42). Alternatively, use of the C_4'-linked (±)-ABA generated antibodies with a strong preference for the (+)-ABA enantiomer (39). MAbs against carboxyl-linked (+)-ABA similarly show a preference for (+)-ABA (21).

As with IAA-C, enzyme tracers for ABA EIAs have been made with carbodiimides. An exception has been the conjugation of AP to a previously synthesized ABA-BSA conjugate using glutaraldehyde to link the two proteins (8).

Cytokinins

Antigens for the production of antibodies to cytokinins and enzyme tracers for cytokinin EIAs are synthesized using cytokinin-ribosides by a periodate reaction which breaks the sugar ring and links amino groups of the carrier protein to the resulting aldehydes. RIAs using PAbs have been reported for zeatin riboside (ZR) (35, 40), and 2-isopentenyladenosine (IPA) (16), and for the synthetic cytokinin benzyladenosine (BA) (6). EIAs have also been developed for ZR with PAbs (13, 24). In addition, MAbs have been raised against several cytokinins (e.g., 4, 47).

Antibodies for cytokinins are fairly specific for their "family" of cytokinins (e.g., zeatin, dihydrozeatin, etc.) and specificity has been only slightly improved by the production of MAbs (Table 4). In this laboratory, ammonium sulfate-precipitated anti-ZR PAbs have been found to be highly active, requiring as little as 0.8 μg protein per assay in an ELISA.

Table 4. Cross-reactivities of a polyclonal antibody and a monoclonal antibody raised against ZR.

Compound	Cross-reactivity: % of Zeatin Riboside	
	Polyclonal[1]	Monoclonal[2]
Zeatin riboside	100	100
Zeatin	88	42
cis-Zeatin	6.6	–
Dihydrozeatin	5.8	0.1
Dihydrozeatin riboside	0.4	0.2
Isopentenyladenine	0.2	–
Isopentenyladenosine	0.8	0.2
Benzyladenine	0.2	0.4
Benzyladenosine	0.3	–
Kinetin	0.05	0.08
Kinetin riboside	0.07	–
Adenine	0.05	0

[1] Antibody used for studies in reference (24).
[2] Reference (4).

A single bleed providing 11.5 ml of serum has supplied enough material to coat over 1500 96-assay plates. Similar high titers have been reported by others for ZR PAbs (13).

Gibberellins

Because of the large number of GAs, the specificity conferred by immunoassays is not always an advantage. Depending on the GA, antibodies raised to a particular GA may be very specific, or they may cross-react with a few related GAs. For example, antibodies which have been raised against GA_3 also recognize GA_7, GA_{20}, and to some extent GA_1 (1, 46) whereas those against GA_9 are quite specific (1). Since the specific GAs present in a tissue may not be known to the researcher, the antibody at hand may or may not be useful for quantifying GAs in a given system. Such antibodies have been used, however, in combination with techniques to separate the GAs in the tissue extract. In one study, after thin-layer chromatography, each peak co-chromatographing with a known GA was quantified with two different GA antibodies which were reactive with the GA thought to be present, but which were also known to have different cross- reactivities (3).

Gibberellin antigens and enzyme tracers have been synthesized via the carboxyl group using the mixed anhydride and carbodiimide reactions as with IAA (46) and samples are methylated with diazomethane before analysis. Since only small amounts of most GAs are available for use, the production of antibodies and development of immunoassays for several GAs has been demonstrated from only a few mg of starting material (1), as has the synthesis of the enzyme tracers for EIA (2).

USE OF THE IMMUNOASSAYS FOR ANALYSIS OF PLANT HORMONES

Immunoassays have the sensitivity to detect very small amounts of antigen, and thus offer the potential for measuring hormones from very small samples of plant tissues. Large numbers of immunoassays can be run in tandem, adding the advantage of speed to the technique. These characteristics have stimulated interest in EIAs and RIAs as methods for identifying differences in plant hormone levels in specific tissues.

Balanced against these expectations has been the realization that, although the antibodies used in immunoassays are highly specific for the antigen, other materials in plant extracts can competitively and non-competitively inhibit the antigen-antibody interaction, giving spurious readings (26, 37). These interactions can vary, not only with extracts of tissues from different species (7), but also with different treatments of the same extract (27), and, most likely, different experimental treatments of the same tissue. Thus, there is a need for validation of the assay with each type of extract examined.

Such validation may indicate the need for further purification of the extract, and, in some cases the extract preparation for an immunoassay may be comparable to that used for a physico-chemical analysis (32). If purification steps are added, larger amounts of tissue must be used initially to accommodate for loss, and an internal standard must be added to calculate the loss. Generally, these have been radiolabeled standards (e.g., 4), although analysis of a number of split samples with cold standard added to half can also give an average estimation of loss (38).

Validation of an immunoassay is best done by comparing the results to those obtained with an independent method, but there are several internal checks which can be used to identify problems in the assay. One of the most widely used is that of parallelism. An extract dilution curve which is not parallel to that of a standard hormone indicates the presence of interfering substances with which the antibody is interacting, and further clean-up of the tissue extract is needed (26). The amount of clean-up needed to obtain parallel dilution curves can differ with the assay. An acidic ether partitioning step eliminated non-parallel interference in extracts of crown gall tumors for the IAA-N PAb ELISA (25), but a crude extract gave a parallel dilution in the ZR ELISA (24) (Fig. 2).

Whereas nonparallel dilutions indicate interference, parallel dilutions cannot assure the absence of interference. Spiking samples with known amounts of standard can indicate interfering interactions, even with extracts which exhibit parallelism (see Table 5 and reference 15). In addition, compounds with binding affinities similar to that of the antigen can give parallel dilutions, as has been shown for NAA in the IAA-N RIA (28). In such cases, further purification of the sample is needed, and with

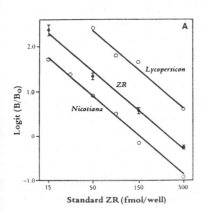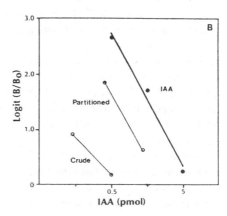

Fig. 2. A. Dilutions of tobacco (*Nicotiana tabacum*) and tomato (*Lycopersicon esculentum*) crown gall tumor tissue crude extracts compared with standard ZR dilutions in the ZR PAb ELISA. B. Dilution of tobacco crowngall tumor tissue extracts, both crude and acidic (pH 3.0) ether-partitioned, compared with standard IAA dilutions in the IAA-N PAb ELISA.

Table 5. Readings, using the IAA-N PAb and ZR PAb ELISAs of added hormone, in the presence of a constant volume of extract which had been tested for parallelism (see Fig. 2).

Hormone	pg Added (A)	pg[1] Read (B)	B/A
IAA Extract 1	88	149	1.69
	263	480	1.83
	875	1313	1.50
	2630	5100	1.94
Extract 2	88	292	3.32
	263	604	2.3
	875	1995	2.28
	2630	7420	2.82
ZR	18	31	1.74
	176	332	1.88

[1] Corrected for endogenous hormone.

compounds such as NAA, the interference may not be apparent until such purification is done.

The technique of successive approximation can be used to establish the amount of purification needed to eliminate interfering materials (30). The extract, with an internal standard, is put through a series of purification steps, and after each step, the amount of hormone in the original extract is estimated. If the amount of hormone estimated after step 2 is different from that estimated after step 1, step 2 is needed to eliminate some interfering materials which were present after step 1. If, however, the estimation of hormone amount does not change with further purification, the lesser amount of purification is sufficient.

Several laboratories have provided validation of immunoassays by independent means, such as gas chromatography-mass spectrometry (GC-MS), for particular tissues. Using a tetradeuterated, ring-labeled analog, a validation of the IAA-N RIA by GC-SIM-MS indicated that, in methanolic extracts of etiolated corn seedling shoots, subjected to both acidic and basic ether partitioning, the levels of free IAA measured by RIA agreed with those obtained by GC-SIM-MS (27). In contrast, levels of IAA in the extract after base hydrolysis, which was used to free conjugated IAA, were higher than GC-SIM-MS determinations. Chromatography on DEAE and LH-20 Sephadex removed the interfering compounds. These results emphasize that not only the tissue, but also treatment of the tissue during extraction, can influence performance in an immunoassay. Comparisons between GC-SIM-MS have also been made in this laboratory using IAA-C PAbs (5 and unpublished).

Good agreement has been observed between RIA and GC-SIM-MS determinations of a cytokinin in a cultured anise cell line (10) and of ABA in developing rapeseed embryos (Table 6) (11) and pepper leaves (31). In the latter case, the crude extract gave consistently higher RIA-determined values, and polyvinylpyrrolidone powder added to the assay mixture was effective in reducing this interference. GC (29) and HPLC (32) have also been used for confirming immunoassay results.

Table 6. ABA concentrations (ng mg^{-1} fresh weight) in developing embryos of rapeseed (*Brassica napus* L. cv. Tower) determined by RIA and GC-SIM-MS[1]. For the latter, a hexadeuterated analog of ABA, synthesized in this laboratory, was used as internal standard, and base peaks m/z 190 and 194 (a 4 carbon fragment with 2 deuteriums is lost during fragmentation) were used to quantitate endogenous ABA and analog, respectively, using the method of isotope dilution. Each of the mass spectrometric values is an average of two different experimental samples (11).

Days Post-anthesis	Assay	
	RIA	*GC-SIM-MS*
33	2.98 ± 0.45	2.83
38	3.28 ± 0.95	3.16

[1.] GC-SIM-MS = Gas chromatography-selected ion monitoring-mass spectrometry.

OTHER USES OF ANTIBODIES

Immunoaffinity Purification of Hormones

Immobilizing antibodies raised against plant hormones offers the possibility of sequestering the antigen from a complex mixture of substances, thus purifying and concentrating the hormone of interest from plant extracts (23). When eluted, the hormone can be quantitated by a variety of methods. Elution of hormone with an organic solvent may denature the antibody, although there have been reports of repeated methanol elutions from PAb-columns (9, 18) and numerous cycles of acetone elution from a MAb-column (44). Sepharose and microcrystalline cellulose, as well as other supports have been used, and columns as small as 1 ml have been effective. The capacity of a prepared column can be tested with radiolabeled antigen.

An immunoaffinity column can purify an antigen several hundred to a thousand-fold. When followed by HPLC and a simple detector device, such as UV, the results can be very impressive in terms of the purity of the samples (see Chapter E20, Figs. 7A and 8).

Immunocytochemical Localization of Hormones

Success has been achieved in using immunocytochemical techniques for the localization of small molecules in animal tissues and recent reports indicate that this is feasible for localization of plant hormones, as well. ABA was localized in *Chenopodium* by first coupling the hormone to cellular proteins with carbodiimide (34). Anti-ABA PAbs were applied to semithin sections and a light microscope was used to detect an insoluble product of the peroxidase-antiperoxidase (PAP) complex. An example of cellular localization of cytokinin was also reported using PAbs against DHZR on corn root tip sections. Specimens were fixed in formaldehyde, fresh frozen, or freeze-substituted (48). Use of the light and electron microscope to detect immunofluorescence and immunogold, respectively, revealed an absence of cytokinin in the quiescent center and the presence of labeling in surrounding meristematic cells. In both these instances of hormone localization, antisera were raised against haptens conjugated to a protein carrier, and thus bound hormone in the cell was recognized. The problems involved in localizing free hormone have not yet been overcome.

There are many variations in the preparatory steps necessary to localize an antigen, but in the simplest terms, it is as follows. Embedded or nonembedded tissue which has been fixed or frozen is sectioned to insure contact between antibody and antigen as well as to visualize the location of the antigen. A *primary antibody* (for example, mouse IgG MAb) with a well defined specificity for the hormone in question is applied to the section. Following washes which remove excess antibody, a *secondary antibody* (for example, goat anti-mouse IgG) conjugated to a fluorophore, such as fluorescein or L-Rhodamine, is applied. The secondary antibody attaches to the primary antibody and, following washes, the preparation is examined with a fluorescence microscope. A similar approach, minus the need for sectioning, was used to visualize the linkage of IAA to an artificial support. IAA was covalently linked at its carboxyl group to ω-aminopentyl-agarose gel beads by means of carbodiimide. Agarose beads with no IAA linked to them were used as a control (Fig. 3).

Nonspecific retention properties of either the primary or secondary antibody pose a major problem in the interpretation of localization of hormone. One obvious control is the omission of primary antibody. Another control is the prior binding of antigen to primary antibody to test for inhibition of the observed localization. Still another control is the use of preimmune serum in place of primary antibody. This control, however, is not practical if one is using MAbs, considering the limited supply of preimmune serum from a single mouse. When MAbs are used, it is probably better to consider using a nonimmune control such as P3-X63/Ag 8, which is a mouse myeloma clone with no known antigen.

FUTURE PROSPECTS

The use of immunological techniques for plant hormone research is likely to increase. Improvements in the methods of detection in the immunoassays will probably be refinements of an already useful

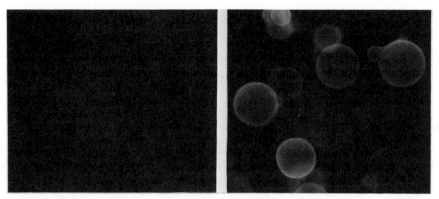

Fig. 3. Localization of IAA to ω-aminopentyl-agarose beads using an immunochemical technique. The control beads (left) received no IAA but did receive primary antibody (anti-IAA MAb; cross- reactivities listed in Table 2) followed by washing (with PBS and 0.1%BSA), then secondary antibody (goat anti-mouse IgG with L-Rhodamine), with more washes as above. For the treated sample (right), IAA was linked to the gel beads by a carbodiimide and treated with antibodies as in control. Gel preparations were examined with epiflouresence using a Zeiss Universal microscope with an excitation filter of 530 nm and a barrier filter of 470 nm. Fluoresced IAA-linked beads are evident (right; ×250).

technology. EIA is especially adaptable to a variety of detection methods, and a new one, electrochemical detection, is already in the early stages of use (14). An electrochemically detectable product is necessary. An example of this is phenol, which is produced from phenyl phosphate, an electroinactive and noninterfering chemical, through the activity of alkaline phosphatase, an enzyme already commonly used in EIA. Electrochemical detection, whether it be linked to liquid chromatography (LC-EC) or to a flow injection analysis (FIA-EC), will probably find increased use because of the low levels of detectability and relatively low cost of instrumentation. Other detection methods will surely appear in the years ahead.

Regarding the antibodies themselves, several are already available commercially, and, as immunoassays become more common, this availability should increase. Since immunoassays allow the analysis of multiple samples simultaneously, require less sophisticated equipment than most physico-chemical methods, and may require less extract purification (depending on the tissue), immunoassays should extend the capability for measuring plant hormones in a number of situations. As more systems are examined and validated, the types of interference presented by plant extracts should be better understood and clean-up procedures better defined and perhaps more standardized. Immobilized antibodies will likely be used more frequently for that clean-up, and immunoaffinity columns will probably be commercially available in the next few years. Similarly, other uses for antibodies raised against hormones will likely increase, including immunocytochemical localization of at least the bound hormones.

Immunological techniques present powerful tools for research into plant hormones. Immunoassays will not replace other methods of analysis but should provide complementary techniques for the quantitation of plant hormones.

Acknowledgements.

The authors wish to thank Timothy Stroup who provided IAA-C MAb cross-reactivity data, Leslie Leverone for the photographs showing immunocytochemical localization of IAA, and G. Douglas Winget for translating the computer disc.

References

1. Atzorn, R., Weiler, E. W. (1983) The immunoassay of gibberellins. I. Radioimmunoassays for the gibberellins A_1, A_3, A_4, A_7, A_9, and A_{20}. Planta 159, 1-6.
2. Atzorn, R., Weiler, E. W. (1983) The immunoassay of gibberellins. II. Quantitation of GA_3, GA_4, and GA_7 by ultra-sensitive solid-phase enzyme immunoassays. Planta 159, 7-11.
3. Atzorn, R., Weiler, E. W. (1983) The role of endogenous gibberellins in the formation of α-amylase by aleurone layers of germinating barley caryopses. Planta 159, 289-299.
4. Badenoch-Jones, J., Letham, D. S., Parker, C.W., and Rolfe, B.G. (1984) Quantitation of cytokinins in biological samples using antibodies against zeatin riboside. Plant Physiol. 75, 1117-1125.
5. Caruso, J. L., Pence, V. C., Stroup, T. L., Nienaber, M. A. (1985) Comparison of enzyme-linked immunosorbent assay (ELISA) and gas chromatography-mass spectrometry for the quantitation of indole-3-acetic acid. Plant Physiol. Suppl. 77, 2.
6. Constantinidou, H. A., Steele, J. A., Kozlowski, T. T., Upper, C. D. (1978) Binding specificity and possible analytical applications of the cytokinin-binding antibody, anti-N^6-benzyladenosine. Plant Physiol. 62, 968-974.
7. Crozier, A., Sandberg, G., Monteiro, A. M., Sundberg, B. (1986) The use of immunological techniques in plant hormone analysis. In: Plant growth substances 1985. pp. 13-21, Bopp. M. ed. Berlin, Springer-Verlag.
8. Daie, J., Wyse, R. (1982) Adaptation of the enzyme-linked immunosorbent assay (ELISA) to the quantitative analysis of abscisic acid. Anal. Biochem. 119, 365-371.
9. Davis, G. C., Hein, M. B., Chapman, D. A., Neely, B. C., Sharp, C. R., Durley, R. C., Biest, D. K., Heyde, B. R., Carnes, M. G. (1986) Immunoaffinity columns for the isolation and analysis of plant hormones. In: Plant growth substances 1985. pp. 44-51, Bopp, M. ed. Berlin, Springer-Verlag.
10. Ernst, D., Schafer, W., Oesterhelt, D. (1983) Isolation and quantitation of isopentenyladenosine in an anise cell culture by single-ion monitoring, radioimmunoassay and bioassay. Planta 159, 216-221.
11. Finkelstein, R. R., Tenbarge, K. M., Shumway, J. E., Crouch, M. L. (1985) Role of ABA in maturation of rapeseed embryos. Plant Physiol. 78, 630-636.
12. Fuchs, S., Haimovich, J., Fuchs, Y. (1971) Immunological studies of plant hormones. Detection and estimation by immunological assays. Eur. J. Biochem. 18, 384-390.
13. Hansen, C. E., Wenzler, H., Meins, F., Jr. (1984) Concentration gradients of trans-zeatin riboside and trans-zeatin in the maize stem. Plant Physiol. 75, 959-963.
14. Heineman, W. R., Halsell, H. B. (1985) Strategies of electrochemical immunoassay. Analytical Chem. 57, 1321A-1331A.

15. Jones, H. G. (1985) Correction for non-specific interference in hormone immunoassay. 12th International Conference on Plant Growth Substances, Heidelberg, Abstract no. 1105.
16. Khan, S. A., Humayun, M. Z., Jacob, T. M. (1977) A sensitive radioimmunoassay for isopentenyladenosine. Anal. Biochem. *83*, 632-635.
17. LePage-Degivry, M. T., Duval, D., Bulard, C., Delaage, M. (1984) A radioimmunoassay for abscisic acid. J. Immun. Methods *67*, 119-128.
18. MacDonald, E. M. S., Morris, R. O. (1985) Isolation of cytokinins by immunoaffinity chromatography and analysis by high-performance liquid chromatography radioimmunoassay. In: Methods in Enzymology, vol. 110. Steroids and isoprenoids, Part A. pp. 347-358, Law, J. H., Rilling, H. C., eds. New York, Academic Press.
19. Madej, A., Haggblom, P. (1985) Radioimmunoassay for determination of indole-3-acetic acid in fungi and plants. Physiol. Plant. *64*, 389-392.
20. Maggio, E. T. (1980) Enzymes as immunochemical labels. *In* Enzyme-immunoassay, pp. 53-70, Maggio, E.T., ed. Boca Raton, Florida, CRC Press, Inc.
21. Mertens, R., Deus-Neumann, B., Weiler, E. W. (1983) Monoclonal antibodies for the detection and quantitation of the endogenous growth regulator, abscisic acid. FEBS Lett. *160*, 269-272
22. Mertens, R., Eberle, J., Arnscheidt, A., Ledebur, A., Weiler, E. W. (1985) Monoclonal antibodies to plant growth regulators. II. Indole-3-acetic acid. Planta *166*, 389-393.
23. Morris, R. O., Akiyoshi, D. E., MacDonald, E. M. S., Morris, J. W., Regier, D. A., Zaerr, J. B. (1982) Cytokinin metabolism in relation to tumor induction by *Agrobacterium tumefaciens*. In: Plant growth substances, pp. 175-183, Wareing, P. F. ed. New York, Academic Press.
24. Pence, V. C., Caruso, J. L. (1986) Auxin and cytokinin levels in selected and temperature-induced morphologically distinct tissue lines of tobacco crown gall tumors. Plant Science (in press)
25. Pence, V. C., Caruso, J. L. (1986) ELISA determination of IAA using antibodies against ring-linked IAA. Phytochemistry (in press).
26. Pengelly, W. L. (1986) Validation of immunoassays. In: Plant growth substances 1985. pp. 35-43, Bopp, M. ed. Berlin, Springer-Verlag.
27. Pengelly, W. L., Bandurski, R. S., Shulze, A. (1981) Validation of a radioimmunoassay for indole-3-acetic acid using gas chromatography-selected ion monitoring- mass spectrometry. Plant Physiol. *68*, 96-98.
28. Pengelly, W., Meins, F., Jr. (1977) A specific radioimmunoassay for nanogram quantities of the auxin, indole-3-acetic acid. Planta *136*, 173-180.
29. Raikhel, N. V., Hughes, D. W., Galau, G. A. (1986) An enzyme-immunoassay for quantitative analysis of abscisic acid in wheat. In: Molecular Biology of Plant Growth Control, Fox, J. E., Jacobs, M., eds. New York, Alan R. Liss, Inc. (in press).
30. Reeve, D. R., Crozier, A. (1980) Quantitative analysis of plant hormones. In: Encyclopedia of plant physiology, new series, Vol. 9, pp. 203-280, MacMillan, J., ed. Heidelberg, Springer-Verlag.
31. Rosher, P. H., Jones, H. G., Hedden, P. (1985) Validation of a radioimmunoassay for (+)-abscisic acid in extracts of apple and sweet-pepper tissue using high-pressure liquid chromatography and combined gas chromatography-mass spectrometry. Planta *165*, 91-99.
32. Sandberg, G, Ljung, K., Alm, P. (1985) Precision and accuracy of radioimmunoassay in the analysis of endogenous 3-indoleacetic acid from needles of Scots pine. Phytochem. *24*, 1439-1442.
33. Siddle, K. (1985) Properties and application of monoclonal antibodies. In: Alternative immunoassays, pp. 13-37, Collins, W. P. ed. New York, John Wiley & Sons.
34. Sotta, B., Sossountzov, L., Maldiney, R., Sabbagh, I., Tachon, P., Miginiac, E. (1985) Abscisic acid localization by light microscopic immunochemistry in *Chenopodium polyspermum* L. J. Histochem. Cytochem. *33*, 201-208.

35. Vold, B. S., Leonard, N. J. (1981) Production and characterization of antibodies and establishment of a radioimmunoassay for ribosylzeatin. Plant Physiol. 67, 401-403.
36. Walton, D., Dashek, W., Galson, E. (1979) A radioimmunoassay for abscisic acid. Planta 146, 139-145.
37. Wang, T. L., Griggs, P., Cook, S. (1986) Immunoassays for plant growth regulators - a help or a hindrance? In: Plant growth substances 1985. pp. 26-34, Bopp, M. ed. Berlin, Springer-Verlag.
38. Weiler E. W. (1979) Radioimmunoassay for the determination of free and conjugated abscisic acid. Planta 144, 255-263.
39. Weiler, E. W. (1980) Radioimmunoassays for the differential and direct analysis of free and conjugated abscisic acid in plant extracts. Planta 148, 262-272.
40. Weiler, E. W., (1980) Radioimmunoassays for trans-zeatin and related cytokinins. Planta 149, 155-162.
41. Weiler, E. W. (1981) Radioimmunoassay for pmol-quantities of indole-3-acetic acid in plant extracts. Planta 153, 310-325.
42. Weiler, E. W. (1982) An enzyme-immunoassay for cis-(+)- abscisic acid. Physiol. Plant. 54, 510-514.
43. Weiler, E. W. (1984) Immunoassay of plant growth regulators. Annu. Rev. Plant Physiol. 35, 85-95.
44. Weiler, E. W., Eberle, J., Mertens, R., Atzorn, R., Feyerabend, M., Jourdan, P. S., Arnscheidt, A., Wieczorek, U. (1986) Antisera- and monoclonal antibody based immunoassay of plant hormones. In: Immunology in plant science, pp. 27-58, Wang, T. L., ed. Cambridge, Cambridge University Press.
45. Weiler, E. W., Jourdan, P. S., Conrad, W. (1981) Levels of indole-3-acetic acid in intact and decapitated coleoptiles as determined by a specific and highly sensitive solid-phase enzyme-immunoassay. Planta 153, 561-571.
46. Weiler, E. W., Wieczorek, U. (1981) Determination of femtomol quantities of gibberellic acid by radioimmunoassay. Planta 152, 159-167.
47. Woodside, M. L., Latimer, L. J. P., Janzer, J. J., McLennan, B. D., Lee, J. S. (1983) Characterization of monoclonal antibodies specific for isopentenyl adenosine derivatives occurring in transfer RNA. Biochem. Biophys. Res. Commun. 114, 791-796.
48. Zavala, M. E., Brandon, D. L. (1983) Localization of a phytohormone using immunocytochemistry. J. Cell Biol. 97, 1235-1239.

E. THE FUNCTIONING OF HORMONES IN PLANT GROWTH AND DEVELOPMENT

E1. Ethylene in Plant Growth, Development, and Senescence

Michael S. Reid

Department of Environmental Horticulture, University of California, Davis, California 95616, USA.

INTRODUCTION

Amongst hormones in both plant and animal kingdoms, ethylene, a gaseous hydrocarbon, is unique. Despite its chemical simplicity, it is a potent growth regulator, affecting the growth, differentiation, and senescence of plants, in concentrations as little as 0.01 μl/l. As recently as twenty years ago, plant physiologists were divided as to whether this gas, which had been shown to have a range of striking effects on plant tissues, could properly be called a hormone. Since then, the advent of gas chromatographic means of detecting and measuring ethylene, the elucidation of its biosynthetic pathway, and the discovery of potent regulators of its production and action, have provided powerful tools for physiologists to explore the role of ethylene in plant growth and development. Ethylene is now considered to be one of the important natural plant growth regulators, and the literature abounds with reports of its effects on almost every phase of the life of plants. Although the majority of studies have concentrated on particular processes, particularly fruit ripening, flower senescence, and abscission, many other reported responses of plants to ethylene may be important parts of normal growth and development. The proceedings of several symposia (8,18,19,38,54,62) and a number of reviews (1,17,39,45,60,68) provide an excellent background in the subject; the pathway of ethylene biosynthesis and the role of ethylene in leaf senescence are discussed in Chapters B4 and E14 respectively. The aim of this chapter is briefly to review present understanding of the role of ethylene in plant growth and development, highlighting effects of this gas that have received less attention in previous reviews.

What are the selective advantages of this gaseous compound that led to its widespread adoption by plants as a hormone? Although the normal pathway of ethylene biosynthesis in plants is a complex, carefully regulated, series of enzymatic processes, ethylene is also a common product of oxidation of many types of organic molecules. In particular, peroxidation of unprotected long-chain fatty acids can generate substantial quantities of ethylene. It is conceivable that there was a selective advantage to primitive plants of the ability to recognize and respond to situations where stresses such as high temperature, high light, or attack by microorganisms caused the generation of ethylene gas. One can imagine that once such a response system was selected, there would be a further advantage to plants of developing a controlled system of biosynthesis that would enable this new regulator to be produced in response to a variety of stimuli, not simply external oxidizing stresses.

The gaseous nature of ethylene may also explain its manifold hormonal actions in plants. Lacking the nervous system of the animal kingdom, plants appear poorly equipped to sense distant stimuli. The ready diffusion of ethylene through the intercellular spaces may, for example, act as a signal of damage, stress, or physical contact. Similarly the diffusion of ethylene may be an important part of the coordination of ripening of quite different tissues in ripening fruits. Another advantage of ethylene's diffusibility might be the fact that its concentrations in the tissue can be altered simply by changing the rate of synthesis. Whereas reducing the concentration of other hormones requires metabolism or detoxification of that already in the tissue, ethylene diffuses away, and the concentration in the tissue rapidly reaches the equilibrium between the rate of production and the rate of diffusion. One can imagine that rapid growth responses such as response to contact stimuli (7) and the inhibition or acceleration of shoot elongation (17) would benefit from such a rapid and sensitive response system.

In contrast to volatile insect pheromones which function as long-distance sex attractants, ethylene does not appear to be utilized by plants as a "long distance" hormone, despite its ability to to be transported over long distances by diffusion or air movements. There is no evidence, for example, that ethylene levels in orchards are sufficiently augmented by the presence of some ripening fruits to initiate the ripening of others. Of course, the diffusibility of ethylene *is* a frequent problem for operators of greenhouses, cool stores, and transportation facilities, who must avoid exposing sensitive crops (such as lettuce and some cut flowers) to products of incomplete combustion, gases from ripening rooms, or ripening fruits.

M. S. Reid

TECHNIQUES

Because ethylene is a common environmental pollutant, plants are often accidentally exposed to it in physiologically active concentrations, sometimes with devastating results (Fig. 1). Such effects are often then studied in the laboratory, where ethylene can easily be applied, either as a gas, or by utilizing Ethephon (2-chloroethyl phosphonic acid), a compound which decomposes to release ethylene at physiological pH. Almost every phase of plant growth and development has been shown to be affected by ethylene in some plant or other. Of course, this does not mean that any particular response is a general effect on all plants, nor that ethylene is a natural regulator of the affected process. A satisfactory demonstration of a natural role for ethylene should also include evidence that ethylene production increases prior to, or concomitantly with, the process being studied, or that ethylene sensitivity of the tissue increases at that time. In addition, application of stimulators and inhibitors of ethylene production and action should, respectively, accelerate and retard the process.

Physiologists working with ethylene are particularly fortunate in the techniques that they have available for testing its role in plant growth and development. The gas chromatograph, which spurred the tremendous advances in our knowledge of ethylene effects during the last twenty

Fig. 1. Plants grown in a greenhouse heated with unvented gas heaters. Ethylene concentrations in the greenhouse air were greater than 0.5 μl/l. (a) defoliation of *Pisonia*; (b) discoloration of *Asplenium*; (c) Yellowing and defoliation of *Schefflera*; (d) Epinasty of *Syngonium*.

years, has been refined by the addition of photo-ionization detectors, which enable even more sensitive measurement of ethylene, permitting determination of resting production rates where before we had to content ourselves with that frustrating concentration, "trace amounts".

Modulators of Ethylene Biosynthesis and Action

The elucidation of the pathway of ethylene biosynthesis (67) included the discovery of the immediate precursor to ethylene, 1-aminocyclopropane-1-carboxylic acid (ACC). This compound is an excellent exogenous precursor for ethylene, and enables application of "ethylene" in liquid media without the disturbing effects of pH which accompany the use of Ethephon (50), our previous best choice. In addition, we have some useful inhibitors of ethylene production. Aminoethoxyvinylglycine (AVG) and its analogues inhibit ACC synthase, the enzyme that converts S-adenosyl methionine (SAM) to ACC. These chemicals are effective at vanishingly low concentrations (3), which is fortunate, since they are very expensive, when available. Amino(oxyacetic) acid (AOA), another inhibitor of pyridoxal phosphate-requiring enzymes like ACC synthase, is also relatively effective (21). Although rather toxic, it is freely available, and inexpensive. Inhibitors of the terminal step in ethylene biosynthesis, the so-called Ethylene-Forming Enzyme (EFE) include free-radical scavengers such as sodium benzoate, high temperature, uncouplers, low oxygen, and Co^{++} (67).

For years, ethylene physiologists used CO_2 as an inhibitor of ethylene action (11) in testing the involvement of ethylene in processes they were examining. Although many recognized the possible other explanations of CO_2 effects (e.g., inhibition of respiration, effects on a range of enzymes, on cellular pH, photosynthesis and stomatal aperture), it was all that was available. The discovery that Ag^+ was a potent and specific inhibitor of ethylene action (5), and formulation of this ion in the stable, mobile and non-phytotoxic thiosulfate complex (STS), provided an unmatched tool for probing the involvement of ethylene in almost any phase of plant growth and development (61). It seems surprising that many workers studying ethylene effects fail to utilize this tool in exploring ethylene responses.

Another inhibitor with some unique advantages is 2,5-norbornadiene (NBD), a cyclic olefin which appears to compete with ethylene at its binding site (57). This chemical has the advantage over Ag^+ of being volatile, and can therefore be applied in experiments where Ag^+ cannot be used, for example because of phytotoxicity, or because of the mass of the tissue under study. Its volatility also allows experiments where it is desirable to treat with an inhibitor only temporarily. The disadvantages of NBD are that relatively high concentrations of this material (ca. 2,000 µl/l) are required to overcome some ethylene responses, that it can be difficult to design a protocol to maintain a steady treatment concentration (53), and that it has an offensive and rather pervasive aroma.

Being volatile itself, ethylene can also be applied for short periods, a property of this hormone of which advantage is all too rarely taken. Frequently, however, the "hypobaric" reduction of tissue ethylene concentrations has been used to test for ethylene involvement in plant responses. In this technique, reduction of the total pressure to 0.1 atm and supplying pure oxygen places the tissue in a normal oxygen partial pressure where the diffusivity of ethylene is greatly increased. Hypobaric ethylene removal is relatively difficult, technically, because of the problems of providing appropriate vessels, of controlling pressures, and of maintaining humidity at low pressures.

ETHYLENE RESPONSES

Release of Dormancy

The ability of many plants to grow in regions with adverse climates, such as freezing winters, or hot dry summers, is closely tied to their ability to suspend operations during adverse weather, and to recommence, often quite rapidly, when the weather improves. The timing and the mechanism of the reactivation of growth and development is still imperfectly understood, but in some cases it appears that ethylene may play some role.

The dormancy of many seeds can be broken by application of growth regulators. In a number of species, ethylene application stimulates the germination of dormant seeds (36,59). A number of authors have demonstrated substantial production of ethylene by germinating seed, but it is probable that this ethylene is the *result* of the stresses of germination, and therefore not associated with release of dormancy.

The flowering of many bulb crops is controlled by a proper schedule of storage at high and low temperatures to break dormancy and initiate flower development. For years it has been known that ethylene pollution of the storage area can cause flower malformations or abortion (termed "blasting") in these crops (16). It has been now shown (30) that exposure of dormant iris, narcissus, tulip, freesia and gladiolus propagules to smoke or to ethylene *at the appropriate time* will hasten shoot and root growth, shorten the time to flowering, and increase the number of small propagules which successfully flower (Fig. 2). The mechanism by which ethylene "breaks" the dormant state of these propagules is still unexplored. The pronounced effect of ethylene on sprouting of potatoes (55) is associated with increased respiration and mobilization of carbohydrates (28). It is conceivable that such mobilization is the basis for the shorter dormant period and improved flowering of bulbous crops treated with ethylene.

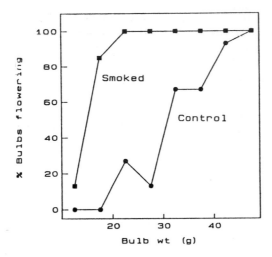

Fig. 2. Effect of ethylene exposure prior to forcing on the flowering of bulb crops. Cured bulbs of *Narcissus tazetta* cv. Soleil d'Or were graded by weight and exposed to .."smoke from smouldering wood and fresh leaves for several hours on each of 10 consecutive days". The bulbs were then stored, and planted in the normal manner. (Redrawn from the data of Imanishi, H. (1983) Scientia Hortic. *21*, 173-80)

Shoot Growth and Differentiation

The period between imbibition of a seed and the unfolding of the first true leaf is the most perilous time in the life of any green plant. Even if the succulent new plant survives the attacks of pests and pathogens, its path to the sun may be frustrated by barriers in the soil. The dramatic effects of ethylene on the growth and development of etiolated seedlings were the basis for the discovery by Neljubov (42) of the physiological action of ethylene. The so-called "triple response" involves a reduction in elongation, swelling of the hypocotyl, and a change in the direction of growth. This response may well be the means by which seedlings grow around obstacles in the soil (22). It presumably results from a redistribution of auxin in response to the ethylene induced by the contact stress between the seedling and the obstacle.

In another "avoidance" mechanism, some aquatic plants, such as star-wort (31,32), and rice (35), grow much faster when exposed to ethylene. In the star-wort, the compressed internodes of the floating rosette that forms the surface portion of the plant start to elongate very rapidly following submergence, or treatment with ethylene (Fig. 3). Not only is the ethylene content of the aerial spaces in the submerged shoots high, but also the effect of submersion can be eliminated by a pretreatment with Ag^+. The data are consistent with the hypothesis that increased intercellular ethylene content is the signal that the plant has been flooded. Rapid diffusion into the air keeps intercellular ethylene low when the plants are above water. Flooding (and the sparing solubility of ethylene in water) results in a rapid rise in intercellular ethylene content, and a growth response directed to returning the leaves to an aerial environment.

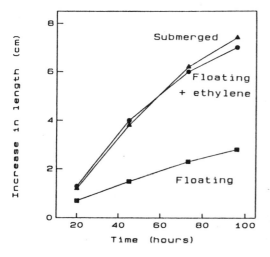

Fig. 3. The effect of ethylene (1.5 µl/l) or submergence on stem extension in the star-wort (*Callitriche platycarpa* L.) Detached rosettes were placed in half-strength Hoagland's solution and treated in various ways. Length of the plant was measured at intervals. From (32).

One of the classic symptoms of ethylene exposure in plants is the downward growth of the petioles, termed "epinasty" (Fig. 4). This symptom is considered to result from a redistribution of auxin in response to the ethylene treatment; increased auxin in the upper part of the petioles causes increased growth there, and a consequent downward bending of the petiole. Although epinasty is typically a symptom of inadvertent exposure of plants to ethylene, it may sometimes function as an avoidance mechanism. Epinasty is a symptom common to plants that are flooded, or stressed in other ways. The epinastic curvature of leaves of flooded plants is now known to be due to the transport of ACC from the roots to the petioles, and its conversion, there, to ethylene (9). It is possible that other epinastic responses to stress are also mediated via enhanced ethylene production in the petiole.

Fig. 4. Epinasty in poinsettia (*Euphorbia pulcherrima*) plants treated with ethylene at different concentrations. The bracts were removed from the plants to demonstrate the curvature of the petioles.

Stress ethylene may also modify shoot growth. The reduced growth and improved trunk diameter of unstaked shade trees is probably due to ethylene produced in response to flexing stress (41). Tree trunks treated for an extended period with ethylene are of markedly greater diameter than air-treated controls (Fig. 5). The site of production of the ethylene induced by flexing, and the way in which it exerts its effects have not yet been determined.

Fig. 5. Effect of ethylene on the growth of pine tree trunks. The portion of the trunk between the arrows was treated with air (A) or 5 to 20 μl/l ethylene (B) for 99 days. (Photo, Neel).

Low concentrations of ethylene applied to plants of the "diageotropica" tomato, a mutant which normally grows horizontally, partially restored the plants to a normal growth habit. Zobel (70) suggested that this mutant was unable to produce ethylene, and hypothesized that basal ethylene production is required for normal growth. Other workers have shown that the mutant can produce ethylene at normal rates (33), and suggest that its response may be the result of insensitivity to ethylene, or possibly auxin.

In our studies, we have found that early application of Ag^+ as STS to potted plants of various taxa has no perceptible effect on growth, development, and flowering, while effectively inhibiting the eventual abscission of flower petals (12). Since normal growth continues even when the action of ethylene is strongly inhibited, it seems that ethylene is not involved in normal growth and development. The curious behavior and ethylene response of the *diageotropica* mutant must relate to modulation of the effects of growth regulators other than ethylene.

Root Growth and Differentiation

Roots, too, respond markedly to the ethylene content of their environment. This is of particular importance, because the soil atmosphere can contain quite high concentrations of ethylene (31). Typically, root growth is stimulated at low concentrations of ethylene, and inhibited at higher concentrations (32, Fig. 6).

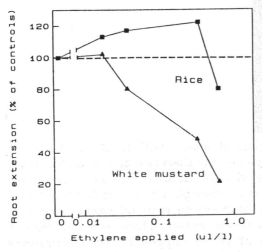

Fig. 6. Elongation of primary roots of rice and white mustard after 24 h treatment with ethylene at various concentrations. Redrawn from (32)

The anatomy of roots can also be affected by their ethylene content. A particularly good example of this response is the development of aerenchymatous roots in flooded maize plants (32). This adaptive response to increased intercellular ethylene can also be induced by ethylene treatment and prevented by Ag^+. Studies of the induction of alcohol dehydrogenase in flooded roots of mutant corn plants have suggested that the response to flooding may be to ACC rather than to ethylene (4), an observation that warrants further examination.

There has been little study of another interesting response of roots of a number of species to ethylene: the massive production of root hairs (Fig. 7). It is not yet known whether ethylene is involved in the natural development of root hairs. However, it is conceivable that in this phenomenon, too, ethylene acts as a signal, indicating that the root is well embedded in the soil.

Responses to Physical Stimuli

Plant physiologists have long been fascinated by the ability of plants to respond to physical stimuli. Although a number of these reponses (for example, those to gravity and light) appear to be mediated through changes in the distribution of auxin, ethylene has been implicated as the active agent in some cases (7).

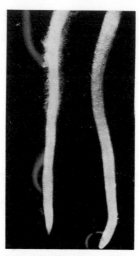

Fig. 7. Effect of treatment for 4 days with 1.5 µl/l ethylene (left) or air (right) on the formation of root hairs on mung bean roots. (Photo, Robbins)

Even quite gentle stimulation of plants will result in increased production of ethylene (26). Ethylene may therefore be a mediator in the response of some plants to tactile stimuli. In many climbing vines, roots attach the vine to supporting walls or other plants by modified root hairs. The stimulation of root hair development on roots of plants exposed to ethylene (Fig. 7) suggests that localized contact-induced ethylene may be the signal for the elaboration of these supports. In preliminary experiments, we have shown that ethylene treatment of *Philodendron* induces a felting of roothairs on the adventitious roots at the plant's nodes.

Adventitious Rooting

Ever since it was demonstrated that application of high concentrations of auxin will stimulate ethylene production, researchers have questioned the relative roles of auxin and ethylene in the formation of adventitious roots during auxin-stimulated rooting of cuttings. The possible involvement of ethylene is strengthened by reports that rooting of hard-to-root plants can sometimes be improved by scarification (wounding) of the base of the cuttings, a treatment that would induce production of wound ethylene. Studies of the effects of ethylene, ethylene producing compounds, and inhibitors of ethylene synthesis and action on rooting of cuttings have led to conflicting interpretations of ethylene's role in the process. Using a rooting chamber which allowed us to apply ethylene to the base of mung bean cuttings and not to the upper portions, we showed (53) that ethylene does indeed stimulate adventitious rooting (Fig 8). Since application of inhibitors of ethylene synthesis and action to similar cuttings reduced rooting, we concluded that ethylene plays a role in adventitious rooting of untreated mung bean cuttings. Its role in rooting of other cuttings is not yet known.

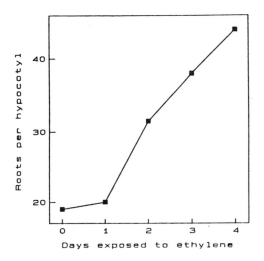

Fig. 8. Effect of period of basal application of dissolved ethylene (equilibrated with 6 µl/l in air) on the number of adventitious roots on the hypocotyl of mung bean cuttings after 4 days. Redrawn from (53)

Abscission

Amongst the most spectactular and commercially important effects of ethylene is the stimulation of abscission. Physiologists are still divided about the role of ethylene in natural abscission, but the majority of the data on hormonal control of abscission are consistent with the following hypothetical scheme (46):

- A gradient of auxin from the subtended organ to the plant axis maintains the abscission zone in a nonsensitive state. This gradient is itself maintained by factors which inhibit senescence of the organ. Thus, auxins, cytokinins, light, and good nutrition tend to reduce or delay abscission.

- Reduction or reversal of the auxin gradient causes the abscission zone to become sensitive to ethylene. Application of auxin proximal to the abscission zone, removal of the leaf blade, or treatments which accelerate its senescence (shading, poor nutrition, ethylene) therefore hasten abscission. Ethylene, or stresses which enhance its production, also may hasten abscission by reducing auxin synthesis and/or interfering with its transport from the leaf. Where ABA stimulates abscission, it may do so by stimulating ethylene production, or by interfering with the production, transport, or action of auxin.

- Once sensitized, the cells of the abscission zone respond to low concentrations of ethylene, whether exogenous or endogenous, by the rapid production and secretion of cellulase and other hydrolytic enzymes, and subsequent shedding of the subtended organ.

This scheme suggests a key role in abscission for ethylene, a view that is still somewhat controversial (56). The role of ABA in abscission is also still in question, but this hormone has been shown to be of primary importance in seed shedding in grasses (44).

Flower Induction

The induction of flowering in pineapple by ethylene or ethylene-producing chemicals is of considerable commercial importance. Other tropical fruits (mango, wax apple) also appear to respond to ethylene. In mango orchards it has long been the practise to hasten flowering by "..keeping smoky fires going for periods of six to seventeen days in such a way that a dense smoke is passing up through the trees" (14). This treatment ensures flowering in climates where trees do not experience drought or chilling stresses (which might be suggested to act through stimulation of stress ethylene production). No information is yet available on the mechanism by which ethylene induces flowering in these species. Application of high concentrations of ethylene during the inductive long night has been shown to *prevent* the induction of flowering in *Pharbitis nil* (58). This response is also unexplained, but we are presently using it as a probe to examine the molecular basis of the induction of flowering in this species.

Sex Expression

Not long after Ethephon became available commercially, it was demonstrated that application of this material to seedlings would dramatically change the ratio of male to female flowers in members of the Cucurbitaceae (1). Such applications were used to accelerate fruit production in lines of cucumber that naturally produced a number of male flowers before any female flowers appeared. It seems as though ethylene may be an important part of the natural determination of flower sex. Application of Ag+ has been shown to induce male flowers on the gynoecious cucumber cultivars used to produce F1 hybrid seed, and is now used commercially to permit propagation of these lines. AVG induces a similar response, but application of Ethephon to AVG-treated plants does not overcome the AVG effect (29). The suggestion that ACC, rather than ethylene, is involved in determination of flower sex requires further investigation.

Flower Opening

Ethylene may be involved in different ways in the opening of flowers. Although low concentrations of ethylene accelerate the opening of carnation buds (13), similar concentrations strongly inhibit the opening of tight rose buds (23, Fig. 9).

De Candolle (15) suggested that the ..."unusual movements (of the sexual organs of several flowers)...interest the physiologist on two counts, firstly as a proof of plant sensitivity, and secondly as an indication of the analogy existing between these organs and the corresponding organs of animals"! The rapid growth of the filaments in intact flowers of

Gaillardia. can also be seen in detached filaments treated with auxin (37). The hypothesis that auxin-induced ethylene synthesis plays a part in this rapid growth needs to be tested by exploring the effects of some of the specific inhibitors of ethylene synthesis and action that are now available.

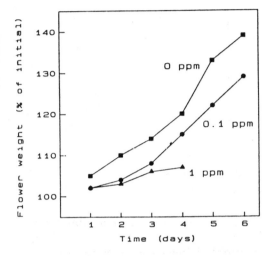

Fig. 9. Effect of ethylene on the opening of tight rose buds (cv. Sonia). Bud opening was measured by changing fresh weight of the flower. Redrawn from (23)

Flower Senescence

The bulk of the research on the physiology of flower senescence has been carried out on a limited number of species; morning glory (*Ipomoea*) and carnation (*Dianthus*) have been the most thoroughly studied (24). In these flowers, senescence is associated with a burst of ethylene production, and can be induced by treatment with exogenous ethylene or ACC and prevented by pre-treatment with inhibitors of ethylene synthesis or action. We have used *Petunia hybrida* L. (cv. Pink Cascade) as a model system for studying some aspects of flower senescence (66). These flowers are an ideal experimental system in that they are easily produced in large numbers, they show a precise timing of the start of senescence, and their color changes from pink to purplish-blue during senescence. Like climacteric fruits, ethylene-sensitive flowers appear to become more sensitive to ethylene as they age. After a 24 h treatment with different concentrations of ethylene, petunia flowers were of markedly different colors, depending on their age and on the concentration of ethylene with which they were treated (Fig. 10). Opening buds were hardly affected by 1 µl/l ethylene, whereas flowers that had been open for 1 day responded nearly as much as they did to 50 µl/l.

The inhibition of flower senescence by Ag+ has formed the basis of an effective practical treatment for cut flowers (49). Commercial growers of

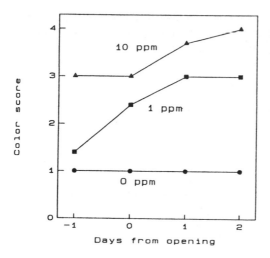

Fig. 10. Changes in the sensivity of aging *Petunia* flowers to ethylene. Flowers of different ages were harvested at the same time, and exposed to different concentrations of ethylene for 24 h. High color scores (normal pink = 0, completely blue = 4) indicate senescence. Redrawn from (66)

carnations commonly treat the flowers with STS after harvest to inhibit ethylene action. This treatment can extend the life of the flowers as much as 4-fold (Fig. 11).

We have studied the control of ethylene synthesis in senescing flowers (10,65); in a pattern similar to that seen in many fruits, the onset of senescence is associated with a coordinated increase in the activities of ACC synthase and of the EFE, the enzyme system converting ACC to ethylene. Detailed study of ethylene production in aging carnation petals has revealed that the biosynthesis of ethylene is separated between the base and the upper portion of the petal (40). Basal portions of the petals produce little ethylene when treated with ACC. Upper portions produce

Fig. 11. Effect of pretreatment with STS (4mM Ag⁺, 10 min) on carnation flowers held in deionized water at 20°C for 14 days.

little ethylene on their own. These results are interpreted to indicate that the majority of the ACC synthase is located in the lower portion of the petal, and that the EFE is largely located in the upper portion.

The spectacular effects of ethylene on carnations and some other flowers, and the benefits obtained by pretreating them with Ag+ have given rise to the erroneous generalization that flower senescence is caused by ethylene. For many flowers this is not true. In our experience, ethylene will not cause petal wilting in many flowers: rose, gerbera, and chrysanthemum, to name some of the most important. Nor will pretreatments with inhibitors of ethylene synthesis or action extend the vase life of these and of many other flowers. These observations suggest that the ethylene-mediated senescence of carnation and morning glory is only one type of flower senescence. There has, as yet, been no study of the control of senescence in those flowers whose longevity is not affected by ethylene.

Pollination

The rapid senescence of some flowers following their pollination has provided an interesting experimental system for examining the control of flower senescence (43). Pollination causes a very rapid increase in ethylene production, first by the gynoecium, and then by the petals (Fig. 12).

A number of workers, intrigued by the rapidity of the pollination response, have sought the nature of the pollination stimulus. Auxin, which is present in high concentrations in pollen, has been suggested as a possible pollination stimulus, but it does not move in the stigmas of carnations (48). We demonstrated that the stimulus was not ethylene *per se* because pollinated flowers wilted just as rapidly when we placed a tube over the pollinated stigmas and aspirated the tube to remove ethylene

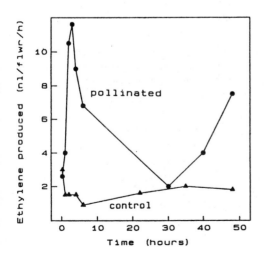

Fig. 12. Effect of pollination on ethylene production in *Petunia* flowers. From (66)

produced by the gynoecium (63). The finding that pollen of many species contains substantial concentrations of ACC (64) led us to hypothesize that ACC might be the stimulus for the pollination response. The stigma of carnation flowers contains a much higher activity of the EFE than the other tissues of the flower, and is therefore well prepared for rapid conversion of ACC to ethylene.

Although there is sufficient ACC in the pollen to explain the initial burst in ethylene production by the gynoecium, the role of pollen ACC in the early ethylene production by pollinated flowers has been questioned (27). Cosmos pollen, which contained very little ACC, greatly stimulated ethylene production by *Petunia* stigmas. Pretreatment of stigmas with AVG prevented the pollen-stimulated burst of ethylene production, suggesting that it comes from newly synthesized ACC (Fig. 13). These data suggest that the early effects of pollination are related to some other interaction between the pollen and the style.

Whatever the role of pollen ACC, there is increasing evidence that synthesis and transport of ACC plays and important part in the post-pollination events. The increased ethylene production by the gynoecium following pollination or wounding indicates an increase in the activity of ACC synthase (since the EFE is constitutive and in high activity in this organ) and we have demonstrated such an increase in pollinated Petunia flowers. Using radioactively labelled ACC, we showed that ACC applied to the stigma is transported as ACC to the petals (48), and have suggested that ACC may be the mobile pollination stimulus. The finding that pre-treatment of the stigma with AVG delayed pollination-induced senescence (27) supports this suggestion. In cyclamen, however, petal wilting could not be induced by ethylene or by ACC until the flowers were pollinated (25), indicating, at least for that species, that another regulatory factor may be involved in the pollination response.

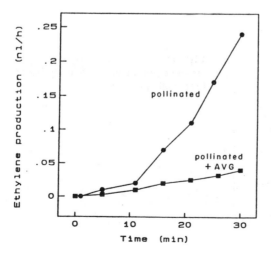

Fig. 13. Effect of 2μl droplets of AVG (2 nmol) or H_2 (control), applied to *Petunia* stigmas 3 h in advance of pollination, on early ethylene evolution from styles excised from freshly opened flowers Redrawn from (27)

Wound Responses

In many plant responses, ethylene appears to function as a wound hormone. Its production rises rapidly when plants are wounded or exposed to stress, and the responses of plants to ethylene often seem to be tailored to reducing stress or averting infection of wounds. Thus, for example, leaf abscission accompanying severe drought stress may be an adaptive response to ethylene which reduces the evaporative surface of the affected plant. Similarly, ethylene can elicit the production of phyto-alexins—compounds which appear to combat fungal or bacterial infection in wounded plants (69).

A spectacular wound response of some plants to ethylene is the stimulation of secretory processes. In some species of *Prunus* application of Ethephon results in the stimulation of gum production—normally a response to infection or other stresses. In *Hevea*, the rubber tree, a similar response is valuable commercially (2). Ethephon treatment maintains the flow of latex from the trees, thereby considerably improving yields, and reducing labor costs. There are some indications that the vascular blockages that accompany "wilt" diseases, and are also found in the stems of some cut ornamentals, may also be an ethylene-mediated wound response. Tracheids in the rachis of cut fronds of maidenhair fern become completely occluded with a pectinaceous gum after two days in the vase (Fig. 14). Such deposits are prevented by a pre-treatment with Ag+, suggesting that they are a response to ethylene produced following wounding of the rachis (20).

Fruit Development and Ripening

The stimulation of fruit ripening was one of the earliest reported effects of ethylene gas, and is the basis for what is probably the most widespread use of ethylene and compounds which liberate it. Commercially, harvested bananas, tomatoes, avocados, melons, and kiwifruit are ripened

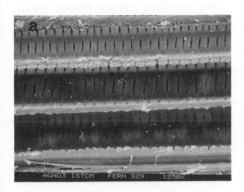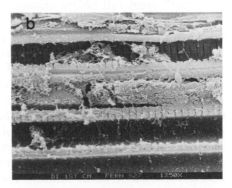

Fig. 14. Tracheids from the rachis of fronds of maidenhair fern held after harvest in 25 µl/l AgNO₃ (a) or deionized water (b) for two days. (Photo, Fujino)

with ethylene gas. Ethephon is sprayed in processing tomato fields late in the season to ensure uniform ripeness of the fruit. As in the case of abscission, there is still considerable discussion of the natural role of ethylene in fruit ripening. For many years the question of whether ethylene production by fruit rose before or after the other events of ripening (usually signalled by the onset of the respiration climacteric) was hotly debated. For most of the fruit that have been examined using single-fruit samples and high-sensitivity gas chromatography, it has been found that increased ethylene production is coincident with, or follows, the onset of the respiration climacteric (Fig. 15).

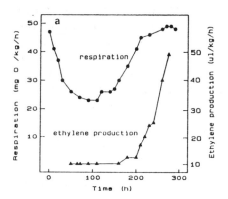 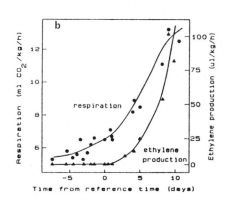

Fig. 15. Respiration and ethylene production of a feijoa fruit (a) and of Cox's orange apples (b) during the onset of ripening. The data on Cox apples are means of many fruit sampled at different times—day 0 is the day for each apple on which ethylene production started to increase. Redrawn from (47 & 52)

The melon, in which a substantial rise in ethylene production occurs *before* the onset of ripening (45), is not typical of climacteric fruits. Most workers studying the control of the onset of ripening now support the hypothesis (47) that the concentrations of an unidentified "ripening inhibitor" must fall before endogenous ethylene can induce the onset of the "climacteric" patterns of respiration and ethylene production that accompany fruit ripening. Perhaps the best evidence for such an inhibitor comes from work with avocados. Some cultivars of avocado will not ripen while attached to the tree, even if treated with high concentrations of ethylene. Shortly after removal they become very sensitive to applied ethylene, and will ripen in the absence of exogenous ethylene after a few days (6). Extraction and identification of the hypothetical inhibitor would provide an important practical tool for controlling fruit ripening that might have far-reaching practical implications.

"Non-climacteric" fruits are those in which the events of ripening are not associated with a sudden increase in respiration or ethylene production, although the fruit may be sensitive to ethylene (for example, ethylene is used to de-green non-climacteric oranges and lemons). In a study of the ripening of cherry fruits (51), we used Ag+ as a probe to determine whether ethylene played any role in natural ripening. Whether sprayed on the tree or applied to detached fruit through their pedicel, Ag+ had no perceptible effect on fruit ripening. Similar results were obtained with grapes, another non-climacteric fruit. It thus appears that ethylene plays no role in natural ripening of non-climacteric fruit. The identification of the mechanisms controlling their ripening could have considerable practical implications.

Seed Dispersal

The fact that abscission of tissues within the fruit precedes explosive dispersal of seeds in some species suggest that here, too, ethylene may play an important role. In the "squirting cucumber", ethylene production by the fruits rises substantially in the period prior to the splitting of the fruit rind (34). This increased ethylene production is presumed to be responsible for inducing the abscission of the turgid fruit tissues which precedes seed ejection. A similar mechanism may be the basis of seed dispersal in many other species.

DISCUSSION

In reviewing the role of ethylene in plant growth and development it becomes clear that this role may be quite different from those of the other plant hormones. Although ethylene appears to be involved in a great many plant responses, it is almost always as a signal of environmental or physiological change. There is scant evidence that ethylene is an important component of the balance of growth regulators that maintains normal vegetative growth; indeed, plants grow quite normally under hypobaric conditions, or after application of sufficient Ag+ to strongly inhibit the action of ethylene. Changes in plant growth are, however, often associated with increased production of or sensitivity to ethylene. Ethylene production increases in response to environmental stresses (physical stimuli, flooding); sensitivity to endogenous ethylene increases in organs reaching physiological maturity (maturing fruits and flowers, or aging leaves).

The induction of any change in plant behavior is the result of an interaction between tissue sensitivity and the concentration of the effective hormone. Ethylene does not act alone in affecting plant behavior; the action of ethylene can almost always be modulated by application of other plant hormones. It may be that the sensitivity of the tissue to ethylene is a function of the concentrations of other hormones.

It is also apparent that information on the response of plant tissues to ethylene, and its role in growth and development far outstrips our

275

understanding of how these responses are mediated. We know much about the biochemical changes associated with ethylene-mediated abscission and fruit ripening, and several groups are now studying the molecular biology of these processes. In contrast, almost nothing is known of the mechanisms by which ethylene mediates such processes as dormancy release, aerenchyma development, growth inhibition or stimulation and stimulation of latex secretion. Elucidation of these mechanisms provides a rich field of research for those interested in the regulation of plant growth and development.

References

1. Abeles, F.B. (1973) Ethylene in Plant Biology. Academic Press, New York, 302 pp.
2. D'Auzac, J., Ribaillier, D. (1969) Ethylene, a new agent stimulating latex production in *Hevea brasiliensis.* C. R. Acad. Sci. Paris, Ser. D. 268, 3046-3049.
3. Baker, J.E., Wang, C.Y., Lieberman, M., Hardenburg, R.E. (1977) Delay of senescence in carnations by a rhizobitoxine analog and sodium benzoate. HortScience 12, 38-39.
4. Bennett, D.C., Freeling, M. (1986) Induction of alcohol dehydrogenase by 1-acetyl-1-cyclopropane carboxylic acid. J. Cellular Biochem. Supplement 10B, 1986. *18.*
5. Beyer, E.M. Jr. (1976) A potent inhibitor of ethylene action in plants. Plant Physiol. *58*, 268-271.
6. Biale, J.B., Young, R.E. (1981) Respiration and ripening in fruits--retrospect and prospect. In: Recent Advances in the Biochemistry of Fruits and Vegetables, pp. 1-39, Friend, J., Rhodes, M.J.C., eds. Academic Press, London.
7. Biro, R.L., Jaffe, M.J. (1984) Thigmomorphogenesis: Ethylene evolution and its role in the changes observed in mechanically perturbed bean plants. Plant Physiol. *62*, 289-296.
8. Blanpied, G.D. (1985) (Moderator) Ethylene in postharvest biology and technology of horticultural crops. Symposium of the American Society for Horticultural Science. HortScience 20, 40-60.
9. Bradford, K.J., Yang, S.F. (1980) Xylem transport of 1-aminocyclopropane-1-carboxylic acid, an ethylene precursor, in waterlogged tomato plants. Plant Physiol. *65*, 322-326.
10. Bufler, G., Mor, Y., Reid, M.S., Yang, S.F. (1980) Changes in the 1-aminocyclopropane-1-carboxylic acid content of cut carnation flowers in relation to their senescence. Planta *150*, 439-442.
11. Burg, S.P., Burg, E.A. (1967) Molecular requirements for the biological activity of ethylene. Plant Physiol. *42*, 144-152.
12. Cameron, A.C., Reid, M.S. (1983) The use of silver thiosulfate to prevent flower abscission from potted plants. Scientia Hortic. *19*, 373-378.
13. Camprubi, P., Nichols, R. (1979) Ethylene-induced growth of petals and styles in the immature carnation inflorescence. J. Hort. Sci *54*, 225-258.
14. Chandler, W.H. (1950) Evergreen Orchards. Lea & Febiger, Philadelphia. 452 pp.
15. De Candolle, A.-P. (1832) Physiologie Vegetale, ou Exposition des Forces et des Fonctions Vitales des Vegetaux. Bechet Jeune, Paris. Vol 2, 517.
16. De Munk, W.J. (1975) Ethylene disorders in bulbous crops during storage and glasshouse cultivation. Acta Hort 51, 321-235.
17. Eisinger, W. (1983) Regulation of pea internode expansion by ethylene. Ann. Rev. Plant Physiol. *34*, 225-240.
18. Friend, J., Rhodes, M.J.C. (1981) (Eds) Recent Advances in the Biochemistry of Fruits and Vegetables. Academic Press, London. 275 pp.
19. Fuchs, Y., Chalutz, E. (1984) (Eds) Ethylene. Biochemical, Physiological, and Applied Aspects. Martinus Nijhoff/Dr W.D. Junk, The Hague. 341 pp.

M. S. Reid

20. Fujino, D.W., Reid, M.S., Vandermolen, G.E. (1983) Identification of vascular blockages in rachides of cut maidenhair (*Adiantum raddianum*) fronds. Scientia Hortic. *21*, 381-388.
21. Fujino, D.W., Reid, M.S., and Yang, S.F. (1981) Effects of aminooxyacetic acid on postharvest characteristics of carnation. Acta Hortic. *113*, 59-64.
22. Goeschl, J.D., Rappaport, L., Pratt, H.K. (1966) Ethylene as a factor regulating the growth of pea epicotyls subjected to physical stress. Plant Physiol. *41*, 877-884.
23. Goszczynska, D.M., Reid, M.S. (1985) Studies on the development of tight cut rose buds. Acta Hortic. *167*, 101-108.
24. Halevy, A.H., Mayak, S. (1979) Senescence and postharvest physiology of cut flowers, Part 1. In: Horticultural Reviews, Vol. 1, pp. 204-36, Janick, J., ed. AVI, Westport.
25. Halevy, A.H., Whitehead, C.S., Kofranek, A.M. (1984) Does pollination induce corolla abscission of cyclamen flowers by promoting ethylene production? Plant Physiol. *75*, 1090-1093.
26. Hiraki, K., Ota, Y. (1975) The relationship between growth inhibition and ethylene production by mechanical stimulation in *Lilium longliflorum*. Plant Cell Physiol. *16*, 185-189.
27. Hoekstra, F.A., Weges, R. (1986) Lack of control by early pistillate ethylene of the accelerated wilting of *Petunia hybrida* flowers. Plant Physiol. *80*, 403-408.
28. Huelin, F.E., Barker, J. (1939) The effect of ethylene on the respiration and carbohydrate metabolism of potatoes. New Phytol. 38, 85-104.
29. Hume, B., Lovell, P. (1982) Female flower production in *Cucurbita pepo* is controlled by ACC (1α amino cyclo propane-1-carboxylic acid) and not by ethylene. Plant Physiol. *69*, 137.
30. Imanishi, H., Fortanier E.J. (1982/83) Effects of exposing Freesia corms to ethylene or to smoke on dormancy-breaking and flowering. Scientia Hortic. *18*, 381-389.
31. Jackson, M.B. (1985) Ethylene and responses of plants to soil waterlogging and submergence. Ann. Rev. Plant Physiol. *36*, 145-174.
32. Jackson, M.B. (1982) Ethylene as a growth promoting hormone under flooded conditions. In: Plant Growth Substances, pp. 291-301, Wareing, P.F., ed. Academic Press, N.Y.
33. Jackson, M.B. (1979) Is the *diageotropica* tomato ethylene deficient? Physiol. Plant. *46*, 347-351.
34. Jackson, M.B., Morrow, I.B., Osborne, D.J. (1972) Abscission and dehiscence in the squirting cubumber (*Ecballium elaterium*). Can. J. Bot. *50*, 1465-1471.
35. Kende, H., Acaster, M.A., Jones, J.F., Metraux, J.-P. (1982) On the mode of action of ethylene. In: Plant Growth Substances, 1982, pp. 269-277, Wareing, P.F., ed. Academic Press, London.
36. Ketring, D.L. (1977) Ethylene and seed germination. In: The Physiology and Biochemistry of Seed Dormancy and Germination, pp. 157-178, Khan, ed. Elsevier, Amsterdam.
37. Koning, R.E. (1983) The roles of auxin, ethylene, and acid growth in filament elongation in *Gaillardia grandiflora* (*Asteraceae*). Amer. J. Bot. 70, 602-610.
38. Lieberman, M. (1983) Postharvest Physiology and Crop Preservation. NATO Advanced Study Institute Series. Series A: Life Sciences, *46*, Plenum, New York. 572 pp.
39. Lieberman, M. (1979) Biosynthesis and action of ethylene. Ann Rev. Plant Physiol., 30, 533-591.
40. Mor, Y., Halevy, A.H., Spiegelstein, H., Mayak, S. (1985) The site of 1-aminocyclopropane-1-carboxylic acid synthesis in senescing carnation petals. Physiol. Plant. *65*, 196-202.
41. Neel, P.L., Harris, R.W. (1971) Factors influencing tree trunk growth. Arborist's News *36*, 115-138.
42. Neljubov, D. (1901) Ueber die horizontale Nutation der Stengel von *Pisum sativum* und einiger anderen Pflanzen. Beih. Bot. Centralbl. *10*, 128-139.

277

43. Nichols, R., Bufler, G., Mor, Y., Fujino, D.W., Reid, M.S. (1983) Changes in ethylene production and 1-aminocyclopropane-1-carboxylic acid content of pollinated carnation flowers. J. Plant Growth Regul. 2, 1-8.
44. Osborne, D.J. (1983) News Bulletin of the British Plant Growth Regulator Group 6, 8-11.
45. Pratt, H.K., Goeschl, J.D. (1969) Physiological role of ethylene in plants. Ann. Rev. Plant Physiol. 20, 541-584 .
46. Reid M.S. (1985) Ethylene and abscission. HortScience 20, 45-50.
47. Reid, M.S. (1974) The role of ethylene in the ripening of some unusual fruits. In: Colloques internationaux C.N.R.S. # 238. Facteurs et regulation de la maturation des fruits. pp. 189-192, Ulrich, R., ed. CNRS, Paris.
48. Reid, M.S., Fujino, D.W., Hoffman, N.E., Whitehead, C.S. (1984) 1-Aminocyclopropane-1-carboxylic acid (ACC)--The transmitted stimulus in pollinated flowers? J. Plant Growth Regul. 3, 189-196.
49. Reid, M.S., Paul, J.L., Farhoomand, M.B., Kofranek, A.M., Staby, G.L. (1980) Pulse treatments with the silver thiosulfate complex extend the vase life of cut carnations. J. Amer. Soc. Hortic. Sci. 105, 25-27.
50. Reid, M.S., Paul, J.L., Young, R.E. (1980) Effects of Ethephon and betacyanin leakage from beet root discs. Plant Physiol 66, 1015-1016.
51. Reid, M.S., Pech, J.C., Latche, A. (1985) Non intervention d'ethylene dans la crise respiratoire chez les cerises. Fruits 40, 197-203.
52. Reid, M.S., Rhodes, M.J.C., Hulme, A.C. (1973) Changes in ethylene and CO2 during the ripening of apples. J. Sci. Fd. Agric. 24, 971-979.
53. Robbins, J., Reid, M.S., Rost, T., Paul, J.L. (1985) The effect of ethylene in adventitious root formation of Mung bean (Vigna radiata) cuttings. J. Plant Growth Regul. 4, 147-157.
54. Roberts, J.A., Tucker, G.A. (1985) Ethylene and Plant Development. Butterworths, London. 416 pp.
55. Rylski, I., Rappaport, L., Pratt, H.K. (1974) Dual effects of ethylene on potato dormancy and sprout growth. Plant Physiol. 53, 658-662.
56. Sexton, R., Lewis, L.N., Trewavas, A.K., Kelly, P. (1985) Ethylene and abscission. In: Ethylene and Plant Development, pp. 173-96, Roberts J.A., Tucker, G.A., eds. Butterworths, London.
57. Sisler, E.C., Pian, A. (1973) Effect of ethylene and cyclic olefins on tobacco leaves. Tobacco Sci. 17, 68-72.
58. Suge, H. (1972) Inhibition of photoperiodic floral induction in Pharbitis nil by ethylene. Plant Cell Physiol. 13, 1031-1038 .
59. Taylorson, R.B. (1979) Response of weed seeds to ethylene and related hydrocarbons. Weed Sci. 27, 7-10.
60. Thimann, K.V. (1980) Senescence in Plants. CRC Press, Boca Raton, Florida. 276 pp.
61. Veen, H. (1983) Silver thiosuphate: an experimental tool in plant science. Scientia Hort 20, 211-224.
62. Wareing, P.F. (1982) Proceedings of the 11th International Conference on Plant Growth Substances. Academic Press, London. 683 pp.
63. Whitehead, C.S., Fujino, D.W., Reid, M.S. (1983) The roles of pollen ACC and pollen tube growth in ethylene production by carnations. Acta Hortic. 141, 229-234.
64. Whitehead, C.S., Fujino, D.W., Reid, M.S. (1983) Identification of the ethylene precursor, 1-aminocyclopropane-1-carboxylic acid (ACC) in pollen. Scientia Hortic. 21, 291-297.
65. Whitehead, C.S., Halevy, A.H., Reid, M.S. (1984) Control of ethylene synthesis during development and senescence of carnation petals. J. Amer. Soc. Hortic. Sci. 109, 473-475.
66. Whitehead, C.S., Halevy, A.H., Reid, M.S., (1984) Roles of ethylene and 1-aminocyclopropane-1-carboxylic acid in pollination and wound-induced senescence of Petunia hybrida flowers. Physiol. Plant. 61, 643-648.
67. Yang, S.F. (1985) Biosynthesis and action of ethylene. HortScience 20, 41-45.

68. Yang, S.F., Hoffman, N.E. (1984) Ethylene biosynthesis and its regulation in higher plants. Ann. Rev. Plant Physiol. *35*, 155-189.
69. Yang, S.F., Pratt, H.K. (1978) The physiology of ethylene in wounded plant tissues. In: Biochemistry of Wounded Plant Tissues, pp. 595-622, Kahl, G., ed. Walter de Gruyter, Berlin.
70. Zobel, R.W. (1973) Some physiological characteristics of the ethylene-requiring tomato mutant *diageotropica*. Plant Physiol. *52*, 385-389.

E2. Polyamines as Endogenous Growth Regulators

Arthur W. Galston and Ravindar Kaur-Sawhney

Department of Biology, Yale University, New Haven, Connecticut 06511, USA.

POLYAMINES AS ESSENTIAL CELLULAR COMPONENTS

It is probable that all cells contain the diamine putrescine (Put; 1,4-diaminobutane) and the triamine spermidine (Spd), while eukaryotic cells contain the tetraamine spermine (Spm) as well (6, 13). In both prokaryotes and eukaryotes (53), including higher plants (30), mutants lacking the ability to biosynthesize polyamines (PAs) are unable to grow and develop normally (53). Since the addition of PAs to these mutants generally restores normal growth and development, it is reasonable to conclude that PAs are essential to all cells. This conclusion is reinforced by the demonstrable effects of "suicide inhibitors" of the main PA-biosynthetic enzymes, ornithine decarboxylase (ODC) and arginine decarboxylase (ADC: Fig. 1). These compounds, α-difluoromethylornithine (DFMO) and α-difluoromethylarginine (DFMA), specifically and irreversibly bind to and inhibit ODC and ADC, respectively. The ensuing decline in cellular PA titers is accompanied by a diminution or cessation of growth and development, which are restored upon the addition of the relevant PA.

Given the biological ubiquity and apparent indispensability of PAs, we need, as plant physiologists, to ask at least the following basic questions:

1. How are PAs biosynthesized and metabolized in plants?
2. What is their location in the cell?
3. What is their probable function?
4. Do PAs normally regulate growth and development of "normal" plants?
5. Can changes in PA titer help explain the action of physical (light, temperature, stress) and chemical (nutrients, hormones) agents affecting plant growth?
6. Are PAs translocated? If so, do they function as hormones?

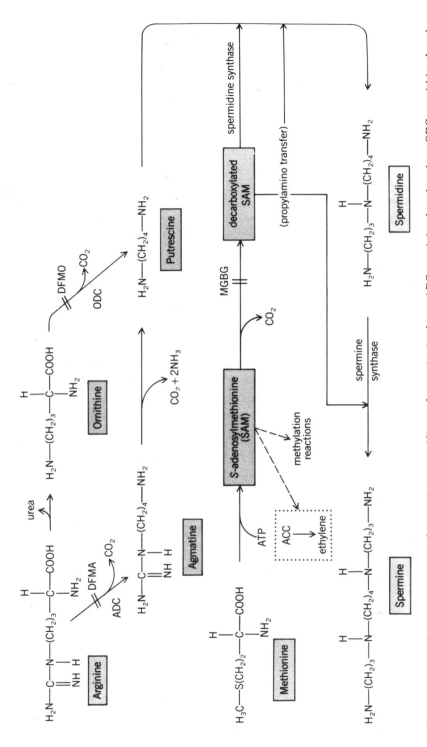

Fig. 1. Pathways for the biosynthesis of putrescine, spermidine and spermine in plants. ADC = arginine decarboxylase; ODC = ornithine decarboxylase; DFMO = α-difluoromethylornithine; DFMA = α-difluoromethylarginine; MGBG = methylglyoxal bis-(guanylhydrazone). From (21a).

7. What is the effect of the administration of exogenous PAs or PA analogs on entire plants, excised organs, individual cells and protoplasts?
8. Are any of these effects potentially important in agriculture?

THE BIOSYNTHESIS AND METABOLISM OF POLYAMINES IN PLANTS

Evidence supporting the biosynthetic scheme in Fig. 1 for plants has recently been extensively reviewed (48). While much of the original evidence came from microorganisms and animals, a recent spate of activity in the plant field has led to considerable documentation of similar systems in plants, especially in the angiosperms. Considerable interest has been displayed in the participation of S-adenosylmethionine (SAM) in the biosynthesis of Spd and Spm, since SAM is also the source of the plant hormone ethylene, by way of the intermediate 1-amino-1-cyclopropane carboxylic acid (ACC). Ethylene is well known as a senescence-inducer (1) and PAs have antisenescence activity, particularly with excised plant parts (23). Thus, any system that controls the flow of carbon atoms through SAM could control the developmental fate of the cell, tissue or organ involved. It is noteworthy that many stress conditions will lead to an increased production of both ethylene and Put (see below), but not other PAs.

Put is oxidized by a diamine oxidase (DAO), yielding 4-aminobutyraldehyde, which then cyclizes spontaneously to form pyrroline: other products of the reaction are NH_3 and H_2O_2 (48). DAO activity is especially high in the *Leguminosae*, where it may represent up to 3% of the total protein of the cell. In some plants, the 4-aminobutyraldehyde is further oxidized to 4-aminobutyric acid (GABA). Polyamine oxidase (PAO), especially abundant in the cereals (47), catalyzes an analogous reaction, yielding pyrroline, 1,3-diaminopropane (Dap) and H_2O_2 (48). When Spm is the substrate, aminopropylpyrroline is formed instead of pyrroline.

The DAO of pea contains two identical subunits, each containing Cu^{2+} and sharing one carbonyl group; DAOs from other sources contain –SH groups in addition to Cu^{2+} and carbonyl (48), while other DAOs contain FAD. PAO in several cereals seems to be especially associated with lignified tissues (25, 47) or guard cells (25), and at least some activity is in the cell wall (25).

Put and other PAs may exist in the form of conjugates with such phenolic acids as cinnamic and ferulic (49). Under certain conditions, these conjugates may constitute up to 90% of the total PAs of the cell. Put is also the source of carbon atoms for the formation of the 5-membered pyrrolidine ring of nicotine and related alkaloids. The fate of Put in alkaloid synthesizing cells appears to be determined by auxin: thus, at low levels of NAA (c. 1μM) tobacco callus grows little but makes lots of alkaloid, while at higher levels (c. 10μM), growth is stimulated and alkaloid synthesis falls, the Put being diverted into conjugate formation (54).

Cadaverine (Cad: 1,5-diaminopentane) is formed from lysine in some plants (48). It is found free in actively-growing leguminous roots and is converted, via oxidation and cyclization, to the 6-membered piperideine ring of anabasine and other alkaloids. A lysine oxidase is apparently responsible for the initial oxidation.

LOCALIZATION OF POLYAMINES AND THEIR BIOSYNTHETIC ENZYMES

Since PAs are relatively small, soluble, diffusible molecules at cellular pH's, their immobilization in the cell for localization purposes is difficult to achieve. Disruption of the cell to achieve particle separation may result in many artifacts, especially if pH changes cause altered distribution patterns. The fact that endogenous and exogenous arginine may have entirely different metabolic fates *en route* to PAs and that endogenous and exogenous PAs are metabolized differently probably indicates some compartmentation of enzymes and substrates in the metabolic pathways related to PAs.

The availability of DFMO and DFMA, which undergo irreversible covalent bond formation with ODC and ADC, respectively, affords a means of localizing these enzymes within tissues, but so far not conclusively within the structure of the cell. Labeling these inhibitors with ^{14}C or ^{3}H permits autoradiographic techniques to be employed, as it has been for ODC in animals (37). In plants, it appears by the use of ^{3}H-DFMO that ODC is localized in the nuclei of tobacco ovules (R.D. Slocum unpublished), a conclusion in agreement with cell fractionation studies based on centrifugal separations (36). On the same basis, ADC seems to be in the cytosol, while Spd synthase, once reported to occur in chloroplasts, can now only be said to be particulate (55). PAO is probably wall-localized, at least in the cereals (25).

Immunochemical methods with labeled antibodies have also been used in animal, but not plant investigations. The availability of natural anti-Spm immunoglobulins in rabbit serum should permit localization of this important PA, at least in its bound forms.

Obviously, the localization of PAs and the enzymes of their metabolism within the plant cell is largely *terra incognita*, and much remains to be done.

FUNCTIONS OF POLYAMINES

At cellular pH's, PAs are polycations (6, 13) and thus bind readily to such important cellular polyanions as DNA, RNA, phospholipids and acidic protein residues. Through such binding, PAs could affect the synthesis of macromolecules, the activity of macromolecules, membrane permeability, and partial processes of mitosis and meiosis. This has been shown convincingly *in vitro*, but the *in vivo* evidence is still scanty.

Membrane Structure and Function

Effects on the cellular plasma membrane are perhaps easiest to demonstrate. PAs can stabilize otherwise labile oat protoplasts against lysis, decrease the leakage of betacyanin from wounded root tissue and preserve thylakoid structure in excised barley leaves. The activity of membrane-localized enzymes in both animals and plants can be affected by PA-mediated changes in membrane fluidity and fine structure. One important example is the ethylene-generating system, whose activity is regulated by temperature-dependent inhibitory effects of Ca^{2+} and Spm. Similarly, PAs can counteract hormone-induced changes in membrane permeability and can affect membrane-localized proton-secreting systems, one of the probable targets for auxin (see reviews 21, 45, 48, 53).

Interactions with Nucleic Acids

Spm-DNA complexes have a regular conformation, stabilizing the DNAs against thermal denaturation *in vitro*. Spd and Spm also facilitate conformational changes, such as the B → Z transition in methylated poly-nucleotides (9), and together with other basic cellular components such as histones, may control DNA conformation important in nucleosome assembly and gene expression. Depletion of intracellular PAs sensitizes DNA to alkylating agents, perhaps by exposing previously protected groups on DNA. These studies have been recently reviewed (45, 48, 53).

The structure and function of microbial tRNA's is affected by PAs (43), a fact that has led investigators with plant tRNA and rRNA to postulate similar action of PAs bound to those entities. PAs are also found as integral components of a plant virus and a bacteriophage, affect the organization of DNA in bacterial nucleoids (18) and control chromosome condensation and nuclear membrane dissolution during pre-prophase. Specific PAs are apparently required for maintenance of several yeast ds RNA plasmids (see reviews 45, 53).

Control of Protein Structure and Enzyme Activity

PAs have been reported to control the phosphorylation of a nucleolar protein in *Physarum polycephalum*, a slime mold (27). This protein was identified as ODC, and the authors suggest that PAs control their own synthesis through phosphorylative inactivation of their main biosynthetic enzyme. Another worker suggests that ODC inactivation is brought about by direct linkage to Put. PAs have also been reported to stimulate various kinases in animals and plants, probably by virtue of their size and charge. Other enzymes whose activities are affected by PAs include an NADPH oxidizing enzyme and fructose 1,6-bisphosphatase. While most of the effects are produced by ionic binding of PAs to the macromolecule, there is evidence of covalent binding of PAs to protein, possibly mediated by transglutaminase (see reviews 21, 45, 53).

Effects on the Synthesis of Macromolecules

In many types of cells, correlations have been reported between PA biosynthesis and titer on the one hand and cellular proliferation on the other (2). This is consistent with the finding that various systems for the *in vitro* synthesis of proteins are stimulated by the addition of PAs. Inhibition of PA biosynthesis retards growth in microbial (53), animal (31) and plant cells (12). In some instances, these inhibitions can be reversed by the addition of PAs (12, 45, 53). The suicide inhibitor of ODC, DFMO, was in fact synthesized in order to control the growth of malignant cells in cancer patients (33). MGBG, an inhibitor of SAM decarboxylase, is also able to inhibit the growth of some types of malignant cells, especially in concert with DFMO or other drugs.

When DFMO is added to rat hepatoma tissue culture cells, Spd titer declines, followed by a decline in polysome content, decreased ^{3}H-leucine incorporation into protein, and a prolongation of the G_1 phase of the cell cycle. It appears that a high titer of Spd or Spm is required for entry into the S phase, when DNA synthesis occurs. Similar results have been reported in a wide variety of prokaryotic and eukaryotic cells, both plant and animal (45, 53).

Buffering of Cellular pH

In plants showing Crassulacean acid metabolism (CAM), Put is synthesized along with secretion of malate into the cytoplasm (35). Put is also increased by increased external acidity (56), by SO_2 (38), or by excess NH_4^+ (38), all of which produce an acid stress on the exterior of the plasma membrane. This raises the possibility that the reversible protonation of the multiple amino groups of PAs may thus serve to buffer cellular pH. Quantitatively, this seems possible, since the titer of Put in stressed cells may reach 0.4mM. Other stress stimuli that also result in the accumulation of Put (see below) may possibly act through a common effect on pH.

POLYAMINES IN PLANT GROWTH AND DIFFERENTIATION

Through the development of a series of mutants in *Escherichia coli* and *Saccharomyces cerevisiae* (53), PAs have been shown to be obligate growth factors for both prokaryotic and eukaryotic cells. In yeast, Put is required for attainment of optimal growth and development, while the higher PAs are required for sporulation. Mammalian cells, especially in tissue culture, have also been shown to require PAs for normal growth and development.

PA titer and plant growth rate are positively correlated in a wide variety of plant growth systems (15, 21, 45, 48), and the interruption of PA biosynthesis by inhibitors (45, 48) or mutation results in altered growth patterns (30), some of which can be reversed by the application of particular PAs. In a few cases, the application of exogenous PAs to protoplasts (24) or cells (22) in tissue culture has resulted in a temporary or sustained increase in cell division.

The ratio of diamines to higher polyamines (Put/Spd + Spm) is generally directly correlated with elongation rate, especially in seedling organs (Fig. 2, 43a). It appears that the Put → Spd transformation is especially important in controlling the rate of cell division. Put appears to be injurious, while high Spd (and Spm) seem essential for the G_1 → S transition.

The development of the ovary and ovules during maturation seems highly sensitive to PA titer, especially in tomato and other solanaceous plants (12, 44). Application of DFMO to young tomato pistils immediately after pollination blocks the subsequent growth and development of the ovary (12, 34). This inhibition can be overcome in part by the subsequent application of Put or Spd, indicating that reduction of PA titer below some critical value inhibits the prolific cellular division that prepares the way for fruit development. Similarly, PA overproducers in tobacco selected from inhibitor-resistant strains or revertants of mutations show altered differentiation patterns, producing staminoid structures instead of ovules in the ovaries (30) and petaliferous stamens (30). Other PA mutant

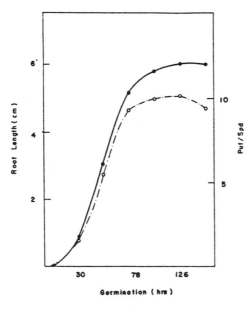

Fig. Relation of Put/Spd ratio (– · – · –) to the length of root (————) in Alaska pea seedling germination. From (43a)

strains show altered floral morphology as well (R. L. Malmberg, personal communication).

Hydroxycinnamic acid conjugates of PAs are found only in reproductive apices of some solanaceous plants (10), suggesting a possible role in floral induction. The different sex organs also have different spectra of PA-containing conjugates (10), indicating a possible relation to sexual differentiation. In corn, a male-sterile line has a very low titer of PA conjugates and a complete absence of feruloyl-Put conjugates (32).

Senescence in many plant organs, both *in situ* and upon their excision from the plant, is correlated with a decline in PA titer (26,50). Since the addition of exogenous PAs in the millimolar concentration range delays or prevents such senescence-related processes as the decline of chlorophyll, protein and RNA content (measured by decreases in protease and RNase activities) in a wide range of monocot and dicot leaves (23, Fig. 3), it has

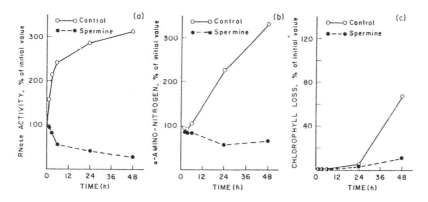

Fig. 3. Kinetics of the effect of 1 mM spermine on (a) RNase activity; (b) free α-amino nitrogen content; (c) chlorophyll loss in excised peeled oat leaf segments incubated in the dark. From (23).

been proposed that PA titer controls senescence, at least in leaves (23). ADC activity is usually well correlated with leaf well-being (26). Thus inhibitors or other agents conducing to a decline in ADC activity generally promote senescence (26), while light, hormones and other treatments that inhibit senescence lead to an increase in ADC activity (26). Exogenous PAs seem to interact with a Ca^{2+}-specific site on the outer membrane (23), and only a brief contact between leaf and PA (a few minutes in a 48 h. period) suffices to prevent senescence. PAs act through an inhibition of ethylene formation (20) or action (4).

REGULATION OF POLYAMINE TITER IN THE PLANT

The synthesis of PAs is sensitive to several externally-manipulable variables, including light, hormones, excision and stress. Control seems to be exerted through ADC, ODC and Spd synthase in different plants.

Light

In excised oat leaves, ADC activity and PA titer fall in darkness and rise in white light (26). In etiolated peas, the $P_r \rightarrow P_{fr}$ phytochrome transition increases ADC activity in buds, while decreasing it in epicotyls (14). Thus, P_{fr} effects on ADC in each organ parallel P_{fr} effects in growth of the organ. This is perhaps the only known case of simultaneous photoinduced induction and repression of the same enzyme in different organs. Changes in PA titer reflect the altered ADC activity and growth, and the altered ADC activity is not simply a consequence of altered growth rate, as shown by kinetic and surgical procedures.

Hormones

Either PAs or auxin will stimulate dormant *Helianthus tuberosus* explants to grow *in vitro*. Since auxin application results in an increase in PA titer and macromolecular synthesis, it has been proposed that auxins act through PAs to produce growth promotion in that tissue. We have already mentioned a similar situation in the tomato ovary, where auxin application that induces parthenocarpy requires active ODC. Auxin-induced rooting in mung bean seedlings is also reduced in the presence of MGBG, an inhibitor of SAM decarboxylation, and the inhibition is reversed by the application of arginine or ornithine (reviewed in 21, 45).

The gibberellin-induced elongation of dwarf pea internodes in the light is accompanied by a rise in ADC activity and PA titer, while application of DFMA partially prevents these effects (R. Kaur-Sawhney *et al*, in press). A similar GA-PA interaction exists in lettuce hypocotyls (18). In the GA-induced elongation of pea internodes, which involves both cell division and cell elongation, PAs appear to play a role in cell division and not cell elongation (46). The GA-induced increase in α-amylase in barley aleurone is inhibited by MGBG, but there is no change in PA titer following GA administration (61). GA and PA both induce a large increase in ODC activity in this same barley system (28 and reviews 21, 45).

Cytokinins increase PA biosynthesis and titer in lettuce (11) and cucumber (51) cotyledons and in red irradiated etiolated pea buds. In cucumbers, cytokinins can also reverse an inhibitory effect of abscisic acid on PA biosynthesis (51).

We have already mentioned the quantitative interplay of PAs and ethylene, possibly resulting in part from competition for SAM, a common precursor. PAs and ethylene also inhibit each other's biosynthesis and action. Thus, exogenous PAs reduce auxin-induced ethylene production in petals (52), leaves and fruits (4) and senescing orange peel (16). PAs apparently block ethylene biosynthesis at the ACC → ethylene step, known also to be Ca^{2+}-sensitive (4, 20) and at the prior ACC synthase step (16, 20). When ACC synthesis is blocked by PAs, there is apparently an increased flow of carbons from SAM into PAs (16). Conversely, when PA synthesis is blocked by MGBG or DFMA, ethylene synthesis is promoted (42). In etiolated pea seedlings, exogenous application of ethylene inhibits ADC activity, while depletion of endogenous PAs increases ADC activity (5). Thus fluctuations in the comparative rates of ethylene and PA production could produce consequent changes in ADC activity.

Stress
Physical. More than thirty years ago, K^+-deficiency was found to increase Put levels in plants (41), and this effect has been widely confirmed (48, 58). The increased Put "replaces" 30% of the cationic loss represented by K^+ deficiency. K^+ deficiency rapidly and reversibly induces higher ADC activities in young oat seedlings grown in sand culture on modified Hoagland's solution (58). The increase requires *de novo* protein synthesis (56), and is correlated with an increased stress-induced incorporation of ^{35}S-methionine into a 39 kilodalton band on a denaturing polyacrylamide gel.

Put accumulation and ADC increase also result from osmotic stress (19, Fig. 4), acid feeding (56, Fig. 5), high NH_4^+ feeding (29) or exposure to

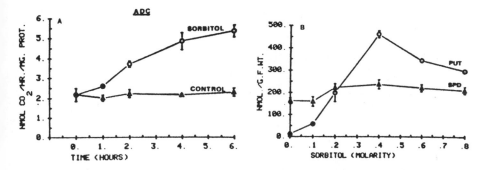

Fig. 4. A. Effect of osmotic stress on arginine decarboxylase activity. Leaf segments were floated over buffer (Δ) or buffer + 0.4 M sorbitol (O) under light, and samples taken at the indicated times and assayed for enzyme activity. Bars represent ± S.E.M.
B. Effect of osmotic concentration on polyamine titer in peeled oat leaf segments. Polyamine titer determined after 4 hr. incubation in osmoticum or control buffer. From (19).

the atmospheric pollutants SO_2 (38) or Cd^{2+} (L.H. Weinstein et al, unpublished). The response to osmotic stress in excised oat leaves occurs within minutes, as shown by timed application of cycloheximide to stressed systems. There is evidence that Put accumulation results not only from ADC activation, but also from a decrease in Spd synthase

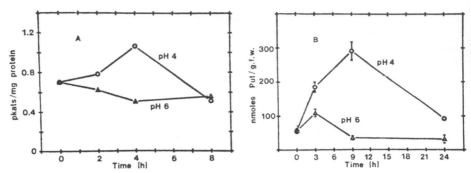

Fig. 5. A. Time-course of changes in arginine decarboxylase activity in peeled oat leaf segments at pH 4.0 or 6.0. Oat leaf tissue was incubated for up to 8 hours at either pH 4.0 or 6.0. At various times throughout this incubation, some tissue was removed and arginine decarboxylase activity determined. O, pH 4.0; Δ, pH 6.0. Enzyme activity is expressed as pkat/mg protein (1 katel = 1 mol/s). Each data point represents the mean of duplicate samples.
B. Time-course of changes in putrescine titer in peeled oat leaf segments at pH 4.0 or 6.0. Oat leaf tissue was incubated for periods of up to 24 hours at the pH indicated. At various times throughout the incubation, tissue was removed and the titer of putrescine determined. O, pH 4.0; Δ, pH 6.0. Each data point represents the mean of triplicate samples ± 1 S.D. From (56).

activity (A.F. Tiburcio *et al*, unpublished). This may also be true with acid feeding and the Put/Spd ratio increases with increase in acidity (Fig. 6).

The stress-increased Put could represent (a) the cause of the injury syndrome (b) the plant's defense against injury, or (c) a metabolic side-effect unrelated to stress. Recent experiments with simultaneous application of stress and DFMA indicate that the first possibility is closest to the truth (A.F. Tiburcio *et al*, unpublished).

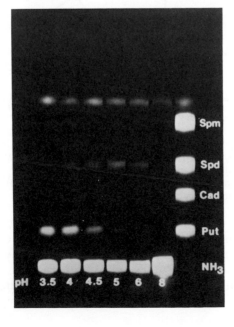

Fig. 6. Effect of external pH on polyamine titers in peeled oat leaf segments. Oat leaf tissue was incubated for 8 hours at various pH values. Perchloric acid extracts were dansylated and separated by thin-layer chromatography along with known polyamine standards. The thin-layer chromatograph was photographed in ultraviolet light to reveal dansylated polyamines. Note the rise in putrescine and decline in spermidine as the pH decreases. From (59).

Biological. Yeast and other fungi have only the ODC pathway for Put biosynthesis (53). This led us to inquire whether phytopathogenic fungi might be selectively inhibited by DFMO. In petri dish culture on Czapek's medium, both DFMO and DFMA produced marked growth inhibitions that were reversible with applied Put or Spd (39, Figs. 7, 8). The DFMA

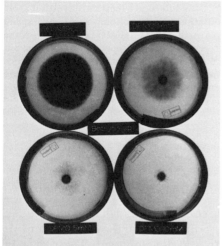

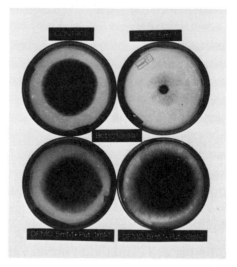

Fig. 7. The response of *Botrytis sp* mycelial growth to increasing levels of DFMO. From (39)

Fig. 8. The reversal of DFMO inhibition of mycelial growth of *Botrytis sp* by putrescine. From (39)

inhibition turned out to be trivial, since DFMA is converted to DFMO by arginase, in an analog of the arginine → ornithine conversion. When DFMO was sprayed on a unifoliolate leaf of bean infected with uredospores of *Uromyces phaseoli*, partial or complete protection against infection was achieved, depending on the concentration and timing (Fig. 9). Both pre-and post-infection sprays were effective, and it appeared that some protective effect moved from sprayed to unsprayed tissues. It appears crucial to inhibit fungal growth with DFMO before hyphal penetration of leaf cells, where the uninhibited leaf ADC could furnish Put to the fungus (40). Since several other diseases of plants are also benefited by DFMO application, this differential and specific inhibition of PA biosynthesis could turn out to be an important new method of disease control.

CONCLUSIONS

While the precise physiological role of the PAs remains unclear, we are compelled to consider them as candidates for active regulators of plant

Fig. 9. Effect of post-treatment with DFMO on development of local lesions on unifoliolate bean leaves inoculated with uredospores of *Uromyces phaseoli* L, race 0. Left column, sprayed with 0.01% Tween 20 in pH 7 buffer. 1 (top), 3 (middle) and 5 (bottom) days after spore inoculation. Right column, same, but spray contained 1 mM DFMO as well. Note protection by DFMO applied 1 and 3 days, but not 5 days after inoculation with uredospores. After five days, fungal hyphae have already penetrated the leaf, from which they can obtain PAs. From (40)

growth (21). Althougn they cannot yet be considered as hormones, because of scant evidence about their translocatability[1] (7, 57), they are present in all cells, and essential to normal growth and development. Their titer is responsive to such physiological controls as light, hormones, injury and stress, and external application can affect important physiological processes. Effects of added PAs and inhibitors of their biosynthesis, both *in vivo* and *in vitro*, are impressive. The next few years should yield much exciting new information, permitting an evaluation of the true role of PAs in the control of plant growth.

References.

1. Abeles, F. B. (1973) Ethylene in plant biology. Academic Press, New York.
2. Abraham, K.A., Alexander, P. (1981) Role of polyamines in macromolecular synthesis. Trends in Biochemical Sciences 64, 106-107.
3. Altman, A., Kaur-Sawhney, R. and Galston, A.W. (1977) Stabilization of oat leaf protoplasts through polyamine-mediated inhibition of senescence. Plant Physiol. 60, 570-574.
4. Apelbaum, A. Burgoon, A.C., Anderson, J.D., Lieberman, M., Ben-Arie, R., Mattoo, A.K. (1981) Polyamines inhibit biosynthesis of ethylene in higher plants. Plant Physiol. 68, 453-456.
5. Apelbaum, A., Goldblust, A., and Icekson, I. (1975) Control by ethylene of arginine decarboxylase activity in pea seedlings and its implication for hormonal regulation of plant growth. Plant Physiol. 79, 635-640.
6. Bachrach, U. (1973) Function of Naturally Occurring Polyamines, Academic Press, New York.

[1.] But see chapter A1, page 1 and 2.

7. Bagni, N., Pistocchi, R. (1985) Putrescine uptake in *Saintpaulia* petals. Plant Physiol. 77, 398-402.
8. Bagni, N. Torrigiani, P., Barbieri, P. (1983) *In vitro* and *in vivo* effect of ornithine and arginine decarboxylase inhibitors in plant tissue culture. In Advances in Polyamine Research Vol 4, pp. 409-417. Bachrach, U., Kaye, A., Chayen R., eds. Raven Press, New York.
9. Behe, M., Felsenfeld, G. (1981) Effects of methylation on a synthetic polynucleotide: The B-Z transition in poly (dG-M^5dC)-poly(dG-M^5dC) Proc. Natl. Acad. Sci. USA 78, 1619-1623.
10. Cabanne, F., Martin-Tanguy, J., Martin, C. (1977) Phénolamines associées à l'induction floral et à l'état reproducteur du *Nicotiana tabacum* var. Xanthi n.c. Physiologie Veg. 15, 429-443.
11. Cho, S-C. (1983) Effects of cytokinin and several inorganic cations on the polyamine content of lettuce cotyledons. Plant Cell Physiol. 24, 27-32.
12. Cohen, E., Arad, S.M., Heimer, Y.M., Mizrahi, Y. (1982) Participation of ornithine decarboxylase in early stages of tomato fruit development. Plant Physiol. 70, 540-543.
13. Cohen, S.S. (1971) Introduction to the Polyamines. Prentice-Hall, Englewood Cliffs, N.J.
14. Dai, Y.-R., Galston, A.W. (1981) Simultaneous phytochrome controlled promotion and inhibition of arginine decarboxylase activity in buds and epicotyls of etiolated peas. Plant Physiol. 67, 266-269.
15. Dumortier, F.M., Flores, H.E., Shekhawat, N.S., Galston, A.W. (1983) Gradients of polyamines and their biosynthetic enzymes in coleoptiles and roots of corn. Plant Physiol. 72, 915-918.
16. Even-Chen. Z., Mattoo, A.K., Goren, R. (1982) Inhibition of ethylene biosynthesis by aminoethoxyvinylgycine and by polyamines shunts label from 3,4-(^{14}C)-methionine into spermidine in aged orange peel discs. Plant Physiol. 69, 385-388.
17. Feirer, R.P., Mignon, G., Litvay, J.D. (1984) Arginine decarboxylase and polyamines required for embryogenesis in wild carrot. Science 223, 1433-1435.
18. Flink, L., Pettijohn, D.E. (1975) Polyamines stabilize DNA folds. Nature 253, 62-63.
19. Flores, H., and Galston, A.W. (1982) Polyamines and plant stress: activation of putrescine biosynthesis by osmotic shock. Science 217, 1259-1261.
20. Fuhrer, J., Kaur-Sawhney, R., Shih, L.-M., Galston, A.W. (1982) Effects of exogenous 1,3-diaminopropane and spermidine on senescence of oat leaves. II. Plant Physiol. 70, 1597-1600.
21. Galston, A.W. (1983) Polyamines as modulators of plant development. Bioscience 33, 382-88.
21a Galston, A.W. (1986) Plant morphogenesis. In McGraw-Hill Yearbook of Science and Technology pp. 351-354.
22. Huhtinen, O., Honkanen, J., Simola, L.K. (1982) Ornithine and putrescine supported divisions and cell colony formation in leaf protoplasts of alders (*Alnus glutinosa* and *A. incana*). Plant Sci. Lett. 28, 3-9.
23. Kaur-Sawhney, R., Galston, A.W. (1979) Interaction of polyamines and light on biochemical processes involved in leaf senescence. Plant Cell Environ. 2, 189-196.
24. Kaur-Sawhney, R., Flores, H.E., Galston, A.W. (1980) Polyamine-induced DNA synthesis and mitosis in oat leaf protoplasts. Plant Physiol. 65, 368-371.
25. Kaur-Sawhney, R., Flores, H.E., Galston, A.W. (1981) Polyamine oxidase in oat leaves: a cell wall localized enzyme. Plant Physiol. 68, 494-498.
26. Kaur-Sawhney, R., Shih, L.M., Flores, H.E., Galston, A.W. (1982) Relation of polyamine synthesis and titer to ageing and senescence in oat leaves. Plant Physiol. 69, 405-410.
27. Kuehn, G.D., Affolter, H.-U., Atmar, V.J., Seebeck, T., Gubler, U., and Braun, R. (1979) Polyamine-mediated phosphorylation of a nucleolar protein from *Physarum polycephalum* that stimulates rRNA synthesis. Proc. Natl. Acad. Sci. USA 76, 2541-2545.

28. Kyriakidis, D.A. (1983) Effect of plant growth hormones and polyamines on ornithine decarboxylase activity during the germination of barley seeds. Physiol. Plant. 57, 499-504.
29. LeRudulier, D., Goas, G. (1975) Influence des ions ammonium et potassium sur l'accumulation de la putrescine chez les jeunes plantes de Soja hispida Moench, privées de leurs cotyledons. Physiol. Veg. 13, 125-136.
30. Malmberg, R.L., McIndoo, J. (1983) Abnormal floral development of a tobacco mutant with elevated polyamine levels. Nature 305, 623-625.
31. Mamont, P.S., Duchesne, M.-C., Grove, J., Bey, P. (1978) Anti- proliferative properties of DL-α-difluoromethyl ornithine in cultured cells. A consequence of the irreversible inhibition of ornithine decarboxylase. Biochem. Biophys. Res. Commun. 81, 58-66.
32. Martin-Tanguy, J., Cabanne, F., Perdrizet, E., and Martin, C. (1978) The distribution of hydroxycinnamic acid amides in flowering plants. Phytochem. 17, 1927-1928.
33. Metcalf, B.W., Bey. P., Danzin, C., Jung, M.J., Casara, P. and Vevert, J.P. (1978) Catalytic irreversible inhibition of mammalian ornithine decarboxylase (E.C.4.1.1.17) by substrate and product analogues. J. Am. Chem. Soc. 100, 2551-2553.
34. Mizrahi, Y., Heimer, Y.M. (1982) Increased activity of ornithine decarboxylase in tomato ovaries induced by auxin. Physiol. Plant. 54, 367-368.
35. Morel. C., Villanueva, V.R., Queiroz, O. (1980) Are polyamines involved in the induction and regulation of the crassulacean acid metabolism? Planta 149, 440-444.
36. Panagiotidis, C.A., Georgatsos, J.G., and Kyriakidis, D.A. (1982) Superinduction of cytosolic and chromatin-bound ornithine decarboxylase activities of germinating barley seeds by actinomycin D. FEBS Lett. 146, 193-196.
37. Pegg, A.E., Seeley, J., and Zagon, I.S. (1982) Autoradiographic identification of ornithine decarboxylase in mouse kidney by means of α-(5-^{14}C) difluoromethyl ornithine. Science 217, 68-70.
38. Priebe, A., Klein, H., and Jager, H.-J. (1978) Role of polyamines in S0$_2$-polluted pea plants. J. Exp. Bot. 29, 1045-1050.
39. Rajam, M.V., Galston, A.W. (1985) The effects of some polyamine biosynthetic inhibitors on growth and morphology of phytopathogenic fungi. Plant Cell Physiol. 26, 683-692.
40. Rajam, M.V., Weinstein, L.H., Galston, A.W. (1985) Prevention of a plant disease by specific inhibition of fungal polyamine biosynthesis. Proc. Natl. Acad. Sci. USA. 82, 6874-6878.
41. Richards, F.J., and Coleman, R.G. (1952) Occurrence of putrescine in potassium-deficient barley. Nature (London) 170, 460.
42. Roberts, D.R., Walker, M.A., Thompson, J.E. and Dumbroff, E.B. (1984) The effects of inhibitors of polyamine and ethylene biosynthesis on senescence, ethylene production and polyamine levels in cut carnation flowers. Plant Cell Physiol. 25, 315-322.
43. Sakai, T.T., Cohen, S.S. (1976) Effects of polyamines on the structure and reactivity of tRNA. Prog. Nucl. Acid Res. 17, 15-42.
43a. Shih, H-J. and Galston, A.W. (1985) Correlations between poly-amine ratios and growth pattens in seedling roots. Plant Growth Regulation 3, 353-363.
44. Slocum, R.D., Galston, A.W. (1985) Changes in polyamine biosynthesis associated with post-fertilization growth and development in tobacco ovary tissues. Plant Physiol. 79, 336-343.
45. Slocum, R.D., Kaur-Sawhney, R., Galston, A.W. (1984) The physiology and biochemistry of polyamines in plants. Arch. Biochem. Biophys. 235, 283-303.
46. Smith, M.A. Davies, P.J., and Reid, J.B. (1985) Role of polyamines in gibberellin-induced internode growth in peas. Plant Physiol. 78, 92-99.
47. Smith, T.A. (1976) Polyamine oxidase from barley and oats. Phytochemistry 15, 1565-1566.
48. Smith, T.A. (1985) Polyamines. Ann. Rev. Plant Physiol. 36, 117-143.

49. Smith, T.A., Negrel, J., Bird, C.R. (1983) The cinnamic acid amides of the di- and polyamines. In Advances in Polyamine Research 3, pp. 347-370, Caldarera, C.M., Zappia, V., Bachrach, U., eds., Raven Press, New York.

50. Srivastava, S.K., Raj, A.D.S., Naik, B.I. (1981) Polyamine metabolism during ageing and senescence of pea leaves. Indian J. Exp. Biol. 19, 437-40.

51. Suresh, M.R., Ramakrishna, S., Adiga, P.R. (1978) Regulation of arginine decarboxylase and putrescine levels in *Cucumis sativus* cotyledons. Phytochemistry 17, 57-63.

52. Suttle, J.C. (1981) Effects of polyamines on ethylene production. Phytochemistry 20, 1477-1488.

53. Tabor, C.W., Tabor, H. (1984) Polyamines. Ann. Rev. Biochem. 53, 749-790.

54. Tiburcio, A.F., Kaur-Sawhney, R., Galston, A.W. (1985) Correlation between polyamines and pyrrolidine alkaloids in developing tobacco callus. Plant Physiol. 78, 323-326.

55. Yamanoha, B., Cohen, S.S. (1985) S-adenosylmethionine decarboxylase and spermidine synthase from Chinese cabbage. Plant Physiol. 78, 784-790.

56. Young, N.D., Galston, A.W. (1983) Putrescine and acid stress: induction of arginine decarboxylase activity and putrescine accumulation by low pH. Plant Physiol. 71, 767-771.

57. Young, N.D., Galston, A.W. (1983) Are polyamines transported in etiolated peas? Plant Physiol. 73, 912-914.

58. Young, N.D., Galston, A.W. (1984) Physiological control of arginine decarboxylase activity in K-deficient oat shoots. Plant Physiol. 76, 331-335.

59. Young, N.D., (1984) Plant stress and polyamine metabolism: induction of arginine decarboxylase in stressed oat leaves. Ph.D. Dissertation, Yale University

E3. Gibberellins and Plant Cell Elongation

Jean-Pierre Métraux

Agricultural Division, CIBA-Geigy Ltd., 4002 Basle, Switzerland.

INTRODUCTION

Gibberellins (GAs) are well known for their spectacular effects in intact plants. They were first discovered in the secretory products of *Gibberella fujikuroi*, a fungus infecting rice seedlings (for an historical account see 47). Diseased plants grow tall and spindly and tend to fall over under their own weight. In 1926, Kurosawa showed that fungal extracts applied to plants could induce the same symptoms as the pathogen. After, two compounds were crystallized from extracts and given the names Gibberellin A and B. It was not until the 1950s that the first chemical structure of GA was characterized. During that period, a number of laboratories reported that extracts of higher plants could induce similar biological responses as those obtained with fungal GA. This opened the way to intensive analytical research and GAs were eventually detected in various taxa of lower and higher plants. The notion that GAs are in fact genuine plant growth regulators gradually emerged. Chemical identification was pursued vigorously: more than 70 different GAs have been discovered so far. The principal metabolic pathways have also been extensively documented (see chapter B2). Besides its effect on stem elongation GAs affect a number of physiological processes such as fruit and flower formation, dormancy of vegetative organs as well as seed germination (23). There is considerable diversity among the processes regulated by GA and the existence of a common mechanism of action at the molecular level remains speculative. Progress at elucidating the mode of action of GA has been slow mainly for the difficulty of separating the primary action from remotely connected effects. The use of simple biological systems consisting mostly of one tissue has greatly helped diminishing such problems. An example of a simple tissue is the aleurone layer of cereals used to study the regulation of seed germination by GA. In this tissue, GA controls the synthesis of hydrolytic enzymes by regulating the expression of their genes (22). Excised stem segments of lettuce or oat are other examples of simple target tissues used to analyze the mode of action of GAs on stem elongation (25, 31).

GA AND ELONGATION IN WHOLE PLANTS

GAs applied to intact plants induce sizable elongation of stem tissue, this effect being more pronounced in rosette or in dwarf species (23). Whole plants and particularly mutants with low GA levels were extensively used to identify the active structures of GA and their metabolisms. It was found that GA_1 is the active form of GA regulating shoot elongation in maize, rice and pea (48). These results were obtained by studying the response of a number of GA-mutants in which specific steps in the biosynthesis of GA_1 are blocked. In all cases normal growth resumes when exogenous GA_1 is added back to the normal level in the tissue. Whether control of shoot elongation by GA_1 is a general phenomenon remains unanswered, though GA_1 has been identified in a number of higher plant species (48). Further analyses of dwarf mutants are currently under way. A detailed account on the control of growth in tall plants by GA_1 is presented by J. Reid in chapter E4 of this Volume.

Since GA affects growth in intact plants, the possibility exists for practical use to increase yield of crop plants. Agricultural applications are exemplified by the use of GA to control stalk elongation in sugarcane (39).

GA AND ELONGATION IN EXCISED STEM TISSUE

The relative complexity of whole plants makes detailed studies on the effect of GA at the cellular and molecular level difficult. Interactions of various kinds (for example hormones, metabolites or tissue interactions) interfere with experimentation designed to monitor the sole effect of GA. An excised stem tissue which responds directly to GA, without participation of other factors, is therefore desirable. The reduction of an elongating plant to an excised stem tissue is often problematic: once excised, most stem tissues loose their ability to respond to exogenously supplied GA. The search for excised stem segments responding to GA proved successful for only a few tissues. Among those, hypocotyls of dark-grown lettuce were found to elongate markedly when treated with GA in the light (Fig. 1A and 2A)(54). Growth of *Avena* stems can be studied in GA-responsive stem sections of light-grown oat plants (Fig. 1B and 2B)(29). Excised stems of deep-water rice elongate in response to GA (Fig. 1C and 2C)(49). This is similar to elongation taking place when whole rice plants or sections are partially submerged by rising flood waters (41). Other examples of tissues retaining responsiveness to GA after excision are cucumber hypocotyls, wheat coleoptiles, and cowpea epicotyls (16, 28, 51).

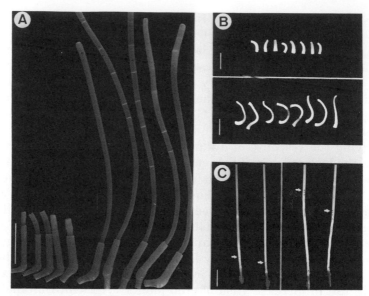

Fig. 1. A) Effect of GA$_3$ on the growth of *Avena* stem sections incubated in darkness for 48 h. Left: controls; right: GA$_3$ (30 μM). Sucrose (0.1 M) was added as a substrate in both treatments. The bar indicates 1 cm. (Courtesy of P. Kaufman, Univ. of Michigan, Ann Arbor). B) Effect of GA$_3$ on the growth of lettuce hypocotyls incubated in the light for 30 h. Top: controls; bottom: GA$_3$ (10 μM). KCl (10 mM) was present in both treatments. The bar indicates 4 mm. C) Effect of GA$_3$ on the growth of deep-water rice stem sections incubated in the light for 48 h. Left: controls; right: GA$_3$ (5 μM). The arrow indicates the position of an ink dot placed 1 cm above the node at time zero. The bar indicates 2 cm. Ink dots were placed 2 mm apart. Note that most of elongation takes place in the first zone above the node. (Courtesy of A. Bleecker, Plant Research Laboratory, Michigan State Univ., East Lansing).

The next sections will mainly focus on progress made with excised systems which are more amenable to simple experimentation.

Cellular Basis for Elongation

The increase in length of an extending shoot results from the increase in length of existing as well as newly divided cells. Cell division, or more simply the formation of new cross-walls does not, in itself, contribute to new volume, unless the recently formed cells expand. The effect of GA on stem extension has been considered with special emphasis on cell elongation. Histological observations on excised *Avena* stem sections indicated that new cells were mainly generated in the intercalary meristem (31). In this tissue GA causes cell divisions to stop while cell elongation is stimulated in the intercalary meristem and above it (29). In excised sections of lettuce hypocotyls cell divisions remain unaffected by GA treatments (Fig. 3). Suppression of cell divisions with either gamma-rays or 5-fluorodeoxyuridine leads to undiminished elongation of GA-treated sections (58). Elongation of newly formed cells is stimulated in

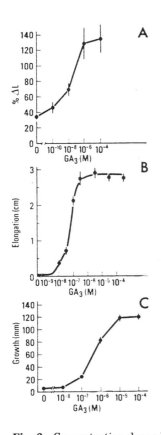

Fig. 2. Concentration dependence of the growth response in excised sections of lettuce hypocotyls (A); *Avena* stems (B); and deep-water rice stems (C). From (1,49,54)

GA-treated stem sections of deep-water rice (Table 1) (49). This is also found when stem sections of deep-water rice are partially submerged (42). These observations support the notion that the effect of GA is based on pronounced cellular elongation.

It should be pointed out that GA-induced stem growth can also be accompanied by an increase in cell number. This is born out by observations of subapical meristems in rosette and caulescent plants where both the rate of cell divisions and the size of the meristem are increased by GA (52). The same effects take place in the intercalary meristems of certain monocots (Table 1)(49). The effect of GA in relation to the mitotic process was studied with respect to the timing of the cell cycle. It was proposed that GA shortens the interphase by inducing cells at the G1 phase to synthesize DNA (38). The mode of action of GA in meristems deserves attention of its own. Perhaps the study of excised systems such as the highly responsive stems of deep-water rice might provide new insights in this process.

Effect of Light and Other Factors

Light

In excised lettuce hypocotyls, blue and far red light inhibit elongation and this effect is reversed by GA (54). Similar observations were made in intact lettuce seedlings (65). These studies indicated that phytochrome and blue light receptors are involved in this response (65). At other wavelengths or in the dark, lettuce sections grow rapidly and remain unaffected by GA (54). Peas, dwarf peas and cucumbers are examples of plants which are inhibited by far red light only. This inhibition is also reversed by GA (28, 37, 55). Inhibition by light reduces the sensitivity of pea stems to endogenous GA rather than changing the level of this hormone in the tissue (34). Light inhibition of growth also occurs in oat stems but application of GA does not fully restore the rate of elongation up to the level obtained in the dark (30). Thus reversal of light inhibition is an important physiological basis of GA-induced growth.

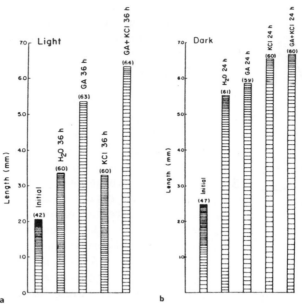

Fig. 3. Mean section length, cell length and number in excised lettuce hypocotyls before and after 36 h incubation in light a) or darkness b). From (58)

Table 1. Number and size of subepidermal cells in the growing regions of internodes of deep-water rice incubated in H_2O or in GA_3 (5 µM) for 2 days. The counts were made in the regions that had elongated during the last 20 h of incubation. From (49)

Treatment	Increase in internodal length	Total number of cells in one file of the newly elongated region	Average cell length
	mm		µm
H_2O	1.1 ± 0.4	18.7 ± 5.7	57.3 ± 2.6
5 µM GA_3	41.3 ± 2.9	308.4 ± 25.1	135.0 ± 3.8

Auxin

Since other hormones such as auxin can modulate elongation of excised stem tissue, GA-induced growth has been explained by an effect of GA on auxin biosynthesis, auxin being the real regulator of growth. Excised stem tissues where auxin and/or GA show pronounced effects on growth should be considered critically when analyzing the mode of action of GA. In such cases, a contribution of auxin to the regulation of growth cannot be ignored a priori and the experimental appropriateness of such materials for the study of GA has to be carefully evaluated. Pea stems are an example of a tissue where GA enhances auxin biosynthesis (36). GA increases the levels of free auxin by an effect on the racemisation of L-tryptophan to D-tryptophan. L-tryptophan has no effect on elongation whereas D-tryptophan stimulates the elongation of excised stem segments in a similar way as auxin does. D-tryptophan is limiting for the synthesis of auxin and is proposed to be its precursor (36). In lettuce hypocotyls, various natural and synthetic auxins have no effect on elongation (24, 43, 54). Moreover antiauxins do not prevent GA-induced growth (24). In *Avena* stem sections, the situation is somewhat different. Elongation of the intercalary meristem is reduced by application of physiological concentrations of auxin in GA-treated stems in the dark or in the light (30). Auxin alone inhibits elongation in the dark and enhances it somewhat in the light (3, 30). GA-stimulated elongation in *Avena* stems in the dark is unlikely to be mediated by auxin since under these conditions auxin is inhibitory (30). Interestingly, when *Avena* stem sections are pretreated with auxin, the response time of a subsequent exposure to GA is shortened considerably. It was hypothesized that a factor required for GA-induced growth is more rapidly produced in auxin-pretreated tissue (3).

Ethylene

Ethylene has been shown to mediate the effect of GA in aquatic and semi-aquatic plants (46, 49). Partially submerged plants produce elongated internodes. Ethylene induces similar internode elongation in air-grown plants (46, 49). Sensitivity of the tissue to ethylene was proposed as a mechanism by which these plants adapt to the water level (46). At the surface of the water, ethylene produced by the plants diffuses rapidly into the air. When the water level rises, ethylene accumulates in the submerged tissue and causes growth. Inhibitors of GA biosynthesis prevent growth of submerged and of ethylene-treated plants (46, 49). Ethylene affects the responsiveness of the tissue to endogenous GA (46, 49). Such a modulatory mechanism could possibly be achieved by modifications of a hypothetical receptor site for GA or of biochemical responses taking place after the primary action of GA. The effect of ethylene was also tested in lettuce hypocotyls. Growth, measured either as increase in length or in fresh weight, remained unaffected by ethylene at concentrations up to 5000 ppm, independently of the presence of GA (24).

Dihydroconiferyl Alcohol

In lettuce hypocotyls GA-induced growth is influenced by the presence of cotyledons (26, 54). Growth is impaired in seedlings with severed cotyledons, while the relative growth increase induced by GA_3 remains the same (54). A factor was isolated from cotyledons and identified as dihydroconiferyl alcohol (DCA) which enhances GA-induced growth but has no effect when applied alone (53). The possibility that GA-induced growth is modulated by a factor produced in the cotyledons has, however, been reevaluated (45). No evidence was found for the involvement of a cotyledon factor, including DCA, in GA-induced growth of lettuce hypocotyls (45). The decrease in growth after excision of cotyledons was explained by the removal of fast growing meristematic cells which are responsive to GA (45, 54).

Growth Kinetics

The comparison of the growth response at different GA concentrations indicates that it is the rate of elongation which is affected and not the duration of the growth response (25, 31). Rapid growth measuring-devices were used to determine the kinetics of changes in elongation rate after GA treatment. In lettuce hypocotyls the rate of elongation is increased 10 to 15 min. after the application of the hormone (Fig. 4A)(43), for *Avena* stems the response-time is 3.5 hours (31). This lag-time can be shortened to about 20 min. by a pretreatment with auxin (Fig. 4B)(3). In lettuce maximal growth rate was obtained one hour after treatment (43). The rapid initiation of growth by GA in lettuce hypocotyls compares well with the timing of auxin-induced growth in other tissues (11). The implication

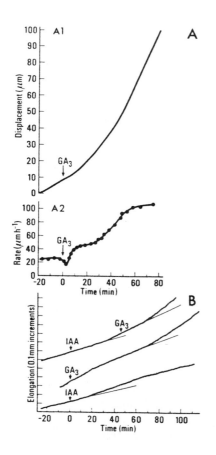

Fig. 4. A) Short-term kinetics of the response of a single section of lettuce hypocotyl to GA_3 (1 µM). Data are presented as displacement (A1) and calculated rates (A2). From (43) B) Fine kinetics of growth in single sections of *Avena* stems in response to IAA (100 µM), GA_3 (100 µM) or IAA followed by GA_3 alone. From (3)

302

of the rapid short term response with respect to possible mode of action will be discussed later. The rapidity of the GA response in lettuce might be partly explained by the dimensions of the tissue: the ratio surface to volume is particularly elevated which favors the penetration of GA to the target cells. The persistance of the GA response after removal of the hormone is also noteworthy: a short exposure to GA is sufficient to induce a sustained growth response, as observed in excised stems of lettuce or oats (25, 31). By comparison, auxin needs to be continuously present in the medium to maintain high growth rate (25). Good retention by the tissue and slow metabolism were proposed to explain this aspect of the GA response (25).

Control of the Rate of Elongation

Growth of plant cells results from the irreversible yielding of the cell wall to the force imposed by turgor. However, yielding of the cell wall is not a purely physical process but is controlled by cell wall metabolism. This is well demonstrated by tests of cell wall extensibility where the input of metabolism can be disconnected from the extensile behaviour. Growth of a living tissue is linear over shorter time intervals while cell walls *in vitro* deform non-linearly when the force of turgor is replaced by uniaxial or multiaxial loads (61). Similarly, maintenance of turgor is also dependent on metabolism. Elongation can theoretically be controlled by the rate of water uptake or by changes in the mechanical properties of the cell wall. The effects of hormones have been repeatedly analysed with respect to changes in turgor pressure or with respect to the rheology of the cell wall. Generally, growth enhancement by GA does not seem to result from an increase in turgor alone. Experiments were undertaken to determine the regulatory role of turgor in GA-treated sections of lettuce hypocotyls. Turgor can be manipulated via the internal solute concentration by incubating sections in KCl solutions of various strengths. Growth rate is not necessarily increased by higher solute concentrations in the tissue, unless GA is present, which shows that GA must affect the yielding properties of the cell wall (59). A time-course of cation uptake and cell sap osmolality shows that, while cell sap osmolality of KCl-treated sections increases as a result of K^+ uptake, growth is not above that of controls (Fig. 5). Addition of GA to KCl-treated sections results in an enhancement of growth without an increase in osmolality when compared to KCl-treated controls. Osmolality in fast-growing GA-treated hypocotyls decreases the same way as in controls (59). These results indicate, that unless the yielding properties of the cell wall are changed, the rate of elongation is unlikely to be affected by changes in turgor due to cell sap osmoticum. Water transport into expanding cells could be another factor limiting the rate of elongation. Using a pressure-probe technique whereby turgor is monitored continuously in a single cell

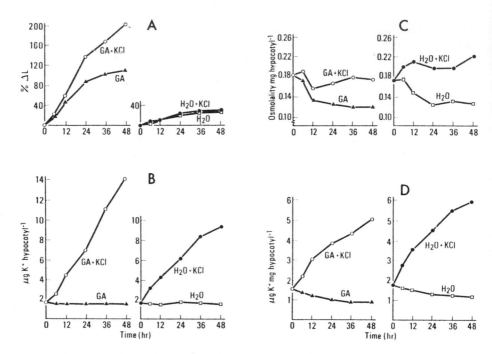

Fig. 5. Analyses of growth (A), K^+ uptake (B, D) and osmolality of expressed cell sap (C) in excised light-grown lettuce hypocotyl sections. From (59)

by a micromanometer, water transport was not found to be limiting for cell expansion (8). Auxin, which induces a rapid increase in the rate of elongation, has no effect on hydraulic conductance in pea internode tissue measured at the cellular or at the tissue level (8). The pressure-probe technique might prove useful to further investigate the biophysics of GA-induced elongation. GA-enhanced growth of lettuce hypocotyls in sections or seedlings is accompanied by an increase in cell wall extensibility as measured with an extensometer or by an osmotic technique (Fig. 6)(27, 59). The physical properties of the cell walls in *Avena* stem sections also change after GA-treatments (Fig, 7)(2). In cucumber seedlings where the response to GA is rather small when compared to that in lettuce hypocotyls or oat stems, elongation is the result of a decrease in osmotic potential while cell wall extensibility remains unaffected (28). In this tissue the situation is somewhat complicated by the high responsiveness to auxin which results in a strong growth response accompanied by an increase in cell wall extensibility (28). To be significant for the onset of the growth response, changes in cell wall properties should at least coincide in time with the increase in growth rate induced by GA. In *Avena* stems the plasticity of the cell walls decreases 1-2 hours after exposure to GA (Fig. 7) (2). This precedes the response time for GA-induced growth reported to be 3.5 hours (31). This result should now be evaluated in the light of the

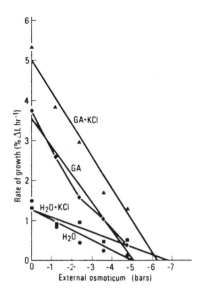

Fig. 6. Determination of extensibility and water potential of H_2O-, GA_3- and KCl-treated hypocotyl sections incubated in light in mannitol for 6 h. GA_3, 5 µg/ml, KCl, 10 mM. Extensibility is measured by the slope of the line relating growth to external osmoticum. From (59)

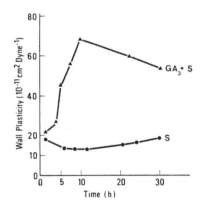

Fig. 7. Changes in extensibility (measured as irreversible plastic deformation) of *Avena* stem sections incubated in sucrose (S) (0.1 M) or sucrose and GA_3 (30 µM). From (2)

shorter lag-times published for auxin-pretreated *Avena* stem sections (3). For technical reasons (dimension of the tissue or rheological method) such measurements are often difficult to carry out in detailed time-lapse as would be ideally desired, and one has to rely on data obtained much after the rate of elongation has increased. This is the case in lettuce where changes in extensibility were recorded several hours after GA treatment. The weight of the evidence, considering the results obtained in the turgor experiments, still strongly suggests that GA affects the yielding properties of cell walls in lettuce hypocotyl sections. In auxin-sensitive tissues, high rates of elongation are also paralelled by increased cell wall yielding and a decrease in turgor (62). These findings have prompted a number of studies on the biochemical basis of GA-induced changes in cell wall extensibility.

Hypotheses on the Mode of Action

The Role of Cell Wall Acidification

Evidence presented above suggests certain analogies between auxin and GA induced elongation. Both hormones rapidly affect elongation as well as cell wall extensibility. It is not surprising that attempts were made to explain rapid effects of GA in terms of the acid growth hypothesis, proposed to explain auxin-induced growth. The theory is based on acid stimulated growth and states that auxin stimulates the activity of a plasma membrane proton pump which acidifies the cell wall space, causing cell wall loosening and consequently growth (7). Usually, four predictions are used to determine if the acid growth theory applies to an auxin-responsive tissue: 1) auxin induces cells to acidify the cell wall space and the lag-time of acidification is similar to that of auxin-induced growth; 2) acidic solutions must be able to substitute for auxin in promoting growth; 3) if acidification is prevented by neutral buffers or inhibitors of proton secretion, auxin-induced growth is also blocked. 4) If acidification is induced by other compounds such as fusicoccin (FC) or α-naphtylacetate, growth is enhanced. In its ability to simply explain auxin-induced growth, this theory has received widespread attention and has generated an impressive amount of research. Attempts were made to explain GA-induced growth by the acid growth hypothesis. In lettuce hypocotyls acidic solutions rapidly enhance elongation (60). However, GA-induced elongation is not accompanied by a sizeable change in medium acidification. This does not result from the lack of detection for technical reasons nor to the inability of the tissue to acidify the medium since: i) the experiments were performed with scrubbed hypocotyls which allows a good diffusion of chemicals in and out of the tissue and ii) during growth enhancement with FC and KCl, the pH of the medium is readily observed to drop (Fig. 8)(60). Interestingly, when sections are incubated in FC without KCl in the medium enhancement of growth is equally strong but medium acidification is absent (60). The presence of salts in the outside medium affects acidification in a crucial way. Measurable acidification in auxin treated pea stems is also dependent on the presence of potassium ions in the outside medium (63). The failure to detect acidification could be attributed to buffering by the cell wall, secreted protons being exchanged for the counterions of negatively charged cell wall components. Cations present in the external medium displace this exchange so that the less retained protons become measurable in the outside medium. This argument may hold for FC but not for GA-treatments, since addition of KCl to GA-induced sections does not result in acidification (60). A shift from pH 6 to pH 4 causes a transient increase in growth rate which

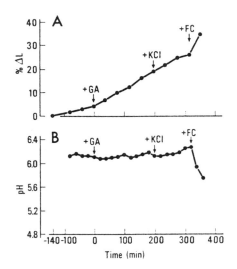

Fig. 8. The effect of GA_3, $GA_3 + KCl$, and $GA_3 + KCl + FC$ on elongation (A) and medium acidification (B) by scrubbed lettuce hypocotyl sections preincubated in H_2O. From (60)

reaches values comparable to those obtained with GA. When GA is added to sections growing at pH 4, the response is similar to that measured at pH 6 (43). If GA promotes proton secretion, then addition of GA to tissue already at low pH should not lead to further growth enhancement. This provides additional evidence that GA-induced growth is mediated by other mechanisms than acidification. *Avena* stem segments also respond to acidic solutions but acidify their medium after GA treatments. However, GA-induced acidification is only increased by 17 % over controls after a period of 90 min (31). This is a small effect comparatively to auxin treatments where acidification increases more than 10 fold (7, 62). As in lettuce hypocotyls, *Avena* stems respond fully to GA, irrespectively of the pH of the external medium (31). Again, this would not be expected should GA induce growth by acidifying the cell wall. In conclusion, the experimental evidence indicates that GA induces stem elongation in lettuce or oats by other mechanisms than acidification of the cell wall space. Since the pH of the cell wall solution is not limiting, other factors have to be considered for the regulation of elongation.

The Role of Calcium

An alternative proposal to explain the rapid effect of GA on elongation has been advanced by Moll and Jones (44). This hypothesis capitalizes on the role of Ca^{++}-ions and is based on the observation that $CaCl_2$ solutions rapidly block GA-induced elongation in lettuce hypocotyls. When $CaCl_2$ is added to untreated controls, elongation decreases within minutes. This inhibition can be reversed by GA but not by acidic solutions (Fig. 9)(44). EGTA or EDTA also reverse Ca^{++}-inhibited growth (44). Pretreatment of

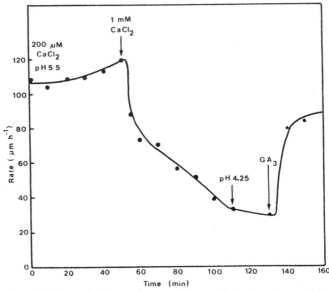

Fig. 9. The effect of CaCl$_2$ and GA$_3$ on the growth rate of a lettuce hypocotyl in a flow-past experiment. The initial medium was 200 µM CaCl$_2$, 1 mM HEPES, 1 mM KCl, 10 µM EDTA, pH 5.5. Changes in the medium are indicated with arrows. At 50 min, the CaCl$_2$ concentration was increased to 1 mM CaCl$_2$ with no other change in the medium. 50 µm EDTA was added at 80 min (not indicated) with no effect on the growth rate. At 105 min, the medium was switched to pH 4.25, 1 mM CaCl$_2$ still present and the other components the same. At 130 min, GA$_3$ (50 µM) was added to this same medium. From (44)

sections with ruthenium red, which prevents movement of Ca^{++}-ions across membranes inhibits GA-induced growth, while leaving controls unaffected (25). The growth rate of ruthenium red-treated sections is increased by EGTA (25). Both EGTA and ruthenium have been shown to remain on the outside of the plasma membrane in animal tissues (25). If these drugs behave similarly in cells of lettuce hypocotyls, then the level of Ca^{++}-ions in the cell wall and the movement of Ca^{++}-ions into the cell might be important for early events in the stimulation of growth by GA. The first component, e.g. the removal of Ca^{++}-ions bound to the cell wall, could have a direct effect on wall extensibility and growth since Ca^{++}-ions reduce the extensibility of cell walls in algae and higher plants (62). The inhibition by Ca^{++}-ions seems to result from non-covalent ionic interactions between Ca^{++} and crosslinked cell wall polymers (62). The other component of the calcium hypothesis deals with intracellular calcium concentrations which might be as critical for the control of growth. Many physiological processes in plants have been proposed to be controlled by calcium levels (9). The passage of calcium into the cytosol either from higher extracellular concentrations or from intracellular compartments leads to activation of the calcium-binding protein calmodulin. Activation of calmodulin is a prerequisite for diverse regulatory responses (9). Gravitropism, phototropism, auxin transport, cytokinin action were all demonstrated to have specific calcium requirements. Auxin-induced growth was proposed to be accompanied by an increase in cytoplasmic Ca^{++} (20). An increase in cytoplasmic calcium concentration is thought to stimulate vesicle flow from the Golgi to the cell wall. This increases the rate of cell wall synthesis and turnover which in turn increases the rate of elongation (20). A precedent for Ca^{++}-controlled vesicle flow exists in animal cells where Ca^{++}-ions trigger presynaptic vesicles. While attractive, the calcium hypothesis warrants further attention. First of all there is a need to work out a clean experimental technique allowing the intracellular calcium concentrations to be critically determined. The regulatory functions of GA-induced changes in calcium concentrations and the mechanisms controlling Ca^{++}-ion release by GA are other questions to be tackled.

The Role of Cell Wall Metabolism

Changes in cell wall extensibility and growth could presumably arise from changes in chemical composition of the cell wall. The effect of GA on cell wall metabolism has been considered largely with respect to cell wall synthesis. *Avena* internodes incorporate ^{14}C-labeled glucose in the cell wall at increased rates after about 1 hour of exposure to GA (31). This corresponds to the lag-time of GA-induced elongation in this tissue. Sustained cell wall synthesis in GA-treated *Avena* stems was also observed microscopically. Meristematic cells maintain their thickness during the rapid extension process (31). Despite a strong correlation, it is difficult to determine if changes in incorporation occurred prior to the increase in rate of elongation. In lettuce seedlings, incorporation of labeled glucose into the cell wall is not changed by treatments with GA for the first 8 hours of incubation, a period after which the rate of incorporation drops in both treatments (56). In the same material, total cell wall mass increases in parallel with the rate of elongation, but the increase in the rate of elongation precedes the increase in production of new wall materials (56). This indicates that GA-induced cell wall synthesis is not causative for growth, but rather functions to maintain the structural integrity of the cell wall undergoing rapid elongation. The carbohydrate content of the pectic, hemicellulosic and of the cellulosic fraction increases in seedlings exposed to GA for 48 hours (27). Such end-point measurements do not allow the causative role of wall synthesis to be evaluated. Synthesis of cell wall materials as a regulatory mechanism of growth should not be completely dismissed. Finer analytical methods are needed to follow compositional changes in cell wall polymers along the time-course of the growth response. Furthermore, experiments with single cells of *Nitella* and with stem sections support the concept that the inner layer of the cell wall predominantly controls growth (50, 61, 62). The analytical approach should therefore be focused on this part of the cell wall which is newly synthesized during GA-treatment.

Inhibitors have been utilized to determine how synthesis of cell wall polymers affects the rate of elongation. 2,6-Dichlorobenzonitrile, an inhibitor of cellulose synthesis, supressed GA-enhanced growth of *Avena* stems (31). That cellulose synthesis is a prerequisite for GA-enhanced growth is very likely. It probably is a part of the synthetic activity sustaining the long-term growth response. It remains to be seen if early events of GA-induced growth, particularly the rapid increase in growth rate, are dependent on cellulose synthesis. In single cells of *Nitella*, cellulose synthesis is not coupled to the control of the rate of elongation (50). Cell wall metabolism should also be considered in terms of turnover of cell wall components During this process, stress-bearing bonds are broken, possibly by cell wall hydrolases, allowing for extension and release of cell wall polymers which are replaced in subsequent synthetic steps (35). Pulse-labelling of cell walls in pea internode sections provided evidence for an effect on the release of xyloglucan after auxin treatment (35). This observation was later confirmed using a centrifugation technique to remove water-soluble components from the cell wall space (63). The release of xyloglucan from stems with acidic solutions was also documented and supported the basic premises of the acid growth hypothesis (21). Cell wall preparations from pea stem homogenates did not release xyloglucan after treatment with low pH buffers (5). Soluble xyloglucan might have leached from the cell walls during the preparation procedure (5). The release of xyloglucan *in vivo* by auxin or low pH, which was thought to result from covalent bond breaking, was reevaluated in the light of such experiments. It was suggested that the binding of highly water-soluble xyloglucans was pH dependent and resulted from a physical rather than chemical change in the cell wall (5). In lettuce hypocotyls, GA has no effect on xylose and glucose content of cell wall materials recovered from the incubation medium or from wall preparations obtained by tissue homogenization which seems to rule out a role for the release of xyloglucan in the control of GA-induced growth (24). This experimental approach might be worthwhile pursuing and emphasis should also be given to other cell wall components.

The Role of Peroxidases and Phenolic Cell Wall Components

The conceptual ground for some of the hypotheses discussed above is that GA enhances growth by increasing the extensibility of the cell wall. The ideas exposed in this section take the opposite view, namely that GA prevents reactions which otherwise cause stiffening of the cell wall. Cell walls of suspension-cultured spinach cells contain esterified ferulic acid which is attached to non-reducing pectinic carbohydrate moieties (15). Adjacent ferulic acid residues are catalyzed into diferuloyl crosslinks by a cell wall peroxidase (Fig. 10A)(12, 13). This is thought to stiffen the cell wall structure and consequently to limit its expansion. The action of cell wall peroxidases is thus growth limiting. While enhancing the growth rate of cultured spinach cells, GA was found to suppress the secretion of a peroxidase into the culture medium (12, 13). GA was proposed to regulate cell expansion by inhibiting peroxidase activity (12, 13). In spinach cells peroxidase also catalyzes the conversion of water-soluble phenolics into hydrophobic quinones or quinone polymers (13). The inhibition of peroxidase activity by GA results in a decrease in hydrophobicity of the cell wall, making the cell wall more available to the plasticizing effect of water and to the loosening action of hydrolases. The control of cell wall hydrophobicity might be another growth-limiting function of peroxidase.

Peroxidase also seems to be involved in the crosslinking of the cell wall protein extensin. This protein is bound to cell wall phenolics rather than polysaccharides. Extensin can only be removed from cell walls by solubilization in $NaClO_3$ while anhydrous hydrogen fluoride which solubilizes cell wall carbohydrates has no effect (14). Extensin molecules are crosslinked by isodityrosil residues (14, 40). These residues are

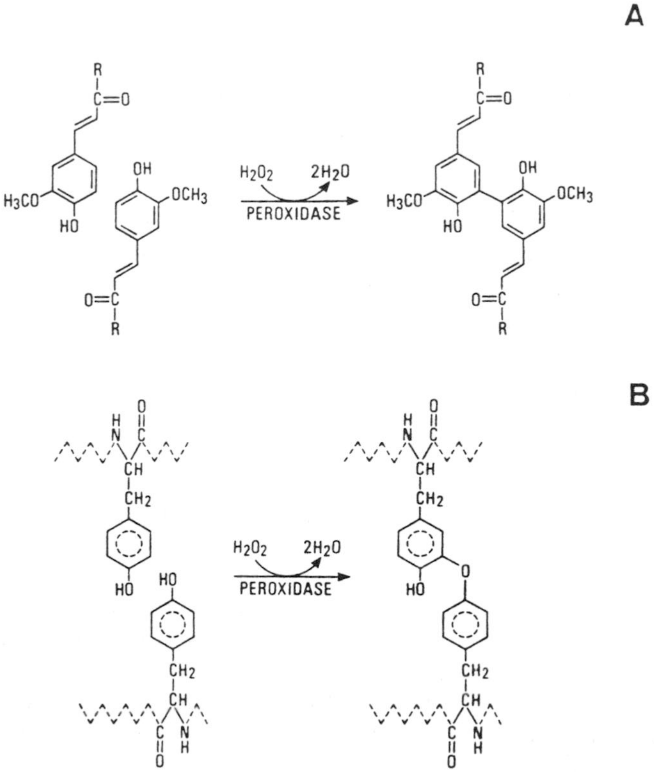

Fig. 10. A) Proposed mechanism for the association of feruloyl esters to form diferulate-polysaccharide crosslinks (From 12). B) Proposed mechanism for the formation of isodityrosyl residues that crosslink extensin molecules. From (40)

interpolypeptide bridges, somewhat analogous to the S-S cyclic cysteine peptides of polypeptide chains. A peroxidase is likely to be involved in the synthesis of isodityrosyl residues (Fig. 10 B)(14). This suggests that peroxidase controls the formation of the extensin network. It indicates another way by which peroxidases could be implicated in the regulation of cell wall extensibility and growth.

The location of the peroxidase in the cell wall is essential for the hypotheses described above. To maintain the peroxidase active, hydrogen peroxide must be available in the cell wall. A number of studies provided evidence for the presence of peroxidases in cell walls (6, 17). In horseradish, hydrogen peroxide was shown to be produced in cell walls at the expense of either NADPH or NADH (10). These cell walls also contain NAD-specific malate dehydrogenase (MDH) which reduces malate to produce NADH (19). Byl and Terry (6) have reported the presence of peroxidase, MDH and glutamate oxaloacetate transferase (GOT) in the cell wall solution of sugarcane leaf spindles as well as in pea stems. In sections of sugarcane leaf spindles, GA enhances elongation with a concomitant decrease in activity of cell wall located peroxidase, MDH and GOT (64). This is consistent with the idea of a malate-oxaloacetate shuttle across the plasma membrane to transport oxaloacetate back into the cytoplasm (19). Such a scheme includes the requirements for a functional peroxidase system in the cell wall. These observations need further confirmation and represent a promising starting point for investigations in other GA-responsive tissues.

The regulation of peroxidase activity by GA remains unknown. The observation that Ca^{++}-ions affect peroxidase secretion in sugarbeet cells could be of interest in this context (18). The effect of GA on calcium movement described above might conceivably be tied to the secretion of peroxidase.

The regulation of cell wall lignification by light and GA has been considered as a possible mechanism controlling elongation in lettuce hypocotyls. The effect of light and GA was studied on the activity of phenylanaline ammonia-lyase (PAL). This enzyme controls the production of phenylpropanoid precursors of ferulic acid and lignin. In the light, lettuce hypocotyls show a high activity of PAL while growth is inhibited. An inverse relationship exists also in GA-treated sections where high rates of elongation are correlated with a low activity of the enzyme (4). An increase in cell elongation would have been expected should phenylpropanoids synthesized via PAL be involved in light-induced inhibition of growth. This was not found to be the case. When sections are incubated in the light with α-aminooxy-β-phenylpropionic acid (AOPP), an inhibitor of PAL, the activity of the enzyme is completely inhibited without influencing elongation (4). Similar experiments with t-cinnamic acid also show that the products of PAL activity are not limiting elongation of GA-treated lettuce hypocotyl sections (4).

GA Binding

Structural similarities between animal steroids and GA's suggest a mode of action of GA analogous to animal steroid hormones. The prominent features in the mode of action of animal steroids are the existence of a receptor protein in the cytosol, migration and association of the receptor-hormone complex to the nucleus and formation of m-RNA which is translated into proteins in the cytoplasm. Up to now there is no good evidence which indicates that GA action conforms to this model. Efforts were repeatedly made to find cytoplasmic binding sites for GA. In homogenates of cucumber hypocotyls, a soluble proteinaceous macromolecular component was found which stably and reversibly binds to [^{3}H]GA$_4$ (32). Those GAs which are active in the cucumber bioassay compete with GA$_4$ for the binding of the protein. These experiments were repeated using another binding assay in which equilibrium dialysis is replaced by filtration through DEAE-cellulose filter discs to separate free GA from GA bound to the receptor. Saturable binding of GA$_4$ to a cytoplasmic protein was found (32). A strong correlation also exists between binding affinity and biological activity in the cucumber bioassay. Binding specificity relating to biological activity is an important criterion for a putative hormone receptor. Another type of GA-binding , somewhat unrelated to the steroid model, has been described by Stoddart and coworkers (57). GA binds stably to a homogenate of lettuce hypocotyl tissue. GA binding is associated with a fraction which sediments at low centrifugal forces and is enriched in cell walls. Binding of [^{3}H]GA$_1$ to this cell wall fraction can be decreased by unlabelled GA$_1$ or GA$_9$ which are active in the elongation test, but not by GA$_8$ which is inactive. No other interactions between tissue components and GA$_1$ were observed. The stability of GA$_1$ binding suggests a covalent interaction. When growth is inhibited by L-2-(4 methyl-2,6-dinitroanilino)-N-methyl propionamide (MDMP), incorporation of [3H]GA$_1$ to the wall fraction is also decreased. The causality of GA-binding and elongation remains to be evaluated, although it might be of significance in this system, where the effect of GA is confined to elongation and is related to the extensibility of the cell wall. Though interesting, current knowledge on binding sites represent only first steps in our approach to the understanding of the perception mechanisms of GA. The major tasks ahead are the elucidation of the function of receptors and an understanding of the events taking place between the binding and the response.

CONCLUDING REMARKS

From the foregoing it appears that the use of excised stem tissues has allowed significant progress in the understanding of the physiological processes underlying GA-induced stem elongation. One must also recognize areas of ignorance, particularly with respect to early events of the GA response. For instance, the search for GA receptors has remained particularly unyielding and will continue to be a cardinal objective for future research. In a recent and refreshing look at the concept of plant hormones, analogies were drawn between plant hormone action and adaptation processes in sensory physiology (33). The broad dose-response curves of many plant hormones as well as the often recorded biphasic growth responses suggest that plant hormones act similarly as sensory stimulants. The biochemical mechanisms underlying adaptation to sensory stimulants in bacteria involve modifications of the receptor for the stimulant (mainly by protein methylation). If adaptation mechanisms can be clearly substantiated in plants then the search for proteins modified during adaptation might provide a new basis to approach the search for receptors. Another obscure area is the often hypothesized activation of hormone-specific genes, the product of which should lead to an action in the cell wall causing it to become extensible under the force of turgor. The credibility of this hypothesis suffered somewhat after the findings of very rapid induction times for hormone-induced elongation These lag periods (10–15 min.) were thought to be too rapid for gene expression to be involved. Recent findings of rapid effects of auxin on translational processes which correlate in time with the onset of elongation, spurred renewed interest in such mechanisms. New insights can now be gained at the regulation and eventually at the localization and characterization of auxin-specific genes involved in the control of elongation. Regulation of gene expression in GA-responsive stem tissues awaits now attention of its own. Encouraging results obtained with studies on the role of calcium in the rapid response to GA suggest new alternatives to probe into the mode of action of GA. At the cellular level, the regulation of cell wall extension in GA-treated tissue deserves further investigations. This applies particularly to the biochemical and physiological significance of various peroxidase-controlled stiffening mechanisms.

References

1. Adams, P.A., Kaufman, P.B., Ikuma, H. (1973) Effects of GA and sucrose on the growth of oat stem segments. Plant Physiol. 51, 1102-1108.
2. Adams, P.A., Montague, M.J., Tepfer, M., Rayle, D.L., Ikuma, H., Kaufman, P. B. (1975) Effects of gibberellic acid on the plasticity and elasticity of *Avena* stem segments. Plant Physiol. 56, 757-760.

3. Adams, P.A., Ross, M.A. (1983) Interaction of indoleacetic acid and gibberellic acid in short-term growth kinetics of oat stem segments. Plant Physiol. 73, 566-568.

4. Barnes, L., Jones, R.L. (1984) Regulation of phenylanaline ammonia-lyase activity and growth in lettuce by light and gibberellic acid. Plant Cell and Environm. 4, 89-95.

5. Bates, G.W., Ray, P.M. (1981) pH dependent interactions between pea cell wall polymers possibly involved in wall deposition and growth. Plant Physiol. 68, 158-164.

6. Byl, T., Terry, M.E. (1985) The cell wall peroxidase system and plant growth. Plant Physiol. 77, suppl.4.

7. Cleland, R.E., Rayle, D.L. (1978) Auxin, H^+-excretion, and cell elongation. Bot. Mag. (Tokyo), Spec. Issue, 1, 125-139.

8. Cosgrove, D.J., Cleland, R.E. (1983) Osmotic properties of pea internodes in relation to growth and auxin action. Plant Physiol. 72, 332-338.

9. Dieter, P. (1984) Calmodulin and calmodulin-mediated processes in plants. Plant Cell Environm. 7, 371-380.

10. Elstner, E.F., Heupel, A. (1976) Formation of hydrogen peroxide by isolated cell walls from horseradish (Armoracia lapathifolia Gilib.) Planta, 130, 175-180.

11. Evans, M.L. (1974) Rapid responses to plant hormones. Ann. Rev. Plant Physiol. 25; 195-223.

12. Fry, S.C. (1979) Phenolic components of the primary cell wall and their possible role in the hormonal regulation of growth. Planta 146, 343-352.

13. Fry, S.C. (1980) Gibberellin-controlled pectinic acid and protein secretion in growing cells. Phytochemistry 19, 735-740.

14. Fry, S.C. (1982) Isodityrosine, a new cross-linking amino acid from plant cell wall glycoprotein. Biochem. J. 204, 449-455.

15. Fry, S.C. (1983) Feruloylated pectins from the primary cell wall: Their structure and possible functions. Planta 157, 111-123.

16. Garcia-Martinez, J.L., Rappaport, L. (1984) Physiology of gibberellin-induced elongation of epicotyl explants from Vigna sinensis. Plant growth regulation 2, 197-208.

17. Gaspar, Th., Penel, Ch., Thorpe, T., Greppin, H. 1982) Peroxidases 1970-1980: A Survey of their Biochemical and Physiological Roles in Higher Plants. Université de Genève, Centre de Botanique, Genève.

18. Gaspar, T., Kevers, C., Penel, C., Greppin, H. (1983) Auxin control of calcium-mediated peroxidase secretion by auxin-dependent and auxin-independent sugarbeet cells. Phytochemistry 22, 2657-2660.

19. Gross, G. (1977) Cell wall-bound malate dehydrogenase from horseradish. Phytochemistry 16, 319-321.

20. Hertel, R. (1983) The mechanism of auxin transport as a model for auxin action. Z. Pflanzenphysiologie 112, 53-67.

21. Jacobs, M., Ray, P.M. (1975) Promotion of xyloglucan metabolism by acid pH. Plant Physiol. 56, 373-376.

22. Jacobsen, J.V. (1983) Regulation of protein synthesis in aleurone cells by GA and ABA. In: The Biochemistry and Physiology of Gibberellins. Vol. 2, pp. 159-187, Crozier, A., Ed. Praeger, New York.

23. Jones, R.L. (1973) Gibberellins: their physiological role. Ann. Rev. Plant Physiol. 24, 571-598.

24. Jones, R.L. (1980) The physiology of gibberellin-induced elongation. In: Plant Growth Substances 1979. pp 188-195. Skoog, F., Ed. Springer Verlag, Berlin, Heidelberg, New-York.

25. Jones, R.L., Moll, C. (1983) Gibberellin-induced growth in excised lettuce hypocotyls. In: The Biochemistry and Physiology of Gibberellins, pp. 95-128, Crozier, A., Ed. Praeger, New York.

26. Kamisaka, S. (1973) Requirements of cotyledons for gibberellic acid-induced hypocotyl elongation in lettuce seedlings. Isolation of the cotyledon factor active in enhancing the effect of gibberellic acid. Plant Cell Physiol. 14, 747-755

27. Katsu, N., Kamisaka, S. (1983) Quantitative and qualitative changes in cell wall polysaccharides in relation to growth and cell wall loosening in *Lactuca sativa* hypocotyls. Physiol. Plant. 58, 33-40.
28. Katsumi, M., Kazama, H. (1978) Gibberellin control of cell elongation in cucumber hypocotyls. Bot. Mag. (Tokyo), Special Issue No. 1, 141-185.
29. Kaufman, P.B. (1965) The effects of growth substances on intercalary growth and cellular differentiation in developing internodes of *Avena sativa*. II. The effects of gibberellic acid. Physiol. Plantarum 18, 703-724.
30. Kaufmann, P.B. (1967) Role of gibberellins in the control of intercalary growth and cellular differentiation in developing *Avena* internodes. Ann. N.Y. Acad. Sci. 144, 191-203.
31. Kaufman, P.B., Dayanandan, P. (1983) Gibberellin-induced growth in *Avena* internodes. In : The Biochemistry and Physiology of Gibberellins, vol 2, pp. 129-157, Crozier, A., Ed. Praeger, New York.
32. Keith, B., Brown, S., Srivastava, L.M. (1982) *In vitro* binding of gibberellin A$_4$ to extracts of cucumber measured by using DEAE-cellulose filters. Proc. Natl. Acad. Sci. USA. 79, 1515-1519.
33. Kende, H. (1983) Some concepts concerning the mode of action of plant hormones. In: Strategies of Plant Reproduction, pp. 147-156, Meudt, W.J., Ed. Allanheld & Osmun, Totowa, New Jersey.
34. Kende, H., Lang, A. (1963) Gibberellins and light inhibition of stem growth in peas. Plant Physiol. 39, 435-440.
35. Labavitch, J.M. (1981) Cell wall turnover in plant development. Ann Rev. Plant. Physiol. 32, 385-406.
36. Law, D.M., Hamilton, R.H. (1985) Mechanism of GA-enhanced IAA biosynthesis. Plant Physiol. 77, s4.
37. Lockhart, J.A. (1956) Reversal of light inhibition of pea stem growth by the gibberellins. Proc. Natl. Acad. Sci. USA, 42, 841-848.
38. Loy, J.B. (1977) Hormonal regulation of cell division in the primary elongating meristems of shoots. In: Mechanisms and Control of Cell Division, pp. 92-110, Rost, T.L., Gifford, E.M., Eds. Dowden, Hutchinson and Ross, Stroudsburg, Pennsylvania.
39. Martin, G.C. (1983) Commercial uses of gibberellins In: The Biochemistry and Physiology of Gibberellins, Vol 2, pp. 395-444, Crozier, A., Ed. Praeger, New York.
40. McNeil, M., Darvill, A.G., Fry, S.C., Albersheim, P. (1984) Structure and function of the primary cell walls of plants. Ann. Rev. Biochem. 53, 625-663.
41. Métraux, J.-P., Kende, H. (1983) The role of ethylene in the growth response of submerged deed-water rice. Plant Physiol. 72, 441-446.
42. Métraux, J.P., Kende, H. (1984) The cellular basis of the elongation response in submerged deep-water rice. Planta 160, 73-77.
43. Moll, C., Jones, R.L. (1981) Short-term kinetics of elongation growth of gibberellin-responsive lettuce hypocotyl sections. Planta 152, 442-449.
44. Moll, C., Jones, R.L. (1981) Calcium and gibberellin- induced elongation of lettuce hypocotyl sections. Planta 152, 450-456.
45. Moll, C., Jones, R.L. (1984) The role of cotyledon factors in gibberellic acid-enhanced elongation of lettuce seedling hypocotyls. Plant Sci. Lett. 34, 283-294.
46. Musgrave, A., Jackson, E., Ling, E. (1972) *Callitriche* stem elongation is controlled by ethylene and gibberellin. Nature New Biol. 238, 93-96.
47. Phinney, B.O. (1983) The history of gibberellins in: The Biochemistry and Physiology of Gibberellins, vol 1, pp. 19-52, Crozier, A., Ed. Praeger, New York.
48. Phinney, B.O. (1984) Gibberellin A$_1$ dwarfism and the control of shoot elongation in higher plants. In: The Biosynthesis and Metabolism of Plant Hormones, Soc. Exp. Biol. Seminar No. 23, pp. 17-45, Crozier, A., Hillman, J.R., Eds. Cambridge University Press.
49. Raskin, I., Kende, H. (1984) The role of gibberellin in the growth response of submerged deep-water rice. Plant Physiol. 76, 947-950.

50. Richmond, P.A., Métraux, J.P. (1984) Cellulose synthesis inhibition, cell expansion and pattern of cell wall deposition in *Nitella* internodes. In: Structure, Function and Biosynthesis of Plant Cell Walls, pp. 475-476, Dugger, W.M., Bartnicki-Garcia S., Eds. Waverly Press, Baltimore.

51. Rose, R.J. (1974) Differential effect of cycloheximide onthe short-term gibberellin and auxin growth kinetics of gamma coleoptiles. Plant Sci. Lett. 2, 233-237.

52. Sachs, R.M. (1965) Stem elongation. Ann. Rev. Plant Physiol. 16, 73-96.

53. Shibata, K., Kubota, T., Kamisaka, S. (1974) Isolation and chemical identification of a lettuce cotyledon factor, a synergist of the gibberellin action in inducing lettuce hypocotyl elongation. Plant Cell Physiol. 15, 191-194.

54. Silk, W.K., Jones, R.L. (1975) Gibberellin response in lettuce hypocotyl sections. Plant Physiol. 56, 267-272.

55. Simpson, G.M., Wain, R.L. (1961) A relationship between gibberellic acid and light in the control of internode extension of dwarf pea (*Pisum sativum*) J. Exp. Bot. 12, 207-216.

56. Srivastava, L.M., Sawhney, V.K., Taylor, E.P. (1975) Gibberellic acid induced cell elongation in lettuce hypocotyls. Proc. Natl. Acad. Sci. USA 72: 1107-1111.

57. Stoddart, J.L. (1983) Sites of gibberellin biosynthesis and action. In: The Biochemistry and Physiology of Gibberellins, vol 2, pp. 1-54, Crozier, A., Ed. Praeger, New York.

58. Stuart, D.A., Durnam, D.J., Jones, R.L. (1977) Cell elongation and cell division in elongating lettuce hypocotyl sections. Planta 135, 249-255.

59. Stuart, D.A., Jones, R.L. (1977) The roles of extensibility and turgor in gibberellin- and dark-stimulated growth. Plant Physiol. 59, 61-68.

60. Stuart, D.A., Jones, R.L. (1978) The role of acidification in gibberellic acid- and fusicoccin-induced elongation growth of lettuce hypocotyls sections. Planta 142, 135-145.

61. Taiz, L., Métraux, J.P., Richmond, P.A. (1981) Control of cell expansion in the *Nitella* internode. In: Cytomorphogenesis in Plants, pp. 231-259, Kiermayer, O., Ed. Springer, Wien, New-York

62. Taiz, L. (1984) Plant cell expansion. Ann. Rev. Plant Physiol. 35, 585-657.

63. Terry, M.E., Jones, R.L. (1981) Effect of salt on auxin-induced acidification and growth by pea internode sections. Plant Physiol. 68, 59-64.

64. Terry, M.E. Personnal communication.

65. Thomas, B., Tull, S.E., Warner, T.J. (1980) Light dependent gibberellin responses in hypocotyls of *Lactuca sativa*. L. Plant Sci. Lett. 19, 355-362.

E4. The Genetic Control of Growth via Hormones

James B. Reid

Botany Department, University of Tasmania, Hobart 7001, Australia.

INTRODUCTION

Genetics has been used extensively as a tool to understand the developmental and biochemical pathways of certain bacteria and fungi. However, this is generally not the case in higher plants where, due to increased complexity, even the genetic control of developmental processes such as internode elongation, flowering and fruiting has only been determined in a limited number of species. Indeed, it is only relatively recently that biochemical techniques have become available to determine the action of these genetic systems in cases where the genes regulate growth via changes in the amounts of, or sensitivity to, various plant hormones. In the present article a limited number of developmental systems are discussed in order to illustrate how an understanding of the genetic control of a process can lead to a more complete picture of the way plant hormones control development.

One of the major advantages of using single gene differences to understand the control of development is the assumption that a single primary action is involved due to the one gene: one protein hypothesis (4). This is perhaps not the case when various inhibitors or environmental treatments are used to alter the development of a plant. In addition, specific inhibitors of hormone biosynthesis or action are not always available (e.g. ABA). In these circumstances, in the absence of mutants, the only avenue available to find what processes are controlled by a hormone are either to examine the effects of its application or to correlate its endogenous level with physiological responses. However, it must be noted that a single gene may cause a multitude of pleiotropic effects (i.e. influence a wide range of seemingly unrelated developmental processes) if the primary site of action is of a basic nature. Furthermore, to make full use of mutants, isogenic lines are required in order to avoid incorrectly attributing a difference to one gene when it is caused by some difference in the genetic background. However, the careful use of segregating progenies or a large number of pure lines has been used successfully in certain circumstances (50,38).

HORMONAL CONTROL OF DEVELOPMENT IN BRYOPHYTES

Bryophytes offer an excellent system in which to show the usefulness of genetics to hormone physiology since they possess a conspicuous haploid phase during which relatively simple changes in cell development occur. In mosses, some of these changes appear to be under the control of the auxins and cytokinins (e.g., 7, 23, 2). Mutants which regulate the level of, or senstivity to, these hormones have been isolated and used extensively in the study of the control of development (for detailed reviews see 8, 79).

The development of the protonema in mosses such as *Funaria hygrometrica* or *Physcomitrella patens* normally commences with the germination of the spore to produce the primary chloronemata. This consists of an irregular branching filament of cylindrical cells which have cross walls perpendicular to the filament axis and contain a large number of round chloroplasts (see 8, 79). Subsequent development leads to the production of cells of a second type, the caulonemata, which differ from chloronemata in possessing cross walls oblique to the filament axis, a reduced number of spindle-shaped chloroplasts and a very regular pattern of branching.

Branching in the caulonemata may then give rise to secondary chloronemata or to buds that develop into leafy gametophores (8, 79). The transition from chloronemata to caulonemata is under strong environmental control and does not occur if the protonema are grown at low light intensities, low temperatures (6) or in a continuous flow of liquid medium (10). The last result suggests that some substance normally responsible for caulonemata production is leached from the system. Auxin (IAA or α-NAA) stimulates the transition from chloronemata to caulonemata (23) and can overcome the inhibitory effect of low light intensities (8) or of growth in a continuous flow of liquid medium (10) suggesting that auxin is the controlling substance. Cytokinins stimulate the production of buds and gametophores in many mosses (e.g., 7, 2). It appears that in *P. patens* auxins are also necessary for this development (10) and that light may also be critical (1).

Mutants influencing these developmental processes were first examined in *F. hygrometrica* (17). Mutants were isolated which only formed primary chloronemata or only chloronemata and caulonemata but no gametophores. Auxin could cause partial reversion of certain of these mutants to the wild-type phenotypes. Cove, Ashton and coworkers have isolated a range of developmental mutants from *P. patens* and successfully divided them into a number of categories based upon their development (1,2) (Table 1). The possible mode of action of the mutations involved has been suggested by comparison of the phenotype produced with that of wild-type plants treated in various ways (e.g., application of auxins and cytokinins, drip-cultures, etc.) and with the effect of auxins and cytokinins

Table 1. Categories of developmental mutants in *Physcomitrella patens* their phenotypes compared with the wild-type (normal), their response to applied auxins and cytokinins and their possible modes of action (1, 2, 10, 77, 79, 80).

Mutant Category	Phenotype on Minimal Medium			Effect of Applied		Possible Mode of Action
	Chloronemata primary + secondary	Caulonemata	Gametophores	Auxin	Cytokinin	
1	more	none	none	little	little	auxin insensitive
2	more	normal	none	little	normal growth	blockage of cytokinin production
3	normal	normal	normal	resistant	wild-type response	defective auxin uptake
4	slightly more	no normal caulonemata	none	normal growth	little	blockage of auxin production
5	many more	slightly fewer	few or none	normal growth	little	reduced auxin production
6	many more	slightly more	few or none	normal growth	little	reduced auxin production
7	more	normal or slightly fewer	few	normal growth	little	reduced auxin production
8	many more	more	normal	sensitive	resistant	unknown but not related to auxin or cytokinin levels
9 (OVE)	few	normal or slightly more	many, abnormal	–	–	cytokinin overproducers
10 (OVE)	normal	normal	more, slightly abnormal	–	–	cytokinin overproducers?

applied to the mutant phenotype (e.g., reversion of the mutant type to the wild type form or insensitivity to applied hormones) (Table 1) (1, 2). Evidence of this type has confirmed the importance of auxin and cytokinin to the normal development of the moss plant. Auxin, in addition to being required for caulonemata production, also appears to be essential for gametophore production. In contrast, cytokinin does not appear to be limiting for caulonemata production but does control the development of gametophores (1).

The division into categories on phenotypic grounds has yielded a larger number of mutant categories than appears to be warranted at the physiological and biochemical levels (Table 1). For example, categories 4, 5, 6 and 7 all appear to be caused by mutations which reduce the level of auxin production to some extent. The mutants are therefore 'leaky' to varying degrees resulting in differing levels of phenotypic disturbance. Such a range of mutants has shown that low levels of auxin are essential to caulonemata production while higher levels are necessary for gametophore production (1). The gametophore (cytokinin) over-producing mutants (categories 9 and possibly 10 in Table 1) have allowed both N^6-(Δ^2-isopentenyl) adenine and zeatin to be identified from the culture medium. The levels of these cytokinins in the mutant cultures (Fig. 1) were 100 fold greater than from wild-type cultures and consequently positive identifications were possible (80). The use of protoplast fusion to carry out somatic hybridization and complementation analysis (necessary since category 9 mutants are sterile) has elegantly shown that at least 3 loci are involved with the category 9 phenotype (see 79). This system is therefore well suited to studies on cytokinin biosynthesis (77).

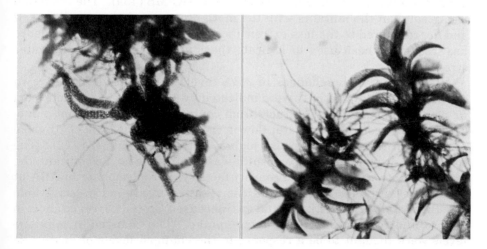

Fig.1 Gametophores from a cytokinin overproducing mutant (left) and the wild type (right) grown in liquid culture (both at same magnification). (Photograph provided by T. L. Wang)

In addition to mutations possibly involved with auxin and cytokinin production (categories 2, 4, 5, 6, 7, 9, and 10), mutations which possibly influence auxin uptake (category 3) or sensitivity (category 1) have been isolated (1). Insensitivity may be attributable to altered receptor properties or some other step between reception and the phenotypic response. This area is perhaps one of the major gaps in our understanding of plant development and the availability of mutations to probe this area may enable these steps to be elucidated. The availability of mutations influencing many aspects of development, such as occurs in the moss *P. patens*, allows the control mechanisms to be taken apart step by step and is potentially able to yield a far greater understanding than where only single mutations are available.

WILTY PHENOTYPES AND ABSCISIC ACID

Mutants have been used with great success to understand the role of abscisic acid (ABA) in water stress (71, 74, 75, 29, 53) and seed germination (26, 66). In tomatoes three non-allelic, recessive, wilty mutants, flacca (*flc*), sitiens (*sit*) and notabilis (*not*) have been isolated (see 69). All three result in a tendency for the mutants to wilt if subjected to the mildest water stress. The mutants also possess much higher rates of transpiration than control varieties because their stomata open wider and resist closure in the dark or in wilted leaves (69). Plants homozygous for the genes *flc*, *sit* and *not* treated with ABA revert to the phenotype of the normal control varieties, Rheinlands Ruhm (for *flc* and *sit*) and Lukullus (for *not*) (74). The endogenous level of ABA was found to be reduced in all three mutants (74), even when measured by GC-MS (43a). The reduction was greater in the mutants with the more pronounced abnormalities, *sit* and *flc*, compared to the less extreme mutant *not* (69, 74). This evidence has frequently been used to indicate the involvement of ABA in water stress.

In addition to the changes in ABA levels quantitative increases in auxin and cytokinin activity were indicated by bio-assays of extracts from *flc* plants compared with extracts from control Rheinlands Ruhm plants (73, 71). The resistance to water flow in the roots and the level of ethylene evolution in the shoots was also higher in the *flc* plants (74, 72). The conclusion drawn from these results was that, at least for mutant *flc*, mutant plants were unable to convert farnesyl pyrophosphate to ABA at the normal rate (44, 74) and that this resulted in altered bio-assayable cytokinin and auxin activity and in increased stomatal resistance to closure and root resistance to water uptake. The changes in ethylene levels were thought to be a response to the changed levels of auxin-like substances and may mediate pleiotropic effects of *flc* such as leaf epinasty, swelling of the upper stem and rooting along the stem (69, 72). However, recent studies have failed to produce a difference in IAA levels between wild-type and *flc* plants and differences in ethylene production may reflect the water status of the plants (43b). Recently a series of double mutant homozygotes has been produced from the mutants *flc*, *sit* and *not* (75) since the individual mutants all appear to be 'leaky' for ABA accumulation (74, 43a). The double homozygous types, *not flc* and *not sit*, led to a more extreme wilty phenotype than any of the single mutant types while *flc sit* produced only a marginally more severe phenotype than the single homozygous mutant types *flc* and *sit* (75). These results support the notion that all three mutants are leaky although direct measurements of endogenous ABA levels are still required. However, the severity of the double mutant types (*not sit* > *not flc* > *flc sit*) is unexpected (75) given the severity of the single mutants (*sit* > *flc* > *not*; 74). The precise biochemical site of action of these mutations is unknown but the results from the double homozygotes suggest *not* may be distinct from *flc* and *sit*. This may allow further insight into how these genes control ABA levels.

Wilty mutants have also been found in potato (64), pea (34), corn (48), *Capsicum scabrous diminutive* (70) and *Arabidopsis thaliana* (28). In potato the droopy mutant results in excessive stomatal opening. Stomatal conductances were reduced by applied ABA and the leaves from droopy plants accumulated less ABA than normal plants when water-stressed (53). These results suggest droopy is similar to the wilty mutants described in tomatoes although the former did not show some associated characters such as leaf epinasty, stem swelling and frequent rooting along the stem. The wilty mutant in corn appears to be caused by an inadequate water supply due to a delay in the differentiation of metaxylem vessels (48) while in *Capsicum scabrous diminutive* (70) an increased concentration of ions in the guard cells may be responsible. In peas a single recessive gene, *wil*, causes wilting (34) and results in a lower percent water content, a lower water potential and a lower diffusive resistance (13). Grafting studies indicate that this is not attributable to the rootstock (78). Analysis of endogenous ABA levels by GCMS-MIM using a deuterated internal standard showed that leaves of droughted *wil* plants possess less ABA than *Wil* types and in this respect are similar to the ABA deficient mutants in tomato and potato (78).

The wilty-like mutants in *Arabidopsis* are of particular interest since both ABA insensitive and ABA deficient mutants have been isolated. The isolation of the mutants used an ingenious method rarely employed in higher plants (28, 29). Previous work by Koornneef and co-workers (30) with *Arabidopsis* had yielded 'non-germinating GA-dwarfs'. These plants failed to germinate unless provided with gibberellin (GA) but if continually provided with GA they germinated and developed into fertile plants with a wild-type phenotype. As in other species, 'germinating GA-dwarfs' were also found (e.g., 45; see next section). Five loci were identified, three (*ga-1*, *ga-2* and *ga-3*) possessing both 'nongerminating' and 'germinating' mutant alleles. It is suggested that the mutations at all five loci control steps in the GA biosynthetic pathway and that many mutations are leaky, the GA requirement for germination being much lower than for stem elongation (30). Since ABA has been implicated in seed dormancy, the selection for ABA mutants began by screening for germination revertants amongst M_2 and M_3 progenies from EMS treated nongerminating types possessing ga-1. This is an elegant selection method since the revertants are self-detecting (i.e., they are the germinating seeds amongst nongerminators) (28). From the range of revertants produced, germinating GA-responsive extreme dwarfs were crossed to parental nongerminating ga-1 types (Table 2). The ability to restore germination was found to be due to a recessive gene (called aba) which, on a wild-type (e.g., Ga-1) background, resulted in a smaller and weaker plant with a slightly yellow-brownish colour and symptoms of withering. The withering symptoms can be reduced by applying ABA and both seeds and leaves of aba types contain reduced ABA levels, suggesting that aba regulates the biosynthesis of ABA (28). Homozygous aba restores germination on a ga-1 background suggesting GA is required for germination only if ABA is present. Work with a range of independently produced ga-1 and aba mutants with differing severities of action suggests that the balance of GA to ABA may be crucial for germination. Mutant aba also leads to reduced seed dormancy and increased rate of water loss compared with comparable wild-type plants (Table 2). A final point raised

Table 2. The genotypes of certain developmental phenotypes of *Arabidopsis*.

Phenotype	Genotype
Wild-type	Ga-1 Aba
Nongerminating, extreme-dwarf	ga-1 Aba
Germinating (revertant) extreme-dwarf	ga-1 aba
Withering, reduced seed dormancy and ABA levels	Ga-1 aba

by the gene aba is that homozygous aba plants segregating from plants heterozygous at the aba locus lacked seed dormancy indicating that the germination behaviour of ripe seeds is determined by the embryo genotype and not by that of the maternal genotype (28). Karssen et al. (26) showed that although both maternal and embryonic ABA occurred in developing seeds it was the embryonic ABA which controlled seed dormancy. The use of mutant types in this way is a powerful tool if it is necessary to determine whether a seed or fruit character is under the control of the maternal or embryo genotype.

ABA-insensitive mutants were selected by their good growth on a medium supplemented with 10 μM ABA (normal plants grow poorly on this medium) (29). Five mutants were isolated (occupying at least three loci) and these, like the *aba* mutants, possessed reduced seed dormancy. Three also exhibited increased withering and water loss like the ABA deficient *aba* types. ABA levels in young seeds of the insensitive types were found to be higher than, or similar to, wild-type seeds. It is suggested that the most likely cause of this ABA insensitivity is due to reduced availability of a receptor or reduced binding-capacity of the receptor for ABA. The three loci involved have been named *abi-1*, *abi-2* and *abi-3*. Mutants *abi-1* and *abi-2* also influence water relations. This may suggest that *abi-1* and *abi-2* operate at a more basic stage in ABA action than *abi-3* providing a genetic tool to further explore the physiology of hormone action. The striking effect of ABA mutants on dormancy in *Arabidopsis* also appears to occur in other species since both the droopy mutant in potatoes (65) and the *sit* mutant in tomatoes (Koornneef unpub. in 28) promote precocious seed germination. The precocious germination of viviparous mutants in maize also appears to involve reduced ABA levels and sensitivity (66).

The work with wilty mutants across all species shows that a wide range of physiological causes are involved. These range from anatomical changes which limit water transport (48) to changes in ionic concentrations within the guard cells which limit their function (70) to mutations which influence the level of, or sensitivity to, ABA (72, 53, 28, 29). The development of a broad range of mutations influencing different aspects of a particular developmental process is of importance since it allows the partial processes to be examined. Production of double or triple mutants may then allow the interaction and interdependence of these processes to be determined so that the limiting or controlling steps may be identified. Unfortunately such recombinants have rarely been produced, limiting the full potential of the physiological-genetic approach. The work with ABA deficient and insensitive mutants outlined above is one of the key pieces of evidence that suggests the importance of this hormone to the normal water relations of the plant, even though many questions and some controversy remain (35).

INTERNODE LENGTH AND THE GIBBERELLINS

The excellent work of Phinney and co-workers (e.g., 45) with maize and of Brian and co-workers (e.g., 9) with garden peas initiated the work on the involvement of internode length genes with the gibberellins. Since this time similar work has been done with many other species including sweet peas (32), *Pharbitis nil* (3), rice (36), wheat (54, 68), tomatoes (81), *Arabidopsis thaliana* (30) and *Phaseolus vulgaris* (16). The internode length genes have received this attention because of their relatively high frequency and their commercial significance with the advent of many dwarf crops.

In maize thirty mutants which influence plant height have been described (47). The early work by Phinney (45) demonstrated that the five nonallelic dwarf mutants, d_1, d_2, d_3, d_5, and an_1, became phenocopies of normal types after treatment with small amounts of GA_3. All contained less GA-like activity than normal types when examined using the d_5 maize bioassay. This led to the conclusion that all five mutations inhibit different steps in the GA biosynthetic pathway. In addition, another five nonallelic dwarf mutants, D_8, pe_1, mi_2, na_1 and na_2 showed little or no response to applied GA_3 (45). The four GA-sensitive dwarfs, d_1, d_2, d_3 and d_5 have recently received considerable attention and their sites of action in the GA biosynthetic pathway have been determined with varying degrees of certainty.

Gene d_5 blocks the cyclisation of copalylpyrophosphate (CPP) to ent-kaur-16-ene (ent-kaurene) by reducing the B activity of ent-kaurene synthetase (Fig. 2). Cell free extracts from normal maize shoots synthesise ent-kaurene as the major diterpene. Extracts from d_5 seedlings showed a marked reduction in ent-kaurene synthesis regardless of whether mevalonic acid, geranylgeranylpyrophosphate or CPP was used as substrate. A concomitant increase in ent-kaur-15-ene (ent-isokaurene) synthesis was observed (18), a metabolite not in the pathway leading to the GAs (20).

Gene d_1 blocks the 3β-hydroxylation of GA_{20} to GA_1 since (^{13}C,^3H) GA_{20} was metabolised to $[^{13}C,^3H]GA_{29}$-catabolite and $[^{13}C,^3H]GA_1$ by shoots of normal seedlings whereas $[^{13}C,^3H]GA_{29}$ and $[^{13}C,^3H]GA_{29}$-catabolite were produced by shoots of d_1 seedlings (67) (Fig. 2). The gene d_2 probably controls the carbon-7-oxidation step since GA_{53}-aldehyde is not active when applied to d_2 seedlings while GA_{53} is active in promoting leaf-sheath elongation. Likewise, d_3 probably controls the C-13-hydroxylation step since GA_{12}-aldehyde lacks activity while GA_{53}-aldehyde is active (47, 46) (Fig. 2). Application data and grafting results also support the site of action of d_5 and d_1 (e.g., 47). These results have been made possible by the determination of the native GAs in maize (19), the construction of a probable biosynthetic pathway (47,46) and the availability of appropriate intermediates. Biochemically, these mutations are possibly the best understood genes controlling plant hormone production.

In peas, an undisputed role for the involvement of certain internode length genes in GA biosynthesis has only recently been established (50, 49, 22, 21), even though early application data led Brian and co-workers to suggest such a relationship (9). Scepticism during the 1960s and 1970s existed because extraction of GA-like substances could not establish a consistent difference between tall (conferred by the gene Le on an otherwise wild-type background) and dwarf (le) varieties of peas (e.g., 24).

Recent work has shown that the shoots of tall (*Le*) plants possess GA_1 (49, 22), which is absent, or at least present below the levels of detection, in the shoots of light-grown dwarf (*le*) plants (50, 11). This suggests that in peas

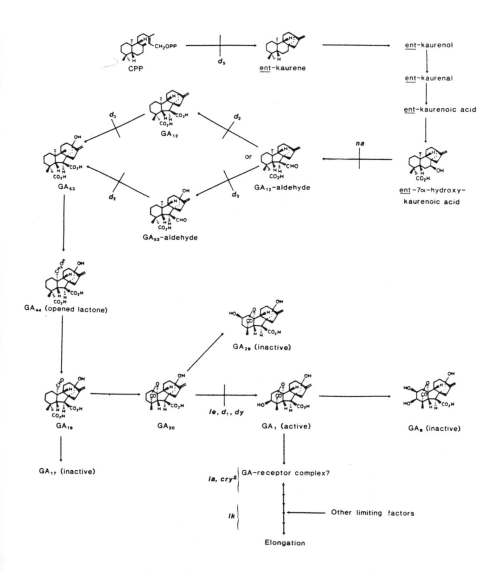

Fig. 2. The proposed sites of action of certain length mutants in maize (d_1, d_2, d_3, d_5), peas (*le*, *na*, *la* cry^s, *lk*) and rice (*dy*) in the early 13-hydroxylation gibberellin biosynthetic pathway and possible subsequent steps leading to elongation. Possible alternative pathways from GA_{12}-aldehyde to GA_{53} are indicated (25, 46).

le blocks the 3β-hydroxylation of GA_{20} to GA_1 (Fig. 2). This was confirmed when the immature shoot tissue of *Le* peas was shown to metabolise $[^3H,^{13}C]GA_{20}$ to $[^3H,^{13}C]GA_1$, $[^3H,^{13}C]GA_8$ and $[^3H,^{13}C]GA_{29}$. In contrast, no $[^3H,^{13}C]GA_1$ and only a trace of $[^3H,^{13}C]GA_8$ (in one line only) was found in comparable *le* plants (21). However, the trace of $[^3H,^{13}C]GA_8$ suggests that gene *le* is a leaky mutation. As would be expected *Le* plants show a more or less equivalent growth response to the application of GA_1 or GA_{20} while plants possessing the leaky gene *le* respond far more to GA_1 than to GA_{20} (22). These results support the suggestion that GA_1 may be the only active GA for elongation in plants possessing the early 13-hydroxylation pathway for gibberellin biosynthesis (e.g. maize, peas and rice; 47, 46). The genes at the *le* locus in peas thus appear to control the same step as d_1 in maize (67) and *dy* in rice (36).

The early difficulties in determining the action of the gene *Le* resulted from a number of causes including poor separation techniques, extraction of plants at an age before the tall (*Le*)/dwarf (*le*) difference was maximal and extraction of inappropriate tissue. It would seem obvious that if the biochemical effect of a gene controlling a developmental response is to be examined, the tissue extracted should show the developmental difference and be of an age where this effect is maximal. However, this has often been overlooked because plants of the appropriate age have been too large and difficult to grow or the hormone concentration has been greater in some other tissue (e.g. developing seeds) and thus easier to measure.

Seven nonallelic internode length genes, in addition to *le*, have been examined in peas (62, 57). Three further dwarfing mutants, *na*, *lh* and *ls* (Fig. 3) appear to influence steps in the GA biosynthetic pathway since they contain little or no biologically active GA-like substances and become phenocopies of tall types after treatment with small amounts of GA_1 (49, 22, 63). Feeds of $[^3H,^{13}C]GA_{20}$ to *na* plants show no dilution of metabolites by endogenous ^{12}C-GAs indicating that *na* prevents the production of C_{19}-GAs at the current levels of detection by GCMS analysis. Treatment with GA-precursors suggests that *na* may block the conversion of *ent*-7α-hydroxy kaurenoic acid to GA_{12}-aldehyde since *na* plants show no response to even 100 µg of *ent*-kaurene, *ent*-kaurenol, *ent*-kaurenal, *ent*-kaurenoic acid or *ent*-7α-hydroxy kaurenoic acid but show a 90 per cent increase in elongation in response to even 1 µg of GA_{12}-aldehyde (Ingram and Reid, unpub.) (Fig. 2). Confirmation of the biochemical site of action of *na* and determinations of the sites of action of *lh* and *ls* must await the availability of appropriately labelled intermediates from the GA-biosynthetic pathway. Like *le*, the mutations *na*, *lh* and *ls* affect elongation in both the light and the dark (56) (Fig. 3) and are probably leaky since the homozygous double recessive plants (e.g., *na ls* and *na lh* plants) are shorter than comparable single homozygotes (57). Examination of these types with more sensitive analytical

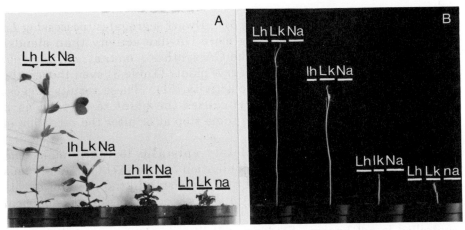

Fig. 3. The effect of the internode length mutants *lh*, *lk* and *na* on the growth of (A) 29 day-old light grown pea plants and (B) 14 day-old dark grown pea plants.

techniques such as radioimmunoassay and enzyme-immunoassay may allow the quantitative effects of these mutations to be examined. Mutants *na*, *lh* and *ls* in peas are similar to the d_2, d_3 and d_5 mutants in maize (47) or the *dx* mutant in rice (36) since they inhibit the production of biologically active GAs by blocking steps in the GA-biosynthetic pathway prior to the production of C_{19}-GAs (63).

The genes at three other internode length loci in peas, *la*, *crys* and *lk* influence GA sensitivity (Fig. 2) (51, 63). Gene lk causes the production of the erectoides phenotype in which the length of the internode is approximately 10 percent of comparable tall plants (Fig. 3). In this respect the effect is similar to mutant *na* which produces the nana phenotype. However, other pleiotropic effects of *lk* (i.e., multiple phenotypic effects of one gene) such as leaflet appearance, lack of lateral branching and brittleness of the stem and petiole make the two phenotypes clearly distinguishable (57). The most dramatic difference is seen after treatment with saturating doses of GA_1. The *lk* plants only show a weak response, elongating less than 10 per cent as much as *na* plants (63) and at no levels of applied GA_1 do they become phenocopies of wild type plants. The level of GA-like substances in *lk* segregates is similar to that of *Lk* segregates (63). Gene lk is therefore a GA-insensitive dwarfing gene having certain similarities to the insensitive dwarfing genes in wheat (e.g., 54, 68) and maize (45).

The gene combination *la crys* (Table 3) results in another type of GA insensitivity in which the plants behave as if continually saturated with GAs (51). This phenotype, slender (12), has also been found in other species (e.g., barley, 15). All available evidence suggests that GA levels are not important in determining the internode length of slender peas (*la crys*). They do not respond to applied GA_3 (9) or GA synthesis inhibitors such as AMO 1618 or PP 333 (e.g., 51). Dwarf segregates (possessing *La* and/or *Cry*) contain quantitatively more GA-like activity than slender segregates and even addition of the GA synthesis mutant, *na*, does not result in a phenotypic change in *la crys* plants (Table 3) even though the plants contain little or no GA-like activity (51). These results suggest that the gene combination *la crys* causes the plant to respond as if saturated with GAs by influencing some step at or after the reception of the GA stimulus.

The gene *lk* is almost completely epistatic to (i.e., masks the expression of) the slender gene combination *la crys* (57) (Table 3). This suggests that the gene *lk* may act at a later step in the chain of events leading from the GA-receptor site to the production of the physiological response than does *la crys* (Fig. 2). However, the nature of the steps being controlled is not known. Stoddart (68) likewise has suggested that the GA-insensitive dwarfing gene *Rht 3* in wheat may concern subcellular associations of the biologically active endogenous-GAs and does not operate through the altered metabolism of active GAs. The nature of plant hormone receptor-sites and their physiological action is an area of uncertainty, especially for the GAs. If a series of mutations influencing these processes can be found and the sequence of their action determined from their genetic interactions a powerful system would be available to further explore the biochemical nature of these steps.

The eighth internode length locus examined in peas, *lm*, results in a general reduction in size and vigour of all parts of the shoot regardless of which other internode length genes are present (31). The first visible effect of gene *lm* is a distortion of the roots (43) which is associated with abnormal cells in the root cortex. This suggests that the primary effect of *lm* is not the control of internode length, a view supported by the fact that no genetic interactions occur with the other internode length loci. This demonstrates the need to separate mutations exerting a direct and controlling influence over a developmental process from those having only a secondary or indirect effect.

Table 3. The phenotype of certain internode length genotypes in peas. Epistasis is indicated when the gene present at a locus does not matter (e.g., *Le* or *le*).

Genotype	Phenotype
Le Na La Cry Lk	Wild-type tall
le Na La Cry Lk	Dwarf, reduced GA_1 levels
Le or *le, na La Cry Lk*	Nana, all C_{19}-GAs reduced
Le or *le, Na* or *na, la crys Lk*	Slender, insensitive to GA levels
Le or *le, Na* or *na, La Cry* or *la crys, lk*	Erectoides, GA insensitive

FLOWERING

Considerable genetic variation exists within most species for the control of flowering and many single gene differences have been reported (see 40). However, unlike the developmental processes so far discussed, the chemical nature of the 'flowering hormone(s)' remains obscure even though considerable indirect evidence from techniques such as grafting support their existence (see 5). Discussion here will be restricted to the control of flowering (and associated characters such as apical senescence) in peas where both the genetics and physiology have been examined (see 41, 43). This work demonstrates that even though the biochemical control of flowering has not been determined, flowering mutants can be isolated and divided into groups influencing the level of the 'hormone(s)', sensitivity to the 'hormone(s)' and indirect or secondary effects due to changes in the growth rate, etc., of the plant.

In peas the genes at six major loci, *lf, e, sn, hr, dne* and *veg*, interact to control the flowering phenotype (for reviews see 41, 43) (Table 4). Phenotypes range from day- neutral types, through quantitative long day types, to types with an almost qualitative requirement for long days (37, 41) (Fig. 4). They vary in the number of vegetative nodes before flowering from as low as 4 to over 150 under warm, short-day conditions (counting from the cotyledons as zero). Certain genotypes possess large vernalisation responses while others show little or no response to vernalisation (42, 41). The change from vegetative to reproductive growth in peas is thought to be controlled by the time at which the balance of a flower promoter to a flower inhibitor passes a threshold (38) even though the biochemical nature of these substances is not known. The two complementary genes, *Sn* and *Dne*, control steps leading to the synthesis of the flower inhibitor in the cotyledons and leaves (38, 27, 41) since stocks homozygous for either *sn* or *dne* lack the flower delaying ability of *Sn Dne* stocks when grafted to *sn Dne* or *Sn dne* scions. The delaying ability of the gene combination *Sn Dne* occurs only under short photoperiods suggesting that long photoperiods promote flowering by retarding the pathway leading to inhibitor production (42, 27). This process is phytochrome controlled with the size of the inhibitory effect depending on the length of the dark period when 24 h cycle lengths are used (60). Consequently, plants possessing *Sn Dne* are photoperiodic while plants possessing *sn* and/or *dne* are essentially day-neutral. The product of the *Sn Dne* pathway, as well as influencing flower initiation also influences a range of other developmental processes including flower development, yield, apical senescence (or more correctly cessation of apical growth), vegetative vigour and the production of basal laterals (33, 37, 61, 43). Consequently, if other genes mask the photoperiodic effect of the *Sn Dne* combination on flower initiation their presence may be traced using one or more of these pleiotropic characters.

Table 4. The phenotypic effect and possible site and mode of action of the genes controlling flowering in peas.

Gene	Phenotypic Effect	Site of Action	Possible Mode of Action
sn	Promotes flowering and senescence by blocking photoperiod responses	Leaves including the cotyledons	Blocks the producton of a graft transmissible inhibitor of flowering and apical senescence in SD
dne	(as above)	(as above)	(as above)
hr	Reduces the size of the photoperiod responses	Leaves excluding the cotyledons	Causes *Sn Dne* activity to be reduced at an early age
e	Delays flower initiation on certain genetic backgrounds	Primarily cotyledons	Reduces the ratio of flower promotory: flower inhibitory substances
lf	Lowers the minimum flowering node	Shoot apex	Increases sensitivity to the ratio of flower promotory: flower inhibitory substances
veg	Blocks flower initiation	Shoot apex	Makes the apex insensitive to even highly promotory ratios of the flowering hormones

The genes E and Hr modify the expression of the $Sn\ Dne$ combination since the E/e and Hr/hr gene differences have little effect in homozygous sn and/or dne plants (37, 39, 41) (Fig. 4). Gene E has its major influence in the cotyledons where it appears to reduce inhibitor production by the Sn Dne pathway resulting, in some circumstances, in early flower initiation (38, 41). Hr exerts its major influence in the foliage leaves (55). The output of the $Sn\ Dne$ pathway appears to decrease rapidly with age in hr plants which therefore behave as quantitative long day plants. This decrease is reduced in Hr plants and they therefore behave as near qualitative long day plants (39, 59, 55) (Fig. 4). Genes E and Hr are therefore interpreted as influencing hormone levels by regulating the ontogenetic stages and tissues in which the 'structural' genes Sn and Dne operate (Table 4).

Genes Lf and Veg both operate at or near the shoot apex and influence the sensitivity of the plant to the putative flowering hormones (38, 61) (Table 4). Four alleles, lf^a, lf, Lf and Lf^d are known at the lf locus (41) and respectively lead to a higher promotor to inhibitor ratio being required for flowering (Fig. 5). Gene veg prevents flower initiation under all conditions so far examined (61). Consequently, in veg plants the shoot apex is unable to initiate flowers even under conditions highly inductive

Fig. 4. The effect of the flowering genes controlling hormone levels, E, Sn and Hr on the flowering node (indicated by tape and arrows), development of pods, cessation of apical growth and senescence in 12 week old pea plants grown in a 10 h photoperiod. Note that Sn and Hr influence many facets of growth (A) even if the flowering node of the plants is similar due to the presence of E (B). All plants are homozygous for $lf\ Dne\ Veg$.

Fig. 5. The effect of the flowering alleles lf^a, lf and Lf, which control the sensitivity of the apex to the flowering hormones, on the growth of 7 week old, day-neutral (*sn Dne hr Veg*) pea plants. The flowering node for each plant is indicated by the arrows. Note the pleiotropic effects of the *lf* alleles on peduncle length, lateral outgrowth and pod development (41, 43). (Photograph provided by Dr. I.C. Murfet, Univ. of Tasmania.)

for flowering in wild-type plants. This gene can therefore be considered analagous to the gibberellin-insensitive dwarf discussed earlier whereas the *lf* alleles exert only a quantitative effect on sensitivity of the apex to the flowering hormones. However, even in *veg* plants the effects of the *Sn Dne* pathway can be observed if vegetative characters under the control of the *Sn Dne* system are examined such as the cessation of apical growth, vegetative vigour and lateral branching (61). This might suggest that *veg* operates at a step well after the primary site of reception of the inhibitor in the apex. Likewise, in Lf^d plants various processes occurring with age and known to be influenced by the *Sn Dne* product proceed regardless of the fact that flower initiation has not occurred (43) suggesting a more primary and fundamental site of perception and action for the *Sn Dne* product (e.g., as a promotor of vegetative growth).

Other genes in peas (e.g., the internode length genes *le, la* and cry^s, 41) have minor pleiotropic effects on flowering. These effects are probably secondary consequences of the altered growth in these genotypes. The involvement of gibberellins with the genes controlling flowering and apical senescence in peas has been investigated extensively using application data, metabolism of labelled gibberellins and examination of the levels of endogenous gibberellins by both bioassays and GC-MS (e.g., 52, 11). These studies provided inconclusive results regarding the direct involvement of any of the flowering genes with the gibberellins. However, recent genetic evidence shows that the expression of the flowering genes is not masked by the presence of altered gibberellin levels caused by genes *le, na, ls* or *lh* (see 41, 58) even though the flowering genes may result in quantitative changes in gibberellin levels and/or metabolism because of their influence on vegetative vigour and their control of apical growth. These results suggest the flowering genes do not operate by directly influencing gibberellin metabolism and demonstrate the usefulness of producing combinations of developmental mutants since they provide firm evidence where physiological and biochemical data have been inconclusive.

Although the nature of the 'flowering hormone(s)' remains obscure an understanding of the genetical control of flowering allows the physiology of the partial processes to be examined and provides a firm basis for biochemical comparisons. The similarity of the categories of mutants involved in flowering to those shown for biochemically better understood developmental responses (e.g., internode length) offers hope that resolution of the control of this fundamental transition may soon be possible.

CONCLUDING REMARKS

The preceding discussion shows that the genetic systems regulating elongation, germination, transpiration and protonema development operate, at least partially, by influencing the levels of, or sensitivity to, plant hormones such as the gibberellins, auxins, cytokinins and abscisic acid. In many of these cases this information provides the most direct evidence for the involvement of a particular class of hormones with a particular developmental process (e.g., 74, 28, 46, 21). However, such multi-disciplinary studies are rare and, unlike the work with bacteria and fungi, few programs to select mutants specifically to examine physiological processes in higher plants have been undertaken. It is to be hoped that over the next few years, through the use of a combined genetical, physiological and biochemical approach, an understanding of the control of development in higher plants comparable to that already established in certain bacteria and fungi will be obtained.

A feature common to all the developmental processes examined is the clear division of mutants into categories. For example, mutations influencing hormone levels exist for all the developmental responses discussed. These mutations normally appear to regulate biosynthesis and provide direct evidence that endogenous hormone levels control aspects of plant development. This can be illustrated by the work on internode length where wild-type maize, wheat and peas are all capable of responding to low levels of applied GAs (45, 54, 22) indicating that elongation is limited by the levels of endogenous GAs. A reduction in the level of endogenous GAs by either GA biosynthesis mutants or applied inhibitors

(45, 22) results in reduced elongation. The ontogenetic changes in GA levels in the shoots of tall peas also show a close correlation between GA levels and elongation with both beginning at low levels, rising to a maximum before decreasing again prior to senescence (14). GA_1 levels are probably the key to this control system in plants possessing an early 13-hydroxylation pathway for the synthesis of GAs (47, 21, 46).

This controlling role for plant hormone levels over development is contrary to recent suggestions that control may reside primarily in tissue sensitivity (e.g., 76). However, the equally important role of tissue sensitivity is stressed by the widespread occurrence of hormone sensitivity mutants which have been suggested to influence the concentration or binding properties of the hormone receptor site. They offer a powerful, but as yet unexploited, tool for examining the primary action of plant hormones.' Many of the developmental processes are also influenced by mutations that appear to have a primary effect unrelated to the developmental process being examined. These mutations serve to illustrate the complex interactions involved in plant development and the many independent factors that may limit a particular developmental process even though one group of hormones may normally be exerting a controlling role. Mutants can profitably be used to identify and examine these partial processes.

In the majority of cases so far examined hormone mutants have been considered to have similar effects through the plant. Few have been examined to determine what changes occur during ontogeny or from one tissue to another. However, where this has been examined the mutants have shown precise ontogenetic and tissue specificity (49, 58). Since hormones are almost certainly involved in controlling ontogenetic and tissue specific developmental changes, detailed examination of the present mutants and the selection of mutants with ontogenetic or tissue specificity may be most beneficial. Synthesis and sensitivity mutants may also provide valuable probes for examining the hormonal control of developmental processes such as flowering and monocarpic senescence where the biochemical nature of the hormones is not yet established.

Acknowledgements

I wish to thank Drs. T.J. Ingram, I.C. Murfet, J.J. Ross and T.L. Wang and Mr. P.J. Dalton for helpful discussions and comments relating to the manuscript, Mrs. T. Grabek for technical assistance and the Australian Research Grants Scheme for financial support.

References.

1. Ashton, N.W., Cove, D.J., Featherstone, D.R. (1979) The isolation and physiological analysis of mutants of the moss, *Physcomitrella patens*, which over-produce gametophores. Planta *144*, 437-442.

2. Ashton, N.W., Grimsley, N.H., Cove, D.J. (1979) Analysis of gametophytic development in the moss, *Physcomitrella patens*, using auxin and cytokinin resistant mutants. Planta *144*, 427-435.
3. Barendse, G.W.M., Lang, A. (1972) Comparison of endogenous gibberellins and the fate of applied radioactive gibberellin A_1 in a normal and a dwarf strain of Japanese morning glory. Plant Physiol. *49*, 836-841.
4. Beadle, G.W., Tatum, E.L. (1941) Genetic control of biochemical reactions in *Neurospora*. Proc. Nat. Acad. Sci. U.S.A. *27*, 499-506.
5. Bernier, G., Kinet, J.M., Sachs, R.M. (1981) The physiology of flowering. vol. 1. p. 149. CRC Press, Boca Raton, Florida.
6. Bopp, M. (1959) Versuche zur Analyse von Wachstum and Differenzierung des Laubmoosprotonemas. Planta *53*, 178-197.
7. Bopp, M. (1963) Development of the protonema and bud formation in mosses. J. Linn. Soc. (Bot.) *58*, 305-309.
8. Bopp, M. (1983) Developmental physiology of bryophytes. In: New manual of bryology, vol. 1, pp. 276-324, Schuster, R.M., ed. The Hattori Botanical Lab., Miyazaki, Japan.
9. Brian, P.W. (1957) The effects of some microbial metabolic products on plant growth. Symp. Soc. Exp. Biol. *11*, 166-181.
10. Cove, D.J., Ashton, N.W., Featherstone, D.R., Wang, T.L. (1980) The use of mutant strains in the study of hormone action and metabolism in the moss *Physcomitrella patens*. *In* The proceedings of the fourth John Innes Symposium, 1979, pp. 231-241, Davies, D.R., Hopwood, D.A., eds. The John Innes Charity, Norwich.
11. Davies, P.J., Emshwiller, E., Gianfagna, T.J., Proebsting, W.M., Noma, M., Pharis, R.P. (1982) The endogenous gibberellins of vegetative and reproductive tissue of G2 peas. Planta *154*, 266-272.
12. de Haan, H. (1927) Length factors in *Pisum*. Genetica *9*, 481-497.
13. Donkin, M.E., Wang, T.L., Martin, E.S. (1983) An investigation into the stomatal behaviour of a wilty mutant in *Pisum sativum*. J. Exp. Bot. *34*, 825-834.
14. Eckland, P.R., Moore, T.C. (1968) Quantitative changes in gibberellin and RNA correlated with senescence of the shoot apex in the 'Alaska' pea. Amer. J. Bot. *55*, 494-503.
15. Foster, C.A. (1977) Slender: an accelerated extension growth mutant of barley. Barley Genet. Newslett. *7*, 24-27.
16. Goto, N., Esashi, Y. (1973) Diffusible and extractable gibberellins in bean cotyledons in relation to dwarfism. Physiol. Plant. *28*, 480-489.
17. Hatanaka-Ernst, M. (1966) Entwicklungsphysiolgische Untersuchungen an strahlen-induzierten Protonemamutanten von *Funaria hygrometrica* Sibth. Z. Pflanzenphysiol. *55*, 259-277.
18. Hedden, P., Phinney, B.O. (1979) Comparison of *ent*-kaurene and *ent*-isokaurene synthesis in cell-free systems for etiolated shoots of normal and dwarf-5 maize seedlings. Phytochem. *18*, 1475-1479.
19. Hedden, P., Phinney, B.O., Heupel, R., Fujii, D., Cohen, H., Gaskin, P., MacMillan, J., Graebe, J.E. (1982) Hormones of young tassels of *Zea mays*. Phytochem. *21*, 391-393.
20. Hedden, P., Phinney, B.O., MacMillan, J., Sponsel, V.M. (1977) Metabolism of kaurenoids by *Gibberella fujikuroi* in the presence of the plant growth retardant, N, N, N-trimethyl-1-methyl-(2',6',6'-trimethylcyclohex-2'-en-1'-yl) prop-2-enylammonium io-dide Phytochem. *16*, 1913-1917.
21. Ingram, T.J., Reid, J.B., Murfet, I.C., Gaskin, P., Willis, C.L., MacMillan, J. (1984) Internode length in *Pisum*. The *Le* gene controls the 3β-hydroxylation of gibberellin A_{20} to gibberellin A_1. Planta *160*, 455-463.
22. Ingram, T.J., Reid, J.B., Potts, W.C., Murfet, I.C. (1983) Internode length in *Pisum*. IV. The effect of the *Le* gene on gibberellin metabolism. Physiol. Plant. *59*, 607-616.
23. Johri, M.M., Desai, S. (1973) Auxin regulation of caulonema formation in moss protonema. Nature new Biol. *245*, 223-224.

24. Jones, R.L., Lang, A. (1968) Extractable and diffusible gibberellins from light- and dark-grown pea seedlings. Plant Physiol. *43*, 629-634.
25. Kamiya, Y., Graebe, J.E. (1983) The biosynthesis of all major pea gibberellins in a cell-free system from *Pisum sativum*. Phytochemistry *22*, 681-689.
26. Karssen, C.M., Brinkhorst-van der Swan, D.L.C., Breekland, A.E., Koornneef, M. (1983) Induction of dormancy during seed development by endogenous abscisic acid: studies on abscisic acid deficient genotypes of *Arabidopsis thaliana* (L.). Heynh. Planta *157*, 158-65.
27. King, W., Murfet, I.C. (1985) Flowering in *Psium*: a sixth locus, *Dne*. Ann. Bot. *56*, 853-846.
28. Koornneef, M., Jorna, M.L., Brinkhorst - van der Swan, D.L.C., Karssen, C.M. (1982) The isolation of abscisic acid (ABA) deficient mutants by selection of induced revertants in nongerminating gibberellin sensitive lines of *Arabidopsis thaliana* (L.) Heynh. Theor. Appl. Genet. *61*, 385-393.
29. Koornneef, M., Rueling, G., Karssen, C.M. (1984) The isolation and characterisation of abscisic acid-insensitive mutants of *Arabidopsis thaliana*. Physiol. Plant. *61*, 377- 383.
30. Koornneef, M., van der Veen, J.H. (1980) Induction and analysis of gibberellin sensitive mutants in *Arabidopsis thaliana* (L.) Heynh. Theor. Appl. Genet. *58*, 257-263.
31. Lindqvist, K. (1951) The mutant 'micro' in *Pisum*. Hereditas *37*, 389-420.
32. Magara, J. (1963) Notes on the possible role of endogenous 'gibberellins' in the determining of monofactorial dwarfism in dwarf sweet peas (*Lathyrus odoratus* L.) and in mutants d_1 and d_5 of maize. Ann. Physiol. Veg. *5*, 249-261.
33. Marx, G.A. (1968) Influence of genotype and environment on senescence in peas, *Pisum sativum* L. Bioscience *18*, 505-506.
34. Marx, G.A. (1976) "Wilty": a new gene of *Pisum*. Pisum Newslet. *8*, 40-41.
35. Milborrow, B.V. (1981) Abscisic acid and other hormones. In: The physiology and biochemistry of drought resistance in plants, pp. 348-88, Paleg, L.G., Aspinall, D., eds. Academic Press, Sydney.
36. Murakami, Y. (1972) Dwarfing genes in rice and their relation to gibberellin biosynthesis. In: Plant growth substances, 1970, pp. 166-174, Carr. D.J., ed. Springer-Verlag, Berlin.
37. Murfet, I.C. (1971) Flowering in *Pisum*. Three distinct phenotypic classes determined by the interaction of a dominant early and a dominant late gene. Heredity *26*, 243-257.
38. Murfet, I.C. (1971) Flowering in *Pisum*: reciprocal grafts between known genotypes. Aust. J. Biol. Sci. *24*, 1089-1101.
39. Murfet, I.C. (1973) Flowering in *Pisum*. *Hr*, a gene for high response to photoperiod. Heredity *31*, 157-164.
40. Murfet, I.C. (1977) Environmental interaction and the genetics of flowering. Ann. Rev. Plant Physiol. *28*, 253-278.
41. Murfet, I.C. (1985) *Pisum sativum* L. In: Handbook of flowering, vol. IV, pp. 97-1126, Halevy, A.H., ed. CRC Press, Boca Raton, Florida .
42. Murfet, I.C., Reid, J.B. (1974) Flowering in *Pisum*: the influence of photoperiod and vernalising temperatures on the expression of genes *Lf* and *Sn*. Z. Pflanzenphysiol. *71*, 323-331.
43. Murfet, I.C., Reid, J.B. (1985) The control of flowering and internode length in *Pisum*. *In* The pea crop: a basis for improvement, pp. 67-80, Hebblethwaite, P.D., Heath, M.C., Dawkins, T.C.K., eds. Butterworths, London.
43a. Neill, S.J., Hogan, R. (1985) Abscisic acid production and water relations in wilty tomato mutants subjected to water deficiency. J. Exp. Bot. *36*, 1222-1231.
43b. Neil, S. J., McGaw, B.A., Horgan, R. (1986) Ethylene and 1-aminocyclopropane-1-carboxylic acid production in *flacca*, a wilty tomato mutant, subjected to water deficiency and pretreatment with abscisic acid. J. Exp. Bot. *37*, 535-541..
44. Nevo, Y., Tal, M. (1973) The metabolism of abscisic acid in *flacca*, a wilty mutant of tomato. Biochem. Genet. *10*, 79-90.

338

45. Phinney, B.O. (1961) Dwarfing genes in *Zea mays* and their relation to the gibberellins. In: Plant growth regulation, pp. 489-501, Klein, R.M., ed. Iowa State College Press, Ames, Iowa.

46. Phinney, B.O. (1984) Gibberellin A_1, dwarfism and the control of shoot elongation in higher plants. In: The biosynthesis and metabolism of plant hormones, Soc. Exp. Biol. Seminar Series 23, pp. 17-41, Crozier, A., Hillman, J.R., eds. Cambridge University Press, London.

47. Phinney, B.O., Spray, C. (1982) Chemical genetics and the gibberellin pathway in *Zea mays* L. In: Plant growth substances 1982, pp. 101-110, Wareing, P.F., ed. Academic Press, London.

48. Postlethwait, S.N., Nelson, O.E. (1957) A chronically wilted mutant of maize. Amer. J. Bot. *44*, 628-633.

49. Potts, W.C., Reid, J.B. (1983) Internode length in *Pisum*. III. The effect and interaction of the *Na/na* and *Le/le* gene differences on endogenous gibberellin-like substances. Physiol. Plant. *57*, 448-454.

50. Potts, W.C., Reid, J.B., Murfet, I.C. (1982) Internode length in *Pisum*. I. The effect of the *Le/le* gene difference on endogenous gibberellin-like substances. Physiol. Plant. *55*, 323-328.

51. Potts, W.C., Reid, J.B., Murfet, I.C. (1985) Internode length in *Pisum*. Gibberellins and the slender phenotype. Physiol. Plant. *63*, 357-364.

52. Proebsting, W.M., Davies, P.J., Marx, G.A. (1978) Photoperiod-induced changes in gibberellin metabolism in relation to apical growth and senescence in genetic lines of peas (*Pisum sativum* L.). Planta *141*, 231-2238.

53. Quarrie, S.A. (1982) Droopy: a wilty mutant of potato deficient in abscisic acid. Plant Cell Envir. *5*, 23-26.

54. Radley, M. (1970) Comparison of endogenous gibberellins and response to applied gibberellin of some dwarf and tall wheat cultivars. Planta *92*, 292-300.

55. Reid, J.B. (1979) Flowering in *Pisum*: the effect of age on the gene *Sn* and the site of action of gene *Hr*. Ann. Bot. *44*, 163-173.

56. Reid, J.B. (1983) Internode length genes in *Pisum*. Do the internode length genes effect growth in dark-grown plants? Plant Physiol. *72*, 759-763.

57. Reid, J.B. (1986) Internode length in *Pisum*. Three further loci, *lh, ls* and *lk*. Ann. Bot. *57*, 577-592.

58. Reid, J.B. (1986) Gibberellin mutants. In: A genetic approach to plant biochemistry, Plant Gene Research, vol. 3, pp. 1-34, King, P.J., Blonstein, A.D., eds. Springer- Verlag, Wien, New York.

59. Reid, J.B., Murfet, I.C. (1977) Flowering in *Pisum*: the effect of genotype, plant age, photoperiod and number of inductive cycles. J. Exp. Bot. *28*, 811-819.

60. Reid, J.B., Murfet, I.C. (1977) Flowering in *Pisum*: the effect of light quality on genotype *lf e Sn Hr*. J. Exp. Bot. *28*, 1357-1364.

61. Reid, J.B., Murfet, I.C. (1984) Flowering in *Pisum*: a fifth locus, Veg. Ann. Bot. *53*, 369-382.

62. Reid, J.B., Murfet, I.C., Potts, W.C. (1983) Internode length in *Pisum*. II. Additional information on the relationship and action of loci *Le, La, Cry, Na* and *Lm*. J. Exp. Bot. *34*, 349-364.

63. Reid, J.B., Potts, W.C. (1986) Internode length in *Pisum*. Two further mutants, *lh* and *ls*, with reduced gibberellin synthesis, and a gibberellin insensitive mutant, *lk*. Physiol. Plant. *66*, 417-426..

64. Simmonds, N.W. (1965) Mutant expression in diploid potatoes. Heredity *20*, 65-72.

65. Simmonds, N.W. (1966) Linkage to the S-locus in diploid potatoes. Heredity *21*, 473-479.

66. Smith, J.D., McDaniel, S., Lively, S. (1978) Regulation of embryo growth by abscisic acid in vitro. Maize Genet. Co-op. Newslet. *52*, 107-108.

67. Spray, C., Phinney, B.O., Gaskin, P., Gilmour, S.J., MacMillan, J. (1984) Internode length in *Zea mays* L. The dwarf-1 mutant controls the 3β-hydroxylation of gibberellin A_{20} to gibberellin A_1. Planta *160*, 464-468.
68. Stoddart, J.L. (1984) Growth and gibberellin-A_1 metabolism in normal and gibberellin-insensitive (*Rht 3*) wheat (*Triticum aestivum* L.) seedlings. Planta *161*, 432-438.
69. Tal, M. (1966) Abnormal stomatal behaviour in wilty mutants of tomato. Plant Physiol. *41*, 1387-1391.
70. Tal, M., Eshel, A., Witztum, A. (1976) Abnormal stomatal behaviour and ion imbalance in *Capsicum scabrous diminutive*. J. Exp. Bot. 27, 953-960.
71. Tal, M., Imber, D. (1970) Abnormal stomatal behaviour and hormonal imbalance in *flacca*, a wilty mutant of tomato. II. Auxin- and abscisic acid-like activity. Plant Physiol. *46*, 373-376.
72. Tal, M., Imber, D., Erez, A., Epstein, E. (1979) Abnormal stomatal behaviour and hormonal imbalance in *flacca*, a wilty mutant of tomato. V. Effect of abscisic acid on indoleacetic acid metabolism and ethylene evolution. Plant Physiol. *63*, 1044-1048.
73. Tal, M., Imber, D., Itai, C. (1970) Abnormal stomatal behaviour and hormonal. imbalance in *flacca*, a wilty mutant of tomato. I. Root effect and kinetin-like activity. Plant Physiol. *46*, 367-372.
74. Tal, M., Nevo, Y. (1973) Abnormal stomatal behaviour and root resistance, and hormonal imbalance in three wilty mutants of tomato. Biochem. Genet. *8*, 291-300.
75. Taylor, I.B., Tarr, A.R. (1984) Phenotypic interactions between abscisic acid deficient tomato mutants. Theor. Appl. Genet. *68*, 115-119.
76. Trewavas, A. (1981) How do plant growth substances work? Plant Cell Envir. *4*, 203-228.
77. Wang, T.L., Beutelmann, P., Cove, D.J. (1981) Cytokinin biosynthesis in mutants of the moss *Physcomitrella patens*. Plant Physiol. *68*, 739-744.
78. Wang, T.L., Donkin, M.E., Martin, E.S. (1984) The physiology of a wilty pea: abscisic acid production under water stress. J. Exp. Bot. *351*, 1222-1232.
79. Wang, T.L., Futers, T.S., McGeary, F., Cove, D.J. (1984) Moss mutants and the analysis of cytokinin metabolism. *In* The biosynthesis and metabolism of plant hormones, Soc. Exp. Biol. Seminar Series 23, pp. 135-164, Crozier, A., Hillman, J.R. eds. Cambridge University Press, London.
80. Wang, T.L., Horgan, R., Cove, D.J. (1981) Cytokinins from the moss Physcomitrella patens. Plant Physiol. *68*, 735-738.
81. Zeevaart, J.A.D. (1984) Gibberellins in single gene dwarf mutants of tomato. Plant Physiol. Suppl. 75, 186.

E5. Auxin Transport

Philip H. Rubery
Department of Biochemistry, University of Cambridge, Cambridge, U.K.

INTRODUCTION

The concept of translocation of chemical messengers in higher plants was expressed in the Nineteenth Century (e.g., 38) and gained concreteness with Went's discovery of auxin (Wuchsstoff) (45) which was eventually shown to be indole-3-acetic acid (Fig. 3, I) (12). The polarity of auxin transport in cereal seedlings was established by the 1930s and later found to be a widespread feature of shoot and root tissues. Good recent accounts of the historical development of auxin polar transport studies are available (16, 19). Polar auxin transport has long been linked, at least in theory, to polar developmental and growth phenomena such as apical dominance and tropisms. It is not a principal task of this Chapter to unravel current controversies and uncertainties in such areas, although scientific interest in polar auxin transport has been sustained and reinforced by the often implicit assumption that auxin concentration is an important physiological and developmental variable. The term 'hormone', denoting the concept of a *transported* chemical messenger, was coined in 1905 and soon applied to plants (40) although the phytohormone concept has rightly been subjected to critical analysis.

Polar auxin transport is the main theme of this Chapter, but it should be emphasised that both endogenous and applied auxins can move (much more rapidly than by polar transport) in the plant's vascular systems. For example, auxin applied to mature leaves is translocated in the phloem like a photoassimilate and has also been reported as moving in the xylem transpiration stream. In these pathways, transport occurs by mass flow down thermodynamic gradients of osmotic potential or water potential. As will be discussed later, the direction of polar transport is determined by distribution of protein carriers rather than by gross potential gradients.

It should also be noted that inactive auxin conjugates such as inositol esters may move in the vascular system (26) and be 'activated' by enzymatic hydrolysis at unloading zones, before being distributed by polar transport. An important but largely unsolved problem is how auxin, a highly permeant molecule, is kept relatively confined and does not exchange freely between the three different transport systems available to it (28).

PHYSIOLOGICAL CHARACTERISTICS OF POLAR AUXIN TRANSPORT

These have been extensively studied for 50 years and frequently reviewed (e.g., 8, 17, 9). Only a summary will be given here. In tissue sections, polar transport is classically detected and analysed by applying auxin to one end, either as a fixed source or as a pulse, and following the movement through the tissue, often into an agar 'receiver' block at the opposite end (Fig. 1). The availability of radiolabelled auxin removed the constraint of bioassy and enabled any concomitant metabolism or immobilisation to be assessed.

In shoot tissues, the polarity of auxin transport is basipetal: that is, auxin moves preferentially from morphologically apical to more basal regions. In roots, it continues in the same physical direction as in shoots, but is now described as acropetal since it is directed towards the root tip. In immature growing portions of the root, the situation is less clear and basipetal transport away from the root tip appears to occur.

Both flux and velocity can be estimated from the kinetics of auxin arrival at the receiver (Fig. 2), or by following the movement of a pulse of radiolabelled auxin through the tissue. The assumptions involved in the application of these and other methods, and their inaccuracies and uncertainties, are extensively discussed in (19). The interested reader should be aware that 'velocity' does not here refer to a transport stream of constant concentration moving at a uniform speed like a conveyer belt, as

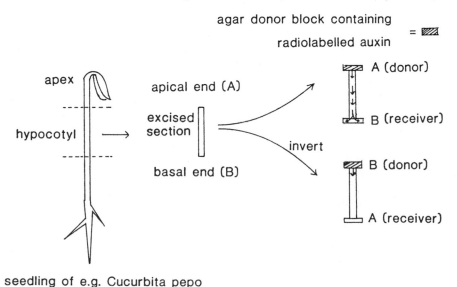

Fig. 1. Basipetal polar auxin transport through a seedling hypocotyl section. Polarity is independent of orientation with respect to gravity.

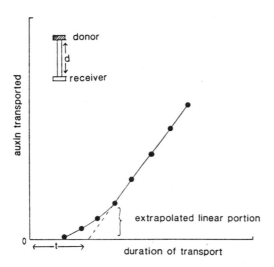

Fig. 2. The velocity of auxin transport is classically measured by the intercept method (19). The linear portion of the arrival curve of auxin into a receiver is extrapolated to zero time. The abscissa intercept (t) is taken as the time required for the front of an auxin stream to reach the end of the section, length d. The velocity is then d/t. Since the early part of the curve is nonlinear, especially with a sensitive method of auxin assay, this 'velocity' is subject to qualification, as discussed in the text.

the method illustrated implicitly assumes (Fig. 2). It may conveniently be regarded as an experimentally determined parameter, with dimensions of length.time^{-1}. The measured velocity of IAA transport is between 5 and 20 mm h^{-1}. This is consistent with the basic mechanism of movement being diffusion rather than a mass flow (8). To anticipate later discussion, although isotropic diffusion does not have a velocity as such (the time taken for an average molecule to traverse a given distance is proportional to the distance squared), the polar movement of auxin is a true transport process and essentially results from a biologically-specified *asymmetric* diffusion. Thus molecules have on average a greater probability of basipetal than of acropetal movement—which is manifested as an overall velocity, for example of a 'front' of auxin moving away from a fixed source. However, some molecules will move in advance of the front (8), accounting for the nonlinear early part of the arrival curve (Fig. 2). It is necessary to be aware of the problem that the more sensitive the method of auxin detection, the sooner the 'advance guard' of auxin arrival can be detected. A pulse of auxin moving by polar transport will have a 'centre of mass' that proceeds with a velocity, although the width of the pulse broadens diffusively.

Polar transport is specific for IAA and synthetic auxins, although 2,4-D (Fig. 3; II) in particular moves about 80% more slowly than IAA. It is an energy requiring process and an uphill auxin concentration gradient can be established along the tissue. Anaerobiosis and metabolic poisons such as cyanide and dinitrophenol inhibit polar transport. The basipetal/

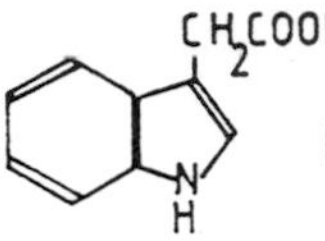

indole-3-acetic acid (IAA)

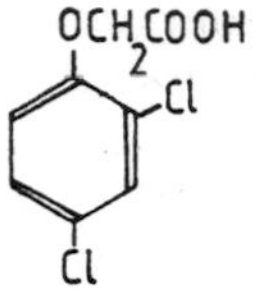

2,4-dichlorophenoxyacetic acid (2,4D)

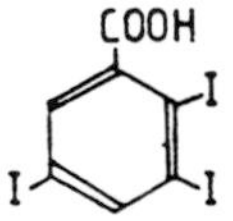

2,3,5-triiodobenzoic acid (TIBA)

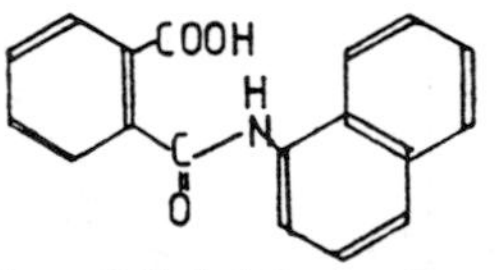

N-1-naphthylphthalamic acid (NPA)

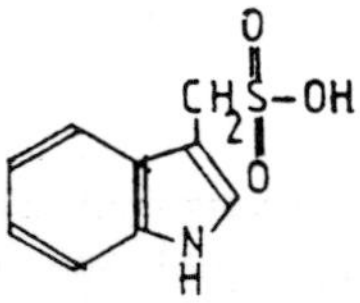

indole-3-methanesulphonic acid

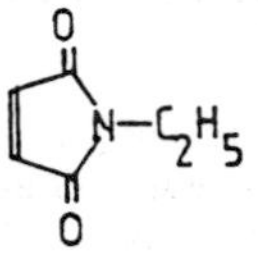

N-ethylmaleimide

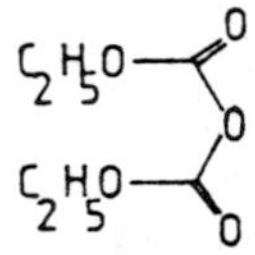

diethylpyrocarbonate

Fig. 3.

acropetal polarity ratio tends to decline as tissues age. Whether this is related to the ability of added auxin to prevent the decay of polar transport in excised sections (8) is unknown. Polar transport appears to occur through parenchymatous cells, particularly those associated with or differentiating into vascular tissue, but not through the vascular elements themselves (25a, 28).

Less work has been done with intact plants than with excised sections, but the same conclusions as to polarity emerge, although the detailed pathways of auxin distribution are not always easily explained. For example, auxin applied to the shoot tip of 5-day old pea seedlings accumulates in lateral root primordia rather than proceeding to the primary root apex (30).

Specific and highly potent inhibitors of polar transport through sections and intact plants, known as phytotropins, have been valuable tools. They include 2,3,5-triiodobenzoic acid (TIBA; Fig. 3, III) and N-1-naphthylphthalamic acid (NPA; Fig. 3, IV)--others are listed and compared in (20). NPA and TIBA are both noncompetitive transport inhibitors (6), and yet TIBA (but not NPA) can itself be polarly transported (42). This apparent paradox will be discussed later. Phytotropins have long been known to work by blocking the 'secretion' of auxin from the basal ends of cells, and a basally-located secretion pump was proposed to account for polarity (8, 19). This model has now been simplified by the chemiosmotic hypothesis of polar auxin transport (8, 27, 33, 37), in which 'secretion' is seen as energetically-downhill auxin anion efflux, and whose discussion now follows.

THE ROUTES BY WHICH AUXIN CAN CROSS MEMBRANES: BACKGROUND TO THE CHEMIOSMOTIC HYPOTHESIS

Before going on to consider current views as to the mechanism of polar auxin transport, we must understand the fundamental processes by which auxin can cross membranes. The feature which dominates auxin transport is the lipid-soluble weakly acidic nature of both naturally occurring IAA and of most synthetic auxins such as NAA and 2,4-D (pK values 4.7, 4.2 and 2.8, respectively). Their hydrophobic nature makes these auxins highly permeant substances in their uncharged forms. The strongly acidic auxin indolemethanesulfonic acid (Fig. 3, V) has been the subject of recent interest (4) and will be discussed later.

The movement of both the hydrophobic undissociated acid molecules (designated AH) and of the more polar anions (A⁻) must be taken into account. The relative concentrations of these species depend on the strength of the acid (dissociation constant = K) and the solution pH and are given by the familiar Henderson-Hasselbalch equation for buffers.

$$pH = pK + \log_{10} \frac{[A^-]}{[HA]} \tag{1}$$

$$\text{or} \qquad \frac{[A^-]}{[HA]} = 10^{(pH-pK)} \tag{2}$$

Consider first diffusion between two buffered aqueous compartments through a biological membrane across which an electrical potential difference is maintained. A convenient and important example is the plasmalemma, membrane potential negative inside, which separates the cytoplasm (pH approx 7) from the cell wall (pH approx 5). For the sake of simplicity, we shall first neglect anion transport and consider the situation where the only form of auxin that crosses the membrane is AH, which is the major permeant species (Fig. 4A). Thus the cells will take up auxin by diffusion of AH down its concentration gradient. As the AH molecules enter the cytoplasm they will ionize to A^- and H^+ in accordance with the dissociation constant. This will maintain the AH gradient while the cytoplasmic concentration of A^- will rise above its external concentration. Equilibrium will be reached when the AH concentrations are the same in each compartment. Provided that the cytoplasm is adequately buffered against the H^+ released by auxin ionization, the total auxin concentration in the cytoplasm will exceed that outside the cell because of the 'trapping' of the A^- anions. The energy for auxin accumulation is expended on maintaining the transmembrane pH gradient via biochemical pH-stats and proton pumping from the cytoplasm.

Thus cells act rather like sponges in the sense that they 'soak up' auxin from a relatively acidic apoplast. This was known in the 1930s (references in (36)) but its significance for polar transport, as discussed in the next section, has only more recently been appreciated. It may also be noted that there need be no spatial restriction on the location of receptors which mediate auxin action and that, unlike non-penetrant animal hormones, there is no second messenger requirement as such. However, coupling messengers such as Ca^{2+} probably are necessary for reasons of amplification and integration with other regulatory signals (28).

The equilibrium condition (scheme of Fig. 4A) for auxin accumulation is expressed most simply by Eqn (3) and more usefully by Eqn (4).

$$[H^+]_i [A^-]_i = [H^+]_o [A^-]_o \qquad (3)$$

$$\frac{[\text{total auxin}]_i}{[\text{total auxin}]_o} = \frac{1 + 10^{(pH_i - pK)}}{1 + 10^{(pH_o - pK)}} = \frac{([H^+]_i + K)[H^+]_o}{([H^+]_o + K)[H^+]_i} \qquad (4)$$

where subscripts i and o refer to inside and outside compartments.
The auxin accumulation found experimentally is below that predicted by Eqn (4) which does not allow for anion permeability (see Eqn (7) below). However, the pH dependence of net auxin uptake follows a characteristic 'titration curve' which reflects the concentration of AH, the major permeant species (31, 36).

Obviously, the more permeant a molecule, the faster it will move across membranes in response to a particular concentration gradient. This is expressed quantitatively by the Permeability Coefficient (P) which is the constant of proportionality relating transmembrane flux to driving force. It is analogous to the Frictional Coefficient in mechanics and has the dimensions of velocity. For IAAH, P is about 10^{-5} m s^{-1}. Some feel for the meaning of this number comes from the simple equation (Eqn 5) which gives the time $(t_{\frac{1}{2}})$ for auxin to reach 50% of its equilibrium concentration within a sphere, radius r.

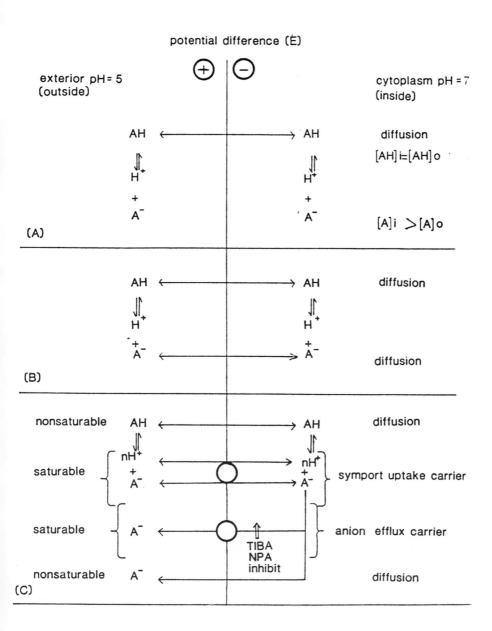

Fig. 4. Progressively more complete illustrations of the routines of transmembrane auxin transport. (A) only the protonated species (AH) move by diffusion; (B) AH and the auxin anion (A⁻) both move by diffusion; (C) diffusive and carrier-mediated components operate.

$$t_{\frac{1}{2}} = \frac{r}{3} \log_e 2 \left(\frac{[H^+]_i + K}{P_{AH}[H^+]_i} \right) \tag{5}$$

For an internal pH = 7 and pK = 4.7, this becomes

$$t_{\frac{1}{2}} = \frac{46.3\,r}{P_{AH}} \tag{6}$$

Now let us consider the effect that taking account of anion permeability will have on the accumulation ratio for auxin (Fig. 4B). Because of its charge, A^- is intrinsically less permeant than AH (with P_{A^-} approximately 10^{-8} m s^{-1}). The cytoplasm will still act as an anion trap but A^- will to some extent leak back into the external medium by passive diffusion down the electrochemical gradient set up by AH uptake and ionization. Both chemical (i.e., the A^- concentration difference across the membrane) and electrical components of this gradient act in the same direction, favoring anion efflux. The full equation for the accumulation ratio (Eqn 7; scheme of Fig. 4B) (8, 33) includes, as the above argument suggests, the membrane potential (E) as well as the permeability coefficients for AH and A^-. It reduces to Eq (4) when P_{A^-} is zero. The larger P_{A^-} becomes, the lower is the auxin accumulation (33) (Fig. 5).

$$\frac{[\text{total auxin}]_i}{[\text{total auxin}]_o} = \frac{([H^+]_i + K)\{P_{AH}[H^+]_o RT(1 - e^{-FE/RT}) + P_{A^-} KFE\}}{([H^+]_o + K)\{P_{AH}[H^+]_i RT(1 - e^{-FE/RT}) + P_{A^-} KFE e^{-FE/RT}\}} \tag{7}$$

So far, we have only considered intrinsic diffusive transport. However, membrane carriers are also intimately involved in polar auxin transport and may be regarded as providing additional saturable components to P_{A^-} and P_{AH} which are susceptible to developmental and biochemical control (Fig. 4C) (28, 33).

Of course when we consider real cells rather than an idealised single compartment, the overall accumulation ratio will mask the separate contributions of cytoplasm and vacuole, but under most conditions the buffered alkaline cytoplasm remains a sink for auxin. However, because it usually occupies most of the cell volume, the vacuole may contain as much or more auxin than the cytoplasm even though it is more acidic.

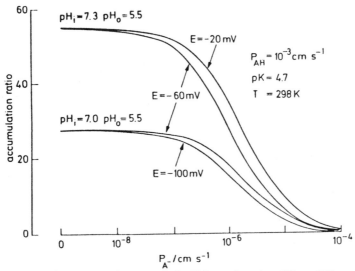

Fig. 5. The behaviour of the accumulation ratio for IAA as a function of P_A at different values of E and internal pH (pH_i), according to Eq. (7).

A MODEL FOR POLAR AUXIN TRANSPORT

Tissue polarity probably reflects an underlying asymmetry of individual cells. Since polar auxin transport is highly substrate specific, it seems likely that differences in membrane permeability, rather than in driving forces, are involved and are determined by membrane proteins rather than by membrane lipids. A simple model has been proposed (8, 33, 37) in which a carrier protein which catalyses the efflux of auxin anions is localized preferentially at the basal ends of cells in the transport pathway. In effect there would be a local increase in P_{A^-} relative to P_{AH}. This would allow a cell taking up auxin from a source to bring its cytoplasmic concentration of auxin to effective equilibrium with a higher external auxin concentration in the cell wall at the basal end of the cell than at the apical end. This is because the auxin accumulation ratio is lower at the basal end with its higher density of anion efflux sites. The argument can then be applied as auxin moves from cell to cell along a linear file to account for the basipetal polarity of the tissue as a whole. The model, dubbed (8) the 'chemisosmotic polar diffusion hypothesis', predicts, in accord with observation, that an uphill concentration gradient of auxin can be established along the tissue in the direction of polar auxin transport (Fig. 6). Thus, the energy stored in an electrochemical gradient set up by cellular accumulation of auxin from an acidic apoplast can be applied to polar transport. Back diffusion will only oppose this to a small degree because of the high membrane permeability to IAAH. Inhibition of polar transport by disruption of energy metabolism is seen as reflecting

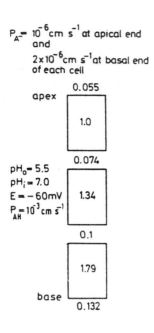

$P_A = 10^{-6} cm\ s^{-1}$ at apical end and $2 \times 10^{-6} cm\ s^{-1}$ at basal end of each cell

apex

0.055

1.0

0.074

$pH_o = 5.5$
$pH_i = 7.0$
$E = -60mV$
$P_{AH} = 10^{-3} cm\ s^{-1}$

1.34

0.1

1.79

base

0.132

Fig. 6. Simulation of IAA equilibrium distribution (Eq. (7)) between cytoplasm and exterior of a file of cells which are relatively more permeable to IAA⁻ at their basal ends. The choice of P_A- values is arbitrary: P_A-/P_{AH} could well be greater than 10^{-3} because of carrier participation. The numbers inside and outside the boxes ('cells') represent total IAA concentration; the vacuole has been omitted for simplicity. From (33).

disruption of metabolically maintained pH and electrical gradients across the plasma membrane. The model requires that auxin move from cell to cell by transport across the plasma membrane and the intervening cell walls rather than primarily through plasmodesmata; symplastic transport is unlikely to be the dominant pathway for highly permeant molecules. This is supported by the observation that polar transport persists in tissues whose plasmadesmatal connections have been largely broken (39) and that auxin in the polar transport stream passes through the free space between cells (2).

Further analysis of the model to ascertain how well it can account for the velocity, profile, and pulse propagation characteristics of polar transport will be discussed after consideration of auxin carriers.

CARRIER-MEDIATED AUXIN TRANSPORT

The General Picture

There is now extensive kinetic evidence for the existence of two carrier-mediated routes for transmembrane auxin transport, in addition to the diffusive movement of IAAH and IAA⁻ discussed above (33, 37, 41) (Fig. 4C). Two basic types of observation suggested the operation of carriers. (1) The uptake of low concentrations of radioactive IAA can be partially inhibited by nonradioactive IAA, showing the presence of a saturable component superimposed on the nonsaturable diffusive uptake due to IAAH transport. (2) A component of efflux is inhibited by TIBA and NPA (polar transport inhibitors). This can be shown directly by inhibition of unidirectional efflux, or by stimulation of *net* uptake. Such observations have been made using suspension-cultured cells (37), more complex tissue systems (stem and root segments of various species) (1, 5, 6, 41), and also *in vitro* preparations of membrane vesicles (3, 14). A simple interpretation of the detailed results, summarised below, is that two carrier systems are present (32, 33, 41): an 'uptake carrier' together with an 'efflux carrier' which is sensitive to inhibition by NPA and TIBA (32, 41). It has been suggested that the uptake carrier is an electroneutral 1:1 symport of H^+ and IAA^- (10, 31) and that the efflux carrier is an IAA^- uniport (32, 33) (Fig. 4C). More complex and speculative proposals are that the symport carrier may be electroimpelled with a $2H^+/1IAA^-$ stoicheiometry (14, 15), and that the efflux carrier may involve counter-transport of IAA^- and Ca^{2+} (15). Resolution of the exact mechanisms involved will require further evidence.

Before dealing in greater depth with the auxin carriers, it will be helpful here to summarise the main reasons for believing that they are indeed involved in *polar* auxin transport in terms of the chemiosmotic hypothesis. (1) Mathematical analysis and computer simulation (11, 23) show that the model accounts rather well for experimental findings, including a need for *saturable* carriers to achieve the observed velocity and concentration profiles characteristic of polar transport. (2) A saturable component to polar auxin transport through tissue segments can be demonstrated (9). (3) The substrate and inhibitor specificity of the carriers matches that of polar auxin transport (14, 33). (4) Use of a fluorescently-tagged monoclonal antibody to an NPA-binding protein has enabled its distribution in pea stem cells to be studied (Fig. 7): although the images are as yet of low resolution, a preferential basal cellular localisation is indicated (16). An attempt to determine how attempted apoplast pH modification may affect polar transport has not given a clear pattern of results (13).

The strongly acidic auxin indolemethanesulfonic acid provides an interesting test of the model since it is fully ionised under all physiological conditions. As might be expected, its uptake is slow relative to IAA and essentially pH-independent (4). It exerts typical auxin responses, but at 10-fold or greater than IAA concentrations (4). This may reflect poor penetration and/or the untested possibility of weak binding to IAA receptors. It appears that the sulfonate can interact with auxin carriers since a degree of basipetal polar transport can be demonstrated, but only at the high (100 μM) concentrations needed for uptake into the cytoplasm. Kinetic studies with zucchini hypocotyl segments and endomembrane vesicles indicate that the sulfonate is a substrate for the auxin uptake carrier (Km approx. 10-fold greater than for IAA) (P. H. Rubery, M. Sabater, unpublished data). Curiously, a strong *acropetal* polarity was

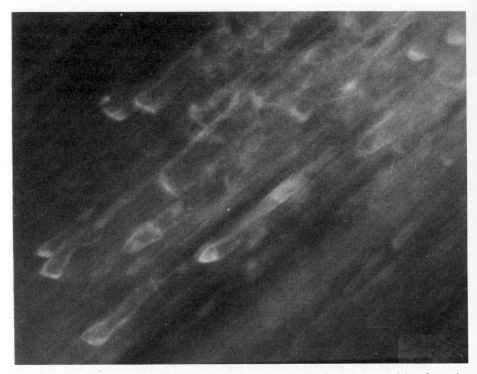

Fig. 7. Immunofluorescence in pea stem cells (in longitudinal section) resulting from the incubation of the tissue with fluorescein-labelled monoclonal antibodies to the NPA binding site. The fluorescence occurs predominantly in the plasma membrane at the basal ends of parenchyma cells sheathing the vascular bundles. (These cells are seen diagonally across the center of the photograph: top–upper right; base–lower left. Below these cells to the right is a pale strip of autofluorescence caused by the vascular bundle.) The occurrence of the NPA binding site at the basal ends of the cells indicates that this is the location of the basal efflux carrier conferring polarity to auxin transport. (Photograph courtesy of Mark Jacobs. From 16).

observed in corn coleoptiles at 1.0 µM indolemethanesulfonate (4), where membrane penetration would be very limited. Studies with phytotropins could help to clarify this situation.

The Uptake Carrier (5, 7, 31, 32, 33, 37, 41)

The carrier responsible for the saturable component of auxin uptake will transport IAA and 2,4-D, but not NAA or benzoic acid. The apparent Michaelis constant for IAA at pH 5.0 is about 1.0 µM—in the same concentration range as physiological auxin levels. The carrier shows counter transport behaviour in that addition of a saturating concentration of nonradioactive auxin after labelled auxin uptake had been proceeding for several minutes resulted in an immediate efflux of labelled auxin. This is because unidirectional carrier-mediated influx of radioactive auxin is immediately halted, but its mediated efflux can continue until passive permeation of cold auxin masks that component also. Since counter flow was seen using radiolabelled 2,4-D in crown gall cells, where 2,4-D is not a substrate for the TIBA-sensitive auxin efflux system, the uptake carrier is likely to operate reversibly. Counter flow also demonstrates that net carrier-mediated auxin uptake still proceeds, in the face of the negative membrane potential, as flux equilibrium is approached. Thus the uptake carrier is unlikely to catalyse electrically uncompensated auxin anion influx. However, an electroneutral symport mechanism is plausible. Uptake showed no dependence on any external cation other than H^+ and the pH optimum of carrier-mediated IAA uptake (pH 5) was higher than that of the stronger acid 2,4-D (pH 4) (31). This would be expected if the carrier were to bind and translocate a proton together with an auxin anion, since the pH optimum for transport will reflect the concentration of a carrier-H^+/A^- species. Such an auxin anion/H^+ symport would be reversible and accumulative from an acidic external medium and would be energetically equivalent to diffusive transport of AH, provided that the symport was symmetrical in its influx and efflux properties and obligatorily coupled so that H^+ and A^- could not be transported separately.

A theoretical analysis of the kinetic behaviour of AH diffusion plus a multiproton symport (responsive to the membrane potential and more accumulative than electroneutral symport) did not fit the uptake data obtained with intact cells and tissue segments (10). However plasma membrane-enriched vesicle preparations, which appear to retain the essential auxin transport features of intact cells, exhibit a greater auxin accumulation from acidic external media than can be explained by IAAH diffusive uptake or the energetically equivalent electroneutral symport (14, 15). This led to a $2H^+/1IAA^-$ symport being proposed (15). Although more recent studies suggest that a saturable *binding* component may contribute to the total apparent auxin 'uptake' by vesicle preparations (3). Thus the simple electroneutral symport hypothesis remains to be disproved but the evidence for it remains rather indirect and it is difficult to reconcile with the effects of valinomycin in the vesicle system (14, 15). There are precedents for varying stoicheiometry of symport carriers and the system may turn out to be rather more complex than the simplest models so far used as working hypotheses.

The Efflux Carrier and its Regulation (5, 7, 32, 33, 37, 41)

There are two main lines of argument and evidence which favor a model in which the component of auxin efflux that can be inhibited by TIBA and NPA is catalysed by a distinct carrier rather than representing a facet of the uptake system discussed above (Fig. 4C). First, it would require a very complex model with many *ad hoc* postulates to incorporate all the substrate- and inhibitor-specificity features in a single 'umbrella' system. In particular, the model would have to be very asymmetric to explain why TIBA effects are only exerted at the cytoplasmic face of the membrane . Second, the uptake carrier system and the TIBA-sensitivity of efflux can be separated experimentally. Thus pretreatment of tissues with low concentrations of N-ethylmaleimide (Fig. 3, VI) (reacts with cysteine side chains in proteins) or diethylpyrocarbonate (Fig. 3, VII) (reacts with histidine side chains) will irreversibly abolish the TIBA-stimulation of net IAA uptake without substantially affecting the uptake carrier (32).

Some further insight has recently been obtained into the relationships between IAA efflux and the actions of TIBA and NPA. First, Depta and Rubery (7) investigated the transmembrane transport of radiolabelled TIBA by *Cucurbita pepo* (zucchini) hypocotyl segments. As well as the expected diffusive uptake of the protonated lipophilic species TIBAH, there was also a carrier-mediated component of TIBA efflux which could be blocked by auxins as well as by NPA. The reciprocal transport interactions of IAA and TIBA are reminiscent of the ability of TIBA, as well as IAA, to be polarly transported—albeit rather slowly. The TIBA-inhibition of polar IAA transport is however non-competitive (6) and cannot result from simple competition for a carrier substrate binding site. Accordingly, a two-site model was proposed (7) in which carrier molecules can specifically bind auxin and TIBA at separate sites. Occupation of either one of the sites would allow translocation of the appropriate ligand across the membrane. In contrast, a carrier complex with both sites occupied would not be active in substrate translocation because of a conformational change, which could embrace the association of loaded separate subunits to form a non-functional heterooligomer. This feature would result in pure non-competitive inhibition of IAA transport by TIBA, but would not allow mutual inhibition of binding. This is not necessarily incompatible with literature reports of TIBA inhibiting auxin binding (see (34)) in microsomal preparations which include endoplasmic reticulum as well as plasma membrane: the majority of auxin binding sites are on the endoplasmic reticulum (31), where the TIBA inhibition could be localised. The transport model of course relates to the plasma membrane. Nevertheless, compromise models can be constructed which do allow partial binding inhibition and mixed transport inhibition (7).

The two site model may be extended to account for the behaviour of NPA which, like TIBA, is a noncompetitive inhibitor of polar auxin transport but which, unlike TIBA, is not itself polarly transported. Although it stimulates net uptake of both IAA and TIBA, NPA is a poor competitor for auxin binding but is an effective inhibitor of TIBA binding to microsomes (6). However, total high affinity NPA binding was only weakly inhibited by TIBA. This pattern of observations suggests that NPA may bind to a *third* site for which TIBA and IAA are poor ligands. This 'NPA site' is envisaged as a regulatory site whose occupation would prevent carrier translocation. If plasma membrane components are responsible for the NPA inhibition of TIBA binding to microsomes, then it could be proposed that NPA also competes at the 'TIBA site' on the auxin carrier, but without being translocated. A highly schematic diagram illustrates these possibilities in a simplified way (Fig. 8).

The concept of a discrete NPA site was first proposed from binding studies (42), and it was later pointed out that the IAA efflux carrier could include catalytic and regulatory subunits, either permanently associated or capable of interacting if the proteins are mobile in the membrane (33). Recent work in my laboratory has provided additional evidence which supports the NPA site being located on a separate polypeptide from the auxin efflux carrier (Morris, D.A., Rubery, P.H., Bowyer, J. unpublished and 25). The effect on the auxin transport characteristics of Zucchini hypocotyl segments caused by pretreating them with with inhibitors of 80S ribosomal protein synthesis was invesigated. Cycloheximide (CH) and 2-(4-methyl-2,6-dinitroanilino)-N-methylpropionamide (MDMP) were used and their potency in Zucchini confirmed by (^{35}S)methionine incorporation. The inhibitor pretreatments did not affect intracellular pH,

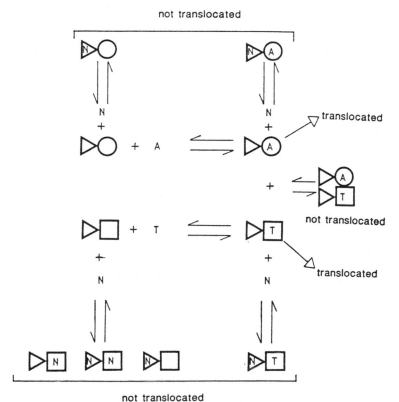

Fig. 8. Diagram to illustrate some possible interactions of IAA (A), TIBA (T), and NPA (N) with catalytic and regulatory binding sites on the plasmamembrane auxin-efflux carrier system. O = IAA site, □ = TIBA site, ▶ = NPA site. See text and (7) for further discussion.

as assessed by distribution of the labelled probe (2-^{14}C)dimethyl-oxazolidine-2,4-dione (1) and did not have a general antagonistic effect on membrane transport since uptake of 3-O-methylglucose, which occurs by proton-symport, was unaffected. The stimulation by NPA of net IAA uptake was abolished after 2-3h pretreatment, suggesting that the receptor for NPA had been lost because of protein turnover—when protein synthesis is inhibited, the normal processes of protein breakdown continue. (In general, proteins can turn over at very different rates. Little is known, especially in higher plants, of the mechanisms which underly such differential stability. This applies with particular force to plant plasma membrane proteins.) When efflux was examined directly, after preincubation with CH, the inhibitory effect of NPA was no longer detectable, but efflux measured in the absence of NPA was unaltered. This confirmed that NPA sensitivity had been lost and also suggested that the efflux carrier itself remained unperturbed under the conditions used (3 h pretreatment with 10 µM CH). Furthermore, conditions can be established whereby loading the segments with nonradioactive IAA stimulates net radiolabelled IAA uptake because of competition for the saturable efflux carrier (5, 7, 41): this stimulation was unaffected by mild CH treatment (1.0 µM for 1.5 h) whereas the NPA effect was about 50% reduced. These observations are consistent with a regulatory NPA-binding subunit which turns over faster than the catalytic subunit for auxin efflux. Endomembrane vesicle preparations from hypocotyls pretreated with CH or MDMP showed a substantially reduced NPA stimulation of net IAA uptake, compared with controls (Sabater, M., Rubery, P.H., unpublished data). Preliminary results indicate that the uptake carrier is also a target for CH, with a similar stability to the NPA binding target.

Some predictions can be tested. For example, NPA *binding* should be reduced by CH and MDMP treatment. Also, it may be possible to establish conditions whereby the polar transport of auxin through tissues can be desensitised to NPA. However, this may prove difficult to demonstrate since CH can itself inhibit polar transport (29) (including in zucchini hypocotyls pretreated with 1.0 µM CH for 3h (unpublished)), but the phenomenon has been little characterised. Attenuation of the symport, or nonspecific effects, could be implicated.

MATHEMATICAL ANALYSES OF THE CHEMIOSMOTIC POLAR DIFFUSION MODEL

Mitchison (23) and Goldsmith et al (11) have adopted rather different mathematical approaches to the question of whether the chemiosmotic theory can account quantitatively for the dynamic aspects of polar transport as well as for its directional and energetic characteristics. Both conclude (11, 23) that the experimentally observed features of auxin pulse propagation and of the distribution profiles and arrival curves found in tissue segments with a continuous auxin supply can be accomodated by the theory—provided that the permeability coefficients are within certain realistic ranges. The original papers must be consulted for a full account of this complex work, but some of their conclusions will be summarised in a simplified form.

Different anion permeabilities at the two ends of the cell do indeed lead to asymmetry of auxin movement; Cytoplasmic streaming makes only a small contribution to transport velocity (23); velocity decreases with increasing cell length (11, 23) (consistent with experimental data (30)); *saturability* of carriers is needed to explain the observed shapes of distribution profiles (23). The predicted values of the permeability coefficients needed to give velocities of the observed order of 10 to 20 mm h^{-1} (2.8 to 5.6×10^{-6} m s^{-1}) depend on assumptions made about the permeability of the tonoplast; if the vacuole is freely accessible as a transcellular auxin transport route, then P_{A^-} in the plasma membrane at the end of the cell should be about 2.6×10^{-6} m s^{-1}—about the same as the auxin transport velocity (23). However, if the tonoplast is (unrealistically) assumed completely impermeant so that auxin flow is restricted to the cytoplasm and chanelled round the vacuole, then P_{A^-} is an order of magnitude smaller (23). Measurements of the permeability of isolated vacuoles to auxin are clearly needed, but have not yet been reported, to help assess the path of auxin through the cell. This could be important since the *position* in cells of a relatively impermeant vacuole could influence the *direction* of auxin fluxes through a tissue, as Mitchison has argued in the context of shoot gravitropism (24). Finally, these analyses suggest that the role of the symport carrier could be in supplementing the intrinsic permeability of membranes to IAAH to permit observed transport velocities. However, since the fundamental requirement of the theory is for a higher P_{A^-}/P_{AH} *ratio* at the basal end of the cell, polar transport could in principle involve an asymmetric distribution of a symport carrier as well as of the efflux carrier (33).

Milborrow and Rubery (22) have pointed out that in tissues which have a *gradient* of carriers, the asymmetry at cellular junctions necessary for polar transport could be achieved if in each individual cell the carrier density were uniform, but decreased in a regular way down a file of cells. This would only achieve effective polarity over short distances, and was proposed in the context of ABA transport in the elongating zone of roots where a gradient of ABA carriers (possibly ABA^-/H^+ symport) is found (1, 22). In such a model polar transport could be controlled developmentally in terms of carrier biosynthesis and turnover rather than by more elaborate mechanisms involving the establishment of polarity in individual cells—the subject of the next section.

SPECULATION CONCERNING THE ESTABLISHMENT AND MAINTENANCE OF POLARITY

How polarity is regulated is completely unknown but various possibilities can be distinguished. Work with single cells, notably *Fucus* eggs, has shown the importance of electric fields and currents in the polarity of growth (18). The idea has been discussed that plasma membrane auxin carriers could congregate basally because of membrane electrophoresis (28). If the cells in the polar transport pathway also responded to auxin by increasing electrogenic H^+ efflux, the changing pH and electrical gradients could (besides feedback modulating auxin uptake) help maintain carrier distribution and perhaps account for the known ability of auxin to maintain and stimulate its own polar transport system (8).

Alternative scenarios are suggested by analogy with lymphocyte capping in which membrane proteins may move directionally by ATP-driven interaction with the cytoskeleton or by being carried along in a membrane flow set up by localised endocytosis. However, such speculation should be dampened by the realisation that there is no direct evidence for plasma membrane protein mobility or for endocytosis in turgid plant cells with intact walls (discussed in 35).

The evidence described earlier for rapid turnover of a regulatory NPA-binding subunit prompts a different line of thought. Namely, the localised insertion or removal of particular proteins could generate an asymmetric distribution, possibly requiring monomer aggregation or anchorage to the cytoskeleton to ensure its stability. The efflux carrier itself could then be uniformly distributed with a basal activity that could be augmented by association with its regulatory subunit, whose localised distribution could be the critical asymmetry in polar transport (16). Carrier down-regulation could be accomplished in the longer term by subunit turnover and in the short term by a naturally-occurring NPA analog. Such endogenous regulators have not yet been identified, but the parallel situation in which endorphins, opiate peptides, were discovered long after their pharmacological analogs like morphine should inspire renewed efforts to attain this important goal.

P. H. Rubery

AUXIN TRANSPORT AND OUR UNDERSTANDING OF AUXIN ACTION

One of the reasons for challenging the idea that auxin concentration changes are a major factor influencing rates of molecular processes (see Introduction) is the apparent contradiction between the relatively large changes in *exogenous* auxin concentration needed to elicit a particular fractional response and the relatively small changes in *endogenous* auxin level associated with the same degree of response. However, our knowledge of transmembrane auxin transport and auxin metabolism suggests that this argument may be too simple. There is not a necessarily simple linear relationship between auxin concentration in a compartment with access to a receptor and either external auxin concentration or its averaged out internal concentration in total tissue water (which is all that current methods can directly measure). As the exogenous auxin concentration is increased, an initially linear rise in concentration at a remote receptor could be damped out by (i) saturation of auxin carriers along the transport pathway; (ii) changes in transmembrane pH gradients brought about by auxin; (iii) metabolic buffering by auxin destruction or conjugation. Similarly, the compartmentation of endogenous auxin could be altered by stimuli (e.g., the gravity vector) which produce growth responses—for instance by modulation of carrier activity or transport driving forces.

It is possible that an IAA receptor involved in growth control is on the *outer face* of the plasma membrane (20b) and so responds to the cell wall IAA concentration. This would have important consequences for auxin action and also for our view of the physiological importance of polar auxin transport. IAA is known to cause cell wall acidification, and this has for some time been widely regarded as the way in which auxin stimulates cell extension (the acid growth hypothesis—see Chapter C1). However recent careful quantitative work indicates that, in maize coleoptile at least, acidification is not a sufficient explanation of auxin action (20a). But whatever is the principal means by which auxin loosens cell walls, concomitant acidification (and increased cell volume) would act to reduce the IAA concentration available to an outward-facing receptor by stimulating uptake into the cells. This would be applicable to the auxin relations of intact plants (as opposed to sections bathed in an auxin reservoir). Indeed, as previously discussed (35), redistributive change of the IAA concentration in the cell wall, which is a low volume/low pH compartment, is substantially more sensitive to alterations in transmembrane pH gradients than is the cytoplasmic concentration. Membrane hyperpolarisation would oppose this to a small extent (33). Such decreases in cell wall auxin concentration would thus tend to act as a feedback damping constraint on auxin action. (In contrast, for an inward-facing receptor, reinforcement would result.) However, the polar transport system could perhaps maintain a supply of auxin to the surface of the responsive cells by a source-sink transfer. Seen in this way, polar auxin transport would be an integral part of a prolonged tissue elongation response to IAA. Since both elongation and polar transport are characteristic of young tissues, the decline in polar transport that occurs during tissue maturation could be a factor (as well as restrictive changes in cell wall chemistry) that limits cell elongation.

Finally, Hertel has speculated (15) that auxin action and auxin transport would be directly linked if the auxin anion carrier could act as a Ca^{2+} gate and allow counter-transport of IAA^- and Ca^{2+} down their respective electrochemical gradients. The carriers are proposed to occur on the tonoplast as well as on the plasma membrane so that auxin could promote Ca^{2+} release from wall or vacuole reservoirs so elevating the cytoplasmic Ca^{2+} concentration which would then act, perhaps via calmodulin, as a coupling messenger for auxin (28). This interesting proposal at present lacks direct evidence, and is a further area in which the study of isolated vacuoles could be invaluable.

References

1. Astle, M.C., Rubery, P.H. (1983) Carriers for abscisic acid and indole-3-acetic acid in primary roots: their regional localisation and thermodynamic driving forces. Planta 157, 53-63.
2. Cande W.Z., Ray, P.M. (1976) Nature of cell-to-cell transfer of auxin in polar transport. Planta 129, 43-52.
3. Clark, K.A., Goldsmith, M.H.M. (1985) A component of pH-driven IAA accumulation in zucchini membrane vesicles is saturable binding. (Abstr.) Plant Physiol. (Suppl.) 77, 443.
4. Cohen, J.D., Baldi, B.G., Bialek, K. (1985) Strongly acidic auxin indole-3-methane-sulfonic acid. Synthesis of (^{14}C)indole-3-methanesulfonic acid and studies of its chromatographic, spectral, and biological properties. Plant Physiol. 77, 195-199.
5. Davies, P.J., Rubery, P.H. (1978) Components of auxin transport in stem segments of Pisum sativum L. Planta 142, 211-219.
6. Depta, H., Eisele, K.H., Hertel, R. (1983) Specific inhibition of auxin transport: action on tissue segments and in vitro binding to membranes from maize coleoptiles. Plant Sci. Lett. 31, 181-192.
7. Depta, H., Rubery, P.H. (1984) A comparative study of carrier participation in the transport of 2,3,5-triiodobenzoic acid, indole-3-acetic acid, and 2,4-dichlorophenoxy-acetic acid by Cucurbita pepo L. hypocotyl segments. J. Plant Physiol. 115, 371-387.
8. Goldsmith, M.H.M. (1977) The polar transport of auxin. Annu. Rev. Plant Physiol. 28, 439-478.
9. Goldsmith, M.H.M. (1982) A saturable site responsible for polar transport of indole-3-acetic acid in sections of maize coleoptiles. Planta 155, 68-75.
10. Goldsmith, M.H.M., Goldsmith, T.H. (1981) Quantitative predictions for the chemiosmotic uptake of auxin. Planta 153, 25-33.
11. Goldsmith, M.H.M., Goldsmith, T.H., Martin, M.H. (1981) Mathematical analysis of the chemiosmotic polar diffusion of auxin through plant tissues. Proc. Natl. Acad. Sci. USA 78, 976-980.
12. Haagen-Smit, A.J., Leech, W.D., Bergren, W.R. (1941) Estimation, isolation and identification of auxins in plant material. Science 93, 624-625.

13. Hasenstein, K-H., Rayle, D. (1984) Cell wall pH and auxin transport velocity. Plant Physiol. 76, 65-67.
14. Hertel, R., Lomax, T.L., Briggs, W.R. (1983) Auxin transport in membrane vesicles from *Cucurbita pepo* L. Planta 157, 193-201.
15. Hertel, R. (1983) The mechanism of auxin transport as a model for auxin action. Z. Pflanzenphysiol. 112, 53-67.
16. Jacobs, M., Gilbert, S.F. (1983) Basal localization of the presumptive auxin transport carrier in pea stem cells. Science 220, 1297-1300.
17. Jacobs, W.P. (1979) Plant hormones and plant development. 339pp. Cambridge University Press.
18. Jaffe, L.F., Nuccitelli, R. (1977) Electrical controls of development. Annu. Rev. Biophys. Bioeng. 6, 445-476.
19. Kaldewey, H. (1984) Transport and other modes of movement of hormones (mainly auxins). In: Hormonal regulation of development II. The functions of hormones from the level of the cell to the whole plant. Scott, T.K., ed. Encyc. Plant Physiol., New Ser. 10, 80-148.
20. Kateckar, G.F., Geissler, A.E. (1980) Auxin transport inhibitors IV. Evidence of a common mode of action for a proposed class of auxin transport inhibitors: the phytotropins. Plant Physiol. 66, 1190-1195.
20a Kutschera, U., Schopfer, P. (1985) Evidence against the acid growth theory of auxin action. Planta 163, 483-493.
20b Löbler, M., Klämbt, D. (1985) Auxin-binding protein from coleoptile membranes of corn (*Zea Mays* L.) II Localization of a putative auxin receptor. J. Biol. Chem. 260, 9854-9859.
21. Mer, C.L. (1969) Plant growth in relation to endogenous auxin with special reference to cereal seedlings. New Phytol. 68, 275-294.
22. Milborrow, B.V., Rubery, P.H. (1985) The specificity of the carrier-mediated uptake of ABA by root segments of *Phaseolus coccineus* L. J. Exp. Bot. 36, 807-822.
23. Mitchison, G.J. (1980) The dynamics of auxin transport. Proc. R. Soc. London Ser. B 209, 489-511.
24. Mitchison, G.J. (1981) The effect of intracellular geometry on auxin transport. II Geotropism in shoots. Proc. R. Soc. London Ser. B 214, 69-84.
25. Morris, D.A., Rubery, P.H. (1985) Effects of translation inhibitors on NPA-sensitive auxin net-uptake and efflux by *Cucurbita pepo* hypocotyl segments. (Abstr.) Plant Physiol. (Suppl.) 77, 18.
25a Morris, D.A., Thomas, A.G. (1978) A microautoradiographic study of auxin transport in the stem of intact pea seedlings (*Pisum sativum* L.). J. Exp. Bot. 29, 147-157.
26. Nowacki, J., Bandurski, R.S. (1980) *Myo*-inositol esters of indole-3-acetic acid as seed auxin precursors of *Zea mays* L. Plant Physiol. 65, 422-427.
27. Raven, J.A. (1975) Transport of indoleacetic acid in plant cells in relation to pH and electrical potential gradients, and its significance for polar IAA transport. New Phytol. 74, 163-172.
28. Raven, J.A., Rubery, P.H. (1982) Coordination of development: hormone receptors, hormone action, and hormone transport. In: Molecular biology of plant development, pp. 28-48, Smith, H., Grierson, D., eds. Blackwell Scientific, Oxford.
29. Riov, J., Goren, R. (1979) Inhibition of polar indole-3-acetic acid transport by cycloheximide. Plant Physiol. 63, 1217-1219.
30. Rowntree, R.A., Morris, D.A. (1979) Accumulation of ^{14}C from exogenous labelled auxin in lateral root primordia of intact pea seedlings. Planta 144, 463-466.
31. Rubery, P.H. (1978) Hydrogen ion dependence of carrier-mediated auxin uptake by suspension-cultured crown gall cells. Planta 142, 203-206.
32. Rubery, P.H. (1979) The effects of 2,4-dinitrophenol and chemical modifying reagents on auxin transport by suspension-cultured crown gall cells. Planta 144, 173-178.

33. Rubery, P.H. (1980) The mechanism of transmembrane auxin transport and its relation to the chemiosmotic hypothesis of the polar transport of auxin. In: Plant growth substances 1979, pp. 50-60, Skoog, F., ed. Springer, Berlin Heidelberg New York.
34. Rubery, P.H. (1981) Auxin receptors. Annu. Rev. Plant Physiol. 32, 569-596.
35. Rubery, P.H. (1984) Auxin binding and membrane receptors in relation to auxin action and transport. In: Membranes and compartmentation in the regulation of plant functions. Annu. Proc. Phytochem. Soc. Eur. 24, 267-282, Boudet, A.M., Alibert, G., Marigo, M., Lea, P.J., eds. Clarendon Press, Oxford.
36. Rubery, P.H., Sheldrake, A.R. (1973) Effect of pH and cell surface charge on cell uptake of auxin. Nature (London) New Biol. 244, 285-288.
37. Rubery, P.H., Sheldrake, A.R. (1974) Carrier-mediated auxin transport. Planta 118, 101-121.
38. Sachs, J. (1880) Stoff und Form der Pflanzenorgane I. Arb. Bot. Inst. Wurzburg 2, 452-488.
39. Sheldrake, A.R. (1979) Effects of osmotic stress on polar auxin transport in Avena mesocotyl sections. Planta 145, 113-117.
40. Soding, H. (1923) Werden von der Spitze der Haferkoleoptile Wuchshormone gebildet? Ber. Dtsch. Bot. Ges. 41, 396-400.
41. Sussman, M.R., Goldsmith, M.H.M. (1981) Auxin uptake and action of N-1-naphthyl-phthalamic acid in corn coleoptiles. Planta 151, 15-25.
42. Thomson, K-S., Hertel, R., Muller, S., Tavares, J.E. (1973). N-1-Naphthylphthalamic acid and 2,3,5-triiodobenzoic acid: In-vitro binding to particulate cell fractions and action on auxin transport in corn coleoptiles. Planta 109, 337-352.
43. Trewavas, A.J. (1982) Growth substance sensitivity: the limiting factor in plant development. Physiol. Plant. 55, 60-72.
44. Wangerman, E., Mitchison, G.J. (1981) The dependence of auxin transport on cell length. Plant Cell Environ. 4, 141-144.
45. Went, F.W. (1928) Wuchsstoff und Wachstum. Rec. Trav. Bot. Neerl. 25, 1-116

E6. The Induction of Vascular Tissues by Auxin

Roni Aloni

Department of Botany, The George S. Wise Faculty of Life Sciences, Tel Aviv University, Tel Aviv 69978, Israel.

INTRODUCTION

The vascular tissues of the plant connect the leaves and other parts of the shoot with the roots. The vascular system is composed of two kinds of conducting tissues: the phloem, through which organic materials are transported and the xylem, which is the conduit for water and soil nutrients. Vascular development in the plant is an open type of differentiation, it continues as long as the plant grows by apical and lateral meristems. Thus, there is a continuous development of new vascular tissues that are in dynamic relationship to one another. In spite of the complexity of structure and development of the vascular tissues, there is evidence that the auxin, indole-3-acetic acid (IAA), is the main limiting and controlling factor for both phloem and xylem differentiation (1, 12, 13). This fact raises a major problem of development, which is, what are the mechanisms by which one stimulus controls the differentiation of complex patterns of phloem and xylem? However, it should be emphasized that other growth regulators may also be involved in vascular differentiation. They are beyond the scope of this article and the reader is directed to recent reviews on this topic (e.g., 19, 21).

A major problem in studying vascular differentiation is the difficulty of observing the phloem. Therefore, it is not surprising that reports on phloem differentiation are confusing and often contradictory. Some of these contradictions have been clarified by using a new clearing technique for phloem (4) and will be discussed below.

The aim of this chapter is to present a summary of current thoughts and evidence on the role of auxin in controlling phloem and xylem differentiation.

STRUCTURE AND DEVELOPMENT OF VASCULAR SYSTEMS

In order to discuss the role of auxin in vascular differentiation, there is need to present some of the basic features of these tissues.

The first organized vascular system found in lower members of the plant kingdom consisted of phloem with no xylem (9). Thus, we find the first developed sieve tubes in members of the brown algae. Much later during the evolution of plants, in the transition from aquatic to terrestrial habitats, the water conducting system developed.

The vascular system of higher plants usually consists of phloem and xylem. Both are complex tissues, as they consist of several types of cells. The conducting cells in the phloem are the sieve elements. These cells, which transport the photosynthetic products, lose their nucleus at maturity. Their cytoplasm is connected through the sieve areas to build the sieve tubes. In the xylem the conducting cells are the tracheary elements which function in the conduction of water, as non-living cells after autolysis of their cytoplasm. The tracheary cells are characterized by their secondary wall thickenings, which keep their shape when they are dead, against the pressure formed by their surrounding cells. Two functional conduits are distinguished in the xylem : the tracheid, which is a non-perforated long cell, characterized by bordered pits, and the vessel which is a long continuous tube built up of numerous vessel elements connected end-to-end by perforation plates. Each vessel is limited in length by imperforate walls in its distal parts. It is important to point out that the vessels and not vessel elements are the operating conduction units of the xylem in angiosperms. Therefore, their dimensions are important parameters for studying long distance transport of water, xylem adaptations or pathology of the xylem.

There is a clear relationship between the pattern of the vascular tissues and that of the leaves. Changes in the phyllotaxis, which is the arrangement of the leaves on the stem, during plant development are associated with similar changes in the pattern of the vascular tissues descending from the leaves towards the roots. In addition, the development of the vascular tissues is correlated to leaf development. Vascular tissues are initiated in the based of the leaf primordium and their course of differentiation is in accordance with leaf expansion.

In the embryo, and in the young portions of the shoots and roots, primary phloem and primary xylem are formed from the apical meristem - the procambium. During the development of gymnosperms and dicotyledons, a lateral vascular meristem, the cambium, differentiates in the older parts of the plant axis and produces the secondary phloem and xylem. The secondary tissues increase the width of the woody parts. The procambium and the cambium are considered as two continuous developmental stages of a single meristem with no clear dividing lines between them (e.g., 15). Accordingly, the primary and secondary vascular tissues are formed along the plant axis in a continuous pattern. Thus, we can find in the stem, vessels whose upper portions consist of primary vessel elements and their lower portions of secondary elements (Aloni and Plotkin, unpublished).

Along the plant axis the vascular tissues are organized in longitudinal strands which are also called vascular bundles, or they are arranged in a continuous cylindrical complex. The xylem is formed in the inner side towards the pith and the phloem towards the outside. In some plant families phloem can be found both in the outer and inner side of the xylem. Xylem does not differentiate in the absence of phloem. On the other hand, phloem often develops with no xylem. Thus, we find phloem anastomoses (Fig. 1A) between the longitudinal bundles in many plant species (4). Bundles of phloem with no concomitant xylem are usually found in the stem of many species. For example, in mature internodes of *Coleus*, on each bundle of phloem with xylem (a collateral bundle), there is another bundle which consists of phloem only (3). In the young internodes there are more phloem- only bundles than collateral bundles. In the course of vascular development, phloem differentiation precedes that of xylem. Therefore, the first elements to differentiate from the procambium are the sieve elements and the differentiation of secondary phloem in the spring precedes that of the secondary xylem (9). In tissue culture the vascular elements differentiate from parenchyma cells by redifferentiation and have a similar sequential pattern in which the phloem appears before the xylem (1).

Along the plant axis there is a general and gradual increase in the diameter of both sieve and tracheary elements from leaves to roots. The narrowest vascular elements are found in the leaves and the largest elements are observed in the roots. The basipetal increase in vascular element size is often associated with basipetal decrease in vascular element density (5). Thus, for example, vessel density is generally greater in branches where vessels are smaller than in roots where they are larger.

THE INDUCTION OF VASCULAR DIFFERENTIATION BY LEAVES AND BY AUXIN

Developing buds and young growing leaves induce vascular tissues below them. In the spring the young leaves stimulate cambium reactivation and formation of new phloem and xylem which extend downwards from the developing buds towards the roots (17). The removal of young leaves from stems reduces or prevents vascular differentiation below the excised leaves (12, 14). The effect of the leaves on vascular differentiation is polar: leaves promote vascular differentiation in the direction of the roots (12) but have no effect, or even a slight inhibitory effect on vascular development in the direction of the shoot tip (2). The stimulation effect of the leaves can be demonstrated by grafting shoot apices with a few leaf primoridia onto callus, which results in the formation of both phloem and xylem below the graft in the callus tissue (23).

It is well known that one of the major signals produced by the young leaves is auxin, which moves in a polar fashion towards the roots. A source of auxin replaces the young leaves in promoting vascular differentiation and regeneration (12, 18), (Fig. 2). Vascular differentiation occurs along the polar flow of auxin from leaves to roots. In their excellent studies, Jacobs (12) and LaMotte and Jacobs (14) have established that indole-3-acetic acid is the limiting and controlling factor

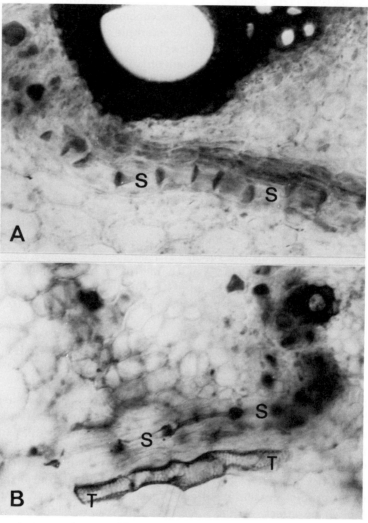

Fig. 1. Transverse sections, taken from internodes of *Luffa cylindrica*, stained with 2% lacmoid in 96% ethanol (1), both x 200. *A*. From an intact untreated plant. The photograph shows a typical phloem anastomosis consisting of sieve elements only. *B*. From a decapitated stem treated for 10 days with 1.0% naphthaleneacetic acid (NAA) applied in a lanolin paste. The high auxin concentration induced the differentiation of tracheary elements in the anastomosis. S, sieve element: T, tracheary element.

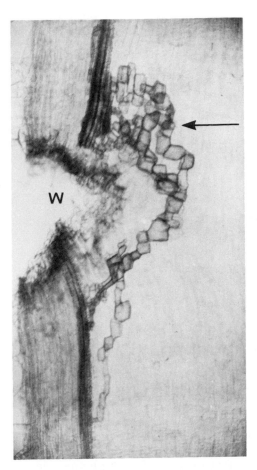

W

Fig. 2. A longitudinal view of xylem regeneration around a wound induced by IAA in a young internode, with the leaves and buds above it excised. The tissue was taken from a decapitated internode of *Cucumis sativus* treated for 7 days with 0.1% indole-3-acetic acid in lanolin, which was applied to the upper side of the internode immediately after the wounding. The tissue was cleared with pure lactic acid and remained unstained, x 50. The photograph shows a typical polar pattern of xylem regeneration around a wound which was induced by the basipetal polar flow of auxin. The polar regeneration is characterized by a dense appearance of many regenerative tracheary elements (arrow) immediately above the wound that differentiated close to the wound surface. Below the wound there are few elements in defined files that connect to the damaged strand at greater distances from the wound. The polar pattern of xylem regeneration reflects the pattern of auxin movement around the wound. As the vascular strands are the preferable pathways for auxin movement, the applied IAA moved basipetally in the damaged strand to where it was interrupted by the wound and was forced to find a new pathway around the obstacle, resulting in a somewhat higher auxin concentration immediately above the wound. This higher auxin concentration resulted in many more regenerative tracheary elements above and close to the wound than below it. W, wound.

in both xylem and phloem regeneration. Auxin alone replaced both qualitatively and quantitatively the effect of the leaves on vascular differentiation in *Coleus* (12, 14, 22).

Phloem is induced at low auxin levels while there is a need for higher auxin levels for xylem differentiation (14, 22). Bruck and Paolillo (8) reported that more phloem than xylem differentiated after leaf excision in *Coleus*. They suggested that the phloem-only bundles represent an auxin deficient stage of potentially collateral bundles, which depends on auxin availability to the bundle for fulfillment. In tissue cultures, phloem differentiates with no xylem at low auxin levels, whereas both phloem and xylem are induced at higher auxin concentrations (1), (Fig. 3). Thus, xylem differentiation only takes place at higher auxin levels. As this also fulfils the requirement for phloem differentiation, it explains why xylem does not differentiate in the absence of phloem and always accompanies the pattern of phloem. These results in tissue culture were made possible

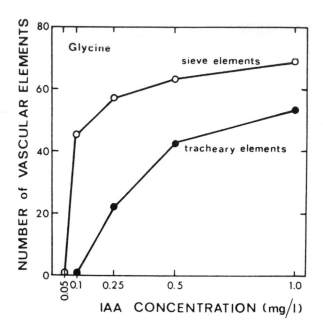

Fig. 3. Effect of indole-3-acetic acid (IAA) on the differentiation of sieve and tracheary elements in tissue culture of *Glycine max* after 21 days. Notice that the low auxin concentration (0.1 mg/1) induced sieve elements with no tracheary elements (1).

by applying the lacmoid clearing technique (4) which enables the study of the three dimensional structure of both phloem and xylem in the whole callus. In previous studies of vascular differentiation in tissue culture (23), the role of auxin level was not detected, probably because of the use of a thin-section technique.

I propose that phloem anastomoses which consist of phloem with no xylem (4), and occur between the longitudinal bundles, are induced by streams of low auxin level. It is therefore expected that high auxin levels would induce xylem in these anastomoses. Figure 1B supports this view, by showing the differentiation of anastomosis of phloem with xylem that was induced by application of high auxin concentration to decapitated young stems of *Luffa*.

An additional factor to be considered in vascular control is the capacity of mature vascular tissues to transport auxin. Auxin from mature leaves moves rapidly in a non-polar fashion in the phloem (10). When the phloem is damaged below mature leaves there is a quantitative increase in vascular differentiation, indicating the replacement of a long non-functional region of damaged tissues (7). It is believed that this promoting effect of the wounding is due to an additional source of auxin in the wound region which arrives from the mature phloem. This is an additional mechanism which enables leaves to regulate their supportive vascular system.

THE CONTROL OF VASCULAR STRAND FORMATION

In his elegant series of experiments done with pea seedlings, Sachs (18, 19) brings evidence supporting his hypothesis that canalization of auxin flux determines the orderly pattern of vascular differentiation from leaves to roots.

According to Sachs (18) the organization of the vascular tissues in ordered bundles is determined by the flow of auxin through the cells prior to vascular differentiation. Auxin movement from an auxin source probably occurs initially by diffusion. As the auxin diffuses through the cells it induces the formation of the polar auxin transport system (see chapter E5.) along a narrow file of cells through which diffusion has been taking place. This occurs in the direction away from the auxin source regardless of its spatial orientation. Thus, the horizontal movement of auxin around a wound induces the cells to transport auxin in a horizontal direction (Fig. 2). The continuous polar transport of auxin induces a further complex series of events which finally result in the formation of a vascular strand. When the vascular strand has developed it remains the preferential pathway for auxin transport, as cells possessing the ability to transport auxin are associated with vascular tissues (11, 16).

As the vascular strands are the fastest channels for auxin flow they become the preferable pathways for further auxin transport. Therefore new streams of auxin from young developing leaves are directed towards the vascular strands, thus continuing the development of a discrete vascular network whose position is determined by the location of the leaf primorida. At the same time the canalization of auxin transport to and in the vascular strands precludes the formation of new vascular strands in the vicinity of pre-existing strands. In addition, the polarized cells in the strand, which are well supplied with auxin, are not subject to the influence of an inductive source of auxin coming from another direction and this will thus prevent the expression of another smaller source of auxin in its neighbourhood. On the other hand, a pre-existing vascular strand which is not supplied with auxin (e.g., descending from an old leaf) acts as a sink for any new stream of auxin (coming from a young leaf) so that a new vascular strand will be formed towards a pre-existing strand which has a low supply of auxin.

The above hypothesis and results which stem from Sachs studies give us the basis for understanding the three dimensional patterns of strands organization, and are useful for studying the special relations between the arrangement of the leaves on the stem (phyllotaxis) and the pattern of vascular bundles descending from them.

THE CONTROL OF VASCULAR ELEMENT SIZE AND DENSITY

A new hypothesis has been proposed in order to explain the general increase in vessel size and decrease in vessel density (i.e., number of vessels per transverse-sectional area) from leaves to roots (5).

The hypothesis is based upon the assumption that the stable and steady polar flow of auxin from leaves to roots controls the polar changes in vessel size and density along the plant axis. Accordingly, the following six-point hypothesis was proposed:

(1) Basipetal polar flow of auxin from leaves to roots establishes a gradient of decreasing auxin concentration from leaves to roots.

(2) Local structural or physiological obstruction to auxin flow results in a local increase in auxin concentration.

(3) The distance from the source of auxin to the differentiating cells controls the amount of auxin flowing through the differentiating cells at a given time, thus determining the cells' position in the gradient.

(4) The rate of vessel differentiation is determined by the amount of auxin that the differentiating cell receives; high concentrations cause fast, low concentrations result in slow differentiation. Therefore, the duration of the differentiation process increases from leaves to roots.

(5) The final size of a vessel is determined by the rate of cell differentiation. Since cell expansion is stopped when the secondary wall is deposited, rapid differentiation results in narrow vessels, while slow differentiation permits more cell expansion and therefore results in wide tracheary elements. Decreasing auxin concentration from leaves to roots therefore results in an increase in vessel size in the same direction.

(6) Vessel density is controlled by the auxin concentration; high concentrations induce greater, low concentrations lower densities. Therefore, vessel density decreases from leaves to roots (5).

Experiments with bean (5), *Acer* stems (6) and *Pinus* seedlings (20), support the six- point hypothesis. The experiments show that the rate of vessel (5, 6) and tracheid differentiation (20) decreases with increasing distance from the auxin source. The rate of vessel formation in bean was found to be constant at any one distance from the auxin source (5).

Various auxin concentrations applied to decapitated bean stems induced substantial gradients of increasing vessel diameter and decreasing vessel density from the auxin source towards the roots (Figs. 4, 5). High auxin concentration yielded numerous vessels that remained small because of their rapid differentiation. Low auxin concentration resulted in slow differentiation and therefore in fewer and larger elements.

Interestingly, auxin concentration was found to influence the pattern of vessels in the secondary xylem of bean plants as seen in a transverse section. Immediately below the auxin source the vessels were arranged in

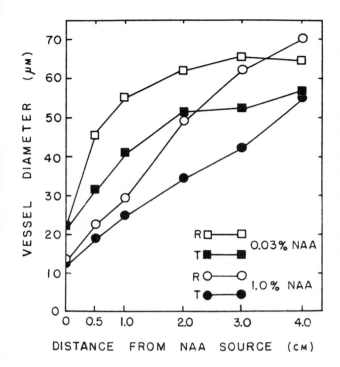

Fig. 4. Effect of low (0.03% NAA; squares) and high (1.0% NAA; circles) auxin concentrations on the radial (R) and tangential (T) diameter of the late-formed vessels along decapitated internodes of *Phaseolus vulgaris*. Both auxin concentrations used induced a substantial gradient of increasing vessel diameter with increasing distance from the site of application. The high auxin concentration yielded the narrowest vessels immediately below the application site (5).

layers (Fig. 5A). Further down, where lower levels of auxin are expected, the vessels were grouped in bundles (Fig. 5B).

CONCLUSION

Auxin, which is the limiting and controlling factor for both phloem and xylem differentiation, induces phloem with no xylem at low auxin levels. Xylem differentiation only takes place at higher auxin levels. As this also meets the requirements for phloem differentiation, it explains why xylem does not differentiate in the absence of phloem and always accompanies the pattern of the phloem.

The polar movement of auxin from leaves to roots induces vascular strands along the flow of auxin. The three dimensional organization of vascular strands along the stem can be explained by the canalization of auxin which causes the development of polar auxin transport, followed by vascular differentiation, in the tissues through which the auxin is initially diffusing. The non-polar transport of auxin in the phloem promotes vascular differentiation in instances of wounding and serves as an additional mechanism which regulates vascularization in wound regions.

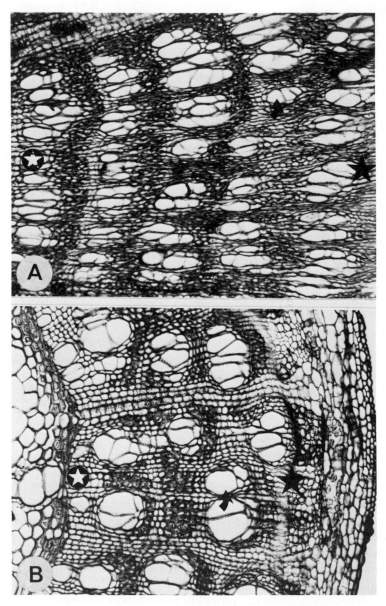

Fig. 5. Transverse sections, taken along the same internode of *Phaseolus vulgaris* treated with 0.1% naphthaleneacetic acid. Both sections are presented in the same direction and in the same magnification x 100. *A*. Half a centimeter below the site of auxin application. *B*. 4.0 cm below the NAA application. The photographs show an increase in vessel diameter and decrease in vessel density with increasing distance from the auxin source, as well as the changes of vessel pattern from layers A to bundles B. More secondary xylem was induced by the auxin at 0.5 cm (A) than at 4.0 cm (B) below the site of application. These changes in secondary xylem width are marked by the stars, which show equivalent locations in the photographs, thus marking the borders of the secondary xylem areas that were induced by the auxin in the 20 day experimental period. *Arrow* marks a late formed secondary vessel (5).

The polar flow of auxin controls the size and density of vascular elements along the plant axis. The general increase in the diameter of vascular elements and decrease in their density from leaves to roots is suggested to be due to a gradient of decreasing auxin concentration from leaves to roots.

References

1. Aloni, R. 1980. Role of auxin and sucrose in the differentiation of sieve and tracheary elements in plant tissue cultures. Planta *150*, 255-263.
2. Aloni, R. and W. P. Jacobs. 1977. Polarity of tracheary regeneration in young internodes of *Coleus* (Labiatae). Amer. J. Bot. *64*, 395-403.
3. Aloni, R. and W. P. Jacobs. 1977. The time course of sieve tube and vessel regeneration and their relation to phloem anastomoses in mature internodes of *Coleus*. Amer. J. Bot. *64*, 615-621.
4. Aloni, R. and T. Sachs. 1973. The three-dimensional structure of primary phloem systems. Planta *113*, 345-353.
5. Aloni, R. and M. H. Zimmermann. 1983. The control of vessel size and density along the plant axis - a new hypothesis. Differentiation *24*, 203-208.
6. Aloni, R. and M. H. Zimmermann. 1984. Length, width and pattern of regenerative vessels along strips of vascular tissue. Bot. Gaz. *154*, 50-54.
7. Benayoun, J., R. Aloni, and T. Sachs. 1975. Regeneration around wounds and the control of vascular differentiation. Ann. Bot. *39*, 447-454.
8. Bruck, D. K. and D. J. Paolillo, Jr. 1984. Replacement of leaf primordia with IAA in the induction of vascular differentiation in the stem of *Coleus*. New Phytol. *96*, 353-370.
9. Esau, K. 1969. The phloem. *In* W. Zimmermann, P. Ozenda and H. D. Wulff, (ed.) Encyclopedia of plant anatomy, Vol. V. pt. 2. Borntraeger, Berlin.
10. Goldsmith, M. H. M., D. A. Cataldo, J. Karn, T. Drenneman, and P. Trip, 1974. The rapid nonpolar transport of auxin in the phloem of intact *Coleus* plants. Planta *116*, 301-317.
11. Jacobs, M. and S. F. Gilbert. 1983. Basal localization of the presumptive auxin transport carrier in pea stem cells. Science *220*, 1297-1300.
12. Jacobs, W. P. 1952. The role of auxin in differentiation of xylem around a wound. Amer. J. Bot. *39*, 301-309.
13. Jacobs, W. P. 1970. Regeneration and differentiation of sieve-tube elements. Int. Rev. Cytol. *28*, 239-273.
14. LaMotte, C. E. and W. P. Jacobs. 1963. A role of auxin in phloem regeneration in *Coleus* internodes. Dev. Biol. *8*, 80-98.
15. Larson, P. R. 1982. The concept of cambium. In : New perspectives in wood anatomy. Bass, P. (ed.) Martinus Nijhoff/Dr. W. Junk Publisher. The Hague. pp. 85-121.
16. Morris, D. A. and A. G. Thomas. 1978. A microautoradiographic study of auxin transport in the stem of intact pea seedlings (*Pisum sativum* L.). J. Exp. Bot. *29*, 147-157.
17. Reinders-Gouwentak, C. A. 1965. Physiology of the cambium and other secondary meristems of the shoot. In : Ruhland, W. (ed.) Handbuch der Pflanzenphysiologie XV/1. Springer, Berlin, Gottingen, Heidelberg, pp. 1077-1105.
18. Sachs, T. 1969. Polarity and the induction of organized vascular tissues. Ann. Bot. *33*, 263-275.
19. Sachs, T. 1981. The control of the patterned differentiation of vascular tissues. Adv. Bot. Res. *9*, 152-255.
20. Saks, Y. and R. Aloni. 1985. Polar gradients of tracheid number and diameter during primary and secondary xylem development in young seedlings of *Pinus pinea* L. Ann. Bot. *56*, 771-778.

21. Savidge, R. A. and P. F. Wareing. 1981. Plant-growth regulators and the differentiation of vascular elements. *In* Xylem cell development. Barnett, J. R. (ed.) Castle House Publications. Kent. pp. 192-235.
22. Thompson, N. P. and W. P. Jacobs. 1966. Polarity of IAA effect on sieve-tube and xylem regeneration in *Coleus* and tomato stems. Plant Physiol. *41*, 673-682.
23. Wetmore, R.H. and J. P. Reir. 1963. Experimental induction of vascular tissues in callus of angiosperms. Amer. J. Bot. *50*, 418-430.

E7. Hormones and the Orientation of Growth

Peter B. Kaufman and Il Song

Division of Biological Sciences, University of Michigan, Ann Arbor, Michigan 48109, USA.

Hormones are chemical messengers which act at target sites to regulate rates and amounts of growth of cells in tissues of roots, stems, leaves, buds, flowers, and fruits. In this chapter, we shall focus on the roles that hormones play in the orientation of growth of plant organs, particularly of roots and shoots. The basic growth-orienting processes that we shall discuss include phototropism—the orientation of shoots toward unilateral light sources; gravitropism—the orientation of roots downwards and of shoots upwards in response to gravistimulation (placement of plants horizontally); and thigmotropism—the change in orientation of growth in stems from one of rapid elongation to one of repressed elongation and promoted lateral expansion as a result of mechanical perturbation.

PHOTOTROPISM

Nature of the Response and Its Adaptive Significance.

It is well-known that unilateral light plays a central role in orienting shoots of plants; that is, it causes them to grow asymmetrically so that shoots become oriented toward sources of higher light intensity, as we have all observed with plants growing in a window. The adapative significance of this response is quite obvious: it allows the leaves and stems of shoots to capture more radiant energy in the process of photosynthesis.

When shoots grow toward unilateral light sources, we call this a positive photoropic response. Roots are generally negatively phototropic and thus bend away from unilateral light sources. Some vines in the tropics, such as *Monstera deliciosa* and *Philodendron* spp., also exhibit negative phototropic responses in their shoots during seedling and early vegetative stages of development (37). Because of this, their shoots, at these stages, tend to grow toward objects such as tree trunks that cast shadows and are loci where light intensities are lower than in surrounding areas where trees are not growing. The shoots of these lianas (vines) attach themselves by means of aerial roots as they grow up the tree trunk. Later, when the shoots of these plants are older and have reached the canopy of the trees upon which they grow, they exhibit a positive phototropic response. The adaptive significance of both phototropic responses is easily seen.

The Three Primary Components of the Phototropic Response Mechanism in Higher Plants.

The overall mechanism of positive phototropic curvature can be divided into three phases: (a) the light perception phase, (b) the transduction phase, and (c) the asymmetric growth response phase. In light reactions such as phototropism, it is inherently assumed that the perception phase involves absorption of particular wavelengths of light by a particular phototropically active pigment. The action spectrum for phototropism in grass seedlings indicates that greatest positive phototropic curvatures occur in blue light (peaks occur at 420, 436, and 475 nanometers) with a second, but lower, peak occuring in the ultraviolet region (peak occurs at 370 nanometers). The nature of the phototropically active pigment has been a matter of controversy, some saying that it is a carotenoid pigment, and others saying that it is a flavinoid pigment (5, 6, 32, 41); most current evidence supports the idea that a flavinoid pigment is the primary photoreceptor pigment for the process (8, 9, 32). During the tranduction phase, the auxin, indole-3-acetic acid, becomes asymmetrically distributed in most plants—with more accumulating on the shaded side than on the illuminated side. We shall have more to say about the experimental evidence for how this IAA assymetry may arise in the next section. During the asymmetric growth response phase, cell elongation is inhibited on the illuminated side and continues at more or less normal rates on the shaded side. As a consequence, the shoot curves toward the light.

How Is Auxin Asymmetry Established in Phototropically Stimulated Shoots?

To answer this question, one can postulate several possible ways by which IAA becomes asymmetrically distributed in shoots of seedlings which are illuminated unilaterally: (a) light causes a net transport of IAA to the shaded side; (b) light causes destruction of IAA on the illuminated side; (c) less IAA is synthesized on the lighted side; and (d) light causes a decrease in rate of IAA transport on the lighted side.

Two experimental approaches have been used to examine these possibilities (40). One involves the use of diffusates collected in agar blocks which are then assayed by the IAA-specific and sensitive *Avena* coleoptile curvature test. The other involves the use of ^{14}C-IAA applied in donor agar blocks placed on grass coleoptile or dicot seedling shoot tips, and the measurement of radioactivity in both tissue samples and in receiver agar blocks following phototropic stimulation of the seedlings. We shall next examine the results of experiments that have employed each of these protocols.

The auxin diffusate—bioassay experiments basically involve excision of *Avena* coleoptile tips, placing them on agar receiver blocks to collect diffusible auxin following various types of light or mechanical treatments, and testing the auxin activity in the diffusates by means of the *Avena* coleoptile curvature bioassay. These experiments were done over the course of several decades, starting in the early 1900s. They showed several important results: (*a*) the coleoptile tip is necessary for phototropic curvature to occur, for if it is absent, no curvature occurs in unilaterally illuminated seedlings; and if the tip is replaced on the cut seedling stump, curvature does occur. Thus, the tip of the grass seedling coleoptile is the primary source of diffusible auxin that mediates the phototropic curvature response. (*b*) The auxin moves basipetally from the tip region, and when collected from the basal ends of coleoptile tips that have been unilaterally illuminated, one finds about 60% of the diffusible auxin in receptor blocks placed below the shaded side and 40% in blocks placed below the illuminated side. (*c*) If a mica or glass barrier is placed vertically through coleoptile tips of unilaterally illuminated *Avena* seedlings, no such asymmetry in diffusible auxin in the receptor blocks is seen; the amount of diffusible auxin in essentially the same in receiver blocks placed below shaded and illuminated sides of the coleoptiles. From this series of experiments, it was concluded that light causes a lateral migration of auxin across the coleoptile tip, and because more auxin accumulates on the shaded side, greater cell elongation occurs here, causing the seedling shoot to display a positive phototropic response. This conclusion is the basis for the famous Cholodny-Went hypothesis, first proposed in 1928 (41). It has been invoked to explain both phototropic curvature responses in shoots and gravitropic curvature responses in roots and shoots (see next section), and until recently, has largely been unchallenged.

Studies (5, 8, 9, 38, 41, 42) based on the use of ^{14}C-IAA to follow IAA distribution and rate of movement in tissue to which it is applied in unilaterally illuminated seedlings of both cereal grasses and the dicot, sunflower, are summarized below as follows: When ^{14}C-IAA is applied to the apical portions of excised *Avena* coleoptiles whose tips have been removed (to remove the primary endogenous IAA source), the ratio of radioactivity obtained in basally applied receiver agar blocks is 25:75 for the illuminated and shaded sides, respectively. The ratio of radioactivity in the tissue halves is 35:65 for the illuminated and shaded portions, respectively. These results indicate that for *Avena* seedlings, under these conditions, the IAA that is applied to the coleoptile apex moves in a basipetal direction, and in doing so, becomes asymmetrically distributed in the tissue itself and this leads to a difference in the amounts of IAA that arrive at the receiver agar blocks. Further, under these conditions, the unilateral light treatment, in comparison with dark controls, has no effect on the total amount of radioactivity obtained in the reciver blocks. Thus, one must rule out the possibility that asymmetry in IAA in the tissue comes about by light causing an inactivation of IAA on the illuminated side.

The above results are not always obtained with other plants. A similar experiment with ^{14}C-IAA was performed on sunflower seedlings, and in this case, no asymmetric distribution of radioactive IAA occurs in the stem tissue of unilaterally illuminated seedlings (42). If IAA does not get asymetrically distributed in such seedlings, it is possible that other hormones, such as native gibberellins, do so, and these could be the effector hormones. It is also possible, that in such seedlings, light has a more direct effect—that of decreasing the rate of cell elongation on the illuminated side.

In studies on the influence of light on the rate of longitudinal transport of ^{14}C-IAA in *Avena* coleoptiles, it was found that light significantly represses the rate of basipetal IAA transport in uniformly illuminated coleoptiles, relative to dark controls (38). The amount of decrease in IAA transport rate is 12 to 17% over a short period following exposure of the coleoptiles to blue light for 15 minutes. Similar results were also obtained for transport of ^{14}C-IAA in maize coleoptiles. It therefore appears that the asymmetry in IAA that develops in tissue of unilaterally illuminated seedlings could be explained by light causing a diminution in the rate of basipetal transport of IAA on the lighted side. How light brings about a decrease in IAA transport rate is unknown and should be explored.

From the above, it is clear that the role of IAA, or of other hormones, in regulating the positive phototropic curvature response needs to be reinvestigated. The Cholodny-Went hypothesis of light inducing IAA transport to the shaded sides of unilaterally illuminated seedlings is open to question, particularly in view of the fact that (a) the rate of IAA transport to the shaded side may be too slow in relation to the kinetics for the bending response, and (b) one could also explain the asymmetry on the basis of light causing a decrease in rate of IAA transport downwards on the illuminated sides of the seedlings. We are further confounded by the fact that no IAA asymmetry has been shown experimentally to occur in some plants, like the unilaterally illuminated sunflower seedlings. Moreover, all the above plants were first grown in the dark (etiolated), then given unilateral light treatments. In de-etiolated, green seedlings, there is no convincing experimental evidence that such IAA asymmetry does, in fact, occur when the seedling are illuminated unilaterally (42). Here, the positive phototropic curvature response could be explained by light directly causing a diminution in rate of cell elongation on the illuminated sides, as compared with that on the shaded sides of such green seedlings.

A reinvestigation of the role of auxin in the phototropic response of maize coleoptiles has in fact been carried out by Briggs and co-workers (3), with the seeming vindication of the Cholodny-Went hypothesis in that system. They have shown that there is a growth stimulation on the shaded side coupled with a growth depression on the irradiated side. There is also a basipetal migration of the growth responses which occurs at the same rate as the growth stimulation caused by exogenous applications of IAA. By applying different concentrations of IAA in lanolin unilaterally or symmetrically to intact coleoptile tips, they demonstrated that the endogenous auxin content was limiting, because applied auxin enhanced growth. The concentration of asymmetrically applied IAA needed to counteract unilateral light was only about twice that which, when applied symmetrically, induced an overall growth rate equivalent to that on the shaded side of a non-auxin-treated, unilaterally illuminated coleoptile. Thus a x2 concentration difference between shaded and illuminated sides could explain phototropic curvature of maize coleoptiles. This is within the range of auxin asymmetry commonly found.

GRAVITROPISM

Nature of the Gravitropic Response in Higher Plants and Its Adaptive Significance.

One of the best, and agriculturally important, manifestations of the gravitropic curvature response is seen in shoots of cereal grasses which have become prostrated by the action of wind or rain. This phenomenon is called lodging. Once lodged, and provided that the shoots are not too heavy with grain, they begin to manifest an upward bending response. In cereal grasses, this curvature response takes place at swollen localized regions of the leaf sheath bases and, in some grasses (mainly C-4 grasses), also at the bases of the internodes.Such regions are referred to, respectively, as leaf sheath and internodal pulvini. It requires almost 48 hours for the shoots of lodged cereal grasses to attain an upright position (90° curvature response) (Fig. 1). At 30°C, they bend upward at the rate of about 1.5° per hour (in oat (*Avena*) shoots). This response is clearly temperature-dependent, as *Avena* shoots, for example, show no such upward bending at 40°C or at 4°C. Those held at the latter temperature "store" growth potential, for if these shoots are now placed at 30°C, rapid upward bending ensues. On the other hand, the response is very sluggish or nil for shoots held at 40°C for two days, then placed at 30°C.

Fig. 1. Gravitropic response in two oat (*Avena*) shoot segments photographed every three hours over a 48-hour period. The segment on the right was held in a normal position with the basal stem portion held firmly to permit the apical portion to bend upwards. The segment on the left has the apical sheath portion held firmly in place, permitting the stem portion to bend upwards. This upward bending response in gravistimulated grass shoots, such as oat (*Avena*), occurs at the swollen leaf-sheath pulvinus. From (21).

    The upward bending response is even more rapid in prostrated (gravistimulated), dark-grown maize seedlings (2). Upward bending is initiated in the coleoptile within one to three minutes and progresses basipetally to the mesocotyl, where bending is first seen within five minutes after shoots are placed horizontally. By 60 minutes, the shoots have attained an upright position after a couple of overshoots to the right and left of vertical. Thus, the rate of upward bending is of the order of 1.5 to 2 degrees per minute. Time-lapse photographs of this upward bending response in maize seedlings are shown in Figure 2.

    What we have just described in older cereal grass shoots and in maize seedlings is termed a negative gravitropic curvature response. It is also manifest in setae of mosses and liverworts, shoots of lower vascular plants and ferns, newly developing shoots of conifers, and shoots of both

Fig. 2. Time-lapse photograph of a seedling of maize or corn (*Zea mays*) during gravitropic curvature. The initial photograph was taken just as the seedling was placed horizontally (gravistimulated). Successive photographs were taken at 15-minute intervals. The India ink mark at "N" denotes the node between the coleoptile (at left) and the mesocotyl or first internode (at right). From (2).

monocots and dicots at all stages of shoot development. With the exception of grasses, *Ephedra*, and scouring rushes (*Equisetum* spp.), which have active intercalary growth systems and show the negative gravitropic response at localized sites associated with their nodal regions, most shoots manifest the response over extensive, elongating portions of their internodes. This can be seen in elongating conifer shoots and in dicots such as mung bean and sunflower seedlings. The adaptive significance of this negative gravitropic curvature response is that it orients the photosynthetically active shoots toward the sun so that more incident solar radiation is captured in photosynthesis. The upright orientation of shoots may also be important for pollen dispersal by wind, attracting pollinators (e.g., insects, birds, bats), and in seed dispersal. For humans, upright growth of shoots is essential for proper harvesting of crops such as grains. Plant breeders have developed semidwarf cultivars of cereals, such as wheat and rice, which are resistant to lodging during heading (flowering and grain-filling) stages of development. More traditional, taller cultivars are very prone to lodging, and when their shoots do not show upward bending (recovery), yield losses can be very severe.

A positive gravitropic curvature response is displayed by primary and most adventitious roots. In prostrated seedlings, it requires about one to two hours for the roots to attain a vertical orientation toward the center of gravity. Lateral roots do not always show such a pronounced curvature response. Some grow outwards at ca. 45°, and others show no downward bending in the normal course of their development.Those roots which show no positive gravitropic response are termed ageotropic. Ageotropic growth is also displayed by shoots of 'lazy' mutants (e.g., ageotropica mutant of tomato and lazy mutants of rice, wheat, and maize), stolons or runners (e.g., strawberries), and underground stems or rhizomes (e.g., bamboos, quackgrass, and potato). The adaptive significance of the ageotropic response in stolons and rhizomes is one of facilitating vegetative propagation. The 'lazy' mutants of maize and tomato, of course, have no agricultural value because of their permanently "lodged" growth habit. The positive gravitropic curvature response in primary and adventitious roots facilitates the growth of these roots into the soil and thus provides anchorage for the plant and facilitates the acquisition of water and mineral nutrients.

The Primary Components of the Gravitropic Response Mechanism in Higher Plants.

As with phototropism, we can conveniently subdivide the gravitropic response mechanism into three sequential components: gravity perception, gravity transduction, and asymmetric growth response. For convenience, we shall discuss the gravitropic responses in roots and shoots of higher plants in terms of these three components.

Gravity Perception in Roots and Shoots.

In both roots and shoots, gravity is considered to be perceived by starch-containing plastids (amyloplasts or chloroplasts), which we call statoliths (1, 12, 15, 17, 26, 41, 42). The cells which contain statoliths are termed statocytes.In dicot and monocot roots, the starch statoliths occur in the root-caps, whereas in primary shoots of these plants, they occur in parenchyma cells of cortex, usually in close proximity to vascular bundles of the primary vascular system. When shoots or roots are gravistimulated, the starch statoliths fall to the bottoms of the statocyte cells within less than one minute, usually impinging on the rough endoplasmic reticulum and the plasma membrane. That the starch grains are essential for the gravitropic response to occur in roots is witnessed by the fact that when root-caps are carefully excised (as is easily demonstrated with maize seedlings, which have a meristematic calyptrogen between the root-cap and the root apex, making it easy to separate the cap from the apex), the roots do not show a positive gravitropic response; however, when the root-caps are replaced on the root apices, the gravitropic curvature response is restored (41, 42). In some roots, such as those of the onion (*Allium*) and *Arabidopsis* mutants (36), no starch statoliths are evident in the root-caps, yet the roots show positive gravitropic curvature when they are placed horizontally.In these plants, we do not know what the statoliths are; possibly, mitochondria or some other type of storage plastid (e.g., proteoplasts that store protein bodies or elaioplasts that store lipids) serve this function.

In the leaf-sheath pulvini of cereal grass shoots, the starch statoliths are actually chloroplasts which contain large starch grains. They occur in statocytes located in U-shaped cell clusters on the inner sides of each vascular bundle in the ground parenchyma (22). In *Avena fatua* the acquisition of gravisensitivity correlates with the development of 14-16 statocytes in association with each vascular bundle (45). When the shoots are gravistimulated, the starch statoliths in the pulvini cascade to the bottoms of the statocytes within two minutes, and most complete their descent within 15 to 30 seconds. Proof of their essentiality for negative gravitropic curvature response to occur in these shoots has not yet been unequivocally established (22, 24).

One of the big mysteries in the gravitropic response mechanism is how perception of the gravitropic signal by organelles such as starch statoliths leads to the next phase of the response, namely, transduction. One idea that is extant is that the starch statoliths serve as information carriers (e.g., of enzymes or of Ca^{2+}) (21, 24, 26). As the starch statoliths cascade to the bottoms of the statocyte cells in gravistimulated cereal grass pulvini, they drag down with them tonoplast membranes. These unit tonoplast membranes, as well as the double plastid membranes, could be sites where hormone-synthesizing or hormone-deconjugating enzymes are located. The starch statoliths of cereal grains show marked esterase activity: when stained with fluoroscein diacetate and excited by UV light, the statoliths light up bright yellow. Such esterase activity could be responsible for deconjugating IAA from its inositol ester conjugate, or of GAs from their glucosyl ester conjugates, at the site of the plasma membrane or the rough ER at the bottom of the statocyte. Even if this mechanism were operative, one still has to explain (a) the apparent physical and/or biochemical asymmetry built into statocyte cells on the top as compared with the bottom sides of gravistimulated pulvini, and (b) how the graviperception signal gets transmitted radially to neighboring cells which do not contain statoliths, but which do show an elongation response in gravistimulated pulvini.

Gravity Transduction in Roots and Shoots.

The second component of the gravitropic response mechanism is transduction. The essential component of this process is the development of hormone asymmetry (16, 18, 19, 35, 36, 41, 42, 43, 44). Basic dogma from the Cholodny-Went hypothesis says that the asymmetry comes about as a result of basipetal transport of IAA from the upper to the lower sides of gravistimulated roots and shoots. Cell elongation is then inhibited in the former and stimulated in the latter (where up to a 10,000-fold difference in sensitivity of root cells, as compared to shoot cells, to IAA is the basis for the difference in the cell elongation response) (41). It will become apparent after reading the material which follows (a) that IAA is not the only hormone involved in regulating the upward bending response in gravistimulated shoots; (b) in some systems, IAA is not basipetally transported in gravistimulated organs, whereas in others, it is; and (c) that nonhormonal messengers, such as Ca^{2+}, may be involved in regulating gravitropic curvature (40). Thus, as with phototropism, the Cholodny-Went hypothesis has been challenged as the sole mechanism for explaining how upward bending occurs in shoots and downward bending occurs in roots. Some investigators (11) have gone so far as to say that hormones may not even be involved in mediating the response. This is highly unlikely in view of what we know about the regulation of gravitropic curvature with exogenously applied hormones such as IAA, GAs, ABA, and ethylene, and endogenous changes which occur in the distribution of one or more of these same hormones in gravistiumulated organs.

Root Gravitropism
The role of hormones in regulating root gravitropism has been controversial for the past decade (41, 42, 43, 44). IAA has been the primary hormone cited as regulating the process. Many experiments have shown that the root cap produces a growth inhibitor and that this inhibitor causes a reduction of growth on the lower side of the root. The

Cholodny-Went hypothesis proposes that this inhibitor is auxin, which moves from the root cap to the lower side of the growing zone of a horizontal root. A primary source of auxin for the root is the shoot, but apparently auxin synthesis can also take place in the root tip itself (39). In addition, IAA applied to the root tip can migrate back to the growing zone (7), and, if placed laterally on the tip it can induce a bending toward that side. When IAA is applied laterally to the growing zone of a horizontal root, three times more IAA moves downward to the lower side of the root than in the reverse direction (23). This polarity of lateral auxin movement is abolished by the removal of the root cap. If there is a higher level of endogenous IAA in the lower side of a horizontal root (a point not yet conclusively settled) then it would probably inhibit growth via the synthesis of ethylene, because if ethylene synthesis is blocked then IAA no longer inhibits root growth but only causes a stimulation (30). ABA (abscisic acid) has also been considered to be important; it was thought that ABA is produced in the root cap and moves to the elongation zone on the lower side of a gravistimulated root where it inhibits cell elongation and causes the root to bend downward. Recently, this regulatory role for ABA in the positive gravitropic response of the root has been challenged (27, 28). At physiological concentrations (0.1 µM), its initial effect is to promote cell elongation, and this occurs during the very time when the root completes its positive gravitropic curvature response (28). Furthermore, primary roots of carotenoid-deficient mutants of *Zea mays* (corn), which contain no ABA, still respond normally to gravity (27). (The lack of ABA in the roots of such plants is consistent with the idea that ABA can be synthesized via the oxidation of carotenoids [see Chapter B5].) (27) Likewise, treatment of primary roots of normal corn seedlings with fluridone (an inhibitor of carotenoid biosynthesis) results in no detectable ABA, yet the roots are strongly graviresponsive. The IAA role is thus still a viable one, and in fact, today, IAA receives most support as being the primary hormone that regulates the root gravitropic curvature response.

When roots are gravistimulated, one of the initial events which takes place is a rapid proton efflux on the upper side of the elongation zone of the root (26, 29). This is easily seen with the proton efflux indicator dye, bromocresol purple, which changes from red at pH 6.5 to yellow at more acid pH's; the shift from red to yellow indicates proton efflux. No such color change occurs on the lower side of the downward bending root. Such a change on the upper side of the root would make the pH in the vicinity of the cell walls lower, and this would favor the enhancement in the activities of one or more cell wall-loosening enzymes. This has not yet been demonstrated experimentally, but it is a likely event. It could explain the rapid increase in cell elongation that occurs on the upper side, in the growing zone, of a gravistimulated root.

We cannot leave the subject of root gravitropism without mentioning an important nonhormonal regulator of the process, namely calcium (24, 25, 26). Exogenously applied [45]Ca typically moves to the lower side of tips of gravistimulated roots. When Ca^{++} is placed on the lower side of the tip, the root will curve downward. EDTA or EGTA, calcium chelators, will block the effect of Ca^{++} in inhibiting root curvature. Thus, its role in regulating the curvature response is undisputed. One of its primary physiological roles is to activate the enzyme modulator, calmodulin; the calmodulin activates plasma membrane bound Ca ATPases in plant cells. Ca^{++} is also essential for the basipetal transport of IAA in both roots and shoots, and it may have important regulatory effects on the cell elongation process itself. It may also be important in gravity perception because the starch statoliths themselves have significant amounts of Ca^{++} associated with them. The descent of the starch grains in the root cap in gravistimulated roots may provide a mechanism for getting Ca^{++} to the lower side of the root where it exerts its regulatory effect on cell elongation back of the root apex. The means by which a Ca^{++} redistribution causes downward bending may be via IAA since the presence of Ca^{++} on the lower side of the root enhances the downward movement of IAA applied to the upper side (23). In addition Ca^{++} has been found to be necessary for the growth-inhibitory action of IAA in roots (14). Ca^{++} redistribution in the root cap may cause the downward movement of auxin which is moving basipetally from the root cap. Alternatively, the Ca^{++} might possibly move backward to the growing zone and there influence the distribution and inhibitory action of auxin. This auxin could have arrived in the growing zone either by acropetal movement from the stem, or by basipetal movement from the root cap. As auxin is only inhibitory in the presence of Ca^{++} the geotropic response could, in fact, be caused by evenly distributed auxin in the presence of an assymetric distribution of Ca^{++} (14). The exact mechanism awaits elucidation.

Shoot Gravitropism.

In shoots, several hormones are implicated in the regulation of negative gravitropic curvature (21, 31, 33, 35, 36, 39, 41, 42, 43, 44). These include IAA and its inositol ester and amide-linked conjugates, GAs and their glucosyl ester and glucoside conjugates, and ethylene. The hormone(s) involved apparently vary with the experimental system (plant) being considered, and as expected, multihormonal control occurs in most of them. We shall first examine hormonal control by IAA in gravistimulated maize seedlings.

As mentioned earlier, the upward bending response in maize seedlings occurs very rapidly, usually within one hour at 30°C (2). It starts within three minutes in the coleoptile and five minutes in the mesocotyl (first internode). The onset of auxin asymmetry is also very rapid. Analyses (2) for free IAA in gravistimulated seedlings indicates that top half / bottom half asymmetry first appears within three minutes in the mesocotyls. This is two minutes before the mesocotyls start to bend upward; thus, IAA could be an important effector hormone in triggering the upward bending response. Full IAA asymmetry is established within 15 minutes after seedlings are oriented horizontally. Fifty-seven percent of the free IAA occurs in the lower halves of the mesocotyls at this time and remains at this level at 30, 60, and 90 minutes after seedlings are first gravistimulated. We can only speculate on how the asymmetry in IAA is established. Further experiments are necessary to determine whether or not basipetal transport is fast enough. It is also possible that the IAA asymmetry arises as a result of release of IAA from its conjugates to a greater extent on the lower side, or that more IAA synthesis begins to occur here soon after seedlings are gravistimulated.

In gravistimulated cereal grass shoots at older stages of development (late vegetative and early reproductive stages), the leaf sheath and internodal pulvini are the primary sites for the upward bending response, as we have mentioned earlier. At least two hormones, IAA and GAs, are important in regulating the response during the transduction stage (21). If *Avena* shoots (45 days old) are gravistimulated for 24 hours, they attain an upward curvature that is of the order of 30°. If the leaf sheath pulvini from these plants are partitioned into top and bottom halves to compare with "left" and "right" halves in vertical control plants, and these tissue fractions are analyzed for free IAA, one obtains the results depicted in Figure 3. The vertical control puvini contain 60 to 70 nanograms of free IAA per gram dry weight of pulvinus tissue. The sum total of free IAA in the gravistimulated pulvini (top plus bottom halves) is 420 nanograms per pulvinus, indicating that enhanced IAA synthesis (a seven times increase over that of controls) is induced by gravistimulation in these shoots. An average of 120 nanograms IAA occurs in the top halves and 300 nanograms IAA occurs in the lower halves, an asymmetry of ca. 1:2.5 top/bottom. In *Avena fatua* a top/bottom assymetry of IAA can be detected

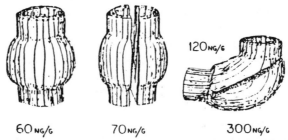

60 NG/G   70 NG/G   300 NG/G

Fig. 3. Amount of free IAA in ng/g dry weight in *Avena* (oat) leaf-sheath pulvinus tissue. At left, pulvini were also left upright, but were divided into 'left' and 'right' halves. At right, pulvini were gravistimulated for 24 hours so that shoots attained a 30° upward bending response; they were then divided into top and bottom halves for the free IAA analysis. From (9)

in as short a time as 15 minutes of gravistimulation, when only 1-10 degrees of upward bending has occurred (45). How and when the IAA asymmetry arises is still a moot question. One thing that is certain is that IAA is not transported downward across gravistimulated pulvini; ^{14}C-IAA applied to the tops of such pulvini does not move basipetally into the tissue to any appreciable extent except in the uppermost 0.5 to 1 mm of pulvinus tissue just under the IAA agar donor block. However, the acquisiton of gravisensitivity in *Avena fatua* correlates with the development of an ability by the leaf sheath base to control the rate of basipetal transport of ^3H-IAA (45). It is possible that the IAA gets conjugated to form amide-linked IAA (the primary IAA conjugate in vegetative *Avena* tissue), and as such, cannot move. Pulvini have been found to contain three times as much conjugate as free IAA. The increase in IAA in the lower halves of gravistimulated pulvini could be explained by either a differential IAA synthesis, occurring more at the bottom than at the top, and/or the release of IAA from its amide-linked conjugate. Gravistimulation also brings about changes in the levels of free GAs, as well as GA conjugates, in *Avena* leaf sheath pulvini (33). With gravistimulation of shoots to 30° (24 hours required), and comparable fractionation of pulvinus tissue into lower and upper halves (compared with "left" and "right" halves of vertical control pulvinus tissue), one finds that GA conjugates predominate in the upper halves, that acidic GA_3, GA_4, and GA_7-like GAs are in greatest abundance in the lower halves, and that total free GAs and GA conjugates decrease slightly as a result of gravistimulation. When the *Avena* shoots are fed ^3H-GA_4, and one "chases" the metabolites formed during gravistimulation, it is clear that the lower halves produce and/or retain more of the free acidic GAs, and the upper halves produce more of the highly water-soluble GA conjugates. Differential GA and/or GA-conjugate synthesis is clearly implicated, but part of the asymmetry could also arise as result of differential movement of free GAs and of GA conjugates to the respective pulvinus halves.

The Role of Ethylene in Shoot Gravitropism. Ethylene has been impl-icated as being another one of the regulatory hormones in shoot gravi-tropism (13, 21, 36). Following gravistimulation of cereal grass shoots, one finds very large increases in ethylene evolution from the tissues where curvature takes place, with more ethylene emanating from the lower halves than the upper halves. In *Avena* shoots that have been gravistimulated, the burst in ethylene production brought about by this treatment occurs some six hours after shoots are first placed horizontally; this is almost five and one-half hours after initiation of upward bending. Thus, one can conclude that ethylene in this system does not play a primary role in the initiation of curvature. Its enhanced production and the greater amount produced on the lower sides of gravistimulated shoots may be brought about, respectively, by the greatly enhanced levels of IAA

387

synthesized in puvini of gravistimulated shoots and the greater amounts of IAA found in the lower halves. Perhaps ethylene plays a role in the differential expansion of cells that occurs later during the course of upward bending.

In conifer shoots, such as those of *Cupressus arizonica*, ethylene, auxin, and gibberellin all appear to be important in the upward orientation (hyponastic growth response) of elongating lateral shoots (4). In these plants, GA_3, high light intensity, decapitation, and certain levels of IAA increase ethylene evolution and induce upturning of the lateral shoots. The ethylene evolution precedes the initiation of shoot upturning, so it could be important in causing this hyponastic response. Additional evidence in support of this idea comes from exogenous ethylene appliations to the shoots; here, such applications cause upward bending. Furthermore, removal of endogenous ethylene from the plants with an external mercuric perchlorate trap causes the branches to grow downwards. Whether ethylene or GAs or IAA or a combination of these hormones initiates the upward bending response in these plants is not yet clear. Studies are needed on the kinetics for ethylene evolution and on changes in levels of endogenous IAA and GAs during the course of hyponastic growth of these lateral branches.

Basis For the Asymmetric Growth Response after the Transduction Phase of Gravitropic Curvature.

We have already alluded to possible mechanisms that could explain how positive gravitropic curvature might occur in roots. Similar mechanisms may be operating in shoots. These basically involve proton pumping (46), activation of Ca ATPases via calcium redistribution and calmodulin synthesis (26, 40), and auxin-induced cell wall-loosening and wall synthesis (22). As for the shoot negative gravitropic curvature response, we shall use as our model the grass leaf-sheath pulvinus (22). The scheme in Figure 4 depicts a cascade of events that could explain the upward bending response mechanism in pulvini of grass shoots. The asymmetric distribution of IAA and of GAs and their conjugates is simply the start of a chain reaction which leads to asymmetric growth. In this chain of events, protein and cellulose synthesis are essential; likewise, differential hormone-stimulated cell wall loosening is involved. Both processes would occur early in the process of upward bending, the wall loosening starting first, followed by cell wall synthesis. Experimental evidence shows that new proteins are synthesized on both upper and lower sides of upward-bending pulvini, and several have been identified. Cellulase synthesis increases in the upper halves; this could allow the cell walls to become folded under the stress created by upward bending in the pulvinus. Invertase activity increases markedly in the lower halves; this could provide hexose substrate from sucrose for the cell wall synthesis process. Future work will indicate when the major protein changes occur, what the major proteins are, and how hormones such as GAs and IAA regulate their synthesis or degradation.

THIGMOMORPHOGENESIS

Thigmonastic growth responses, or thigmomorphogenetic responses, refer to the effect of mechanical perturbation on plant growth (19). It is manifested in roots which grow away from stones or other barriers in the

MODEL TO EXPLAIN UPWARD BENDING RESPONSE IN GRAVISTIMULATED CEREAL GRASS SHOOTS

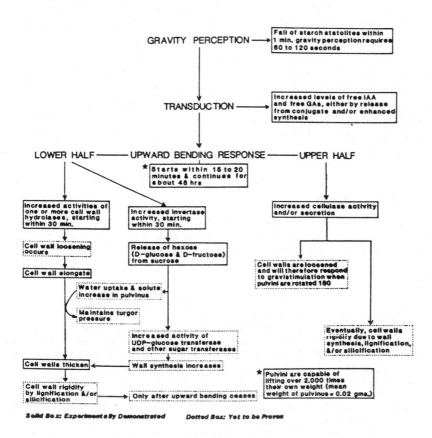

Fig. 4. Cascade of events which could explain the biochemical basis for the upward bending response in gravistimulated cereal grass pulvini. Modified from (22).

389

soil, the contact coiling of tendrils, the closure of leaf traps of *Drosera* (sundew) and *Dionaea* (Venus' fly-trap) upon contact by insects or other contact stimuli, and the diminution in stem length with concommitant thickening caused by mechanical perturbation.

The last-mentioned example, growth retardation by mechanically perturbed plants, has probably been studied most extensively, especially in connection with its regulation by hormones. Several hormones appear to be involved in the response (10). It is well-known that stress induced by various stimuli, such as flooding, or mechanical perturbation (MP), will induce ethylene formation in plants. In beans, MP elicits the production of the ethylene precursor, ACC (1-aminocyclopropane-1-carboxylic acid) from its precursor, SAM (S-adenosyl methionine), peaking at 30 and 90 minutes; ethylene levels then rise, peaking at about two hours and ceasing by four hours. This ethylene production typically causes the internodes to thicken and show reduced linear extension growth. However, the hormonal regulation of this response may be yet more complicated. Recent evidence indicates that MP lowers the levels of endogenous GAs and increases the levels of endogenous IAA and ABA. ^{14}C-IAA basipetal polar transport is also reduced significantly by MP. In this bean system, exogenously applied GA_3 decreases the levels of endogenous IAA and ABA. Most recent evidence, then, suggests that MP of bean internodes induces $CH_2=CH_2$ synthesis, which in turn, induces the accumulation of high levels of IAA and the production of ABA, both of which contribute to a diminution in internodal extension. The thickening of the internodes may be due to the effect of ethylene in promoting lateral cell expansion.

Acknowledgements

We thank Richard Pharis of the University of Calgary for his helpful discussions in connection with the writing of this chapter and the NASA Space Biology Program under the direction of Thora Halstead for financial support under NASA grant, NAGW-34.

References

1. Audus, L.J. (1969) *In* Physiology of Plant Growth and Development, pp. 201-242, Wilkins, M.B. ed. McGraw-Hill Book Co., N.Y.
2. Bandurski, R.S., Schulze, A., Dayanandan, P., Kaufman, P.B. (1984) Response to gravity by *Zea mays* seedlings. Time course of the response. Plant Physiol. *74*, 284-288.
3. Baskin, T. I., Briggs, W. R., Iino, M. (1986) Can lateral auxin redistribution account for photopropism of maize coleoptiles. Plant Physiol. *81*, 306-309.
4. Blake, T.J., Pharis, R.P., Reid, D.M. (1980) Ethylene, gibberellins, auxin and the apical control of branch angle in a conifer, *Cupressus arizonica*. Planta *148*, 64-68.
5. Briggs, W.R. (1963) The phototropic responses of higher plants. Ann. Rev. Plant Physiol. *14*, 311-353.
6. Curry, G.M., Thimann, K.V. (1961) Phototropism: The nature of the photoreceptor in higher and lower plants. *In* Progress in Photobiology, pp. 127-34, Christensen, B.C., Buchmann, B., eds. Elsevier Pub. Co., Amsterdam, Netherlands.

7. Davies, P. J., Doro, J. A., Tarbox, A. W. (1976) The movement and physiological effect of indoleacetic acid following point applicaionts to root tips of *Zea mays*. Physiol. Plant *36*, 333-337.

8. Dennison, D.S. (1979) Phototropism. *In* Physiology of Movements, Encyclopedia of Plant Physiology, New Series. Vol. 7. pp. 506-566, Haupt, W., Feinleib, M.E., eds. Springer-Verlag, New York.

9. Dennison, D.S. (1984) Phototropism. *In* Advanced Plant Physiology, pp. 149-162, Wilkins, M.B., ed. Pitman Pub. Co., Inc., Marshfield, Mass.

10. Erner, Y., Jaffe, M.J. (1982) Thigmomorphogenesis: the involvement of auxin and abscisic acid in growth retardation due to mechanical perturbation. Plant and Cell Physiol. *23*, 935-941.

11. Firn, R.D., Digby, J. (1980) The establishment of tropic curvature in plants. Ann. Rev. Plant Physiol. *31*, 31-48.

12. Haberlandt, G. (1928) The statocytes of stems. pp. 606-609. *In* Physiological Plant Anatomy. Macmillan Pub. Co., N.Y.

13. Harrison, M., Pickard, B.G. (1984) Burst of ethylene upon horizontal placement of tomato seedlings. Plant Physiol. *75*, 1167-1169.

14. Hasenstein, K-H., Evans, M.L. (1986) Calcium dependence of rapid auxin action in maize roots. Plant Physiol. *81*, 439-443.

15. Heathcote, D.G. (1981) The geotropic reaction and statolith movements following geostimulation of mung bean hypocotyls. Plant Cell Environ. *4*, 131-140.

16. Hertel, R., DelaFuente, R.K., Leopold, A.C. (1969) Geotropism and the lateral transport of auxin in the corn mutant amylomaize. Planta *88*, 204-214.

17. Iversen, T.H. (1969) Elimination of geotropic responsiveness in roots of cress (*Lepidium sativum*) by removal of statolith starch. Physiol. Plant. *22*, 1251-1262.

18. Jacobs, M. (1983) The localization of auxin transport carriers using monoclonal antibodies. What's New In Plant Physiology *14*, 17-20.

19. Jacobs, M., Gilbert, S.F. (1983) Basal localization of the presumptive auxin transport carrier in pea stem cells. Science *220*, 1297-1300.

20. Jaffe, M.J. (1981) Thigmomorophogenesis and thigmonasty. *In* McGraw-Hill Yearbook of Science and Technology, pp. 394-395. McGraw-Hill Book Co., New York.

21. Kaufman, P.B., Dayanandan, P. (1984) Hormonal regulation of the gravitropic response in grass shoots. *In* Hormonal Regulation of Plant Growth and Development, Purohit, S.S. ed. Vol. I. Agro Botanical Pub., Bikaner, India.

22. Kaufman, P.B., Song, I., Pharis, R.P. (1985) Gravity perception and response mechanism in graviresponding cereal grass shoots. *In* Hormonal Regulation of Plant Growth and Development, Vol. II. pp. 189-200, S S. Purohit, ed. Agro Botanical Pub., Bikaner, India.

23. Lee, J.S., Evans, M.L. (1985) Polar transport of auxin across gravistimulated roots of maize and its enhancement by calcium. Plant Physiol. 77, 824-827.

24. Lee, J.S., Mulkey, T.J., Evans, M.L. (1983) Gravity-induced polar transport of calcium across root tips of maize. Plant Physiol. *73*, 874-876.

25. Lee, J.S., Mulkey, T.J., Evans, M.L. (1983) Reversible loss of gravitropic sensitivity in maize roots after tip application of calcium chelators. Science *220*, 1375-1376.

26. Moore, R., Evans, M.L. (1986) How roots perceive and respond to gravity. American Jour. Bot. *73*, 574-587.

27. Moore, R., J.D. Smith, Fong, F. (1985) Gravitropism in abscisic acid deficient seedlings of *Zea mays*. Amer. J. Bot. *72*, 1311-1313.

28. Mulkey, T.J., Evans, M.L., Kuzmanoff, K.M. (1983) The kinetics of abscisic acid action on root growth and gravitropism. Planta *157*, 150-157.

29. Mulkey, T.J., Kuzmanoff, K.M., Evans, M.L. (1981) Correlation between proton-efflux patterns and growth patterns during geotropism and phototropism in maize and sunflower. Planta *152*, 239-241.

391

30. Mulkey, T.J., Kuzmanoff, K.M., Evans, M.L. (1982) Promotion of growth and hydrogen ion efflux by auxin in roots of maize pretreated with ethylene biosynthesis inhibitors. Plant Physiol. 70, 186-188.
31. Naqvi, S.M., S.A. Gordon. (1966) Auxin transport in Zea mays L. V'coleoptiles. I. Influence of gravity on the transport of indoleacetic acid-2-^{14}C. Plant Physiol. 41, 1113-1118.
32. Ninnemann, H. (1980) Blue light photoreceptors. BioScience 30,166-170.
33. Pharis, R.P., Legge, R.L., Noma, M., Kaufman, P.B., Ghosheh, N.S., LaCroix, J.D., Heller,K. (1981) Changes in endogenous gibberellins and the metabolism of GA$_4$ after geostimulation in shoots of the oat plant (Avena sativa). Plant Physiol. 67, 892-897.
34. Pickard, B.G., Thimann, K.V. (1966) Geotropic response of wheat coleoptiles in absence of amyloplast starch. J. Gen. Physiol. 49, 1065-1086.
35. Pickard, B.G. (1985) Roles of hormones in geotropism. In Hormonal Regulation of Development III. Role of Environmental Factors. Encyclopedia of Plant Physiology, new series, Vol. 11, pp. 193-281, Pharis, R.P., Reid, D.M. eds. Springer-Verlag, Berlin.
36. Pickard, B.G. (1985) Early events in geotropism of seedling shoots. Ann. Rev. Plant Physiol. 36, 55-75.
37. Ray, T. R., Jr. (1979) Slow motion world of plant behavior visible in rain forest. Smithsonian 9 (12), 121-130.
38. Shen-Miller, J., Cooper, P., Gordon, S.A. (1969) Phototropism and photoinhibition of basipetal transport of auxin in oat coleoptiles. Plant Physiol. 44, 491-496.
39. Scott, T.K., Matthyse, A. (1984) Function of hormones at the whole plant level of organization. In Hormonal Regulation of Development II. Encyclopedia of Plant Physiology, New Series, Vol.10, pp. 217-243, Scott, T.K., ed. Springer-Verlag, New York.
40. Slocum, R.D., Roux, S.J. (1983) Cellular and subcellular localization of calcium in gravistimulated oat coleoptiles and its possible significance in the establishment of tropic curvature. Planta 157, 481-492.
41. Thimann, K.V. (1977) Hormone Action in the Whole Life of Plants. Univ. of Mass. Press, Amherst, Mass.
42. Wareing, P.F., Phillips, I.D.J. (1981) Growth and Differentiation in Plants. Third Edition. Pergamon Press, New York.
43. Wilkins, M.B. (1966) Geotropism. Ann. Rev. Pl. Physiol. 17, 379-408.
44. Wilkins, M.B. (1984) Gravitropism. In Advanced Plant Physiology, pp. 163-185, Wilkins, M.B. ed. Pitman Publishing, Inc., Marshfield, Mass.
45. Wright, M. (1986) The acquisition of gravisensitivity during development of nodes of Avena fatua. J. Plant Growth Regulation 5, 37-47.
46. Wright, L.Z., Rayle, D.L. (1983) Evidence for a relationship between H$^+$ excretion and auxin in shoot gravitropism. Plant Physiol. 72, 99-104.

E8. HORMONAL REGULATION OF APICAL DOMINANCE

Imre A. Tamas

Biology Department, Ithaca College, Ithaca, NY 14850, USA.

INTRODUCTION

Plants develop from zygote to the adult form through a series of cellular and subcellular events. As young leaves, fruits and other structures emerge, they are elaborated into forms which are characteristic for each species. Moreover, they tend to appear in a highly ordered spatial relationship relative to each other. As circumstances vary, morphology may be altered or the arrangement of organs could change. The nature of these responses clearly indicates that the process of development is closely coordinated and that there is a hierarchical relationship among the interacting plant structures. Coordination of developmental activity among different parts is assumed to be achieved through a process of correlative control. In this view, dominant structures exert regulatory influence over others through a correlative signal which inhibits or otherwise alters development in the target structure.

The dominance of the growing shoot apex over other structures in the plant has been the subject of intensive study for the past century. A number of reviews have appeared covering various aspects of apical dominance (see 11, 16, 25, 29). For a broad overview of the problem, the reader is referred to these. In addition, some closely related topics are discussed in detail elsewhere in this volume (see: bud induction mutants in mosses, Chapter E4; control of polarity by auxin, Chapter E6; stolon development, Chapter E14; hormone directed transport, Chapter E12).

MANIFESTATIONS OF APICAL DOMINANCE

The growing shoot apex is known to exert correlative influence over a range of developmental events including axillary bud growth, the orientation of laterals, and the development of rhizomes and stolons. Of these, the inhibition of axillary bud growth by the shoot apex has been most intensively studied and thus became most closely associated with the concept of apical dominance. The present review will primarily draw on information related to axillary bud growth and its hormonal regulatory mechanism. Other aspects of apical dominance will also be treated on a selected basis, particularly as they pertain to the question of regulation.

Correlative Regulation of Axillary Bud Growth

Control by the Vegetative Shoot Apex

Axillary buds in several plant species have been shown to originate directly from the growing shoot apex. A small group of cells in the axil of a leaf primordium becomes isolated from the apical meristem and develops into the apex of the axillary bud (35). After further development, a visible bud is formed. In *Phaseolus*, a species with incomplete apical dominance, axillary bud growth continues at a slow rate even in the presence of an intact shoot apex (see 16). Both mitosis and cell expansion could be observed in these buds. In plants with complete apical dominance, such as *Tradescantia*, bud development is arrested at an early stage and mitotic activity is inhibited (26).

If the shoot apex is detached, or its dominant influence rendered ineffective, growth and mitotic activity in the lateral buds resumes after some delay. There is considerable variation in the length of the lag period. This may depend, in part, on the stage of the cell division cycle at which the inhibited cells are held. For example, mitosis can be observed within an hour after release in the axillary buds of *Cicer*, followed by the resumption of bud growth and DNA synthesis in that order (11). This has been interpreted to mean that the cells, at the time of release, have already duplicated their DNA (i.e., they are at the G_2 stage) and thus are able to undergo mitosis without much delay. On the other hand, *Tradescantia* requires over four days following shoot decapitation for bud growth to resume (26). The cells of inhibited *Tradescantia* buds are held at the G_1 stage, and thus DNA needs to be synthesized before mitosis could occur.

The length of the lag period preceding bud growth may also depend on the degree of inhibition imposed by the plant. The buds of *Phaseolus*, which are incompletely suppressed in the intact plant, can show increased internodal expansion four hours after decapitation of the main shoot (see 16). This is in contrast to the much longer time required by *Tradescantia* in which bud growth is totally supressed.

The degree of bud inhibition may also influence the sequence of early events after the removal of the dominant shot apex. In *Tradescantia*, bud outgrowth is preceeded by the resumption of mitotic activity in the bud apex (26). In the less suppressed buds of *Phaseolus*, however, initial growth after release results entirely from the enlargement of internodal cells of the bud (see 16). Cell division begins about a day after the onset of growth.

These results point up the problem of locating the site of primary events involved in bud growth control. Axillary bud response to correlative inhibition could conceivably involve specific target cells within the bud. So far, none have been identified. However, differential response to inhibition has been noted in some cells of the axillary bud meristem in *Tradescantia*. In the inhibited buds of this plant, a restricted group of cells, the "zone of inhibition", has been identified at the tip of the axillary bud meristem (7). After decapitation of the shoot, DNA and histone synthesis was initiated in these cells which was followed by the resumption of mitosis. In *Phaseolus* on the other hand, where cell expansion appears to be the first response to bud release, the putative target cells may very well differ in location and structure from those in *Tradescantia*.

Control by Reproductive Structures

The correlative effect of the shoot apex over axillary buds is most highly expressed early in plant development. Consequently, much of our knowledge of this phenomenon is derived from work with plants in their juvenile state. As plants mature, the emergence of reproductive structures initiates a major rearrangement in the plant's overall morphology. This is likely to shift the correlative relationships among the various organs and could alter the dominant position of the shoot apex. Flower induction in *Perilla* demonstrates this point. After photoperiodic flower induction, an increased number of axillary shoots can be found in these plants (5). In oats, the emergence of the inflorescence releases the lateral buds from inhibition and allows them to develop into tillers (12). Flowering, therefore, can decrease the level of dominance expressed by the shoot apex in some plants. On the other hand, reproductive development in *Phaseolus* severely restricts axillary bud growth. When axillary bud growth was investigated in adult *Phaseolus* plants, the growth rate was found to be unaffected by the appearance of flowers (40). However, as the growing fruits were about to attain their full size, axillary bud growth was suddenly terminated. A rapidly developing cultivar, 'Redkloud', showed faster fruit growth and earlier bud inhibition compared to a slower cultivar 'Redkote'. In both cultivars, the cessation of bud growth coincided with the appearance of fully grown fruits. Complete fruit removal allowed the resumption of axillary bud growth in both cultivars and caused up to two-fold increase in the combined total length of axillary shoots (40). It seems, therefore, that fruits of bean plants are able to exert "reproductive dominance" over the growth of axillary buds. In peas, defloration caused both the continuation of apical growth (which stops otherwise) and the emergence of axillary shoots (23). The effect on lateral growth was most pronounced near the region from which the flowers were detached. It is interesting to note that decapitation, a common method of releasing axillary buds from dominance, had no effect in these plants. Therefore, dominance over bud growth was exerted not by the shoot apex but by the reproductive structures. At present, it is not known how common this type of dominance is among plant species.

Control of Bud Development in Mosses

The gametophores, or leafy shoots, of mosses develop from buds which are initiated on the protonema. Before buds can be induced, the protonema has to progress from the chloronema to the caulonema stage, e.g., only the caulonema stage is capable to form buds (see 8). When the growth of the caulonema apex is experimentally interfered with, the filament reverts back to the chloronema stage. The site of bud induction (i.e., the position of the target cell) is at a specific distance from the caulonema apex (8). As the gametophore itself develops, the growth of its laterals is controlled by the developing apex (27). Removal of the gametophore apex causes the outgrowth of laterals. These results indicate that bud development in mosses, as in seed plants, is under close apical control.

In the study of moss development, significant advances have been made regarding the cellular effects of hormones during bud initiation. Because of the relative simplicity of the caulonema structure, biochemical and structural changes in individual cells can be followed during bud formation with a degree of precision not readily attainable with more complex plants. As will be seen later, these cellular events provide important clues about the mechanism of hormonal control and thus are highly relevant to the problem of apical dominance in general.

HORMONAL EFFECTS

Auxin

The discovery of auxin was closely linked to the notion that the tips of growing structures, such as coleoptiles and shoots, were rich souces of auxin, and were capable of affecting growth elsewhere in the plant through an auxin signal. The ability of growing shoot tips to release auxin seemed to parallel their ability to repress axillary bud growth. This observation led Thimann and Skoog to ask whether bud inhibition is caused by auxin (41). They tested their idea by decapitating Vicia plants and treating the stump with auxin. They found that the treatment prevented the outgrowth of axillary buds, just as the intact apex did. Since this initial demonstration, the ability of auxin to replace the shoot apex in apical dominance has been confirmed repeatedly (see 16). The active substance responsible for the inhibition of axillary buds has been isolated from Phaseolus shoot tips (particularly from the young, developing leaves) and identified by gas chromatography-mass spectrometry to be indoleacetic acid (48).

The question of whether a relationship can be demonstrated between IAA transport and apical dominance has been investigated by a number of workers. When a ring of lanolin containing triiodobenzoic acid (TIBA), an IAA transport inhibitor, was placed around the stem between the apex and the bud, bud growth inhibition was relieved (44). Furthermore, when branching (lacking apical dominance) and non-branching lines of tomato were tested for their ability to export radiolabeled IAA from the shoot apex, only the latter was able to do so (33). This suggests that the branching character is due to the failure of the shoot apex to export IAA. It may be concluded from the foregoing results that the release of IAA from the apex, and subsequent IAA transport, is an essential component in axillary bud growth inhibition.

A similar conclusion may be reached about the possible role of IAA in reproductive dominance over axillary bud growth. When the pedicels of individual fruits on *Phaseolus* plants were treated with naphthylphthalamic acid (NPA), an IAA transport inhibitor, resting axillary buds lower down the stem resumed growth, as did the buds on defruited plants (37). The buds on intact plants remained inhibited. Resumption of bud growth was not the result of decreased competition for nutrients between fruits and buds because on NPA treated plants fruit development was not inhibited (in fact, fruit growth was substantially enhanced). Deseeding experiments on *Phaseolus* plants showed that the source of the inhibiting signal is the seed (e.g., buds resumed growth when fruits were deseeded). When the deseeded pods were treated with IAA, the buds did not grow (38). Thus, the inhibiting signal from fruits, as well as from shoot apices, can be duplicated by IAA treatment. Whether IAA from fruits is responsible for axillary bud growth regulation in mature plants has not been confirmed but can be inferred from the available evidence. Developing fruits and seeds have been recognized to be rich in IAA (4). When the deseeded fruits of *Phaseolus* or *Glycine* were treated with ^{14}C-IAA, ^{14}C-labeled material was recovered from neighboring leaves, fruits and axillary buds (37, 39). In *Glycine*, IAA in both its free and esterified form was identified in phloem exudate (14). After the plants were depodded, the IAA ester level in the exudate decreased to about one-fifth of its previous level. Therefore, fruits seem to be able to release IAA which could take part in regulating bud growth activity.

Cytokinins

In many species, cytokinin treatment of the plant releases buds from correlative inhibition. In soybeans for example, when benzyladenine was applied directly to inhibited buds, growth was initiated (3). A relationship between endogenous cytokinin concentration and axillary bud growth has been demonstrated by comparing two genetic lines of tomato with differing degrees of apical dominance (24). The 'torosa-2' mutant with strong apical dominance contained less cytokinin than the normal line. No difference in auxin activity was found between these plants. In etiolated pea seedlings (where decapitation and removal of one cotyledon confers dominance of the cotyledon-less bud over the other) the application of ^{14}C-benzyladenine (^{14}C-BA) to the roots resulted in substantially greater accumulation of ^{14}C in the dominant bud compared to the other (30). These results show that axillary bud growth and cytokinin activity in the bud are closely correlated.

The suggestion has been made that apical dominance may be maintained by regulating cytokinin distribution in the plant (see 29). It was assumed that shoot growth requires root-produced cytokinin which accumulates preferentially in the shoot apex, thus depriving the axillary buds of their supply of cytokinin. In decapitated *Solanum* cuttings, when supplied with ^{14}C-BA at the base, IAA application to the apical stump caused a decrease in the amount of ^{14}C transported to the axillary buds (49). These results suggest that cytokinin transport to, and accumulation in, the axillary buds may be under apical control and could conceivably contribute to the overall controlling influence on bud growth. Simultaneous application of benzyladenine together with IAA to the stump in decapitated soybean plants enhances the bud suppressing action of IAA (3). An effect of cytokinin on the shoot apex was also demonstrated using grafting experiments with two isogenic lines of tomato differing in their ability to suppress lateral growth (45). The cultivar 'Craigella', expressed less apical dominance than 'Craigella Lateral Suppressor', and contained a higher level of a cytokinin (tentatively identified as N^6-[Δ^2-isopentenyl] adenosine) in the roots. Grafting a 'Lateral Suppressor' scion on 'Craigella' rootstock maintained apical dominance and increased the IAA level in the apex and the ABA level just below the apex. The reverse graft had no effect on hormonal levels. The results were interpreted as indicating an indirect inhibitory effect of root-derived cytokinins on axillary bud growth *via* an increase in the levels of IAA and ABA in the apex. Thus while cytokinins from the roots can enhance lateral bud growth, if the apical bud is highly dominant then the root-derived cytokinins are probably transported principally to the apical bud enhancing the apical dominance.

Either a redirection of the cytokinin supply from the roots may be necessary to initiate axillary bud growth or the decline in auxin supply from the apex enables the axillary buds to commence cytokinin synthesis. Although added cytokinin in rootless *Solanum* cuttings was able to stimulate axillary bud growth (49), it was subsequently shown that shoots are competent to produce their own cytokinin (46). However, the repressed lateral buds do not do so. Experiments with a purine synthesis inhibitor, hadacidin, suggest that synthesis of endogenous cytokinins within the buds themselves is required for bud growth to occur. Following decapitation, hadacidin treated axillary buds remain inhibited, and the inhibitory effect can be reversed by cytokinin treatment (22).

Gibberellins

There are few reports on the relationship between axillary bud development and endogenous gibberellin levels, and the evidence appears to be contradictory (42). However, studies with applied substances suggest a possible role for gibberellins in the regulation of bud development. Generally, the outgrowth of axillary buds can be substantially enhanced with the external application of GA_3. The effect does not seem to depend on the location of the treatment. Bud growth stimulation has been achieved with GA_3 application to the decapitated shoot stump (3, 19), or to the buds themselves (28). The growth response can be altered by plant age and other experimental variables (see 29).

Plant response to GA_3 can be influenced by the presence of other growth substances. In peas, the treatment of the decapitated stump with GA_3 alone was shown to stimulate lateral bud growth. However, GA_3 enhanced the inhibitory effect of IAA when the two were applied together (19). The latter effect was thought to result from the more efficient release of IAA by the dominant apex because GA_3 was found to stimulate the basipetal transport of radiolabeled IAA in these plants (19). GA_3 can also show synergistic interaction with cytokinins. When GA_3 was applied to axillary buds of intact pea plants it had no effect, but it strongly enhanced the growth promoting action of kinetin when both were present simultaneously (32). Furthermore, kinetin elicited a response earlier than GA_3, and it had considerable effect even without GA_3. From these results the authors concluded that the two substances affected different phases of growth, namely kinetin released the buds from inhibition and GA_3 acted during the post-release phase.

Abscisic Acid

According to several reports, axillary bud growth inhibition is correlated with the ABA content of the buds. If the shoot is decapitated or otherwise treated to release the buds from inhibition, the ABA content of the buds declines rapidly. In nonbranching mutants of the tomato, the ABA content of the axillary buds is much higher than in normal plants (42). Whereas ABA treatment of the axillary buds inhibits their growth, the application of ABA to the shoot apex releases axillary buds from inhibition (43). The latter response is probably caused by the inhibition of apical bud growth with the consequent loss of its dominant effect.

In mature *Phaseolus* plants, the growth suppressing effect of fruits on axillary buds was shown to be correlated with increased levels of ABA in the buds (40). Fruit removal or deseeding of the fruits released the buds from growth inhibition and caused a drop in bud ABA levels (38, 40). Bud growth inhibition was restored by either ABA treatment of the buds or the application of IAA to the pod cavity. However, ABA application to the deseeded pods failed to inhibit bud growth suggesting that the bud suppressing effect of fruits was not casused by ABA release from the fruits. Rather, ABA accumulating in inhibited buds probably originated from another source such as the leaves, or was synthesized within the buds under the control of the dominant fruits.

There is, therefore, no indication that ABA functions as the correlative signal released by dominant organs such as growing fruits or shoot apices. The available evidence suggests instead that ABA acts within the axillary buds as a second factor whose level can be regulated by the dominant organ. In the tomato, TIBA treatment of the plant was shown to release the axillary buds from inhibition, but inhibition was restored when the buds were treated with ABA (44). Also, decapitation lowered the ABA level in the axillary buds of *Phaseolus* plants, but this was prevented by the application of IAA to the cut stump (20). It appears therefore that IAA, moving down from the shoot apex, can maintain a high ABA level in the axillary buds. Whether this is responsible for the suppression of bud growth by the apex is an interesting possibility that needs to be explored.

There are some results that seem to contradict this suggestion. The ABA content of axillary buds in decapitated *Phaseolus* plants was shown to decline several hours after bud growth resumed (20). In another study with *Phaseolus*, decapitation of the stem and the release of buds from inhibition failed to cause a substantial change in the ABA content of the buds (47). It remains to be resolved, therefore, why high levels of ABA are generally associated with growth inhibition in buds yet changes in these levels do not coincide with the release of buds from inhibition. One possibility is that decapitation first increases in the buds the level of a growth promoting substance, such as cytokinin, followed by a decline in ABA. Thus the effect of ABA could be subject to modification by other growth regulating substances.

Ethylene

It has been known for some time that IAA is capable of stimulating the synthesis of ethylene. A possible consequence of this is that a localized ethylene buildup, triggered by an IAA signal, could cause growth inhibition in specific structures such as axillary buds. Whether ethylene participates in axillary bud growth regulation has been tested in a number of studies. The results, so far, are contradictory. In etiolated pea plants, IAA treatment caused inhibition of axillary bud growth and an increase in ethylene production by nodal tissue (9). Applied ethylene inhibited lateral buds on decapitated plants showing that the buds were sensitive to ethylene. These findings were confirmed in green, light-grown pea plants as well (6). In *Phaseolus*, however, the application of IAA to the cut stem stump failed to increase ethylene emanation by the node below (see 16).

Under particular circumstances, ethylene seems to make a specific contribution to bud growth regulation. In oats, ethylene is involved in tiller swelling after the tiller has been released from apical dominance (13). When the growing shoot apex of a plant, such as *Phaseolus*, is physically confined in a tube, apical growth is prevented and lateral buds are released (see 16). These responses are attributed to ethylene action because confined apices produce more ethylene than untreated ones and ethylene application to the apex stimulates lateral growth.

TISSUE POLARITY AND THE MECHANISM OF CONTROL

Early in their pioneering work on the hormonal control of apical dominance, Thimann and Skoog noted that little auxin production occurred in inhibited axillary buds of *Vicia* but it increased substantially when growth resumed (41) (this has been confirmed by accurate physicochemical methods [see 16]). Thus it was recognized at the outset that bud growth is correlated with auxin production within the bud, even though auxin from the shoot apex represses axillary bud growth. To explain this apparent paradox, it was suggested that the conversion of a precursor to auxin in the axillary bud is inhibited by auxin arriving from the shoot apex (41). Because the latter needs to pass through tissues of the bud without enhancing growth therein, the implication of this idea is that auxin moving either with or against the normal polarity has different effects, e.g., only basipetally moving auxin in the bud stimulates growth. Just such a dependence of the auxin effect on tissue polarity is indicated by the old report that apically applied IAA promotes elongation in *Pisum* stem segments, whereas treatment of the base causes the opposite result (21). It would be interesting to reinvestigate this phenomenon in light of our current knowledge.

Polarity and Auxin Action

The role of polarity in the control of axillary bud growth has not been explored adequately, but its potential significance is illustrated by studies on xylem differentiation and abscission. In both of these, apical control over the process is ensured because of the intimate interplay between tissue polarity and auxin action. On the one hand, auxin is a key factor in promoting and maintaining polarity. On the other hand, the existing polarity of the tissue often determines the nature of response to auxin action. Xylem vessel differentiation is a case in point. Sachs (31, and Chapter E6) noted in wounded plants that xylem differentiation is oriented toward a source of auxin such as a lateral bud or the site of applied IAA. By varying the pattern of incisions and the location of IAA treatment relative to the wound, the direction of differentiation can be altered, but it always follows the path of IAA transport. These results are interpreted by Sachs to mean that auxin flux through the tissue determines tissue polarity by inducing the formation of auxin transport channels. Furthermore, continued auxin flux is required to maintain and reinforce the existing polarity. Eventually, differentiation of vascular tissue is induced along the auxin transport channels. The observation that monoclonal antibodies against the NPA receptor (the presumed auxin carrier) will specifically bind to cells sheathing the vascular bundles (17) lends strong support to the notion that vascular tissue and special auxin transport cells are closely associated.

It follows from Sachs' hypothesis that apical dominance is an expression of the prevalent polarity of the growing shoot. The continued polar transport of auxin along the shoot prevents the formation of a new polar axis (or the activation of an existing one) in the direction of a lateral organ. This view is supported by results of an experiment in which the effect of two competing sources of auxin on xylem differentiation was studied. If auxin is applied laterally on a decapitated stem, it induces the treated area to differentiate xylem tissue which eventually connects to the existing vascular strand. If the latter is provided with a source of auxin (e.g., a young leaf or auxin applied to the stump) from the apical region above, the site of lateral IAA application fails to form vascular connection with the existing vascular bundle (31). Therefore, the induction of new auxin transport channels (i.e., formation of a new polar axis) requires that auxin flux along the original axis diminishes. If this is to be applicable to the control of lateral bud growth, two assumptions must be made, first that the induction or reactivation of auxin transport channels along the bud axis is the key step in releasing buds from correlative inhibition, and second that this step can be prevented by polar auxin transport along the stem axis. Although these ideas have not been critically tested, some support is provided by the observation that the auxin transport capacity of buds increases after they have been released from inhibition (21).

An important contradiction to the foregoing hypothesis is that in some species, such as *Phaseolus* and *Glycine*, inhibited buds have well developed vascular tissues (see 16) and therefore probably also possess auxin transport channels. It is possible, however, that in these buds the auxin transport channels are nonfunctional and need to be activated before bud growth can resume.

The response of leaf abscission to auxin is distinctly polar which seems to be related to the competing effects of the leaf blade and the shoot apex on the cells of the abscission zone. The presence of the leaf blade prevents abscission by virtue of its continuing release of auxin. The shoot apex has the opposite effect, and this too seems to be mediated by auxin. The removal of the shoot apex delays the abscission of petioles (18), and the abscission enhancing effect of the apex can be duplicated with IAA treatment of the stump (see 2). Thus leaf abscission and axillary bud growth show a fundamental similarity in their response to auxin. First, both are subject to the correlative effect of the shoot apex (which is mediated by auxin), and second the effect of auxin secreted by the shoot apex is opposed by auxin from the appropriate lateral organ (e.g., leaf blade and axillary bud respectively).

The polarity of IAA effects on abscission can be demonstrated in petiole explants which include the abscission zone in the middle. If IAA is applied to the distal end and in sufficiently high concentration, it prevents abscission. But if IAA application is proximal to the abscission zone, or at a low concentration to either end, abscission is stimulated (2). Therefore, IAA arriving from the two ends could elicit opposite responses in the cells of the abscission zone.

The way cells in the abscission zone respond to IAA may in part depend on their sensitivity to, and ability to produce, ethylene. Auxin treatment of *Phaseolus* petiole explants (at the distal end) shortly after excision prevented abscission whereas the same treatment given more than 12 hours later accelerated it. During the initial phase (stage 1), the petiole was found to be insensitive to ethylene treatment but after 12 hours (stage 2) abscission was stimulated by ethylene. Furthermore, auxin action in stage 2 was prevented by the removal of evolved ethylene (1) indicating that the abscission-enhancing effect of auxin could be attributed to increased ethylene production. Therefore, different and contrasting effects of IAA (including those modified by polarity) may be due to changes in the cells' ability to produce or respond to ethylene.

Polarity in the Control of Axillary Bud Growth

Recent work with bud-bearing, isolated stem sections of *Phaseolus* showed polarity to be a factor in the response of axillary buds to auxin. When the apical end of the section was inserted into a sterile agar medium containing sucrose and mineral nutrients, the bud at the middle of the section (ca. 15 mm from the agar) resumed growth. However, if IAA or NAA was also present in the medium, bud growth was prevented (Fig. 1). If the section was implanted with its basal end, auxin in the medium either had no effect or caused a slight stimulation of bud growth. IAA applied to the apical end also caused bud abscission and prevented the appearance of new lateral buds. Basal application did not have these effects (Fig. 1). Because of the basipetal tendency of auxin transport, more auxin could be expected to reach the bud from the apical end compared to the basal end. This, however, is not the explanation of the foregoing data. When IAA transport was measured using ^{14}C-IAA in bud-bearing stem sections, the amount of ^{14}C transported to the bud from the apical cut

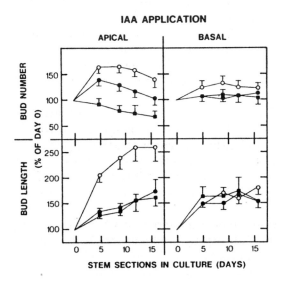

IAA APPLICATION

Fig. 1. Effect of IAA on axillary bud growth, development and abscission in isolated stem sections of *Phaseolus*. Either the apical or the basal end of each section was inserted into the IAA containing medium. IAA concentrations were: O, 0 µM; ●, 1 µM: ■, 10 µM. Results are expressed as the percentage of the value on day 0. Vertical bars equal standard error (Tamas, unpublished).

surface did not exceed that transported from the basal end (Table 1). Furthermore, NPA treatment of the stem sections relieved the inhibition of buds caused by apically applied IAA (Fig. 2). NPA also inhibited ^{14}C-IAA transport from the apical cut surface to the base but not to the bud (Table 2). According to these results, transport of IAA is necessary for the correlative inhibition of axillary buds, but only basipetal transport seems to be effective. Furthermore, there is no difference between basipetal and acropetal IAA transport regarding the amount of IAA accumulation in the

Table 1. Distribution of ^{14}C in bud-bearing, 1 cm long stem sections of *Phaseolus* after 8 hours of basipetal or acropetal ^{14}C-IAA transport. Agar blocks were used as donors and receivers. The sections were cut into five equal parts and were numbered consecutively from apex to base. Results are expressed as percent of total activity in all parts including the donor and reciever (Lim and Tamas, unpublished).

Part		Application of ^{14}C–IAA	
		Apical	Basal
		% radioactivity ± SE	
1.	Apical end	17.6±1.0	0.22±0.05
2.		2.1±0.3	0.34±0.1
3.	Node	0.75±0.1	1.1±0.2
3a.	Axillary bud	0.13±0.05	0.14±0.04
4.		0.63±0.07	4.9±0.9
5.	Basal end	1.2±0.1	28.5±1.9
	Donor	76.7±2.9	64.0±3.8
	Receiver	0.86±0.05	0.76±0.06

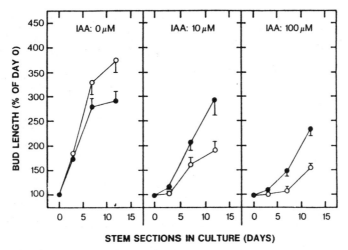

Fig. 2. Effect of NPA and IAA on axillary bud growth. The stem sections were pretreated by floating on distilled water or on NPA solution for six hours prior to the insertion of the apical end into the IAA containing medium. NPA concentrations were O, 0 μM; ●, 50 μM. For other details, see Fig. 1. From (36).

bud, yet the two have opposite effects on bud growth. It is concluded that basipetal IAA transport in the stem, rather than IAA accumulation in the bud, is necessary for bud growth inhibition.

At present, it is not known how a difference in the direction of auxin transport is perceived by the bud so that different developmental responses could be expressed. In an attempt to answer this question regarding the abscission zone, it has been suggested that the relative rates of acropetal and basipetal transport regulate the abscission response by setting up an auxin gradient in the petiole (see 2). The hypothesis was tested by applying IAA in varying amounts to both ends of petiole explants. It was found that abscission was delayed if the amount applied distally exceeded the amount applied proximally. The reverse arrangement stimulated abscission. These same results could be obtained even when the total amount of IAA was varied, as long as the ratio of the two applications did not change. The data show that simultaneous IAA fluxes in opposite direction can cancel each other's effect and the outcome (promotion or inhibition) depends on the ratio of the two.

The question whether this is also applicable to axillary bud growth regulation was recently tested. IAA was applied in varying amounts to both ends of isolated stem segments bearing axillary buds. It was found that basal IAA treatment relieved the bud inhibiting effect of IAA applied to the apex (Lim and Tamas, unpublished). Furthermore, treatment of the base with nonradioactive IAA reduced the amount of ^{14}C-IAA transported from the apex to all parts of the segment. This was interpreted to mean that basal IAA application relieves the growth

Table 2. The effect of NPA on the distribution of ^{14}C in *Phaseolus* stem segments after 8 hours of basipetal ^{14}C-IAA transport. Donor blocks included either 0 or 50 µM NPA. No receivers were used. For other details, see Table 1 (Lim and Tamas, unpublished).

Part		NPA concentration (µM)	
		0	50
		% radioactivity ± SE	
	Donor	52.6±11.1	57.0±2.4
1.	Apical end	34.0±2.0	37.2±5.7
2.		4.2±0.6	2.8±1.0
3.	Node	2.4±0.5	1.3±0.6
3a.	Axillary bud	0.06±0.02	0.13±0.04
4.		1.9±0.6	0.54±0.3
5.	Basal end	4.8±0.4	0.13±0.1

inhibiting effect of IAA moving from the apex by altering the ratio of basipetal-to-acropetal IAA transport.

The Role of Calcium in Hormone Action and Polarity

It is now becoming increasingly evident that calcium ions have a unique regulatory function in a wide variety of processes such as polarized growth, mitosis, nuclear migration, gravitropism and cytoplasmic streaming (15). In many instances, Ca^{2+} was found to mediate specific hormonal effects. Of particular interest is the role Ca^{2+} plays in cell polarity (34). It is believed that the positioning of Ca^{2+} pumps and channels in the plasma membrane at opposite poles of the cell are among the early events of cell polarization, which results in an apex-to-base gradient of Ca^{2+} distribution. There are numerous indications that in tip-growing cells such as pollen tubes, and rhizoids of *Funaria*, the Ca^{2+} level is elevated at the tip (15). Such polarization of Ca^{2+} distribution could be enhanced by IAA transport. Recent experiments suggest that in stems the polar transport of IAA is intrinsically linked to the movement of Ca^{2+} in the opposite direction. IAA treatment of *Helianthus* hypocotyl segments at the apical end was found to stimulate the extrusion of Ca^{2+} at that end (10). Conversely, basal treatment with IAA stimulated Ca^{2+} release at the base. The latter effect was considerably smaller than the former one. Furthermore, Ca^{2+} extrusion was prevented by TIBA, and the absence of Ca^{2+} inhibited IAA transport. The results suggest first that the Ca^{2+} status of the cell is directly linked to IAA transport and second that a change in the direction and rate of IAA movement is likely to alter both the internal concentration and polar distribution of Ca^{2+} in the cell. These changes, in turn, could set off different developmental effects in the cell. There are many experiments showing that the disruption of polarity in Ca^{2+} distribution drastically alters cell structure and development (15). The foregoing data, therefore, provide a plausible model for the interpretation of polar IAA effects.

Recent evidence suggests that Ca^{2+} also has a role in the effect of cytokinin on bud development. In the moss *Funaria*, bud formation in the growing caulonema is induced specifically by cytokinin (see 8, and Chapter E4). Bud induction takes place in the third cell from the caulonema tip. The predictable position of the prospective bud relative to the apex, and the enhancement of growth at the caulonema tip, are two important expressions of the strong polarity that characterizes the caulonema structure. IAA seems to play a key role in the control of polarity because it serves to induce, as well as to maintain, the caulonema stage (see 8). The bud first appears as a localized lateral outgrowth near the apical end of the target cell. After cytokinin treatment, accumulation of membrane-bound calcium is observed in the prospective bud region of the cell (see 15), followed by cell division that produces the first cell of the bud. If sufficient Ca^{2+} is provided but its polar distribution is experimentally disrupted, buds can be induced in all target cells even in the absence of cytokinin. On the other hand, the lack of Ca^{2+} prevents the bud inducing effect of cytokinin. The results indicate that the effect of cytokinin is mediated by Ca^{2+}. The data further suggest that the early events of induction involve the establishment of a new polar axis by the accumulation of Ca^{2+} at the prospective bud site. This raises the interesting question whether axillary bud growth stimulation by cytokinin in higher plants could be viewed as an act of setting a new polar axis. The implication of this idea is that, in addition to stimulating cell division in the bud, cytokinin would also act to reorient IAA transport and cell differentiation in the bud region thus shifting the balance away from the prevailing dominant polarity of the shoot.

CONCLUDING COMMENTS

The basic mechanism of apical dominance remains unresolved even though extensive information is available on specific hormonal effects. Although the correlative signal has not been conclusively identified, IAA is, by all indications, the prospective candidate. All the major classes of growth substances have at least some effect on axillary bud growth, but their interaction is largely undefined. There is strong evidence that cytokinin is a key factor in promoting bud growth. In general, a hormonal regime that enhances the vigor of the apical bud enhances apical dominance, while, when the apical bud is less vigorous, lateral growth may ensue under the influence of growth promotive hormones in the axillary bud.

To understand the mechanism of hormonal control, it will be necessary to explore the fundamental cellular events involved in axillary bud growth inhibition and release. Important progress has been made in recent years regarding the mechanism of auxin transport and its role in tissue polarity. It can be now inferred that auxin dependent cell polarity is involved in the control of axillary bud growth. How auxin controls polar cellular responses needs to be resolved before the mechanism of apical dominance can be elucidated.

References

1. Abeles, F.B., Rubinstein, B. (1964) Regulation of ethylene evolution and leaf abscission by auxin. Plant Physiol. *39*, 963-969.
2. Addicott, F.T. (1982) Abscission. Univ. California Press, Berkeley.
3. Ali, A., Fletcher, R.A. (1971) Hormonal interaction in controlling apical dominance in soybeans. Can. J. Bot. *49*, 1727-1731.
4. Bandurski, R.S., Schulze, A. (1977) Concentration of indole-3-acetic acid and its derivatives in plants. Plant Physiol. *60*, 211-213.
5. Beever, J.E., Woolhouse, H.W. (1975) Changes in the growth of roots and shoots when *Perilla frutescens* L. Britt. is induced to flower. J. Exp. Bot. *26*, 451-463.
6. Blake, T.J., Reid, D.M., Rood, S.B. (1983) Ethylene, indoleacetic acid and apical dominance in peas: A reappraisal. Physiol. Plant. *59*, 481-487.
7. Booker, C.E., Dwivedi, R.S. (1973) Ultrastructure of meristematic cells of dormant and released buds of *Tradescantia paludosa*. Exptl. Cell Res. *82*, 255-261.
8. Bopp, M. (1980) The hormonal regulation of morphogenesis in mosses. *In*: Plant growth substances 1979, pp. 351-361, Skoog, F., ed. Springer-Verlag, Berlin.
9. Burg, S.P., Burg, E.A. (1968) Ethylene formation in pea seedlings; its relation to the inhibition of bud growth caused by indole-3-acetic acid. Plant Physiol. *43*, 1069-1074.
10. De Guzman, C.C., Dela Fuente, R.K. (1984) Polar calcium flux in sunflower hypocotyl segments I. The effect of auxin. Plant Physiol. *76*, 347-352.
11. Guern, J., Usciati, M. (1972) The present status of the problem of apical dominance. *In*: Hormonal regulation in plant growth and development, pp. 383-400, Kaldewey, H., Vardar, Y., eds. Verlag Chemie, Weinheim.
12. Harrison, M.A., Kaufman, P.B. (1980) Hormonal regulation of lateral bud (tiller) release in oats (*Avena sativa* L.). Plant Physiol. *66*, 1123-1127.
13. Harrison, M.A., Kaufman, P.B. (1982) Does ethylene play a role in the release of lateral buds (tillers) from apical dominance in oats? Plant Physiol. *70*, 811-814.
14. Hein, M.B., Brenner, M.L., Brun, W.A. (1984) Effects of pod removal on the transport and accumulation of abscisic acid and indole-3-acetic acid in soybean leaves. Plant Physiol. *76*, 955-958.
15. Hepler, P.K., Wayne, R.O. (1985) Calcium and plant development. Ann. Rev. Plant Physiol. *36*, 397-439.
16. Hillman, J.R. (1984) Apical dominance. *In*: Advanced plant physiology, pp. 127-148, Wilkins, M.B., ed. Pitman, London.
17. Jacobs, M., Short, T.W. (1986) Further characterization of the presumptive auxin transport carrier using monoclonal antibodies. *In*: Plant growth substances 1985, pp. 218-226, Bopp, M., ed. Springer-Verlag, Berlin.
18. Jacobs, W.P. (1955) Studies on abscission: the physiological basis of the abscission-speeding effect of intact leaves. Am. J. Bot. *42*, 594-604.
19. Jacobs, W.P., Case, D.B. (1965) Auxin transport, gibberellin and apical dominance. Science *148*, 1729-1731.
20. Knox, J.P., Wareing, P.F. (1984) Apical dominance in *Phaseolus vulgaris* L.: The possible roles of abscisic and indole-3-acetic acid. J. Exp. Bot. *35*, 239-244.
21. Le Fanu, B. (1936) Auxin and correlative inhibition. New Phytol. *35*, 205-220.

22. Lee, P.K.-W., Kessler, B., Thimann, K.V. (1974) The effect of hadacidin on bud development and its implications for apical dominance. Physiol. Plant. *31*, 11-14.
23. Malik, N.S.A., Berrie, A.M.M. (1975) Correlative effects of fruits and leaves in senescence of pea plants. Planta *124*, 169-175.
24. Mapelli, S., Lombardi, L. (1982) A comparative auxin and cytokinin study in normal and to-2 mutant tomato plants. Plant Cell Physiol. *23*, 751-757.
25. McIntyre, G.I. (1977) The role of nutrition in apical dominance. Symp. Soc. Exp. Biol. *31*, 251-273.
26. Naylor, J.M. (1958) Control of nuclear processes by auxin in axillary buds of *Tradescantia paludosa*. Can. J. Bot. *36*, 221-232.
27. Nyman, L.P., Cutter, E.G. (1981) Auxin-cytokinin interaction in the inhibition, release and morphology of gametophore buds of *Plagiomnium cuspidatum* from apical dominance. Can. J. Bot. *59*, 750-762.
28. Panigrahi, B.M., Audus, L.J. (1966) Apical dominance in *Vicia faba*. Ann. Bot. *30*, 457-473.
29. Phillips, I.D.J. (1975) Apical dominance. Ann. Rev. Plant Physiol. *26*, 341-367.
30. Prochazka, S., Jacobs, W.P. (1984) Transport of benzyladenine and gibberellin A$_1$ from roots in relation to the dominance between the axillary buds of pea (*Pisum sativum* L.) cotyledons. Plant Physiol. *76*, 224-227.
31. Sachs, T. (1986) Cellular patterns determined by polar transport. *In*: Plant growth substances 1985, pp. 231-235, Bopp, M., ed. Springer- Verlag, Berlin.
32. Sachs, T., Thimann, K.V. (1964) Release of lateral buds from apical dominance. Nature *201*, 939-940.
33. Salerno, D.C., Brenner, M.L. (1983) Apical dominance: IAA mobility in the tomato isogenic lines Craigella and Blind. Plant Physiol. *72* (Suppl.), 27.
34. Schnepf, E. (1986) Cellular polarity. Ann. Rev. Plant Physiol. *37*, 23-47.
35. Sussex, I.M. (1955) Morphogenesis in *Solanum tuberosum* L.: Apical structure and developmental pattern of the juvenile shoot. Phytomorphology *5*, 253-273.
36. Tamas, I.A., Barone, C.C., Lim, R. (1985) The relationship between IAA transport and axillary bud growth inhibition. Abstracts, p. 81, 12th Internat. Conf. Plant Growth Subst., Heidelberg.
37. Tamas, I.A., Davies, P.J., Mazur, B.K., Campbell, L.B. (1985) Correlative effects of fruits on plant development. *In*: World soybean research conference III: Proceedings, pp. 858-865, Shibles, R., ed. Westview Press, Boulder.
38. Tamas, I.A., Engels, C.J., Kaplan, S.L., Ozbun, J.L., Wallace, D.H. (1981) Role of indoleacetic acid and abscisic acid in the correlative control by fruits of axillary bud development and leaf senescence. Plant Physiol. *68*, 476-481.
39. Tamas, I.A., Koch, J.L., Mazur, B.K., Davies, P.J. (1986) Auxin effects on the correlative interaction among fruits in *Phaseolus vulgaris* L. *In*: Proceedings, Plant growth regulator society of America, Cooke, A.R., ed. (In press).
40. Tamas, I.A., Ozbun, J.L., Wallace, D.H., Powell, L.E., Engels, C.J. (1979) Effect of fruits on dormancy and abscisic acid concentration in the axillary buds of *Phaseolus vulgaris* L. Plant Physiol. *64*, 615-619.
41. Thimann, K.V., Skoog, F. (1934) On the inhibition of bud development and other functions of growth substance in *Vicia faba*. Proc. Roy. Soc. B *114*, 317-339.
42. Tucker, D.J. (1976) Endogenous growth regulators in relation to side shoot development in the tomato. New Phytol. 77, 561-568.
43. Tucker, D.J. (1977) Hormonal regulation of lateral bud outgrowth in the tomato. Plant Sci. Lett. *8*, 105-111.
44. Tucker, D.J. (1978) Apical dominance in the tomato: the possible roles of auxin and abscisic acid. Plant Sci. Lett. *12*, 273-278.
45. Tucker, D.J. (1981) Axillary bud formation in two isogenic lines of tomato showing different degrees of apical dominance. Ann. Bot. *48*, 837-843.

46. Wang, T.L., Wareing, P.F. (1979) Cytokinins and apical dominance in *Solanum andigena*: lateral shoot growth and endogenous cytokinin levels in the absence of roots. New Phytol. *82*, 19-28.
47. White, J.C., Mansfield, T.A. (1977) Correlative inhibition of lateral bud growth in *Pisum sativum* L. and *Phaseolus vulgaris* L.: studies of the role of abscisic acid. Ann. Bot. *41*,1163-1170.
48. White, J.C., Medlow, G.C., Hillman, J.R., Wilkins, M.B. (1975) Correlative inhibition of lateral bud growth in *Phaseolus vulgaris* L. Isolation of indoleacetic acid from the inhibitory region. J. Exp. Bot. *26*, 419-424.
49. Woolley, D.J., Wareing, P.F. (1972) The interaction between growth promoters in apical dominance. I. Hormonal interaction, movement and metabolism of a cytokinin in rootless cuttings. New Phytol. *71*, 781-793.

E9. Hormones as Regulators of Water Balance

Terry A. Mansfield

Department of Biological Sciences, University of Lancaster, Lancaster,
Lancashire LA1 4YQ, U.K.

The development of strategies which enable growth to continue
without excessive consumption of limited water resources has played a
vital part in the evolution of plants which can survive in terrestrial
environments. Research over the last two decades has established a clear
role for plant hormones in governing the water economy of plants. By
influencing stomatal behavior they can control the expenditure of water,
and by regulating the growth and activities of roots, they can exert some
control over the uptake of water. Our knowledge of the role of hormones in
relation to stomatal functioning is now progressing rapidly and it is
appropriate to devote most of this chapter to this topic. Studies of roots
have not progressed so rapidly, but nevertheless we can also recognise an
important participation of plant hormones in governing activities in roots,
which may often complement effects on stomata to provide an integrated
strategy for improving water balance.

HORMONES AND STOMATAL BEHAVIOR

Abscisic Acid

*Formation and distribution of abscisic acid in relation to the functioning of
stomata.*
   Wright and Hiron (62) found that the ABA content of wheat leaves
increased forty-fold within 30 minutes when they were subjected to a
water deficit severe enough to cause wilting. Subsequent studies have
shown that there is a fairly abrupt rise in the ABA content of the leaves of
many different species as the water potential falls below -1.0 MPa
($= -10.0$ bar). The water potential at which there is an accelerated
production of ABA is very similar in several different species (Fig. 1) (3).
In nutrient deficient plants the production of ABA seems to occur at less
negative water potentials (41).
   Stomatal opening is strongly inhibited by ABA in many different
species. Some typical dose-response curves are shown in Figure 2, from
which it will be seen that the concentration of potassium in the medium
surrounding the epidermis has a major determining influence. This is not

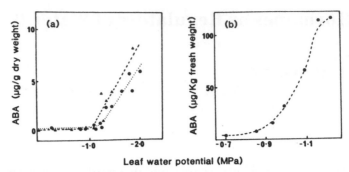

Fig. 1. Relationship between leaf water potential and ABA content. (a) Intact plants of *Ambrosia trifida* (▶) and *Ambrosia artemisifolia* (●). (b) Excised leaves of wheat (ABA produced in 330 min). From (63) and (61).

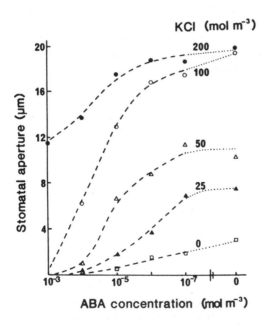

Fig. 2. Effects of ABA on the stomata of *Commelina communis* in different concentations of KCl. Abaxial epidermis was detached from plants that had been grown carefully to avoid water deficits, i.e., there was a minimum endogenous ABA content. From (47).

surprising in view of the apparent mechanism of action of ABA on the guard cells (see below). It is clear from these curves that rises in the endogenous levels of ABA in leaves could readily inhibit stomatal opening, and it is now widely believed that such inhibition plays an important part in water conservation mechanisms.

A period of severe water stress can have a more prolonged effect on the stomata of many species than would be predicted from the state of turgor of the leaf. Thus the stomata may remain closed long after the water supply has been renewed and full turgor restored (1, 49). Attempts to explain this "after-effect" in terms of ABA levels have been reasonably successful with detached leaves allowed to wilt for a few hours, then rewatered (Fig. 3). However, when water stress is imposed on an intact plant, the stomata may remain closed for some time after the level of ABA in the tissue has declined (3). This might indicate that bulk-tissue contents of ABA are an unreliable guide to the amount in the guard cells themselves, and/or that other endogenous factors play a part. Both these possibilities find some support from recent research.

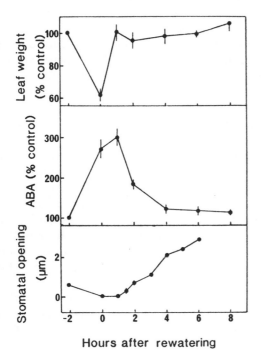

Fig. 3. Changes in leaf weight, ABA content and stomatal opening in *Pisum sativum* during and after a 2-hour period of wilting. The ABA content of unwilted leaves was 0.42 ng mg^{-1} fresh weight. From (12).

An important question which remains unanswered is whether the ABA which acts upon the guard cells is manufactured there, or is transported from other locations in the leaf. Some recent research has indicated that guard cells do have the ability to synthesize ABA (58), but nevertheless there is strong evidence that transported ABA plays a large part in the control of stomatal aperture.

Nearly all the ABA in the leaves of well-watered plants of spinach is contained in the mesophyll chloroplasts (29). ABA is probably synthesized in the cytoplasm before moving to the chloroplasts (17). ABA becomes concentrated in the chloroplasts because of the pH gradients within the cell. The pH of the chloroplast stroma is higher than that of the cytoplasm, particularly in light. ABA is a weak acid (pKa 4.5), and the dissociated ABA molecule is able to penetrate the chloroplast membrane, but the anion formed in the alkaline stroma is not able to escape. If the pH of the chloroplast stroma is 7.5 and that of the cytoplasm 6.5, the distribution ratio for ABA molecules, chloroplast: cytoplasm, will be 10:1. The chloroplast thus functions as a "pH trap" for ABA (37).

The chloroplasts from the mesophyll of wilted leaves contain a much smaller proportion of the total ABA than the chloroplasts from turgid leaves (29). Osmotic stress causes the release of ABA from chloroplasts (17). The relocation of the ABA in the chloroplasts provides a mechanism for its rapid appearance in the cytoplasm, without the need for synthesis of new ABA.

These observations may be important for our understanding of the responses of stomata in leaves under water stress. A reduced water potential in the mesophyll, sufficient to impair the functioning of the chloroplasts, will cause the release of some ABA into the cytoplasm. Anatomical studies have shown that the pathway for movement of such ABA to the guard cells is short, and need not involve the vascular system (36). There is thus a basis for a rapid response of the stomata when the activities of the mesophyll chloroplasts are unfavourably affected by water shortage. Further research is necessary to show how important this mechanism may be compared with the synthesis of ABA in the guard cells themselves. Since the guard cells undergo considerable changes in water potential during their normal functioning, they appear to be unsuitably placed for detecting small changes in the water potential of the bulk of the leaf tissue. This means that a method of communicating information from the mesophyll to the stomata may be of greater consequence for the protection of the tissues against water stress.

Another possible source of the ABA reaching the epidermis is the xylem, which could achieve transport over long distances. ABA occurs in xylem sap, and is at quite high concentrations in stressed plants (18). The export of ABA from the leaves to the roots, and eventual transport back to the leaves, provides a method of controlling stomata which may operate alongside other mechanisms for regulating the turgor of the plant (30, 31). Further evidence of root-derived signals to the stomata comes from studies of cytokinins (see below).

Mechanisms of Action of Abscisic Acid on Stomata

Consideration of the effects of ABA on a variety of processes in plants suggests that its mode of action may be on membrane function, with effects on the permeability to, and active transport of, potassium and also on hydraulic conductivity (55). Early studies of the action of ABA on guard cells showed that it prevents their accumulation of potassium (20, 35). It was later suggested that this was a result of the inhibition by ABA of the proton pump within the plasmalemma (43, 44). This would mean that a primary effect of ABA would be on the influx of potassium into the guard cells, not on its rate of efflux. Other experiments indicate, on the contrary, that the reaction of the guard cells to ABA is a stimulation of ion efflux (32, 59). MacRobbie (32) found that during the first 20 minutes after treatment with ABA, there were very high rates of efflux of rubidium and bromide (analogous to potassium and chloride). The inhibition of stomatal opening by ABA is much smaller when epidermis is supplied with NaCl as a substitute for KCl (24). Both salts are equally effective in supporting increases in the turgor of the guard cells, but the ability to close in response to various stimuli (including ABA) seems to be largely absent in NaCl. This suggests that the efflux mechanisms may show a strong specificity for K^+ ions. Metabolic inhibitors block ABA-induced stomatal closure, and thus it seems that the solute efflux stimulated by ABA is dependent on metabolic energy (60).

External ABA induces stomatal closure as effectively at pH 8.0, at which pH it is not taken up into the guard cells, as at pH 5.0 when it is taken up readily (16). Thus it appears that ABA need not enter the cytosol of the guard cells in order to induce an efflux of ions. This would mean that the site of action of ABA must either be at the outer surface of the plasmalemma, or at a location easily accessible from the outside. Hornberg & Weiler (19) have reported the presence of proteins located in the plasmalemma of the guard cells which bind ABA with a high affinity. This is the first report of proteins which might be involved in the recognition of ABA at its apparent site of action in plant cells.

Evidence has recently accumulated that calcium plays an important part in the regulation of many different cellular activities in animals, through its role as a "second messenger" (46). The discovery of the small protein calmodulin in plants raised the possibility that comparable mechanisms may occur in plant cells (2). Calmodulin, after complexing with calcium ions, stimulates the activity of several enzymes in plants, and in the last few years evidence has begun to emerge that Ca^{2+}-calmodulin may participate in the sequence of cellular events between the initial action of plant hormones and the eventual overt expression of their effects (27).

Recent studies of the action of ABA and calcium on stomatal guard cells have strongly suggested the involvement of calmodulin (11). Calcium has long been known to inhibit stomatal opening in some plants, and studies of the range over which such inhibition occurs (Fig. 4) suggest that it corresponds to the range of free calcium that may exist in the apoplastic space around the guard cells. Calcium ions can exceed 1 mM ($=$ mol m^{-3}) in xylem sap (28) and solutes in the transpiration stream are known to move readily to the region around the guard cells.

At low concentrations of ABA (10^{-9} to 10^{-8}M or 10^{-6} to 10^{-5} mol m^{-3}) there is a strong dependence on the presence of calcium for the inhibition of stomatal opening (Fig. 5), and if the calcium-chelator EGTA is used to reduce the free-space calcium concentration, there is no detectable effect of ABA on the guard cells (Fig. 6). The action of ABA is also inhibited in

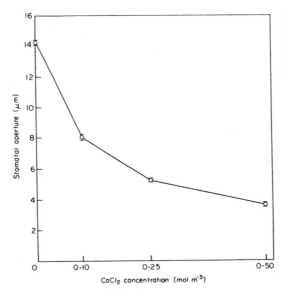

Fig. 4. Inhibition of stomatal opening in *Commelina communis* after incubation of abaxial epidermis for 3 h in a range of concentrations of $CaCl_2$. From (11).

the presence of lanthanum ions (which block calcium channels), and the stomatal closure induced by ABA can be mimicked by a calcium ionophore which is known to increase the flux of calcium from the free space into the cytosol (12).

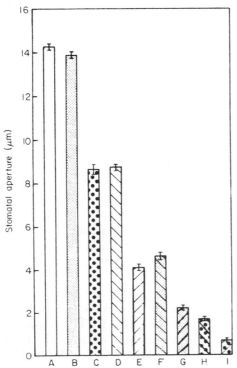

Fig. 5. Inhibition of stomatal opening after incubation of isolated epidermis of *Commelina communis* for 3 h in various combinations of ABA and $CaCl_2$. A, control; B, 10^{-6} mol m^{-3} ABA; C, 10^{-5} mol m^{-3} ABA; D, 0.1 mol m^{-3} $CaCl_2$; E, 0.25 mol m^{-3} $CaCl_2$; F, 10^{-6} mol m^{-3} ABA + 0.1 mol m^{-3} $CaCl_2$; G, 10^{-6} mol m^{-3} ABA + 0.25 mol m^{-3} $CaCl_2$; H, 10^{-5} mol m^{-3} ABA + 0.1 mol m^{-3} $CaCl_2$; I, 10^{-5} mol m^{-3} ABA + 0.25 mol m^{-3} $CaCl_2$. 1 mol m^{-3} = 1 mM. From (11).

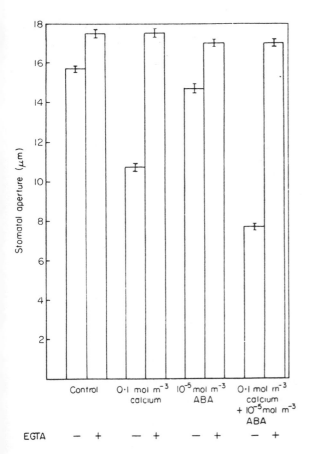

Fig. 6. Effects of calcium and ABA on stomatal opening of isolated epidermis of *Commelina communis* with or without the chelating agent EGTA. The concentration of EGTA was $2 \, mol \, m^{-3} \, (= 2 \, mM)$. From (11).

These indications of the mode of action of ABA on the guard cells are important because they should eventually lead us to a better understanding of the fine-control mechanisms that enable plants to continue to function under varying conditions of water availability in the field. Such detailed understanding will, however, also depend on a knowledge of the contribution of at least two other groups of hormones, the auxins and cytokinins.

Auxins

Control of Stomatal Behavior by Auxins

Early studies with IAA failed to reveal clear effects on stomata. For example, Boysen-Jensen in 1936 (8) supplied excised leaves with IAA via their petioles, but could detect no response. Later experiments in the 1940s and 1950s with synthetic auxins did, however, produce positive results and a number of workers found that compounds such as naphth-1-ylacetic acid (NAA) and naphth-2-yloxyacetic acid (NOXA) caused stomatal closure. These discoveries meant that when modern workers looked again for effects of the natural auxin, IAA, their experiments were often designed to show stomatal closure, not opening. Thus IAA was usually applied to stomata that were already open fully, and when there was no response it was concluded that IAA had no effect. We now know that this conclusion was wrong, and that IAA does exert important controls on stomata. Its action is to stimulate opening, and the closing responses to the synthetic auxins bear no resemblance to the effect of the natural auxin.

The first indication of a substantial role for IAA came from the work of Pemadasa (39) on the factors controlling the opening of adaxial and abaxial stomata. In many herbaceous plants there are stomata on both sides of the leaves, but it is common to find fewer stomata and smaller individual apertures on the adaxial surfaces. Pemadasa showed that externally applied IAA caused increased openings of adaxial stomata, but there was very little effect on abaxial stomata under conditions favourable for opening (Fig. 7). However, when apertures of the abaxial stomata were restricted by a reduced supply of K^+ in the medium used for incubating the epidermis, IAA did cause enhanced opening.

Interactions between IAA and Other Factors
The dose-response relationships between ABA concentration and stomatal closure in epidermal strips are greatly modified by the addition of IAA to the incubation medium (47, 48) (Fig. 8). The inhibitory effect of

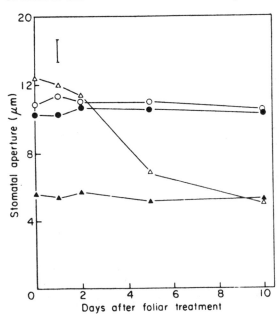

Fig. 7. The effect of application of 1 mol m^{-3} IAA on abaxial (O, ●) and adaxial (□, ■) stomatal opening in *Commelina communis*. The closed symbols are for the controls untreated with IAA. IAA was applied to the surface of the intact leaves on day 0, and epidermis was then removed at the times shown and incubated for 3 h under conditions favorable for stomatal opening. From (39).

ABA is virtually absent in the presence of a high concentration of IAA (10^{-4}M). Another important interaction is found with CO_2. Stomata normally close as the CO_2 concentration in the vicinity of the guard cells increases, but the expression of this response is dependent on the concentration of IAA (Fig. 9). High concentrations of IAA drastically reduce the closing reaction to CO_2, but it is restored partially when ABA is supplied along with IAA. These interactions are thought to be very important in the control of the water balance of the plant, and we shall return to them later.

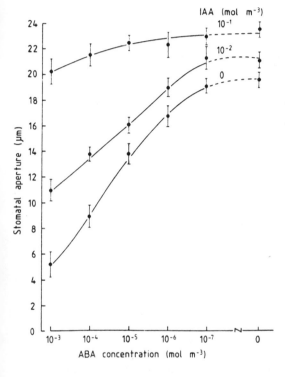

Fig. 8. Influence of IAA on the response of stomata in abaxial epidermis of *Commelina communis* to ABA. A favorable concentration of KCl ($100\ mol\ m^{-3}$) was present in all treatments. From (47).

Cytokinins

Since 1977, much evidence has been reported that cytokinins act as important regulators of stomatal movements. When kinetin is applied to isolated epidermis from a grass, *Anthephora pubescens*, it causes stomatal opening (21). At first it was thought that these responses applied only to the Graminae, but subsequent work has shown that they are more widespread. Age of the leaves may be a determining factor, for in the *Argenteum* mutant of *Pisum sativum* the stomata show no response to kinetin when the leaves have just reached full expansion, but 15 days later there is a significant stimulation of opening (22). Similarly the

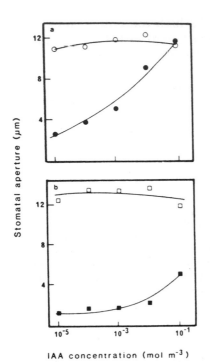

Fig. 9. Stomatal opening on detached abaxial epidermis of *Commelina communis* after incubation for 3 h in light in different IAA concentrations. o = zero CO_2, ● = 700 µl l ^{1} CO_2. Points are means of 60 measurements. b. As in a. but with 10^{-5} mol m^{-3} ABA in the medium. □ = zero CO_2, ■ = 700 µl l^{-1} CO_2. From (48).

stomata of young leaves of *Zea mays* do not respond, but in ageing leaves they open more widely when treated with kinetin (5).

Thus it seems that an effect of endogenous cytokinins might be to determine the extent of stomatal opening on leaves of different ages. Another function may be to communicate information about changing water status from the root to the shoot. There is still some controversy about the location of all the sites of cytokinin synthesis in plants, but there seems little doubt that the roots are a major source of the cytokinins arriving in leaves (54). Over many years, evidence has accumulated that synthesis and/or transport from the roots may be influenced by soil drying, chilling and flooding. Mizrahi (38) has suggested that adaptation to a whole range of stresses (e.g. drought, flooding, nutrient deficiency, salinity, chilling) involves reduced concentrations of cytokinins in the transpiration stream. All of these agents can lead to the same symptoms of stress in the shoot because of the way they can affect uptake of water by the roots.

It has been suggested that for the most efficient, long-term exploitation of water in the soil, a plant would benefit from being able to respond to the water potential in different parts of the root system (25). Contrasting types of stomatal behavior would affect the pattern of water use (26). Four classes of behavior were identified: (1) *pessimistic, non-responsive*, in which the stomata open daily in a fixed but restrained manner so as to use initial soil water by the end of the season; (2) *optimistic, non-responsive*, in which the fixed daily opening routine uses water faster than justified from the initial supply in the soil; (3) *pessimistic, responsive,* in which the stomata react to changing conditions to regulate the use of water so that the available supply is used by the end of the season; (4) *initially optimistic, responsive,* in which water consumption begins as in (2) but there is an ability to reduce the daily degree of stomatal opening to prevent very serious water deficits.

Studies utilizing split root systems have pointed to a possible mechanism by which water consumption, determined by the stomata, may be linked to the availability of water to the root system (7). Roots of individual maize plants were split and divided equally between two containers. After a period of establishment, water was withheld from one container while the other continued to receive a regular supply of water. This supply to one half of the plant's root system was sufficient to maintain adequate water potential and turgor, but nevertheless stomatal opening was reduced, apparently because of the influence of the roots that were in the dry soil (Fig. 10). Leaf sections taken from the "half-dry" plants and placed under favorable conditions for stomatal opening (floated on water in CO_2-free air under high irradiance) continued to display reduced apertures. Thus it seemed that the stomata were affected either by an inhibitor of opening, or by the absence of a promotory agent. Changes in endogenous concentrations of ABA did not adequately explain the data, but application of cytokinins reopened the stomata of the "half-dry" plants (Fig. 11). It is an attractive hypothesis that a plant can use cytokinins from the root system as a chemical signal to the shoot to modify stomatal behavior according to the availability of soil water. This would provide a mechanism for at least a partial achievement of the responsiveness of stomata required by categories (3) and (4) as outlined above. In practice it seems likely that such effects will often combine with those of ABA, and also perhaps IAA, so that there is an integrated response to the three hormones.

Another study using *Helianthus annuus* has given support to the view that stomatal behavior may be directly affected by conditions in the soil (53). There was a poor relationship between stomatal opening and the water potential in the leaves, but a good correlation between opening and soil water potential. Stomatal closure occurred when about two-thirds of the extractable water in the soil had been used, irrespective of leaf water potential at the time. This strongly suggests that information about the availability of soil moisture can in some way be communicated from the roots to the shoot.

Interactions Between Cytokinins and Other Factors
The characteristics of stomatal responses to cytokinins have several features in common with the responses to IAA. For example, there are sometimes no effects when cytokinins are supplied to leaves on their own

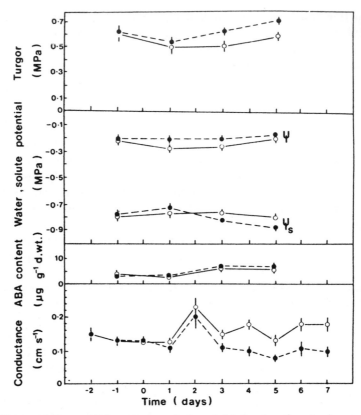

Fig. 10. Water relations, ABA content, and stomatal behavior of maize leaves in plants whose roots were split into two separate rooting containers; water was withheld from part of the root. Control plants have both parts of the root system well watered (○); plants where part of root system was dried (---) were last watered on day 0. Points are means of eight observations. From (7).

(42). This may be because there is already an adequate supply of endogenous cytokinins and/or because the experiments are conducted under conditions favourable for stomatal opening. It is only when there are factors restricting opening that a clear response to cytokinins can be seen.

Kinetin and zeatin have no distinct effect when they are applied to young leaves of *Zea mays* (4). However, if they are supplied together with ABA, they can overcome the strong inhibitory effect of that hormone (Fig. 12). In *Z. mays* the nature of the CO_2-response is determined both by zeatin and by ABA (5). The data in Fig. 13 can be compared with those in Fig. 9, and it will be seen that there are some similarities in the ABA/IAA and ABA/zeatin interactions, though the latter is more complex.

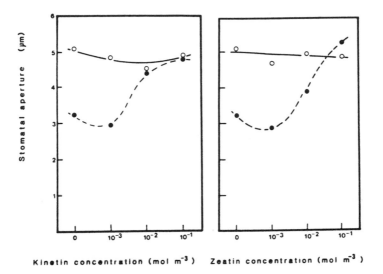

Fig. 11. Stomatal apertures of maize leaves incubated in the light in CO_2-free air on distilled water, kinetin, or zeatin solutions. Leaves taken from well-watered (○) or partly dried plants (---) were sampled 8 h into the light period on day 5 after water was first withheld from part of the root system of half of plants. Points are means of 60 observations. From (7).

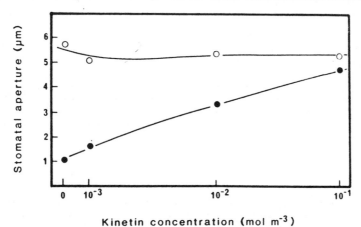

Fig. 12. Stomatal apertures from maize leaf pieces incubated on a range of kinetin concentrations, with (○) or without (---) ABA (10^{-1} mol m^{-3}). Points are means of 60 observations. From (4).

Hormones and the Fine Control of Stomatal Movements

The discovery that the CO_2 responses of stomata are variable is not new, but recent observations have led to some reassessments of their role in controlling the water relations of the plant (34). It is suggested that a major advantage resulting from the response of stomata to CO_2 is their

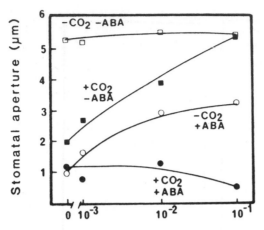

Fig. 13. Stomatal aperture from maize leaf pieces incubated on a range of zeatin concentrations with (+) or without (−) CO_2 (350 µl l^{-1}) and with (+) or without (−) ABA (10^{-1} mol m^{-3}). Points are means of 60 observations. From (5).

partial closure as wind speed increases. Wind takes water vapor away from the leaf surface and brings CO_2 towards it. This means that if the stomata close as the amount of CO_2 in their vicinity rises, some control over the rate of transpiration may result. The importance of such control will depend on changes in leaf temperature brought about by the increased air movement: sometimes increasing wind speed can reduce transpiration because cooling reduces the vapor pressure of water in the intercellular spaces. Thus under certain conditions wind itself may cause a reduction in transpiration for purely physical reasons and in these circumstances a response of the stomata may be superfluous. In many situations, however, such as when the input of solar radiation is moderate to low, wind does increase transpiration and without a partial closure of the stomata, the water consumption of the plant may rise considerably. It is under such conditions that a closing reaction to CO_2 is seen as an advantage.

Most control mechanisms not only bring gains, but also impose penalties, and that is true in this case. If the stomata remain open when wind increases the CO_2 concentration at the leaf surface, there will be an increase in the intercellular space CO_2 concentration, C_i. This will potentially benefit photosynthesis (depending on irradiance) and such benefit, however small, will be lost if the stomata close as wind brings more CO_2 to the leaf surface. Thus it can be suggested that the stomata should not always close in response to CO_2. If the plant is supplied with sufficient water, it is desirable that C_i should rise with increased wind speed, so that the plant can benefit from a small increase in photosynthesis. It is when the plant is running short of water that stomatal closure in wind will become important. Thus the discovery that three plant hormones, ABA, IAA and cytokinins can become involved in regulating the CO_2 responses of stomata is intriguing (Figs. 9 and 13).

This suggests that hormones may provide a much more precise control of stomatal movements than previously recognised. They may contribute to a control system that is extremely elegant—one which changes its characteristics according to the external environment of the plant, especially water supply in the soil. This would mean that hormones are responsible for minute-by-minute control of gaseous diffusion, helping the growing plant to optimise its water consumption relative to the gain of CO_2 for photosynthesis. The existence of a variable CO_2 sensor in the guard cells means that the plant can vary its priorities according to the water supply available (34).

A challenge to plant physiologists in the immediate future will be to determine changes in endogenous levels of cytokinins and auxins in water-stressed plants, to discover how they are related to stomatal movements. ABA has a well-established role in the physiological responses to water stress, and the behavior of ABA during stress/recovery cycles is fairly well documented (cf Figs. 1 and 3). There is little comparable information for cytokinins and auxins, and it will be as difficult to obtain the required precision in analyses (e.g. amounts in the epidermis rather than the leaf as a whole) as has proved to be the case with ABA.

HORMONES AND PHOTOSYNTHESIS

Chemicals which bring about changes in stomatal aperture must also affect rates of net photosynthesis in the intact leaf. Only careful analyses of cause-effect relationships can reveal whether a compound first inhibits mesophyll photosynthesis, and thereafter causes stomatal closure (because of an increased internal CO_2 concentration), or whether stomatal closure precedes a reduction in net photosynthesis. In the former case, however, the internal CO_2 concentration (C_i) will rise, and in the latter case it will fall. This means that measurements or estimations of C_i can prove valuable in helping to interpret the causes of observed effects.

Cummins *et al* (10) decided that ABA caused a reduction in photosynthesis as a result of stomatal closure, and not because there was a direct action on the photosynthetic capacity of mesophyll chloroplasts. Further studies over the years have appeared to confirm this conclusion, but nevertheless some doubts remain and there is a possibility that ABA has a direct action on mesophyll photosynthesis (45). A possible reason for this is that when ABA is applied to the whole leaf, it may affect the action or distribution of other hormones. IAA has been found to stimulate photosynthetic CO_2 uptake, probably by increasing the coupling between electron transport and phosphorylation (51). ABA often antagonizes the effects of IAA in plant tissues (cf. Fig. 8).

Hormone-induced changes in the sensitivity of stomata to CO_2 may act to counteract excessive transpiration in windy conditions, as discussed above. Such effects would lead to a temporary interference with photosynthesis. When hormonal changes bring about long-term adjustments of stomatal aperture, however, such as when the plant has suffered a severe water deficit, it may be desirable for the photosynthetic capacity of the mesophyll to be adjusted accordingly. It seems possible, therefore, that changes in IAA in the leaf could be involved both in adjusting the behavior of stomata to the prevailing conditions, and in altering the activities of the mesophyll to correspond with the modified diffusive conductance of the epidermis.

HORMONES AND INTEGRATED RESPONSES OF THE PLANT TO WATER STRESS

We have much more information on the effects of hormones on stomatal conductance than on other activities in the plant which may be relevant to water conservation. This imbalance in the available information partly explains the emphasis given to stomata in this chapter, but such emphasis is probably justified for other reasons because the stomata provide the major point at which a plant can control its rate of water loss. It is, nevertheless, appropriate to make brief mention of some other responses which may assist the plant in its overall adjustment to conditions of water stress.

Roots

ABA increases the permeability of carrot root tissue to water (15), and a similar conclusion was drawn from studies of exudation from isolated roots of maize (9). Effects on exudation rate could be the result of changes in ion flux and/or water permeability (hydraulic conductivity), and a detailed investigation of the effects of ABA on the exudation process in roots of sunflower indicated separate actions on both of these.

A change in root pressure is unlikely to be of much consequence for the bulk flow of water in the xylem of many plants. However, many herbaceous plants are susceptible to cavitation of the sap in their xylem conduits when they are water-stressed, and root pressure at night may be important in refilling the affected vessels. It thus seems possible that ABA formed in shoots under water stress may be transported to the roots where it can stimulate root pressure.

Root growth and development may also be affected by hormonal changes induced by water deficits. There have been indications from some studies (e.g., 57) that ABA limits the growth of the shoot and enhances the growth of the roots. This could explain the reduced shoot:root ratio that is often found in water-stressed plants.

Leaves

ABA inhibits the active efflux of protons thought to be responsible for the co-transport of sucrose in phloem loading in *Ricinus* (33). A reduction in the rate of transport of sugars out of the leaf would enable more solutes to be retained for the maintenance of turgor. Turgor is necessary for the continuation of growth, but under water stress the continued production of a large area of leaves is clearly undesirable. ABA has been found to inhibit the light-stimulated cell enlargement in leaves of *Phaseolus vulgaris* (56). This could provide an important mechanism for limiting expansion of leaves when plants experience water shortage.

Overall Effects of Hormonal Changes

Mutants of several plant species are known which possess different balances of the hormones discussed in this chapter. A deficiency of ABA production produces 'wilty' plants which are unable to maintain turgor under normal conditions (40, 50). Such mutants provide an important practical demonstration of the importance of ABA for the maintenance of normal water relations in the plant. Higher concentrations of auxin-like substances and cytokinins accompany reduced ABA concentrations in some mutants (50). In the future, physiological investigations of mutants with different capacities for the production of ABA (52) should contribute greatly to our understanding of the ways in which control of water relations is achieved in plants.

Some promising studies of the water relations of different cultivars in relation to ABA production have been carried out in the last few years. Varieties of spring wheat with a high capacity for drought-induced ABA formation have been shown to produce higher yields than low-ABA cultivars, probably because of a higher efficiency of water use (23). By contrast, cultivars of sorghum which showed excessive stomatal closure, apparently because of high ABA production, showed a reduced grain yield in response to drought (14). This suggests that the effects of hormonal balance on CO_2 exchange as well as water vapor loss must be fully considered in future studies.

References

1. Allaway, W.G., Mansfield, T.A. (1970) Experiments and observations on the after-effect of wilting on stomata of *Rumex sanguineus*. Can.J.Bot. 48, 513-521.
2. Anderson, J.M., Cormier, M.J. (1978) Calcium-dependent regulator of NAD kinase. Biochem. Biophys. Res. Comm., 84, 595-602.

3. Beardsell, M.F., Cohen, D. (1975) Relationship between leaf water status, abscisic acid levels and stomatal resistance in maize and sorghum. Plant Physiol. 56, 208-212.
4. Blackman, P.G., Davies, W.J. (1983) The effects of cytokinins and ABA on stomatal behaviour of maize and Commelina. J.Exp.Bot. 34, 1619-1626.
5. Blackman, P.G., Davies, W.J. (1984) Modification of the CO_2 responses of maize stomata by abscisic acid and by naturally occurring and synthetic cytokinins. J.Exp.Bot. 35, 174-179.
6. Blackman, P.G., Davies, W.J. (1984) Age-related changes in stomatal responses to cytokinins and abscisic acid. Ann.Bot. 54, 121-125.
7. Blackman, P.G., Davies, W.J. (1985) Root to shoot communication in maize plants of the effects of soil drying. J.Exp.Bot. 36, 39-48.
8. Boysen Jensen, P. (1936) Growth Hormones in Plants. McGraw-Hill Book Co., New York.
9. Collins, J.C., Kerrigan, A.P. (1983) Hormonal control of ion movements in the plant root. In: Ion Transport in Plants, pp 589-594. Anderson, W.P, ed. Academic Press, London.
10. Cummins, W.R., Kende, H., Raschke, K. (1971) Specificity and reversibility of the rapid stomatal response to abscisic acid. Planta 99, 347-351.
11. De Silva, D.L.R., Hetherington, A.M., Mansfield, T.A. (1985) Synergism between calcium ions and abscisic acid in preventing stomatal opening. New Phytol. 100, 473-482.
12. De Silva, D.L.R., Cox, R.C., Hetherington, A.M., Mansfield, T.A. (1985) Suggested involvement of calcium and calmodulin in the responses of stomata to abscisic acid. New Phytol. 101, 555-563.
13. Dörffling, K., Streich, J., Kruse, W., Muxfeldt, B. (1977) Abscisic acid and the after-effect of water stress on stomatal opening potential. Z. Pflanzenphysiol. 81, 43-56.
14. Durley, R.C., Kannangara, T., Seetharama, M., Simpson, G.M. (1983) Drought resistance of Sorghum bicolor. 5. Genotypic differences in the concentrations of free and conjugated abscisic acid, phaseic acid and indole-3-acetic acid in leaves of field grown drought stressed plants. Can.J.Plant Sci. 63, 131-145.
15. Glinka, Z., Reinhold, L. (1972) Induced changes in the permeability of plant cells to water. Plant Physiol. 49, 602-606.
16. Hartung, W. (1983) The site of action of abscisic acid at the guard cell plasmalemma of Valerianella locusta. Plant, Cell & Environment 6, 427-428.
17. Hartung, W., Heilmann, B., Gimmler, H. (1981) Do chloroplasts play a role in abscisic acid synthesis? Pl.Sci.Lett. 22, 235-242.
18. Hoad, G.V. (1975) Effect of osmotic stress on abscisic acid level in xylem sap of sunflower (Helianthus annuus). Planta 124, 25-29.
19. Hornberg, C., Weiler, E.W. (1984) High affinity binding sites for abscisic acid on the plasmalemma of Vicia faba guard cells. Nature 310, 321-324.
20. Horton, R.F., Moran, L. (1972) Abscisic acid inhibition of potassium flux into stomatal guard cells. Z. Pflanzenphysiol. 66, 193-196.
21. Incoll, L.D., Whitelam, G.C. (1977) The effect of kinetin on stomata of the grass Anthephora pubescens. Planta 137, 243-245.
22. Incoll, L.D., Jewer, P.C. (1985) Cytokinins and stomata. In Stomatal Function, Zeiger, E., Farquhar, G.D., Cowan, I.R., eds. Stanford University Press.
23. Innes, P., Blackwell, R.D., Quarrie, S.A. (1984) Some effects of genetic variation in drought-induced abscisic acid accumulation on the yield and water use of spring wheat. J.Agric.Sci, Camb. 102, 341-351.
24. Jarvis, R.G., Mansfield, T.A. (1980) Reduced stomatal responses to light, carbon dioxide and abscisic acid in the presence of sodium ions. Plant, Cell & Environment 3, 279-283.

25. Jones, H.G. (1980) Interaction and integration of adaptive responses to water stress: the implications of an unpredictable environment. *In* Adaptation of Plants to Water and High Temperature Stress, pp. 353-365, Turner, N.C., Kramer, P.J., eds. John Wiley & Sons, London.
26. Jones, H.G. (1983) Plants and Microclimate. Cambridge University Press.
27. Kelly, G.J. (1984) Calcium, calmodulin, and the action of plant hormones. Trends in Biochemical Sciences, January 1984, 4-5.
28. Kirby, E.A., Pilbeam, D.J. (1984) Calcium as a plant nutrient. Plant, Cell & Environment 7, 397-405.
29. Loveys, B.R. (1977) The intracellular location of abscisic acid in stressed and non-stressed leaf tissue. Physiol. Plant. 40, 6-10.
30. Loveys, B.R. (1984a) Diurnal changes in water relations and abscisic acid in field-grown *Vitis vinifera* cultivars. III The influence of xylem-derived abscisic acid on leaf gas exchange. New Phytol. 98, 563-573.
31. Loveys, B.R. (1984b) Abscisic acid transport and metabolism in grapevine (*Vitis vinifera* L.). New Phytol. 98, 575-582.
32. MacRobbie, E.A.C. (1981) Effects of ABA in 'isolated' guard cells of *Commelina communis* L. J.exp.Bot. 32, 563-572.
33. Malek, T., Baker, D.A. (1978) Effect of fusicoccin on proton co-transport of sugars in the phloem loading of Ricinus communis L. Plant Sci. Lett. 11, 233-39.
34. Mansfield, T.A., Davies, W.J. (1985) Mechanisms for leaf control of gas exchange. BioScience 35, 158-164.
35. Mansfield, T.A., Jones, R.J. (1971) Effects of abscisic acid on potassium uptake and starch content of stomatal guard cells. Planta 101, 147-158.
36. Mansfield, T.A., Wellburn, A.R. and Moreira, T.J.S. (1978) The role of abscisic acid and farnesol in the alleviation of water stress. Phil.Trans.Roy.Soc., Lond., B284, 471-482.
37. Milborrow, B.V. (1984) Inhibitors. *In* Advanced Plant Physiology, pp 76-110, Wilkins, M.B., ed. Pitman, London and Massachusetts.
38. Mizrahi, Y. (1980) The role of plant hormones in plant adaptation to stress conditions. 15th Colloquium of International Potash Institute, pp 75-86.
39. Pemadasa, M.A. (1982) Differential abaxial and adaxial stomatal responses to indole-3-acetic acid in Commelina communis L. New Phytol. 90, 209-219.
40. Quarrie, S.A. (1982) Droopy: a wilty mutant of potato deficient in abscisic acid. Plant, Cell & Environment 5, 23-26.
41. Radin, J.W. (1984) Stomatal responses to water stress and to abscisic acid in phosphorus-deficient cotton plants. Plant Physiol. 76, 392-394.
42. Radin, J.W., Parker, L.L., Guinn, G. (1982) Water relations of cotton plants under nitrogen deficiency. V. Environmental control of abscisic acid accumulation and stomatal sensitivity to abscisic acid. Plant Physiol. 70, 1066-1070.
43. Raschke, K. (1977) The stomatal mechanism and its responses to CO_2 and abscisic acid: observations and hypothesis. *In* Regulation of Cell Membrane Activities in Plants pp. 173-183, Marre, E., Ciferri, O., eds. Elsevier/North Holland, Amsterdam.
44. Raschke, K. (1979) Movements of stomata. *In* Encyclopedia of Plant Physiology (New Series) vol 7, pp. 383-441, Haupt, W., Feinleib, M.E., eds. Springer-Verlag, Berlin.
45. Raschke, K., Hedrich, R. (1985) Simultaneous and independent effects of abscisic acid on stomata and the photosynthetic apparatus in whole leaves. Planta 163, 105-118.
46. Rasmussen, H., Barrett, P.Q. (1984) Calcium messenger system: an integrated view. Physiological Reviews, 64, 938-984.
47. Snaith, P.J., Mansfield, T.A (1982a) Stomatal sensitivity to abscisic acid: can it be defined? Plant, Cell & Environment 5, 309-311.
48. Snaith, P.J., Mansfield, T.A. (1982b) Control of the CO_2 responses of stomata by indol-3-ylacetic acid and abscisic acid. J.exp.Bot. 33, 360-365.
49. Stålfelt, M.G. (1955) The stomata as a hydrophotic regulator of the water deficit of the plant. Physiol. Plant. 8, 572-593.

50. Tal, M., Nevo, Y. (1973) Abnormal stomatal behaviour and root resistance, and hormonal imbalance in three wilty mutants of tomato. Biochem. Genet. 8, 291-300.
51. Tamas, I.A., Schwartz, J.W., Hagin, J.W., Simmonds, R. (1974) Hormonal control of photosynthesis in isolated chloroplasts. In: Mechanisms of Regulation of Plant Growth, pp. 261-268. Beileski, R.L., Ferguson, A.R., Cresswell, M.M., eds. Royal Society of New Zealand, Wellington.
52. Taylor, I.B., Tarr, A.R. (1984) Phenotypic interactions between abscisic acid deficient tomato mutants. Theor. Appl. Genet. 68, 115-119.
53. Turner, N.C., Schulze, E.-D., Gollan, T. (1985) The responses of stomata and leaf gas exchange to vapour pressure deficits and soil water content. II In the mesophytic herbaceous species *Helianthus annuus*. Oecologia (Berl.) 65, 348-355.
54. Van Staden, J., Davey, J.E. (1979) The synthesis, transport and metabolism of endogenous cytokinins. Plant, Cell & Environment 2, 93-106.
55. Van Steveninck, R.F.M., Van Steveninck, M.E. (1983) Abscisic acid and membrane transport. *In* Abscisic Acid, pp 171-235, Addicott, F.T., ed. Praeger Publishers, New York.
56. Van Volkenburg, E., Davies, W.J. (1983) Inhibition of light-stimulated leaf expansion by abscisic acid. J.exp.Bot. 34, 835-845.
57. Watts, S., Rodriguez, J.L., Evans, S.E., Davies, W.J. (1981) Root and shoot growth of plants treated with abscisic acid. Ann.Bot. 47, 595-602.
58. Weiler, E.W., Schnabl, H., Hornberg, C. (1982) Stress-related levels of abscisic acid in guard cell protoplasts of *Vicia faba* L. Planta 154, 24-28.
59. Weyers, J.D.B., Hillman, J.R. (1980) Effects of abscisic acid on [86]Rb$^+$ fluxes in *Commelina communis* L. leaf epidermis. J.exp.Bot. 31, 711-720.
60. Weyers, J.D.B., Paterson, N.W., Fitzsimons, P.J., Dudley, J.M. (1982) Metabolic inhibitors block ABA-induced stomatal closure. J.exp.Bot. 33, 1270-1278.
61. Wright, S.T.C. (1977) The relationship between leaf water potential and the levels of abscisic acid and ethylene in excised wheat leaves. Planta 134, 183-189.
62. Wright, S.T.C., Hiron, R.W.P. (1969) (+) abscisic acid, the growth inhibitor induced in detached wheat leaves by a period of wilting. Nature (London) 224, 719-720.
63. Zabadal, T.J. (1974) A water potential threshold for the increase of abscisic acid in leaves. Plant Physiol. 53, 125-127.

E 10. Hormones and Reproductive Development

James D. Metzger
USDA/ARS Metabolism and Radiation Research Laboratory, Fargo, North Dakota 58105, USA.

INTRODUCTION

To farmers, horticulturalists, and others whose livelihood depends on growing plants, it is obvious that the transition from vegetative to reproductive development is a critical phase in the life cycle of higher plants. It is difficult to overestimate the impact the direct products of flowering have on human endeavors. Because they are an integral part of the human diet, production of seeds and fruits form the foundation of all nations' economies. Thus, it becomes equally evident that the ability to manipulate and control flowering with simple treatments has enormous potential both from an economic standpoint as well as increasing food production for an ever growing human population. However, the development of such cultural techniques is predicated on a thorough understanding of the physiological, biochemical, and molecular aspects of reproductive development. Although a great deal of descriptive information exists for many species about the influence of environmental factors on reproductive development, knowledge of the mechanisms by which the floral transition takes place is almost totally lacking.

Flowering, or more precisely, reproductive development is composed of many independent but highly coordinated processes. To discuss the role of hormones in flowering *per se* would be pointless; it is much more convenient to arbitrarily divide flowering into several temporally related sequences. *Flower initiation* is the production of flower (or inflorescence) primordia. *Evocation*, on the other hand, is the term used to describe those processes that occur in the apex prior to, and are required for the formation of flower primordia. Since one cannot precisely determine the exact point at which a flower primordium is formed, it is necessary to wait for the appearance of true floral structures; this is usually referred to as *flower formation*. *Flower development* encompasses the processes occurring between flower formation and anthesis.

In many species, the onset of reproductive development is regulated by environmental factors. Many environmental factors such as daylength and temperature vary regularly during the year. Numerous species use

431

these seasonal variations as cues to coordinate the initiation of reproductive development with the growing season.

Photoperiodism

Plants in which flowering occurs only under certain daylength conditions are said to be photoperiodic. Photoperiodically sensitive plants fall into several response types. Short-day plants (SDP) flower only when the photoperiod is less than some critical length. Actually, it is more accurate to state that SDP flower only when the dark period is greater than a certain critical length since it is the night length which is measured by the plant. Plants that flower when the daylength is greater than a certain time are called long-day plants (LDP). The requirement can be absolute (obligate) or facultative (quantitative). Plants with a facultative requirement for certain photoperiodic conditions will eventually flower under unfavorable photoperiods, although much later than in inductive conditions. Although LDP and SDP comprise the majority of photoperiodically sensitive plants, three other response types are known. Short-long-day plants (SLDP) are plants that flower only when subjected first to SD followed by LD. Conversely, plants that flower only with the sequence LD then SD are called long-short-day plants (LSDP). Finally, day neutral plants (DNP) are plants with no photoperiodic requirements.

Perception of daylength occurs in the leaves. The primary photochemical event is the absorption of a photon by the chromoprotein phytochrome. The role of the phytochrome in the photoperiodic timing mechanism is discussed in detail in several recent books and reviews (e.g., 64,65), and will not be considered further here.

Vernalization

In numerous species from temperate regions, exposure to the low temperatures of winter promotes the initiation of reproductive development upon the return of warmer temperatures the following spring. This phenomenon is known as vernalization, and is unique because the perception and transduction of the environmental cue occurs during the cold period while floral development is manifested later when the plants are under warmer (growth-promoting) temperatures. An exception is brussel sprouts, *Brassica oleracea*, in which flower initiation occurs during the thermoinductive treatment

Analogous to the various photoperiodic response types, plants can be classified according to their response to low (thermoinductive) temperatures. Summer annuals (or simply annuals), which have no requirement for a thermoinductive treatment, complete their entire life cycle in one growing season. Seeds of winter annuals germinate during late summer or early autumn and overwinter as seedlings. These plants then flower the following growing season. Usually the cold requirement is facultative; that is, flowering will eventually occur without vernalization, but takes considerably longer than plants subjected to thermoinductive temperatures. Furthermore, winter annuals as a group tend to be sensitive to thermoinductive temperatures at most stages of development. Indeed, imbibed seeds of many winter annuals can be vernalized (seed vernalization). Biennials, in contrast, require one full season of vegetative growth, have an obligate requirement for a thermoinductive treatment, and exhibit a period of juvenility in which plants cannot be vernalized. Numerous perennials also have a cold requirement for flowering.

Vernalization temperatures can also promote flowering in species not normally considered cold requiring. The LDP *Spinacia oleracea* responds to thermoinductive temperatures with a shortening of the critical daylength requirement. In other LDP as well, low night temperatures can compensate for a long dark period (i.e., SD) resulting in flowering under normally noninductive conditions.

In contrast to photoperiodic plants, the site of perception of temperature lies in the apical region of the plant. Often, cold requiring plants also have a requirement for certain photoperiodic conditions following thermoinduction. Most winter annuals and biennials behave as LDP once thermoinduction is complete. On the other hand, the cold requiring perennial *Chrysanthemum morifolium* needs SD after thermoinduction.

Table 1 shows a list of representative plants in each of the various response groups. Both photoperiodism and vernalization have important ecological consequences in allowing plants to fill special niches. Biennials and winter annuals are often the first plants to set seed in spring and early summer. Many LDP flower in midsummer while SDP produce seed toward autumn. Moreover, precise timing of flowering also ensures that many individuals from a given population will be flowering at the same time thereby maximizing outcrossing. Although one usually thinks of vernalization and photoperiodism as adaptations that evolved to avoid initiation of reproductive development right before the onset of winter and possibly not being able to produce seed, they are also important to time reproductive development to coincide with other seasonal environmental factors. Some tropical plants are strongly photoperiodic (certain species can respond to changes in daylength as little as 15 minutes) and this apparently is to time flowering in relation to seasonal variations in rainfall (i.e., monsoon or rainy season).

Table 1. Representative Plants of Each Flowering Response Type

Species (common names)	Family
LONG DAY PLANTS	
Rudbeckia bicolor (cone flower)	Compositae
Anethum graveolens (dill)	Umbelliferae
Spinacia oleracea (spinach)	Chenopodiaceae
Hyoscyamus niger(henbane, annual strain)	Solanaceae
Nicotiana sylvestris (Nicotiana)	Solanaceae
Agrostemma githago (corn cockle)	Caryophylaceae
Silene armeria (Sweet William campion)	Caryophylaceae
Lolium perrene (perennial ryegrass)	Gramineae
SHORT DAY PLANTS	
Xanthium strumarium (cocklebur)	Compositae
Kalanchoë blossfeldiana (Kalanchoë)	Crassulaceae
Perilla crispa (Perilla)	Labiatae
Nicotiana tabacum (tobacco, var Maryland Mannoth)	Solanaceae
Euphorbia pulcherrima (poinsetta)	Euphorbiaceae
Crysanthemum morifolium (Crysanthemum var. Honeysweet)	Compositae
Pharbitis nil (Japanese morning glory)	Convolvulaceae
Glycine max (soybean var. Biloxi)	Leguminoseae
LONG SHORT DAY PLANTS	
Bryophyllum daigremontianum (Bryophyllum)	Crassulaceae
Cestrum nocturnum (night jessamine)	Solanaceae
SHORT LONG DAY PLANTS	
Coreopsis grandiflora (tickseed)	Compositae
Echeveria harmsii (Echeveria)	Crassulaceae
DAY NEUTRAL PLANTS	
Zea mays (maize or corn)	Gramineae
Cucumis sativus (cucumber)	Cucurbitaceae
Glycine max (soybean var. Wilkins)	Leguminoseae
Lycopersicon esculentum (tomato)	Solanaceae
Nicotiana tabacum (tobacco var. Wisconsin 38)	Solanaceae
Gossypium hirsutum (cotton)	Malvaceae
COLD-REQUIRING PLANTS	
Althaea rosea (hollyhocks)	Malvaceae
Thlaspi arvense (field pennycress)	Cruciferae
Brassica oleracea (brussel sprouts)	Cruciferae
Hyoscyamus niger (henbane, biennial strain)	Solanaceae
Triticum aestivum (winter wheat)	Gramineae
Beta vulgaris (beet)	Chenopodiaceae
Daucus carota (carrot)	Umbelliferae
Crysanthemum morifolium (Crysanthemum var. Shuokin)	Compositae

Juvenility

In many plants, flowering cannot be induced despite being subjected to the proper inductive conditions until a certain size or age is obtained. This refractory period is known as the *juvenile phase*. The length of the juvenile period can be as short as a few days or weeks in some herbacious plants or as long as forty years in some species of trees (Table 2). Juvenility is a serious obstacle in breeding programs for economically important forest trees.

Table 2. Comparison of the duration of the juvenile phase in various species

Species (common name)	Duration of Juvenile Phase
Chenopodium rubrum (coast blite)	0
Pharbitis nil (Japanese morning glory)	0
Perilla crispa (Perilla)	1–2 months
Bryophyllum daigremontianum (Bryophyllum)	1-2 years
Malus pumila (apple)	6-8 years
Citrus sinensis (orange)	6-7 years
Citrus paradisi (grapefruit)	6-8 years
Pinus sylvestris (Scotch pine)	5-10 years
Betula pubescens (birch)	5-10 years
Pyrus communis (pear)	8-12 years
Larix decidua (European larch)	10-15 years
Pseudotsuga menziesii (Douglas-fir)	15-20 years
Fraxinus excelsia (ash)	15-20 years
Acer pseudoplatanus (sycamore maple)	15-20 years
Picea abies (Norway spruce)	20-25 years
Abies alba (white fir)	25-30 years
Quercus robur (English oak)	25-30 years
Fagus sylvatica (European beech)	30-40 years

The transition from the juvenile phase to one permissive of flowering (adult or mature phase) is called *phase change*. It is important to understand that the transition to the adult state does not necessarily mean that the plant has been induced to flower; certain environmental conditions such as photoperiod or thermoinductive temperatures may still be required. Plants that have become mature but have not flowered because of improper conditions are termed *ripe-to-flower*. At present, there is no good method to distinguish a juvenile plant from a mature one, except by the ability to form flowers.

Another feature exhibited by the two phases is a marked stability through cell division. Reversion to the juvenile phase is not observed in cuttings from flowering (mature) plants. Generally speaking, grafting scions from plants in one state to receptors of the other does not usually alter the phase of either graft partner. Re-juvenation normally occurs only with sexual reproduction.

Often other morphological characteristics differ in juvenile and mature plants. For example, the shape and thickness of leaves, phyllotaxis, or the growth habit of stems may be different. What relationships, if any, these changes have with the ability to form flowers is not known since none of these characteristics are always associated with a transition to maturity. A comparison of correlative morphological characteristics in juvenile and adult plants is shown in Table 3. Rooting ability of cuttings

Table 3. Juvenile and adult characteristics in selected species. Adapted from (15)

Characteristic	Species	Juvenile form	Adult form
Growth Habit	Hedera helix	plagiotropic	orthotropic
	Ficus punula	plagiotropic	orthotropic
	Metrosideros diffusa	plagiotropic	orthotropic
	Euonymus radicans	plagiotropic	orthotropic
Leaf Shape	Cupressus spp.	acicular	scale-like
	Acacia spp.	pinnate	phyllodes
	Eucalyptus spp.	oval, sessile	lanceolate with petioles
	Pinus spp.	flat, glaucous	scale- and bract-like
	Hedera helix	palmate	ovate, entire
Phyllotaxis	Eucalyptus spp.	opposite	alternate
	Hedera helix	alternate	spiral
Anthocyanin	Malus pumila	+	−
Pigmentation	Carya illinoisiensis	+	−
in Leaves	Acer rubrum	+	−
	Hedera helix	+	−
Thorniness	Robinia pseudoacacia	thorns	no thorns
	Malus robusta	thorns	no thorns
	Cirtus	thorns	no thorns
Autumn Leaf	Fagus sylvatica	keep leaves	abscise
Abscission	Quercus spp.	keep leaves	abscise
in Deciduous	Robinia pseudoacacia	keep leaves	abscise
Trees	Carpinus spp.	keep leaves	abscise
Rooting	Hedera helix	+	−
Ability of	Quercus spp.	+	−
Cuttings	Fagus sylvatica	+	−
	Pinus spp.	+	−
	Pyrus malus	+	−

is one of these characteristics which has some economic importance. Often cuttings from forest trees lose the ability to root following phase change to the adult state, thus hindering mass introduction of genetically improved clones.

It appears that juvenility provides a mechanism whereby flowering is prevented until the plant is large enough to survive supplying growing reproductive structures and subsequent developing seeds with assimilates. This is particularly critical with perennials which also have demands for those same assimilates from storage organs.

Floral stimulus

In photoperiodically sensitive plants, the site of perception of daylength is the leaf, whereas the apex is where the morphological change occurs. This suggests that some message is transferred from the leaf to the apex causing the transition to flower formation. This signal is termed the *floral stimulus, florigen,* or *flower hormone.* Further evidence for the existence of the floral stimulus are the observations that non-induced plants flower when a leaf from an induced plant is grafted onto them. Successful flowering in the non-induced receptor has also been obtained with several interspecific grafts between different photoperiodic response types. Likewise, receptor plants with a cold requirement can be made to flower without vernalization by grafting flowering donors that are either SDP, LDP or DNP. A summary of successful flower induction in receptor plants under non-inductive conditions by grafting flowering donors is shown in Table 4.

It is tempting to conclude from these experiments that the floral stimulus is very similar or even identical in all plants. However, numerous examples exist in which successful intra- and interspecific graft unions have been made but no apparent transmission of the floral stimulus occurred (63,76). These negative results have been interpreted as evidence against a unique or ubiquitous floral stimulus (4,5). This has led a number of investigators to consider the possibility that flower formation is controlled by the known classes of plant hormones acting in concert (5).

Floral inhibitors

When a receptor of the DNP *Nicotiana tabacum* cv. Trapezond was grafted to the LDP *N. sylvestris* and maintained in SD, flowering of the receptor was essentially suppressed. Noninduced *Hyoscyamus niger* (annual strain, LDP) also served as an inhibitory donor (42). This suggests the existence of a graft-transmissable floral inhibitor. Another good example of active (and presumably chemical) inhibition of flowering is in the SDP *Fragaria* x *ananassa.* Subjecting daughter plants to SD

Table 4. Selected examples of successful transmission of the floral stimulus between graft partners. Adapted from Lang (41) and Zeevaart (75)

DONOR Species	Flowering Type	Photoperiod/Temp.	RECIPIENT Species	Flowering Type	Photoperiod/Temp.	
Glycine max	SDP	SD	Glycine max	SDP	LD	Leguminoseae
Perilla crispa	SDP	SD	Perilla crispa	SDP	LD	Labiatae
Xanthium strumarium	SDP	SD	Xanthium strumarium	SDP	LD	Compositae
Beta vulgaris	LDP	LD	Beta vulgaris	LDP	SD	Chenopodiaceae
Bryophyllum daigremontianum	LSDP	LD→SD	Bryophyllum daigremontianum	LSDP	LD or SD	Crassulaceae
Crysanthemum morifolium var. Shuokin	CRPa	TIb→LD	Crysanthemum morifolium var. Honeysweet	SDP	LD	Compositae
Anethum graveolens	LDP	LD	Daucus carota	CRP	LD	Umbelliferae
Sinapis alba	LDP	LD	Brassica oleracae	CRP	LD	Cruciferae
Brassica nigra	CRP	TI→LD	Brassica oleracae	CRP	LD	Cruciferae
Blitum capitatum B. virgatum	LDP	LD	Chenopodium rubrum	SDP	LD	Chenopodiaceae
Petunia hybrida	LDP	LD	Hyoscyamus niger	CRP	LD	Solanaceae
Nicotiana tabacum var. Maryland Mammoth	SDP	SD	Hyoscyamus niger	CRP	SD	Solanaceae
Nicotiana tabacum var. Maryland Mammoth	SDP	SD	Nicotiana sylvestris	LDP	SD	Solanaceae
Gossypium hirsutum	DNP	LD	Gossypium davidsonii	SDP	LD	Malvaceae
Bryophyllum daigremontianum	LSDP	LD→SD	Echeveria harmsii	SLDP	SD	Crassulaceae
Pharbitis nil	SDP	SD→LD	Pharbitis nil	SDP	LD	Convolvulaceae
Lunaria annua	CRP	TI→LD	Lunaria annua	CRP	LD	Cruciferae
Kleinia articulata	SDP	SD→LD	Kleinia repens	LSDP	LD	Compositae
Silene armeria	LDP	LD→SD	Silene armeria	LDP	SD	Caryophyllaceae

a CRP = Cold-requiring plant; b TI = Thermoinduced

438

while still attached by stolons to the parent plants that are maintained under non-inductive LD resulted in a great reduction of flowering in the daughter plants. Severing the stolon connection removed the inhibition. Maximum inhibition was observed when assimilate flow from the parents to the daughters was the greatest (27). The most logical interpretation of these results is that a phloem-mobile compound is produced in non-induced leaves, and then transported to the apex where it actively prevents flower initiation.

As with the floral stimulus, the identity of the floral inhibitor(s) remains a mystery. Extractions have not provided any clues to its nature. The floral inhibitor remains a physiological concept.

Active floral inhibitors do not appear to be the basis for all cases of flower inhibition by non-induced leaves. In *Perilla crispa*, the flower promoting ability of induced donor leaf is nullified if non-induced leaves of the receptor lie between the donor leaf and the apex. But this inhibition can be explained entirely by the alteration of assimilate translocation patterns and, hence, movement of the floral stimulus (38). Likewise, non-induced leaves of the SDP *Xanthium strumarium* apparently do no more than to serve as sources of assimilates devoid of the floral stimulus, thereby diluting the floral stimulus and reducing its effective concentration at the apex (80).

Measurement of flowering

A plant is either flowering or it is not; thus, at first glance, measurement should be only a simple qualitative determination. However, such an analysis is too simplistic. A plant may be induced to flower by two different treatments, but in one treatment it produced only one or a couple of flowers, while many flowers are produced following the other treatment. Alternatively, the time required for the completion of some aspect of reproductive development might be affected by different inductive treatments. Clearly, these examples show that there are quantitative aspects to flowering.

As in the study of other developmental processes, it is important to be able to quantitate flowering in relation to various treatments. The methods used to quantitate flowering are numerous, but they basically fall into five categories depending on the experimental design. The simplest is the percentage of plants that have flowered following a particular treatment. Another way to measure the flowering response is to determine the number of buds, flowers, or flowering nodes on an individual plants. An experimenter might choose to determine the number of leaves produced from the beginning of a treatment until flowering is observed. This technique gives temporal data, with the fewer number of leaves, the faster the rate of development. Similar data could be obtained by determining the time (usually in days or weeks) for flowering to occur.

As a final alternative, there are scales with numbered values assigned to different stages of apical development. At various times after the start of a treatment, the apex is examined microscopically and assigned a number, based on an arbitrary scale developed for that particular species. This measure differs from the others in that it is destructive and therefore requires significantly more plants, but it does provide data on flower initiation. The relative advantages and disadvantages of each of these techniques have been discussed in considerable detail (4,41).

HORMONES AND JUVENILITY/PHASE CHANGE

The hormonal role in the control of phase change is not well established, although gibberellins (GAs) may be involved. It does not appear that a specific hormone is required for the maintenance of either the juvenile or adult phase. Reviews of various aspects of juvenility have appeared over the past 20 years (15,58,83). The following discussion on juvenility will be limited to a comparison of evidence for the hormonal basis of juvenility in three greatly different types of plant.

Bryophyllum daigremontianum is a LSDP that has a juvenile phase lasting until the development of 10 to 12 pairs of leaves (72). It has been shown that the leaves perceive the transfer from LD to SD, but in juvenile plants, they are incapable of producing the floral stimulus (72). Application of GA_3 to the leaves of juvenile plants will promote flowering only under SD. When GA_3 is applied in LD, the plants will subsequently flower only if transferred to SD. This period of time between the treatment in LD until the transfer to SD when GA_3 is still effective can be remarkably long: up to four months (72). Since it has been shown that in adult plants flower formation following the transfer from LD to SD requires GA biosynthesis (82), the biochemical basis for juvenility in *Bryophyllum* could reside in an inability for GA biosynthesis following the sequence LD to SD (72).

In conifers, where "flowering" is not normally observed until plants are 10-20 years old, exogenous GAs can induce flowering in 3-12 month old plants (51,52) (Fig. 1). Polar GAs such as GA_3 are effective in members of Cupressaceae and Taxodiaceae families but not in species from Pinaceae. Less polar GAs, such as GA_9 and $GA_{4/7}$, are necessary for promotion of precocious flowering in Pinaceae species (51,52).

It has been postulated that in juvenile conifers, GAs are the limiting factor preventing phase change. However, cessation of GA application to juvenile plants results in a decline in the flowering response and concomitant abscission of newly formed cones (51). Thus, exogenous GA does not cause a true phase change since the mature state should be stable through many mitotic cycles. It may be that in conifers, GAs are limiting for flower formation and development, and overcoming this limitation

Fig. 1. Exogenous GAs will cause precocious "flowering" in juvenile conifers. This four-year-old rooted cutting of *Tsuga heterophylla* was subjected to six weekly spray treatments of a solution containing 200 mg/1GA$_{4/7}$. Untreated plants of the same age failed to flower. Photograph courtesy of Dr. Stephen Ross.

may be one of many alterations resulting from the phase change to the adult state.

In contrast with *Bryophyllum* and conifers, exogenous GA$_3$ causes reversion of many adult woody plants to the juvenile state (79,83) (Fig. 2). These results suggest a positive relationship between GAs and the maintenance of the juvenile phase. Consistent with this is the observation that the levels of GA-like substances were higher in extracts of apical buds from juvenile plants of *Hedera helix* than adult apical buds (20). If it is true that high levels of endogenous GAs are required to maintain the juvenile phase, one would predict that a decline in GA levels in juvenile plants would lead to the transition to the adult phase. However, attempts to artificially reduce endogenous GA levels with growth retardants (inhibitors of GA biosynthesis) have produced equivocal results. Application of CCC (2-chloroethyltrimethyl ammonium chloride) to juvenile *Hedera* plants resulted in dwarf plants but paradoxically also caused an increase in the levels of endogenous GA-like substances (21).

Under certain conditions, mature *Hedera* plants will revert to plants that are either juvenile or show several juvenile characteristics. This fact has been used to study the possible role of hormones in phase maintenance. Rejuvenation of adult shoots was observed when grafted on juvenile stocks. However, the adult leaves had to be removed (14).

Fig. 2 In contrast to conifers, exogenous GAs cause rejuvenation of mature *Hedera helix* (English ivy) plants. Left to to right: Juvenile control. GA-induced reversion to the juvenile form of a mature shoot treated with 5 nmole of GA_3. ABA (5 µmole) prevented rejuvenation caused by 5 nmole GA_3. Mature control shoot. Note the distinct differences in leaf shape and internode length between juvenile and mature forms. From (55).

Rejuvenation of adult plants has also been reported when adult and juvenile plants were grown together in the same culture solution (19). Together, these results suggest that juvenile plants produce a substance(s) that maintains the apex in the juvenile state. Since exogenous GA_3 can also cause rejuvenation (20), it is possible that GAs are involved. Under conditions of low intensity light, adult *Hedera* plants spontaneously revert to the juvenile state; this reversion can be prevented by inhibitors GA biosynthesis (55).

In total the above results are consistent with the notion that in *Hedera*, the ability to maintain high GA levels is necessary for the maintenance of the juvenile state, and, conversely, a diminution of this capacity is associated with the phase change to the adult state. Thus, rather than controlling phase change *per se*, GAs (or lack of GAs) may be involved in phase stabilization (55).

In conclusion, no unified picture for the role of hormones in juvenility and phase change can be presented. Although there is evidence that GAs play a role, it appears to be different in the case of *Bryophyllum* and *Hedera*. This indicates that the physiological and biochemical basis for juvenility and subsequent phase change can be distinct in different species. In the *Bryophyllum*, for example, the physiological basis for the failure of juvenile plants to flower following transfer from LD to SD lies in an inability of the leaves to produce the floral stimulus (72). A similar conclusion was reached for the SDP *Perilla crispa* (68). In other instances, it appears that the apex itself is insensitive to the floral stimulus. Juvenile apices of both *Larix* and *Hedera* failed to flower when grafted to mature stock (14,54). This interpretation assumes that the floral stimulus is produced in the leaves and then transported to the apex akin to photoperiodically sensitive herbacious plants. Even if this is not so, the experiments with *Larix* and *Hedera* do demonstrate that phase change is a property of the apex rather than the leaf as was shown for *Bryophyllum* and *Perilla*. Thus, the concept of phase change is in reality only an operational definition, useful to describe the ability of plants to flower under inductive conditions rather than implying common mechanisms.

HORMONES AND FLOWER FORMATION

This section will examine the role of the known classes of plant hormones in flower formation. Each class will be dealt with separately. In general, the pattern has been to first apply the hormone to the plant under non-inductive conditions (or inductive conditions if inhibition of flower formation is being scrutinized). The hormone is then quantitated as a function of flower formation. Unfortunately, most quantitative work was done before modern analytical techniques using physico-chemical methods were available to physiologists. Instead, relating hormone levels to flower formation was performed with bioassays. Although a good first approximation can be made with bioassays, much of this work must be viewed with caution. The pitfalls of bioassays have been discussed elsewhere (11 and Chapter D1). Wherever possible, I have attempted to use examples in which the investigators have employed physico-chemical techniques for quantitative analysis.

Another useful technique is to use inhibitors of hormone biosynthesis and/or action. In this way, the effect of reducing endogenous hormone levels on flower formation can be assessed. In a similar vein, it has been possible to select hormone-deficient mutants. In certain instances, such mutants have proven extremely valuable in assessing the role of a particular hormone in flower formation.

Auxins

Members of the Bromeliaceae are unique among plants in that they exhibit a strong flowering response following auxin application. This effect is due to auxin-induced ethylene production (77) and will be discussed in greater detail later when the role of ethylene in flower formation is considered. For the most part, however, application of various auxins (either indole acetic acid or naphthalene acetic acid) tend to be inhibitory to flower formation under inductive conditions. This is true for the various response types including: the SDP *Pharbitis nil* and *Chenopodium rubrum*; the LDP *Lolium temulentum* and *Sinapis alba*; and the cold-requiring plants *Lunaria annua* and *Cichorium intybus* (5,77).

These results have been interpreted to mean that the role of auxins in flowering is to prevent flower formation under noninductive conditions. However, quantitative analyses of endogenous auxin levels have been equivocal. In any event, such a proposed role for auxin in the control of flower formation is probably an oversimplification (5,77). It now appears these inhibitory effects are also due to auxin-induced ethylene production (77).

In certain instances, application of low amounts of auxin levels promote flower formation (5), but only under conditions of marginal or partial induction. The significance of these results are uncertain. Nevertheless, it is possible that auxins play some role in some of the processes associated with evocation such as a loss in apical dominance and an alteration in phyllotaxis (5), but this remains to be shown conclusively. The role(s) of auxin in the reproductive transition might be clarified immensely if a) the endogenous compounds were rigorously identified and b) modern physico-chemical methods were used for quantitation.

Ethylene

In many plants, exogenous ethylene, applied either as the gas or by the use of ethylene-releasing agents such as ethrel, inhibits or delays flower formation. As yet, there is no evidence that this inhibition is part of the natural regulating mechanism (77).

In contrast to most plants, species from the Bromeliaceae flower in response to exogenous ethylene. A common and economically important horticultural practice is to induce flowering of bromeliads, particularly pineapples, at will with ethylene-releasing agents (5,77).

Little is known on the role of ethylene in flower formation in bromeliads. In a recent report, treatments that resulted in flower induction were invariably associated with increased ethylene production. Flower induction by one of the treatments, namely mechanical perturbation, could be suppressed by AVG (aminoethoxyvinyl glycine), an inhibitor of ethylene biosynthesis. This inhibition could be reversed by adding ethylene. Furthermore, the immediate precursor to ethylene biosynthesis, aminocyclopropane carboxylic acid, could also induce flowering (13). These results strongly suggest that ethylene is indeed an important regulator of flower formation in bromeliads.

Cytokinins

Flower formation is promoted in a number of species by exogenous cytokinins (5,77). In many of these cases, the promotive effect of cytokinins is seen only in induced or marginally induced plants. Furthermore, in the SDP *Pharbitis nil*, the promotion of flowering by cytokinin was shown to be indirect, i.e., due to enhanced translocation of the floral stimulus and assimilates from the induced leaves (50).

Quantitative analyses of cytokinins in relation to flower formation have not clarified the picture either. In the LDP *Sinapis alba,* an increase in the endogenous cytokinin levels was observed 16 hours after the beginning of LD (1). This positive correlation is in sharp contrast to the negative relationship between endogenous cytokinins and photoperiodic induction observed in the SDP *Xanthium strumarium* (31).

Thus, at present, there is no convincing evidence linking cytokinins and flower induction in a cause and effect relationship. However, a single low dose of cytokinin applied to the apex of noninduced *Sinapis* plants mimicked the early stimulation of mitotic activity caused by an inductive LD (3). Other features of this mitotic activity were identical in the two treatments. The cytokinin effects corresponded very well with an increase in cytokinin-like activity in the leaves following transfer to LD, indicating that cytokinins may play a role in regulating evocational processes (1,3).

Abscisic Acid

As discussed earlier, physiological evidence exists for the presence of a graft-transmissable flower inhibitor. With the discovery of ABA as a potent, naturally occurring growth inhibitor that was possibly involved in the control of bud dormancy, it was reasonable to suspect that ABA might be the flower inhibitor. Indeed, application of ABA to the LDP *Spinacia oleracea* and *Lolium temulentum* repressed LD-induced flower formation (17,18). However, no causal relationship was observed between endogenous ABA levels and the ability to flower; ABA levels in *Spinacia* were in fact higher under LD than SD (39,74). Moreover, the apparent inhibition of flower formation might have been the result of a delay in inflorescence development rather than inhibition of flower initiation (77). Thus, it does not appear likely that ABA plays a role in the regulation of flower formation as a graft-transmissible inhibitor.

In several SDP, ABA promotes flowering (16). However, this promotive action is usually observed only under partially inductive conditions. In other SDP, ABA has no effect or is inhibitory (5,77). As in LDP, there is no apparent role for ABA in flowering formation in SDP.

Gibberellins

Of all the plant hormones that have been applied to plants under strictly noninductive conditions, only GAs have been shown to effectively cause flower formation in a wide variety of species. In general, LDP and plants with a cold requirement are responsive to exogenous GAs while SDP and DNP are not. GA sensitive LDP and cold-requiring plants usually grow as rosettes in noninductive conditions. These generalizations do not always hold, however. For example, the LDP *Hieracium aurantiacum* and *Blitum virgatum,* and the cold-requiring plants *Geum urbanum* and *Lunaria annua* do not form flowers under non-inductive conditions in response to exogenous GA$_3$ (79). On the other hand, application of GA$_3$ caused flower formation in the SDP *Impatiens balsamina* and *Zinnia elagans* maintained in LD (5,79). In still other plants GA inhibits flower formation, particularly woody fruit trees such as cherry, peach, apricot, almond, and lemon (79). These species often exhibit a period of juvenility and as discussed earlier, the basis for GA inhibition of flowering in these species may reside in rejuvenation of adult plants. A comprehensive list of the effects of exogenous GAs on flower formation can be found in (5,79).

In view of the large number of species in which exogenous GAs cause vegetative plants to flower under noninductive conditions, it is logical to conclude that GAs have a critical role in the regulation of flower formation. As yet, however, such a role cannot be defined. This is partly due to the analytical difficulties encountered when investigating GA physiology. At present there are over 70 different GAs, although only a fraction of these appear to be present in any given species. The endogenous GAs appear to be metabolically related. It is likely that only one is responsible for mediating the physiological process under GA control with the others being either precursors or deactivation products (30). Thus, when attempting to correlate endogenous levels of GAs to certain physiological processes, it is important to know the identity of those GAs. However, the instrumentation and analytical expertise necessary to conclusively identify and quantitate endogenous GAs in vegetative tissues has, until recently, been largely unavailable to plant physiologists. Nevertheless, the picture beginning to emerge indicates that GAs may have different roles in the various response types. In the remainder of this section, specific examples will be used to illustrate this point.

As noted earlier, flower formation can be induced in many rosetted LDP with exogenous GAs suggesting that GAs may be limiting in SD. Consistent with this idea are the numerous observations that transfer to LD results in an increase in the levels of one or more GAs or GA-like substances (e.g., 10,37,47). Associated with LD induction are both increased GA biosynthesis and metabolism (10,37,71). Nevertheless, reduction of endogenous GA levels with inhibitors of GA biosynthesis completely suppressed LD-induced stem elongation (bolting) but had no effect on flower formation in *Spinacia oleracea* (71), *Silene armeria* (10) and *Agrostemma githago* (36). In other rosetted LDP, application of GA$_3$ to vegetative plants induces stem elongation but not flowering (63). Thus it appears that in most rosetted LDP, GAs are not limiting for flower formation but do mediate the photoperiodic control of stem elongation. This aspect of reproductive development will be dealt with in greater detail in a later section.

A similar situation has been observed in many rosetted cold requiring plants. Changes in endogenous GA-like substances following thermo-induction appear to be more important for thermoinduced stem elongation than for flower formation *per se* (79). Growth retardants inhibit thermo-induced stem elongation but not flower formation in *Raphanus sativus* (61), *Daucus carota* (33) and *Thlaspi arvense* (48). Application of GA_3 to nonthermoinduced *Lunaria annua* elicits only stem elongation (70).

Even though it does not appear that GAs are directly involved in the transition to flowering in many LDP and cold-requiring plants, the question still remains as to the mechanism by which GAs induce flowering in vegetative plants. One possible answer is that exogenous GA acts indirectly through the production of the floral stimulus. Evidence for this idea comes from grafting experiments with two lines of the LDP *Silene*—one in which exogenous GA_3 induces flowering and stem elongation and another that responds to exogenous GA_3 only with stem elongation. Flowering in the latter can be induced in SD by grafting it to the other line following induction of flowering with GA_3 (63). This must mean that the exogenous GA induced the production of the floral stimulus. A similar conclusion can be drawn from the cold requiring plant *Chrysanthemum morifolium*. Non-vernalized plants induced to flower with GA_3 can cause flowering in SD (non-cold-requiring) lines maintained under LD. Exogenous GA_3 does not induce flowering in the SD lines when the plants are in LD (29).

The vast majority of SDP do not respond to exogenous GA; this fact leads to the conclusion that GAs are usually not limiting for flower formation in LD. But in an ironic contrast to rosetted LDP and cold-requiring plants, growth retardants prevent flower formation in the SDP *Pharbitis nil* if applied before or during the inductive dark period (60,69). About 100 times more GA_3 was required to reverse the inhibition of internode elongation than was needed for flower formation (60). Moreover, the growth retardants acted at the shoot apex and not in the cotyledons, the site of perception of photoperiod. This indicates that the reduction of endogenous GAs reduced the ability of the apex to respond to the floral stimulus (60,69). Thus, in *Pharbitis*, GAs are required for floral initiation, but not for production of the floral stimulus (69).

The opposite conclusion can be made for the LSDP *Bryophyllum daigremontianum*, namely that GAs are necessary for production of the floral stimulus but not initiation. *Bryophyllum* can be induced to flower when GA is applied in SD but not LD, indicating that GA can substitute for the LD portion of the inductive photoperiod (72). GA_{20} has been identified as the major biologically active GA in leaves of *Bryophyllum*; however, it was 20 times less effective than GA_3 in inducing flower formation under SD (23). The transfer of adult plants from LD to SD caused a dramatic increase in the level of GA_{20}. In contrast, plants maintained permanently in SD had no detectable GA_{20} (75). The growth retardant CCC completely inhibited flower formation in *Bryophyllum* when applied during the inductive SD period. This inhibition could easily be reversed with GA_3 (81).

Taken together these results strongly suggest high levels of endogenous GA (presumably GA_{20}) are required for flower formation in *Bryophyllum*. The site of GA action in *Bryophyllum* is in the leaves, not the shoot apex where flower formation occurs (73). Zeevaart (81,82) concluded that the LD part of photoinduction leads to GA production which is absolutely necessary for the production of the floral stimulus in SD. Thus in *Bryophyllum*, GAs regulate the production of the floral stimulus.

To summarize, it was shown that GAs have a variety of roles in reproductive development depending on the species and the response type. In many rosetted LDP and cold requiring plants, GAs are not required for flower formation, but do regulate the closely related phenomenon of bolting. GAs also can play an essential role in the regulation of the production of the floral stimulus as was shown for *Bryophyllum* or in controlling the events associated with the response of the apex to the floral stimulus (i.e., *Pharbitis*).

HORMONES AND THE FLORAL STIMULUS

To date, there is a massive collection of evidence indicating the existence of a stimulus which initiates the transition of an apex from a vegetative state to one committed to reproductive development. From grafting experiments (Table 4), it is logical to conclude that the floral stimulus is very similar or even identical in all response types. Despite many attempts to isolate the floral stimulus, its chemical nature remains as much a mystery today as it was 50 years ago when its existence was first proposed by Chailakhyan (8). Since its conception, the floral stimulus has been envisioned as a single or at most a few substances different from the known plant hormones (4,76). The failure to identify the chemical nature of the floral stimulus has led some to question this hypothesis. As noted earlier, while there are numerous reports in which floral induction can be transferred between graft partners of different response types, there have also been many failures. Furthermore, except for two examples, all successful grafts have been within families. The exceptions are the induction of flowering in the LDP *Silene* (Caryophyllaceae) following grafting to the SDP *Perilla* (Labiatae) and *Xanthium* (Compositae) (63,66). These experiments were criticized, however, for not having enough non-induced control grafts (76).

Bernier and colleagues have suggested that instead of a unique and specific flower-inducing compound, the stimulus is primarily a certain balance of the known plant hormones and assimilates arriving at the apex in a specific sequence (3,4,5). Support for this hypothesis comes from the observations that exogenous hormones can often mimic one or several evocational processes. For example, application of a single low dose of cytokinin to the apex of the LDP *Sinapis alba* under non-inductive conditions caused several evocational events including an increase in the mitotic index and the sub-division of vacuoles into smaller ones (3). Since cytokinin application cannot completely substitute for LD in inducing flowering, only "partial evocation" occurred (3,4). Based on defoliation experiments, transport of a mitotic stimulus from the leaves began about 16 h after the beginning of the LD treatment (1). This corresponded with an increase in cytokinin-like activity in the leaves (2). Other evocational processes relating to energy metabolism seemed to be regulated by soluble sugars (5).

It is difficult, however, to reconcile this theory with some important experimental observations. First, if flower initiation is the sum of partial evocational events, then one should be able to induce flowering by using a combination of treatments. To date, this has not yet been successfully accomplished. Second, it is unlikely that a specific ratio of the known hormones arriving at the apex is the critical factor in floral initiation. It has been shown that a single leaf under inductive conditions can induce flower formation. In the presence of non-induced leaves, a special ratio of hormones emanating from the induced leaf would almost certainly be altered by the time it reached the apex (76). Third, it is difficult to see how a certain specific sequence of essential compounds could be an overriding factor in flower initiation. Leaves from induced *Perilla* plants continue to export the floral stimulus in non-inductive conditions for up to three months (68). Under these conditions how could the sequential arrival of hormones and intermediary metabolites at different apicies be maintained? Fourth, there is always a danger in over-interpreting the results of experiments in which substances are applied to plants and assuming that the observed effect is identical to a certain aspect of evocation. The action of exogenous compounds can be indirect and this should be considered before formulating a hypothesis concerning the role of a hormone in flower formation. A case in point is the instance discussed earlier where application of cytokinins to the cotyledons of suboptimally induced plants of the SDP *Pharbitis nil* promoted flowering (50). It was shown that the action of cytokinin was in the cotyledons, not the apex. Furthermore, it was believed that the enhanced flowering response by cytokinin was due to greater export of assimilates along with a concomitant increase in the amount of floral stimulus moving from the cotyledons to the apex (50). Finally, the results of grafting experiments between different response types are more easily explained on the basis of a unique floral stimulus. The site of action of GA when applied to non-induced *Bryophyllum* plants is in the leaves not the apex (72). In this case, the action of exogenous GA seems to be linked to the production of a stimulus in the leaves, which is different from GA, and is capable of initiating the transition to floral development (72,81). Further evidence for the uniqueness of the floral stimulus was obtained in grafting

Fig. 4. Application of GA$_3$ to *Spinacia* plants under SD causes the plants to assume the LD growth habit (elongated petioles and stem). Ten micrograms of GA$_3$ were applied to the plant on alternate days for 10 days. Photograph courtesy of Dr. Jan A. D. Zeevaart.

(Fig. 7). Following transfer to LD, the level of GA$_{19}$ progressively declined. The GA$_{20}$ content, on the other hand, increased in proportion to the lowering of the GA$_{19}$ level (Fig. 7). An increase in the level of GA$_{29}$ was also observed following the transfer to LD; however, the start of the increase lagged a day or so behind the increase in GA$_{20}$ levels. The rise in

Fig. 5. Inhibitors of GA biosynthesis retard the apearance of morphological characteristics associated with the transfer of *Spinacia* plants to LD. Approximately 17 mg of AMO 1618 [2 isopropyl-4trimethylammoinum chloride)-5-methylphenyl piperidine-1-carboxylate] in 10 ml of water was applied daily to the roots. The inhibition caused by the growth retardant was reversed with exogenous GA$_3$. Application of GA$_3$ was indentical to that described in Fig. 4. Photograph courtesy of Dr. Jan A. D. Zeevaart.

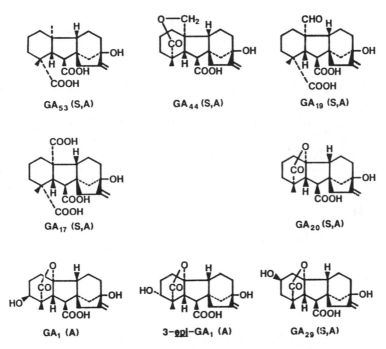

Fig. 6. The structure of the endogenous GAs identified in shoots of *Spinacia oleracea* L. and *Agrostemma githago* L. by combined gas chromatography–mass spectrometry. Letters in parentheses indicate in which species the GA was identified (S = *Spinacia*; A = *Agrostemma*). From (37) and (45).

the levels of both GA_{20} and GA_{29} occurred just prior to the onset of stem elongation (Fig. 7). The other endogenous GAs (GA_{17} and GA_{44}) remained fairly constant under either photoperiodic regime (47).

The rise in the level of GA_{20} and the concomitant decline in the level of GA_{19} implies a product-precursor relationship between these two GAs. Inasmuch as GA_{20} can cause stem elongation in *Spinacia* in SD, but GA_{29} is biologically inactive, it was proposed that the conversion of GA_{19} to GA_{20} was under photoperiodic control and stem elongation was regulated by the endogenous level of GA_{20} (47).

That the conversion of GA_{19} to GA_{20} is under photoperiodic control has been subsequently confirmed. $[^2H]$-GA_{53} was metabolized by *Spinacia* shoots to $[^2H]$-GA_{44} and $[^2H]$-GA_{19} in SD, while in LD, $[^2H]$-GA_{20} was also formed (24). When plants were allowed to accumulated $[^2H]$-GA_{44} and $[^2H]$-GA_{19} in SD and then transferred to LD, a decline was observed in the levels of both $[^2H]$-GA_{44} and $[^2H]$-GA_{19} while at the same time the amount of $[^2H]$-GA_{20} increased (24). In later work, photoperiodic regulation of the conversion of GA_{19} to GA_{20} was directly demonstrated using $[^{14}C]$-GA_{19} synthesized enzymatically(25). Cell-free preparations from *Spinacia* leaves were made which could also complete this conversion. Enzyme activity was proportional to the length of the light

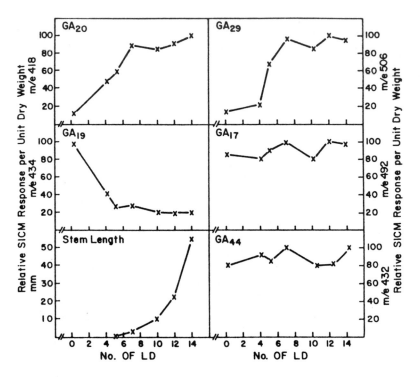

Fig. 7. Changes in the relative levels of five GAs and stem length in *Spinacia* shoots as affected by different durations of LD treatment. GA levels measured by combined gas chromatography selected ion current monitoring. The highest concentration (ion current response/unit dry wgt.) of each GA was arbitrarily assigned a value of 100, and the other concentrations were epxressed in proportion to this value. The ion curent response of the molecular ion was used in all cases except for GA_{19}, in which case the base peak was used. From (47).

period that the plants were exposed prior to preparation of the cell-free extract. Once the plants were returned to darkness, extractable enzyme activity declined rapidly (25). The nature of photoperiodic regulation of this enzyme is unknown.

Knott (40) has shown that the site of perception of photoperiod in *Spinacia* for both flowering and stem elongation is the leaf. This raises the possibility that GAs move in tandem with the floral stimulus from the leaf to the apex where they initiate stem elongation and flower formation. Photoperiodic treatment also affected the levels of GAs in phloem exudate from *Spinacia* leaves in a similar way to that observed in whole leaves (46). Thus, the GA_{20} content in the phloem exudate increased with LD treatment. This is consistent with the notion that GA_{20} is the substance hypothesized by Knott (40) to be synthesized in the leaves under LD, translocated to the apex where it promotes stem elongation (46).

The picture in the LDP *Agrostemma githago* is not so clear as in *Spinacia*. Shoots of *Agrostemma* were shown by GC-MS to contain GA_{53}, GA_{44}, GA_{19}, GA_{17}, GA_{20}, GA_1, and 3-*epi*-GA_1 (37) (Fig. 6). Of the endogenous GAs tested, GA_1 elicited more stem growth in SD than GA_{20}, while both GA_{17} and 3-*epi*-GA_1 had virtually no activity (37). Since GA_{20} is the immediate precursor to GA_1, and in other plants GA_1 appears to be the GA responsible for biological activity (the other endogenous GAs are presumably either precursors to, or deactivation products of GA_1), it is reasonable to suspect that GA_1 mediates photoperiodic control of stem growth in *Agrostemma* (37). Nevertheless, quantitative analysis of endogenous GA levels as a function of photoperiod showed complex changes when plants were transferred to LD (Fig. 8). The levels in shoot extracts of GA_{17}, GA_{19}, GA_{20}, and GA_{44} increased rapidly just prior to the onset of stem elongation, but fell sharply during the most rapid growth phase. The level of GA_1 rose somewhat well after stem elongation commenced (37). Thus, in *Agrostemma*, there does not appear to be a good correlation between photoperiodically induced stem growth and endogenous GA levels. However, these measurements were from whole shoots. Since GA acts at the shoot apex, the really important factor is the level of GA there; GA levels elsewhere in the shoot would be irrelevant or, at most, indirectly relevant to stem elongation.

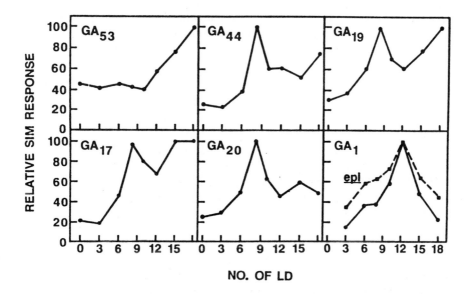

Fig. 8. Changes in the relative levels of seven endogenous GAs in *Agrostemma* shoots as a function of LD treatment. GA levels determined by combined gas chromatography—selected ion current monitoring as described in Fig. 6. From (37).

Hormones and Sex Expression

There are two basic groups of flowers: they are either *perfect*, in which the flower contains both stamens and pistils, or they are *imperfect*, in which case either the pistils or stamens are present. A *monoecious* plant has both the staminate and the pistilate flowers on the same plant (e.g., *Cucumis sativus* and *Zea mays*) while in *dioecious* plants, the male and female flowers are on separate plants (e.g., *Spinacia* and *Cannibus sativa*). The sex of imperfect flowers has a genetic basis, but environmental factors such as photoperiod, temperature, and nitrogen status have an influential role (32,58). Exogenous hormones can can modify the sexuality of flowers suggesting that hormones mediate genetic and environmental control of sex expression (9). The role of hormones in sex expression of three species is discussed below.

The role of hormones in sex expression is perhaps best understood in cucurbits. In *Cucumis*, perfect flowers are initiated but one sex organ fails to develop. Application of auxins to flower buds at the bisexual stage leads to the formation of female flowers, while exogenous GAs results in male flower formation (37,52). It now appears that IAA acts through ethylene (59). Treatments which reduce endogenous ethylene levels promote maleness (7). Inhibitors of GA biosynthesis, on the other hand, cause a tendency towards feminization (28,52).

Endogenous hormone levels are also consistent with their postulated role in sex expression. Shoots of a hermaphroditic line of *Cucumis* had a higher auxin content than an andromonoecious line (22). Endogenous levels of ethylene and GA-like substances were also correlated with sex expression: high levels of ethylene production were associated with plants containing pistilate flowers (7,56) and staminate plants contained more GA-like substances than their female counterparts (57). In total, the results strongly suggest that sex expression in *Cucumis* and cucurbits in general is regulated by the internal balance of auxins acting through ethylene and GAs.

Nevertheless, the hormonal balance may not be the sole factor in determining sex expression in *Cucumis*. In a *C. sativus* cultivar in which the number of pistillate flowers increases in response to SD, the endogenous levels of GA-like substances were higher and ethylene production lower in SD than LD, the opposite to what one would predict (62). This suggests that environmental control of sex expression in this species is not necessarily mediated through a balance of GAs and ethylene.

Cannabis sativa is a dioecious plant in which, like *Cucumis*, flower primorida are uncommitted at the time of flower initiation. Application of auxins, ethylene, and GAs affect sex expression in similar fashion as in *Cucumis* (9,32,52). Growth retardants feminize *Cannabis* plants, which can be reversed by GA3 (9). In addition, cytokinins also promote femaleness (9). Following experiments in which plants were defoliated or derooted, it was proposed that leaves play an essential role in sex expression in *Cannabis* by supplying GAs to the flower bud (9). Male plants tend to have higher levels of GA-like substances as well. An increased cytokinin content was associated with female plants (9).

In *Spinacia*, sex expression is under genetic control. There are two sex chromosomes designated as X and Y. Female plants have a genotype of XX and male plants are XY. Photoperiod also influences sex expression. LD increases a tendency towards femaleness despite a genotype of XY. Again, as in many other species, auxins, ethylene, and cytokinins promote female flowers while treatment with GA_3 increases the tendency for the formation of male flowers (9,12).

The qualitative pattern of GA-like substances was different in extracts from male and female plants. Female plants contained high amounts of a GA-like substance that was probably GA_{19} and low amounts of a GA-like substance that was probably GA_{20}. Male plants showed the opposite pattern (12). Since GA_{20} is the postulated GA responsible for biological activity in *Spinacia* (47), the increased level of GA_{20} in male plants is consistent with a role of GAs in sex expression. However, the trend towards feminization observed when plants are subjected to LD is obviously inconsistent with this hypothesis. In addition, male plants bolt sooner in LD than female plants, but both types flower at about the same time. This might be an indication that the higher levels of GA_{20} in male plants are a reflection of more rapid stem elongation and may not have anything to do with sex expression.

FINAL COMMENTS

In countless experiments, plant hormones have been applied to plants in a pharmacological approach to determine the basis for the control of the transition from a vegetative to the reproductive state. Needless to say, the fruits of thousands of hours of labor do not provide us with any certain mechanisms. Perhaps the overriding impediment is the paucity of information about the floral stimulus. Although some success in isolating active fractions has been reported (43,44) the results have not been reproduced in other laboratories (75). Until we know the nature of the floral stimulus—whether it is a unique morphogen or a combination of many factors—it is unlikely that much will be learned about the internal mechanisms controlling reproductive development.

Why has it been so difficult to isolate and characterize the floral stimulus? Fundamental to this problem is that there is no suitable assay. Without such an assay the difficulties simply compound themselves into a vicious circle: how can the presence of an active principle be ascertained if there is no sure way of detecting it? On the other hand, perhaps the lack of an assay is more apparant than real because the active stimulus was not present in any of the fractions that were applied. It should be noted that the majority of extraction experiments assumed *a priori* that because many other plant hormones are acidic compounds soluble in organic solvents, the floral stimulus has similar chemical properties. But what if the floral stimulus were a peptide, carbohydrate, or a volatile lactone? Certainly, these possibilities were precluded by the extraction methods employed in the earlier work. Other uncertainties are lability and method of application to assay plants. While this bewildering maze of obstacles may cause most experimenters to shrug their shoulders and move onto other areas, it would be wise to bear in mind the comment by Zeevaart (76) that with the exception of ABA none of the other hormones were discovered by extracting higher plant tissue.

Flowering is almost certainly the result of selective gene expression—both turning on and switching off of genes. In a number of species, mutants for various aspects of reproductive development have been isolated, (e.g., 53,67). Use of recently developed techniques in molecular biology to analyze the structure and function of genes could be quite useful in determining the molecular basis for these mutations. Already this kind of approach has allowed advances in the understanding how genes control, on a molecular level, development in insects and vertebrates (6). While this avenue would not directly lead us to the identification of the floral stimulus, it certainly would tell us the molecular details of what the floral stimulus does. Such a vantage point could provide the insights necessary for a breakthrough.

References

1. Bernier, G., Bodson, M., Kinet, J-M., Jacqmard, A., Havelange, A. (1974) The nature of the floral stimulus in mustard. *In* Plant growth substances 1973, pp. 980-986. Hirokawa, Tokyo.
2. Bernier, G., Kinet, J-M., Claes, A. (1979) The role of cytokinin in floral evocation in *Sinapis alba*. Tenth International Conference on Plant Growth Substances, Madison, Wisconsin, pg. 30 (abstract).
3. Bernier, G., Kinet, J-M., Jacqmard, A., Havelange, A., Bodson, M. (1977) Cytokinin as a possible component of the floral stimulus in *Sinapis alba*. Plant Physiol. *60*, 282-285.
4. Bernier, G., Kinet, J-M., Sachs, R.M. (1981) The physiology of flowering, Vol. I, The intiation of flowers. CRC Press, Boca Raton.
5. Bernier, G., Kinet, J-M., Sachs, R.M. (1981) The physiology of flowering, Vol. II, Transition to reproductive growth. CRC Press, Boca Raton.
6. Brown, D.D. (1984) The role of stable complexes that repress and activate eucaryotic genes. Cell *37*, 359-365.
7. Byers, R.E., Baker, L.R., Sell, H.M., Herner, R.C., Dilley, D.R. (1972) Ethylene: A natural regulator of sex expression in *Cucumis melo* L. Proc. Nat. Acad. Sci. USA *69*, 717-720.
8. Chailakhyan, M.Kh. (1936) On the hormonal theory of plant development. Dokl. Acad. Sci. USSR *12*, 443-447 (In Russian).

9. Chailakhyan, M.Kh., Khryanin, V.N. (1980) Hormonal regulation of sex expression in plants. *In* Plant growth substances 1979, pp. 331-344, Skoog, F., ed. Springer-Verlag, New York.
10. Cleland, C.F., Zeevaart, J.A.D. (1970) Gibberellins in relation to flowering and stem elongation in the long day plant *Silene armeria*. Plant Physiol. *46*, 392-400.
11. Crozier, A., Durley, R.C. (1983) Modern methods of analysis of gibberellins. *In* The biochemistry and physiology of gibberellins, Vol. II, pp. 485-538, Crozier, A., ed. New York: Praeger.
12. Cúlafic, L., Neskovic, M. (1974) A study of auxins and gibberellins during shoot development in *Spinacia oleracea* L. Arh. Biol. Nauka. Beograd. *26*, 19-27.
13. DeProft, M., Van Dijck, R., Philippe, L., DeGreet, J.A. (1985) Hormonal regulation of flowering and apical dominance in bromeliad plants. Twelfth International Conference on Plant Growth Substances, Heidleberg, Fed. Rep. of Ger., pg. 93 (abstract).
14. Doorenbos, J. (1954) "Rejuvenation" of *Hedera helix* in graft combinations. Proc. Koninkl. Ned. Akad. Wetenschap., Ser. C *57*, 99-102.
15. Doorenbos, J. (1965) Juvenile and adult phases in woody plants. In: Encyclopedia plant physiol, Vol. XV/1, pp. 1222-1235, Ruhland, W., ed. Springer, Berlin.
16. El-Antably, H.M.M., Wareing, P.F. (1966) Stimulation of flowering in certain short-day plants by abscisin. Nature *210*, 328.
17. El-Antably, H.M.M., Wareing, P.F., Hillman, J. (1967) Some physiological responses to d.l. abscisin (dormin). Planta *73*, 74-90.
18. Evans, L.T. (1966) Abscisin. II. Inhibitory effect on flower induction in a long-day plant. Science *151*, 107.
19. Frank, H., Renner, O. (1956) Uber verjungung bei *Hedera helix* L. Planta *47*: 105-14.
20. Frydman, V.M., Wareing, P.F. (1973) Phase change in *Hedera helix* L. I. Gibberellin-like substances in the two growth stages. J. Exp. Bot. *24*, 1131-1138.
21. Frydman, V.M., Wareing, P.F. (1974) Phase change in *Hedera helix* L. III. The effects of gibberellins, abscisic acid and growth retardants on juvenile and adult ivy. J. Exp. Bot. *25*, 420-429.
22. Galun, E., Izhsar, S., Atsmon, D. (1965) Determination of relative auxin content in hermaphrodite and adromonoecious *Cucumis sativus* L. Plant Physiol. *40*,, 321-326.
23. Gaskin, P., MacMillan, J., Zeevaart, J.A.D. (1973) Identification of gibberellin A$_{20}$, abscisic acid, and phaseic acid from flowering *Bryophyllum daigremontianum* by combined gas chromatography-mass spesctrometry. Planta *111*, 347-352.
24. Gianfagna, T., Zeevaart, J.A.D., Lusk, W.J. (1983) The effect of photoperiod on the metabolism of deuterium-labeled GA$_{53}$ in spinach. Plant Physiol. *72*, 86-89.
25. Gilmour, S.J., Zeevaart, J.A.D., Schwenen, L., Graebe, J.E. (1985) The effect of photoperiod on gibberellin metabolism in cell-free extracts from spinach. Plant Physiol. *S77*, 92.
26. Goodwin, P.B. (1978) Phytohormones and fruit growth. *In* Phytohormones and related compounds: A comprehensive treatise, Vol. II, pp. 175-249, Letham, D.S., Goodwin, P.B., Higgins, T.J.V. eds. Elsevier, Amsterdam.
27. Guttridge, C.G. (1959) Evidence for a flower inhibitor and vegetative growth promoter in the strawberry. Ann. Bot. *23*, 351-360.
28. Halevy, A.H., Rudich, J. (1967) Modification of sex expression in muskmelon by treatment with the growth retardant B-995. Physiol. Plant. *20*, 1052-1058.
29. Harada, H. (1962) Etude des substances naturelles de croissance en relation avec la floraison-isolement d'une substance de montaison. Rev. Gén. Bot. *69*, 201-297.
30. Hedden, P., MacMillan, J., Phinney, B.O. (1978) The metabolism of the gibberellins. Annu. Rev. Plant. Physiol. *29*, 149-192.
31. Henson, I.E., Wareing, P.F. (1977) Cytokinins in *Xanthium strumarium* L.: Some aspects of the photoperiodic control of endogenous levels. New Phytol. *78*, 35-45.

32. Heslop-Harrison, J. (1972) Sexuality of angiosperms. In Plant Physiology: A treatise. Vol. VIC, Physiology of development: From seeds to sexuality, pp. 133-290, Steward, F.C., ed. Academic Press: New York.

33. Hiller, L.K., Kelly, W.C., Powell, L.E. (1979) Temperature interactions with growth regulators and endogenous gibberellin-like activity during seedstalk elongation in carrots. Plant Physiol. 63, 1055-1061.

34. Jacques, M. (1969) Les différents aspects morphologiques florcusion chez les *Blitum capitatum* et *virgatum* en rapport avec les modalités des processus dinduction floral. C.R. Acad. Sci. 268D, 1045-1047.

35. Jacques, M. (1973) Transfert par voie de greffage du stimulus photoperiodique. C.R. Acad. Sci. 276D, 1705-1708.

36. Jones, M.G., Zeevaart, J.A.D. (1980) Gibberellins and the photoperiodic control of stem elongation in the long-day plant, *Agrostemma githago* L. Planta 149, 269-273.

37. Jones, M.G., Zeevaart, J.A.D. (1980) The effect of photoperiod on the levels of seven endogenous gibberellins in the long-day plant *Agrostemma githago* L. Planta 149, 274-279.

38. King, R.W., Zeevaart, J.A.D. (1973) Floral stimulus movement in *Perilla* and flower inhibition caused by noninduced leaves. Plant Physiol. 51, 727-738.

39. King, R.W., Evans, L.T., Firn, R.D. (1977) Abscisic acid and xanthoxin contents in the long-day plant *Lolium temulentum* L. in relation to photoperiod. Aust. J. Plant Physiol. 4, 217-223.

40. Knott, J.E. (1934) Effect of a localized photoperiod on spinach. Proc. Amer. Soc. Hortic. Sci. Suppl. 31, 152-154.

41. Lang, A. (1965) Physiology of flower initiation. In Encyclopedia of plant physiology, pp. 1380-1536, Ruhland, W., ed. Springer-Verlag, Berlin.

42. Lang, A., Chailakhyan, M.Kh., Frolova, I.A. (1977) Promotion and inhibition of flower formation in a dayneutral plant in grafts with a short-day plant and a long-day plant. Proc. Natl. Acad. Sci. USA 74, 2412-2416.

43. Lincoln, R.G., Cunningham, A., Hamner, K.C. (1964) Evidence for a florigenic acid. Nature 202, 559-561.

44. Lincoln, R.G., Mayfield, D.L., Cunningham, A. (1961) Preparation of a floral initiating extract from *Xanthium*. Science 133, 756.

45. Metzger, J.D., Zeevaart, J.A.D. (1980) Identification of six endogenous gibberellins in spinach shoots. Plant Physiol. 65, 623-626.

46. Metzger, J.D., Zeevaart, J.A.D. (1980) Comparison of the levels of six endogenous gibberellins in roots and shoots of spinach in relation to photoperiod. Plant Physiol. 66, 679-683.

47. Metzger, J.D., Zeevaart, J.A.D. (1980) Effect of photoperiod on the levels of endogenous gibberellins in spinach as measured by combined gas chromatography-selected ion current monitoring. Plant Physiol. 66, 844-846.

48. Metzger, J.D., (1985) Role of gibberellins in the environmental control of stem growth in *Thlaspi arvense* L. Plant Physiol 78, 8-13.

49. Nitsch, J.P. (1965) Physiology of flower and fruit development. In Encyclopedia of plant physiology, Vol. 15, Part 1, pp. 1537-1647, Ruhland, W., ed. Springer-Verlag, Berlin.

50. Ogawa, Y., King, R.W. (1979) Indirect action of benzyladenine and other chemicals on flowering of *Pharbitis nil* Chois. Plant Physiol. 63, 643-649.

51. Pharis, R.P., Morf, W. (1968) Physiology of gibberellin induced flowering in conifers. In Biochemistry and physiology of plant growth substances, pp. 1341-1356, Wightman, F., Setterfield, G., eds. Runge Press, Ottawa, Canada.

52. Pharis, R.P., King, R.W. (1985) Gibberellins and reproductive development in seed plants. Annu. Rev. Plant Physiol 36, 517-568.

53. Pierik, R.L.M. (1967) Regeneration, vernalization and flowering in *Lunaria annua* L. in vivo and in vitro. Meded. Landbouwhogescn. Wageningen 67, 1-71.

54. Robinson, L.W., Wareing, P.F. (1969) Experiments on the juvenile-adult phase change in some woody species. New Phytol. *68*, 67-78.
55. Rogler, C.E., Hackett, W.P. (1975) Phase change in *Hedera helix*. Stabilization of the mature form with abscisic acid. Physiol. Plant. *34*, 148-152.
56. Rudich, J., Halevy, A.H., Kedar, N. (1972) Ethylene evolution from cucumber plants as related to sex expression. Plant Physiol. *49*, 998-999.
57. Rudich, J., Halevy, A.H., Kedar, N. (1972) The level of phytohormones in monoecious and gynoecious cucumbers as affected by photoperiod and ethephon. Plant Physiol. *50*, 585-590.
58. Schwabe, W.W. (1971) Physiology of vegetative reproduction and flowering. *In* Plant physiology: A treatise. Vol. VIA, Physiology of development: Plants and their reproduction, pp. 233-411, Steward, F.C., ed. Academic Press, New York.
59. Shannon, S., De LaGuardia, M.D. (1969) Sex expression on the production of ethylene induced by auxin in the cucumber (*Cucumis sativum* L.). Nature *223*, 186.
60. Suge, H. (1980) Inhibition of flowering and growth in *Pharbitis nil* by ancymidol. Plant Cell Physiol. *21*, 1187-1192.
61. Suge, H., Rappaport, L. (1968) Role of gibberellins in stem elongation and flowering in radish. Plant Physiol. *43*, 1208-1214.
62. Takahashi, H., Saito, T., Suge, H. (1983) Separation of the effects of photoperiod and hormones on sex expression in cucumber. Plant and Cell Physiol. *24*, 147-154.
63. Van de Pol, P.A. (1972) Floral induction, floral hormones and flowering. Meded. Landbouwhogesch. Wageningen *72*, 1-89.
64. Vince-Prue, D. (1975) Photoperiodism in plants. McGraw Hill and Co., London.
65. Vince-Prue, D. (1983) Photomorphogenesis and flowering. *In* Encyclopedia of plant physiology (N.S.), Vol. 16b, pp. 457-490, Shropshire Jr., W., Mohr, H., eds. Springer-Verlag, Berlin.
66. Wellensiek, S.J. (1970) The floral hormones in *Silene armeria* L. and *Xanthium strumarium* L. Z. Pflanzenphysiol. *63*, 25-30.
67. Wellensiek, S.J. (1973) Genetics and flower formation of annual *Lunaria*. Neth. J. Agric. Sci. *21*, 163-166.
68. Zeevaart, J.A.D. (1968) Flower formation as studied by grafting. Meded. Landbouwhogesch. Wageningen *58*, 1-88.
69. Zeevaart, J.A.D. (1964) Effects of the growth retardant CCC on floral initiation and growth in *Pharbitis nil*. Plant Physiol. *39*, 402-408.
70. Zeevaart, J.A.D. (1968) Vernalization and gibberellin in *Lunaria annua* L. *In* Biochemistry and physiology of plant growth substances, pp. 1357-1370, Wightman, F., Setterfield, P., eds. Runge Press, Ottawa.
71. Zeevaart, J.A.D. (1971) Effects of photoperiod on growth rate and endogenous gibberellins in the long-day rosette plant spinach. Plant Physiol. *47*, 821-827.
72. Zeevaart, J.A.D. (1969) *Bryophyllum*. *In* The induction of flowering: Some case histories, pp. 435-456, Evans, L.T., ed. Cornell Univ. Press, Ithaca, New York.
73. Zeevaart, J.A.D. (1969) The leaf as a site of gibberellin action in flower formation in *Bryophyllum daigremontianum*. Planta *84*, 339-347.
74. Zeevaart, J.A.D. (1971) (+)-Abscisic acid content of spinach in relation to photoperiod and water stress. Plant Physiol. *49*, 86-90
75. Zeevaart, J.A.D. (1973) Gibberellin GA_{20} content of *Bryophyllum daigremontianum* under different photoperiodic conditions as determined by gas-liquid chromatography. Planta *114*, 285-288.
76. Zeevaart, J.A.D. (1976) Physiology of flower formation. Annu. Rev. Plant. Physiol. *27*, 321-348.
77. Zeevaart, J.A.D. (1978) Phytohormones and flower formation. *In* Plant hormones and related compounds, Vol. II, pp. 291-327, Letham, D.S., Goodwin, P.B., Higgins, T.J.V., eds. Elsevier/North Holland, Amsterdam.

78. Zeevaart, J.A.D. (1982) Transmission of the floral stimulus from a long-short-day plant, *Bryophyllum daigremontianum*, to the short-long-day plant *Echeveria harmsii*. Ann. Bot. *49*, 549-552.
79. Zeevaart, J.A.D. (1983) Gibberellins and flowering. *In* The biochemistry and physiology of gibberellins, Vol. *2*, pp. 333-374, Crozier, A., ed. Praeger, New York.
80. Zeevaart, J.A.D., Brede, J.M., Cetas, C.B. (1977) Translocation patterns in *Xanthium* in relation to long day inhibition of flowering. Plant Physiol. *60*, 747-753.
81. Zeevaart, J.A.D., Lang, A. (1962) The relationship between gibberellin and floral stimulus in *Bryophyllum daigremontianum*. Planta *53*, 531-542.
82. Zeevaart, J.A.D., Lang, A. (1963) Suppression of floral induction in *Bryophyllum daigremontianum* by a growth retardant. Planta *59*, 509-517.
83. Zimmerman, R.H., Hackett, W.P., Pharis, R.P. (1985) Hormonal aspects of phase change and precocious flowering. *In* Encyclopedia of plant physiology (NS), Vol. II, pp. 79-115, Pharis, R.P., Reid, D.M., ed. Springer-Verlag, New York.

E 11. Hormones and Heterosis in Plants

Stewart B. Rood[1] and Richard P. Pharis[2]

[1]Department of Biological Sciences, University of Lethbridge, Alberta T1K 3M4, Canada.
[2]Plant Physiology Research Group, Department of Biology, University of Calgary, Alberta T2N 1N4, Canada.

INTRODUCTION

Heterosis, or hybrid vigor, occurs when hybrid performance exceeds that of either parent. Heterosis occurs in animals and is particularly common in plants. It is probably important in evolution since heterozygote superiority may ensure the maintenance of genetic polymorphism in natural populations. Further, heterosis is extremely important to agriculture and horticulture and is of increasing importance in forestry. The dramatic yield improvements in cereal crops in the twentieth century have been, in large part, due to the utilization of heterosis. Inbreeding depression refers to the reduction in phenotypic vigor following self-fertilization and is the complementary process to heterosis.

Although heterosis has been unknowingly utilized to improve crop productivity for centuries, the recognition of this phenomenon was not formalized prior to Shull's (36) work in the beginning of the twentieth century. Shull coined the term "heterosis", a truncated version of "heterozygosis" and was able to apply this process in a more predictable fashion to maize (*Zea mays* L.) breeding. Subsequent to Shull's (36) work, maize has remained the model system (38) in the study not only of the genetic basis for heterosis, but also in its physiological basis. Maize is attractive in this regard since: (i) it is diploid rather than polyploid, (ii) the spatial separation of the male (tassel) and female (ear) inflorescences simplifies controlled pollinations, (iii) it is an economically important crop, and most importantly, (iv) maize hybrids generally display considerable heterosis. However, the heterosis phenomenon has also been studied in a range of other economically important plants including wheat (43), barley (26), tomato (45), other vegetable crops (22), forages (14), ornamentals (27), and both coniferous and angiospermous forest trees (15).

New techniques for the production of plant haploids, such as anther culture and the use of mentor pollen (40) may simplify the production of homozygotes and consequently, the controlled production of hybrids may accelerate. This should accelerate research into heterosis in economically important plants.

Heterosis has had three relatively recent reviews (8, 9, 37). As outlined therein, two principal theories have been proposed to explain the genetic basis of heterosis. Complementation, the first theory, describes the interaction between a number of genes to produce the heterotic phenotype. Each gene independently displays incomplete or complete dominance but the resultant polygenic character may display heterosis. For example, hybrid vigor for increased total leaf area in field beans is the result of dominance for increased leaflet number and incomplete dominance for increased leaflet size (7). Thus, the observed heterosis is simply the result of the complementation of genes which separately display simple Mendielian inheritance. Complementation may involve not only genes coded by a given genome but may also involve intergenomic complementation of genes in mitochondria and chloroplasts as well as the nucleus (39).

The second theory for heterosis is overdominance *per se*. The heterozygous condition for a given gene is proposed to confer a superior phenotypic potential of the individual. At the molecular level, overdominance may involve enzymic polymorphism (35). For example, parental maize inbreds have been recognized to contain two different subunits of alcohol dehydrogenase (35). One subunit form is stable but relatively inactive whereas the other form is unstable but very active. A hybrid may contain the two different subunits and consequently will contain a potent and stable type of enzyme. This type of enzymic polymorphism has been shown for a number of enzyme systems, particularly in maize.

HORMONES AND HETEROSIS

The involvement of hormones (e.g., plant growth substances) in heterosis in plants is probable. Heterosis involves almost all apsects of plant growth and development and hormones are intimately involved in, and probably control many of these processes. To demonstrate the involvement in a regulatory manner, of a given hormone, three approaches may be utilized. The exogenous application of the hormone may be expected to influence the growth or developmental process differentially in hybrids and in the inbreds. In particular, if the inbreeding depression is due to a hormone deficiency, the exogenous hormone application should, in part, cure the lesion of inbreeding depression. Conversely, the hybrid may be expected to respond proportionally less to the hormone (e.g., there is no deficiency, and indeed the exogenous plus endogenous levels of hormone may be supra-optimal). Second, a blockage of hormone biosynthesis or transport should influence the growth or developmental process in the opposite direction to that obtained by the exogenous application of the hormone. Third, and perhaps most importantly, there should be a positive correlation in the endogenous levels of the hormone and the growth or developmental processin question. For example, if hormones act in a regulatory manner in hybrid vigor, then hybrids displaying heterosis should contain enhanced levels of promotory hormones, relative to their parental inbreds. However, the establishment of correlation may not, by itself, prove a causal involvement. Further, a simple arithmetic ratio between hormone concentration and growth rate may be misleading. For example, a hybrid which is growing twice as fast as either of its inbred parents may actually be expected to contain many times the endogenous hormone level. Logarithmic relationships between applied hormone concentration and plant growth or development are well established in hormone bioassay systems.

In 1940, Robbins (28) showed that a heterotic maize hybrid WF9 x 38-11 contained higher levels of substances which promoted the growth of the fungus *Phycomyces* than did its parental inbreds. The growth substances were thought to be hormone- and vitamin-like. Similarly, heterotic maize hybrid seedlings have been shown to contain higher levels of substances which promoted the growth of bakers yeast *Saccharomyces cerevisiae* (17). Interestingly, a mixture of extracts from the two parental inbreds was similar in yeast bioassay potency to an extract from the heterotic hybrid, although each alone was less active (17).

Gibberellins

Of the known hormone groups, the involvement of gibberellins (GAs) in the regulation of heterosis in maize appears most likely. Phinney (personal communication) notes that d_5 dwarfs produced from a maize inbred as female parent are noticeably more dwarfed than d_5 seedlings produced from a heterotic genotype. Maize inbreds are more responsive to the exogenous application of GA_3 than hybrids and further, the sensitivity to GA_3 is correlated with the degree of inbreeding (18, 19). This interesting correlation led to further investigations into the possible role of GAs in the regulation of heterosis in maize. The greater responsivity of inbreds to the exogenous application of GA_3 or GA_1 was confirmed for a group of related Canadian maize genotypes (31). Although the exogenous application of GA_3 generally induces a precocious shoot elongation, and subsequently growth is slowed, carefully timed and regulated doses of GA_3 can dramatically accelerate the growth rate of maize. This acceleration is particularly observed for maize inbreds (Fig. 1). This observation that the inbreeding depression can be partly overcome through the exogenous application of GA_3 suggests that an endogenous GA 'deficiency' may underly the inbreeding depression.Using the GA-specific Tan-ginbozu dwarf-rice assay, a heterotic maize hybrid was

Fig. 1. Two maize inbreds (left and right) and their F_1 hybrid (center) 14 days after the application of 0.5 mg gibberellin A_3 (GA) per plant or a control solution (31).

indeed found to contain higher levels of endogenous GA-like substances (for the cone of apical meristematic tissue) (33) (Table 1). Further, under environmentally unfavorable conditions (low temperature) in which heterosis was not displayed (same genotypes), the hybrid and vigorous inbred contained approximately equal levels of endogenous GA-like substances (32). Consequently, potence ratios, which are a quantitative measure of the degree of dominance, are similarly reduced for growth rates and level of GA-like substances by lowering the temperature. Thus, a good correlation between heterosis for growth rate and heterosis for endogenous GA-like substances has been established.

Many of the commonly used inhibitors of GA biosynthesis may also inhibit synthesis of a variety of other substances, most notably sterols (5). Consequently, the use of these GA synthesis inhibitors as an experimental tool in growth studies can be questioned. In maize, however, a number of single gene dwarf mutants exist in which specific points in the GA biosynthetic pathway are inhibited (24, 25). These GA deficient mutants are characterized by a dwarf phenotype, and thus, the consequence of a block in GA biosynthesis is a reduction of growth rate.

Collectively, evidence from: (i) exogenous GA application, (ii) reduction of endogenous GA level (via dwarfing genes), and (iii) evaluation of endogenous GA level in a hybrid and its parental inbreds are consistent in suggesting a role for GAs as a regulatory agent in the

Table 1. Shoot dry weights and levels of GA-like substances in shoot apical meristem cylinders of two maize inbreds and their F_1 hybrid grown under 25/20°C (day/night) temperature conditions (33).

	Days from emergence					
	15 days		*21 days*		*28 days*	
	D. wt (g)	*GA-like activity (μg GA_3 equiv./kg tissue)*	*D. wt (g)*	*GA-like activity*	*D. wt (g)*	*GA-like activity*
Inbred						
CM7	0.30a*	78a	0.72a	6.2a	2.16a	19a
CM49	0.37b	100a	0.85b	8.2a	2.28a	20a
Hybrid						
CM7 X CM49	0.58c	109a	1.97c	22.9b	3.84b	29b

* Within a column, values followed by the same letter do not differ ($P<0.05$).

control of heterosis in maize. At least a part of the reason maize hybrids are vigorous and high yielding may thus be due to increased production of GAs, relative to their inbred parents.

Jacobs (16) has outlined a series of criteria useful in the establishment of a regulatory role for a hormone, namely the PESIGS rules: Parallel variation, Excision, Substitution, Isolation, Generality, and Specificity. These can be applied to test the hypothesis that GAs are involved in the regulation of heterosis in maize. However, some of the criteria are better suited to evaluating the more fundamental hypothesis that GAs control growth rate in maize. Gibberellins are known to be native to maize (10) and, as already outlined, there is parallel variation between heterosis for endogenous level of GA-like substances and growth rate in maize (32, 33). Excision, or removal of the substance is best achieved using the dwarfing genes which block GA biosynthesis (24, 25). Reduced growth accompanies the reduction in endogenous GA level, indicating that GAs probably control growth rate and hence, satisfying in part, Jacobs' (11) excision criterion. Gibberellin A_3 restores normal growth in the dwarf mutants (24, 25) satisfying the substitution criterion. The hormonal response and status of specific cells, tissues or organs has not been adequately tested and hence, the isolation criterion of Jacobs (11) cannot be evaluated yet. Future *in vitro* studies may clarify this aspect.

The hypothesis that GAs control growth rate, and hence, are involved in heterosis is probably generalizable to at least certain other plant systems. For example, low GA level or blockage of C-3 hydroxylation and reduced growth are observed in mutants of rice, barley and peas (see 25). Other less direct correlations between GAs and growth rate have also been reported (see 23). The differential response of inbreds versus hybrids to exogenous GA$_3$ in maize is similar to a differential response of black spruce genotypes (42). Whereas, inbred (selfed) spruce respond to exogenous application of GA$_{4/7}$ with increased height growth and dry matter production, outcrossed genotypes (hybrids) are not very responsive (Table 2). Very rapidly growing outcrossed genotypes are even less responsive (Table 2). The higher level of endogenous GA-like substances in a hybrid versus parental inbreds of maize is probably also generalizable to poplar (30). Fast-growing interspecific poplar hybrids contain higher levels of endogenous GA-like substances than do their slower-growing parental clones (Table 3). Thus, preliminary data indicates that GA involvement in heterosis is generalizable to at least some other plants. Although further research is required to satisfy the PESIGS rules, early results have satisfied certain of Jacobs (11) criteria and hence, support a role for GAs in the regulation of growth rate in general, and heterosis, in particular.

Other research also suggests a role for GA in the regulation of heterosis in maize. Parental pollen tube growth rates of germinating maize are correlated with seedling dry weight, ear weight and grain yield of the resultant hybrids (20). Pollen tube growth is promoted through the exogenous application of GA$_3$ (4, 13), and increases in endogenous GA of untreated pollen is positively correlated with increased pollen germination and rapid pollen tube growth (2, 12). Thus, the correlation between pollen tube growth rate and hybrid vigor of the progeny also appears interrelated with GA physiology.

Table 2. Height, stem dry weight and total seedling dry weights of control and gibberellin A$_{4/7}$-treated[a] black spruce seedlings (adapted from 42).

Family	Height (cm) (age 5 months)		Stem D. wt (g) (age 6 months)		Shoot D. wt (g) (age 6 months)	
	Control	GA$_{4/7}$	Control	GA$_{4/7}$	Control	GA$_{4/7}$
Outcrosses						
59 × 62	15.4	15.5	0.32	0.27	1.09	0.88
52 × 62	15.4	16.1	0.26	0.23	0.87	0.77
62 × 29	14.6	15.6	0.32	0.21	1.14	0.71
59 × 63	14.2	13.1	0.23	0.21	0.85	0.74
52 × 59	13.9	14.3	0.26	0.22	1.00	0.76
63 × 62	13.6	15.1**	0.20	0.22	0.69	0.75
63 × 52	12.7	14.5**	0.18	0.16	0.67	0.55
Selfed						
63 × 63	10.8	12.0**	0.06	0.09	0.23	0.26
65 × 65	8.7	12.3**	0.08	0.16	0.30	0.54*

* Significantly different from corresponding control value (P<0.05).
** Four slowest growers were significantly different from corresponding control values when grouped prior to comparison (P<0.05).
a root drench, 50 ml, 200 mg/l, once weekly from age 3 months.

Table 3. Levels of GA-like substances in bark scrapings from stem internodes of poplar as determined by the dwarf-rice bioassay.

	clone	GA-like activity ($\mu g\ GA_3\ equiv./g\ d.\ wt.\ tissue$)
Populus deltoides		
	D 36	0.47 ± 0.23[a]
	D 56	0.61 ± 0.27
P. deltoides × *P. nigra* (rapid growing interspecific hybrids)		
	DN 138[b]	0.79 ± 0.34
	DN 103	1.23 ± 0.55
	DN 128	2.32 ± 0.42
	DN 160	5.66 ± 1.89

[a] Mean ± s.e. Values presented are total GA-like activities of all chromatographic regions.
[b] D 36 was a parent of DN 103 and DN 160 while D 56 was a parent of the other hybrids.

In an early review, Paleg (21) suggested that GAs play a regulatory role in endosperm mobilization and that this mobilization is probably correlated with the expression of heterosis. Although hybrid vigor cannot simply be attributed to an initial advantage of larger hybrid embryos (1), there is evidence (44), suggesting that early maize seedling performance is positively correlated with subsequent growth rate. Previously, germinants and seedlings of a heterotic maize hybrid were shown to produce higher levels of α-amylase than parental inbreds (34). The regulation of α-amylase in cereal grains is one of the best understood examples of a plant hormone-mediated regulation of gene expression (see Chapter C3). The increased production of α-amylase by the hybrid seedling is consistent with an increased level of endogenous GAs, again implicating GAs in the regulation of heterosis in maize (34).

The greater level of endogenous GAs in maize hybrids might involve either increased rates of biosynthesis or decreased rates of catabolism. Two studies were consistent (29, 31) in showing that maize hybrids oxidatively metabolize [^{3}H]GA_{20} and [^{3}H]GA_1 [both are native in maize (10), and GA_{20} is metabolized to GA_1 (25)] more rapidly than do their inbred parents (Fig. 2). Thus, metabolism in the hybrid is not slower, but rather, is apparently more rapid. This suggests that biosynthesis and turnover of GAs is more rapid in maize hybrids. Interestingly correlations between growth rate and oxidative metabolism of GAs have been reported in other plant systems and in general it appears that rapid metabolism and rapid GA turnover is indicative of rapid growth, and in turn is correlated with high endogenous GA level (see 29).

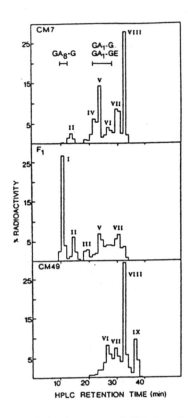

Fig. 2. Elution profiles of radioactivity from reverse-phase C_{18} HPLC loaded with the GA glucosyl-conjugate fraction of extracts from maize inbreds (top and bottom) and their F_1 hybrid (center) following a $[^3H]GA_{20}$ feed. In maize, GA_{20} is converted to GA_1 (and GA_{29}) and then GA_8. Authentic GA_{20}, GA_1, and GA_8 elute at 34, 25, and 13 minutes, respectively, and GA glucosyl conjugates elute coincidental with, or just prior to, the acidic GA. Thus, increasing oxidative metabolism is reflected by movement of radioactivity to the left in the chromatograms (31).

The increased GA level found in heterotic maize hybrids (33) might involve complementation and/or overdominance effects. Gibberellin biosynthesis is a multi-step pathway and consequently, polygenic. It is logical that the complementation of a number of the genes involved in the biosynthetic pathway and/or enzymic polymorphism of isozymes for specific genes in the pathway may contribute to the resultant increased production of GAs in the maize hybrid.

Other Hormones

The roles of other hormones and hence, Jacobs (11) specificity rule, have not been adequately investigated. It has been reported that maize hybrids contain higher levels of indole acetic acid (IAA) (41). However, the previously used methodology, UV absorption following thin layer chromatography, is suspect, and may or may not have represented IAA. The possible involvement of auxin in heterosis is worthy of further study. ABA does not appear to be involved (34), but neither rate of ethylene synthesis nor amounts of cytokinins have been adequately investigated.

Given the number of aspects of maize growth and development which display heterosis, it is unlikely that the regulation can be attributed to a single hormone. Rather, other hormones will probably be shown to be involved. These other hormones may be involved independently or in association with GAs. For example, exogenous application of GA_3 increases the level of IAA in peas (16). Whether it involves GAs and/or other hormones, the mechanism of increasing hormone production in a hybrid offers a potent means of gene amplification. A change in a number of genes which result in enhanced hormone production could in turn, act via the hormone induction of many other enzymes. Consequently, a metabolic cascade will result, influencing many aspects of growth and development.

SUMMARY

In summary, evidence from experiments involving the exogenous application of GA_3, the observed phenotypes of the GA deficient dwarf mutants, analysis of endogenous GA levels, and GA metabolism are consistent in indicating a role for GAs in the regulation of heterosis. The involvement of other hormones is possible, indeed probable, but remains to be experimentally verified. The economic importance of hybrid vigor justifies an increased research effect into this area. Finally, the observation that inbreds are probably depressed in part, due to a GA deficiency (defined very broadly), may provide a mechanism for improving seed yield from inbreds and consequently have an economic application in hybrid seed production.

References

1. Ashby, E. (1930) Studies in the inheritance of physiological characters. I. A physiological investigation of the nature of hybrid vigour in maize. Ann. Bot. *44*, 457-467.
2. Barendse, G.W.M., Rodrigues Pereira, A.S., Berkers, P.A., Dreissen, F.M., Van Eyden-Emons, A., Linskens, H.F. (1970) Growth hormones in pollen, styles and ovaries of Petunia hybrida and *Lolium* species. Acta Bot. Neerl. *19*, 175-186.
3. Berger, E. (1975) Heterosis and the maintenance of enzyme polymorphism. Am. Nat. *110*, 823-829.
4. Bose, N. (1959) Effect of gibberellin on the growth of pollen tubes. Nature *184*, 1577.
5. Coolbaugh, R.C. (1983) Early stages of gibberellin biosynthesis. *In* The biochemistry and physiology of gibberellins, Vol. 1, pp. 53-98. Crozier, A., ed. Praeger Publ., N.Y.
6. Donaldson, C., Blackman, G.E. (1974) The initiation of hybrid vigor in *Zea mays* during the germinating phase. Ann. Bot. *38*, 515-527.
7. Duarte, R. and Adams, M.W. (1963) Component interaction in relation to expression of a complex trait in a field bean cross. Crop Sci. *3*, 185-186.
8. Frankel, R. (1983) Heterosis: Reappraisal of Theory and Practice. Berlin: Springer-Verlag.
9. Gowens, J.W. (1952) Heterosis. Iowa State Coll. Pr., Ames, Iowa.
10. Hedden, P., Phinney, B.O., Heupel, R., Fujii, D., Cohen, H., Gaskin, P., MacMillan, J., Graebe, J.E. (1982) Hormones of young tassels of *Zea mays*. Phytochem. *21*, 390-393.

11. Jacobs, W.P. (1979) Plant Hormones and Plant Development. Cambridge Univ. Press, Cambridge, U.K. 339 pp.
12. Kamienska, A., Pharis, R.P. (1975) Endogenous gibberellins of pine pollen. II. Changes during germination of *Pinus attenuata, P. coulteri,* and *P. ponderosa* pollen. Plant Physiol. *56,* 655-659.
13. Kato, Y. (1955) Responses of plant cells to gibberellin. Bot. Gaz. *117,* 16-24.
14. Kobabe, G. (1983) Heterosis and hybrid seed production in fodder grass. In: Heterosis: Reappraisal of Theory and Practice, pp. 124-37, Frankel, R., ed. Berlin: Springer-Verlag.
15. Larson, C.C. (1956) Genetics in Forest Trees. (English transl. by M.L. Anderson). Essential Books Inc., Fair Lawn, N.J. 224 pp.
16. Law, D.M., Hamilton, R.H. (1984) Effects of gibberellic acid on endogenous indole-3-acetic acid and indoleacetyl aspartic acid levels in a dwarf pea. Plant Physiol. *75,* 255-256.
17. Matskov, F.F., Manzyuk, S.G. (1961) On the role of phsyiologically active substances of the phytohormone and vitamin types in heterosis phenomena in corn. Soviet Plant Physiology *8,* 65-70.
18. Nickerson, N.H. (1959) Sustained treatment with gibberellic acid of five different kinds of maize. Annals Missouri Botanical Gardens *46,* 19-37.
19. Nickerson, N.H., Embler, T.N. (1960) Studies involving the sustained treatment of maize with gibberellic acid I: Further notes on responses of races. Annals Missouri Botanical Gardens *47,* 227-242.
20. Ottaviano, E., Sari-Gorla, M., Mulcahy, D.L. (1980) Pollen tube growth rate in *Zea mays*: Implications for genetic improvement of crops. Science *210,* 437-438.
21. Paleg, L.G. (1965) Physiological effects of gibberellins. Annual Review Plant Physiol. *16,* 291-322.
22. Pearson, O.H. (1983) Heterosis in vegetable crops. *In* Heterosis: Reappraisal of Theory and Practice, pp. 138-88, Frankel, R., ed. Berlin: Springer-Verlag.
23. Pharis, R.P., King, R.W. (1985). Gibberellins and reproductive development in seed plants. Ann. Rev. Plant Physiol. *36,* 517-568.
24. Phinney, B.O. (1961) Dwarfing genes in *Zea mays* and their relation to the gibberellins. *In* Plant Growth Regulation, pp. 489-501, Klein, R.M., ed. Iowa State University Press, Ames.
25. Phinney, B.O. (1984) Gibberellin A_1, dwarfism and the control of shoot elongation in higher plants. *In* The Biosynthesis and Metabolism of Plant Hormones, pp. 17-41. Crozier, A., Hillman, J.R. eds. Cambridge Univ. Pr., Cambridge,
26. Ramage, R.T. (1983) Heterosis and seed production in barley. In: Heterosis: Reappraisal of Theory and Practice, pp. 71-93, Frankel, R., ed. Berlin: Springer-Verlag.
27. Reimann-Philipp, R. (1983) Heterosis in ornamentals. In: Heterosis: Reappraisal of Theory and Practice, pp. 234-259, Frankel, R., ed. Berlin: Springer-Verlag.
28. Robbins, W.J. (1940) Growth substances in a hybrid corn and its parents. Bulletin of the Torry Botanical Club *67,* 565-574.
29. Rood, S.B. (1986) Heterosis and the metabolism of [^{3}H] gibberellin A_1 in maize. Can. J. Bot. (in press).
30. Rood, S.B., Bate, N.J., Blake, T.J., Pharis, R. P. (1986) Heterosis and gibberellins in hybrid poplar. Plant Physiol. *80,* 116 (supp.).
31. Rood, S.B., Blake, T.J., Pharis, R.P. (1983) Gibberellins and heterosis in maize. II. Response to gibberellic acid and metabolism of [^{3}H]GA$_{20}$. Plant Physiol. *71,* 645-651.
32. Rood, S.B., Major, D.J., Pharis, R.P. (1985) Low temperature eliminates heterosis for growth and gibberellin content in maize. Crop Sci. *25,* 1063-1068.
33. Rood, S.B., Pharis, R.P., Koshioka, M., Major, D.J. (1983) Gibberellins and heterosis in maize. I. Endogenous gibberellin-like substances. Plant Physiol. *71,* 639-644.
34. Sarkissian, I.V., Kessinger, M.A., Harris, W. (1964) Differential rates of development of heterotic and nonheterotic young maize seedlings. I. Correlation of differential

morphological development with physiological differences in germinating seeds. Proc. Nat. Acad. Sci. *51*, 212-218.

35. Schwartz, D., Laugher, W.J. (1969) A molecular basis for heterosis. Science *166*, 626-627.

36. Shull, G.H. (1952) Beginnings of the heterosis concept. *In* Heterosis, pp. 14-48. Gowen, J.W., ed. Iowa State Coll. Pr., Ames, Iowa,

37. Sinha, S.K., Khanna, R. (1975) Physiological, biochemical and genetic basis of heterosis. Advances in Agronomy *27*, 123-174.

38. Sprague, G.F. (1983) Heterosis in maize: theory and practice. *In* Heterosis: Reappraisal of Theory and Practice, Frankel, R., ed. Berlin: Springer-Verlag.

39. Srivastava, H.K. (1983) Heterosis and intergenomic complementation: mitochondria, chloroplast, and nucleus. *In* Heterosis: Reappraisal of Theory and Practice, pp. 260-286, Frankel, R., ed. Berlin: Springer-Verlag.

40. Stettler, R.F. (1968) Irradiated Mentor Pollen: Its use in remote hybridization of Black Cottonwood. Nature *219*, 746-747.

41. Tafuri, F. (1966) IAA determination in the kernels of four lines of corn and their hybrids. Phytochemistry *5*, 999-1003.

42. Williams, D.J., Dancik, B.P., and Pharis, R.P. (1986) Early progeny testing and evaluation of controlled crosses of black spruce. Can. J. For. Res. (submitted).

43. Wilson, P., Driscoll, C.J. (1983) Hybrid wheat. *In* Heterosis: Reappraisal of Theory and Practice, pp. 94-123, Frankel, R., ed. Berlin: Springer-Verlag.

44. Woodstock, L.W. (1965) Initial respiration rates and subsequent growth in germinating corn seedlings. BioScience *15*, 783-784.

45. Yordanov, M. (1983) Heterosis in tomato. *In* Heterosis: Reappraisal of Theory and Practice, pp. 189-219, Frankel, R., ed. Berlin: Springer-Verlag.

E 12. The Role of Hormones in Photosynthate Partitioning and Seed Filling

Mark L. Brenner

Department of Horticultural Science and Landscape Architecture, University of Minnesota, St. Paul, Minnesota 55108, USA.

INTRODUCTION

The movement of photoassimilates from sites of synthesis in leaf tissue (source) to the sites of net accumulation in a different tissue (sink) potentially can be regulated at numerous points. Regulation of the net flow of photoassimilates is an integrated process. It is generally accepted that the concentration gradient of photoassimilates between the source and sink is the primary determinant of the current rate of transport and pattern of partitioning (14, 19, 60). However, close examination of the various components involved in the overall process of partitioning indicates that endogenous plant hormones may serve as modulators of many of the specific rate limiting components. This chapter will focus on the involvement of plant hormones as natural regulators of partitioning of photoassimilates especially to developing seeds.

THE PATHWAY OF PHOTOSYNTHATE PARTITIONING

In simplest terms, regulation of photosynthate partitioning can occur within the leaf, along the transport pathway, or within the seed. For clarity, each of these components will be discussed separately (Fig.1).

The extent of partitioning within the leaf may be controlled by the availability of recently fixed carbon which is determined by the rate of photosynthesis itself (19). The recently fixed carbon can first be partitioned to starch for storage within the chloroplast or to triose–phosphates (triose–P) available for export through the chloroplast envelope to the cytosol. Formation of sucrose from triose–phosphates involves both cytosolic fructose–1,6–bisphosphatase (FBPase) and sucrose–phosphate–synthase (SPS). The release of inorganic phosphate (Pi) from triose–P during sucrose synthesis stimulates triose–P export by the phosphate transporter (PT in Fig. 1) in the chloroplast membrane (51).

In many plants sucrose is the prime sugar exported from source tissue to sinks. Sucrose produced within mesophyll cells can be partitioned to

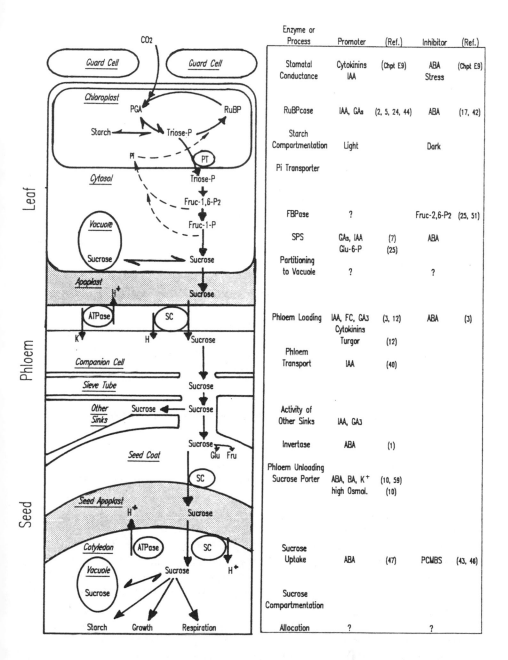

Fig. 1. Schematic representation of the path and possible control points of photosynthesis and partitioning of sucrose to developing seeds. See text for details including definition of abbreviations. SC = sucrose carrier.

either the vacuole for temporary storage (14) or released to the apoplast where it subsequently moves to the sieve elements and companion cells where it is loaded for transport in the phloem (19). Once in the phloem, sucrose is transported and partitioned to other parts of the plant such as roots, expanding leaves, and axillary buds (sinks) where it is either utilized for growth maintenance or storage. Sucrose import by developing seeds differs from that of other parts of the plant in that phloem transport terminates in the maternal tissue and photoassimilates must move apoplastically to the developing zygotic tissue where they are absorbed and utilized (see 54). The unloading of phloem for seed development occurs in the testa of dicotyledonous species while in monocotylenonous plants unloading occurs either in placenta–chalazal tissue of plants like maize (*Zea mays* L.) or in pericarp tissue of plants like wheat (*Triticum aestivum* L.), barley (*Hordeum vulgaris* L.) and rice (*Oryza sativa* L.).

CORRELATION OF ENDOGENOUS PLANT HORMONES WITH SOURCE–SINK RELATIONS

Hormone Content and Distribution in Developing Seeds

To consider hormonal regulation of source–sink relations, it is important to first understand where the hormones occur and if possible to understand their respective sites of origin. While the occurrence of plant hormones in specific tissues does not prove hormonal involvement in source–sink relations, patterns of occurrence of hormones in relation to changes in source–sink processes might indicate the involvement of hormones. In general, developing seeds have higher concentrations of plant hormones than all other plant parts. There are numerous reports of high concentrations of auxins, cytokinins, gibberellins (GAs) and abscisic acid (ABA) in seeds. As will be described below, several authors have reported that hormone content changes distinctly during seed and fruit development.

In early works, hormonal changes were measured only on a total seed basis (11, 35). With the possible exception of cytokinins, the maximum amount of hormones generally occurs during the time of rapid dry matter accumulation (rapid filling period). From later work, it has become evident that hormonal content may be vastly different in the various seed tissues. On a whole tissue basis, the greatest quantity of ABA is generally found in the prime storage tissue of seeds or grain such as cotyledons of soybeans (*Glycine max* L.[47]) and broad bean (*Vicia faba* L.[13]), or embryo and endosperm of maize (28). However, on a concentration basis, a measure presumably of greater physiological significance, ABA is found to be present in higher concentrations in the embryonic axis of soybean (20) and in the embryo of maize (28). ABA is also found in high concentrations in the tissues where phloem unloading occurs, in the seed coats of soybeans (20, 47) and broad bean (13). Likewise, in maize, ABA is found in high concentrations in the pedicel/placento-chalazal tissue (28), the tissue in which unloading occurs in this species. In general the highest concentrations of ABA found in these sites of unloading are observable during the rapid filling stage of seed development.

The occurrence of indoleacetic acid (IAA) in developing soybean seeds on a total seed basis is similar to that for ABA except the maximum amount of IAA is observed several days before the maximum for ABA. IAA is found in highest concentration in the embryonic axis at the time of maximum pod wall elongation, while in the seed coat, IAA reaches a distinct maximum coincident with maximum seed filling (20). In peas (*Pisum sativum*), high concentrations of auxin-like[1] materials are found in the liquid endosperm at the time of maximum pod elongation, and in the embryo (cotyledon and embryonic axis) a small increase in auxin-like material is observed coincident with rapid filling of the seed (16).

Similar to the occurrence of auxin-like material, gibberellin-like material is highest in liquid endosperm of pea at the time of rapid pod elongation (16). In whole pea seeds, GAs (GA_9, GA_{17}, GA_{20}, GA_{29}, and GA_{29}-catabolite) occur maximally during the period of rapid seed growth compared to very low quantities during early seed development or seed maturation (50). In barley, the levels of GA-like material within the grain parallel the pattern for dry weight increase of the grain, with the highest quantities recoverable from the endosperm (36).

In maize and peas, cytokinins are found in highest concentrations in the liquid endosperm of developing seeds (56). In *Phaseolus coccineus*, high amounts of zeatin-like material occur in the suspensor early in seed development, while later in development more polar cytokinins occur (33). Lorenzi et al. (33) propose that the suspensor is the source of cytokinins for the embryo.

Movement of Hormones in Relation to Source-Sink Partitioning

The occurrence of high concentrations of hormones in developing seeds may indicate that 1) they function at that site or in the surrounding tissue; 2) they are produced at that site to be exported and function in some other site such as source tissue; or, 3) they are accumulating at that site thus relieving some other site of excess levels of a given hormone. The latter two require movement of hormones between seeds and source tissue.

---

[1] Auxin-like or GA-like is used to designate that analyses of hormones were based on bioassays rather than on physico-chemical determinations. The substance detected responded similarly to a hormone standard (IAA for auxin-like, GA_3 for GA-like).

Indoleacetic Acid

The classic work of Nitsch on developing achenes providing auxin for strawberry fruit growth led to his suggesting (37) that achenes release auxin that affect processes in leaves. Hein et al. (21) used an EDTA enhanced exudation technique (Fig. 2) to estimate the amount of IAA moving between seeds and source leaves of soybeans. IAA, primarily in the form of ester conjugate(s), was found to be moving acropetally (toward laminae) in petioles (Fig. 3). The highest amount of IAA ester(s) was found in petiole exudate from the mid to late stages of seed filling. Removal of fruits 36 hours prior to exudation reduced the amount of IAA ester recovered in exudate, indicating that fruits were a major source of the IAA conjugate observed in the petiole exudate. A small amount of labeled material that co-chromatographs with IAA can be recovered in source leaves following application of ^{14}C–tryptophan to soybean pods,

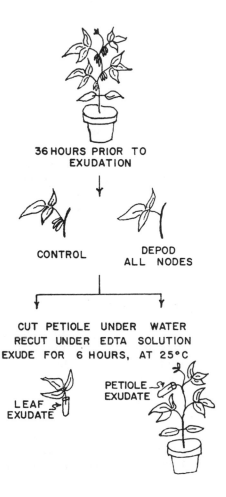

Fig. 2. Diagram of the experimental protocol used to obtained EDTA enhanced exudation of hormones transported in soybean leaf petioles. From (21).

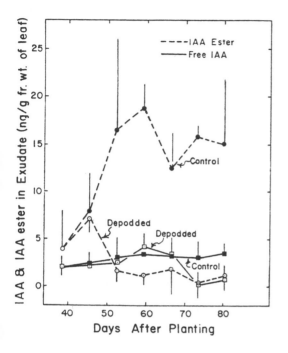

Fig. 3. Amount of free and ester IAA exuded acropetally into 20 mM EDTA from soybean petioles as detailed in Fig. 2. Values represent means of six replicates and their SE. From (21).

and thus provides further evidence that seeds may export IAA to leaves (22). Though depodding nearly eliminated exudation of the IAA esters, the level of IAA in leaves was unaffected 36 hours after the depodding (21). It is possible that the IAA transported to the leaf accumulates and functions in specialized cells of the leaf. Sampling the entire leaf would have masked detection of a hormonal change in a localized area. The fact that IAA will promote stomatal opening (Chapter E9) is especially relevant.

Gibberellins

Defruiting grape plants resulted in leaves containing less GA-like material, suggesting that developing fruit also export GAs to source leaves (24). However, considerable further research is required to verify if fruit export GAs to affect processes in the leaves.

Abscisic Acid

Developing sinks may control the level of ABA in leaves indirectly by serving as sites for accumulation of ABA produced within the leaves (48, 49). Trace quantities of radiolabeled ABA will rapidly move from mature leaves to all other parts of vegetative plants and will generally accumulate in sink tissues (see 7). Labeled–ABA can be found in soybean roots within 15 minutes following application to a leaf, and the ABA is recycled via the xylem back to the shoot apex during the next 30 minutes. ABA applied to source leaves of fruiting plants is exported to developing seeds (15, 49); while ABA applied to filling seeds is immobile (15). Setter et al. (48) found that when translocation from soybean leaves is obstructed or the plant is totally depodded, ABA accumulates in the leaves resulting in stomatal closure and depressed photosynthesis within one hour. Thus it appears that sinks can regulate processes in the source by drawing away ABA which may be inhibit some processes in the source.

ABA movement about the plant is dynamic. In soybean plants the pattern of ABA distribution changes diurnally and with stage of plant development (unpublished data, Cheikh and Brenner). Developing soybean seeds are the major site for ABA accumulation at the time of mid-pod fill (21), and it is only at this stage of plant development that defruiting the plant will result in ABA accumulation within leaves (Fig. 4). Presumably at this stage of development the filling seeds are the only active sinks, while at other stages there are multiple sinks for ABA accumulation.

Further proof that the maternal plant exports ABA to developing seeds is obtained by defoliating a soybean plant and then examining the seeds 48 hours later for ABA content. Sampling seeds of soybean (cv. Clay) resulted in a one-third reduction of the concentration of ABA recovered in the seeds 48 hours after plants were defoliated (8). However, similar treatment to another soybean genotype had no affect on ABA

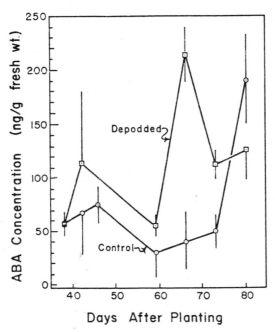

Fig. 4. Concentration of ABA in the second most recently expanded trifoliolate leaf of soybean during fruit development. Depodding treatments consisted of removing all reproductive tissue 32 hours sampling. Values represent means of six replicates and their SE. From (21).

concentration in the seeds from the defoliated plants (Schussler and Brenner, unpublished data). Since defoliation reduced ABA only in the Clay cultivar, it seems that the treatment eliminated a source of ABA rather than removing a substrate for ABA biosynthesis in the seeds. Thus we have hypothesized that the maternal contribution of ABA to developing soybean seeds can vary genotypically from a minimal contribution (the seeds appear to be autonomous for ABA) to a substantial one. Additional proof that seeds are capable of synthesizing ABA is demonstrated by the culturing of isolated soybean cotyledons on a range of sucrose concentrations and observing elevated levels of ABA in the tissue (6) presumably due to osmotic stress. It appears maize kernels synthesize most of their own ABA since culturing kernels *in vitro* in the absence of added hormones contain similar concentrations and distribution patterns of ABA to those kernels removed from field grown ears (28). Finally, isolated wheat ears can synthesize ABA which accumulate in the grain (29).

In summary, as shown in Fig. 5, it appears that at least for soybeans, ABA moves from leaves to filling pods. ABA also is transported from leaves to roots via the phloem and then recycled back to other sinks in the xylem stream (7). IAA-ester is transported from pods to the leaves (21, 22). Cytokinins produced in the roots are carried in the xylem stream to the shoot (9). Though there is no direct evidence that these cytokinins

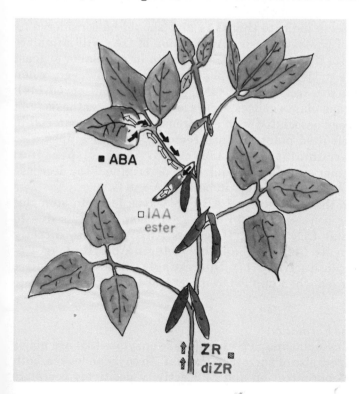

Fig. 5. Diagram of the apparent direction of hormone movement in soybeans in relation to control of source–sink relations.

accumulate in developing fruit, it is known that they do influence fruit set of soybean (9) and presumably subsequent fruit growth. Developing seeds may also synthesized their own cytokinins, though most of the available data do not support this idea (56). Since there is minimal evidence of GAs moving between source and sink tissues, GAs were not included in Fig. 5.

HORMONAL REGULATION OF PROCESSES RELATED TO PHOTOSYNTHATE PARTITIONING

Regulation of Processes in the Leaf

Regulation of Photosynthesis

There is also evidence that IAA enhances stomatal opening (see Chapter E9). Early reports described IAA promotion of photosynthesis within one hour of application to leaves (5). Follow-up experiments indicated IAA imparted this effect by increasing photophosphorylation and CO_2 fixation when tested on isolated chloroplasts (52). However, when similar experiments were repeated (44) IAA only reduced the aging of isolated chloroplasts but seemed to have no direct effect on photophosphorylation of active intact chloroplasts maintained in the presence of BSA (bovine serum albumen).

A number of reports indicate that GAs also should be considered as possible promoters of photosynthesis. Application of GA_3 will stimulate leaf photosynthesis (2) if treatments are applied to intact plants at least several hours before measurement of photosynthesis. However, when isolated chloroplasts are treated with GA_3, no effect is observed (44). Fruit removal on grape plants causes decreased leaf photosynthesis (24), and this effect has been associated with a decrease in GA-like material in the leaves of the defruited plants. Since removing seeded grapes also results in ABA accumulation in leaves (B. Loveys, personal communication), further studies are required to determine if the decrease in photosynthesis is related to changes in both ABA and GAs.

It is well known that increases of ABA in leaves leads to stomatal closure (see Chapter E9) resulting in decreased photosynthesis due to depressed intercellular CO_2 levels. However, ABA may also directly depress photosynthesis (42), perhaps by reducing of ribulose 1,5-bisphosphate carboxylase activity (17). Interestingly, the inhibitory effect of ABA is only observed when applied to intact leaf tissue rather than to isolated chloroplasts.

Regulation of Sucrose Formation

The activities of cytoplasmic SPS and FBPase enzymes (25) are major determinants of the pool size of sucrose in the leaf. In soybean leaves, both endogenous sucrose concentration and SPS activity positively correlate with assimilate export when measured at midday (25). However, when similar comparisons are made on a diurnal basis, these relationships do not hold up during the dark periods, making it clear that there are additional important factors controlling the export of carbon beside SPS activity and net photosynthesis. One possibility is that SPS may be under endogenous hormonal control. We (7) have observed that applying GA_3 or $GA_{4/7}$, at $10^{-6}M$ and $10^{-5}M$ respectively, to the transpiration stream of soybean leaves increases SPS activity. Fructose-2,6-bisphosphate (Fru-2,6-P_2) appears to be a potent regulator of sucrose formation by inhibiting cytosol FBPase activity. When Fru-2,6-P_2 increases, cytosol FBPase activity is suppressed resulting in less Fru-1-P synthesis and a buildup of triose-P in chloroplasts. This leads to starch synthesis and a concomitant decrease in formation of sucrose in the cytosol (25, 51). Regulation of concentration of Fru-2,6-P_2 represents another potent point for hormonal regulation of partitioning but as yet, we do not know if hormones affect the level of Fru-2,6-P_2 in plants as is the case in animal liver cells (23).

Regulation of Phloem Transport

Regulation of Phloem Loading
The capacity of a tissue to load photoassimilates into phloem tissue for long distance transport functionally establishes that tissue as a source tissue. Factors that determine the extent of phloem loading include the availability of sucrose for export, the distribution of the sucrose within the leaf (intra- and intercellular), transfer of sucrose (generally believed to be in the apoplast) towards the sites of phloem loading, and the rate of phloem loading (3, 19). Loading of phloem involves transfer of sucrose through the sieve-element plasma membrane from the vascular boundary. Alternatively, a total symplastic pathway from mesophyll cells to the phloem can be argued (3) with loading occurring across the plasma membrane of mesophyll cells (34).

The transfer process into the sieve element–companion cells of the phloem involves both a high affinity system and a linear system (14, 34). The high affinity system involves a proton/sucrose cotransport directed into the phloem (symport) and is pH dependent. It is seemingly coupled with an outward proton translocating ATPase (3). To compensate for the proton influx, there is a transient efflux of K^+, but there is also a H^+/K^+ exchanging ATPase (19) which maintains the high K^+ content of the phloem. The linear transport system is considered pH independent (14).

Based on experiments where hormones were exogenously applied (see 3) it appears that IAA or the fungal toxin fusicoccin will enhance phloem loading. Both of the later compounds promote membrane proton extrusion (see Chapter C1). Treatment with ABA inhibits sucrose loading (3). In addition, benzyladenine or kinetin also enhanced sucrose loading in stems of *Ricinus*. This is based on experiments where petioles were perfused with cytokinins and exudate from the petioles was sampled for sucrose and K^+. The loading of sucrose into isolated phloem tissue of celery occurs at greater rates when the tissue is placed in a medium adjusted to 200 to 300 m osmolality (with a nonpenetrating solute, PEG 3350 [12]). IAA and also GA_3 (each at 1 μM) promote greater rates of uptake in 200 m osmolal medium but not in media at 100 or 400 m osmolal. Since prior studies have not examined the potential interaction of hormones and turgor on phloem loading, it appears that this area deserves further examination.

Regulation of Phloem Transport

The rate and magnitude of phloem transport is generally believed to be determined by the hydrostatic pressure gradient between the source end and the sink end of the phloem, which is regulated by the rate of phloem loading in the source and by the rate of phloem unloading in the sink. These basic elements describe the Munch pressure hypothesis (14 and references therein). This generally accepted theory also states that there is minimal control exerted along the transport pathway. Thus, if the Munch pressure concept is correct, there would not be any role for plant hormones to directly regulate flow along the phloem transport pathway. Indirectly, however, plant hormones may regulate phloem transport by regulating loading and/or unloading.

A refinement of the explanation describing phloem transport invokes turgor regulation (31). This hypothesis envisages a gradient of solute concentration consisting primarily of sucrose and K^+. While sucrose is preferentially accumulated, K^+ concentration is modulated to maintain the turgor pressure gradient down the phloem pathway. A distinct aspect of this premise is pathway control of phloem transport obtained by adjusting the K^+ concentration through radial exchange of K^+ along the pathway. Data of Vruegdenhil (58) provide evidence that such gradients of K^+ do occur in the phloem of both cassava and castor bean. However, more rigorous experiments are required in which source/sink conditions can be manipulated to determine if K^+ gradients correlate with changes in phloem transport.

Recognizing that several hormones are known to effect K^+ transmembrane movement, future studies should consider if hormones modulate K^+ exchange between sieve elements and surrounding tissue. Though it is known that hormones affect K^+ exchange in guard cells (see Chapter E9) existing data on the application of hormones to directly affect phloem transport are inconclusive (14). Perhaps the most interesting observation is by Patrick (40) who indicated that IAA promotes mobilization of photoassimilate to the point of application on the cut stump of decapitated bean plants by acting upon transfer processes along the transport pathway. Since IAA has been implicated as a regulator of membrane ATPases, it is reasonable to consider IAA as a possible regulator of the K^+ concentration in the phloem. This would mean that IAA might be regulating transport by controlling turgor of the phloem through adjustment of K^+ concentration.

Regulation of Phloem Unloading

Withdrawal of photoassimilates from the phloem plays a major role in establishing the osmotic gradient between source and sink tissues, thereby influencing the rate of phloem transport. As mentioned above, photoassimilates are unloaded in the testa of dicotyledonous plants and in the placenta–chalazal tissue of monocotyledonous plants. Since there are no vascular connections between these sites of unloading and the developing zygotic tissue, it has been experimentally feasible to surgically remove the developing embryo and sample substances released into a bathing medium added to the empty maternal "cup." Most of these experiments have been performed on developing legume seeds. One approach to examine phloem unloading systems from seed cups is to preload developing seed coats with radiolabeled photoassimilates (via transport from labeled leaves), then excise the seeds, remove the embryos, and then sequentially wash the seed coats with various bathing media (10, 56, 60). A second approach is to sample the unloading of substances from attached seed coats. This is accomplished by cutting through the pod wall to surgically remove the distal half of a seed. The remaining embryo tissue is removed and sampling media is added to the seed cup, which is still attached via the funiculus to the pod wall (54).

The addition of a high osmolality solution (400 mmol consisting of sucrose and mannitol) to attached, empty seed cups of *Vicia faba* and peas permits obtaining unloading of radiolabeled sucrose into the trapping medium within the cups at rates equivalent to that which moves into intact seeds (59). Filling empty seeds cups with solutions of lower osmotic potential resulting in reduced unloading. It is proposed (59) that the solutions of high solute concentration lowers the osmotic potential of the solution in the seed–coat apoplast. Consequently, water flows from the phloem system of the seed coats and reduces turgor pressure in the sink end of the phloem. This reduced phloem turgor is transmitted along the transport pathway to the source tissue, where phloem loading is increased. Alternatively, the high turgor treatment may have imparted its effect by altering the activity of membrane bound ATPase/proton carriers, as seems to be the case with sugar beet taproot tissue (60).

Inclusion of metabolic inhibitors in the sampling medium with these seed cup techniques indicates that the unloading of sucrose requires metabolic energy (54). Addition of fusicoccin promotes ^{14}C–photo-assimilate efflux from attached soybean seed coats. As mentioned above, the fungal toxin fusicoccin promotes membrane proton extrusion (see Chapter C1). In contrast to fusicoccin's effect on empty soybean seed coats, treatment of both excised and attached *Phaseolus vulgaris* seed coats with fusicoccin inhibits ^{14}C-photoassimilate efflux even though it does cause acidification of the medium (57). These effects of fusicoccin can be reduced by including orthovandate or ABA in the *Phaseolus vulgaris* seed cups. One model which might explain the above results describes phloem unloading from *Phaseolus vulgaris* seed coats via two independent (and simultaneous) proton carrier pathways located in the plasmalemma (57). One carrier is an outward (into the apoplast) ATPase proton pump electrochemically balanced by K^+ uptake into the phloem. The second carrier is an outward proton/sucrose symporter. It is suggested that fusiccosin inhibits sucrose release because it stimulates the first proton carrier which reduces the proton/sucrose symport carrier activity by reducing the number of protons available for symport with sucrose. Additional data indicate that both the synthetic cytokinin, 6-benzylaminopurine (BAP), or ABA added to excised seed cups stimulate photoassimilate unloading from excised bean seed coats (10). The cytokinin effect is almost immediate, while ABA imparts its effect within 12 minutes of application to the seed cup. IAA, NAA, GA_3, and ACC (1-aminocyclopropane-1-carboxylic acid) are inactive in this unloading system. The action of ABA may be through restricting the ATPase driven proton carrier, thereby allowing the sucrose/proton symport carrier activity to be enhanced. The observation that both ABA and cytokinin act in a similar manner seems quite different to their opposing action on guard cell turgor (Chapter E9).

Application of ABA to filling grain of wheat (15) and barley (55) has been found to enhance the mobilization of recently fixed photoassimilates to the filling grain. It has been proposed that the promotive action of ABA is upon the unloading process (53). The promotive effect of applied ABA on increasing the import of assimilates to intact barley grains appears to be inversely related to the endogenous ABA content of the grain. Treatments with ABA were only promotive when applied to young ears (2 weeks after anthesis). High concentrations of ABA (10^{-3}M) inhibited assimilate import when applied to older ears (3 weeks after anthesis) when their endogenous ABA content had increased five fold (55). This might indicate that while ABA can promote mobilization in cereals, high concentrations might be inhibitory. It should be noted, however, that the promotive effect of ABA on assimilate import to wheat has not been repeatable by other scientists (30).

Active unloading of photoassimilates is not a ubiquitous process for all plant species. For example, unloading into empty maize kernels appears to be a passive process (see 54). Currently, there is no evidence for hormonal regulation of phloem unloading in maize.

Regulation of Assimilate Accumulation

For both legumes and temperate cereals, the sucrose that is unloaded into the apoplast is directly accumulated by the respective cotyledons or endosperm tissues as they develop (54). In contrast, sucrose is inverted to glucose and fructose in the apoplast of the maize pedicel–parenchyma tissue and these are accumulated by the basal endosperm tissue (28).

In vitro uptake of sucrose by developing embryos of legumes occurs by both saturable and nonsaturable mechanisms (32). Follow-up experiments (46) on protoplasts prepared from soybean cotyledons revealed three distinct uptake mechanisms for sucrose: a) a saturable (<10 mM) carrier that is energy-dependent and sensitive to the nonpenetrating sulfhydryl inhibitor p-chloromercuribenzene sulfonate (PCMBS); b) a nonsaturable carrier (at least up to 50 mM) that is suppressed when high concentrations of PCMBS are added; c) a simple diffusive mechanism. The sucrose concentration in the interfacial (apoplastic) region between the seed coat and cotyledon is in the range of 150 to 200 mM and appears to be relatively stable during both the day and night (54). This means that uptake of sucrose in peripheral regions of developing cotyledons by the saturable carrier-mediated component may be of little physiological significance. However, if sucrose moves primarily apoplastically through soybean cotyledons, as it appears to do in bean (41), then it is reasonable to suggest that the sucrose concentration in the central portion of soybean cotyledons may be lower than the saturation point of the sucrose carrier. This would be due to substantial withdrawal of sucrose by the peripheral cells first exposed to the unloaded sucrose.

Sucrose appears to be the prime sugar taken up by wheat endosperm tissue. In addition to a diffusional component, the uptake of sucrose also occurs by a energy-dependent carrier that is sensitive to PCMBS inhibition (43). In contrast, hexoses are the principal sugars taken up by maize endosperm tissue, and their uptake appears to be a nonsaturable passive process (28). Maize embryos take up hexoses in part by a facilitative, metabolically driven process which may be how the embryo can functionally compete with endosperm for available hexose (28).

Exogenous ABA can enhance the accumulation of sucrose by isolated soybean cotyledons (when tested with low sucrose concentrations, 10 mM; [47]), though we have only been able to demonstrate this effect on a specific genotype (cv. Clay) grown in nonstressed environments. Tests with several genotypes grown in the field showed no effect of ABA on *in vitro* sucrose uptake, probably due to the high endogenous ABA in these field derived tissues. However, endogenous ABA content of cotyledons is positively correlated to the sucrose uptake. The concept that ABA may function to enhance sucrose accumulation is further supported by the following manipulative study. Subjecting certain soybean genotypes (cvs. Clay and Evans) to brief periods of drought stress followed by rewatering, results in an accumulation of stress-produced ABA in the developing seeds, presumably from the leaves. The cotyledons of these seeds have significantly greater sucrose uptake compared to cotyledons derived from nonstressed plants. Stressing a different genotype (PI 416.845) does not alter the ABA content of its seeds and also has no effect on subsequent *in vitro* sucrose uptake. We have found that the seeds of this genotype appear to synthesize their own ABA and do not depend on the maternal plant to supply ABA (Schussler and Brenner, unpublished data).

Unlike the promotive effect of ABA on *in vitro* sucrose uptake by soybean embryos (47), neither ABA or BAP (each promotes unloading from *Phaseolus vulgaris* seed cups) affect *in vitro* sucrose uptake by *Phaseolus vulgaris* embryos (38). A unifying hypothesis proposed by Offler and Patrick (38) to explain the difference between these two legumes is that regulation of the substrate concentration (osmotic potential) in the interface between seed coat and embryo tissue is a critical factor affecting photosynthate exchanges from seed coats to embryos. They suggest that alteration of osmotic potential in this interfacial zone affects the turgor of the these tissues, and turgor is one of the key determinants of the rate of unloading. Consequently, in *Phaseolus vulgaris* unloading from seed coats may be the major determinant for assimilate accumulation, while in soybean, uptake of assimilates by embryos may be more significant component compared to unloading by seed coats.

The observation that ABA increases the accumulation of photo-assimilates is not unique to developing seeds. Application of ABA to root tissue discs of sugar beet causes as much as a three-fold increase in sucrose accumulation and the promotive effect is observable within one hour (45). Both IAA and K^+ inhibit the accumulation of sucrose by the sugar beet tissue. ABA also may promote the accumulation of photoassimilate by young developing soybean pods through an increase in invertase activity within the pod tissue (1). Since glucose is superior to sucrose as a substrate for *in vitro* culture of young soybean seeds, it is an interesting idea that ABA may promote the growth of young seeds by increasing the pool size of glucose through increasing invertase activity in the pod tissue.

When pea seeds are grown *in vitro*, GAs do not appear to be required since treatment with GA biosynthesis inhibitors blocks the accumulation of GAs, but has no effect on seed growth (4, 18). These data indicate that GAs do not seem to act within the seed for seed growth and therefore it seems that GAs are not involved with sucrose uptake by seeds.

Indirect Effects of Hormones on Assimilate Partitioning

Compartmentation of photoassimilates is a likely mechanism for sink tissue to maintain the steepness of the concentration gradient from the source tissue. In maize and temperate cereals, the number of endosperm cells and starch granules formed in endosperm cells is highly correlated with kernel growth rate and size at maturity (26, 27, 28). In cereals, endosperm cell number and amyloplast number are established during the early phases of grain development. Since these parameters are not controlled by the levels of available carbohydrates during grain filling (26), it may be speculated that cytokinins might affect the numbers of both endosperm cells and amyloplasts formed within those endosperm cells (28). Cytokinins generally are considered to play a major role promoting cell division, but there is no information on their effecting endosperm cell number. It has been shown that cytokinins can promote chloroplast development (39). Since chloroplasts and amyloplasts are both derived from proplastids, it is possible that amyloplast development also might be altered by cytokinins (28). As mentioned earlier, cytokinin content is high in developing seeds at this critical period (56) but it is not known if any source–sink manipulations that lead to reduced seed development can be related to reduced cytokinin activity.

Hormones, especially cytokinins, also may indirectly affect partitioning by altering the ratio of source to sink tissue by regulating seed set (9). Most of our major crop plants produce only a fraction of their potential seeds compared to the number of flowers produced on the respective plant types. The failure of fertilized flowers to produce seeds has been attributed to insufficient photoassimilates at the time of early fruit development or to a hormonal imbalance at the this time.

Finally, partitioning may be may indirectly altered by regulating the duration of seed fill (27). This can occur by either delaying leaf senescence or extending the seed-filling period by delaying the onset of maturation of seeds. Though minimal information is available, it is reasonable to speculate that hormones may also play a role regulating this process.

SUMMARY

As depicted in Fig. 1, there are experimental data indicating that hormones may act a number of steps involved with the assimilation and transport of photoassimilates to developing seeds. The numerous manipulative studies done to investigate regulation of this overall process have amply demonstrated that overall regulation is extensively integrated (18, 35). Hormones are logical candidates to play key roles in coordinating the respective processes. For those plants that are limited by source activity, transport of promotive hormones from sinks to source tissue may regulate the processes associated with source activity, namely photosynthesis, production of sucrose, and phloem loading. Auxins and gibberellins exported from sinks are reasonable candidates to serve as the promotive signals (2, 3, 5, 21, 22, 24, 37, 52). Relieving source tissue of excess ABA may be a further mechanisms by which sinks can attenuate source leaf activity (17, 21, 42, 48, 49).

Hormones certainly may function to enhance sink activity by increasing the net amount of photoassimilates transferred from the maternal tissue to developing seeds. Action may occur at the sites of phloem unloading (10, 38) or at sites of assimilate accumulation (45, 47).

Turgor also seems to be an important factor regulating phloem loading and unloading (12, 31, 54, 59, 60). It is reasonable to consider that a change in hydrostatic pressure within the phloem is all that is necessary to facilitate communication between source and sinks (59, 60). Based on the observations of Daie (12) that both IAA and GA_3 promote phloem loading when osmotic pressure of the bathing medium is adjusted to 200 to 300 m osmolality, it is obvious that considerable attention should be directed at examining the possible interaction of turgor and hormones in regulating partitioning of photosynthates.

Acknowlegments

Supported in part by the Unites States Department of Agriculture under grant 84-CRCR-1-1484 from the Competitive Reesearch Grants Office. Also supported in part by a grant from the Minnesota Soybean Research and Promotion Council. Contribution from the University of Minnesota Agricultural Experiment Station, St. Paul, MN 55108. Paper No. 2045 of the miscellaneous journal series.,

The author wishes to thank W.A. Brun, R.J. Jones, S.L. Maki and J.R. Schussler for their thoughtful discussions and critical comments during the preparation of this manuscript.

References

1. Ackerson, R.C. (1985) Invertase activity and abscisic acid in relation to carbohydrate status in developing soybean reproductive structures. Crop Sci. 25,615-618.
2. Arteca, R.N., Dong, C.N. (1981) Increased photosynthetic rates following gibberellic acid treatments to the roots of tomato plants. Photosynth. Res. 2,243-249.
3. Baker, D.A. (1985) Regulation of phloem loading. British Plant Growth Regulator Group, Monograph 12,163-176.
4. Baldev, B., Lang, A., Agatep, A.O. (1967) Gibberellin production in pea seeds developing in excised pods: Effect of growth retardant AMO-1618. Science 147,155-156.
5. Bidwell, R.G.S., Levin, W.B., Tamas, I.A. (1968) The effects of auxin on photosynthesis and respiration. In: Wightman, F and G Setterfield (eds): Biochemistry and Physiology of Plant Growth Substances, pp 361-376. The Runge Press, Ottawa.
6. Bray, E.A., Beachy, R.R. (1985) Regulation by ABA of β-Conglycinin expression in cultured developing soybean cotyledons. Plant Physiol. 79,746-750.
7. Brenner, M.L., Brun, W.A., Schussler, J., Cheikh, N. (1986) Effects of endogenous and exogenous plant growth substances on development and yield of soybeans. In: M. Bopp (ed): Plant Growth Substances 1985. Springer-Verlag, Berlin Heidelberg pp 380-386.
8. Brenner, M.L., Hein, M.B., Schussler, J., Daie, J., Brun, W.A. (1982) Coordinate control: The involvement of ABA, its transport and metabolism. In: P.F. Wareing (ed): Plant Growth Substance 1982. Academic Press, New York. pp. 343-352.
9. Carlson, D.R., Dyer, D.J., Cotterman, C.D. (1986) The physiological basis for cytokinin induced increases in pod set in IX 93-100 soybeans. Plant Physiol. (in press).

10. Clifford, P.E., Offler, C.E., Patrick, J.W. (1986) Growth regulators have rapid effects on photosynthate unloading from seed coats of *Phaseolus vulgaris* L. Plant Physiol. *80*,635-637.
11. Crane, J.C. (1969) The role of hormones in fruit set and development. HortSci. *4*,108-111.
12. Daie, J. (1986) Turgor-mediated transport of sugars. Plant Physiol. *80* (Sup),98.
13. Dathe, W., Sembdner, G. (1984) Endogenous plant hormones of the broad bean, *Vicia faba* L. VI Contents of abscisic acid and gibberellins in funicle, pericarp, and seed during fruit development. Biochem. Physiol. Pflanzen. *179*,289-294.
14. Delrot, S., Bonnemain, J.L. (1985) Mechanism and control of phloem transport. Physiol. Veg. *23*,199-220.
15. Dewdney, S.J., McWha, J.A. (1979) Abscisic acid and the movement of photosynthetic assimilates towards developing wheat (*Triticum aestivum* L.) grains. Z. Pflanzen. *92*,183-186.
16. Eeuwens, C.J., Schwabe, W.W. (1975) Seed and pod wall development in *Pisum sativum* L. in relation to extracted and applied hormones. J. Exp. Bot. *26*,1-14.
17. Fisher, E, Stitt, M., Raschke, K. (1985) Effects of abscisic acid on photosynthesis in whole leaves: Changes in CO_2 assimilation, levels of carbon reduction cycle intermediates, and activity of ribulose-1,5-bisphosphate carboxylase. Abstracts of 12th International Plant Growth Substance Conference–1985. pp 28.
18. Garcia-Martinez J.L., Gaskin, P., Sponsel, V.M. (1985) Do endogenous gibberellins control fruit set and seed development in pea? Abstracts of International Plant Growth Substance Conference–1985, pp 103.
19. Giaquinta, R.T. (1983) Phloem loading of sucrose. Ann. Rev. Plant Physiol. *34*,347-387.
20. Hein, M.B., Brenner, M.L., Brun, W.A. (1984) Concentrations of indole-3-acetic acid and abscisic acid in soybean seeds during development. Plant Physiology *76*,951-954.
21. Hein, M.B., Brenner, M.L., Brun, W.A. (1984) Effect of fruit removal on concentration in leaves as measured in petiole exudate of indole-3-acetic acid and abscisic acid during reproductive growth of soybean. Plant Physiology *76*,955-958.
22. Hein, M.B., Brenner M.L., Brun, W.A. (1986) Accumulation of ^{14}C-radiolabel in leaves and fruits after injection of ^{14}C-tryptophan into seed of soybean. Plant Physiol. (in press).
23. Hers, HG, E van Schaftingen (1982) Fructose-2,6-bisphosphate 2 years after its discovery. Biochem J. *206*,1-12
24. Hoad, G.V., Loveys, B.R., Skenek, G.M. (1977) The effect of fruit-removal on cytokinins and gibberellin-like substance. Planta *136*, 25-30.
25. Huber, S.C., Kerr, P.S., Kalt-Torres, W. (1985) Regulation of sucrose formation and movement. In: R.L. Heath, J. Preiss (eds): Regulation of Carbon Partitioning in Photosynthetic Tissue. pp 199-214 Amer. Soc. Plant Physiol.
26. Jenner, C.F. (1985) Control of the accumulation of starch and protein in cereal grains. British Plant Growth Regulator Group, Monograph *12*,195-209.
27. Jenner, C.F. (1986) End product storage in cereals. In: Cronshaw, J., Lucas, W.J., Giaquinta, R.T. (eds.): Phloem Transport. pp 561-572. A.R. Liss, Inc., New York.
28. Jones, R.J., Griffith, S.M., Brenner, M.L. (1986) Sink regulation of source activity: Regulation by hormonal control. In: J. Shannon (ed): Regulation of Carbon and Nitrogen Reduction and Utilization in Maize. Martinus Nijhoff The Hague (in press).
29. King, R.W. (1979) Abscisic acid synthesis and metabolism in wheat ears. Aust. J. Plant Physiol. *6*,99-108.
30. King, R.W., Patrick, J.W. (1982) Control of assimilate movement in wheat. Is abscisic acid involved? Z. Pflanzenphysiol. 106,375-380.
31. Lang, A. (1983) Turgor-regulated translocation. Plant, Cell and Environ. *6*,683-689.
32. Lichner F.T., Spanswick, R.M. (1981) Electrogenic sucrose transport in developing soybean cotyledons. Plant Physiol. *67*,869-874.

33. Lorenzi, R., Bennici, A., Cionini, P.G., Alpi, A., D'Amato, F. (1978) Embryo-suspensor relations in *Phaseolus coccineus*: Cytokinins during seed development. Planta *143*,59-62.
34. Lucas, W.J. (1985) Phloem-loading: A metaphysical phenomenon? *In*: R.L. Heath, J. Preiss (eds): Regulation of Carbon Partitioning in Photosynthetic Tissue. pp 254-271 Amer. Soc. Plant Physiol.
35. Michael, G., Beringer, H. (1980) The role of hormones in yield formation. In: Physiological Aspects of Crop Productivity. Proceedings of the 15th Colloquium of the International Potash Institute pp 85-116.
36. Mounla, M.A.Kh. (1978) Gibberellin-like substances in parts of developing barley grain. Physiol. Plant *44*,268-272.
37. Nitsch J.P. (1959) Auxines et croissance des fruits, II. In: Recent Advances in Botany. Univ. Toronto Press *2*,1089-1093.
38. Offler, C.E., Patrick, J.W. (1986) Cellular pathway and hormonal control of short-distance transfer in sink regions. In: Cronshaw, J., Lucas, W.J., Giaquinta, R.T. (eds.): Phloem Transport. pp 295-306. A.R. Liss, Inc., New York.
39. Parthier, B. (1979) The role of phytohormones (cytokinins) in chloroplast development. Biochem. Physiol. Pflanzen *174*,173-214.
40. Patrick, J.W. (1979) Auxin-promoted transport of metabolites in stems of *Phaseolus vulgaris* L. J. Exp. Bot. *30*,1-13.
41. Patrick, J.W., McDonald, R. (1980) Pathway of carbon transport within developing ovules of *Phaseolus vulgaris* L. Aust. J. Plant Physiol. *7*,671- 684.
42. Raschke, K., Hedrich, R. (1985) Simultaneous and independent effects of abscisic acid on stomata and the photosynthetic apparatus in whole leaves. Planta *163*,105-118.
43. Rijven, A.H.G.C., Gifford, R.M. (1983) Accumulation and conversion of sugars by developing wheat grains. 3. Non-diffusional uptake of sucrose, the substrate preferred by endosperm slices. Plant, Cell and Environ. *6*,417-425.
44. Robinson, S.P., Wiskich, J.T., Paleg, L.G. (1978) Effects of Indoleacetic acid on CO_2 fixation, electron transport and phosphorylation in isolated chloroplasts. Aust. J. Plant Physiol. *5*,425-431.
45. Saftner R.A., Wyse, R.E. (1984) Effect of plant hormones on sucrose uptake by sugar beet root tissue discs. Plant Physiol. *74*,951-955.
46. Schmitt, M.R., Hitz, W.D., Lin, W., Giaquinta, R.T. (1984) Sugar transport into protoplasts isolated from developing soybean cotyledons. Plant Physiol. *75*,941-946.
47. Schussler, J.R., Brenner, M.L., Brun, W.A. (1984) Abscisic acid and its relationship to seed filling in soybeans. Plant Physiol. *76*,301-306.
48. Setter, T.L., Brun, W.A., Brenner, M.L. (1980) Effect of obstructed translocation on leaf abscisic acid, and associated stomatal closure and photosynthesis decline. Plant Physiol. *65*,1111-1115.
49. Setter, T.L., Brun, W.A., Brenner, M.L. (1981) Abscisic acid translocation and metabolism in soybeans following depodding and petiole girdling treatments. Plant Physiol. *67*,774-779.
50. Sponsel, V.M. (1983) The localization, metabolism, and biological activity of gibberellins in maturing and germinating seeds of *Pisum sativum*, cv. Progress No. 9. Planta *159*,454-468.
51. Stitt, M. (1986) Regulation of photosynthetic sucrose synthesis: Integration, adaptation, and limits. In: Cronshaw, J., Lucas, W.J., Giaquinta, R.T. (eds.): Phloem Transport. pp 331-347. A.R. Liss, Inc., New York.
52. Tamas, I.A., Schwartz, J.W., Breithaupt, B.J., Hagin, J.M., Arnold, P.H. (1973) Effect of indoleacetic acid on photosynthetic reactions in isolated chloroplasts. *In*: Plant Growth Substances pp 1159-1168; Hirokawa Publ. Co.
53. Tanner W. (1980) On the possible role of ABA on phloem unloading. Ber. Deutsch Bot. Ges. *93*, 349-351.

54. Thorne, J.H. (1985) Phloem unloading of C and N assimilates in developing seeds. Ann. Rev. Plant Phys. *36*,317-343.
55. Tietz, A., Ludwig, M., Dingkuhn, M., Dorffling, K. (1981) Effect of abscisic acid on the transport of assimilates in barley. Plant *152*,557-561.
56. van Staden, J. (1983) Seeds and cytokinins. Physiol. Plant. *58*,340-346.
57. van Bel, A.J.E., Patrick, J.W. (1984) No direct linkage between proton pumping and photosynthate unloading from seed coats of *Phaseolus vulgaris* L. Plant Growth Regul. 2,319-236.
58. Vreugdenhil, D. (1985) Source-to-sink gradient of potassium in the phloem. Planta 163,238-240.
59. Wolswinkel, P., Ammerlaan, A. (1986) Turgor-sensitive transport in developing seeds of legumes: the role of the stage of development and the use of excised vs. attached seed coats. Plant, Cell and Environ. 9,133-140.
60. Wyse, R. E. (1986) Sinks as determinants of assimilate partitioning: Possible sites for regulation. *In*: Cronshaw, J., Lucas, W.J., Giaquinta, R.T. (eds.): Phloem Transport. pp 197-209. A.R. Liss, Inc., New York.

E13. The Role of Hormones During Seed Development

Ralph S. Quatrano[1]

Department of Botany and Plant Pathology, Oregon State University, Corvallis, Oregon 97331-2902, USA.

INTRODUCTION

The formation of a seed in the life cycle of higher plants is a unique adaptation. It incorporates embryo development with various physiological processes that are meant to insure the survival of the plant in the next generation. These adaptations include the accumulation of nutritive reserves, an arrest of tissue growth and development, and finally desiccation. To survive long periods of time in this dry state until environmental conditions are favorable to resume development into a seedling, numerous seeds have also acquired different mechanisms of dormancy. All of these traits are also of considerable agronomic importance (e.g. nutritive value, yield, germination).

From a more basic viewpoint, seed development has represented a convenient experimental system for the study of the underlying mechanisms of physiological and molecular regulation of cell and tissue development (11,38,42). The stages of seed development and germination involve both spatial and temporal regulation of cell and tissue growth and function. The sequence of events begins with rapid endosperm and embryo growth and differentiation after fertilization, followed by the transition from a state of high metabolic activity and growth to a quiescent state, and finally, a switch back to active growth of the embryo to form a seedling (Fig. 1).

Hormones are thought to play an important role in these processes, since the levels and activities of various hormones change dramatically during this developmental sequence (5,29,37). Cytokinins, auxins, gibberellins (GA) and abscisic acid (ABA) are found in relatively high concentrations in extracts from seeds of different developmental stages. In fact, a large part of our knowledge of hormone biosynthesis and metabolism has been obtained using young seeds (c.f. 26). One must critically ask, however, are the changes in hormones levels/activity

---

[1] Present address: Central Research and Development, E. I. duPont deNemours and Co., Wilmington, Deleware 19898, USA.

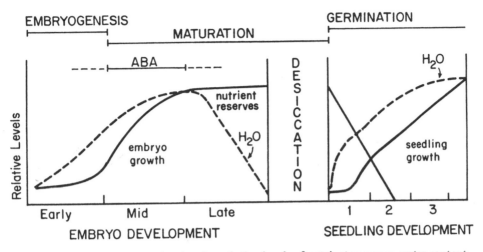

Fig 1. A generalized graph showing the relative levels of nutrient reserves, water content and growth during the embryogenic, maturation and germination stages of embryo and seedling development. Time periods of embryo development vary with species and are not included. Times stated for seedling development are in days. Desiccation separates the end of maturation from the initiation of seedling growth, and ABA levels are temporally correlated with the onset of maturation and prevention of precocious germination during mid-embryo development.

correlated with changes in embryo growth, and, do they play a causal role in embryo development? Let's first review the main points of embryo development in angiosperms.

EMBRYO DEVELOPMENT

What is unique about higher plant embryogenesis when compared with embryo development in other plant groups? During embryo maturation, there is a buildup of nutrient reserves, an arrest of tissue growth, development of desiccation tolerance and the acquisition of dormancy mechanisms (6,51,54). All of these events occur in the latter half of embryo development in most higher plants. Following desiccation, activation of the embryo results in initiating a meristematic type growth pattern forming a seedling, and the utilization of nutrient reserves to support this development (6). All are associated with the "seed strategy," i.e. mechanisms to insure survival of the young seedling. Of course, there are natural exceptions, such as the mangrove (*Rhizophora mangle*) which forms seedlings on the plant (viviparous) without any arrest, desiccation or germination (50). Also, a large number of seeds do not have any significant dormancy and will germinate upon water imbibition and physiological temperatures.

Since these processes are unique to seed plants, embryos from lower plants complete embryogeny without these intervening processes. Experimentally one can ask of the higher plant embryogeny sequence, which of the processes that occur late in seed development are required for normal seedling development? A number of experiments indicate that these processes are not obligatory for completion of the life cycle.Studies with mutants, and *in vitro* culture techniques (to artificially provide the embryo with the required nutrient and environmental conditions) support this conclusion.

Normal seedlings (notwithstanding some somaclonal variation which is more pronounced under certain conditions in some species/varieties than others c.f. 46) are formed *in vitro* from callus cultures via somatic embryogenesis or organogenesis.These developmental sequences occur in the absence of developmental arrest. Also, a few of the viviparous mutants isolated from maize can form seedlings in the absence of desiccation (16,44). Finally, after a period of seed development, isolated embryos are capable of precocious germination into seedlings if cultured on a nutrient medium containing salts, reduced nitrogen (e.g., glutamine) and sucrose (53). Now the question arises, when you compare seedlings from embryos which have, and have not undergone developmental arrest, desiccation, etc., are they identical? What role, if any, do these "maturation" processes and desiccation play in subsequent stages in the life cycle of the plant? Are the function(s) of these processes confined to the seed and germination stages, or do they have effects later in the next generation?

Soybean mutants that lack seed lectin (SBL) and lipoxygenase-l (LXG), both of which normally appear late in development, germinate normally and the resulting seedlings grow, flower and set seed just as wild type. A number of physiological roles proposed for SBL (regulation cell division; pathogen/pest inhibitor) and LXG (lipid oxidation, ethylene biosynthesis) do not seem to critical for completion of plant development. Also, roots that lack SBL nodulate just as well as plants with SBL (21). Preliminary evidence from my lab indicates that wheat seedlings grown to maturity from normally arrested and precociously germinated embryos, showed no differences in a number of morphological and physiological traits in mature vegetative and reproductive plants. Although these maturation events and specific gene products may be of critical importance for some physiological process, agronomic trait or survival in nature, they do not appear to be *required* for completion of the life cycle. They can be thought of as a set of physiological processes that comprise an important but not obligatory pathway of seed development, and whose consequences seem to be restricted to the seed stage of plant development. However, because of these experimental manipulations and mutants available to investigators, approaches can be taken in an attempt to understand the regulation of this *maturation sequence* (Fig. 2).

As a start we can ask, at which stage is the embryo capable of germinating if removed from the seed and cultured in a nutrient medium? To describe specific times after fertilization, or a developmental stage unique to a particular plant (e.g., 45), would be too detailed for our purposes here. Let's divide seed development into three periods, *early*, *middle* and *late*, each representing about a third of the time between

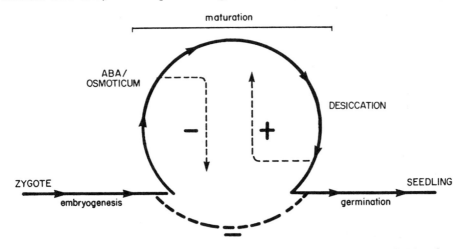

Fig. 2. A generalized representation of embryo development in angiosperms, showing three major stages from zygote to seedling; embryogenesis, maturation and germination. The maturation "loop" can be by-passed in culture by precocious germination of embryos that have undergone some development (i.e., embryogenesis). This occurs in the absence of exogenous hormones but with sources of reduced nitrogen (e.g. glutamine) and energy (e.g., sucrose), and in some cases by premature desiccation. In the presence of high osmoticum and/or ABA, maturation is promoted and precocious germination inhibited. Upon desiccation, these processes are finalized, and normal germination into a seedlings results in response to imbibition and the proper environment. In culture, embryos can reversibly enter or leave the maturation loop by application of high osmoticum/ABA (+) or their removal (−), respectively.

fertilization and desiccation. Using this standard, embryos from wheat (53), rice (49), rapeseed (14,15), soybean (1,2), bean (27), and probably most others, can precociously germinate in culture during the latter half of *early* development.

If embryos are capable of germinating by the start of the middle period, what is present in the seed environment, around/in the embryo, that normally prevents precocious germination *in vivo*? Cytokinins, IAA, GA and ABA levels have been shown generally to be high during these stages (c.f. 29). What do we know about the role of these hormones in seed development and their possible effect on the prevention of precocious germination and the initiation of the maturation pathway? Many of the studies which show changes in hormones levels during seed development and associations of hormones with various developmental events are reviewed elsewhere (e.g., 26,38). This chapter is not meant to be an exhaustive review of these studies but will focus on recent progress in our understanding of embryo development with special emphasis on the role of hormones, specifically ABA.

CHANGES IN LEVELS/ACTIVITY OF HORMONES DURING SEED DEVELOPMENT

Early Embryogenesis

Cytokinins are found in relatively high concentrations in the liquid endosperm stage of early seed growth and their presence coincides with the highest rate of mitosis (52). Studies with isogenic mutants of barley that vary in grain· weight, demonstrate that large-grain lines contain higher amounts of cytokinin at this very early stage of seed development than small-grain lines (47). Based on results such as these, it has been suggested that cytokinin activity at this stage is responsible for enhanced seed size by increasing cell number, resulting in larger storage capacity (30).

In studies of embryo development in interspecific hybrids of *Phaseolus*, the timing of arrest can occur from early to *mid*-maturation, depending on the species combination and direction of the cross (32).When arrest occurs in *early* embryos, they are not able to be excised for rescue by embryo culture. It has recently been shown that when the female parent is grown in hydroponic medium with the cytokinin N^6 benzyladenine (10 µM), embryos that normally arrest at four cells enlarge to more than 100 cells, sufficiently large enough for excision (32). Studies such as these implicate cytokinins in early events of seed and embryo growth.

A number of studies suggest a role of GA in the function of the suspensor during this early stage of embryo growth (7,57). Higher concentrations of GA are found in the suspensor compared to the embryo, and GA can substitute for the suspensor in supporting embryo growth in culture. Protein levels are decreased in embryos of *Phaseolus* when embryos are cultured detached from the suspensor. GA at 0.1–1.0µM restored the protein content to that of freshly excised embryos. The timing and concentration of GA supplied to the embryo from the suspensor appears to be critical for early embryo growth, and points to a possible role of the suspensor in supplying GA to the embryo.

Use of mutants will undoubtedly provide an approach to pursue questions dealing with early embryo function.For example, *dek* mutants of corn are chemically induced mutations that affect both embryo and endosperm development. Some of these mutants are blocked at specific stages early in corn embryo development (48). Similar embryo mutants of *Arabidopsis* have been identified (28).The morphological and genetic traits in both systems are quite well described, and should be valuable in pursuing questions concerning the role of hormones in rescuing these defective embryos.

Mid-embryogenesis

In general, high auxins (IAA) and GA levels have been associated with two phases of reproductive development; active seed growth by cell expansion and fruit growth (13). The GA's found, however, are often the type that have moderate to very low bioactivity (37). Both GA and auxins are highest during *early* to *mid*-embryo development in a number of plants, at a stage when cytokinins are decreasing rapidly and there is little or no ABA detectable (29). The timing of these increases in peas and wheat are correlated with increases in pod and grain length. Using the same isogenic mutants of barley discussed above, investigators found that high single grain weight correlated with higher IAA content (33).

Whereas early pod/grain growth seems to be correlated with the very early increases in GA, the role, if any, in the middle and late stages of seed development is more obscure (37).There does not seem to be a causal relationship between the amount of bioactive GA and late pea seed growth, nor is there an increased degradation of bioactive GA's to inactive ones at this stage (37). Zeevaart (58) showed in *Pharbitis*, that if the later peak of GA accumulation (three weeks after anthesis) was blocked with CCC, seed fresh and dry weights were unaffected. However, progeny of seeds treated with CCC had both a dwarfed growth habit and reduced GA levels. Studies with the *nana* and *Rht* drawfs of pea and wheat clearly show altered GA synthesis/metabolism in vegetative tissues, but the amounts of GA and the qualitative spectrum are unchanged in the seeds of these dwarfs (37,39).Also, in dwarf mutants of *Arabidopsis*, an absolute GA requirement for both germination and elongation growth of seedlings was demonstrated, yet seeds of wild type and mutants reached similar dry weights. These plants, then, can dissociate the effects of GA on growth in vegetative tissue from the same physiological processes in the developing seed or grain. Similarly, even at high GA concentrations during seed development, induction of α-amylase mRNA and protein in aleurone tissue are not detected at these stages (24). Therefore, although GA is present in relatively high concentrations in seeds, its major role seems to be confined to early embryo development and subsequently to germination and seedling growth. The same appears to be true for cytokinins and auxin.

Based on these studies, the role of cytokinins, GAs and auxin in the *mid* and *late* stages of emryogenesis is not clear. However, caution should be exercised in such interpretations. Perhaps the local concentration and/or change in tissue sensitivity to endogenous levels of these hormones may be very critical. Future studies should concentrate on identifying spatial and sensitivity differences during seed development (e.g., 3). In addition, investigators should be more aware of the ratios of hormones as controlling factors. For example, in most plants the *middle* and *late* periods of embryo growth are generally characterized by decreasing amounts/sensitivity of GA with a corresponding increase in amounts/ sensitivity to ABA (37).

Mid to Late Embryogenesis

In a number of different species of monocots and dicots, ABA levels begin to rise and reach their highest levels during the *middle* and *late* period of seed development (c.f. Fig. 1), at a time of decreasing GA and auxin levels, and the initiation of maturation events. It then decreases rapidly to very low levels in the dry seed. In general, ABA reaches a maximum at the same time as seed growth (i.e., dry and wet weight maxima), and declines sharply after the accumulation of reserves and the beginning of the loss of water. In several cases (*Phaseolus* (20), wheat (23), barley (35), *Arabidopsis* (22)) there appears to be a dual peak of ABA accumulation that occurs during the *middle* and *late* period (Fig. 3). In the case of *Arabidopsis*, one peak is associated with a maternal origin whereas the second is with a zygotic origin (22). The levels of ABA that accumulate in these seeds are all in the physiological range, varying roughly between 1-10 μM. Hence, the levels and timing of ABA accumulation *in vivo* is consistent with its having a role in physiological events occurring at this time, i.e., the maturation events of embryo development.What evidence, other than a temporal correlation between high ABA levels and the initi- ation of embryo growth, maturation and prevention of germination, links ABA with the role of a natural regulator of this developmental pathway?

In many species there is an inverse relationship between endogenous ABA and the ability of immature embryos to germinate after removal from the seed. A close correlation exists between ABA levels and embryo growth rates *in situ* in three cultivars of soybeans (1,2). Also, in certain varieties of barley, there is a correlation between reduced ABA content and a shortened dormancy and low sprout resistance (c.f. 29).Additional data centers around two types of studies; effects of ABA on isolated embryos in culture, and, analysis of several types of mutants.

ABA Effects On Embryos In Culture

As stated above, immature embryos placed in culture at the start of mid-embryo development will precociously germinate.The germination specific enzyme, carboxypeptidase C, appears in cotton embryos within one day of precocious germination (11). The same is true of the germination marker in wheat embryos, the small subunit of ribulose bisphosphate carboxylase (40), and for isocitrate lyase in germinating castor bean embryos (10).Although most embryos switch completely from embryogenic to germination processes, *Brassica* embryos seem to be an

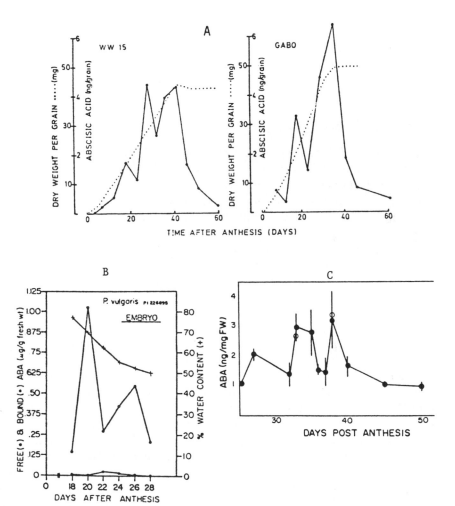

Fig 3. Changes in the ABA levels in wheat grains (A), bean embryos (B), and rapeseed embryos (C). From (24), (20) and (14) respectively.

exception. They concurrently express characteristics of both, gradually acquiring the seedling traits (15).

In the late 1960s, several investigators added ABA to cultures of immature embryos to determine if it prevented precocious germination (c.f. 11). The rationale was that ABA was originally isolated and characterized as an inducer of dormancy and known for its inhibitory effects on seed germination, and as discussed above, because of its high levels during the *mid* to *late* stages of embryo development *in vivo*. In many species, if embryos are removed at the beginning of the *middle* period, and cultured in 0.1-10 μM ABA, precocious germination is inhibited (1,2,14,27,49,53).Not only does ABA prevent germination, but in many cases it results in embryo growth and an accumulation of storage reserves (Fig. 4). ABA stimulates growth and protein accumulation in soybean embryos isolated during the early part of mid embryogenesis, but if embryos from the later stages are cultured in ABA, growth is suppressed but not the protein increases (1,2,12). There is also a clear relationship between embryo ABA concentration and ability to

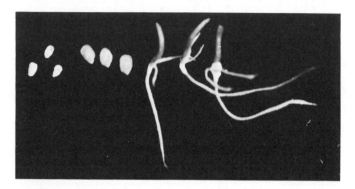

Fig. 4. Stage III wheat embryos (1.5 mm length) at the beginning of incubation (left), after 5 days in culture +ABA (middle) and after 5 days in culture –ABA (right). Note embryo growth and scutellum enlargement in +ABA embryos. From (53).

germinate.Soybean embryos cannot precociously germinate before 21 days (i.e., until the mid stage of embryo development) at which time the ABA concentration is approximately 10 µg/g fresh weight. If the endogenous ABA concentration is experimentally reduced in embryos during mid maturation by washing-out or drying slowly within or removed from detached pods, germination occurs. The extent of germination is correlated with length of washing or drying treatments which in turn, reduces endogenous ABA (1,2).

Drying, however, is not an obligatory requirement for soybean seed maturation or germination—depletion of ABA does seem to be sufficient. Levels below 3-4 µg ABA/g fresh weight is necessary for precocious germination, but during normal maturation of the soybean embryo this level is not reached until the late stage of development. This time corresponds to the onset of rapid dehydration which probably precludes germination in the seed even though ABA levels have declined below the threshold level. There is evidence in wheat, soybean and in rapeseed embryos that with declining ABA levels in late development, there is also a corresponding decrease in tissue sensitivity to ABA, indicating that dehydration and not ABA is probably responsible for inhibiting germination at this stage (c.f. 14,42). Desiccation is probably the normal trigger *in situ* not only to prevent germination of late embryos but to switch the developmental program from maturation to germination. Premature drying of *Phaseolus* and castor bean seeds alters the developmental potential such that, upon rehydration, they germinate and express germination markers rather than resuming maturation (31). Although dehydration is not required in culture for most embryos to switch, it may be necessary for certain embryos such as rapeseed that continue to express certain maturation traits during precocious germination (14).

Finally, previous studies showed that osmotic stress (e.g., sucrose or sorbitol) can both inhibit germination and maintain embryos of wheat (41) and rapeseed (15) in the maturation pathway. It has been assumed that osmotic stress results in increased ABA levels in embryos, similar to its effect on other plant organs such as leaves. Excised cotyledons of soybean cultured separate from the embryo axis increased ABA levels in the presence of 10% sucrose (8). However, ABA levels in rapeseed embryos did not increase significantly (15).The fact that inhibition of germination and maintenance of storage protein synthesis can be uncoupled from high ABA levels suggests that ABA is probably not the primary effector regulating these processes in rapeseed. Other possibilities include changes in ABA catabolism or compartmentation of ABA (c.f. 14,42).

ABA Mutants

Mutants with reduced sensitivity to ABA, or, altered ABA metabolism resulting in abnormal tissue levels of ABA have been studied (c.f. 36) in potato (*droopy*), pea (*wilty*), tomato (*flc, sit, not*), Arabidopsis (*abi, aba*) and maize (*vp*). Karssen (22) has characterized ABA-deficient single-gene mutant lines of *Arabidopsis* which possess symptoms of withering, increased transpiration and a lowered ABA content in ripe seeds and leaves. Dormancy of the mature seeds of ABA mutants was strongly reduced and in high humidity, vivipary was observed. Reciprocal crosses between wild type and ABA mutants showed both embryo and maternal genotypic origins of endogenous ABA. The onset of dormancy correlated with the ABA level regulated by the embryo genotype. More recently in *Arabidopsis*, Koornneef (25) selected and characterize five ABA-insensitive mutants from at least three different loci (*abi1, abi2, abi3*). These three mutants resembled the phenotype of ABA-deficient mutants. All five had the same or higher levels of ABA in developing seeds and fruits, but only two mutants showed only a reduction in seed dormancy. Exogenous ABA was ineffective in restoring the wild type phenotype in *abi1* and *abi2*.

Various viviparous mutants of corn have also been tested for the ability to accumulate and respond to ABA during kernel development. It has been found that *vp1* embryos have normal amounts of ABA, while *vp2*, *vp5*, *vp7* and *vp9* all have reduced levels. Carotenoid and ABA biosynthesis probably share some early steps, so it seems consistent that those *vp* mutants that accumulate low amounts of ABA are also those that show reduced levels of carotenoids. An inhibitor of carotenoid synthesis, fluridone, produces a phenocopy of this class of mutant when applied to wild type ears at a specific time in embryogeny. The vivipary produced is partially reduced by the application of exogenous ABA (16). ABA response has been measured in *vp* mutants by removing the immature

503

embryos from the seed at different times in development and measuring their ability to grow on various concentrations of ABA in culture (44). *Vp8* and *vp5* gave a growth inhibition curve that was identical with wild type embryos. Sensitivity to ABA was reduced about 7 fold in *vp1*, 5.4 fold in *vp7* and in *vp9* a 2 fold less-sensitive response was observed. ABA levels and responses were not measured in these *vp* mutants in stages other than the seed as Karssen (22) reported for *Arabidopsis*. As mentioned earlier, vivipary is a normal occurrence in mangroves, and embryos of this species are insensitive to ABA at concentrations that are effective on those of nonviviparous plants (50).

All of the above results strongly link the initiation of the maturation pathway and inhibition of precocious germination with ABA.Can we make a similar correlation between molecular events and ABA? Are there unique gene products associated with these events that can be increased or decreased by ABA, and if so, at what level does ABA exert this effect?

REGULATION OF GENE EXPRESSION

Gene expression has been extensively studied during embryo and seedling development in a number of monocots and dicots.Changes in the levels of specific mRNA's and proteins have been conveniently catagorized into sets by Dure (11) for cotton embryos, but can be used as a model for our general discussion (Fig. 5A). Based on this model and the discussion above, our interests in this chapter focus on groups 4 and 5, since they represent gene products that correspond to the time of embryo maturation and high ABA levels, and group 7, which represents gene products associated with germination. Also shown in Figure 5 are examples of the changes in the levels of specific mRNA's in cotton (Fig. 5B) and pea (Fig. 5C), as assayed by hybridization of radioactive cDNA sequences to RNA from different stages of embryo and seedling development. Given this background of the developmental changes in gene expression during seed development, can ABA be shown to affect the expression of specific genes and gene sets? Does it stimulate the expression of maturation gene products (e.g., groups 4 and 5), and suppress the expression of germination products (e.g., group 7)?

ABA Effects Specific Protein and mRNA Levels

Ten years ago, evidence was obtained that showed unique but uncharacterized proteins were synthesized in barley aleurone layers in response to ABA (19). More recently, it was demonstrated that one such protein, whose synthesis is increased in barley aleurone in the presence of ABA, is an α-amylase inhibitor (34). In addition, numerous studies with the cereal aleurone system have shown that the stimulation of hydrolases

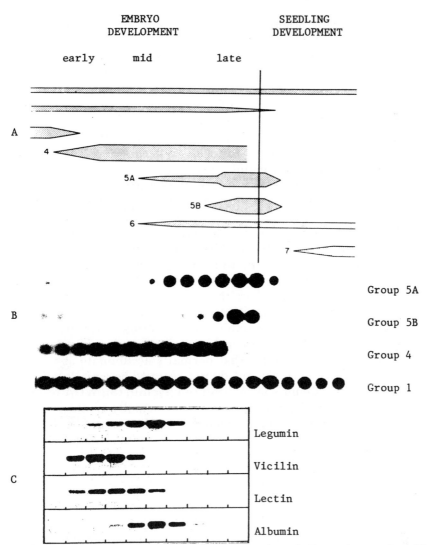

Fig 5. (A) Diagrammatic representation showing the times during embryo and seedling development when different groups of gene products from cotton embryos are detected (11). Changes in mRNA levels during embryo and seedling development: (B) four groups of genes in cotton (11), and, (C) four specific proteins in pea (18). Specific cDNA probes were used on filters containing mRNA from different developmental stages to demonstrate the steady-state amounts of RNA.

by GA can be reversed by ABA, and that this inhibitory effect occurs at both the transcriptional and translational levels (4,9). Hence, ABA affects gene expression by promoting and inhibiting the expression of specific genes. Given this data, and the studies discussed above showing that the endogenous level of ABA reaches its highest levels during *mid* to *late* development in most seeds, raised the likely possibility that the accumulation of gene products at this stage of seed development may be linked to an ABA effect.

In a number of species, specific gene sequences are affected by ABA. In *Brassica* (rapeseed), it has been shown that the *in situ* levels of mRNA for the storage protein cruciferin closely parallels the levels of endogenous ABA (Fig. 6). The same is true for the βsubunit of conglycinin from soybean (8). If embryos of *Brassica* and cotyledons from soybean are removed from the seed during mid maturation when these proteins are not detectable, and cultured in ABA, the levels of both proteins and their corresponding mRNA's are maintained or increased. This ABA effect, however, its not the same in late embryo development (c.f. 15). When an inhibitor of ABA accumulation, fluridone, is given to soybean cotyledons, the response in culture to ABA is reduced or eliminated (8). In wheat, an abundant embryo protein, Em, and wheat germ agglutinin (WGA) localized in the coleoptile have also been shown to accumulate in *mid* to *late* stages of embryogenesis, the time of high endogenous ABA levels in wheat (53,56). Em and WGA dramatically accumulate in a few days when wheat embryos are cultured in the presence of ABA (41,55,56). A corresponding increase in Em mRNA has also been observed (56).

Although these selected examples show a strong correlation between high levels of ABA (both exogenous and endogenous), and gene products *in situ* and *in vitro*, all proteins that normally accumulate at this mid to late stage of embryo development do not seem to be equally affected by ABA. The α' and α subunits of soybean conglycinin (8), WGA localized in the wheat root (43), and napin (*Brassica*) are normally synthesized earlier than the increase in levels of ABA (see Fig. 6). In the case of the α' and α

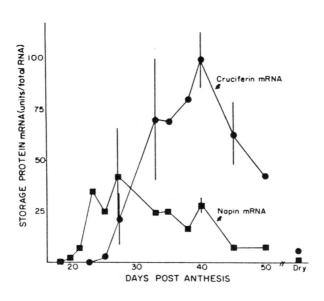

Fig. 6. mRNA levels of the storage proteins from *Brassica*. Compare the accumulation of these mRNA's with ABA levels in *Brassica* (Fig. 3C). The levels of only cruciferin mRNA match the pattern of ABA accumulation.

506

subunits of conglycinin, ABA and fluridone do not have any appreciable effect on their levels. The initial increase of napin and root WGA occurs prior to the increase in ABA, but the levels of both are enhanced by ABA. In the case of napin mRNA and an uncharacterized wheat mRNA sequence (pW511), ABA certainly does increase the levels of these messages, even though the levels *in situ* do not correspond very well with the pattern of ABA accumulation (14,55). In general, ABA does not seem to have a stimulatory effect on endosperm storage proteins in grains (unpublished observations). Hence, during *mid* to *late* maturation, ABA may have a major qualitative effect on certain gene products, only a quantitative effect on some, and no effect on still others.

It should be mentioned again, that an osmoticum such as sucrose can mimic the effects of ABA.These effects are also seen at the molecular level, most extensively described in *Brassica*. In the *Brassica* embryo, this appears to be independent of any changes in endogenous ABA levels, whereas in soybean cotyledons the molecular effects of sucrose are accompanied by increases in endogenous ABA (see p. 503). What now appears to be the case, is that ABA may be one of several triggers, or effectors, that together or independently (in parallel?) coordinately regulate the accumulation of embryo proteins at a particular stage in seed development (c.f. 14,42).

ABA Effects Gene Sets in the Maturation and Germination Pathways

Although the evidence is clear that mRNA and protein levels can be affected by ABA, can we identify a set of genes whose mRNA and protein products are a part of the "maturation loop" (c.f. Fig. 2), directly regulated by ABA? The gene products would have the following characteristics:

(1) accumulate in mid-late development at the time of increasing and/or high levels of ABA and can be inhibited by treatment with an inhibitor of ABA synthesis such as fluridone.
(2) are not essential for germination, i.e., if embryos are precociously germinated in culture, these proteins would not appear.
(3) can precociously accumulate in early embryos if cultured in ABA or an osmoticum; if increased by osmoticum treatment, is the accumulation accompanied by an increase in endogenous ABA and prevented by fluridone?
(4) present in the mature, dry embryo, but are lost and not accumulated during normal germination or in other stages in the normal life cycle.

If so, the products of these genes would be part of a unique maturation set and likely to comprise products from both groups 4 and 5 of Dure (c.f. Fig. 5) and be a subset of the total genes expressed during mid-late embryo development.

A set of proteins whose synthesis normally occurs late in embryogenesis but which are prematurely synthesized in young embryos when they are excised and incubated in ABA have been identified in cotton (*Lea*, c.f. 17) and wheat (acid soluble proteins, c.f. 40). These represent the closest group of gene products that fit the characteristics listed above. Let's take a look at the set described in wheat embryos.

Proteins in an acid-soluble fraction from wheat embryos are normally accumulated in *mid* to *late* development, but can precociously appear in early embryos when cultured in the presence of ABA. These proteins are identical to those normally synthesized later in embryo development, and after five days in culture, embryos accumulate approximately the same levels as what is found after 2-3 additional weeks *in situ*.These proteins are not found in embryos that precociously germinate, and are quickly lost when mature embryos germinate (Fig. 7). Included in this set are the Em protein, coleoptile WGA and what has been tentatively identified as a globulin storage protein.

By differential hybridization we have also identified a set of cDNA clones whose sequences are expressed in mid-late embryo development and found to dramatically increase in ABA-treated embryos (40). Use of the *dek* and *vp* mutants of corn and the embryo-lethal mutants of *Arabidopsis*, described earlier, should help identify and catagorize similar genes that fit these characteristics, and present another "handle" to help establish the role ABA in maturation.

Another part of this approach is to identify gene products that are synthesized and accumulated only during germination (both normal and precocious) and are suppressed by ABA.We have shown in wheat that ribulose bisphosphate carboxylase, a chloroplast-specific protein, fits these requirements (40). Similar resultswere obtained in castor bean embryos using the germination-specific marker, isocitrate lyase (10).Within 24 hours after germination, either precocious or normal, with or without light, wheat embryos begin to accumulate carboxylase mRNA and protein.In the presence of ABA, carboxylase protein is not found in the light or dark, and although its mRNA is detectable, it does not accumulate above the very low levels (40). Hence, ABA treatment of wheat embryos, simultaneously stimulates increases in mRNA levels of certain gene sets, while preventing increases in others. At what level does ABA exert this effect?

Levels Of Genetic Control Affected By ABA

Changes in the levels of mRNA during seed development strongly suggests a control component at the level of transcription, either directly at mRNA synthesis or at mRNA stability (c.f. 42). Preliminary evidence clearly suggests an effect at the rate of mRNA synthesis. Nuclei from ABA-treated *Brassica* embryos and preliminary evidence in soybean

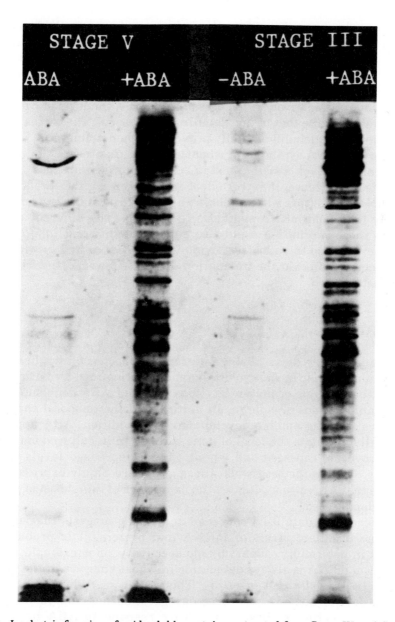

Fig. 7. Isoelectric focusing of acid soluble proteins extracted from Stage III and Stage V wheat embryos cultured for 3 days in +ABA and –ABA. The pH gradient in the gel stained with Coomassie Blue is from 3.5 (top) to 9.5 (bottom). The same number of embryos was used in each extract. These results demonstrate that ABA stimulates the precocious appearance of a set of maturation proteins in Stage III embryos. In –ABA, seedlings form from these embryos without accumulating this protein set. Stage V embryos in +ABA retain the maturation set accumulated while on the plant. These are quickly lost upon germination (–ABA). The proteins accumulated in Stage III embryos in response to ABA in culture are identical to those that are present in Stage V embryos that matured on the plant (compare patterns of Stage III and V +ABA). From (41).

cotyledons show an enhanced synthesis *in vitro* of napin mRNA and βsub-unit of conglycinin respectively. These nuclear "run-on" experiments show a transcriptional component involved in the ABA control mechanism. In wheat, the Em and putative globulin mRNA's are stored in the dry embryo and upon imbibition both mRNA's are lost and not detected after 48 hours. However, in the presence of ABA, the level of Em mRNA is maintained while that of the globulin is increased. The increase in the mRNA level for globulin is totally prevented when α-amanitin, a specific inhibitor of mRNA synthesis in wheat embryos, is present. In contrast, the Em mRNA level is maintained in ABA even in the absence of mRNA synthesis (55). Hence, the ABA effect in maintaining the levels of these two mRNA's in mature wheat embryos appears to be at the level of mRNA stability (Em) and mRNA synthesis (globulin). The different effect of ABA on these two mRNA could also reflect the different inherent half-lives.There is also some indirect evidence in *Brassica* and in wheat that ABA exerts an effect at the translational level. However, the data is not well substantiated.

SUMMARY

A number of studies utilizing different approaches all implicate ABA playing a major role in embryo development during seed formation.These include the analysis of mutants (viviparous and ABA-deficient/insensitive), manipulations of embryos in culture and the temporal changes in ABA levels during embryo development. In addition, ABA appears to specifically modulate the level of gene expression in cultured embryos of cotton, soybean, rapeseed and wheat. Since increased levels of ABA normally occur in the seed when these genes are highly expressed, it is probable that ABA also modulates the expression of this subset of genes in the developing seed.

However, the data from rapeseed and wheat suggests that although ABA may modulate levels of mRNA and protein, it is probably not involved in determining embryogenic specificity. For example, a basal level of Em mRNA is always detectable in wheat embryos in the absence of ABA, both *in situ* and *in vitro*. This is not seen, however, during seedling development, i.e., in three-day germinating embryo tissue. In *Brassica*, immature embryos in culture respond to ABA by making storage proteins. Immature embryos cultured in the absence of ABA continue to exhibit a basal level of storage protein synthesis while mature tissues do not. It appears, therefore, that a signal other than ABA may be involved in determining the embryogenic specificity of Em and storage protein expression.The embryogenic signal may activate an embryo specific gene set (such as Em, *Lea* sequence and some dicot storage proteins) early in development and result in a basal level of expression. ABA could then affect in the up-modulation of expression within this group later in embryogenesis (14,42).

Higher levels of ABA are also required to trigger continued expression of Em in cultured stage V wheat embryos compared to stage III. In *Brassica* also, levels of ABA required to elicit continued expression of storage proteins continues to rise after the onset of germination until they become toxic (14). These results suggest that some germination signal, such as desiccation inactivates the embryo gene set on which ABA had one modulatory effect. In fact, recent evidence supports this idea.

Results with *Phaseolus* show that premature drying not only redirects metabolism from an embryogenic to a germination program, but it does so permanently. The pattern of proteins synthesized *in vitro* by the mRNA fraction from fresh and prematurely dried axes show strong similarities. The mRNA population from rehydrated axes exhibit a different set of proteins when translated *in vitro*. Also, the message for the storage protein phaseolin, is preserved following the normal maturation process and premature desiccation, but following rehydration, this message is no longer detectable in the embryo axes (c.f. 31).

A hypothesis worth considering then, is as follows. At the end of earliest third of seed development (c.f. Figs. 1 and 2), the increasing levels of endogenous ABA promotes the maturation phase of embryo development and represses the seedling phase by modulating gene sets; positively affecting the maturation gene set and negatively affecting the germination gene set. *Both* sets were programmed soon after fertilization for expression during embryo and seedling development by some as yet undefined cue, i.e., the sets were "opened". When embryos are removed from the seed, prematurely, their ABA level drops and seedling development is favored, i.e., precocious germination. The maturation set *slowly* loses its ability to be expressed under these conditions, and desiccation is not an absolute requirement for this switch. If exogenous ABA is added to cultures to restore the high ABA levels found *in situ*, maturation is maintained and seedling development is inhibited. If the embryo continues development in the seed, the sensitivity of the embryo to ABA, and the absolute amount of ABA decreases at the time when the embryo loses water. Hence, desiccation and not ABA prevents germination under normal conditions during these later stages of embryo development. Desiccation also appears to be involved with permanently closing the maturation gene set and making it unavailable for expression after germination. Upon hydration, with low levels of ABA and low sensitivity to ABA, and a "closed" maturation gene set, the embryo begins seedling development by expressing the germination gene set. Gene products remaining from maturation on the plant are lost.

Hopefully this hypothesis will be tested in the next few years so that a more complete understanding of the physiological and molecular controls of embryogenesis can be obtained.

References

1. Ackerson, R. C. (1984) Regulation of soybean embryogenesis by abscisic acid. J. Exp. Bot. *35*, 403-413.
2. Ackerson, R. C. (1984) Abscisic acid and precocious germination in soybeans. J. Exp. Bot. *35*, 414-421.
3. Albone, K. S., Gaskin, P., MacMillan, J. Sponsel, V. M. (1984) Identification and localization of gibberellins in maturing seeds of the curcurbit *Sechium edule*, and a comparison between this curcurbit and the legume *Phaseolus coccineus*. Planta *162*, 560-565.
4. Baulcombe, D., Lazarus, C., Martienssen, R. (1984) Gibberellins and gene control in cereal aleurone cells. J. Embryol. Exp. Morph. *83* (Suppl.), 119-135.
5. Bewley, J. D., Black, M.(1978) Physiology and Biochemistry of Seeds. Springer, N.Y.
6. Black, M. (1983) Abscisic acid in seed germination and dormancy. *In* Abscisic Acid, Chapt. 10, pp. 331-363, Addicott, F. T., ed., Praeger Publishers, New York, NY.
7. Brady, T., Walthall, E. D.(1985) The effect of the suspensor and gibberellic acid on *Phaseolus vulgaris* embryo protein content. Develop. Biol. *107*, 531-536.
8. Bray, E. A., Beachy, R. N. (1985) Regulation by ABA of βconglycinin expression in cultured developing soybean cotyledons. Plant Physiol. *79*, 746-750.
9. Chandler, P. M., Zwar, J. A., Jacobesen, J. V., Higgins, T. J. V., Inglis, A. S. (1984) The effects of gibberellic acid and abscisic acid on α-amylase mRNA levels in barley aleurone layers. Studies using an α-amylase cDNA clone. Pl. Mol. Biol. *3*, 407-418.
10. Dommes, J., Northcote, D. H. (1985) The action of exogenous abscisic and gibberellic acids on gene expression in germinating castor beans. Planta *165*, 513-521.
11. Dure, L. (1985) Embryogenesis and gene expression during seed formation. *In* Oxford Surveys in Plant Molecular and Cell Biology, vol. *2*, pp. 179-197, Miflin, B. J., ed., Oxford Univ. Press, Oxford, UK.
12. Eisenberg, A. J., Mascarnhas, J. P. (1985) Abscisic acid and the regulation of synthesis of specific seed proteins and their messenger RNAs during culture of soybean embryos. Planta *166*, 505-514.
13. Eeuwens, C. J., Schwabe, W. W. (1975) Seed and pod wall development in *Pisum sativum*, L. in relation to extracted and applied hormones. J. Exp. Botan. *26*, 1-14.
14. Finkelstein, R., Crouch, M. (1985) Control of embryo maturation in rapeseed. *In* Plant genetics, in press, Freeling, M., ed., Liss, N.Y.
15. Finkelstein, R. R., Tenbarge, K. M., Shumway, J. E., Crouch, M. L. (1985) Role of ABA in maturation of rapeseed embryos. Plant Physiol. *78*, 630-636.
16. Fong, F., Smith, J. D., Koehler, D. E. (1983) Early events in maize seed development. Plant Physiol. *52*, 350-356.
17. Galau, G. A., Hughes, D. W., Dure, L. (1986) Abscisic acid induction of cloned cotton late embryogenesis-abundant (*Lea*) mRNAs. Pl. Mol. Biol. In Press.
18. Higgins, T J. V. (1984) Synthesis and regulation of major proteins in seeds. Annu. Rev. Plant Physiol. *35*, 191-221.
19. Ho, D.TD-H., Varner, J E. (1976) Responses of barley aleurone layers to abscisic acid. Pl. Physiol. *57*, 175-178.
20. Hsu, F. (1979) Abscisic acid accumulation in developing seeds of *Phaseolus vulgaris* L. Pl. Physiol. *63*, 552-556.
21. Hymowitz, T. (1983) Variation in and genetics of certain antinutritional and biologically active components of soybean seed. *In* Better crops for food, pp. 49-56, Nugent, J., O'Connor, M., eds., Ciba Foundation Symp. *#97*, Pitman, London.
22. Karssen, C. M., Brinkhorst-Van Der Swan, D. L. C., Breekland, A. E., Koornneef, M. (1983) Induction of dormancy during seed development by endogenous abscisic acid: studies on abscisic acid deficient genotypes of *Arabidopsis thaliana* (L.) Heynh. Planta *157*, 158-165.

23. King, R. W. (1976) Abscisic acid in developing wheat grains and its relationship to grain growth and maturation. Planta *132*, 43-51.
24. King, R. W., Salminen, S. O., Hill, R. D., Higgins, T. J. V. (1979) Abscisic acid and gibberellin action in developing kernels of Triticale (c.v. 6A190). Planta *146*, 249-255.
25. Koornneef, M., Reuling, G., Karssen, C. M. (1984) The isolation and characterization of abscisic-acid-insensitive mutants of *Arabidopsis thaliana*. Physiol. Plant. *61*, 377-383.
26. Letham, D. S., Goodwin, P. B., Higgins, T. J. V. (ed) (1978) Phytohormones and Related Compounds: A Comprehensive Treatise. *648* pp. Elsevier, N.Y.
27. Long, S. R., Dale, R. M. K., Sussex, I. M. (1981) Maturation and germination of *Phaseolus vulgaris* embryonic axis in culture. Planta *153*, 405-415.
28. Meinke, D. W. (1986) Use of embryo lethal mutants to study plant embryo development. *In* Oxford Surveys in Plant Molecular and Cell Biology, Vol. *3* in press, Miflin, B. J. ed., Oxford Univ. Press, Oxford, UK.
29. Michael, G., Beringer, H. (1980) The role of hormones in yield formation. In: Physiological Aspects of Crop Productivity, pp 85-116, 15th Colloquium Int. Potash Inst., Bern, Switzerland.
30. Michael, G., Seiler-Kelbitsch, H. (1972) Cytokinin content and kernel size of barley grains as affected by environmental and genetic factors. Crop Sci. *12*, 162-165.
31. Misra, S., Kermode, A., Bewley, J. D. (1985) Maturation drying as the "switch" that terminates seed development and promotes germination. *In* Molecular Form and Function of the Plant Genome, pp. 113-128, van Vloten-Doting, L., Groot, G. S. P., Hall, T. C. eds., Plenum, N.Y.
32. Mok, D. W. S., Mok, M. C., Rabakoarihanta, A., Shii, C. T. (1985) Phaseolus-Wide hybridization through embryo culture. *In* In Vitro Improvement of Crops, in press, Bajaj, Y. P. S. ed., Springer-Verlag, Berlin.
33. Mounler, M. A. Kh., Bangerth, F., Story, V.(1980) Gibberellin-like substances and indole type auxins in developing grains of normal- and high-lysine genotypes of barley. Physiol. Plant. *48*, 568-753.
34. Mundy, J. (1984) Hormonal regulation of α-amylase inhibitor synthesis in germinating barley. Carlsberg Res. Commun. *49*, 439-444.
35. Naumann, R. K., Dorffing, K.(1982) Variation of free and conjugated abscisic acid, and dihydrophaseic acid levels in ripening barley grains. Plant Sci. Lett. *27*, 111-117.
36. Neill, S. J., Horgan, R. (1985) Abscisic acid production and water relations in wilty tomato mutants subjected to water deficiency. J. Exp. Botan. *36*, 1222-1231.
37. Pharis, R. P., King, R. W. (1985) Gibberellins and reproductive development in seed plants. Ann. Rev. Plant Physiol. *36*, 517-568.
38. Phillips, R., Green, C. E., Gengenback, B. G. (eds) (1979) The Plant Seed: Development, Preservation and Germination. Academic Press, N.Y.
39. Potts, W. C., Reid, J. B. (1983) Intermode length in *Pisum*. II. The effect and interaction of Na/na and Le/le gene differences on endogenous gibberellins-like substances. Physiol. Plant. *57*, 448-454.
40. Quatrano, R. S., Ballo, B. L., Williamson, J. D., Hamblin, M. T., Mansfield, M. (1983) ABA controlled expression of embryo-specific genes during wheat grain development. *In* Plant molecular biology, pp. 343-353, Goldberg, R., ed., Liss, New York.
41. Quatrano, R. S., Hopkins, R., Raikhel, N. V.(1983). Control of the synthesis and localization of wheat germ agglutinin during embryogenesis. *In* Chemical taxonomy, molecular biology, and function of plant lectins, pp. 117-130, Goldstein, I., J., Etzler, M. E., eds., Liss, New York.
42. Quatrano, R. S.(1986) Regulation of gene expression by abscisic acid during angiosperm embryo development. *In* Oxford Surveys in Plant Molecular and Cell Biology, vol. 3 in press, Miflin, B. J. ed., Oxford Univ. Press,Oxford, UK.
43. Raikhel, N. V., Quatrano, R. S. (1986) Location of wheat germ agglutinin in developing wheat embryos and those cultured in abscisic acid. Planta. in press.

44. Robichaud, C. S., Wong, J., Sussex, I. M. (1980) Control of in vitro growth of viviparous embryo mutants of maize by abscisic acid. Develop. Genet., *1*, 325-330.
45. Rogers, S. O., Quatrano, R. S.(1983) Morphological staging of wheat caryopsis development. Am. J. Bot. 70, 308-311.
46. Scowcroft, W. R., Larkin, P. J. (1983) Somaclonal variation and genetic improvement of crop plants. *In* Better crops for food, pp. 177-88, Nugent, J., O'Connor, M., eds., Ciba Foundation Symp. *#97*, Pitman, London.
47. Seiler-Kelbitsch, H., Michael, G., Hauser, H., Fischbeck, G. (1975) Cytokiningehalt und Kornentwicklung von Gerstenmutanten mit unterschiedlicher Korngrosse. Z. Plfanzenzuchtung. 75, 311-316.
48. Sheridan, W. F., Neuffer, M. G. (1982) Maize developmental mutants. J. Heredity *70*, 318-329.
49. Stinissen, H. M., Peumans, W. J., DeLanghe, E. (1984) Abscisic acid promotes lectin biosynthesis in developing and germinating rice embryos. Plant Cell Rep. *3*, 55-59.
50. Sussex, I. (1975) Growth and metabolism of the embryo and attached seedling of the viviparous mangrove *Rhizophora mangle*. Amer. J. Bot. *62*, 948-953.
51. Sussex, I. M., Dale, R. M. K. (1979) Hormonal control of storage protein synthesis in Phaseolus vulgaris. *In* The plant seed: development, preservation and germination, pp. 129-141, Phillips, R. L., Green, C. E., Gengenbach, B. G., eds., Academic Press, New York.
52. Tollenaar, M. (1977) Sink-source relationship during reproductive development in maize. A review. Maydica XXII, 49-75.
53. Triplett, B. A., Quatrano, R. S. (1982) Timing, localization, and control of wheat germ agglutinin synthesis in developing wheat e mbryos. Dev. Biol. 91, 491-496.
54. Walbot, V. (1978) Control mechanisms for plant embryogeny. *In* Dormancy and Developmental Arrest, pp. 113-166, Clutter, M., ed., Academic Press, N.Y.
55. Williamson, J. D. (1985) The effects of abscisic acid on gene expression during embryogenesis in wheat. Ph.D. dissertation. Oregon State University. pp. 95.
56. Williamson, J. D., Quatrano, R. S., Cuming, A. C. (1985) E_m polypeptide and its messenger RNA levels are modulated by abscisic acid during embryogenesis in wheat. Eur. J. Biochem. *152*, 501-507.
57. Yeung, E. C., Sussex, I. M. (1979) Embryogeny of *Phaseolus coccineus*: the suspensor and the growth of the embryo-proper in vitro. Z. Pflanzenphysiol. 91, 423-33.
58. Zeevaart, J. A. D. (1965) Reduction of the gibberellin content of Pharbitis seeds by CCC and after-effects in the progeny. Plant Physiol. *41*, 856-62.

E 14. The Role of Hormones in Potato (*Solanum tuberosum* L.) Tuberization

Elmer E. Ewing

Department of Vegetable Crops, Cornell University, Ithaca, New York 14853, USA.

INTRODUCTION

Tuber initiation in the potato plant is accompanied by extensive morphological and biochemical changes above and below ground. It has long been postulated that the changes are mediated hormonally. Before considering the evidence, I would like to review what is known about the tuberization process, its environmental control, and the techniques that have been employed to study it.

Description of Tuberization

A potato tuber is a modified stem, with nodes (the "eyes") and internodes. Its leaves are tiny and scale-like, often dehiscing during harvest and handling. The axis of the tuber is shortened and greatly thickened, and its tissues are packed with starch, but the anatomy of the tuber resembles that of a typical stem. Normally tubers form underground on rhizomes, more commonly called stolons. Tuber initiation begins in the subapical zone of the stolon, and swelling of the tuber occurs acropetally in internodes that were already present at the time of initiation (7). At about the same time that the swelling occurs, the meristematic activity in the stolon apex ceases. Longitudinal sections of stolons during early stages of tuber formation indicate that cell enlargement precedes increases in cell division (26). Starch deposition occurs very early in the ontogeny of the tuber, and it is accompanied by the development of high levels of patatin, a glycoprotein unique to tubers (48).

Tubers form most readily underground. Why this is so is not entirely clear. The most important factor is probably darkness, but the physical resistance of the soil particles may also play a role (P. Struik, personal communication). Under unusual circumstances tubers form above ground, usually at axillary buds. Such "aerial" tubers are much smaller and contain chlorophyll. Aerial tubers illustrate the point that although the tuber most often forms on a stolon, any bud or shoot apex of the potato is capable of tuberizing, given the proper conditions. For example,

515

axillary buds cut from plants that are tuberizing will usually develop tubers if the buds are buried in the soil, particularly if there is a leaf attached to the bud cutting. This ability of leaf-bud cuttings to tuberize affords a useful tool for the study of tuberization.

By taking leaf-bud cuttings from plants exposed to various environmental conditions, it is possible to study how such conditions affect the extent to which the plant is induced to tuberize. For this purpose it is helpful to classify the various growth responses of the buried bud as to the degree of induction which they represent. The range of responses is illustrated for apical cuttings in Fig. 1. Leaf-bud cuttings show a similar progression (10, 36).

Control of Tuber Initiation

Environment

The degree to which a particular potato plant is induced to tuberize is controlled by many factors. The environmental factor which has been most investigated is the effect of photoperiod. It is well known that long nights favor the induction of tuberization: with respect to tuber formation, potatoes are short day plants. (Although the controlling factor is the length of the dark period, the convention of referring to the daylength will be followed in the rest of this chapter.) Presumably the tuberization response is mediated by phytochrome, since five minutes of red light interrupting the daily dark period will reduce tuberization, and far-red light tends to reverse the effect of the red light (1). Temperature also has a pronounced effect on the level of induction to tuberize. Cool temperatures (day temperatures below 30°C and night temperatures below 20°C) favor tuber induction. Both soil temperatures and air temperatures seem to play a role in controlling tuberization. The effects of photoperiod and temperature on tuber induction depend on the irradiance under which the plant is growing. Effects of long days or high temperatures are exaggerated at low levels of irradiance (4, 41). A fourth environmental factor affecting tuber induction is the amount of nitrogen available to the plant. Heavy applications of nitrogen fertilizer reduce the level of tuber induction. Under hydroponic conditions, withdrawal of nitrogen from the nutrient solution may produce immediate tuberization on stolons; addition of nitrogen again may cause tubers to revert to stolons (27, 55).

The mother tuber

Although potato plants can be propagated from seeds, ordinarily they are grown from tubers. The condition of the planted tuber exercises considerable influence over the degree to which the potato plant will respond to the environment. Tubers planted soon after they have passed through their normal dormant period give rise to plants which are less

Fig. 1. Response pattern of apical cuttings, illustrating the typical progression of responses to increasing levels of induction prior to cutting. The most basal node of each cutting was buried in potting mix, as indicated by the dotted line. Cuttings were kept in a mist bench for 11 days. A) Buried bud fails to grow, while roots and the shoot apex grow rapidly. B) Buried bud develops into a leafy shoot or stolon. C) Tuber forms at the tip of the stolon. D) Buried bud forms a sessile tuber. Note that as induction to tuberize increases from A) to D), the growth of the shoot apex and of adventitious roots decreases. From (9).

readily induced to tuberize than are plants developing from old tubers. The physiological age of the tubers appears to be controlled not only by their chronological age, but also by the conditions under which they were grown and stored. Warm growing conditions (e.g., mean temperatures >23°C) and relatively high storage temperatures (e.g., >5°C) result in tubers that behave as if they were physiologically older. Tubers that are extremely old, whether because of prolonged storage at cold temperatures or at warmer temperatures with repeated sprout removal, will not develop normal sprouts. Instead, new tubers will form directly at the eyes and there will be no production of new stems and leaves—a disorder variously referred to as "sprout tubers" or "little potato."

Genetics

There are great genetic differences among potatoes in their responses to all of the above factors. The potato was brought under cultivation in the Andean highlands, where photoperiods are about 12 hours and temperatures are cool. Varieties adapted to these conditions are classified as *Solanum tuberosum* L. group Andigena. When grown under the higher temperatures of the lowland tropics or the long summer days of temperate zones, Andigena potatoes will tuberize weakly or not at all. The potatoes adapted to summers in the temperate zone—the potatoes of southern Chile, Europe, and North America—are classified as *S. tuberosum* group Tuberosum. Tuberosum potatoes give low yields in the highland tropics because tuberization is "turned on" so early and so intensely by the cool temperatures and short days that there is too little shoot growth to support good tuber yields. There are also substantial genetic differences in response to photoperiod within each group.

Leaf area

A final consideration is plant size. When genetic and environmental conditions are highly favorable for tuberization, tubers may form on very small plants. For example, Tuberosum plants grown from seeds sometimes tuberize when only one leaf above the cotyledons has developed, especially if photoperiods are short. At the other extreme, Andigena plants may grow for more than six months without tuberizing when exposed to long photoperiods. Thus there is no particular stage of development that necessarily coincides with tuber initiation.

Experiments with cuttings having different leaf areas (24) demonstrate that when other conditions are equal, cuttings with a large leaf area are more likely to develop tubers than are those with a smaller leaf area. Whether a given set of environmental conditions is favorable for tuber induction may therefore depend upon the leaf area of the plant. A small plant may fail to tuberize, yet tubers may form as the plant attains a greater leaf area even though the environment does not change. It is as though every leaflet were contributing its part toward tuber initiation, with initiation occurring only when the sum of the contributions from all of the leaflets on the plant has attained the necessary level. The threshold will be reached at a small leaf area if conditions strongly favor induction, at a correspondingly greater leaf area if conditions are less favorable for induction, and not at all under very unfavorable conditions (24).

Partitioning of Assimilate and Tuber Induction

Strongly induced cuttings appear to channel all of their assimilate into tuber production. There will be little or no new growth of shoots or roots, even if the cuttings include the shoot apex. The same effect may be seen in whole plants. Within normal ranges shorter photoperiods, cooler temperatures, higher irradiance, lower levels of nitrogen fertilization, physiologically older mother tubers, and earlier maturing varieties will increase the ratio of dry matter partitioned to tubers versus the dry matter in the remainder of the plant. It appears that the reduced growth of shoots and roots accompanying tuber induction is not explained solely by "starvation" of the other organs through diversion of assimilate to tubers. Shoot and root growth rates on cuttings decline with strong inducing conditions even if tuber formation is prevented through excision of buried buds that would produce stolons and tubers.

The overall changes in plant morphology accompanying induction raise the question whether tuber induction should be considered as an indirect effect of the restricted plant growth. Could it be that under inducing conditions the amount of assimilate required by shoots and roots is so diminished that the surplus of assimilate stimulates tuberization? Such an explanation may apply to the dahlia, but apparently not to the potato. There is evidence that short days promote tuberization in dahlia through inhibition of top growth and consequent diversion of assimilate to the tubers; leaf cuttings of dahlia tuberize regardless of photoperiod once competing sinks have been removed (3). The situation in potato is completely different--tubers form on cuttings only if the plants from which the cuttings were taken or the cuttings themselves were exposed to inductive photoperiods. Excision of the above-ground bud from two-node potato cuttings has no effect on the tuberization of the buried bud (12).

We may conclude that in potato the decreased growth of shoots and roots accompanies, but does not directly cause, tuber initiation; nor is the diversion of assimilate to tubers entirely responsible for the restricted growth of shoots and roots. How, then, are the effects of environmental factors on tuber initiation and on other morphological changes mediated?

Grafting Experiments

Intraspecific grafts

Grafting experiments led Gregory (16) to hypothesize that tuber induction is triggered by a stimulus that is produced in the leaf. If a leaf of a potato plant that has been exposed to short photoperiods is grafted to a plant that has been exposed to long photoperiods, tuberization will occur (6, 16, 28). When the leaf of a special Andigena clone that tuberizes under all photoperiods was grafted to a stem of an ordinary Andigena plant in long days, tubers were produced at the buried bud (Fig. 2). In the reciprocal graft, no tubers formed. Thus it appears that in the leaf of the special clone a stimulus is produced under long photoperiods that can be translocated through a graft union, causing tuberization on a clone which ordinarily would not tuberize under long photoperiods.

Fig. 2. Grafted cuttings taken from plants exposed to 20-h photoperiods. The lower leaf on each stock was excised at time of cutting. The portion below the graft union (dotted line) was inserted in potting mix. Cuttings were maintained in a mist chamber under a 20-h photoperiod for 12 days after cutting and grafting. A) Scion taken from a selected clone of Andigena that is able to tuberize under continuous light; stock from ordinary Andigena, requiring short days to tuberize. Three of four stocks tuberized. B) Reciprocal graft from that shown in A. No tuberization on six grafted cuttings. From (12).

Interspecific grafts

More intriguing results have been produced by grafting other species. The stimulus will pass through leafless eggplant or tomato stem segments intergrafted between induced potato scions and noninduced potato stocks; but leaves of tomato or eggplant diminish tuberization when grafted to potato even if such leaves are exposed to short photoperiods (32, 46). On the other hand, it was shown by Nitsch (43) that the Jerusalem artichoke—the tuberization of which is also favored by short photoperiods—would tuberize when grafted to sunflower leaves that had been exposed to short days, but not when grafted to sunflower leaves exposed to long days. This implied that leaves of the sunflower were able to receive and transmit to the Jerusalem artichoke a tuberization signal. These interesting results and their implications for the potato lay fallow in the literature for many years. Then, more or less simultaneously, workers in France (34) and the USSR (5) independently carried out almost identical experiments stimulated by Nitsch's reports. Inasmuch as flowering of the sunflower is controlled by photoperiod, these researchers decided to find a species the flowering of which would be sensitive to photoperiod and which could be grafted to potato. Both groups chose tobacco. Several species of tobacco were utilized, including plants that flower only under long days, plants that flower only under short days, and ones that flower under either long or short days. Results are summarized in Table 1. It will be seen that a tobacco requiring short days for flowering would induce tuberization on potato only if the tobacco leaves received short days; whereas a tobacco that requires long days for flowering caused tuberization on potato when the tobacco leaves were exposed to long days. The experiments seem to show that the stimulus for flowering of the tobacco is graft-transmissible to the potato, where it induces tuberization.

Table 1. Summary of grafting experiments in which potato or tobacco scions were grafted to potato stocks from which all leaves were excised. Plants were kept under long or short photoperiods. From (5 and 34)

Scion	Short Days		Long Days	
	Flowers on tobacco scion[y]	Tubers on potato stock[y]	Flowers on tobacco scion	Tubers on potato stock
Mammoth tobacco	+	+	0	0 or +[z]
Xanthi tobacco	+	+	+	+
Trapezond tobacco	+	+	+	+
Sylvestris tobacco	0	0	+	+
Andigena-type potato[x]	n.a.	+	n.a.	0
Tuberosum potato	n.a.	+	n.a.	+

x   Andigena (5) or a hybrid of Tuberosum X Demissum that required short days for tuberization (34).
y   + = present, 0 = absent, n.a. = not applicable.
z   Results were inconsistent between experiments (5).

Site of Production and Movement of the Stimulus

Role of roots
    Shoot cuttings of Andigena plants were grown hydroponically with the developing adventitious roots continually removed. Leaf-bud cuttings taken from the de-rooted plants still tuberized, provided the shoots had been given short photoperiods (10). Therefore roots are not essential for the production of the induced condition.

Gravity and bud age

Cuttings also permit us to study the effects of gravity on the translocation of the signal to tuberize. If a potato shoot is subjected to inducing conditions, excised, and inserted into soil in such a manner that the three most basipetal buds are buried, tuberization will occur principally at the oldest bud (Fig. 3). The youngest of the three will

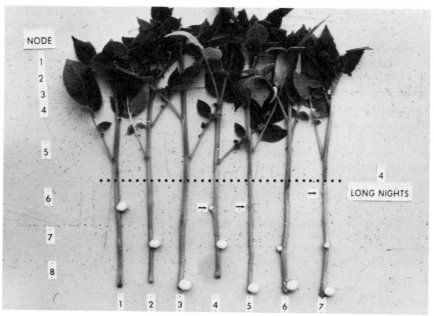

Fig. 3. Shoots containing eight nodes plus the apex were cut from Tuberosum plants (cv. 'Chippewa') that had been exposed to four 8-h days. The leaves at nodes 6, 7, and 8 were excised; and the stems below node 5 were buried, as indicated by the dotted line. From left to right, buds were excised from nodes 7 and 8; 6 and 8; 6 and 7; 8; 7; 6; none. Arrows point to buds or small tubers. Each of the three nodes was capable of tuberizing in the absence of buds at the other two nodes; but if more than one bud was present, tuberization was always strongest at the most basal node. Unpublished figure from data reported in (23).

tuberize very rarely, and then with much smaller tubers than at the other two nodes. We may ask whether the basipetal pattern is produced by the effects of gravity or by a tendency for older buds to tuberize more strongly. By inverting the cuttings and by comparing the effects of different types of cutting, we can see that neither explanation is satisfactory (Fig. 4). Under some conditions the physically upper bud and the younger bud will tuberize more strongly (Fig. 4D). We do not yet understand fully the explanation for the tuberization patterns, but there is some indication that tuberization is expressed most strongly at the bud which is most distant from an illuminated leaf or stem (23).

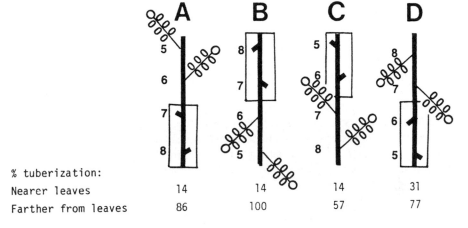

% tuberization:	A	B	C	D
Nearer leaves	14	14	14	31
Farther from leaves	86	100	57	77

Fig. 4. Four-node cuttings were taken from Tuberosum plants (cv. 'Katahdin') that had previously received eight cycles of 8-h days. Cuttings included nodes 5 through 8, counting down from the apex. In treatments A and B, leaves at nodes 7 and 8 were excised, and their buds were buried in containers of potting mix. In treatments C and D, nodes 5 and 6 were the buried buds. Treatments B and D had cuttings inverted in the mist bench; treatments A and C had cuttings upright. The tuberization at the buried buds was always stronger at the bud more distant from the leaves (node 8 in A and B; node 5 in C and D). From (23).

Leaf age

Other experiments with cuttings indicate that both young leaves and old leaves are effective in producing the induced condition, but younger leaves are more effective per unit of leaf area (24). Even the apical bud promotes tuberization on induced plants (24), although there is evidence that under noninductive conditions, tuberization is favored by bud excision (17, 39, 40, 45).

Girdling

When one stem of a two-branched plant was exposed to short days and the other to long days, girdling of either stem blocked the effect of that stem on tuber induction. That is, tuberization was improved if the stem receiving long days was girdled, and tuberization was inhibited if the stem receiving short days was girdled (45).

THE HORMONAL CONTROL OF TUBERIZATION

Although other explanations cannot be excluded, much of the information already presented in this chapter suggests that tuberization is under hormonal control: the alteration of the total morphology of the plant during the induced state, even if tuberization is surgically prevented; transmissibility of the stimulus across grafts; the effects of girdling; and the contribution of the mother tuber to induction would all seem consistent with a hormonal role. Is there a hormone that is produced in the leaves in response to long nights, that is transmissible across grafts, and that induces changes in the morphology and physiology of the plant including tuberization? If we accept the evidence from tobacco/potato grafts, then it does not seem that the stimulus for tuberization is unique to tuberizing species of *Solanum*. Moreover, the stimulus could be a single compound or a balance in concentrations of a group of compounds—some, but not necessarily all, of them hormonal.

As the number of short days to which a plant is exposed increases, the response at buried buds of cuttings taken from the plant changes from no growth, to a shoot, to a stolon, to a tuber (12). Is the stimulus for shoot or stolon formation at the underground bud the same stimulus that when present at a higher concentration induces tuberization, or are there two separate stimuli? Tuber induction is also associated with changes in leaf morphology (59). Leaves are larger, thinner, and have a flatter angle to the stem; axillary branching is suppressed; flower buds abort more frequently; and senescence is hastened when tubers are more strongly induced. Any theory that would explain the hormonal control of tuberization should take into account the manifold effects of tuber induction. Let us now consider the evidence for the involvement of particular hormones.

Endogenous Hormones

Gibberellins

There have been many attempts to measure changes in endogenous hormones as correlated with changes in the degree of induction of the potato plant. One of the first changes that was noted was a decrease in gibberellin-like activity (29, 44, 51, 53). As few as two short days produced a decline in the gibberellin-like activity extracted from Andigena leaves that were previously grown under long photoperiods, and even the chloroplast fraction of Andigena leaves showed reductions in gibberellin-like substances as photoperiods shortened (54). Low irradiance, which tends to inhibit tuberization, greatly increased the gibberellin-like substances in leaves exposed to short days; while long-day leaves were high in such substances regardless of irradiance levels (62). Similarly, high temperatures increased gibberellin-like activity of shoots; but in this respect buds appeared to be much more affected by high temperatures than were leaves (Table 2). If potato plants were grown at high temperatures and buds were excised or chemically inhibited, then the deleterious effect of high temperature on tuberization was largely ameliorated (39). Not only do long photoperiods, high temperatures, and low irradiance have similar effects on tuberization, they all produce effects on shoot morphology that are consistent with known effects of

Table 2. Effect of temperature on gibberellin-like activity in apical buds of potato plants.
From (40).

Day/Night Temperatures °C	Gibberellin activity* (ug GA₃ equivalent/kg f.wt.)	
	Buds	Leaves
20/15	4.0	1.1
35/30	71.8	4.3
LSD (P = 0.05) 15.17 (log transformation)		

*   Lettuce hypocotyl bioassay (means of three plants)

gibberellins (41). Other evidence comes from growing plants in nutrient
solution: a continuous supply of N, which inhibits tuberization, causes
higher gibberellin activity in shoots than does discontinuing the N supply,
which favors tuberization (27). Assays of gibberellin-like activity in
stolons and newly initiated tubers (26, 56) indicate that there is a
substantially lower activity associated with the conversion from stolons to
tubers (Fig. 5). Taken together, there is convincing evidence of a negative

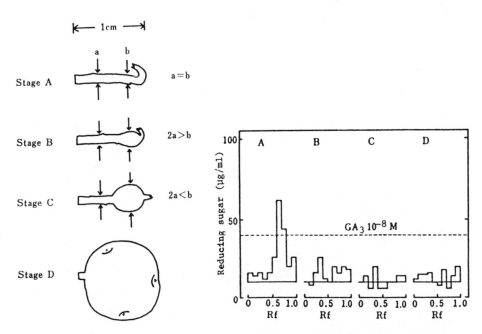

Fig. 5. Changes in the levels of gibberellin-like substances at different stages of tuber development. Stolons at stage A showed no indication of tuberization. At stage B, tuber initiation
had started, but the diameter of the swelling was less than twice the diameter of the stolon
immediately behind it. Stage C tubers were more than twice the diameter of the stolon.
Tubers at stage D were 1 to 2 cm in diameter. Each extract, equivalent to 10 g fresh weight,
was chromatographed on paper and assayed by the Avena endosperm test. Broken line represents the amount of reducing sugar liberated from the endosperms by 10^{-8} M GA₃. From (26).

correlation between the degree of induction to tuberize and the gibberellin-like activity in shoots and stolons.

Inhibitors

What causes the gibberellin activity to decrease during induction? Is there simply a decline in its net production, or is a gibberellin inhibitor produced during tuber induction? An unidentified inhibitor has been reported (29, 51, 56), but more attention has been paid to the possibility that ABA performs the role of gibberellin inhibition. There is some indication that ABA activity increases under inducing conditions (27), but in most cases differences reported have not been very great or have been in the opposite direction (29, 61). There was little increase in ABA activity of stolons at the earliest stage of tuber initiation (Fig. 6).

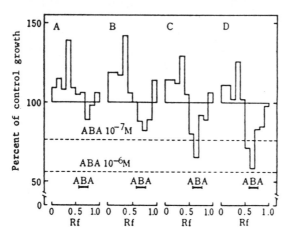

Fig. 6. Changes in the levels of ABA-like substances at different stages of tuber development. Stages and sampling were as explained in Fig. 5. Each extract was chromatographed and assayed by inhibition of Avena coleoptile straight growth induced by 10^{-7} M IAA. Broken lines represent the percent of control growth obtained by 10^{-7} and 10^{-6} M ABA. Markers in the lower part of figure indicate the position of ABA. From (26).

Cytokinins

Another group of hormones that reportedly changes under inducing conditions is cytokinins. The major cytokinin in potato leaves is zeatin riboside, which has been found to be 29% higher in extracts from induced tissues (35). Maximal levels of cytokinins in shoots and underground tissues occurred four and six days, respectively, after exposure to inducing conditions (13, 30). The decline in cytokinin activity of shoots after four days of inducing conditions was attributed to transport to the stolon tips, where metabolic sinks were created that led to tuberization (13, 30). If cytokinin is responsible for induction, and if the levels in the shoots decrease after four days of inducing conditions, then one would expect that leaf-bud cuttings taken after inductive periods progressively longer than four days would tuberize less and less. In our experience the contrary pattern applies; the more days of inducing conditions, the stronger the tuberization on cuttings, until all form sessile tubers. We find no evidence that tuberization depletes the supply of tuberization stimulus in the leaf (10).

The association between tuberization and increased cytokinin activity in shoots has also been noted when the nitrogen supply was removed from hydroponically grown potatoes, though a different interpretation was given (55). In this case it was speculated that some factor other than the cytokinins caused tuber initiation and that once tubers were present, the sink effect increased the photosynthetic activity of the leaf, which in turn produced increased cytokinin activity of the shoot. We have, then, two opposing explanations for the association between the increased cytokinin activity of leaves and tuber initiation: 1) the cytokinins in the leaves are translocated to the stolon where they induce tuberization (13,30); or 2) the initiation of tubers stimulates cytokinin production in the leaves (55). If the first hypothesis is correct, then cytokinins should increase in stolon tips before tubers are initiated. However, cytokinin assays of stolons and small tubers (22, 26) indicated no substantial increase in cytokinin-like activity until tubers were well into their enlargement phase (Fig. 7).

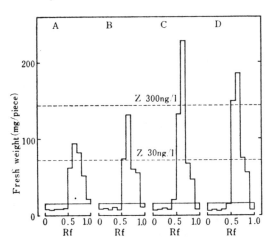

Fig. 7. Changes in the levels of butanol-soluble cytokinins at different stages of tuber development. Stages and sampling were as explained in Fig. 5. Each extract was chromatographed and assayed using soybean callus. Broken lines represent the callus yields with 30 and 300 mg/l zeatin. From (26).

Other hormones

Tuberization in dahlia is associated with changes in evolution of ethylene (2), but similar findings from potato have not been reported. Auxin, the other major category of plant hormone, has received little attention with respect to changes in endogenous content during tuber induction, and one review of the topic has suggested that this may be a serious omission (37). An association between tuber growth rate and auxin content has been noted (33), but, as shown in Fig. 8, only a relatively small increase in auxin activity was found in the first stage of tuberization (26).

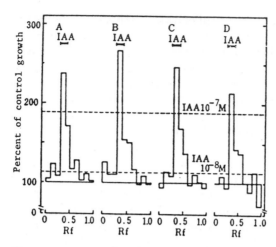

Fig. 8. Changes in the levels of auxin during the course of potato tuberization. Stages and sampling were as explained in Fig. 5. Each extract was chromatographed and assayed by Avena coleoptile straight growth test. Broken lines represent the percent of control growth by 10^{-8} and 10^{-7} M IAA. Markers in the upper part of figure indicate the position of authentic IAA. From (26).

Hormone Applications to Whole Plants

Assays for activities of endogenous hormones have revealed no unequivocal answer as to the role of hormones in the control of tuberization, with the possible exception that tuber initiation is associated with a decline in gibberellin activity. A second approach to the question is through application of exogenous hormones.

Gibberellins

It is abundantly clear from a great number of such experiments that gibberellic acid applications have a strong inhibitory effect on tuberization of whole plants (44). Gibberellins A_1, A_3, A_4, A_5, A_7, and A_9 all inhibited tuberization of leafed stem cuttings (60). Gibberellin applications also promote shoot growth on whole plants or leaf-bud cuttings (38). In both respects exogenous gibberellins appear to mimic the effects of noninductive conditions. There is also a biochemical similarity between plants grown under noninductive conditions and plants grown under inductive conditions but treated with gibberellins. Buried petioles of leaf cuttings taken from either kind of plant had lower levels of the glycoprotein patatin than petioles of cuttings from induced plants (18). Treatment with CCC, which blocks gibberellin synthesis, affords another way of examining the effects of these compounds. The general effect of CCC applications has been an improvement in tuberization, especially where noninductive conditions prevailed (38).

Other hormones

Promotion of tuber initiation on whole plants through the application of other classes of growth substances—including auxin and related compounds, ABA, cytokinins, 2-chloroethylphosphonic acid (ethephon), and a large group of miscellaneous compounds—has been relatively ineffective. In evaluating such experiments it should be noted that almost any chemical applied in dosages that are slightly phytotoxic to the shoot may increase the number of tubers initiated under inductive conditions. The result is typically a larger number of tubers, but smaller average tuber size, producing no increase in total tuber yield. Of much greater interest would be a chemical that could lead to tuber initiation under environmental conditions that would ordinarily prevent all tuberization.

Hormone Applications to Sprout Tubers

The application of the ethylene producing compound, ethephon, to potato tubers that are so old physiologically as to be producing small tubers rather than normal sprouts, produced sprouts that elongated normally and that contained somewhat higher levels of endogenous gibberellins than were present in the sprout tubers (8). Cytokinin activity was similar in the sprouts and small tubers. It is not clear whether ethylene plays a role in normal sprout growth to prevent tuberization, or whether this is a special effect in very old tubers.

Hormone Applications to Cuttings

Abscisic acid

When single node cuttings were taken from Andigena plants that had received 20 short days, tuberization resulted (61). Removal of the leaf from a cutting at the time it was taken led to an orthotropic, elongated shoot rather than a tuber; *but* application of ABA *or* grafting of a leaf back to the cutting gave a tuber. The leaf grafted to the cutting did not need to be from a plant that had been exposed to short days, as long as the bud to which it was grafted came from a plant that had received the short days. In fact, even a tomato leaf produced a tuber when grafted to such a stock. A key point is that the ABA applied to the bud, or the leaf grafted to the cutting, produced a tuber only if the bud had been taken from a plant that had received the short days. Neither the ABA nor the noninduced leaf resulted in tubers when the original cutting came from a plant exposed only to long days. (61)

In interpreting the above results, it is important to note that if leafless cuttings of the type mentioned had come from plants that had been exposed to several weeks of strongly inductive photoperiods, then tubers would have formed even without ABA or a grafted leaf. Thus the buds on the cuttings used in the experiments were somewhere near the verge of being able to tuberize alone. The ABA, or whatever substance(s) were provided by the grafted leaves, "tipped" them toward tuberization. The ABA would not have caused buds on leafless cuttings from long day plants to tuberize; thus by itself it cannot be considered to be the "tuberization stimulus," although it could be one component of the stimulus.

Cytokinin

A second set of experiments with cuttings deals with dipping strongly induced leaf-bud cuttings in cytokinin (36). Repeated treatment with 6-benzyladenine interfered with normal development of sessile tubers. Tuberization was delayed; and instead of the formation of sessile tubers, thickened stolons preceded tuberization. The reason for the interference is not known; perhaps the cytokinins created a sink in the detached leaves that hindered movement of metabolites to the developing tubers.

Hormone Applications to Tissue *In Vitro*

Procedures

Two papers, which in other respects have had an enormous impact on tuberization research, reported a technique that has been largely ignored. Gregory (16) and Chapman (6) cut small pieces of stem containing axilliary buds from plants that had been strongly induced to tuberize, disinfected them, and placed them on an agar medium in the dark. Within about four days after cutting, tubers formed at the buds. Nodes similarly treated from noninduced shoots produced only leafy shoots until cultures were more than three weeks old. To obtain good tuberization it was necessary to add sucrose to the agar medium, and increasing the sucrose concentration up to 10% gave larger tubers (16). Gregory pointed out that the technique might form the basis of a kind of bioassay for the hypothetical tuberization stimulus. If *in vitro* cuttings from, for example, Andigena plants exposed to long photoperiods tuberized only after addition to the agar medium of a substance isolated from induced plants, that would constitute evidence for the role of the isolated compound in controlling tuberization. Unfortunately, since the original papers there has been relatively little published work utilizing the technique, perhaps because—as Gregory himself mentioned (16)—microbial contamination is often difficult to overcome (10).

Much subsequent work on tuberization (15, 49, 57, 58) has been done utilizing a very different *in vitro* system. This latter technique, instead of taking nodes directly from shoots exposed to varying photoperiods, calls for repeated subculturing of sprouts or stolons in the dark prior to the *in vitro* test for tuberization. In this case there is no obvious way to study the effects of photoperiod--the plant material being investigated is cultured under continuous darkness. Another question that may be raised about the technique is that subculturing might deplete the supply of a particular nutrient normally present in the tissue to the point where it would be difficult to distinguish between substances merely required for normal growth and those that are uniquely associated with tuberization. Still another objection is that very long periods are often required before tuberization takes place, and eventually the control treatments may also develop tubers (25).

There is yet a third form of *in vitro* culture. During the last decade seed certification agencies have become very interested in growing potato plantlets *in vitro* as a method of preserving stock cultures free from diseases. Plantlets are subcultured from nodal cuttings under aseptic conditions in agar media containing sugar and a variety of other organic and inorganic nutrients. Various photoperiods are employed, ranging from 12 hours to continuous light. The usual method employed to propagate such cultures is to dissect the plantlets into nodal cuttings every few weeks; but if plantlets are left for longer periods, two months or more, it is common to find tubers under the agar or at the aerial nodes. The small tubers may be convenient for long term storage or as a means of mailing germ plasm to other locations, so attention has been given to factors controlling the tuberization (20, 21). Again, we must be cautious in extrapolating the results of such research to whole plants. The *in vitro* plantlets are relatively insensitive to photoperiod, even when Andigena types are grown. The tubers are forming in the light on plants growing in a highly artificial medium. It sometimes appears that the tuberization is what happens when the growth of the *in vitro* plantlets has been slowed by a limitation of nutrients or by the development of toxic factors. Tuberization as a response to exhaustion may provide a poor model of what happens in the whole plant. With these caveats in mind, let us review some of the results from *in vitro* studies.

Results with subcultured stolons

Subcultured stolons tuberized only if they were supplied cytokinin and sucrose. Tuberization was not complete until about 25 days, considerably later than in the Gregory system (16). Inhibitors of nucleic acid synthesis and protein synthesis did not block the tuberization. Gibberellin strongly depressed tuberization (25), and IAA and NAA promoted it weakly (25, 57). There is some disagreement on the effects of ethylene (15), but most studies have shown that it decreases or eliminates the tuberization of subcultured stolons (42). Additions of ABA to agar media containing 2% sucrose caused a slight swelling of the subapical region of the stolon, but no further development to tubers (25). On 8% sucrose media, ABA alone did not increase tuberization, although it partially overcame the deleterious effects of gibberellic acid (25). Phenolic acids (50) and coumarin also (57) favored tuberization. The enhancement of tuberization by coumarin was blocked by actinomycin D and chloramphenicol (58), a fact which is somewhat surprising considering that these inhibitors of nucleic acid and protein synthesis had little effect on tuberization promoted by cytokinin (49).

Results with light-grown plantlets

The temperature and photoperiod under which plantlets are grown has been shown to affect their size and morphology (20), but it is unclear to what extent this is an indirect photosynthetic response. Shortening the photoperiod of plantlets grown under long days did not favor tuberization (21). Tuberization tended to be promoted by 6-benzylaminopurine, high sucrose, and CCC. It was inhibited by gibberellic acid and ethylene. Additions of ABA had little consistent effect (21).

Results with stem pieces

The few experiments employing the *in vitro* system proposed by Gregory (16) have produced little indication that addition of the known plant hormones to the agar medium can substitute for exposing plants to short photoperiod before bud excision. The best evidence for such a substitution is with kinetin (14), but even in this case results were not entirely clear-cut; the long photoperiod treatment also gave partial tuberization. Other experiments have shown no benefit from kinetin (11). As expected, gibberellic acid decreases tuberization (19). ABA has no effect (10), in contrast to its effect in promoting tuberization of leafless Andigena cuttings from moderately induced plants (61). Perhaps the difference in results is to be explained by the difference in cultivars used or by the presence of sucrose and ammonium nitrate in one case (10) and not in the other (61).

Genetic Studies

Many genetic variants of the potato may be related to hormonal differences. For example, a mutant producing "wilty potato" lacks ABA (52). It nevertheless tuberizes, which further weakens the case for ABA being the tuberization stimulus. Mutants called "giant hills" or "bolters," which tuberize late and have the appearance of high gibberellin content, are fairly common. The "topiary" gene gives a rosette plant totally lacking in apical dominance, with short stolons and strong tuberization (31). A similar morphological response has been achieved by the incorporation into a Tuberosum clone of T-DNA genes from *Agrobacterium tumefaciens* (47). Shoots, which contained a 10-fold increase in cytokinins, were characterized by small leaves, reduced apical dominance, and aerial tubers. Subterranean tubers were smaller and more numerous than on the original clone. The larger number of small tubers and the smaller leaf size are reminiscent of symptoms that accompany foliar applications of cytokinins (or various other growth regulators) to potato plants. In considering whether this should be taken as evidence for the role of cytokinins as the tuberization stimulus, it is worth remembering that photoperiodic induction to tuberize is associated with larger rather than smaller leaf size. Cytokinins may be present at phytotoxic levels whether produced internally or applied externally; in either case, the phytotoxicity may increase the number of tubers initiated.

The difference between the photoperiodic reactions of early and late cultivars of Tuberosum is substantial, and within Andigena populations developed by several plant breeders are clones that cover a broad range of critical photoperiods. Some have an absolute requirement for photoperiods less than 12 hours, even at the most favorable temperatures. Others are able to tuberize under high temperatures and continuous light. Physiologists would do well to consider this wealth of genetic diversity when investigating the role of hormones in tuberization.

SUMMARY AND CONCLUSIONS

Gibberellins

Among the known hormones, the most convincing case for a critical role in control of tuberization is for gibberellins. 1) The environmental changes that increase induction to tuberize cause decreased gibberellin activity of shoots. 2) Gibberellin activity of stolon tips declines at the earliest stage of tuber initiation. 3) High temperatures cause gibberellin content to increase far more in buds than in leaves, and excision of the buds alleviates much of the deleterious effect of high temperature on tuberization. 4) Exogenous gibberellins are highly effective in reducing or eliminating tuberization, and also mimic other effects of the noninductive environment. 5) The effects of exogenous gibberellin application are ameliorated by the gibberellin inhibitor, CCC, which also shows some ability to promote tuberization in noninductive environmental conditions.

The problem, of course, is that the effects of gibberellin are all negative in terms of looking for a tuberization stimulus. To argue that tuberization is under the control of gibberellin, we must turn the concept of a tuberization stimulus on its head and consider that potato buds and stolon tips are programmed to tuberize unless their apical meristems are stimulated to develop into shoots or stolons. Short photoperiods would then promote tuberization by decreasing the supply of gibberellins, permitting the tubers to form. Sprout tubers would form on old tubers after the supply of gibberellins in the mother tuber was depleted.

There is some appeal to this hypothesis, if one can accept the concept of buds being programmed to tuberize as the "normal" state of affairs. However, it is necessary to explain how an induced leaf grafted to a noninduced cutting can promote tuberization. Presumably one could argue that the induced tissue acts as a sink for the gibberellins in the rest of the plant, or transmits a signal that shuts down the gibberellin production.

Another problem would be explaining the continuum of bud responses to varying degrees of induction (Fig. 1). It is easy to accept that a moderate supply of gibberellin (associated with fairly noninductive conditions) would lead to stolon or shoot growth rather than to tuberization; it is more difficult to see why a higher level of gibberellin (associated with extremely noninductive conditions) would lead to a dormant buried bud, the response at the opposite end of the continuum from tuberization. This is especially true when one considers that even leaf-bud cuttings (in which the only bud present is buried) have increasing bud dormancy as induction to tuberize decreases (12).

Inhibitors

An obvious alternative is to invoke a gibberellin inhibitor as the tuberization stimulus. This would be consistent with the observed changes in endogenous gibberellins and the effects of exogenous gibberellins. There have been attempts to isolate such an inhibitor, but to date they have not had much success. Or instead of a specific inhibitor of gibberellins, a more generalized growth inhibitor has been sought. ABA has attracted considerable attention as a logical candidate in this category, but the balance of the evidence is against it. Various phenolic compounds have been suggested, but no data have been published to show that any particular compounds are actually involved.

Cytokinins

There is no doubt that cytokinins must be present in order for tuberization to take place, inasmuch as cell division is one of the early events following tuber initiation. This is not to say, however, that a change in the level of cytokinins in whole plants is the event that triggers tuberization—the switch that shuts off cell division at the apical meristem, and turns on cell enlargement, cell division, patatin synthesis, and starch deposition in the subapical meristem. Cytokinins increase in shoots as plants are induced to tuberize, but the increases are smaller than might be expected for hormonally controlled processes. The reported decline of cytokinin concentrations in shoots after four days of inductive conditions seems inconsistent with the pattern of tuberization on cuttings taken at various periods after induction begins. Cytokinins increase in stolon tips during tuberization, but the increase is relatively small until well after tubers have been initiated (Fig. 7). There have been numerous attempts to improve tuberization of whole plants by foliar applications of cytokinins. Yield differences have generally been marginal, and not strikingly different from results obtained with a great variety of growth substances similarly applied. There is no report that cytokinin application to a noninduced Andigena plant caused it to tuberize.

The other evidence for cytokinins as the tuberization stimulus comes from *in vitro* experiments. Where these have been performed on subcultured stolons, it is difficult to know whether the cytokinin is the unique stimulus for tuberization, or whether it, perhaps like sugar, is a necessary ingredient for tuber formation once tuber induction has occurred. There have been only a few tests of the ability of cytokinin added *in vitro* to substitute for exposure of plants to short days when buds are excised from stems of plants grown under noninducing conditions. Results were not conclusive.

Considering all the data, it seems premature to identify cytokinin as the sole component of the tuberization stimulus.

Hormonal balance

There is no reason why we must seek a single compound to be the tuberization stimulus. Inducing conditions might lead to simultaneous changes in the concentrations of a number of compounds, the balance of which may control tuberization and the many associated morphological changes. The preponderance of thinking at present seems to favor the ratio of gibberellins to cytokinin, or to ABA, as the stimulus. However, there is no justification for ruling out other compounds. Auxin could be a component, especially in controlling the degree of plagiotropism of stolons or in other morphological responses associated with intermediate levels of induction. Nor should we limit our thinking to hormones. Earlier investigators attributed control of tuberization to carbohydrate levels or to the carbon/nitrogen ratio. Hormonal theories by and large have supplanted such explanations; but it may be that sugar concentrations play a role along with hormone levels. As already noted, *in vitro* tuberization occurs very poorly in the absence of sugar. Furthermore, neither glucose nor fructose is as effective as sucrose in meeting the sugar requirement, even though either is superior to sucrose in promoting shoot growth *in vitro* (10).

The hypothetical tuberization stimulus has proved to be as elusive as the analogous flowering stimulus; yet there is good evidence that induction to tuberize produces complex hormonal changes in the potato plant. As more sophisticated and sensitive analytical methods for hormones become available, as gene probes for patatin and related proteins are more fully developed, and as researchers begin to take better advantage of the vast genetic variation available for physiological studies, we can hope to learn whether there is a unique compound controlling tuberization and all its attendant changes, or whether many compounds operate in concert.

References

1. Batutis, E. J., Ewing, E. E. (1982) Far-red reversal of red light effect during long night induction of potato (*Solanum tuberosum* L.). Plant Physiol. *69*, 672-674.
2. Biran, I., Gur, I., Halevy, A.H. (1972) The relationship between exogenous growth inhibitors and endogenous levels of ethylene and tuberization of Dahlias. Physiol. Plant. *27*, 226-230.

3. Biran, I., Leshem, B., Gur, I., Halevy, A.H. (1974) Further studies on the relationship between growth regulators and tuberization of Dahlias. Physiol. Plant. *31*, 23-28.
4. Bodlaender, K. B. A. (1963) Influence of temperature, radiation, and photoperiod in development and yield. In: Growth of the Potato, pp. 199-210, J.D. Ivins and E. L. Milthorpe, ed. Butterworths, London.
5. Chailakhyan, M. Kh., Yanina, L. I., Devedzhyan, A. G., Lotova, G. N. (1981) Photoperiodism and tuber formation in grafting of tobacco onto potato. Doklady Akademic Nauk S. S. S. R. *257*, p. 1276.
6. Chapman, H. W. (1958) Tuberization in the potato plant. Physiol. Plant. *11*, 215-224.
7. Cutter, E. G. (1978) Structure and development of the potato plant. *In* The Potato Crop, pp. 70-152, P.M. Harris, ed., Halsted Press, John Wiley & Sons, New York.
8. Dimalla, G. G., van Staden, J. (1977) Effect of ethylene on the endogenous cytokinin and gibberellin levels in tuberizing potatoes. Plant. Physiol. *60*, 218-221.
9. Ewing, E. E. (1978) Shoot, stolon, and tuber formation on potato (*Solanum tuberosum* L.) cuttings in response to photoperiods. Plant Physiol. *61*, 348-353.
10. Ewing, E. E. (1985) Cuttings as simplified models of the potato plant. *In* Potato Physiology, pp. 153-207, P.H. Li, ed. Academic Press.
11. Ewing, E. E., Senesac, A. H. (1981) *In vitro* tuberization on leafless stem cuttings. *In* 8th Triennial Conference of the European Association for Potato Research (Abs.) pp. 7-8.
12. Ewing, E. E., Wareing, P. F. (1978) Shoot, stolon, and tuber formation on potato (*Solanum tuberosum* L.) cuttings in response to photoperiods. Plant Physiol., Lancaster *61*, 348-353.
13. Forsline, P. L., Langille, A. R. (1975) Endogenous cytokinins in *Solanum tuberosum* as influenced by photoperiod and temperature. Physiol. Plant *34*, 75-77.
14. Forsline, P. L., Langille, A. R. (1976) An assessment of the modifying effect of kinetin on *in vitro* tuberization of induced and non-induced tissues of *Solanum tuberosum*. Can. J. Bot. *54*, 2513-1256.
15. Garcia-Torres, L., Gomez-Campo, C. (1973) *In vitro* tuberization of potato sprouts as affected by ethrel and gibberellic acid. Potato Res. *16*, 73-79.
16. Gregory, L. E. (1956) Some factors for tuberization in the potato. Ann. Bot. *41*, 281-288.
17. Hammes, P. S., Beyers, E. A. (1973) Localization of the photoperiodic perception in potatoes. Potato Res. *16*, 68-72.
18. Hannapel, D. J., Miller, J. C., Jr., Park, W. D. (1985) Regulation of potato tuber protein accumulation by gibberellic acid. Plant Physiol. *78*, 700-703.
19. Harmey, M. A., Crowley, M. P., Clinch, P. E. M. (1966) The effect of growth regulators on tuberization of cultured stem pieces of *Solanum tuberosum*. Eur. Potato J. *9*, 146-151.
20. Hussey, G., Stacey, N. J. (1981) *In vitro* propagation of potato (*Solanum tuberosum* L.). Ann. Bot. *48*, 787-796.
21. Hussey, G., Stacey, N. J. (1984) Factors affecting the formation of *in vitro* tubers of potato (*Solanum tuberosum* L.). Ann. Bot. *53*, 565-578.
22. Jameson, P. E., McWha, J. A., Haslemore, R. M. (1985) Changes in cytokinins during initiation and development of potato tubers. Physiol. Plant. *63*, 53-57.
23. Kahn, B. H., Ewing, E. E. (1983) Factors controlling the basipetal patterns of tuberization in induced potato (*Solanum tuberosum* L.) cuttings. Ann. Bot. *52*, 861-874.
24. Kahn, B. A., Ewing, E. E., Senesac, A. H. (1983) Effects of leaf age, leaf area, and other factors on tuberization of cuttings from induced potato (*Solanum tuberosum*) shoots. Can. J. Bot. *61*, 3193-3201.
25. Koda, Y., Okazawa, Y. (1983a) Influences of environmental, hormonal and nutritional factors on potato tuberization *in vitro*. Japan. Jour. Crop Sci. *52*, 582-591.
26. Koda, Y., Okazawa, Y. (1983b) Characteristic changes in the levels of endogenous plant hormones in relation to the onset of potato tuberization. Japan. Jour. Crop Sci. *52*, 592-597.

27. Krauss, A., Marschner, H. (1982) Influence of nitrogen nutrition, daylength and temperature on contents of gibberellic and abscisic acid and on tuberization in potato plants. Potato Res. 25, 13-21.
28. Kumar, D., Wareing, P. F. (1973) Studies on tuberization in *Solanum andigena*. II. Growth hormones and tuberization. New Phytol. 73, 833-840.
29. Kumar, D., Wareing, P. F. (1974) Studies on tuberization of *Solanum andigena*. New Phytol. 73, 833-840.
30. Langille, A. R., Forsline, P. L. (1974) Influence of temperature and photoperiod on cytokinin pools in the potato *Solanum tuberosum* L. Plant Sci. Lett. 2, 189-191.
31. Leue, E. F., Peloquin, S. J. (1982) The use of the topiary gene in adapting *Solanum* germ-plasm for potato improvement. Euphytica 31, 65-72.
32. Madec, P., Perennec, P. (1959) Le role respectif du feuillage et du tubercule-mère dans la tubèrization de la pomme de terre. Eur. Potato J. 2, 22-49.
33. Marschner, H., Sattelmacher, B., Bangerth, F. (1984) Growth rate of potato tubers and endogenous contents of indolylacetic acid and abscisic acid. Physiol. Plant. 60, 16-20.
34. Martin, C., Vernay, R., Paynot, N. (1982) Physiologie vègètale. Photopèriodisme, tubèrisation, floraison et phènolamides. C. R. Acad. Sc. Paris 295, 565-568.
35. Mauk, C. S., Langille, A. R. (1978) Physiology of tuberization in *Solanum tuberosum* L.: cis-zeatin riboside in the potato plant - its identification and changes in endogenous levels as influenced by temperature and photoperiod. Plant Physiol. 62, 438-442.
36. McGrady, J. J., Struik, P. C., Ewing, E. E. (In press) Effects of exogenous cytokinin on bud development of potato cuttings. Potato Res.
37. Melis, R. J. M., van Staden, J. (1984) Tuberization and hormones. Z. Pflanzenphysiol. 133, 271-283.
38. Menzel, B. M. (1980) Tuberization in potato *Solanum tuberosum* cultivar Sebago at high temperatures: responses to gibberellin and growth inhibitors. Ann. Bot. 46, 259-266.
39. Menzel, B. M. (1981) Tuberization in potato at high temperatures: promotion by disbudding. Ann. Bot. 47, 727-733.
40. Menzel, B. M. (1983) Tuberization in potato at high temperatures: interaction between shoot and root temperature. Ann. of Bot. 52, 5-69.
41 Menzel, B. M. (1985) Tuberization in potato at high temperatures: interaction between temperature and irradiance. Ann. Bot. 55, 35-39.
42. Mingo-Castel, A. M., Negm, F. B., Smith, O. E. (1974) Effect of carbon dioxide and ethylene on tuberization of isolated potato stolons cultured *in vitro*. Plant and Cell Physiol. 53, 798-801.
43. Nitsch, J. P. (1965) Existence d'un stimulus photopèriodique non spècifique capable de provoquer la tubèrisation chez *Helianthus tuberosus* L. Bull. Soc. Bot. Fr. 112, 333-340.
44 Okazawa, Y. (1960) Studies on the relation between the tuber formation of potato plant and its natural gibberellin content. Proc. Crop Sci. Soc. Japan 29, 121-124.
45. Okazawa, Y., Chapman, H. W. (1962) Regulation of tuber formation in the potato plant. Physiol. Plant 15, 413-419.
46. Okazawa, Y., Chapman, H. W. (1963) Effects of tomato scion on tuberization of potato stock. Physiol. Plant. 16, 623-629.
47. Ooms, G., and J.R. Lenton. (1985) T-DNA genes to study plant development: precocious tuberisation and enhanced cytokinins in *A. tumefaciens* transformed potato. Plant Molecular Biology 5: 205-212.
48. Paiva, E., Lister, R. M., Park, W. D. (1983) Induction and accumulation of major tuber proteins of potato in stems and petioles. Plant Physiol. 71, 161-168.
49. Palmer, C. E., Smith, O. F. (1970) Effects of kinetin on tuber formation on isolated stolons of *Solanum tuberosum* L. cultured *in vitro*. Plant Cell Physiol. 11, 303-314.
50. Paupardin, C., Tizio, R. (1970) Action de quelques composès phènoliques sur la tubèrisation de la Pomme de terre. Potato Res. 13, 187-198.

51. Pont-Lezica, R. F. (1970) Evolution des substances de type gibbèrellines chez la pomme de terre pendant la tubèrisation, en relation avec la longueur du jour et la temperature. Potato Res. *13*, 323-331.
52. Quarrie, S. A. (1982) Droopy: a wilty mutant of potato deficient in abscisic acid. Plant, Cell and Environment *5*, 23-26.
53. Racca, R. W., Tizio, R. (1968) A preliminary study of changes in the content of gibberellin-like substances in the potato plant in relation to the tuberization mechanism. Eur. Potato J. *11*, 213-220.
54. Railton, I. D., Wareing, P. F. (1973) Effects of daylength on endogenous gibberellins in leaves of *Solanum andigena*. I. Changes in levels of free acidic gibberellin-like substances. Physiol. Plant. *28*, 88-94.
55. Sattelmacher, B., Marschner, H. (1978b) Relation between nitrogen, cytokinin activity and tuberization in *Solanum tuberosum* L. Physiol. Plant *44*, 65-68.
56. Smith, O. E., Rappaport, L. (1969) Gibberellins, inhibitors, and tuber formation in the potato, *Solanum tuberosum*. Am. Potato J. *46*, 185-191.
57. Stallknecht, G. F., Farnsworth, S. (1982a) General characteristics of coumarin-induced tuberization of axillary shoots of *Solanum tuberosum* L. cultured *in vitro*. Am. Potato J. *59*, 17-32.
58. Stallknecht, G. F., Farnsworth, S. (1982b) The effect of the inhibitors of protein and nucleic acid synthesis on the coumarin-induced tuberization and growth of excised axillary shoots of potato sprouts (*Solanum tuberosum* L.) cultured *in vitro*. Am. Potato J. *59*, 69-76.
59. Steward, F. C., Moreno, V., Roca, W. M. (1981) Growth, form and composition of potato plants as affected by environment. Ann. of Bot. *48*, (Supplement No. 2) 45 pp.
60. Tizio, R. (1971) Action et rôle probable de certaines gibbèrellines (A_1, A_3, A_4, A_5, A_7, A_9, et A_{13}) sur la croissance des stolons et la tubèrisation de la Pomme de terre (*Solanum tuberosum* L.) Potato Research *14*, 193-204.
61. Wareing, P. F., Jennings, A. M. V. (1980) The hormonal control of tuberization in potato. *In* Plant Growth Substances, pp. 293-300, ed., F. Skoog. Springer-Verlag, New York.
62. Woolley, D. J., Wareing, P. F. (1972) Environmental effects on endogenous cytokinins and gibberellin levels in *Solanum tuberosum*. New Phytol. *71*, 1015-1025.

E15. The Hormonal Control of Bud and Seed Dormancy in Woody Plants

Loyd E. Powell

Department of Pomology, Cornell University, Ithaca, New York 14853, USA.

INTRODUCTION

Perennial woody plants of the temperate zone have evolved a dormancy mechanism which helps them to survive winter cold. This mechanism exerts a major influence on their growth and development, to the point where it is probably the single most important factor preventing the adaptation of most temperate zone woody plants to tropical areas. Much is known about the physiology of dormancy, but the basic biochemical mechanism through which it operates is still a mystery. It is commonly assumed to be hormonally controlled, and indeed there is considerable evidence that hormones may play a role in its regulation.

The yearly train of events for shoot development for most trees and shrubs of the temperate zone is shown in Figure 1. Each step in this cycle has received attention with respect to the control mechanisms involved (35). In this chapter we are primarily interested in the events of the last part of the cycle—those related to the type of dormancy often called "rest." (See reference 26 for a recent discussion of terminology.) For prospective, we will begin by briefly discussing events leading up to rest, then rest itself and the related mechanism in seeds of many temperate zone woody plants, and finally the hormonal aspects of this chill-related type of dormancy.

THE ANNUAL SHOOT GROWTH CYCLE IN TEMPERATE-ZONE WOODY PLANTS

Bud break, the first easily visible event in the annual growth cycle, typically takes place in temperate zone woody plants in spring though, of course, bud burst can occur at other times during the growing season as well. These buds in early spring are commonly thought to be free of any physiological dormancy, and need only the warm temperatures and associated growth conditions of spring to activate the buds into growth. The growth of these buds depends on a host of orderly biochemical events

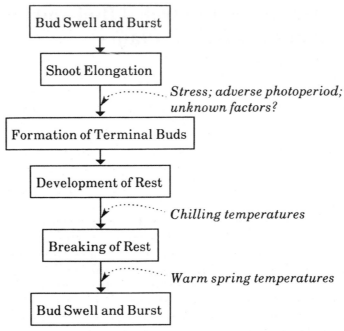

Fig. 1. Annual growth cycle for shoots of temperate zone woody plants.

for proper development. At this early stage they have little or no photosynthetic capacity, and thus depend on stored reserves for energy and building units until the young leaves which issue from them have developed their own photosynthetic capability.

After bud burst in spring there is typically a period of rapid shoot elongation, except on the very short shoots commonly referred to as spurs. In the latter, little growth occurs annually, often amounting to only a few millimeters. In long shoots, many centimeters of elongation may take place under conditions of good nutrition and light exposure, but eventually elongation ceases.

The cessation of terminal growth of many temperate zone woody plants commonly takes place at some point during the growing season, rather than at the end of the growing season with the onset of cold weather. It is not uncommon to read that shoot elongation is a function of daylength, and that short photoperiods are responsible for bringing an end to shoot elongation of many species, with the concomitant formation of terminal buds. It has been shown that many woody species do, in fact, respond to short day conditions in controlled laboratory situations by ceasing shoot elongation (30). However, under the natural conditions that exist in many areas of the temperate zone, woody plants commonly stop shoot elongation and form terminal buds in early summer when daylength is greatest—not least. Clearly, short photoperiod is not the determining factor for these plants. What then, is? A clue is to be had from observations that young vigorous woody plants, particularly those growing under invigorating conditions such as luxury amounts of water and nutrients (especially nitrogen), commonly continue growth well past the time of their older, less vigorous counterparts. In fact, many species growing under nursery conditions can easily be kept growing until cold weather arrives, at the risk, of course, of subsequent cold damage. It is a reasonable assumption that it is the various stresses put upon an older plant that lead to cessation of shoot elongation in early summer. A tree or shrub has many growing points, including developing fruits, all supplied by only a limited input of resources. When these are exhausted, growth ceases. The complete cessation of shoot elongation, rather than its assumption of a slow but steady growth rate, indicates that continued growth will only occur above a certain threshold, whether this threshold is induced by stresses directly or via some secondary message.

Clearly, stresses do occur. In tree fruit orchards it is not uncommon to see temporary wilting of the young tender shoots on hot days when conditions are right for high rates of evapo-transpiration, even when the soil water content is near field capacity. The trees are simply unable to take up water as rapidly as it is lost. Fruit growers also commonly find it necessary to thin or remove large numbers of apples and peaches (and certain other fruits) from the trees in order to obtain optimum fruit size. Failure to thin can result in large numbers of very small fruits at maturity. These two examples demonstrate that the growing points or sinks on a woody plant can be under intense competition. Furthermore, it is easy to demonstrate in a laboratory situation that early termination of shoot growth can result from shortages of water and certain nutrients.

The Rest Period

At some point after shoot growth ceases, a type of physiological dormancy, often called rest, develops in the shoot. It is important to distinguish between this type of dormancy and the type of dormancy displayed by axillary buds which are dominated and prevented from growing by the presence of the shoot tip. When the buds on a shoot have not developed rest, removal of the shoot tip will commonly allow the uppermost 1 or 2 remaining axillary buds to grow. However, if rest has developed sufficiently in the shoot, removal of the shoot tip has no effect on axillary bud growth; axillary bud burst does not occur.

Bud rest varies seasonally in intensity. Based on the amount of gibberellin required on any particular date to activate bud burst, it has been shown that rest intensity increases in peach from the point when it can first be detected in September to a maximum in November; it then subsides to a low intensity by late December (17). Determinations of rest intensity in other species show similar results, though the calendar dates may vary.

Chilling temperatures are required for the elimination of rest. The optimum temperature is about 5-7°C. Lower and higher temperatures are less effective. As the temperature approaches 0°C , it becomes too cold for much physiological chilling to occur. Conversely, temperatures of about 12°C appear to be too warm for much chilling to take place. Higher temperatures can actually reverse the process. Chilling unit accumulation for Geneva, N.Y. is shown in Figure 2, as calculated by a popular model (37). The amount of chilling required to satisfy the chilling requirement varies widely by species, and varieties within species. Generally the equivalent of several weeks of chilling at optimum temperatures are required. In the cold northeastern U.S., many of the common commercial apple cultivars have this requirement fulfilled by February. Some low chilling apple cultivars adapted to warmer climates have a chilling requirement of half or less this amount. There are also a number of high chilling apple cultivars, of research interest only, which appear to have chilling requirements half again as large as the "normal" cultivars. Once the chilling requirement has been satisfied, growth may resume again. However, since the chilling requirement is generally satisfied in winter, when it is still quite cold in most geographical areas, climate-imposed dormancy continues. When the warmer temperatures of spring return, bud burst and shoot elongation can occur once more, to begin again the annual growth cycle.

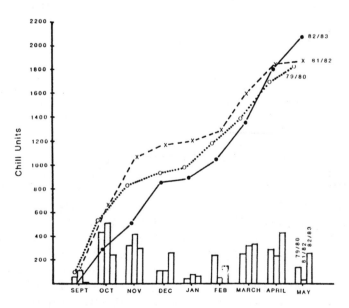

Fig. 2. Chill unit monthly totals and accumulation for 3 years at Geneva, New York. A chill unit is one hour between 2.5 and 9°C (37).

An interesting variant of the above is seen in apple cultivars which have very high chilling requirements. These high chill-requiring cultivars receive part of their chilling in fall and early winter. In cold climates such as exist in the northeastern U.S., it is too cold during the coldest winter months for much physiological chilling to take place. It is not until the milder temperatures of late winter and early spring that they receive the additional chilling required to fulfill their requirement. This high chilling requirement is believed to cause these trees to begin growth later in spring, resulting in late blooming (44). They commonly flower as much as 2-3 weeks later than the "normal" commercial cultivars.

The late growth and blooming characteristics shown by these high chill-requiring cultivars are believed due to an interaction of the chilling requirement and heat unit requirement. Apple trees which are insufficiently chilled require more heat units for bud burst than trees with sufficient chilling (Table 1) (44). The model in Figure 3 is believed to

Table 1. Heat units, in growing degree hours, above 4.5°C (38) required for 50% bud break in two apple cultivars following different periods of chilling.

Cultivar	Chilling Units Received (Hours at 5°C)	Heat Units Required for 50% Bud Burst
Redfield	1,500	5,476
(early blooming)	2,900	4,524
Shear	1,500	15,238
(late blooming)	2,900	6,904

Chilling was at a constant 5°C. Constant temperatures are less efficient than fluctuating temperatures in the chilling process. Therefore, these chill unit values should not be compared with those elsewhere where the trees were subjected to fluctuating orchard temperatures during the chilling season.

represent the situation for delayed growth of apple and similar reacting plants in cold climates. By point D, rest intensity has abated sufficiently, due to chilling, to allow warm temperatures to exert an effect on promoting growth. Were no further chilling to be experienced beyond D, it would require a substantial number of heat units to cause bud burst. However, with increasing amounts of chilling, fewer and fewer heat units are required to promote bud burst. At point E, when rest is completely eliminated, a minimum number of heat units is required. It appears that many late blooming apples begin growth late in spring because of this interaction between chilling and heat unit requirement (10, 44).

Another possibility exists to explain delayed growth in spring. Some species appear to be able to initiate growth at lower temperatures than

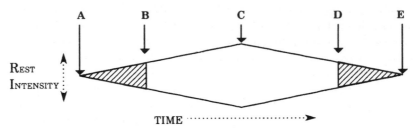

A–Beginning of rest
B–Early rest no longer subject to external stimuli
C–Maximum rest intensity
D–Late rest becomes susceptible to external stimuli
E–End of rest

Fig. 3. Rest period diagram.

other species. For example, *Populus tremuloides* and *Betula papyrifera* are trees capable of initiating growth when daily minimum low temperatures are below 0°C, while *Acer rubrum* and *Fraxinus nigra* are said to have much higher minimum temperature requirements (25). It can be argued, then, that woody plants which exhibit late growth characteristics in spring, such as late blooming apples, simply have a higher temperature requirement for initiation of growth. However, since there is an interaction between heat unit requirement and chilling, it may well be that some species listed as having a high temperature requirement for initiation of growth in spring may in reality have a high chilling requirement, and if such species were more adequately chilled, much of the so-called high heat requirement would disappear.

Studies have been made on the site of rest within the bud. The bud scales have some influence on bud dormancy, but evidently it is not a commanding influence. Removal of the bud scales when rest is not strongly developed may permit bud growth, but when buds are in deepest rest, scale removal has little influence (45). The rest influence apparently resides in the meristematic part of the bud.

SEED DORMANCY

Many temperate zone plants (both woody and herbaceous) ripen their fruits and disperse their seeds during summer and fall. Physiologically it would often be disadvantageous to these seeds for them to germinate shortly after dispersal. Various mechanisms have evolved which delay germination. Seeds of many temperate zone woody plants have evolved a dormancy mechanism which requires several weeks of stratification (i.e., chilling in a moist medium) before germination can occur. This mechanism delays germination until the following spring.

Is the chilling mechanism for seeds the same as that for buds? The similarities between the chilling requirements for these two organs suggest that it is. The temperature optimum for buds and seeds is similar, i.e., about 5-7°C. Temperatures of 0°C are not very effective in satisfying the chilling requirement for either organ, and high temperatures can have a negative effect in breaking dormancy in both organs. Certain growth regulators are capable of substituting for at least part of the chilling requirement in both organs. Perhaps the most persuasive argument is that plants which have high bud chilling requirements produce seeds with high chilling requirements, and those with low bud chilling requirements tend to produce seeds with low chilling requirements (23, 33, 53). Since seeds are far more amenable to manipulation than buds, it would seem that basic studies on the chilling mechanism would best be done on seeds.

Apple seeds have been studied extensively, and seem to be typical of a number of chill-requiring seeds. Mature apple seeds consist of an embryo surrounded by a thin membraneous layer of fused endosperm and nucellus, which in turn is surrounded by the seed coat (28). Unchilled intact apple seeds will not germinate when placed in a moist medium at common room temperatures. They will germinate, however, if the seed coat and endosperm/nucellar membrane are removed, but not if only the seed coat is removed. Thus, the endosperm/nucellar membrane appears to play an important role in preventing germination of unchilled apple seeds. While the naked (no seed coat or endosperm membrane) unchilled embryos of such seeds will commonly germinate, the resulting seedlings are not normal. They form rosettes of leaves (there being little internode elongation), remain dwarfed indefinitely, and are known as physiologic dwarfs (2).

HORMONAL RELATIONSHIPS OF BUD AND SEED DORMANCY IN WOODY PLANTS

It is often assumed that hormones play a major role in regulating rest. Attention has been paid primarily to abscisic acid (ABA), the gibberellins, and cytokinins with respect to their roles in dormancy. There is little evidence that indoleacetic acid or ethylene participate in any commanding way in the "rest" type of dormancy.

Abscisic Acid

ABA has been followed in buds of several species from the onset of rest in summer and fall through the breaking of rest in autumn and winter (35). In some cases, positive correlations between ABA content and depth of dormancy have been established, but in others the relationship is less certain. In these kinds of studies, warm controls are seldom run, but in a few cases where they were included, ABA declined in both warm and cold conditions, though only the latter alleviated dormancy (6, 31). Furthermore, in some instances ABA decline began well before the onset of cold temperatures (42). In a study of 20 apple cultivars, consisting of both low and high chill-requiring types, many of them showed a decline in ABA content in the buds during the cold winter months, but again, no warm controls were included (36). Based on our present state of knowledge, it seems safe to conclude that ABA in buds of many woody plants declines during the cold winter months. However, it is not at all certain that cold temperatures contribute to this decline.

Is there really any evidence that an inhibitor is involved in rest? It is well established that growth promoting hormones tend to fall to low levels in shoots late in the growing season. It could be argued that a low level of growth promoting hormones would, in itself, be sufficient to result in a dormant-like condition. Though on the surface this may sound reasonable, there are arguments for invoking an inhibitor in rest. If it were simply a case of insufficient growth hormones, then the application of growth promoting substances might be expected to promote growth once more. This is often not the case. Also, the fact that rest gradually intensifies long after growth promoting hormones have reached very low levels suggests the build-up of some kind of inhibitory influence. Whether this is ABA or something else remains to be shown. In theory, this ever increasing dormancy need not be due to the gradual increase of an inhibitory compound at all, but rather to something more obscure, e.g., a gradual loss of some kind of essential metabolic function.

The case can be made for involvement of some kind of growth promotive force or substance being generated as a result of chilling. Much of the supporting evidence comes from work with seeds, but since the chilling mechanism in seeds and buds appears to be similar, it seems safe to extrapolate many of the conclusions from one organ to the other. Luckwill (28) long ago found evidence that growth promoting substances appeared during the chilling of apple seeds. Employing the wheat coleoptile bioassay, he suggested that the promoting response was due to auxin. Today we know that wheat coleoptile sections can also respond in a limited way to gibberellins, so at least some of his response may have come from this group of substances.

Rudnicki's research with apple seeds strongly points to the generation of some kind of growth promoting force during chilling (40). He found that much of the ABA had disappeared after 3 weeks of stratification, and that limited germination was possible at this time. As stratification proceeded, germination continued to improve. From time to time during stratification he added various concentrations of ABA to the seeds, and found that it required ever increasing amounts of exogenous ABA to block the germination process. Some kind of promotive force was clearly being generated during stratification which opposed the inhibitory effects of ABA. Though hormone researchers are prone to attribute such effects to the appearance of growth promoting substances, other possibilities should not be overlooked. For example, Lewak and his colleagues have published many papers dealing with enzymatic and other types of changes in apple embryos during stratification; respiratory pathways may change qualitatively and quantitatively during stratification (9); and ATPase activity increases, possibly signalling a change in membrane permeability (54).

Gibberellins

The gibberellins constitute a hormone family for which there is evidence of possible involvement in the rest period. The picture that seems to have gradually emerged is that chilling temperatures are important in the gibberellin status of seeds and buds, though the precise details are far from clear and may vary somewhat with species. During chilling of peach seeds, increases in several gibberellins take place, but these changes seem to be associated more with normal growth and development of the seedling than the release from dormancy (14). Of the several gibberellins undergoing changes during stratification of apple seeds, the most dramatic are changes in GA_4, in which marked shifts in free and bound forms have been detected (16, 19). In hazel, on the other hand, substantial changes in gibberellin activity seem to occur when the seeds are subjected to warm temperatures after chilling is complete (7). Presumably chilling promotes the gibberellin synthesizing system in hazel, but the formation of biologically active gibberellins becomes operative only at higher temperatures.

A further suggestion that gibberellin content may be enhanced by chilling comes from work on physiologic dwarfs. If embryos are removed from mature unchilled seeds which have a chilling requirement, and if these excised embryos are placed in a suitable growth environment, they will often develop into seedlings referred to as physiologic dwarfs. Such seedlings form a rosette of leaves with little or no internode expansion, a situation known to be typical of certain gibberellin deficient plants. These stunted seedlings can be made to develop into more normal appearing plants by either chilling them for a period of time, or by giving them repeated applications of gibberellins (2, 4, 12). Exogenous applications of gibberellins can also replace at least part of the chilling requirement in buds of peach and apricot, this being the basis for determining rest intensity in these fruit trees (17). Gibberellin regulation is also linked to the chilling mechanism in certain biennials and over-wintering annuals which require chilling temperatures for seed stalk elongation (29).

Cytokinins

The evidence for the possible involvement of cytokinins is based on the fact that exogenous applications of these substances may stimulate bud growth/seed germination, and that their endogenous occurrence may correlate well with those same events.

The effectiveness of exogenous applications of cytokinins in stimulating resting buds is uncertain. There are numerous reports that these substances, when applied to nongrowing axillary buds on intact shoots, stimulate these buds to grow, often dramatically. Unfortunately, the type of dormancy which characterized these buds, and its intensity at time of treatment, is seldom stated. It appears that in some cases what was stimulated to grow were axillary buds which were apically dominated. There are reports, though, where applied cytokinins did appear to have a stimulating effect on buds in rest-like dormancy (8, 34, 56). Applications of cytokinins to dormant or partially chilled seeds have alleviated dormancy in some instances, but were unsuccessful in others (3, 27, 41, 50). It has been suggested that some unsuccessful attempts may have been due to poor penetration of the hormone into the embryo, to the wrong concentrations, or to the wrong cytokinin. For example, kinetin and benzyladenine stimulated the germination of apple embryos isolated from dormant seeds to a different extent (27).

If the chilling mechanism which is controlling dormancy is operating through the cytokinin system, then changes in endogenous cytokinin content might be expected during the chilling period. This assumption holds, of course, only if the system responds to changes in concentration, rather than to changes in tissue sensitivity as has been suggested by Trewavas (49). There have been several reports which suggest that there may be a relationship between chilling and the appearance of cytokinins. These include buds and seeds of *Acer pseudoplatanus* (21), *Acer saccharum* (52), apple (5, 24), *Rosa rugosa* (20) and *Populus* (18). Other reports reveal little or no change in cytokinins in response to chilling (see 3 for a more detailed discussion).

Auxin

There is little if any convincing information that auxins play an important role in regulating rest in buds or its counterpart in chill-requiring seeds. Especially noteworthy are several recent reports from Tillberg's laboratory in which studies were made on IAA in seeds of *Pinus silvestris*, *Acer platanoides* and *Rosa rugosa* (46, 47, 48). Her conclusion is that IAA is without a regulatory role in seed dormancy.

Ethylene

There are numerous reports which indicate that ethylene can stimulate germination of seeds and growth of buds, but most of these investigations deal with plant material which does not have a chilling requirement. Even in those instances where a chilling mechanism is involved in the release from dormancy, the positive response obtained by ethylene treatment is generally attributed to ethylene action on events after partial or full release from dormancy by chilling (22, 32, 51, 57). It does not appear that ethylene plays an important regulatory role in the dormancy mechanism in seeds and buds for which chilling is required for release.

CONCLUSION

In the minds of many, ABA is intimately associated with chill-related dormancy. It was, of course, the short day effects on *Acer pseudoplatanus* that led Wareing and his colleagues (11) to the discovery of ABA as a potential regulator of shoot growth. Though later work has failed to confirm that ABA is the chemical signal linked to short days (35), a linkage between ABA and rest is often made. Common assumptions are that ABA is responsible for rest, and chilling causes a diminution of ABA. It remains to be proved that either assumption is correct. The best evidence seems to suggest that while ABA often declines during the winter months, chilling temperatures probably contribute little if anything to the decline. Furthermore, as chilling accumulates, ABA becomes less and less effective as an exogenous inhibitor. If it is effective as an endogenous inhibitor in this type of dormancy, it is surely in the early stages of the process, but even here, the evidence is far from definitive.

On the other hand, chilling does contribute to the generation of a promoting force with respect to bud burst and seed germination. Arguments can be mounted for the importance of both gibberellins and cytokinins in these processes, but after several decades of research it is still not possible to say with absolute certainty that either the gibberellins or cytokinins play primary roles in the regulation of this chill-requiring dormancy mechanism.

References

1. Alscher-Herman, R., Musgrave, M., Leopold, A.C., Khan, A.A. (1981) Respiratory changes with stratification of pear seeds. Physiol. Plant. *52*, 156-160.
2. Barton, L.V. (1956) Growth response of physiologic dwarfs of Malus arnoldiania Sarg. to gibberellic acid. Contrib. Boyce Thompson Institute *18*, 311-317.
3. Black, M. (1980/81). The role of endogenous hormones in germination and dormancy. Israel J. Bot. *29*, 181-192.

4. Blommaert, K.L.J., Hurter, N. (1959) Growth response of physiologic dwarf seedlings of peach, apricot and plum to gibberellic acid. S. African J. Agric. Sci. 2, 409.

5. Borkowska, B. (1976) Variations in activity of cytokinin-like compounds in apple buds during release from dormancy. Biochem. Physiol. Pflanzen 170, 153-157.

6. Borkowska, B., Powell, L.E. (1982/83) Abscisic acid relationships in dormancy of apple buds. Scientia Hort. 18, 111-117.

7. Bradbeer, J.W. (1968). Studies in seed dormancy. IV. The role of endogenous inhibitors and gibberellin in the dormancy and germination of Corylus avellana L. seeds. Planta 78, 266-276.

8. Broome, O.C., Zimmerman, R.H. (1976). Breaking bud dormancy in tea crabapple [Malus hupehensis (Pamp.) Rehd.] with cytokinins. J. Amer. Soc. Hort. Sci. 101, 28-30.

9. Cole, M.E., Solomos, T., Faust, M. (1982) Growth and respiration of dormant flower buds of Pyrus communis and Pyrus calleryana. J. Amer. Soc. Hort. Sci. 107, 226-231.

10. Couvillon, G.A., Erez, A. (1985) Influence of prolonged exposure to chilling temperatures on bud break and heat requirement for bloom of several fruit species. J. Amer. Soc. Hort. Sci. 110, 47-50.

11. Eagles, C.F., Wareing, P.F. (1964) The role of growth substances in the regulation of bud dormancy. Physiol. Plant. 17, 697-708.

12. Flemion, F., Beardow, J. (1965) Production of seedlings from non-chilled seeds. II. Effect of subsequent cold periods on growth. Contrib. Boyce Thompson Institute 23, 101-107.

13. Gianfagna, T.J., Rachmiel, S. (1983) Changes in gibberellin-like activity and seed germination. Plant Physiol. Supplement 72, p. 99.

14. Gianfagna, T.A., Rachmiel, S. (1986) Changes in gibberellin-like substances of peach seed during stratification. Physiol. Plant 66, 154-158.

15. Grochowska, M.J., Karaszewska, A., Jankowska, B., Maksymiuk, J., Williams, M.W. (1984) Dormant pruning influence on auxin, gibberellin, and cytokinin levels in apple trees. J. Amer. Soc. Hort. Sci. 109, 312-318.

16. Halinska, A., Lewak, S. (1978) The presence of bound gibberellins in apple seeds. Bull. Acad. Polonaise Des Sciences. Biol. Sci. Ser. Cl. II, 26, 119-121.

17. Hatch, A.H., Walker, D.R. (1969) Rest intensity of dormant peach and apricot leaf buds as influenced by temperature, cold hardiness and respiration. J. Amer. Soc. Hort. Sci. 94, 304-307.

18. Hewett, E.W., Wareing, P.F. (1973) Cytokinins in Populus x robusta: Changes during chilling and bud burst. Physiol. Plant. 28, 393-399.

19. Isaia, A., Bulard, C. (1978) Relative levels of some bound and free gibberellins in dormant and after-ripened embryos of Pyrus malus cv. Golden Delicious. Z. Pflanzenphysiol. 90, 409-414.

20. Julin-Tegelman, A. (1983) Levels of endogenous cytokinin-like substances in Rosa rugosa achenes during dormancy release and early germination. Z. Pflanzenphysiol. 110, 89-95.

21. Julin-Tegelman, A., Pinfield, N. (1982) Changes in the level of endogenous cytokinin-like substances in Acer pseudoplatanus embryos during stratification and germination. Physiol. Plant. 54, 318-322.

22. Kepczynski, J., Rudnicki, R.M., Khan, A.A. (1977) Ethylene requirement for germination of partly after-ripened apple embryo. Physiol. Plant 40, 292-295.

23. Kester, D.E. (1969) Pollen effects on chilling requirements of almond and almond-peach hybrid seeds. J. Amer. Soc. Hort. Sci. 94, 318-321.

24. Kopecky, F. Sebanek, J. Blazkova, J. (1975) Time course of the changes in the level of endogenous growth regulators during the stratification of the seeds of the 'Panenske ceske' apple. Biol. Plant 17, 81-87.

25. Kozlowski, T.T. (1971) Growth and development of trees, Vol. 1, Academic Press. pp. 314-316.

26. Lang, G.A., Early, J.D., Arroyave, N.J., Darnell, R.L., Martin, G.C., Stutte, G.W. (1985) Dormancy: Toward a reduced, universal terminology. HortScience 20, 809-812.
27. Lewak, S., Bryzek, B. (1974) The influence of cytokins on apple embryo photo-sensitivity and acid phosphatase activity during stratification. Biol. Plant 16, 334-340.
28. Luckwill, L.C. (1952) Growth-inhibiting and growth-promoting substances in relation to the dormancy and after-ripening of apple seeds. J. Hort. Sci. 27, 53-67.
29. Metzger, J.D. (1985) Role of gibberellins in the environmental control of growth in Thlaspi arvense L. Plant Physiol. 78, 8-13.
30. Nitsch, J.P. (1957) Photoperiodism is woody plants. Proc. Amer. Soc. Hort. Sci. 70, 526-544.
31. Orlando, B., Dennis, F.G. (1977) Abscisic acid and apple seed dormancy. J. Amer. Soc. Hort. Sci. 102, 633-637.
32. Paiva, E., Robitaille, H.A. (1978) Breaking bud rest on detached apple shoots: Effects of wounding and ethylene. J. Amer. Soc. Hort. Soc. 103, 101-104.
33. Pasternak, G.P., Powell, L.E. (1980) Chilling requirements of apple seeds from cultivars having low and high chilling requirements for shoot growth. HortScience 15, 408.
34. Pieniazek, J. (1964) Kinetin induced breaking of dormancy in 8-month old apple seedlings on 'Antonovka' variety. Acta Agrobotanica 16, 157-169.
35. Powell, L.E. (1982) Shoot growth in woody plants and possible participation of abscisic acid. In Plant Growth Substances 1982, pp. 363-372, Wareing, P. F., ed., Academic Press.
36. Powell, L.E., Maybee, C. (1984) Changes in abscisic acid during chilling in buds of twenty apple cultivars. HortScience 19, 584.
37. Richardson, E.A., Seeley, S.D., Walker, D.R. (1974) A model for estimating the completion of rest for 'Redhaven' and 'Elberta' peachtrees. HortScience 9, 331-332.
38. Richardson, E.A., Seeley, S.D., Walker, D.R., Anderson, J.L., Ashcroft, G.L. (1975) Pheno-climatography of spring peach bud development. HortScience 10, 236-237.
39. Ross, J.D. (1983) Metabolic control of dormancy breakage in hazel. Plant Physiol. Supplement 72, 98.
40. Rudnicki, R. (1969) Studies on abscisic acid in apple seeds. Planta 86, 63-68.
41. Rudnicki, R., Saniewski, M., Millikan, D.F. (1973) The effect of exogenous plant hormones upon the stratification of apple seeds. Proc. Res. Inst. Pomology Ser. E. No. 3, 539-552.
42. Seeley, S.D., Powell, L.E. (1981) Seasonal changes of free and hydroloyzable abscisic acid in vegetative apple buds. J. Amer. Soc. Hort. Sci. 106, 405-409.
43. Shaltout, A.D., Unrath, C.R. (1983) Rest completion prediction model for 'Starkrimson Delicious' apples. J. Amer. Soc. Hort. Sci. 108, 957-961.
44. Swartz, H.J., Powell, L.E. (1981) The effect of long chilling requirement on time of bud break in apple. Acta Hort. 120, 173-178.
45. Swartz, H.J., Geyer, A.S., Powell, L.E., Lin, S.C. (1984) The role of bud scales in the dormancy of apples. J. Amer. Soc. Hort. Sci. 109, 745-749.
46. Tillberg, E. (1977) Indoleacetic acid levels in Phaseolus, Zea and Pinus during seed germination. Plant Physiol. 60, 317-319.
47. Tillberg, E. (1984) Levels of endogenous indole-3-acetic acid in achenes of Rosa rugosa during dormancy release and germination. Plant Physiol. 76, 84-87.
48. Tillberg, E., Pinfield, N.J. (1981) The dynamics of indole-3-acetic acid in Acer platanoides seeds during stratification and germination. Physiol. Plant 53, 34-38.
49. Trewavas, A.J. (1982) Growth substances sensitivity: the limiting factor in plant development. Physiol. Plant. 55, 60-72.
50. Tzou, D., Galson, E., Sondheimer, E. (1973) The metabolism of hormones during seed germination and release from dormancy. Plant Physiol. 51, 894-897.

51. Wang, S.Y., Faust, M., Steffens, G.L. (1985) Metabolic change in cherry flower buds associated with breaking of dormancy in early and late blooming cultivars. Physiol. Plant. *65*, 89-94.
52. Webb, D.P., Van Staden, J., Wareing, P.F. (1973) Seed dormancy in *Acer*. Changes in endogenous cytokinins, gibberellins and germination inhibitors during breaking of dormancy in *Acer saccharum* Marsh. J. Expt. Bot. *24*, 105-106.
53. Westwood, M.N., Bjornstad, H.O. (1968) Chilling requirements of dormant seeds of 14 pear species as related to their climatic adaptation. Proc. Amer. Soc. Hort. Sci. *92*, 141-149.
.4. Williams, K.M. (1985) ATPase activity and water uptake in apple seeds during stratification. Ph.D. Thesis, Cornell University.
55. Wood, B.W. (1983) Changes in indoleacetic acid, abscisic acid, gibberellins, and cytokinins during budbreak in pecan. J. Amer. Soc. Hort. Sci. *108*, 333-338.
56. Young, E., Werner, D.J. (1986) 6-BA applied after shoot and/or root chilling and its effect on growth resumption in apple and peach. HortScience *21*, 280-281.
57. Zimmerman, R.H., Lieberman, M., Broome, O.C. (1977) Inhibitory effect of a rhizobitoxine analog on bud growth after release from dormancy. Plant Physiol. 59 158-160.

E 16. Hormones in Plant Senescence

Jonathan J. Goldthwaite

Department of Biology, Boston College, Chestnut Hill, Massachussets 02167, USA.

INTRODUCTION

Senescence, as the terminal phase of development of a biological structure, follows characteristic patterns of decomposition and metabolism, that are modulated by a host of internal and environmental cues. All of the major classes of plant hormones, as well as cellular regulators such as nucleotides, calcium, and polyamines alter the rate and/or pattern of senescence in diverse plant materials. Light, temperature, as well as nutritional status also strongly influence this aging process.

Physiologists continue to make progress toward an understanding of senescence regulating mechanisms, how these controls interrelate in the intact organism, and in what way stressful environments and diseases influence aging. This chapter provides an orientation to experimental work in senescence regulation, and will summarize physiological work in selected areas of recent research activity. Related reading in this volume can be found in chapters E1, E2, and E17. Several previous articles will help provide a more comprehensive picture of this field (6,44,49,50)

Some Technical Considerations

Excised organs or organ sections are often employed for senescence experiments to facilitate tissue sampling and solute uptake, as well as to avoid the correlative complexity of the whole plant. When excised organ sections are to be used, the desired and now increasing practice is to sterilize the plant material and subsequently handle the sections or discs aseptically. A less satisfactory alternative is to use non-sterile tissue, and limit the effects of microbial contamination by frequent replacement of sterile incubation media. Replacement of test solutions containing hormones or other solutes can improve treatment effectiveness, even when sterile tissue and solutions are used (Fig. 1). When tissue is enclosed in containers permitting limited gas exchange, the investigator should be aware that high tissue densities can actually diminish the oxygen tension enough to inhibit senescence.

Another potential pitfall which the phytogerontologist must avoid is to mistake death for senescence inhibition. Metabolic inhibitors,

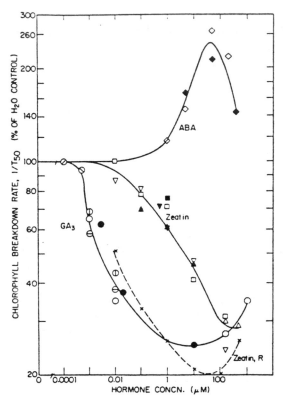

Fig. 1. Senescence rate regulation in discs of *Rumex* leaf tissue by GA₃, zeatin, or ABA. Several experiments are compiled in this figure. Data from the same experiment have the same symbol: GA_3, O; zeatin, □ and Δ; ABA, ◇. Experiment in which zeatin solutions were replaced every 2 days (---). From (30).

anaerobiosis, tissue extracts, and other solutes may in some treatments kill cells prior to their autolysis, thus artifactually preserving high levels of protein and chlorophyll. Effects of such treatments should be checked for reversibility, as well as corroborated with other evidence before firm conclusions can be drawn. Other nonphysiological senescence retardations can result from sublethal cellular stress, such as that resulting from excessive concentrations of osmotically active solutes.

Valid interpretation of hormone effectiveness, tissue hormone responsiveness, and hormone interaction depends upon quantification of senescence *rate* (30); more widespread appreciation of this matter is desirable. A related point of experimentation is the value of carrying out senescence time courses to completion. Although this procedure is more time-consuming, ordinarily requires asepsis, and may involve replacement of treatment media, it helps reveal lethal treatments, and provides data that sometimes can be usefully integrated over the developmental process (18).

Excision necessarily kills a layer of cells at the sample boundary. Cutting also triggers wound ethylene, a period of accelerated ethylene production which, in leaf tissue, typically subsides 6 to 12 hours later (e.g., 32). Assuming asepsis and no further damage, tissue samples are often capable of performing a normal degree of chlorophyll and protein degradation, the customary criteria of senescent autolysis, even while failing to export the products of proteolysis. Although excised material also carries out many of the metabolic activities, and shows many of the hormone and light responses of intact tissue, caution is to be exercised when inferring that control systems shown to be operative in such material are necessarily regulatory under a given set of conditions in the whole plant (e.g., 33).

A study of the effect of deliberately increased wounding upon senescence of oat leaf sections showed that wounding decreased the chlorophyll and protein losses observed in nearly all the wounding and senescence regulator treatments employed, and this decrease was roughly proportional to, and at the location of, the area wounded (17). These results may possibly be explicable as a case of the death-or-senescence artifact described above. Complete recording of senescence time courses under aseptic conditions could permit evaluation of this explanation.

EXCISED TISSUE AND INDIVIDUAL ORGANS

Role of Cytokinins, Gibberellins, and Abscisic Acid

Since the late 1950s, a large number of publications have appeared which demonstrate the widespread and sometimes powerful effects of applied hormones on senescence in leaves, petioles, shoots, and epigeous cotyledons of more than 20 angiosperms, often employing agronomically important cultivars of annuals, but also including a few weedy or woody species. Any one of these excised organs or tissues usually responds to more than one, but not necessarily all, of the major hormone groups. When present, there is qualitative consistency in hormone response: cytokinins, gibberellins, and low concentrations of auxins slow down or delay senescence; abscisic acid and ethylene have an accelerating affect. Such effects, when obtained with a naturally-occurring compound applied at an appropriately low level, indicate the presence of a physiologically coupled receptor system in the aging cells, and provide evidence suggestive of a regulatory role for this hormone. However, further evidence must be developed using the traditional experimental approach (22), or improvements thereon, such as the judicious use of chemical inhibitors or mutants which selectively block the hormone biosynthesis/response system. Beyond this is the need to experimentally unravel the knot of hormone interaction by combining an array of exogenous hormone and inhibitor treatments with measurements of changes in both endogenous hormones and senescence on a careful time-course basis. This is necessary to distinguish cause from effect, and causation from mere temporal association. We also must remain aware that feedback loops as well as interactions between the response to different hormone classes may exist in the tissue under study.

In this section let us examine some progress made regarding three of the hormone groups. The more abundant recent literature on ethylene is considered in the subsequent two sections, along with some studies of auxin and its interplay with ethylene.

The endogenous levels of several hormones have been measured in lettuce leaves (2,5). Dessication of detached leaves, especially when enhanced in the light rapidly caused a dramatic decrease in bioassayed and radioimmunoassayed GAs, a large increase in ABA, determined by gas chromatography, and a drop in bioassayed CKs (2). A 42 hour rehydration period returned these hormone levels at least to that of appropriate control leaves. It was suggested that these hormone changes helped mediate a water stress response by decreasing stomatal aperture. Exogenous GA , ABA, or kinetin affected leaf transpiration and stomatal aperture in a manner qualitatively consistent with this hypothesis. Aharoni, *et al.* (2, 5) pointed out the complexity of this control system, in view of the two-way nature of the water status/hormone status interaction, because of differing short run and long term effects, and because of the ability of applied hormones to override control by water status. It was suggested that water status itself can have a bearing upon leaf metabolism and senescence.

During the senescence of excised lettuce leaves, (5, and Fig. 2) the level of free GAs in exised leaves decreased with the progress of senescence. The GA decrease appeared to be well underway prior to chlorophyll breakdown, consistent with a causative role, although the timing of other senescence processes was not recorded. An inverse correlation of GA level with the pace of senescence was observed when the leaves were treated with kinetin or GA , which delayed chlorophyll breakdown, or ethephon or mild water stress, which accelerated senescence.

In these excised leaves, ABA levels increased somewhat prior to chlorophyll breakdown, and rose more strongly during yellowing (Fig. 2). A positive correlation of ABA level and senescence rate was observed in each of the senescence-altering treatments. Both GA and ABA levels were lower in mature and old attached leaves, than in young leaves from the same plants. While complex hormonal interactions are apparently taking place, these studies have been interpreted to suggest that leaf senescence first is triggered by an early decline in gibberellins, and probably also cytokinins, and once underway is predominantly regulated by ABA (5). It is possible that cytokinins, gibberellins, ethylene, or water

556

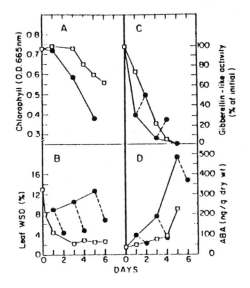

Fig. 2. Effects of water stress on chlorophyll level (A), water saturation deficit (B), GA-like activity (C), and ABA content (D) during senescence of detached lettuce leaves. Mature leaves were kept in darkness at 25°C and 100% RH (☐—☐) or at 25°C and 80 to 85% RH (●—●). Broken lines represent the recovery period of stressed leeaves at 100% RH. Initial content of GA was 9.8 ng GA_3 equiv/g dry weight. From (5).

stress each could in a direct way regulate senescence or *via* an alteration of levels of endogenous ABA, GAs and/or CKs. We cannot yet exclude the possibility that some of the endogenous hormone changes may be associated with, but not causative of, the progress of senescence. Also the role of endogenous auxins and ethylene was not evaluated, nor was the endogenous cytokinin analysis done in detail.

ABA content of attached bean cotyledons during senescence showed little change, other than a small transitory increase approximately half-way through the lifespan (13). ABA increased approximately four-fold in oat leaf sections, and doubled in attached leaves (15). Both increases coincided with the loss of chlorophyll. Under senescence-retarding treatments such as light or kinetin, this ABA increase was not observed in the oat leaf sections. It was concluded that ABA was regulating senescence in this species, although the evidence again was simply of a correlative nature. It was further suggested that treatments which led to stomatal closure could directly, i.e., independent of water potential change, cause ABA accumulation and consequently result in faster senescence.

The Role of Ethylene

In recent years ethylene has received increasing attention as a potential regulator of senescence in vegetative tissues. The question of whether ethylene is involved in the regulation of leaf senescence has been approached in a number of ways: measurement of endogenous ethylene production during senescence, treatment of leaf tissue with exogenous ethylene, changing the rate of ethylene biosynthesis, or absorption of endogenously produced ethylene to diminish its accumulation in cells or their environment. As each approach individually has is own limitations, we are especially interested in studies that have combined these approaches to answer the question.

Differing patterns of ethylene production have been reported in senescing leaf tissue. In some cases, ethylene production peaked prior to the phase of rapid chlorophyll breakdown. This pattern has been reported in detached leaves of *Nymphoides*, *Vicia faba* variety CH4, and *Phaseolus* as well as in tomato leaf segments. In a second pattern, ethylene production increased during the course of senescence and peaked after the onset of rapid chlorophyll breakdown. This pattern has been observed in chilled orange leaves, excised tissue from *Phaseolus* petioles and pulvini, disks from tobacco and beet leaves, and in detached leaves of *Vicia faba* variety CH182, oat, *Ecballium* and *Prunus*. We have observed a third pattern: dock (*Rumex*) leaf disks exhibited two peaks of ethylene production (19). The first peak occurred on the second day after tissue excision, just prior to the rapid phase of chlorophyll and protein breakdown. The second peak of production occurred near the end of the rapid breakdown phase. The physiological significance of these different patterns has yet to be clarified.

It has been confirmed, with tobacco leaves senescing on the plant, that the ethylene production rate increases more than four-fold from the moderate yellowing stage to the complete yellowing stage(4). In both tobacco and bean leaves, ethylene production was found to correlate with internal ethylene levels.

Applications of ethylene or ethylene-releasing compounds have resulted in modest acceleration of leaf tissue senescence in a number of species including tobacco, oat, dock, and bean. Futhermore, when intact bean plants were subjected to reduced pressure to dissipate ethylene, with O_2 and CO_2 levels raised to their atmospheric partial pressures, chlorophyll loss from leaves was inhibited, reinforcing the view that endogenous ethylene was acting as a senescence promotor(34).

Treatments with 1-aminocyclopropane-1-carboxylic acid (ACC), the immediate precursor of ethylene, stimulated ethylene production in dock leaf disks up to 90 times the basal rate though it only accelerated the senescence rate by a 13% (19). The discrepancy between the large acceleration of ethylene production and the smaller acceleration of senescence in response to ACC suggests some limit on the senescence-accelerating effect of elevated rates of ethylene production. ACC accelerated the senescence of detached oat leaves in the dark to a similar degree (16).

Retardations of senescence and ethylene production by some of the ethylene biosynthesis inhibitors have been reported. $CoCl_2$ inhibited both ethylene production and senescence in oat leaf sections (16), although it is not known whether the effect was reversible by ethylene. Aminoethoxy-vinylglycine (AVG) inhibited ethylene production and senescence in tobacco leaf disks; the senescence inhibition was partially reversed by exogenous ethylene (4, and Fig. 3 data at zero Ag^+). AVG also inhibited chlorophyll loss from oat leaf segments (16). The ethylene antagonist, silver ion, also inhibited senescence in tobacco (Fig. 3).

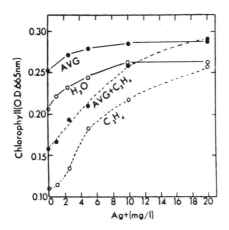

Fig. 3. Effect of increasing concentration of $Ag^{+($ which counteracts the effects of ethylene) on chlorophyll retention by tobacco leaf discs treated with ethylene and AVG. Chlorophyll content at zero time was at 0.520 O.D. units. Leaf discs were pretreated by floating for 30 min on $AgNO_3$ solutions and thereafter were allowed to senesce for 5 days in 25-ml flasks. Ethylene at 5 μl/l was applied during the first 4 days. AVG at 0.1 mM was applied continuously from day 0. Measurements of chlorophyll were made on the 5th day. From (3).

In our work on dock, we tested three inhibitors of ethylene biosynthesis, $CoCl_2$, aminooxyacetic acid (AOA), and AVG. The rate of senescence was inhibited 21 to 39% by these inhibitors at 0.1mM, and somewhat stronger inhibitions were obtained at higher inhibitor levels (20). The inhibitions of ethylene production and senescence by $CoCl_2$ were only partially reversible by exogenous ethylene. This suggested that $CoCl_2$ may have been substantially retarding senescence by mechanisms additional to its inhibition of ethylene production, i.e., to a degree non-specifically. In the cases of both AOA and AVG, ACC or exogenous ethylene completely reversed the inhibitions of ethylene production and senescence. This indicated a specific inhibition of senescence in dock by these two compounds through their suppression of ethylene production.

At the ultrastructural level 2 ppm ethylene for 48 hr induced an apparent transformation of thylakoids into macrograna, an enlarged chloroplast substructure not observed in attached tomato leaves senescing in charcoal-purified air (14). Additional microscopic studies under physiological conditions of hormone level perturbation would be a valuable aid to mechanistic interpretation.

Hormone Interactions That May Involve Ethylene

A number of investigators have raised the question of whether other senescence-regulating hormones act though a regulation of (endogenous) ethylene production or action. Results at least consistent with this possibility are those in which ABA (16) or high auxin (1, and Fig. 4) stimulated, and cytokinin (23) inhibited, both ethylene evolution and senescence. Results that appear to run counter to this hypothesis include some in which cytokinin (1, and Fig. 4), especially when combined with auxin stimulated ethylene evolution while inhibiting senescence. In one study, ABA treatment led to inhibited ethylene production, while senescence was stimulated (24). It may be that the above discrepancies could in some cases be resolved by more careful kinetic analysis over a complete time course.

Ethylene antagonists have been only occasionally been employed in an attempt to show whether the senescence regulation by another hormone can be blocked. Thus, the acceleration of senescence by ABA in oat leaf sections was only partially overcome by addition of Ag^+, AVG, or $CoCl_2$ (16). Since ABA treated tissue evolved ethylene faster within 1 day, it was inferred that ABA acted both by stimulating ethylene production and, to a larger degree, by an independent senescence-accelerating route.

Even though kinetin accelerated ethylene production somewhat in tobacco leaf disks, one could imagine its action mechanism on senescence

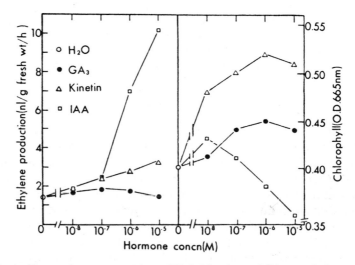

Fig. 4. Effects of increasing concentrations of IAA, kinetin, and GA_3 on ethylene production and chlorophyll retention by tobacco leaf discs. Leaf discs were allowed to senesce for 96 h in 50-ml flasks. Ethylene production was measured after the first 48 h of incubation, and the average hourly rate was calculated. Chlorophyll was extracted after additional 72 h of incubation in sealed flasks which were ventilated daily. Chlorophyll content at zero time was 0.647 O.D. units. From (1).

to be an *via* the inhibition of the binding or action of ethylene. This possibility would seem to be somewhat discredited by the observation in the same study that kinetin inhibited chlorophyll loss more strongly than did Ag^+, CO_2 or a combination of these two ethylene antagonists (1). It was also concluded that ethylene bound to its site of action somehow regulated its evolution by negatively feeding back on its biosynthesis (1), or by direct competition for the free ethylene pool between ethylene action (perhaps involving metabolism of the hormone) and diffusion of the hormone out of the tissue (16). IAA is a key promotor of ethylene biosynthesis, while the synergistic action of kinetin may have been effected *via* maintaining the free IAA level through an inhibition of conjugate formation (1). The simultaneous senescence inhibition by kinetin probably occurred by another, parallel, action which more directly delayed senescence *via* promotive effects on protein synthesis and RNA stability.(1)

In subsequent studies on excised tobacco leaf tissue, a number of sugars, most notably D-galactose, sucrose, and lactose stimulated ethylene production several fold with or without exogenous IAA (32,41). The combined effect of sugar with IAA was more than additive. Radiolabeled IAA was taken up and was decarboxylated, or conjugated as an ester or an amide, or catabolized by an unknown route within 1 day of incubation. While sucrose treatment had no obvious effect on formation of these metabolites, it greatly enhanced the subsequent metabolism of the IAA ester over 3 to 5 days. It was hypothesized that sugar stimulated the release of free IAA from the hydrolysis of conjugate(s). This could help explain reported effects of sugars on several auxin-mediated phenomena, including stem growth, embryogenesis from callus, and leaf senescence. Auxin levels increase in senescing oat, bean, and tobacco leaves. An increase in endogenous carbohydrates might control the IAA level, thereby influencing ethylene production, as well as other auxin-mediated processes (32,41).

In detached rice leaves, ethylene production again appears to accelerate senescence (24). While the cytokinin benzyladenine (BA) at higher concentrations led to accelerated ethylene production after 3 days of incubation, it delayed senescence. ABA after 2 days resulted in accelerated senescence and inhibited ethylene production. It was suggested that BA and ABA, despite the apparently contrary effects on ethylene evolution, may nevertheless control senescence *via* ethylene action, perhaps by inducing a decrease or an increase, respectively, in tissue sensitivity to ethylene. However it appears unnecessarily dogmatic to conclude that these hormones, as well as light, cold, and various metabolic inhibitors all influence senescence by way of the ethylene hormone response.

Some of the discrepancies considered above may be due to the use of different species, some of which involve hormonal regulation of senescence *via* ethylene, and others in which ethylene-independent routes of senescence regulation are activated by cytokinins, gibberellins, auxins or ABA. It may also be true that both ethylene-mediated and ethylene-independent senescence regulation are simultaneously effected by one of the other hormones.

HORMONES IN THE SENESCENCE OF ORGAN SYSTEMS OR WHOLE PLANTS

A more complex experimental system inevitably confronts us when we attempt to understand hormone regulatory systems as they operate in shoots or entire plants. The complexity of interorgan and intertissue correlative effects during reproductive development and vegetative senescence in the whole soybean has been described by Nooden, including a list of 14 correlative control signals, or influences (35). Indeed as is argued elsewhere, there is no reason to anticipate simplicity in such a biological system (37), though it is possible for one signal to have a cascade of effects.

Another form of complexity, over which the careful investigator can exert influence, reflects the difficulties of designing controlled experiments upon larger systems. Thus, one must attempt to control water stress, nutritional status and vigor of the root system, stomatal aperture with its attendant effects on photosynthesis and xylem sap flux, and age heterogeneity of material in an experimental comparison, etc. Failure to account for such variables can potentially lead to artifacts and often hinders interpretation. A few examples where these concerns have been addressed are included in some of the literature discussed in the remainder of this chapter.

Soybean

This annual crop cultivated for its protein and energy rich seeds exhibits a dramatic overall senescence, or monocarpic senescence, of the entire plant body. This senescence also includes the nitrogen-assimilating root nodules and the process is closely related to flowering and the development of seeds in the many axillary fruits borne on the shoot. A summary of the phenomenology of this process, and of a number of experimental manipulations thereof, can be found in reference 35.

Flower induction, and hence the eventual senescence of the plant which follows reproductive development, is very sensitive to environmental conditions, being induced by nights longer than a critical value, which has been agronomically selected in North America to optimize yield in a given latitudinal zone ("maturity group"). The rate of seed and fruit development is further stimulated by additional short days and is also sensitive to the day and night temperature regime (47). It concluded that substantial yield increases in soybean will require genetic or chemical modification of photoperiodic sensitivity to suppress the rate of seed-fill (47).

Soybeans homozygous recessive for two loci controlling photoperiodic response and for two other loci influencing indeterminateness of shoot growth habit after flowering showed a dramatically delayed leaf senescence (DLS), even though seed development continued. These plants yielded 75 to 86% of the seed dry weight per hectare compared to that given by two cultivars, while nitrogen fixation was not clearly increased in the DLS lines (40). This showed that complete leaf senescence was not required for moderate seed yield, although maximum yields were thought to require complete leaf senescence. Also, the seed yield in DLS was somehow limited, even though plant nitrogen was not limiting. Workers interested in agronomic application of DLS were cautioned that expression is strongly limited by field environment and by genetic background (40).

A number of studies have employed surgical or genetic suppression of seed development to manipulate the senescence porcess. Plants from which all young pods were manually removed showed a pattern of shoot accumulation of dry matter and reduced nitrogen in the shoot that was similar to controls (8,9). Likewise, the timing of the rise and fall of chlorophyll and RuBPcarboxylase (RuBPCase) in leaves throughout the canopy, and of nitrogenase in roots was affected little by depodding (Fig. 5 and 6). However, the rate of senescent decline in these parameters was somewhat inhibited by depodding, with the result that leaves of depodded plants retained some chlorophyll for up to two weeks longer than controls. Similar results were obtained in a comparison of genetic male-sterile and fertile isolines (44). Surprisingly, the presence of nodules did not delay the senescence of podded or depodded plants (Fig. 5). The initiating event in shoot senescence was, therefore, not the seed-filling process (9). Loss of photosynthetic function (RuBPCase and chlorophyll) of depodded plants was not accompanied by depletion of phosphorous or reduced nitrogen of the leaves, neither was nodule senescence accompanied by carbohydrate depletion in the roots. Thus the accumulation of macronutrients by developing seeds does not cause senescence of vegetative plant parts *via* a depletion of their nutrient supply (9,35).

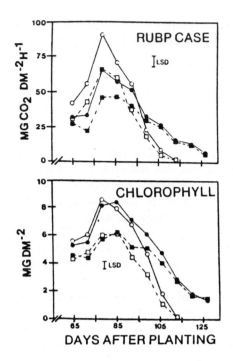

Fig. 5. Effect of pod removal on RuBPCase activity, chlorophyll concentration, and chlorophyll of upper leaves of soybean plants. Nodulated podded (——o), nodulated depodded (——●), nonnudulated podded (– –□), and nonnodulated depodded (– –■). From (9).

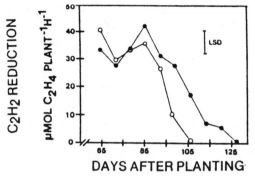

Fig. 6. Effect of pod removal on nitrogenase activity as measured by acetylene reduction of nodulated soybean roots. Nodulated podded (——o), and nodulated depodded (——●). From (9).

Once the photosynthetic decline has begun, depodding can influence the rate of senescence. Pod removal caused a slight inhibition of chlorophyll loss, a slight acceleration of RuBPCase loss and photosynthesis decline, and a surprising stabilization or even an increase in leaf soluble protein (48). The protein increase was shown to be largely due to accumulation of at least two putative storage polypeptides of 80 and 28 kilodaltons molecular mass.

This study nicely illustrates the sometimes unappreciated need for measurement of multiple and specific markers in senescence studies. It appears that a minimum set of parameters to define senescence in attached leaves should consist of quantitative measurements of the chlorophylls and of RuBPCase to monitor compositional senescence of the chloroplast as well as CO_2 exchange rates to monitor functional (photosynthetic) senescence, plus appropriate compositional and functional measures of cytoplasmic senescence. Also, the content of total protein should be measured, since its decrease usually correlates with senescence in this organ. It is suggested that workers in senescence regulation should define such a set of multiple markers for proper assay of the progress of senescence in other organs and tissues also.

Several surgical manipulations involving pod and/or seed removal from chamber-grown soybeans have demonstrated a localized behavior of the accelerating effect of the fruits upon leaf yellowing (29). This undefined corelative influence has been referred to as an hypothetical "senescence signal" without prejudice as to its nature (35). Although an effect could be clearly observed when fruit development was eliminated just prior to the seed growth phase (29), a further delay of entire plant death has been obtained by defloration, indicating a cumulative effect of reproductive structures over the course of their development (26). These studies show that the flowers and fruits accelerate but do not cause senescence, consistent with the studies discussed above. The disagreement of interpretation on this point between the more recent (9) and the earlier (29) study may be attributable to an overstatement in the latter of the strength of the senescence-retarding effect of defruiting. This could have come about because of the dependence of this work on a senescence parameter, visible leaf yellowing, which was insensitive to most of the chlorophyll degradation time-course (28,43), and which needed to be complemented by the assay of additional senescence markers.

Steam-girdling of the petiole had no effect on the time of leaf yellowing in fruiting soybeans, although this treatment resulted in higher nitrogen content at the time of yellowing, presumably due to inhibited leaf nitrogen export in the heat-treated phloem (35,36). This suggested that the aforementioned accelerating effect of the fruits is somehow communicated to the leaf *via* the xylem. Because developing fruits on defoliated basal nodes did not much accelerate terminal yellowing of leaves at upper defruited nodes, it was concluded that the seeds probably do not act by diversion of senescence inhibitors (cytokinins, nutrients, etc.) which may be coming from the roots to the leaves in the xylem sap stream (35,36). Based upon their series of indirect experiments, Nooden and coworkers promoted the hypothesis that the leaf senescence accelerating signal in soybean was an increased level of a senescence accelerating hormone produced in growing seeds which moved to the subtending (or nearby) leaf; ABA was considered a prime candidate (35).

In a study of sink regulation of source photosynthesis, it was concluded that sink inhibition by depodding or steam girdling of the petiole inhibited leaf photosynthesis rates through ABA-induced stomatal closure (45). Leaf ABA level increased 10-fold two days after depodding, and increased 25-fold one day after girdling. Radioactive ABA, injected into the petiole, and moving *via* the transpiration stream was used to label the leaf ABA pool (46). This permitted an analysis of the dynamics of ABA catabolism, synthesis, and export in control and sink-inhibited plants. Although girdling apparently caused a transient increase in ABA synthesis, the majority of the ABA increase in both girdled and depodded plants was attributed to an inhibition of export of the hormone from the leaf blade (Fig. 7). One day after injection of control plants, about 60% of the radioactivity that had left the leaf was in the pods and 40% was in the rest of the shoot. Another study using radioactive ABA on intact soybeans likewise failed to support the idea that ABA was the senescence signal: ABA injected into pods at various developmental stages showed very little movement to the leaves (39).

Although the ABA content of leaves did increase 8-fold during their senescence, the increase came late in the process, after the majority of protein and chlorophyll was lost (43). Thus, increased levels of endogenous ABA may help accelerate the latter portion of leaf senescence

Fig. 7. Effect of translocation-obstructing treatments on total ABA-derived radioactivity in leaf blades and pods of soybean. The ($\pm$)[2-^{14}C]ABA was injected into petioles at T = -4t and the treatments (control [O], depodding [$\triangle$], and girdling [$\square$]) were applied at T = 0 h. From (46).

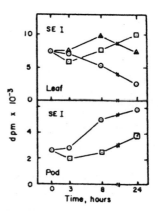

in this species. What initiates the process, and what delays senescence in deflowered or defruited plants is unclear.

The cytokinins are another hormone class which has been long hypothesized to be causally involved in senescence of vegetative organs in shoots. These compounds might influence senescence both through a depletion in their biosynthesis and export in the root xylem sap as plants age, and by being diverted from shoot vegetative tissues under the influence of floral induction and fruit development (35,49). The output of four cytokinins in soybean root pressure exudate (xylem sap) underwent a severe decrease which occurred after flowering, prior to rapid seed and nodule growth, and was overlapped by the period of yellowing and abscission of leaves at the more basal nodes (21). It was concluded that this cytokinin peak was primarily related to shoot reproductive development, while the decline in hormone flux could have influenced leaf senescence. Because exogenous applications of cytokinin can delay soybean leaf senescence, these hormones may indeed contribute to the control of monocarpic senescence in this species.

Radioactive zeatin and zeatin riboside were transpirationally supplied to excised soybean nodes bearing a leaf and one or more pods at the pod elongation or early podfill stage (38). After 1 day, the leaf contained 65% of the radioactivity, extensively metabolized. The pods contained only 8% and the seeds, 2% of the supplied label, which showed that growing pods did not divert much cytokinin from the leaves, at least at these developmental stages and under these experimental conditions. Many of the studies in this area have presumed the root to be the sole, or predominant, source of cytokinins for the rest of the plant. Because this assumption has been challenged in the pea (7), it will be important to reexamine the senescence studies in this light.

Pea

Another senescence phenomenon apparent in whole plants is known as apical senescence. It occurs in peas following the start of reproduction as the first stage of overall senescence. The process results in arrested development and localized senescence (chlorosis) of the terminal bud and associated leaves (12). This senescence pattern is reversible by various treatments at any point prior to apical death.

The G2 line of pea, unlike most common cultivars, has dominant alleles at both the *Sn* and *Hr* genetic loci. Flowering and fruit development in this variety occur regardless of photoperiod, but apical senescence only develops under long days and it absolutely depends upon the simultaneous presence of developing fruit.

The earliest symptoms of apical senescence in G2 peas precede the period of maximum seed fresh weight growth, and the presence of apical senescence under different photoperiods cannot be simply accounted for by a given threshold of seed growth rate. Although nutrient levels were not directly measured, it has been suggested that fruits do not act by a nutrient drain on the apex, since this might be proportional to the collective fruit growth rate (16a). Morphologically, however, G2 plants grown in long days, in comparison with those grown in short days, have fewer nodes with smaller leaves located between the shoot apex and the growing fruits. This was a consequence of both faster fruit growth and slowing apical node production under long days. Bearing in mind this morphological difference, senescence of the apex in long days, as contrasted with short days, could accordingly result from the combined influence of locally decreased levels of hypothetical leaf-derived senescence-inhibitory hormones (and photosynthate or other nutrients), and/or locally increased levels of hypothetical fruit-derived senescence promotive hormones.

Evidence in support of the former hypothesis has been obtained from studies on GAs in relation to senescence in G2 peas (42). Bioassayed levels of GAs, mainly GA_{20} and GA_{19}, in leaves and stipules of G2 peas decreased about 90% when plants grown in short days were moved to long days. There was no change in GA levels with photoperiod in I2 peas which are recessive at the Sn locus and undergo apical senescence regardless of daylength. GA_{12} aldehyde, from which all the other GAs are derived, is metabolized in G2 shoots to GA_{53}, GA_{44}, GA_{19} (and/or GA_{17}) and GA_{20}. There was considerably more metabolism of ^{14}C from GA_{12} aldehyde through this pathway in short days as compared to long days (Fig. 8)(11), supporting the earlier finding that shoots in short days contained more GA_{19} and GA_{20}. These experiments also showed that photoperiod controls GA metabolism in G2 peas at one or both of the steps between GA_{12} aldehyde and GA_{53}.

When transport of photosynthate from the upper leaves to the apical bud was examined it was found that more photosynthate moved to the apical bud in short days than long days. Moreover, there was a 25-fold shift of partitioning of photosynthate within the apical bud in favor of the reproductive structures in long days as compared to short days. Thus apical senescence may be attributable to a nutrient starvation of the developing vegetative structures caused by the strong sink activity of the very young developing reproductive structures (Kelly and Davies, personal communication). Kelly and Davies suggest that a nutrient diversion to reproductive structures early in their development may be a general phenomenon in the induction of plant senescence. Such a nutrient shift may well be hormonally mediated since, in G2 peas, it follows a sharp decline in GA activity in the shoot, associated with less vigorous vegetative tissue development. This is supported by the fact that exogenous GA_{20} applied to the shoot tip delayed apical senescence in a way that nearly duplicated the effect of short days (42).

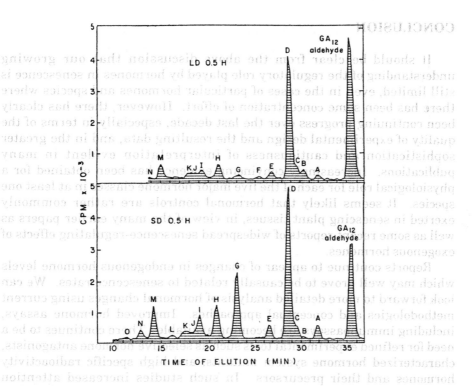

Fig. 8. Representative elution profiles of radiolabeled [^{14}C]GA$_{12}$ aldehyde metabolites from a C$_{18}$HPLC column with a water to acetonitrile gradient. The peaks are lettered in the order of increasing polarity. The metabolites were extracted from G2 pea shoots grown in short days (under which growth is indeterminate) or long days (under which apical senescence takes place) and allowed to take up [^{14}C]GA$_{12}$ aldehyde in water for 0.5h through the cut base of the stem in the light. (Peak G was identified as GA$_{53}$. Peaks I, J, K received label from metabolism of peak G, and I, J, and K were identified as GA$_{44}$, GA$_{20}$, and GA$_{19}$ plus GA$_{17}$, respectively.) From (11).

To examine the second hypothesis that growing fruits send some senescence accelerating hormone(s) to the apex, radioactive CO$_2$ was supplied to developing fruits of G2 peas under long day conditions. The shoot apex was extracted and shown to contain radioactivity derived from the fruit. Less than 1% of the fruit radiolabel was exported to the apex and this material was shown not to be ABA. Because the movement from fruit to apex was lower in long days than short days, it seems unlikely that apical senescence can be attributed to a fruit-derived senescence promoter (10, Hamilton and Davies, in preparation). The likelihood that the hypothetical senescence hormone was ABA had been previously diminished by an experiment which showed that exogenous ABA applications to the apex could not restore the effectiveness of a surgically reduced fruit load in long days (12).

CONCLUSION

It should be clear from the above discussion that our growing understanding of the regulatory role played by hormones in senescence is still limited, even in the cases of particular hormones and species where there has been some concentration of effort. However, there has clearly been continuing progress over the last decade, especially in terms of the quality of experimental design and the resulting data, and in the greater sophistication and cautiousness of interpretation evident in many publications. Increasingly convincing evidence has been obtained for a physiological role for each of the five major hormone classes in at least one species. It seems likely that hormonal controls are rather commonly exerted in senescing plant tissues, in view of the many earlier papers as well as some recent reports of widespread senescence-regulating effects of exogenous hormones.

Reports continue to appear of changes in endogenous hormone levels which may well prove to be causally related to senescence rates. We can look forward to more detailed analysis of hormonal changes using current methodologies and conceptual approaches. Improved hormone assays, including immunoassays are becoming available. There continues to be a need for refined experimental tools such as selective hormone antagonists, characterized hormone system mutants, and high specific radioactivity hormones and their precursors. In such studies increased attention should be paid to location and timing of such changes in the plant, to possible changes in tissue responsiveness, and to differential metabolism and relative activity of various chemical forms of the hormones (*e.g.*, 27). Additional classes of hormones may exist which might also regulate senescence.

Another area in which we can anticipate further study to be fruitful is in the clarification of the relationship between nutritional and hormonal control of senescence. The literature discussed in this article abounds with argumentation which often attempts to discredit such a role for the mineral macronutrients and photosynthates. It appears probable that some interaction, perhaps localized, between nutrient fluxes and the hormonal controls is occurring *in vivo* in monocarpy and in other forms of senescence.

A final point worthy of emphasis is the recognition that many studies of senescence regulation deal with the progress of, rather than the initiation of, this aging process. Thus, the onset of vegetative senescence may occur very early, possibly even as a direct result of the completion of leaf expansion (9). An early major effect of daylength to help initiate apical senescence was clearly observed in the G2 pea. The possibility exists that some causative events in monocarpic senescence may also be traced back as far as the photoperiodic induction step itself, as had been indicated by the earlier studies on cocklebur (25), and other species.

Acknowledgment

I thank Marilyn Grant, Patrice Kennedy Coughlin, and, Margo Webber and for their bibliographic assistance.

References

1. Aharoni, N., Anderson, J.D., Lieberman, M. (1979) Production and action of ethylene in senescing leaf discs. Effect of indoleacetic acid, kinetin, silver ion, and carbon dioxide. Plant Physiol. 64,805-809.

2. Aharoni, N., Blumenfeld, A., Richmond, A.E. (1977) Hormonal activity in detached lettuce leaves as affected by leaf water content. Plant Physiol. 59,1169-1173.

3. Aharoni, N., Lieberman, M. (1979) Ethylene as a regulator of senescence in tobacco leaf discs. Plant Physiol. 64,801-804.

4. Aharoni, N., Lieberman, M., Sisler, H.D. (1979) Patterns of ethylene production in senescing leaves. Plant Physiol. 64,796-800.

5. Aharoni, N., Richmond, A.E. (1978) Endogenous gibberellin and abscisic acid content as related to senescence of detached lettuce leaves. Plant Physiol. 62,224-228.

6. Biswal, U.C., Biswal, B. (1984) Photocontrol of leaf senescence. Photochem. Photobiol. 39,875-879.

7. Chen, C.M., Ertl, J.R., Leisner, S.M., Chang, C.C. (1985) Localization of cytokinin biosynthesis sites in pea plants and carrot roots. Plant Physiol. 78,510-513.

8. Crafts-Brandner, S.J., Below, F.E., Harper, J.E., Hageman, R.H. (1984a) Effects of pod removal on metabolism and senescence of nodulating and nonnodulating soybean isolines. I. Metabolic constituents. Plant Physiol. 75,311-317.

9. Crafts-Brandner, S.J., Below, F.E., Harper, J.E., Hageman, R.H. (1984b) Effects of pod removal on metabolism and senescence of nodulating and nonnodulating soybean isolines. II. Enzymes and chlorophyll. Plant Physiol. 75,318-322.

10. Davies, P.J. (1981) Do developing fruits export a senescence factor? Quantitative analysis of export from developing pea fruits. Plant Physiol. 67 (4 Suppl.), 115.

11. Davies, P.J., Birnberg, P.R., Maki, S.L., Brenner, M.L. (1986) Photoperiod modification of [^{14}C] gibberellin A_{12} aldehyde metabolism in shoots of pea, line G2. Plant Physiol. 81, 191-196.

12. Davies, P.J., Proebsting, W.M., Gianfagna, T.J. (1977) Hormonal relationships in whole plant senescence. In Plant Growth Regulation. pp 273-280. Pilet, P.E., ed. Springer-Verlag, Berlin.

13. Dumbroff, E.B., Brown, D.C.W., Thompson, J.E. (1977) Effect of senescence on levels of free abscisic acid and water potentials in cotyledons of bean. Bot. Gaz. 138,261-265.

14. Fukuda, K., Toyama, S. (1982) Electron microscope studies on the morphogenesis of plastids. XI. Ultrastructual changes of the chloroplasts in tomato leaves treated with enthylene and kinetin, Cytologia 47,725-736.

15. Gepstein, S., Thimann, K.V. (1980) Changes in the abscisic acid content of oat leaves during senescence. Proc. Natl. Acad. Sci. USA 77,2050-2053.

16. Gepstein, S., Thimann, K.V. (1981) The role of ethylene in the senescence of oat leaves. Plant Physiol. 68,349-354.

16a Gianfagna, T.J., Davies, P.J. (1981) The relationship between fruit growth and apical senescence in the G2 line of peas. Planta 152,356-364.

17. Giridhar, G., Thimann, K.V. (1985) Interaction between senescence and wounding in oat leaves. Plant Physiol. 78,29-33.

18. Goldthwaite, J. (1974) Energy metabolism of Rumex leaf tissue in the presence of senescence-regulating hormones and sucrose. Plant Physiol. 54,399-403.

19. Hackett, B.P., Goldthwaite, J. (1982) Ethylene: Role in Rumex leaf tissue senescence. Plant Physiol. 69 (4 Suppl.),8.

20. Hackett, B.P., Goldthwaite, J. (1983) On the role of ethylene in *Rumex* leaf tissue senescence. Plant Physiol. *72* (1 Suppl.),22.

21. Heindl, J.C., Carlson, D.R., Brun, W.A., Brenner, M.L. (1982) Ontogenetic variation of four cytokinins in soybean root pressure exudate. Plant Physiol. *70*,1619-1625.

22. Jacobs, W.P. (1959) What substance normally controls a given biological process? I. Formulation of some rules. Devel. Biol. *1*,527-533.

23. Jana, S., Choudhuri, M.A. (1982) Ethylene production and senescence in submerged aquatic angiosperms. Aquatic Botany *13*,359-365.

24. Kao, C.H., Yang, S.F. (1983) Role of ethylene in the senescence of detached rice leaves. Plant Physiol. *73*,881-885.

25. Krizek, D.T., McIlrath, W.J., Vergara, B.S. (1966) Photoperiodic induction of senescence in *Xanthium* plants. Science *151*,95-96.

26. Leopold, A.C., Niedergang-Kamien, E., Janick, J. (1959) Experimental modifications of plant senescence. Plant Physiol. *34*,570-573.

27. Letham, D.S., Gollnow, B.I. (1985) Regulators of cell division in plant tissues. XXX. Cytokinin metabolism in relation to radish cotyledon expansion and senescence. J. Plant Growth Regul. *4*,129-145.

28. Lindoo, S.J., Nooden, L.D. (1976) The interrelation of fruit development and leaf senescence in 'Anoka' soybeans. Bot. Gaz. *137*,218-223.

29. Lindoo, S.J., Nooden, L.D. (1977) Studies on the behavior of the senescence signal in Anoka soybeans. Plant Physiol. *59*,1136-1140.

30. Manos, P.J., Goldthwaite, J. (1975) A kinetic analysis of the effects of gibberellic acid, zeatin, and abscisic acid on leaf tissue senescence in *Rumex*. Plant Physiol. *55*,192-198.

31. Meir, S., Philosoph-Hadas, S., Aharoni, N. (1984) Role of IAA conjugates in inducing ethylene production by tobacco leaf discs. J. Plant Growth Regul. *3*,169-181.

32. Meir, S., Philosoph-Hadas, S., Epstein, E., Aharoni, N. (1985) Carbohydrates stimulate ethylene production in tobacco leaf discs. I. Interaction with auxin and the relation to auxin metabolism. Plant Physiol. *78*,131-138.

33. Miller, B.L., Huffaker, R.C. (1985) Differential induction of endoproteinases during senescence of attached and detached barley leaves. Plant Physiol. *78*,442-446.

34. Nilson, K.N., Hodges, C.F. (1983) Hypobaric control of ethylene-induced leaf senescence in intact plants of *Phaseolus vulgaris* L. Plant Physiol. *71*,96-101.

35. Nooden, L.D. (1984) Minireview: Integration of soybean pod development and monocarpic senescence. Physiol. Plant *62*,273-284.

36. Nooden, L.D. (1985) Regulation of soybean senescence. World Soybean Research Conference, 1984. Westview Press, Boulder, in press.

37. Nooden, L.D., Leopold, A.C. (1984) Hormonal regulatory systems in plants. In: Scott, T.K., ed., Encyclopedia of plant physiology, new series, *10*,4-22.

38. Nooden, L.D., Letham, D.S. (1984) Translocation of zeatin riboside and zeatin in soybean explants. J. Plant Growth Regul. *2*,265-279.

39. Nooden, L.D., Obermeyer, W.R. (1981) Changes in abscisic acid translocation during pod development and senescence in soybeans. Biochem. Physiol. Pflanzen *176*,859-868.

40. Phillips, D.A., Pierce, R.O., Edie, S.A., Foster, K.W., Knowles, P.F. (1984) Delayed leaf senesconce in soybean. Crop Sci. *24*,518-522.

41. Philosoph-Hadas, S., Meir, S., Aharoni, N. (1985) Carbohydrates stimulate ethylene production in tobacco leaf discs. II. Sites of stimulation in the ethylene biosynthesis pathway. Plant Physiol. *78*,139-143.

42. Proebsting, W.M., Davies, P.J., Marx, G.A. (1978) Photoperiod-induced changes in gibberellin metabolism in relation to apical growth and senescence in genetic lines of peas (*Pisum sativum* L.). Planta *141*,231-238.

43. Samet, J.S., Sinclair, T.R. (1980) Leaf senescence and abscisic acid in leaves of field-grown soybean. Plant Physiol. *66*,1164-1168.

44. Schweitzer, L.E., Harper, J.E. (1985) Leaf nitrate reductase, D-ribulose-1, 5-bisphosphate carboxylase, and root nodule development of genetic male-sterile and fertile soybean isolines. Plant Physiol. 78,61-65.

45. Setter, T.L., Brun, W.A., Brenner, M.L. (1980) Effect of obstructed trans- location on leaf abscisic acid, and associated stomatal closure and photosynthesis decline. Plant Physiol. 65, 1111-1115.

46. Setter, T.L., Brun, W.A., Brenner, M.L. (1981) Abscisic acid translocation and metabolism in soybeans following depodding and petiole girdling treatments. Plant Physiol. 67,774-779.

47. Thomas, J.F., Raper, C.D. Jr. (1976) Photoperiodic control of seed filling for soybeans. Crop Sci. 16,667-672.

48. Wittenbach, V. (1983) Effect of pod removal on leaf photosynthesis and soluble protein composition of field-grown soybeans. Plant Physiol. 73,121-124.

49. Woolhouse, H.W. (1983) Hormonal control of senescence allied to reproduction in plants. In Strategies of plant reproduction. pp. 201-233. Meudt, W.J., ed. Allanheld-Osmun, Totowa.

50. Yang, S.F., Hoffman, N.E. (1984) Ethylene biosynthesis and its regulation in higher plants. Ann. Rev. Plant Physiol. 35,155-189.

E17. Postharvest Hormone Changes in Vegetables and Fruit

Pamela M. Ludford

Department of Vegetable Crops, Cornell University, Ithaca, New York 14853, USA.

INTRODUCTION

At the time of harvest, there is a large potential for change in physiological processes going on in edible plant tissue. On removal from the parent plant, vegetables are deprived of their normal supply of water, minerals, and organic molecules including hormones, which normally would be supplied by translocation from other parts of the plant. Although little new photosynthesis is being carried out, there is active transpiration and tissues can transform many of the constituents already present. While postharvest changes in fresh vegetables cannot be stopped, they can be slowed down within certain limits.

The kind and extent of physiological activity in detached plant parts determine their storage longevity to a large extent. Thus some, such as seeds, tubers, bulbs, and fleshy roots, are morphologically and physiologically adapted to maintain the tissue in a dormant state, both innate and imposed, until environmental conditions are favorable for germination or growth. Metabolic activity is depressed, but not halted, in such organs. Regrowth is triggered in the spring, probably by a change in the hormone balance. Most vegetables stored over extended periods in the fresh state are biennials that break rest and eventually sprout during storage, hence terminating their usefulness for commercial purposes. The cells of most other plant parts, such as fruit, leaves, stems, petioles, etc., differ in that they are physiologically primed for senescence rather than dormancy. Fruit ripening is usually associated with the development of optimal eating quality and constitutes the final stages of maturation, and thus can be regarded as a typical senescence phenomenum. Many plant materials classified as fruits botanically are considered to be vegetables for commercial or legal purposes. The cucurbits are consumed in both the immature and the fully mature states as cucumbers or zucchini and melons.

Most cultivated vegetable parts are removed suddenly from the natural environment, and often held for short-term transportation or long-term storage in stressful environments, including low temperatures, artificial atmospheres, or both in combination to reduce respiration rates. Storage of fruit and vegetables can be prolonged by ethylene removal using ethylene scrubbers (e.g., Purafil), flushing with nitrogen gas, or by hypobaric storage. Transient ethylene production is also triggered by stress or injury, so wounding during harvest and transport has an obvious effect on storage, added to which many bacterial pathogens have the capacity to synthesise ethylene. Quite apart from ethylene effects in fruit ripening, the effect on leaf abscission has an immediate result on leafy vegetables. The commercial storage of cabbage along with apples can be disastrous (33).

The few research reports available on stored plant materials have suggested that endogenous hormones continue to function and appear to control physiological events. This conclusion is apparent from correlative evidence of hormonal balances in detached plant organs and easily observed physiological events such as rest, dormancy and compulsive regrowth. The term "dormancy" in this chapter is used as a state in which growth is temporarily suspended due to unfavourable conditions, i.e., imposed dormancy, while innate dormancy (true dormancy or rest) occurs when growth cannot take place even under favourable conditions due to the condition of the plant material itself. Exogenous methods using applied growth substances on plants or their excised parts have been widely used to extrapolate to endogenous hormonal responses. In many cases there are problems with this approach, often because of difficulties in uptake and distribution into bulky tissue such as fruit or tubers. All the commonly identified endogenous hormones, i.e., auxins, gibberellins (GAs), cytokinins, abscisic acid (ABA), and ethylene, appear to be present, as well as polyamines.

Interaction and balance between opposing promotory and inhibitory hormonal factors is the idea behind the control of metabolism in postharvest storage, be it of rest and regrowth in vegetable storage organs or of maturation and ripening in vegetable fruit. The interesting difference is in what constitutes these opposing factors. In fruit, ethylene is one important promoter of ripening, and ABA may be another (36), while auxins, GAs and cytokinins are possible candidates for the role of ripening inhibitors. The latter are high in young seeds and developing fruit and may affect the changing sensitivity of maturing climacteric fruit to ethylene. In leafy vegetative tissue, ethylene causes leaf abscission and cytokinins retard senescence. In storage organs, ABA acts more as an inhibitor of regrowth, while auxins and GAs are likely to promote it.

FRUIT RIPENING

Under normal conditions, fruit ripening occurs as an integrated sequence of changes including softening, colour change, accumulation of sugars and aromatics, coupled with a decline in organic acids. These active metabolic processes are accompanied in many fruits by an increase in respiration termed the climacteric. Coincident with, or just prior to, the increase in respiration in climacteric fruit there is a pronounced increase in the production of ethylene. At least some of the ripening changes can be separated from each other experimentally including the respiration climacteric and the ethylene peak (36), but ripening is inhibited in climacteric fruit under conditions that inhibit ethylene synthesis. Ethylene application initiates the respiration response even in non-climacteric citrus fruit but here continued application is necessary. The main difference between climacteric and nonclimacteric fruit is seen in their ability to produce ethylene autocatalytically in response to threshold levels of ethylene. Commonly referred to as the ripening hormone, ethylene has a cascade effect in climacteric fruit leading to the "one rotten apple in the barrel" syndrome. Thus ethylene plays a significant role in the changes that occur with the climacteric in fruit ripening, and its production is intimately involved in fruit ripening changes.

The most desirable fresh state for consumption of the climacteric tomato fruit, *Lycopersicon esculentum* Mill., is at the termination of ripening and the beginning of senescence, the red ripe stage. However, for commercial purposes to facilitate shipment fruit are frequently harvested at the "mature green" (MG) stage and then ripened by ethylene application. A difficulty in commercial harvesting is the determination of the precise MG stage, as some immature fruits respond to exogenous ethylene but do not undergo a normal ripening. This may account for complaints of poor quality in winter-shipped tomatoes. The bell pepper, *Capsicum annuum* L., cannot be ripened to a satisfactory red color if removed from the plant in the immature or green stage. On the other hand avocado will not undergo ripening or the climacteric rise in respiration while still attached to the plant, with IAA suspected to be the inhibitor.

The rate of protein synthesis increases during the early stages of ripening in several climacteric fruit (36), although the way in which such synthesis is controlled is not clear. While this may just reflect an increase in protein turnover, it is also related to *de novo* synthesis of ripening-specific enzymes. For instance, the enzyme polygalacturonase (PG) is absent from green tomato pericarp tissue, is first detected when fruit begin to colour, and increases progressively during ripening, along with acid invertase. Until the first appearance of ethylene in tomato fruit neither the respiratory climacteric, appearance of PG nor the increase in polysomes and cytoplasmic mRNA takes place (38).

Ethylene

Parts of the ethylene synthetic pathway via 1-aminocyclopropane)-1-carboxylic acid (ACC) were established in fruit (52). Aminoethoxyvinyl-glycine (AVG) is effective in inhibiting ethylene synthesis in slices of green tomatoes, but relatively ineffective in pink and red tomato fruit. However, isolated ACC synthase from pink and red fruit is sensitive to low levels of AVG, and the ineffectiveness with fruit tissue may simply reflect relatively high endogenous levels of ACC at the pink and red stages of fruit development (7). At the preclimacteric MG stage, ACC synthase activity, ACC content, and ethylene-forming enzyme (EFE) activity are low, but increase markedly on ripening following the breaker stage. Addition of exogenous ACC to many vegetative tissues results in greatly increased ethylene production, but this is not the case with preclimacteric fruit of apple and cantaloupe, i.e., both the conversion of SAM to ACC and that of ACC to ethylene are restricted in these preclimacteric fruit (52). In preclimacteric MG tomato fruit, ACC slightly enhances the ripening process and ethylene production, and exogenous ethylene treatment increases the capability to convert ACC to ethylene, i.e., increases EFE activity before ACC synthase activity (52). This is also true for cantaloupe, *Cucumis melo* L., where wounding increases the activity of both ACC synthase and the EFE. Tomato fruit tissue is also capable of conjugating ACC to N-malonyl-ACC (MACC). The malonylation enzyme is promoted by exogenous ethylene treatment and the content of ACC conjugate increases as ripening progresses (28). Levels of ACC and ACC synthase are higher in the placenta of freshly harvested fruit than in the outer pericarp (27).

The role of ethylene as the ripening hormone has been brought into question and the investigation of ripening has been aided by several non-ripening tomato mutants: *Nr* - never ripe; *rin* - ripening inhibitor; and *nor* - non-ripening. Both *rin* and *nor* fruit fail to ripen (with the exception of seed maturation) and do not display a climacteric rise in CO_2 or ethylene evolution (43). They have reduced lycopene levels, their chlorophyll content remains high, with very low, if any, PG levels. External ethylene applications have little effect in inducing ripening though it will bring about a temporary stimulation of CO_2 evolution which is repeatable (43). Wounding of the fruit, by contrast, causes an increase in both CO_2 and ethylene production, so the capability to produce ethylene is not lacking (19), and ACC and ACC synthase are present in the *rin* mutant (7). Furthermore, in wild tomato species that ripen on the vine but remain green, two species show ethylene production correlated to fruit softening, while in two others external ripening changes are not correlated to ethylene production (17). Ripening may thus be determined by changes in sensitivity to ethylene rather than the amount produced.

When shipped over long distances, fruit can be subjected to low temperature stress. Ethylene production in a number of chilling-sensitive vegetables is stimulated by chilling temperatures of 0 to 15°C. This may occur in tissue which does not usually produce significant amounts of ethylene, such as preclimacteric fruit. In chilled cucumber (*Cucumis sativus*), increased ethylene production is due to increased capacity to make ACC, but the increase in ACC, ACC synthase, and ethylene is not apparent until subsequent warming. An increase in ACC synthase activity during the warming period can be inhibited by cycloheximide treatment but not by cordycepin or α-amanitin, suggesting stimulated production of mRNA coding for ACC synthase during chilling, i.e., mRNA is transcribed during the chilling stage but translation is not completed until transfer to warmer temperatures (50). Prolonged chilling exposure damages the conversion of ACC to ethylene by the EFE. Ethylene production in zucchini squash (*Cucurbita pepo*) also increases rapidly on transfer from chilling temperatures to 10°C, especially those held at low oxygen levels (34). In 'Honey Dew' melon (*C.melo* var.*inodorous*), however, ACC accumulates during the chilling period without waiting for subsequent warming (Wang, personal communication), so it seems that not all sensitive fruit respond in the same way to chilling. Under most commercial conditions, chilling is not likely to exceed four days, and the ethylene generated upon warming probably is responsible for the chlorophyll loss and pitting noted in chilled cucumbers.

Other Hormones

Abscisic acid.

In a number of fruit (e.g., pear, avocado), the level of free ABA is constant during maturation and increases during ripening, with implications of a ripening promoter (36). However, in kiwi fruit the ABA content decreases rapidly during ripening (45), and in tomato the pattern of change is a little different. ABA levels increase during the growth of the tomato fruit reaching a peak at the MG stage, and declining prior to the respiration climacteric and ripening (Fig.l), even in the non-ripening mutant *rin*. The ABA peak comes a little later in the mutant *Nr*, and about 20 days later in *nor*, where ABA levels are lower (32). Only in two Japanese tomato cultivars,'Kyoryoku-goko' (22) and 'Fukuju' (45) is a 3-fold increase in free ABA seen during ripening after the MG stage to a much higher level than reported in other studies (over 300 vs.175 ng/g fresh weight), and which could be increased still further by harvesting at the turning or pink stages. Changes in bound ABA reflect those of free ABA but at about one-seventh of the levels, similar to the 10:1 ratio of free/bound ABA throughout avocado ripening. Thus the increase in free ABA must represent net synthesis rather than release of the bound form. Free ABA accumulates to similar levels in both attached and detached

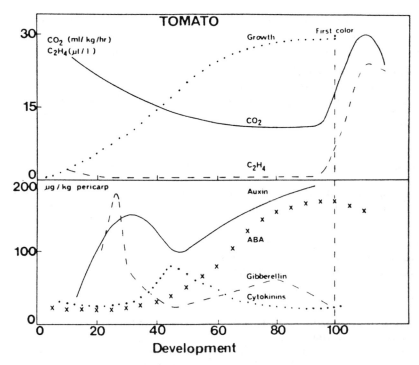

Fig. 1. Trends in free hormone levels in tomato pericarp tissue during development and ripening. From (31).

fruit, unless the fruit is detached very early (15 days after anthesis) when levels are lower, showing that ABA is synthesised in the detached fruit and is not dependent only on translocation from the plant (32).

Auxins.

It is known that indoleacetic acid (IAA) influences ethylene formation through the induction of ACC synthase. Endogenous auxin concentrations are thought to be highest after pollination during the early stages of fruit development and lowest during maturation. The early work of Nitsch showed control by developing seeds and/or applied auxin over early development of strawberry receptacles, and recent work has quantitated free and conjugated IAA along with free ABA in strawberry achene and receptacle tissue during development (5). It has been suggested that ripening may be related to a decline in auxins, oxygen and ethylene interactions, and changes in hydroperoxide levels (16). It is difficult to show this using applied exogenous auxin, since these can stimulate ethylene production. Contrasting results obtained by different types of application (e.g., dipping or spraying fruit vs. vacuum infiltration) can be explained by limited penetration of auxin into the tissue when applied by dipping. It was shown, using dipped whole fruit and vacuum-infiltrated cut discs, that 2,4-D caused a dual effect in tomato fruit tissue, namely an increase in ethylene production which promoted ripening, but also a delay in ripening (49). The last effect prevailed, but depended on the uniformity of the auxin distribution and its concentration. No close correlation has yet been found between fruit ripening and endogenous auxin content (31). However, in strawberry receptacle auxin appears to control stage-specific formation of polypeptides during fruit development (48).

Gibberellins.

Similarly with GAs, their endogenous levels are thought to be high in very young fruit and they may play a role in retarding senescence. However, exogenous applications sometimes stimulate ethylene production and confuse the issue. One interesting aspect of applied GAs has been found with colour changes. Fruit ripening is associated with the conversion of chloroplasts to chromoplasts. Rind of Valencia oranges reaches maximum orange colour during winter months but tends to regreen during spring and summer, chromoplasts revert back to chloroplasts, and this is enhanced by GA (42). GA$_3$ delays appearance of lycopene during ripening of tomatoes (14), and preharvest treatment of citrus and apricot fruit with GA$_3$ retards chlorophyll loss and carotenoid synthesis, but again no close correlation has yet been found between fruit ripening and endogenous gibberellin content (31).

Cytokinins.

Senescence in leaves can generally be delayed by cytokinin treatment, and fruit ripening can be regarded as a senescence phenomenum. High levels of endogenous cytokinin may also delay fruit ripening, and levels may decline as ripening proceeds. Cytokinin activity is high for two weeks after anthesis in cherry tomato when cell division is most active (1), and declines with their ribosides from MG to the red ripe stage in standard varieties (12, 13). The non-ripening mutant *rin* not only contains higher levels than normal fruit, but the rate of decline is less and high levels of zeatin glucoside are found which could act as a storage form to ensure containued high zeatin levels. Increased cytokinin levels in seeded tomato fruit can be obtained by reducing the ratio of foliage to fruit and hence lowering sink competition (47), and this delays the rate of ripening after the breaker stage. Although applied cytokinins are not necessarily effective, kinetin also delays ethylene-induced degreening in peel of banana slices and loss of chlorophyll in other fruit (31).

Thus, cytokinins, GAs, and possibly IAA are implicated in delay of ripening, while ABA levels may increase until ripening is in progress.

UNDERGROUND STORAGE ORGANS

Wounding Responses

Ethylene effects on bulky storage organs are of interest because of possible wound ethylene production after harvest. Treatment with ethylene results in a sharp rise in respiration, especially in the presence of O_2 rather than air. These organs can be separated in their respiration response to cyanide (CN). One group yields CN-sensitive fresh slices (e.g., potato, turnip, rutabaga, Jerusalem artichoke), while the other gives CN-resistant slices using the alternative path of respiration, as in carrot and parsnip. After ethylene treatment, the former group produces CN-resistant fresh slices, rather than CN-sensitive. Other changes resulting from ethylene treatment include increased numbers of polysomes and changes in gene expression. However, it was shown for carrot that the ethylene-stimulated changes are not necessarily coupled (35).

Many papers have investigated the wound reactions and aging of cut slices from storage organs such as potato tuber, carrot root or sugar beet. Wounding induces the synthesis of mRNA and an increase in nucleolar size and protein synthesis along with RNase activity. Inactive ribosomes from resting potato tubers can be activated by added mRNA from polysomes of wounded potato.

The idea of a wound-hormone has evolved from the hypothetical "Wundreiz" of Wiesner and Haberlandt's "leptohormone", through traumatic acid of bean pericarp, and the "β-inhibitor complex", of which a major component was ABA rather than a complex of phenolic compounds, to the endogenous hormone balance usually present (37). Endogenous hormones have been found in all plant storage organs investigated, including potato tuber, the swollen stem of kohlrabi, carrot root, beet, sweet potato, and Jerusalem artichoke tuber. They increase after wounding, as with IAA and cytokinin activity in potato, although this could indicate differences in sensitivity or form (bound vs. free) rather than concentration differences. Exogenous applications of GA_3 can amplify endogenous hormone activities and, most important, hormones can affect metabolic activity of cut slices. The addition of GA_3 stimulates protein synthesis, RNA synthesis, nucleolar size and RNA polymerase activity still further in wounded tissue, although cells of uninjured tubers are not responsive (37). Polyamines inhibit RNase activity in cut potato, and spermidine and spermine also inhibit the rise in betacyanin leakage that normally takes place from cut discs of beet root, *Beta vulgaris*, (4). Polyamines may affect wound-induced or senescence-induced destabilization of cell membranes in plant storage organs. Thus hormones may work together to counteract the catabolic processes caused and to stimulate the synthetic activity necessary to "heal a wound".

Hormonal Changes During Storage

Onion.

The onion, *Allium cepa* L., most important of the bulb crops, demonstrates innate dormancy or rest, dormancy, and regrowth, whose morphological changes correlate with changes in endogenous hormonal balances. In common commercial practice, onion bulbs are cured, i.e., air-dried for about 10 days after harvest, then stored at low temperatures for an extended time period, e.g., 2°C for nine months, September through May. Preharvest foliar spraying with maleic hydrazide (MH) delays sprouting of onion (as well as potato and garlic) during storage and is regularly used. While onion storage in the U.S.is usually around 2°C, MH treatment also is effective with higher storage temperatures (18-26°C) common in the Middle East (30).

Hormonal changes which induce the rest period are initiated while the onion is maturing and still in the field. Morphological changes consist of a rapid lateral growth of the bulb combined with a steady senescence of the leaves. Auxin activity decreases in both the green foliage and bulb apices as the leaves senesce from lush green to the weakened soft-neck stage (24). When all foliage has become procumbent on the ground, auxin activity disappears while ABA activity is present in the tops. At this developmental stage the crop is usually lifted from the soil (undercut) and set on top of it and then harvested within a few days. Inhibitors appear to be synthesized in the leaf and translocated to the bulb apex, since ABA activity in harvested bulbs is high. Also bulbs whose green tops were prematurely dried by a desiccator spray, or removed while green, show early sprouting (41). Premature defoliation should be avoided in practice since movement of inhibitor to the bulb apex appears necessary to establish dormancy during storage.

Bioassays of endogenous hormones in bulbs (central plugs containing apical tissue) stored at 2°C in the dark show a hormonal balance high in ABA activity and low in IAA, GA, and cytokinin activities from harvest to mid-winter (24). This is the rest period and no sprout growth or cell division occurs for several weeks even under the most favorable conditions. In January, the beginning of imposed dormancy is shown by a rapid rise in sprouting and apical cell division on transfer to favourable conditions. There is a decline in ABA activity during dormancy and the following regrowth stage, accompanied by an increase in growth promoter activity (Fig. 2). By April, when some bulbs sprout even in storage and most have well-developed internal leaves (Fig. 3), most hormone activity declines with the balance in favour of growth promoters. Application of exogenous growth substances to excised onion apices shows that only kinetin can break rest in non-temperature induced apices, while ABA prolongs the innate dormancy period. Thus cytokinins may be part of the breaking of innate dormancy, and ABA may be part of the inhibitor complex maintaining this state.

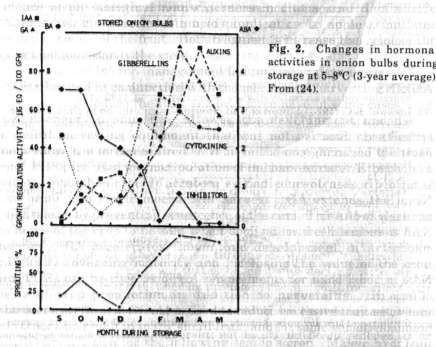

Fig. 2. Changes in hormonal activities in onion bulbs during storage at 5–8°C (3-year average). From (24).

Potato.

The potato tuber, *Solanum tuberosum* L., is a swollen underground stem or stolon that can overwinter in the soil or under proper storage conditions. It also shows the biennial postharvest features of innate dormancy, dormancy and regrowth. Although investigation of tuberization under inducing conditions has involved much hormone work (see Chapter E 14), postharvest hormone studies are more concerned with the induction and termination of the innate dormancy period and the possible role of an endogenous inhibitor. The latter, found in potato peel extract, was shown to decline naturally with the termination of rest (18). A rise in endogenous GA activity and decline in ABA activity seem to be involved in the dormancy break of potato tubers (23), and the application of GA₃ to promote sprouting in seed potatoes is an approved commercial practice in some countries. Increases in cytokinin activities are also found during the transition to dormancy, with conjugation being of possible importance in cytokinin movement from storage tissue to active meristems (46). The polyamines putrescine, spermidine and spermine are found equally distributed in all parts of dormant potato tubers (25), along with their biosynthetic enzymes, arginine decarboxylase, ornithine decarboxylase, and SAM decarboxylase. With breaking of dormancy and beginning of

Fig. 3. Sprouting of onion in storage.

sprout growth, polyamine levels increase in apical buds, but not in dormant lateral buds or non-bud tissue. Thus the break of dormancy may also involve changes in polyamine levels, although it is uncertain whether these changes are the cause or the result of the breaking of dormancy.

Root Crops

Most commercially stored root crops are biennial plants, and low temperatures along with high humidities are generally used for storage. Although no rest period has been demonstrated, all show some degree of dormancy and regrowth and, with cold-induction, flowering. Treatment with GA₃ can cause carrot (*Daucus carota* L.), among other species, to bolt following non-vernalizing temperatures, and endogenous GA-like activity increases during vernalization at 5°C (21). However, stem elongation precedes floral differentiation and no flowering takes place without cold induction, so that endogenous GAs are not implicated in cold induction during storage. Roots are also sites of synthesis for cytokinins, which are exported in the xylem. Cambium tissue was found to be the site for cytokinin synthesis in carrot root (9). ABA in root tissue may derive from leaves and build up during vegetative growth, and in sugar beet is preferentially catabolized to phaseic acid and dihydrophaseic acid in the

584

root, although ABA conjugation could take place in both the taproot sink and source leaves (11).

LEAFY CROPS

Cabbage

Within the Cruciferae, white cabbage (*Brassica oleracea* L., Capitata group) is an important fresh vegetable crop that can be stored up to six months. The traditional method involves harvesting during low field temperatures and storing with ventilation of cold winter air (common storage). This can result in quality loss, typified by loss of chlorophyll and weight loss, to the point of making it unmarketable, a condition to which ethylene may be a contributing factor. Use of technical improvements in storage such as refrigeration, ethylene removal, and controlled atmosphere (CA) has made possible longer storage times. Cabbage can be satisfactorily stored at 2°C under CA conditions (5% CO_2, 2.5% O_2), extending the storage life for several months. There is considerable diversity in inherent keeping quality among cultivars.

The long storing cultivar 'Green Winter' apparently has an inhibitor-controlled dormant period for several months after harvest, since ABA-activity rises rapidly for about eight weeks after harvest and then declines (23). After three months IAA and GA activities are slightly higher in heads stored in refrigerated air than in CA-stored heads (5% CO_2, 2.5% O_2). Cytokinin activity is not affected by CA conditions (23), being high at harvest, declining to a low point during early dormancy, and increasing when regrowth begins in the apices. Cytokinin profiles show varietal differences between cv. 'Excel' and three cabbage breeding lines of different storage capabilities (41). High levels (100 ppm) of ethylene in air at 0°C for five weeks storage of fresh cabbage result in 6-fold increases in endogenous auxin activity and 12-fold in GA activity over the air control, while ABA activity is undetectable (24). The external leaves of all heads become bleached, desiccated and abscised from the main stalk (Fig. 4). Even at 1 ppm, ethylene in air has a detrimental effect magnifying or accelerating changes taking place over the long storage period, whether these are in degreening and leaf abscission, weight loss, sugar loss, changes in organic acid content, or increased respiration rates (20). The increased CO_2 and reduced O_2 of CA conditions seem to counteract the ethylene effects seen in air storage.

When decapitated trimmed stems from sampled heads of cabbage are rooted in potting mix and grown in the greenhouse, regrowth of axillary buds takes place with time, sometimes followed by flowering. The rate and type of regrowth varies depending on the state of the cabbage head, possibly reflecting the hormonal state (29). Regrowth is slow for cabbage straight from the field, and also after four weeks of storage. All plants

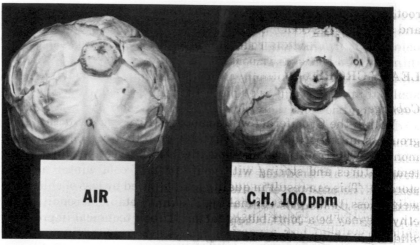

Fig. 4. Cabbage stored in ethylene atmosphere showing abscised leaves.

form heads with no flowering (Fig. 5). This could represent the dormant period with a preponderance of ABA. After four months in refrigerated air storage, regrowth on planting is fast and most plants bolt and flower, with differences between cultivars in the rate of regrowth (Fig. 5). Cabbage stored under CA conditions show no flowering even after five months. By this time ABA levels may have dropped, as in air storage, but possibly the ratio between IAA, GA and cytokinins is not optimal for flowering. However, after a storage time of ten weeks in air, some cultivars bolt and flower, while others remain vegetative, indicating a possible borderline hormone balance between either IAA, GA, cytokinins, or ABA and these promoters. In cv.Bartola regrowth and flowering is speeded up by the presence of ethylene during air storage (Fig.6), while 5 ppm ethylene even in CA storage allows flowering at this ten week stage as well as after 5 months (29). Actual measurements made on these heads indicate that ABA still tends to be high at 10 weeks, and IAA levels remain high in the non-flowering CA-stored heads. However, these analyses were done on large "apical samples" which included stem, leaf bases, and axillary buds as well as apices. Recent work shows that axillary buds contain twice as much free ABA, for instance, as the rest of the "apical" tissue, and chance inclusion of more axillary buds could disproportionately slant the hormonal results (Ludford, unpublished data).

Other Leafy Vegetables

While cabbage cultivars can be stored for months, others in the *B.oleracea* L.species, e.g., Brussels sprout (Gemmifera group), cauliflower (Botrytis group) and broccoli (Italica group) can be stored from only a few days to a few weeks due to their rapid loss of edible quality. With broccoli

Fig. 5. Regrowth of cabbage stem axillary buds. A. Vegetative heads. B. Flowering.

and Brussels sprout, this involves desiccation, yellowing, and abscission of leaves, while cauliflower also shows curd growth called "riciness". Quality loss is accentuated by ethylene. Treatment of broccoli with AVG, the ethylene synthesis inhibitor, reduces ethylene production and respiration and retards yellowing and senescence so that it is still in saleable condition with green colour retention and compactness (6). Ethylene, known to promote loss of chlorophyll in fruit ripening, was used to promote the blanching of celery as far back as 1924 (39).

Applied cytokinins are often effective in prolonging the shelf life of leafy vegetables by slowing down senescence. Synthetic and natural cytokinins extend the storability of both Brussels sprout and broccoli (41). Good quality in broccoli is maintained by benzyladenine (BA) treatment and low temperatures of 2°C, reflected in sensory evaluations of the cooked broccoli (6). Application of BA as a postharvest dip delays

Fig. 6. Regrowth and flowering of stem axillary buds from "Bartola" cabbage heads stored for 10 weeks. A. Air. B. Controlled atmosphere.

senescence and maintains green colour and fresh appearance in many other leafy vegetables besides crucifers, including endive, escarole, spinach, parsley, green onion, and asparagus. However, application for the use of BA and other synthetic cytokinins on leafy vegetables has not been approved in the U.S. Senescence in celery, *Apium graveolens* L., is also delayed by BA treatment, with respiration rates decreased by 27%

and total weight loss reduced by 47% over 22 days (51). Pithiness of the edible celery petiole can be stimulated by flooding and nutrition-deficiency after a prolonged period, but much faster by water stress. Not too surprisingly, the latter is associated with an increase in endogenous free ABA. However, exogenous ABA application also stimulates petiole pithiness of detached leaves (3), and shortens the storage life of broccoli and Brussels sprout (41).

It was suggested that these opposing effects of applied BA and ABA could be due to a relationship between senescence and stomatal aperture (39), with ABA enhancing closure of stomata in the light, and cytokinins maintaining opening. While this may be reasonable in intact plants, harvested leafy vegetables are detached plant parts, and transpiration with the resulting wilting is a major factor in initial deterioration. The older view that cytokinins delay senescence by the maintenance of RNA and protein synthesis, delaying the degradation of protein and chlorophyll, and by slowing down the rate of respiration, seems more applicable. A microsomal enzyme fraction from fresh cauliflower heads hydroxylates the cytokinins isopentenyl adenine and isopentenyl adenosine to zeatin and zeatin riboside, again demonstrating that cytokinin biosynthesis, or rather conversion, is not limited to roots and seeds (10). Ethylene treatment of the enzyme system reduces the conversion by 28-43%. Part of the senescence promoting activity of ethylene could thus be through preventing cytokinin conversions or even synthesis.

Although most reports discuss exogenous hormone treatment, endogenous cytokinin activities decrease during accelerated aging of outer leaves of Brussels sprout while GA and inhibitor activities increase (41). English gardeners remove the terminal buds after the axillary buds or sprouts have begun to form to increase the size of sprouts, but decapitation at a younger stage results in axillary shoots. Lateral buds of the younger plants contain more GA-like activity than those of older plants which would not show shoot extension (40). The "riciness" of cauliflower or elongation of the floret peduncles may be GA controlled, and cauliflower curds treated with growth inhibitors including inhibitors of GA synthesis such as chlormequat (CCC) frequently inhibit curd growth (41). Good quality sprouts would therefore appear to need high cytokinin content along with low GA, with the latter improving cauliflower quality also.

Seedlings of lettuce, *Lactuca sativa*, were used for one of the earliest GA bioassays, but little is known about the postharvest endogenous hormone content. Two cytokinins have been isolated from butterhead lettuce, mostly from the innermost developing leaflets (26). Again exogenous cytokinin treatments enhance chlorophyll retention in lettuce (2), but is of little practical value since only outer leaves are affected, and these would be trimmed by the time they reach the consumer because of damage in handling.

CONCLUSIONS

Plant hormones obviously play an important role in the postharvest physiology of vegetables, but little is solidly known about it. As is usually the case, the effects observed are due more to hormonal balance than to the activity of any one hormone, although sensitivity to growth substances must be considered as a possible controlling factor rather than hormone level itself (44). It has been claimed that our known hormones do not have the requisite variety for a signalling system to match the variety of the situation that has to be controlled (8). This is based on the belief that hormone molecules carry complex information, whereas they may be agents with one simple bit, on or off, with concentration dependence for control of magnitude of response (15). There may instead be a multiplicity of target cells, with complexity resulting after the hormonal reaction.

References

1. Abdel-Rahman, M., Thomas, T.H., Doss, G.J., Howell, L. (1975) Changes in endogenous plant hormones in cherry tomato fruits during development and maturation. Physiol. Plant 34, 39-43.
2. Aharoni, N., Back, A., Ben-Yehoshua, S., Richmond, A.E. (1975) Exogenous gibberellic acid and the cytokinin isopentyladenine retardants of senescence in romaine lettuce. J. Amer. Soc. Hort. Sci. 100, 4-6.
3. Aloni, B., Pressman, E. (1979) Petiole pithiness in celery leaves: induction by environmental stresses and the involvement of abscisic acid. Physiol. Plant. 47, 61-65.
4. Altman, A. (1982) Polyamines and wounded storage tissues - Inhibition of RNase activity and solute leakage. Physiol. Plant. 54, 194-198.
5. Archbold, D.D., Dennis, F.G. (1984) Quantification of free ABA and free and conjugated IAA in strawberry achene and receptacle tissue during fruit development. J. Amer. Soc. Hort. Sci. 109, 330-335.
6. Batal, K.M., Heaton, E.K., Beuchat, L.R., Granberry, D.M. (1981) Effects of N^6-benzyladenine and storage temperatures on shelf-life and quality of raw and cooked broccoli (Brassica oleracea L. var. Italica). HortScience 16, 284.(Abstract).
7. Boller, T., Herner, R.C., Kende, H. (1979) Assay for and enzymatic formation of an ethylene precursor, 1-aminocyclopropane-1-carboxylic acid. Planta 145, 293-303.
8. Canny, M.J. (1985) Ashby's law and the pursuit of plant hormones: a critique of accepted dogmas using the concept of variety. Aust. J. Plant Physiol. 12, 1-7.
9. Chen, C-M., Ertl, J.R., Leisner, S.M., Chang, C-C. (1985) Localization of cytokinin biosynthetic sites in pea plants and carrot roots. Plant Physiol. 78, 510-513.
10. Chen, C-M., Leisner, S.M. (1984) Modification of cytokinins by cauliflower microsomal enzymes. Plant Physiol. 75, 442-446.
11. Daie, J., Campbell, W.F., Seeley, S.D. (1981) Temperature-stress-induced production of abscisic acid and dihydrophaseic acid in warm- and cool-season crops. J. Amer. Soc. Hort. Sci. 106, 11-13.
12. Davey, J.E., Van Staden, J. (1978) Endogenous cytokinins in the fruits of ripening and nonripening tomatoes. Pl. Sci. Lett. 11, 359-364.
13. Desai, N., Chism, G.W. (1978) Changes in cytokinin activity in ripening tomato fruit. J. Food Sci. 43, 1324-1326.
14. Dostal, H.C., Leopold, A.C. (1967) Gibberellin delays ripening of tomatoes. Science 158, 1579-1580.

15. Firn, R.D. (1985) Ashby's law of requisite variety and its applicability to hormonal control. Aust. J. Plant Physiol. *12*, 685-687.
16. Frenkel, C., Eskin, M. (1977) Ethylene evolution as related to changes in hydroperoxides in ripening tomato fruit. HortScience *12*, 552-553.
17. Grumet, R., Fobes, J.F., Herner, R.C. (1981) Ripening behavior of wild tomato species. Plant Physiol. *68*, 1428-1432.
18. Hemberg, T. (1947) Studies of auxins and growth inhibiting substances in the potato tuber and their significance with regard to its rest period. Acta Hort. Berg. *14*, 133-220.
19. Herner, R.C., Sink, K.C. 1973. Ethylene production and respiratory behavior of the *rin* tomato mutant. Plant Physiol. *52*, 38-42.
20. Hicks, J.R., Ludford, P.M. (1981) Effects of low ethylene levels on storage of cabbage. Acta Hort. *116*, 65-73.
21. Hiller, L.K., Kelly, W.C., Powell, L.E. (1979) Temperature interactions with growth regulators and endogenous gibberellin-like activity during seedstalk elongation in carrots. Plant Physiol. *63*, 1055-1061.
22. Inaba, A., Yamamato, T., Ito, T., Nakamura, R. (1980) Comparison of ripening characteristics and sensory evaluation in attached and detached tomato fruits. J.Japan. So c. Hort. Sci. *49*, 132-138.
23. Isenberg, F.M.R. (1979) Controlled atmosphere storage of vegetables. Hortic. Rev. *1*, 337-394.
24. Isenberg, F.M.R., Thomas, T.H., Abdel-Rahman, M., Carroll, J.C., Howell, L., 1974. The role of natural growth regulators in rest, dormancy and regrowth of vegetables during winter storage. Proc. XIX Intern. Hort. Congress, Warszawa, Vol. 2, pp. 129-138.
25. Kaur-Sawhney, R., Shih, L., Galston, A.W. (1982) Relation of polyamine biosynthesis to the initiation of sprouting in potato tubers. Plant Physiol. *69*, 411-415.
26. Kemp, T.R., Knavel, D.E., Hamilton, J.L. (1979) Isolation of natural cytokinins from lettuce leaves. HortScience *14*, 635-636.
27. Kende, H., Boller, T. (1981) Wound ethylene and 1-aminocyclopropane-1-carboxylate synthase in ripening tomato fruit. Planta 151, 476-481.
28. Liu, Y., Su, L-Y, Yang, S.F. (1985) Ethylene promotes the capability to malonylate 1-aminocyclopropane-1-carboxylic acid and D-amino acids in preclimacteric tomato fruits. Plant Physiol. 77, 891-895.
29. Ludford, P.M., Hillman, L. (1985) Regrowth of cabbage axillary buds. Acta Hort. *157*, 219-226.
30. Matlob, A.N. (1979) The effect of preharvest foliar sprays of maleic hydrazide on the sprouting of onion bulbs during storage. Mesopotamia J. Agric. *14*, 145-152.
31. McGlasson, W.B. (1978) Role of hormones in ripening and senescence. *In* Postharvest Biology and Biotechnology, pp 77-96, Hultin, H.O. and Milner, M. eds., Food and Nutrition Press, Inc., Westport, CT.
32. McGlasson, W.B., Adato, I. (1976) Changes in the concentrations of abscisic acid in fruits of normal and *Nr, rin* and *nor* mutant tomatoes during growth, maturation and senescence. Aust. J. Plant Physiol. *3*, 809-187.
33. McKeon, A.W., Lougheed, E.C., Murr, D.P. (1978) Compatability of cabbage, carrots and apples in low pressure storage. J. Amer. Soc. Hort. Sci. *103*, 749-752.
34. Mencarelli, F., Lipton, W.J., Peterson, S.J. (1983) Responses of 'zucchini' squash to storage in low-O_2 atmospheres at chilling and non-chilling temperatures. J. Amer. Soc. Hort. Sci. *108*, 884-890.
35. Nichols, S.E., Laties, G.G. (1985) Differential control of ethylene-induced gene expression and respiration in carrot roots. Plant Physiol. 77, 753-757.
36. Rhodes, M.J.C. (1980) The maturation and ripening of fruits. *In* Senescence in Plants, pp. 157-205, Thimann, K.V., ed., CRC Press, Inc., Boca Raton, FL.
37. Rosenstock, G., Kahl, G. (1978) Phytohormones and the regulation of cellular processes in aging storage tissues. *In* Biochemistry of Wounded Plant Tissues, pp. 623-671, Gunter K., ed., Walter De Gruyter, Berlin, New York.

38. Speirs, J., Brady, C.J., Grierson, D., Lee, E. (1984) Changes in ribosome organization and messenger RNA abundance in ripening tomato fruits. Aust.J. Plant Physiol. *11*, 225-233.
39. Thimann, K.V. (1980) The senescence of leaves. *In* Senescence in Plants, pp. 85-115, Thimann, K.V., ed., CRC Press, Inc., Boca Raton, FL.
40. Thomas, T.H. (1972) The distribution of hormones in relation to apical dominance in Brussels sprouts (*Brassica oleracea* var.gemmifera L.) plants. J. Exp. Bot. *23*, 294-301.
41. Thomas, T.H. (1981) Hormonal changes during senescence, ripening and regrowth of stored vegetables. *In* Quality in Stored and Processed Vegetables and Fruit, pp. 253-265, Goodenough, P.W. and Atkin, R.K., eds., Academic Press, London.
42. Thompson, W.W., Lewis, L.N., Coggins, C.W. (1967) The reversion of chromoplasts to chloroplasts in Valencia oranges. Cytologia *32*, 117-124.
43. Tigchelaar, E.C., McGlasson, W.B., Buescher, R.W. (1978) Genetic regulation of tomato fruit ripening. HortScience *13*, 508-153.
44. Trewavas, A.J. (1982) The regulation of development and its relation to growth substances. What's New Plant Physiol. *13*, 41-44.
45. Tsay, L.M., Mizuno, S., Kozukue, N. (1984) Changes of respiration, ethylene evolution, and abscisic acid content during ripening and senescence of fruits picked at young and mature stage. J. Japan. Soc. Hort. Sci. *52*, 458-463.
46. Van Staden, J., Dimalla, G.G. (1978) Endogenous cytokinins and the breaking of dormancy and apical dominance in potato tubers. J. Exp. Bot. *29*, 1077-1084.
47. Varga, A., Bruinsma, J. (1974) The growth and ripening of tomato fruit at different levels of endogenous cytokinins. J. Hort. Sci. *49*, 135-142.
48. Veluthambi, K., Poovaiah, B.W. (1984) Auxin-regulated polypeptide changes at different stages of strawberry fruit development. Plant Physiol. *75*, 349-353.
49. Vendrell, M. (1985) Dual effect of 2,4-D on ethylene production and ripening of tomato fruit tissue. Physiol. Plant. *64*, 559-563.
50. Wang, C.Y. (1982) Physiological and biochemical responses of plants to chilling stress. HortScience *17*, 173-186.
51. Wittwer, S.H., Dedolph, R.R., Tuli, V., Gilbart, D. (1962) Respiration and storage deterioration in celery (*Apium graveolens* L.)as affected by post-harvest treatments with N[6]-benzylamino-purine. Proc. Amer. Soc. Hort. Sci. *80*, 408-416.
52. Yang, S.F., Hoffman, N.E. (1984) Ethylene biosynthesis and its regulation in higher plants. Ann.Rev.Plant Physiol. *35*, 155-189.

E 18. Hormones in Tissue Culture and Micro-Propagation

Abraham D. Krikorian, Kevin Kelly, and David L. Smith
Department of Biochemistry, Division of Biological Sciences, State University of New York at Stony Brook, Stony Brook, New York 11794, USA.

INTRODUCTION

Aseptic culture techniques have figured prominently in the study of plant growth and development. The identity of hormones and role of growth regulators came about in large measure as a result of these studies. The first successful aseptic cultures were those of excised root tips. Somewhat later it became possible to grow callus cultures derived from storage organ explants or the cambial region of woody species. These cultures grew slowly and the stimuli to cell division usually entailed the addition of auxins such as indole-3-acetic acid (IAA) or naphthaleneacetic acid (NAA) to otherwise simple media comprised of mineral salts, sucrose and a few vitamins (16).

A major advance was the discovery that addition of liquid endosperms such as coconut water to culture media dramatically increased the amount and rate of cell division. This was significant because the explants used did not include cambium. In fact, the carrot root plugs were small (order of 2.5 mg fresh weight) and excision was far enough away from the cambium to guarantee that the component cells would have stopped dividing (33).

In tobacco it was shown that stem explants containing bits of vascular cylinder as well as pith required only auxin as an additive to the culture medium to achieve significant callus growth. When pith explants without vascular tissue were tested, they would not grow well unless complex additives such as yeast or malt extract, or coconut water were supplied in addition to an auxin. This work opened the way to the discovery of kinetin, the first chemically identified plant cell division factor. It soon followed that kinetin in the presence of IAA or a synthetic substitute could bring about cell division in tobacco and a number of other species in a completely defined medium (32).

Further advances in our appreciation of the complexity of growth requirements of excised tissues *in vitro* came about from studies on crown gall tumors. Proliferative growths on stems of plants such as tobacco,

593

kalanchoe and periwinkle could be initiated by inoculation with *Agrobacterium tumefaciens*. The time that the crown gall bacterium was allowed to remain in contact with the tissue prior to its elimination by an appropriate heat (or antibiotic) treatment determined the complexity of the additives required *in vitro* to sustain the proliferation when explanted and cultured. The more fully transformed the plant tissue, the more autotrophic it was. In other words, crown gall tumors could be maintained on a simple medium of mineral salts and a carbon source because they were synthesizing their own growth substances (4).

Minimal media without hormonal additives rarely serve, however, as a vehicle to sustain growth of normal tissues. Growth stimulating substances may be supplied in the form of natural fluids like coconut water which are known to contain auxins, cytokinins, gibberellins and sugar alcohols like myo-inositol or as known chemical compounds (33). In either case, the guiding principle is that *in vitro* (and *in situ*) one requires at least two systems to be operative. One involves auxins and the other cytokinins. The extent to which this classic and simplistic view now needs to be modified or amplified will emerge in the course of this chapter.

ASEPTIC CULTURE SYSTEMS

The term tissue culture is being used nowadays in a generic sense to include the *in vitro* culture of virtually any plant part at any level of organization (12, 41). This means protoplasts, cells, callus, tissues, organs and even whole plants. Even so, it is increasingly becoming evident that only precise designation of what is being cultured and why will more effectively communicate what strategies a tissue culturist might adopt (7, 12). Since this necessarily impacts the way and which growth regulators are used, it will be useful to review the range of culture techniques in use.

Embryo Culture

It was shown as far back as the early 1920s that one could sometimes stimulate growth, otherwise unobtainable or erratic, of certain embryos in aseptic culture. In some cases, embryos with poorly developed food reserves do not germinate because they are dependent on external nutrient sources. For instance the seeds of orchids contain a very small embryo comprised only of a simple mass of cells (25). The embryo is totally dependent upon exogenous sugar to germinate. In nature this sugar is provided by a symbiotic mycorrhizal relationship. Another example where embryos fail to germinate involves the formation of inhibitors in the seed. Here embryos often germinate only after an appropriate period of dormancy. In some plants (e.g., iris) one can eliminate both the dormancy requirement and the effect of germination inhibitors present in the seed of some hybrids by excising embryos and rearing them in sterile culture in the absence of the chemical constraints to a size sufficient for independent growth. Aseptic culture has become a widely used and routine procedure for rescuing embryos that would not normally grow into plantlets. The growth regulators used must foster normal growth if the strategy is to be effective. This means no callus (31).

Meristem, Shoot, or Stem-Tip Culture

The discovery of the ability of explanted meristems (apical growing points) and shoot tips of the orchid *Cymbidium* grown in aseptic culture to produce, when appropriately cut, protruberances which resemble normal protocorms which can grow to plantlets provided in the 1960s the most dramatic impetus for the further development of procedures for multiplying and maintaining plants in aseptic culture (31). Shoot tip cultures from many other plants have since been exploited in the obtaining, maintaining and multiplication of stocks. In some cases, one generates one plant from one cultured shoot tip; in others, multiple shoots can be stimulated to form by addition of appropriate levels of cytokinin. As long as development of shoots emerging from the proliferated area at the base of a shoot tip explant can be maintained at a rate consistent with their removal by excision, one has an "open-ended" system. One merely maintains a hormone balance, usually higher in auxin, which favors the continued formation of undifferentiated growth that will organize in culture by adjusting the medium -usually increasing cytokinin or lowering auxin. As these shoots (with or without roots) are removed from the proliferating mass and are transferred to an environment or different medium conducive to further root development, say auxin-rich, new proliferations grow to replace them (7, 12, 26).

A variation on this theme involves the stimulation of some plants to form precocious axillary shoots in profusion when stimulated by exogenous cytokinins. Since axillary shoots can, in turn, produce additional axillary branches as each newly formed shoot is subcultured, the method is a good one for rapid clonal multiplication. It has been applicable to a great variety of species ranging from herbaceous foliage plants to bulbous monocotyledons and woody species (12, 19).

Other organ cultures

When we fully understand what developing organs receive in terms of stimuli and nutrients and are able to recreate the environment in which they originate *in situ*, we should be able to grow them separately, and if not from their initiating cells, then from their primordia. At the present the achievements of classical organ culture apply more to roots than other organs. Leaves and fruits rarely reach significant size *in vitro* (33).

Anthers and Ovules as Sources of Haploid Cells

In the mid-1960s, cultured anthers of several plants were shown to be capable of yielding haploid plantlets. Cultures can be initiated from anthers containing immature pollen grains, actually microspores prior to the development of the mature male gametophyte. In tobacco, for instance, the vegetative nucleus divides to give rise to the proembryo while still within the original wall of the pollen grain. The process, called androgenesis, has potential for raising haploid plants and is a breakthrough with consequences for genetics, plant breeding and agriculture. This is because homozygous diploids can be raised by the use of colchicine or by taking advantage of the fact that many cultured cells and tissues spontaneously undergo endopolyploidization. Androgenesis generally involves no 'use of exogenous hormones to initiate the system. This implies that they produce what they need. It is a matter of catching them at the right stage (28). Gynogenesis, the process whereby haploid plants are produced in vitro by induction of haploid tissues from the female gametophyte is not as far technically advanced as androgenesis and generally involves rather elaborate culture media (44).

Protoplast Cultures

Enzymatic isolations, *en masse*, of still viable wall-less cells (protoplasts) have been familiar since 1960. Cellulases and pectinases, generally derived from certain wood-degrading fungi, capable of dissolving the intercellular components and cell wall are now widely used to produce protoplasts from different plants and organs and even their aseptically derived and cultured tissues and cells (30). Having reconstituted walls around them, they are able to divide, proliferate and in some instances, eventually give rise to plants. In still other cases, having obtained naked protoplasts, and having overcome in this way the barrier to cell fusion inherent in the cell wall, fusion of protoplasts may lead to fused nuclei and production of reconstituted cells from which new plants can grow (13). If the protoplasts were from haploid cells (as from anther or ovule culture) new diploid cells could be produced in a sort of artificial fertilization or syngamy. This is callled parasexual or somatic hybridization. In select cases one can envision the exploitation of the methods in the production of novel plants, even between evolutionarily disparate organisms, by fusion breeding techniques. Successful production and growth of protoplasts into plantlets will be a key feature in the implementation of genetic transformation or engineering studies (39). The hormonal requirements, if any, for regeneration of wall are not understood and regeneration media are traditionally complex. Once walled cells are formed, they are grown in the way specific for the plant (36).

USE OF AUXINS, CYTOKININS, AND OTHER GROWTH REGULATORS FOR CALLUS INDUCTION AND MAINTENANCE

Auxins

Auxins share in common their ability to cause cell enlargement and elongation. In cultured systems they may promote cell division. Despite the fact that there are a number of naturally occurring auxins (14), most of these are not generally available for routine use other than IAA. Because of their stability, synthetic auxins are extensively employed. The most commonly available and used are 2,4-dichlorophenoxyacetic acid (2,4-D), 1-naphthaleneacetic acid and indole-3-butyric acid (IBA). There are also many compounds which are derivatives of the chloro substituted phenylacetic or phenoxyacetic acids that have found widespread use (33). In some cases compounds which are not strictly auxins, such as dicamba (3, 6-dichloro-o-anisic acid) or picloram (4-amino-3,5,6-trichloropyridine-2-carboxylic acid) (both of which are herbicides at higher concentrations), have been used as auxin substitutes (3,6,8,39).

In practice, the use of auxins is an art. It is not possible to dictate a particular concentration to be used in any single case. There is an extensive literature on the initiation of callus cultures of various plants or cultivars (3, 6, 7, 8) and it is beyond the scope of this chapter to go into this. It may be of interest to note, however, that a concentration of auxin substantially lower in concentration may be necessary to maintain cultures once they are initiated-whether on semi-solid media or in liquid. One of the best examples of a nominally permanent change in auxin requirement of a culture is that of "habituation" to auxin. Here, tissues that originally required an exogenous auxin for growth, gradually or suddenly lose this requirement (23). (The same applies to cytokinins (23).) It may also be useful to point out that lowering auxin level to $\frac{1}{4}$ or $\frac{1}{2}$ that used in a semisolid medium is useful when one grows materials in liquid. Presumably this is due to better availability of auxin when immersed in a bathing fluid.

In many instances, addition of any one of the auxins to a basal medium may be enough to initiate and sustain callus growth. Therefore, one ordinarily uses a single auxin at a time. However, since there may be different sites of action or target molecules, it can be helpful to use more than one auxin simultaneously when a tissue is recalcitrant (12). Tissue culture of monocotyledons, particularly cereal grains and palms has been achieved in some cases through the use of rather high levels of synthetic auxins like 2,4-D. These levels would ordinarily be considered to be herbicidal but cell proliferation in the absence of exogenous cytokinin is frequently achievable, and morphogenesis (either formation of somatic embryos or adventitious organs) ensues when the auxin is removed or lowered in the medium (18, 37).

Cytokinins

Kinetin, the prototype molecule for the synthetic adenyl cytokinins and zeatin which is about 10 times more potent and generally considered the prototype of the naturally-occurring cytokinins, are both widely used in tissue culture. Dihydrozeatin, also naturally occurring, is not widely used. N^6-benzylaminopurine, a synthetic, is perhaps more widely used than either kinetin or zeatin (7, 12). N^6-Δ^2-isopentenyl adenine is also widely used. Not surprisingly, whatever is active, readily available, and inexpensive is used. When tested in the carrot root phloem or tobacco pith assay, any of these cytokinins are active, but only in the presence of an auxin. In those cases where an exogenous auxin supply is not needed, it is assumed that the system is synthesizing its own auxin (9).

The ability of N,N'-diphenylurea, first isolated from coconut water (though there is a likelihood that it was actually a contaminant picked up during the purification) and related compounds to substitute for cytokinin-active adenine derivatives has been demonstrated in several callus culture bioassay systems (33). Particular phenylurea derivatives such as N-phenyl-N'-4-pyridylurea are every bit as active as zeatin or even more so but because they are not commonly available, they are not extensively used (24). Therefore, adenyl cytokinins are but one class of substance active in cell division. (Even here, azakinetin was long ago shown to have substantial activity (33) even though it is not an "adenyl" compound.) The substituted phenylureas comprise another class. That many other substances with high level of cytokinin activity will emerge if looked for, is emphasized by the finding that dihydroconiferyl alcohol found in bleeding sap of *Acer pseudoplatanus* and maple syrup is active in various callus assays (22). Leucoanthocyanins and other phenolics provide yet other examples of substances with cell division stimulating activity (33). Liquid endosperms when active in specific culture systems can rarely be substituted for with completely defined media. Also, the device of using "conditioned" medium (i.e., in which prior growth has occurred) to facilitate growth of protoplasts, cells and other tissues implicates various other growth promoters (16). Similarly, the device of using "nurse" cultures and "feeder" layers of cells which are alive but cannot divide (because of X-ray inactivation etc.) to feed protoplasts and/or cells all indicate we do not have a full understanding of the full chemical needs (16, 38).

Gibberellins

Despite the wide ranging physiological effects of gibberellins, their addition, primarily of GA_3, to tissue culture media has only been occasional (10). In some systems like the carrot root bioassay, they result in cell divisions rather than enlargement (33).

Abscisic Acid (ABA)

ABA can be used to slow down growth and to moderate the effects of cytokinins and auxins. In somatic embryogenesis it may prevent precocious germination and bring about a more normal embryo development especially when the hormone balance has been perturbed and there is too much callus etc. (2, 16).

CYTOKININ–AUXIN RATIOS AND ADVENTITIOUS ORGANO-GENESIS FROM INTACT ORGANS AND CALLUS CULTURES

In the course of investigating the interaction of kinetin and auxin in cell division in the tobacco pith system, it was learned that manipulating auxin:cytokinin ratios could affect organogenesis. When the level of auxin relative to that of cytokinin is high, roots form; when the cytokinin relative to that of auxin is high, shoots form. When the ratios are about the same, a callus mass is produced (26, 32). The device of adjusting auxin:cytokinin ratios in an attempt to induce shoots and roots (Fig. 1) is now well-established in tissue culture (9).

In the so-called "indirect organogenesis" route for multiplication *in vitro* (12), the strategy is to induce a callus by whatever means, stimulate shoots on the callus, and then remove the shoots and root them (cf Table 1). Each of these steps may not be discrete or synchronous and the levels of the exogenous hormone are different. For example an auxin, or an auxin and a cytokinin might be needed to establish a callus; then a higher level of cytokinin relative to auxin might be needed to stimulate shoot buds and then only an auxin might be used to induce roots on the separated shoot. The same principles apply in "direct organogenesis" (Table 1). The difference here is that minimal, if any, callus is involved in the

Fig. 1. Organized vs. unorganized growth in cultured callus of *Nicotiana langsdorfii* as influenced by cytokinin:auxin ratios. a) unorganized callus; b) roots formed from callus when added auxin ratio was higher relative to added cytokinin; c) shoots formed from callus when added cytokinin ratio was higher than added auxin.

599

proliferation of organized structures. These form adventitiously directly on an explanted organ (12). The shoots are removed and root formation is fostered with auxin. Unfortunately, there are many instances where morphogenesis cannot be induced despite use of elaborate protocols. Even in *Nicotiana* manipulation of auxin and cytokinin ratios is not equally effective among the many species and varieties. The Wisconsin 38 strain of the 'Havana' cultivar of *N. tabacum* is particularly responsive and is a "model" system. An increasing number of plants are responding to manipulation, however, and commercialization has even been possible in the case of quite a few horticultural cultivars. Obviously, those many plants which do not respond to the methods do not figure in these activities (16).

The hormonal regimens for generation of plants from cells grown in suspension or from protoplasts fit into the principles outlined here. Free cells, or cell clusters, directly produced in liquid or by first producing a callus on semi-solid medium and then disrupting this or permitting it to continue growth in liquid, can be plated in or on semi-solid media and provided with a sequence of hormones to elicit morphogenesis(16).

Table 1. Strategies for Multiplication of Higher Plants *In Vitro*

• Stimulation of indirect organogesesis
    adventitious shoot and/or root formation on a callus

• Stimulation of direct organogenesis
    adventitious shoot and/or root formation on an organ or tissue explant without an intervening callus

• Use of shoots from terminal, axillary, or lateral buds for precocious branching and multiple bud formation
    shoot apical meristems (no leaf primordia present)
    shoot tips (leaf primordia or young leaves present)
    buds
    nodes
    shoot buds on roots

• Stimulation of somatic embryogenesis
    direct formation on a primary explant
    indirect formation from cells grown in suspension or semisolid media

• Stimulation of direct plantlet formation via an organ of perennation formed *in vitro*

• Implementation of micrografting

• Ovule culture

• Embryo rescue

• Mega- and microspore culture

• Infection with a crown gall plasmid genetically altered to give teratoma-like tumors

---

HORMONES AND SOMATIC EMBRYOGENESIS

Generally speaking, though not universally, explanted tissues may now be induced to proliferate in liquid media, to multiply more or less indefinitely, and to release from their callus free cells and small units which, via small pro-embryonic globules, can produce somatic embryos and thence form plantlets that develop and reproduce normally. This is especially true of carrot and other umbellifers (2, 34) and of an ever increasing number of other species (2). Thus, all the living, mature normally diploid cells of the plant body retain the full information of the zygotic nucleus. Moreover, such cells retain in their cytoplasm the ability to make the information fully effective. Cultured tissues can thus provide unique material for the investigation of hormonal stimuli and factors that control development starting from free cells.

Somatic embryos of carrot were first encountered in suspension cultures supplemented with coconut water (16, 33, 34). Later it was shown that coconut water was not required for the process and that an auxin such as 2,4-D or NAA could be used (2, 16). Most protocols now involve exposure of a tissue explant to a synthetic auxin, usually 2,4-D. This initial stage of culture is critical for it is here that cells which are programmed or programmable to express their totipotency respond by dividing. The cell division rate is generally slow at first but populations of cells are conserved and their numbers increased. Competent cells are easily recognized by their generally dense cytoplasmic contents and poor vacuolization. These cells are for all practical purposes the equivalent of young zygotes (16). They can be maintained and sustained in this totipotent state for varying periods. The total removal or drastic lowering of the level of the exogenous auxin leads to their fuller expression of competence by the formation of large proembryonic globules (sometimes called proembryonic masses, PEMs). The proembryos can further pass through varying stages of development, e.g., in a dicotyledonous species the heart stage, the torpedo stage, the cotyledonary stage. All stages but especially the cotyledonary forms are generally readily reared to fully mature plants. Figure 2 provides a summary scheme of the definitive events associated with somatic embryogenesis from suspension cultures of competent cells. The evidence that normal development in such cultures depends on the presence of a variety of growth factors is quite strong (2, 33, 34). More striking is the evidence that the type of growth—by cell division or enlargement, organized or unorganized—is dependent upon their balanced effects and interaction with the external environment. In some cases the spontaneous formation of somatic embryos within a suspension occurs at high frequency, but in others imperfectly or not at all. Occasionally, when the reading of the developmental program is not accurate, one gets abnormal forms resembling embryos but these are not capable of further growth. These are called neomorphs. That the cells of neomorphs may be reprogrammed to correct this reading is suggested by the fact that neomorphs can be induced to callus and when the proper permissive sequence and environment (both chemical and otherwise) is supplied, normal somatic embryos develop (21).

Long continued culture of totipotent cells in suspension is frequently accompanied by a loss of their competence. Similarly, the smaller the unit size of cell clusters that are tested for embryogenic potential, the greater is their requirement for exogenous growth-promoting substances and environmental stimuli such as darkness (34).

There are many species, however, in which the theoretically-existent embryogenic (or even morphogenic) competence has not been demonstrated despite extensive efforts. Even when a system works, there may be a problem of yield, for many cells do not develop but merely proliferate. Others do not grow at all. But even in those cases where expression of competence is realized relatively easily, in no case is the growth of the embryogenic forms uniform, neither is the yield 100%. While many small globular forms might proceed directly to the heart-

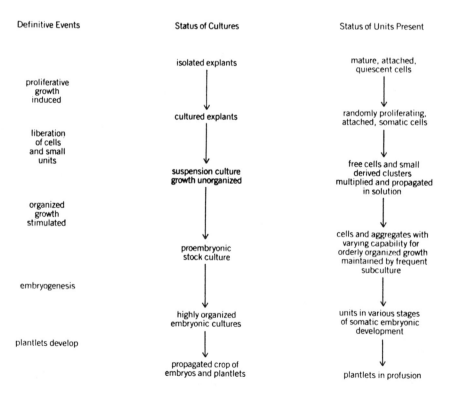

Fig. 2. Summary scheme of the critical events in developing a suspension culture system competent to form somatic embryos (20).

shaped stage, and on to a plant, there are many variations on this pattern. Nevertheless, one can anticipate that somatic embryos generated from suspensions will have an increasingly important place in research and *in vitro* propagation (20).

Some investigators distinguish between an *indirect* route of somatic embryo formation (the process described in some detail here) and a *direct* route. The direct route normally involves direct formation of embryos on a primary explant, usually grown on semi-solid media, without intervention of a callus stage (Table 1). The significance of this for our discussion of hormones is that induction and expression of competence are apparently not necessarily readily separable into two discrete phases. In suspension cultures, one usually uses an auxin, removes it, and somatic embryos develop in an essentially basal medium (2). In *direct* somatic embryogenesis, auxin is added and the embryos develop and may even continue to mature in the presence of the hormone (2). The direct somatic embryogenesis route may provide an avenue for true cloning of plants since anything that minimizes callus might reduce the chance for changes to occur (16, 20).

MICROPROPAGATION OR MULTIPLICATION USING PRE-FORMED OR ORGANIZED PROPAGULES

Meristem, shoot- or stem tip, node, axillary, or lateral bud culture

Micropropagation was originally defined as any aseptic culture procedure involving the manipulation of plant organs, tissues or cells that produces a population of plantlets by bypassing the normal sexual process or non-aseptic vegetative propagation (15). In practice, however, stem tips and lateral buds have been the most commonly used starting point or origin of the primary explant (3, 6, 7, 12). The use of callus, free cells or other of the more demanding approaches covered above rarely comes to mind when the word micropropagation is used.

It stands to reason that a good familiarity with the biology of any given plant is the first of several prerequisities to successful manipulation of organized propagules in aseptic culture. Botanists appreciated long before axenic culture came into laboratory practice that the development of the higher plant body involved the suppression of many, and the development of relatively few, actual or potential primordia. Production of new growing regions, tissues, organs and even new plants were held in check by correlative influences and inhibitions (16). IAA or IBA were early used to induce root formation on cuttings (15). The key feature of micropropagation via precocious axillary branching, however, directly arose out of the early work on kinetin. Wickson and Thimann (43) were taken by the possibility that kinetin might well exert an effect on the development of buds, not just their initiation as had been suggested by the work of Skoog and Miller (32). Figure 3 summarizes that essentials of the observation that cytokinin can antagonize the inhibitory effect of auxin on lateral bud elongation. One is essentially releasing the tendency for a shoot apex to exhibit apical dominance over the lateral buds.

Murashige (26) and others have found it helpful to segregate the sequence of events associated with the multiplication process into stages as follows: I – establishment stage, II – shoot multiplication, III – rooting.

Frequently, specific media or aseptic culture conditions are associated with each of these stages and viewing the problem in this orderly fashion can be helpful in developing a strategy, or attempting interpretation of data. However, one ought to avoid the implication that these stages are always associated with a particular hormonal regimen. Also, the stages are not necessarily temporally discrete and separable. Murashige (12, 26) initially reduced the stages of *in vitro* propagation to three: I, II, III. But a fourth stage, IV or final transfer to the natural environment stage now is an integral part of the procedure. Moreover, an initial stage O has also been added. Stage O involves selection of the mother plant and adoption of a program of pretreatment to render the final strategy to be workable (12).

The fact that many papers have been written on *in vitro* multiplication should be indication enough that the field is largely empiric (6, 7, 9, 12, 17). Even when plants can be multiplied by forcing of precocious axillary branching with cytokinins (Fig. 3) not all species respond equally well and there are many species where an acceptable methodology is still unavailable (12, 47).

It is generally true that the smaller the primary explant, the greater the difficulty encountered in the establishment of growth and the

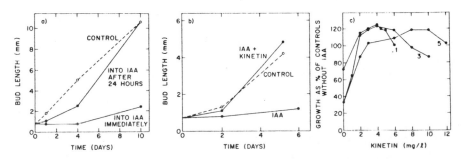

Fig. 3. Interactions of auxin and kinetin in lateral bud growth of pea. a) bud elongation is immediately inhibited by 4mg/1 IAA. If buds are exposed to 4mg/IAA 24 hours after excising them from the parent plant, the elongation is slightly retarded but thereafter growth is the same or slightly faster than the control. This indicates that if auxin supply is interrupted the laterals begin growth; b) the inhibition of bud elongation by IAA is completely reversed by 4mg/1 kinetin; c) the level of kinetin needed to overcome inhibition of lateral bud growth by auxin increases as the concentration of auxin increases. Note that the level of kinetin needed to overcome inhibition when auxin is provided at 1, 3 and 5 mg/1 is roughly equal. From (43).

promotion of shoots. One can, of course, resort to using a large shoot tip explant to minimize contraints brought on by starting with a small strict apical meristem, but there are some situations such as obtaining of virus-free stock where the meristematic dome is the preferred starting material (35).

The points of strategy in terms of hormones in each of the stages of micropropagation are as follows. Stage O involves means to bring the donor plant to a vigorous stage of growth. This means selection of material at the right time of year so that endogenous inhibitors provide no obstacles. Breaking of dormancy by hormone treatment (12) may be necessary. Increasing light or even etiolation in some cases is useful. Stage I involves design of media to stimulate bud or shoot primordia initiation and development, or to stimulate precocious axillary bud break and development. Here cytokinins and auxins may be supplied but if the system is synthesizing adequate auxin, then only a cytokinin may be necessary. Occasionally, a low level of GA_3 may be useful since it will permit some extension growth and allow the system to be managed better (12, 19). Stage II involves increasing or sustaining the level of bud production or precocious branching. This is generally achieved by increasing the cytokinin level but it may be preferable to opt for a lower multiplication rate by use of low levels of cytokinin since too rapid an increase may lead to genetic change. Indeed, it is sometimes helpful to pulse a system intermittently with cytokinin between subcultures to modulate the branching patterns more effectively—namely hormone – "blank" medium – hormone – blank, and so on. There is also the process of "vitrification" to be taken into account. Here the tissue *in vitro* becomes "glassy" and succulent. While some consider it to be the result of overexposure to high cytokinin, it is not clear what the cause(s) are (3, 47). It may be that a compact multiple branched plant is desired. In that case there may be little to be gained by pulsing. In plants where a uniform single stem is preferred, such as forest trees, a system may profit by pulsing combined with use of a gibberellin to foster stem elongation. High light will tend to keep internodes short. Stage III involves use of auxins to stimulate roots. While this stage is generally done under aseptic conditions, it is possible in many cases to carry this out *ex vitro* (3, 7, 47). The long established principles that apply to rooting of cuttings in conventional propagation procedures (15) apply here as well. Sometimes difficult-to-root species are easier to root after *in vitro* shoot multiplication. Removal or leaching of "excess" cytokinin from shoots is sometimes critical since they generally antagonize rooting. "Flushing" of microcuttings can be carried out in a "blank" medium prior to exposure to auxin (12). Systems known to be hard-to-root because of endogenous inhibitors also profit by such flushing or any other means to destroy or remove the inhibitor(s). Sometimes addition of activated charcoal to the medium to remove excess hormone etc. is helpful but the level of auxin must be increased to account for adsorption. Striking or rooting of cuttings is a demanding art in some cases; very easy in others. Some investigators subcategorize Stage III by distinguishing between a Stage III MC (microcutting) and Stage III MS (multiple shooted) (7). In III MC a single microcutting is rooted; in III MS a multiple shooted cutting is rooted and this gives a bushy plant. Stage IV involves full establishment of the plant *ex vitro*. This requires skillfull management but rarely involves exogenously added hormones outside of what may be normally used for a particular crop, e.g., growth retardants to keep plants short (27). See Fig. 4.

The horticultural nature of the micropropagation process involving precocious axillary branching is emphasized when seen in the context that plants so multiplied originate from small cuttings. Indeed, certain conifers within the genera *Pinus, Picea, Tsuga, Pseudotsuga, Thuja, Juniperus, Sequoia* and *Araucaria* have been manipulated with a combination of traditional and aseptic culture strategies. This involves repetitive cytokinin treatment of the intact tree at very high levels (order of 200 mg/l) during periods of optimum growth so as to foster buds or shoots (usually from latent axillary bud meristems at the base of a needle

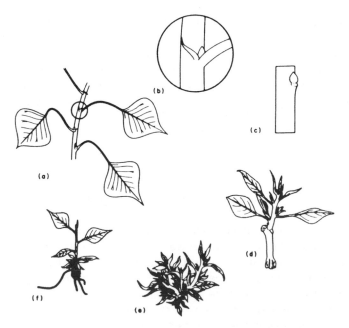

Fig. 4. Diagrammatic scheme of multiplication of *Sapium sebiferum*, a woody member of the Euphorbiaceae. The strategy is similar for many other plants where precocious axillary branching is to be stimulated. a) a young branch with lateral buds from a mature, tested tree; b) close-up of the lateral bud; c) primary explant comprising a single bud; d) precocious branches formed by bud break and; e) a culture with a well established but compact branching system; f) a rooted plantlet (19).

cluster or at the apex of the fascicles). These forced shoots are juvenile in morphology and in some cases resemble those of newly germinated seedlings. They may be excised and further forced *in vitro* to produce additional buds which in turn can be separated and rooted (1).

Another example derives from the use of a normally micropropagational strategy without the aseptic component (12). Leaf bud cuttings from cassava prepared from field grown plants and rooted under mist have been calculated as being able to yield 4,000,000 plants from an initial, mature stock plant with 500 healthy leaves. As each newly propagated plant grows and produces additional leaves with buds, new material is available for propagation (29).

MULTIPLICATION VIA ORGANS OF PERENNATION FORMED IN ASEPTIC CULTURE

Some plants form organs of perennation *in vitro*. When this occurs, one has the means for multiplication at another level and it may well turn out that direct planting or germ plasm storage of plants can be implemented by this means. Potatoes can form miniature tubers, gladioli can form cormlets, bulbils have been encountered in certain lilies, onion, narcissus, hyacinth, *Dioscorea*, etc. and, of course, protocorms are produced by orchids. None of these means other than the orchid protocorm system has been so controllable that it has been seriously adopted as a means of mulitplication. Even so, these organs may provide yet another strategy in select cases. Studies on the control of tuber formation of *in vitro* multiplied plantlets of potato from nodal origin suggest that at the moment the hormonal controlling factors are elusive (12).

OVERALL DISCUSSION AND COMMENTARY

It becomes at once apparent that tissue culturists frequently seem to be bound to an empiric approach to achieve a given end. We have attempted, however, to show that the practical goals frequently aimed for are not modest. One is, in essence, aiming to control growth and development! Figure 5 provides a picture of the many points of potential response in a system and perhaps justifies, in part, why one frequently has to resort to extensive, and seemingly arbitrary testing.

A number of attempts have been made over the years to draw relationships between the nature and level of endogenous growth substances, culturability, and morphogenetic competence. From our perspective, these attempts have not been very instructive either theoretically or helpful practically (45). The multiplicity of effects encountered in varying systems is, as implied, a consequence of interactions and various levels of sensitivity on the part of essentially heterogeneous cell populations in tissue and organ explants (37). There is little doubt that the origin and physiological state of the initial explant is critical to responsiveness. This is tacitly recognized when one elevates Stage 0 of micropropagation to a separate level. It is also underscored in anther culture work or spore work (42) where little more than a basal medium is used. But the wide range of successfully cultured explants from a single test organism, moderates arguments that one tissue source is always necessarily better than another. A rule of thumb among experienced workers is to utilize whatever information may be available as to the normal zones of rapid growth in selecting a primary explant. General analysis of growth substance content whether of auxins, cytokinins, gibberellins or abscisic acid and other inhibitors will never be, at best, more than a rough guide to physiological status. There are many

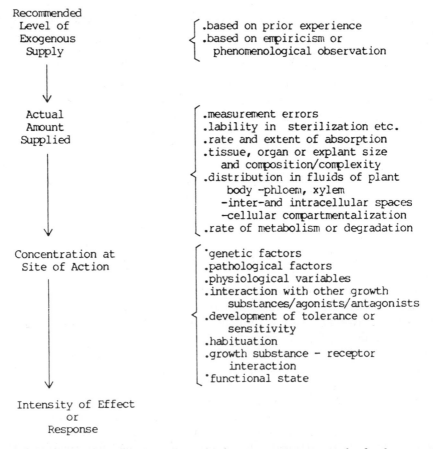

Recommended
Level of
Exogenous
Supply

{
.based on prior experience
.based on empiricism or
  phenomenological observation
}

Actual
Amount
Supplied

{
.measurement errors
.lability in sterilization etc.
.rate and extent of absorption
.tissue, organ or explant size
  and composition/complexity
.distribution in fluids of plant
  body —phloem, xylem
  —inter-and intracellular spaces
  —cellular compartmentalization
.rate of metabolism or degradation
}

Concentration at
Site of Action

{
·genetic factors
.pathological factors
.physiological variables
.interaction with other growth
  substances/agonists/antagonists
.development of tolerance or
  sensitivity
.habituation
.growth substance – receptor
  interaction
·functional state
}

Intensity of Effect
or
Response

Fig. 5. Some factors that affect the relationship between exogenous supply of a plant growth substance and the response it elicits.

interrelationships that can and are set in motion when explanted tissues and organs are placed in culture. Surely the location of growth substances within a cell or tissue complex is as important as the overall level. Figure 6 provides a simplistic scheme of some of the many possible interactions in a tissue culture environment. The chemical form, level and sequence of exposure to growth regulators all can play a role. Use of agents such as activated charcoal etc. to modify the rate of release, delivery or absorption of a growth regulator are all readily explainable from such a perspective. These are part of the tissue culturists armamentarium and can play a decisive role (6, 7, 12). Some workers feel that they can modify response by carefully selecting material to be cultured. Certain tissues show a strong seasonal component as to culturability and this in part may be due to accumulation of inhibitors (12). There is no doubt some investigators are better than others in getting cultures to grow.

Radiommune assays and monoclonal antibodies may well play a role in permitting correlations to be made in morphogenctically competent systems as they express their potential (40). But a fundamental problem here is that the responsive systems have often already been determined as to their responsiveness and the later events in the culture process merely facilitate the expression. A good experimental control for comparison may, therefore, not be that easy to achieve. Newer and more sensitive chemical tests (40) for quantifying plant growth substance levels can play an important role here but it is critical to have well defined biological test

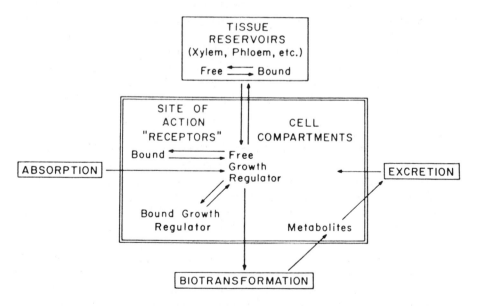

Fig. 6. Diagram showing some of the interrelationships of absorption, distribution, binding, release, metabolism, excretion and re-absorption of a growth substance by an aseptically cultured system.

systems on which to do such analyses. An example which emphasizes the difficulty of such an approach involves a carrot culture grown in this laboratory. It produces somatic embryos in profusion, and the stemline callus can undergo a wide range of morphogenetic expression. Yet this culture has been grown on a basal salt medium supplemented only with sucrose. Exogenous growth regulators have never been used in the initiation or maintenance of the culture. Even so, exogenously added growth regulators can modify the growth and expression of the totipotency. All this underscores a fundamental problem—namely that culture systems ordinarily lack important temporal, spatial and directional information that is ordinarily provided in a controlled geometric setting in the intact plant body. A complex and very responsive cytoplasm can respond to external and internal cues and it is able to compound these informational cues to determine its pace and course of its development.

As the field of Molecular Biology continues to develop, new techniques are becoming available to explore basic mechanisms involved in the action of plant hormones. Currently most prominent is the crown gall tumor system which permits the insertion of a specific segment of DNA from the pathogenic bacterium *Agrobacterium tumefaciens* into the genome of a higher plant (5, 46). Defined genes on the transferred DNA have been shown to code for enzymes involved in the metabolism, by unusual pathways, of auxins and cytokinins. Mutations in the genes which code for auxin synthesis result in tumors that spontaneously produce shoots while those for cytokinin production give rise to tumors with roots. The obvious inference is that these genes cause the plant cells to produce these hormones in sufficient quantities to cause undifferentiated growth without need for an exogenous source of these growth substances. The crown gall system thus allows the study of phytohormone regulation at a level that involves but a handful of genes. To disclose early control mechanisms by sorting through observations of complicated phytohormone actions of the sort described in this chapter would seem to be a most difficult and frustrating task. Using genetically manipulated crown gall DNA, it should now be possible to approach this complex regulation scheme. By infecting plant cells with a strain of bacteria that has a mutation in a gene which controls the auxin metabolism of the plant, for example, one could follow the step-by-step alterations of RNA, protein or hormones which occur as the shoot-tumor morphology develops.

The difficulty still remains, however, of assaying for these relatively minor events among the many other reactions that are part of a normal plant system. To address this problem, newly developed techniques involving the use of antibodies may prove most useful. Although most molecules are not able to elicit antibody production, they can be chemically bound to larger molecules and antibodies against this joint complex can be raised. These antibodies will also be able to bind the specific phytohormone alone, even if it is present in otherwise undetectable quantities (40). The application of these developing techniques to plant systems will require copious groundwork before useful results can be expected. With the proper perspective and patience, however, valuable information will be uncovered which could allow control of phytohormone metabolism so as to elevate further the usefulness of tissue culture systems for basic research and commercialization.

References

1. Abo El Nil, M. (1980) Method for asexual reproduction of coniferous trees. U. S. Patent. 4,353,184. 14pp.
2. Ammirato, P. (1983) Embryogenesis. In: Handbook of plant cell culture. Vol 1. Techniques for propagation and breeding. pp. 82-123. Evans, D., Sharp, W., Ammirato, P. Yamada, Y., eds. Macmillan Publishing New York.
3. Bajaj, Y.P.S. (ed) (1985) Biotechnology in agriculture and forestry 1. Trees I. Springer-Verlag, Berlin Heidelberg.
4. Braun, A. (1969) Abnormal growth in plants. In: Plant physiology - A treatise. pp. 370-420. vol VB, Steward, F. ed. Academic Press, New York.
5. Chilton, M., Drummond, M. Merlo, D., Sciaky, D., Montoya, A., Gordon, M., Nester, E. (1977) Stable incorporation of plasmid DNA into higher plant cells: The molecular basis of crown gall tumorigenesis. Cell 11, 263-271.
6. Conger, B. (ed) (1981) Cloning agricultural plants via in vitro techniques. CRC Press, Boca Raton.
7. DeFossard, R. (1984) Tissue culture for plant propagators. 2nd ed. University of New England, Armidale.
8. Evans, D., Sharp, W., Ammirato, P., Yamada, Y. (eds.) (1983) Handbook of plant cell culture. vol. 1. Techniques and applications. Macmillan Publishing, New York.
9. Flick, C., Evans, D., Sharp, W. (1983) Organogenesis pp. 13-81. In: Evans, D., Sharp, W., Ammirato, P., Yamada, Y., eds. vol. I. Techniques for propagation and breeding. Macmillan Publishing, New York.
10. Fry, S., Street, H. (1980) Gibberellin-sensitive cultures. Plant Physiol. 65, 472-477.
11. Gelvin, S. (1984) Plant tumorigenesis In: Plant-microbe interactions. Molecular and genetic perspectives. In: Kosuge, T., Nester, E. eds. vol. 1. pp. 343-377. Macmillan Publishing, New York.
12. George, E., Sherrington, P. (1984) Plant propagation by tissue culture. Handbook and directory of commercial operations. Exegetics Ltd., Eversley, Basingstoke.
13. Gleba, Y., Sytnik, K. 1984) Protoplast fusion. Genetic engineering in higher plants. Springer-Verlag, Berlin.
14. Hangartner, R., Peterson, M., Good, N. (1980) Biological activities of indolacetylamino acids and their uses as auxins in tissue culture. Plant Physiol. 65, 761-767.
15. Hartmann, H., Kester, D. (1984) Plant propagation. Principles and practices. Prentice Hall, Englewood Cliffs.
16. Krikorian, A. (1982) Cloning higher plants from aseptically cultured tissues and cells. Biol. Rev. 57, 151-218.
17. Krikorian, A., Cronauer, S. (1984) Banana. In Handbook of plant cell culture. vol. 2, pp. 327-48. Sharp, W., Evans, D., Ammirato, P., Yamada, Y. eds. Macmillan Publishing, New York.
18. Krikorian, A., Kann, R. P. (1986) Oil palm improvement via tissue culture. Plant Breeding Reviews 4, 175-201.

19. Krikorian, A., Kann, R. (1986) Multiplication of the Chinese tallow, *Sapium sebiferum*. *In* Tissue culture in forestry. vol. II. case histories. Bonga, J., Durzan, D. eds. Martinus, Nijhoff/W. Junk, The Hague .

20. Krikorian, A. D., Kann, R. P., O'Connor, S. A., Fitter, M. S. (1986) Totipotent suspensions as a means of multiplication. *In* Tissue culture as a plant production system for horticultural crops. pp. 61-72. Zimmerman, R. ed. Martinus Nijhoff/Dr. W. Junk Publishers, The Hague.

21. Krikorian, A., Kann, R., O'Connor, S., S., Fitter, M., Cronauer, S., Smith, D. L. (1986) The range of morphogenetic responsiveness in aseptically cultured daylily tissues and cells. *In* Progress and prospects in forest and crop biotechnology. Valentine, F. ed. Springer-Verlag, New York.

22. Lee, T. S., Purse, J. G., Price, R. J., Horgan, R., Wareing, P. (1981) Diconiferylalcohol. A cell division factor from Acer species. Planta *152*, 571-577.

23. Meins, F., Binns, A. (1979) Epigenetic clonal variation in the requirement of plant cells for cytokinins. *In* The clonal bases for development. pp. 185-201. Subtelny, S., Sussex, I. eds. Academic Press, New York.

24. Mok, M., Mok, D., Armstrong, D., Shudo, K., Isogai, Y., Okamoto, T. (1982) Cytokinin activity of N-phenyl-$N^1$1,2,3,-thiadiazol-5-ylurea (Thidiazuron). Phytochemistry 21, 1509-1511.

25. Morel, G. (1974) Clonal multiplication of orchids. *In* The orchids. Scientific studies. pp. 169-222. Withner, C. ed. John Wiley & Sons, New York.

26. Murashige, T. (1977) Manipulation of organ initiation in plant tissue cultures. Botan. Bull. Acad. Sinica *18*, 1-24.

27. Nickell, L (1983) Plant growth regulating chemicals. CRC Press, Boca Raton.

28. Nitsch, C., Godar, M. (1979) The role of hormones in promoting and developing growth to select new varieties in sterile culture. *In* Plant regulation and world agriculture. pp. 49-62. Scott, T. ed. Plenum Publishing, New York.

29. Patena, L. F., Barba, R. C. (1979) Rapid propagation of *Cassava* by leaf bud cuttings. Phil. Jour. Crop Sci. *4*, 53-62.

30. Pilet, P. E. ed. (1985) The physiological properties of plant protoplasts. Springer-Verlag, Berlin.

31. Raghavan, V. (1976) Experimental embryogenesis in vascular plants. Academic Press, New York.

32. Skoog, F., Schmitz, R. (1972) Cytokinins. *In* Plant physiology: A treatise. vol. 6B, pp. 181-213. Steward, F., ed. Academic Press, New York.

33. Steward, F., Krikorian, A. (1971) Plants, chemicals and growth. Academic Press, New York.

34. Steward, F., Israel, H., Mott, R., Wilson, H., Krikorian, A. (1975) Observations on growth and morphogenesis using cultured cells of carrot. Phil. Trans. Roy. Soc. London *273B*, 33-53.

35. Styer, D., Chin, C. (1983) Meristem and shoot tip culture for propagation, pathogen elimination, and germplasm preservation. Hort. Rev. *5*, 221-277.

36. Thorpe, T. ed. (1981) Plant tissue culture methods and applications in agriculture. Academic Press, New York.

37. Tran Thanh Van, K. (1981) Control of morphogenesis in *in vitro* cultures. Ann. Rev. Plant Physiol. *32*, 291-311.

38. Tran Than Van, K., Toubart, P., Cousson, A., Darvill, A. G., Gollin, D. J., Chelf, P., Albersheim, P. (1985) Manipulation of morphogenetic pathways of tobacco by oligosaccharins. Nature *314*, 615-617.

39. Vasil, I., ed (1984) Cell culture and somatic cell genetics of plants. vol. 1. Laboratory procedures and applications. Academic Press, New York.

40. Weiler, E. (1984) Immunoassay of plant growth regulators. Ann. Rev. Plant Physiol. *35*, 85-95.

41. Wetter, R., Constable, F. eds (1982) Plant tissue culture methods. 2nd ed. National Research Council of Canada, Ottawa.
42. White, R. (1979) Experimental investigations of fern sporophyte development. pp. 505-549. In: The experimental biology of ferns. Dyer, A., ed. Academic Press, London.
43. Wickson, M., Thimann, K. (1958) The antagonism of auxin and kinetin in apical dominance. Physiol. Plant. *11*, 62-74.
44. Yang, H., Zhan, C. (1982) *In vitro* induction of haploid plants from unpollinated ovaries and ovules. Theor. Appl. Genet. *63*, 97-104.
45. Zaerr, J., Mapes, M. (1982) Action of growth regulators. pp. 231-255 In: Tissue culture in forestry. Bonga, J., Durzan, D., eds. Martinus Nijhoff/Dr. W. Junk Publishers, The Hague.
46. Zambryski, P., Holsters, M., Kruger, K., Depicker, A., Schell, J., Van Montagu, M., Goodman, H. (1980 (1980) Tumor DNA structure in plant cells transformed by A. *tumefaciens*. Science 209, 1385-1391.
47. Zimmerman, R. ed. (1986) Tissue culture as a plant production system for horticultural crops. Martinus Nijhoff/Dr. W. Junk Publishers, The Hague.

E 19. Natural and Synthetic Growth Regulators and Their Use in Horticultural and Agronomic Crops

Thomas J. Gianfagna

Department of Horticulture and Forestry, Rutgers University, New Brunswick, New Jersey 08903, USA.

INTRODUCTION

Plant growth regulators (PGRs) have been an important component in agricultural production even prior to the identification of plant hormones. For example, in order to synchronize flowering in mango or in pineapple, fires were lit adjacent to fields (43) in which these crops were grown. The ethylene generated as a result of incomplete combustion stimulated flowering, although this explanation was not understood at the time. Heat treatment of lemons was also used to stimulate ripening and degreening (12), the ethylene generated from inefficient heating units again promoting the ripening process. PGRs are now used on over one million hectares worldwide on a diversity of crops each year (52). Most of these applications are confined to high-value horticultural commodities rather than field crops, although there are several significant exceptions, such as chlormequat chloride (2-chloroethyltrimethyl-ammonium chloride) for lodging control in wheat, glyphosine (N,N-bis(phosphonomethyl)glycine) for increased sugarcane yield, and DEF (S,S,S-tributylphosphorotrithioate) for cotton defoliation.

There have also been several promising results with PGRs for soybean and corn. Triiodobenzoic acid was registered briefly to increase yield in soybean, presumably by stimulating branching and thus allowing light penetration into the canopy for fruit set. In corn, dinoseb, used primarily as a preemergence herbicide, has been found to increase grain yield by 10-15%. Dinoseb treatment stimulated earlier tasseling and improved ear filling. Nevertheless, both compounds have failed to provide consistent yield increases, and there was considerable genotype variation in response to treatment. Significant opportunities certainly still exist for the development of PGRs to increase yield in the major crops.

Most of the current uses for PGRs in the high-value horticultural crops are not, however, for compounds that increase crop yield directly, whether by increasing either the total biological yield or the harvest index. Compounds that provide economic benefit by enhancing crop

614

quality, or those that aid in more efficient crop management are more common instead. For example, gibberellic acid (GA) is used to reduce the incidence of physiological rind disorders in citrus, and daminozide (N,N,dimethylaminosuccinamic acid) application to apple stimulates color development of the fruit. Both treatments increase the value of the crop, but not necessarily the yield.

PGRs that aid in crop management fall into several catagories. First, PGRs can be used in conjunction with the mechanical harvesting of crops. Compounds that reduce the fruit removal force have in some cases, such as sweet cherry production, allowed the development and use of mechanical harvesting equipment that was not particularly effective alone because the force required to remove the crop damaged the trees. Second, PGRs are used for manipulation of the harvest date. Ethephon (2-chloroethylphosphonic acid) accelerates and concentrates ripening of tomato prior to a single mechanical harvest. GA extends the lemon harvest season by preventing senescence of the rind. This allows a greater percentage of the crop to be sold when demand for fresh lemons is high. Conversely, GA accelerates flower bud development in artichoke, significantly advancing maturity and increasing the value of the crop. Third, PGRs can be used as direct replacements for hand labor in other aspects of crop production in addition to harvesting. In apple, naphthalene acetic acid (NAA) is used to reduce excessive fruit set, which would otherwise result in many small fruit, and for some varieties, inhibit flower bud production for the following season's crop. Maleic hydrazide (1,2-dihydro-3,6-pyradazinedione) is used in tobacco to inhibit the growth of lateral buds, which would otherwise reduce leaf quality. For both cases—fruit thinning in apple and sucker control in tobacco—a labor-intensive cultural practice can be accomplished efficiently with PGRs.

In most of the examples described above, PGRs are useful because they can in some way modify plant development. This may occur by interfering with the biosynthesis, metabolism, or translocation of plant hormones, or PGRs may replace or supplement plant hormones when their endogenous levels are below that needed to change the course of plant development. The plant growth retardants such as chlormequat chloride and ancymidol (α-cyclopropyl-α(4-methoxyphenyl)-5-pyrimidine-methanol), which are used to control internode elongation in lily, pointsettia, and other floricultural crops, act by inhibiting gibberellin biosynthesis. Ethephon, on the other hand, accelerates ripening and abscission by releasing ethylene at a time when endogenous levels of this hormone are low.

Other compounds may more directly affect plant metabolism, circumventing the hormonal control system. For example, lateral bud development can be prevented in tobacco by inhibiting cell division with maleic hydrazide.

While the biochemical mechanism of action of PGRs is not well understood, it is still possible to classify compounds on the basis of their similarity in action to the naturally occurring plant hormones. This approach will be taken in the following description of the chemistry, physiology, and use of PGRs in agriculture.

AUXINS

The auxin-type PGRs comprise some of the oldest compounds used in agriculture. Shortly after indole acetic acid (IAA) was identified as a naturally occurring compound, it was synthesized and became readily available. IAA is not found in itself to be useful in agriculture because it is rapidly broken down to inactive products by light and microorganisms. Nevertheless, a number of synthetic compounds were found to act similarly to IAA in the auxin bioassay tests. Indolebutyric acid (IBA) and NAA were found to increase root development in the propagation of stem cuttings. 2,4-dichlorophenoxyacetic acid (2,4-D) stimulates excessive, uncontrolled growth in broadleaf plants for which it is used as a herbicide. NAA and naphthalene acetamide (NAAM) are used to reduce the number of fruit that set in apple, whereas 4-chlorophenoxyacetic acid (4-CPA) is used to increase fruit set in tomato. The auxins 2,4,5-trichlorophenoxy-propionic acid (2,4,5,-TP) and the dichlorophenoxy analog (2,4-DP) are used to prevent abscission of mature fruit in apple.

Stimulation of fruit set
One of the first recorded effects of auxins was the stimulation of fruit set in unpollinated ovaries of solanaceous plants. Gustafson (20) demonstrated that pollen was a rich source of auxin, and that in some species pollination alone was all that was required for fruit set to occur. In tomato, chemical stimulation of fruit set is all that is needed for fruit growth to take place. In addition, compounds that block the transport of auxin from the ovary to the pedicel of the flower also stimulate fruit set (4). In California, the early spring crop of tomatoes is treated with 4-CPA at 25-50 ppm to stimulate fruit set at a time of the year when cool night temperatures, which inhibit fruit set in tomato, are likely. This treatment results in an increase in yield and earlier harvest.

Chemical thinning
Removal of excessive numbers of young fruit from apple and pear trees is a common orchard management practice, although this results in drastic reductions in the total biological yield. There are two main reasons for removing as many as 80% of the flowers: first, to increase the total marketable yield by increasing the size of the remaining fruit; and second, to reduce the phenomenon of biennial bearing in order to maintain

production levels from year to year. The effect of fruit thinning on fruit size is probably related to the leaf/fruit ratio. As this ratio is reduced below 30/1, fruit size is reduced as well (32). The time at which fruit thinning is done is as important to fruit size as the amount of fruit thinning. In order to improve flower bud production and fruit size in apple, thinning should take place within 30 days after full bloom. In apple, the period of cell division in the fruit is brief, ending approximately 20 days after full bloom (57). Removing excess fruit during this period can stimulate cell division within the remaining fruit. This early period is also of critical importance for floral initiation. The extent to which new flower buds are produced will partially determine yield the following season. The presence of fruits tends to inhibit floral initiation. The two auxin-type compounds used in chemical thinning of apple and pear are NAA and NAAM. NAA is applied at 2-5 ppm, 7-20 days after full bloom, whereas NAAM is effective during the same time period, but higher rates (17-34 ppm) are used (58).

Several mechanisms have been proposed to explain the effect of auxins on fruit abscission. Early observations that auxin application reduced the early drop of flowers and fruit soon after the bloom period led to the suggestion that auxin first stimulated fruit set, and that then because of increased competition between fruits for nutrients and assimilates, a greater percentage of fruit abscised during the June-drop period (51). This early increase in fruit set, however, is not always observed. Luckwill (30) proposed that fruit abscission occurs because NAA and NAAM induce embryo abortion. Without seed growth, fruit senescence and abscission take place prematurely. While the number of viable seeds is often correlated with fruit abscission, this is not always the case, suggesting that embryo abortion may not be the primary factor influencing abscission. NAA does cause increased ethylene evolution from apple fruit within one day after application (56). Ethylene is known to reduce auxin transport from the leaf blade to the petiole (3) and to induce the synthesis of enzymes that degrade the abscission zone (2). Perhaps the induction of fruit abscission by NAA is mediated by ethylene, which stimulates abscission of fruit in a manner similar to its effect on the abscission of leaves (Fig. 1).

Prevention of fruit drop

Frequently, the mature fruit of apple, pear, lemon, and grapefruit will abscise prior to the time of commerical harvest. This obviously reduces the potential crop yield, and may cause growers to begin harvesting the crop earlier than is desirable, resulting in lower quality fruit. Under natural conditions, there seems to be an inverse relationship between the auxin content of the fruit and the tendency toward abscission (31). The role of auxin in abscission is complicated. Clearly, application of auxin soon after fruit set results in an acceleration of abscission; however, when

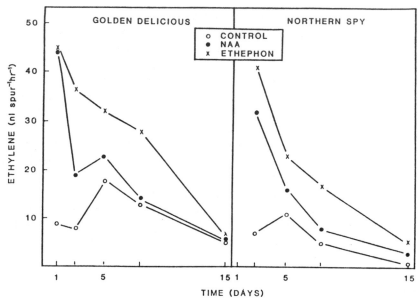

Fig. 1. Ethylene evolution from fruit spurs in response to post-bloom treatment of Golden Delicious' and Northern Spy' apple trees with NAA and ethephon. From(56)

auxins such as 2,4,5-TP, NAA, and 2,4-D are applied during the mid-stages of fruit growth, abscission is delayed or prevented (48). In addition, auxin application may decrease the response of fruit abscission zones to exogenously applied ethylene. NAA and 2,4,5-TP are used at 10-20 ppm just prior to the beginning of fruit drop in apple. Repeat applications may be necessary with NAA; 2,4,5-TP prevents fruit drop for a longer period. In citrus, 2,4-D at 25 ppm prevents premature fruit drop and allows an extension of the harvest season into the summer (50).

Propagation

Rooting of stem cuttings was one of the first uses of auxins. The most commonly used compound is IBA, which has only weak auxin activity, but is relatively stable and insensitive to the auxin-degrading enzyme systems. It is also not readily translocated. Other compounds such as NAA and 2,4-D will also promote root development, but they are more easily translocated to other parts of the stem cutting where they may have toxic effects (24). The auxins stimulate root development by inducing root initials which differentiate from cells of the young secondary phloem, cambium, and pith tissue.

Herbicidal action

2,4-D and picloram (4-amino-3,5,6-trichloropicolinic acid) are two auxin-type herbicides that at low concentrations bring about growth responses similar to IAA. At higher concentrations they are herbicidal:

2,4-D is commonly used to control broadleaf weeds in grasses, and picloram is used for vegetation control on noncrop land because of its high activity and soil persistence. Both compounds cause epinastic bending in leaves, a cessation of growth in length, and increased radial expansion. After several days tumors may form, followed by a softening and collapse of the tissue. Epinastic bending and stem swelling are characteristic of ethylene effects on plants, and auxin-induced ethylene biosynthesis may partially account for the effect of these compounds on plant growth. Treatment of plants already exposed to 2,4-D with compounds that inhibit ethylene synthesis or action does not reverse the herbicidal effects of the auxins, however (1). Auxin herbicides cause an increase in DNA, RNA, and protein levels in treated tissue, but the greatest effect, is on RNA levels (7). In addition, the level of RNAase activity is higher in resistant plants than in sensitive plants. One aspect of the herbicidal activity of the synthetic auxins seems clearly to be a disturbance in RNA metabolism.

GIBBERELLINS

Despite considerable enthusiasm for the potential uses of GA in agriculture that existed when this compound was rediscovered by US and British scientists in the 1950s, major GA use remains limited to the management of fruit crops, the malting of barley, and the extension of sugarcane growth in certain production regions.

There are over 70 gibberellins found in both higher plants and the *Gibberella* fungus, although only two commercial products are available- GA_3 and a mixture of GA_4 and GA_7. Both are produced by fermentation cultures of the fungus. A formulation of $GA_{4/7}$ and benzyladenine is also available, which is currently being used to induce fruit elongation in apple and also to increase the extent of lateral branching in young trees.

Increasing fruit size in grape

GA is used extensively on seedless grape varieties to increase the size and quality of the fruit (8). Prebloom sprays of 20 ppm induce the rachis of the fruit cluster to elongate. This creates looser clusters that are less susceptable to disease during the growing season. GA not only reduces pollen viability, but also decreases ovule fertility in grape. Application of GA at bloom, therefore, results in a decrease in fruit set, which in turn reduces the number of berries per cluster and increases the weight and length of the remaining fruit. An additional application of GA during the late bloom to fruit set period will further increase berry size. It has been suggested that this later application of GA increases the mobilization of carbohydrates to the developing fruit (46) (Table 1).

Seeded varieties generally do not respond favorably to GA treatment. However, in Japan, the seeded variety 'Delaware', is cluster-dipped in 100

Table 1. Effects of GA$_3$ applied at bloom and/or fruit set on fruit growth in 'Thompson Seedless' grapes. From (8)

Treatment	Berry Number	Cluster compactness*	Berry Wt. (g)	Berry Length (cm)
Control	91	3.9	2.7	1.8
GA at bloom	61	2.8	3.3	2.3
GA at fruit set	90	4.1	4.0	2.3
GA at bloom + fruit set	64	3.0	4.8	2.6

* Cluster compactness increases with higher values.

ppm GA to induce parthenocarpic fruit development and to increase berry size.

Stimulating fruit set

Not all crops respond as positively as the tomato does to auxin-induced fruit set. However, a number of deciduous fruit tree species such as apple and pear, as well as some citrus species, can be induced to set fruit with GA, or a combination of GA and auxin. In Europe, poor fruit set in apple and pear can drastically reduce crop yield. Consistently unfavorable weather during the pollination period has led to the development of a hormone mixture to induce parthenocarpic fruit set in apple (27). In pear, spring frost injury to the ovules or style can prevent fertilization and the stimulation for fruit set. Application of GA$_3$ at 15-30 ppm can induce parthenocarpic fruit development and salvage what would have been a lost crop.

In citrus, fruit set of mandarin oranges is often light. Application of GA during full bloom can increase fruit set. Girdling branches will also increase fruit set as well as cause an increase in endogenous GA-like substances in the region above the branch girdle (55).

Modifying fruit and tree shape

Promalin, a mixture of benzyladenine and GA$_{4/7}$ is used to control fruit shape in 'Delicious' type apples. High temperatures during the bloom period will often reduce the length-to-diameter ratio, resulting in uncharacteristicly round fruit instead of the elongated apples that consumers have come to expect. Promalin applied at bloom will increase length-to-diameter ratio of the fruit (59) (Fig. 2). Increased fruit size may also result from treatment.

Promalin is also being used to increase lateral branching in non-bearing apple trees. Young trees typically have a strong, vigorously growing central leader, with a few upright growing branches. For fruit production, this is an undesirable tree shape, so mechanical devices are used to force the lateral branches to grow more horizontally. Promalin

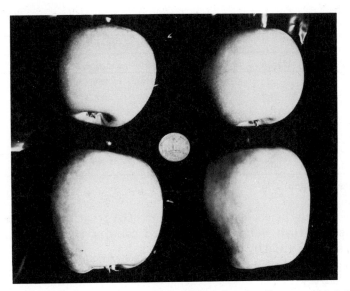

Fig. 2. Effect of Promalin (below) on fruit shape of 'Mutsu' apple. (Courtesy of J. A. Hopfinger)

will stimulate branching and increase the branch angle, as well as increase shoot elongation, all of which aid in the development of a scaffold branching system more suitable for fruit production.

Effects on fruit ripening

GAs are used to delay fruit ripening in lemon in order to increase the availability of fruit during the months of May to August when demand is high, but production is low. GA is applied in November or December in order to delay the harvest date and increase storage life of the fruit (10) (Table 2).

Delaying harvest is also important for a number of other citrus species including 'Navel' oranges and grapefruit. While fruit abscission can be controlled by 2,4-D, changes associated with the senescence of the rind will decrease the quality of the fruit. This reduces the usefulness of holding the fruit on the tree longer in order to allow harvest to take place during the period of high consumer demand. GA application will reduce the occurrence of physiological rind disorders such as water spot, creasing, rind staining, and softening by delaying senescence of the rind tissue (9).

Increasing yield in sugarcane

Sugarcane growth is very sensitive to the reductions in average daily temperature normally experienced during the winter months in many cane-producing regions of the world, especially Hawaii. GA application is used to overcome the reduced growth of the 3-5 internodes undergoing

Table 2. Influence of gibberellic acid on lemon tree harvest patterns one year after treatment. (From *The Citrus Industry* vol. 2)

GA Treatment*	Harvest Dates (1960–61)					
(ppm)	March	May	June	August	October	January
	percentage of crop harvested					
0	9.2	46.5	23.7	12.3	5.8	2.5
5	10.0	40.6	24.1	16.9	5.8	2.6
10	10.1	39.4	23.7	17.0	7.2	2.6
20	11.2	37.7	22.7	17.9	8.0	2.5
40	10.3	36.2	22.8	18.6	9.5	2.6

* GA treatment applied on Nov. 17, 1959.

expansion during the cooler winter season. GA treatment has resulted in an increase in fresh weight of harvested cane of 10.9 ton/ha, and has increased sucrose yield by 1.1 ton/ha (34).

Malting of barley

GA is used to increase the yield of barley malt and to decrease the time required for this process to occur. Embryo growth and yield of malt extract are competitive processes; by increasing the rate of malting relative to embryo growth, a greater yield of malt extract occurs (38). Application of GA to germinating barley supplements the endogenous content of this hormone and accelerates the production and release of hydrolytic enzymes that degrade the storage proteins and carbohydrates of the endosperm into the sugars and amino acids that comprise the malt extract.

Controlling flower bud production

Spring application of GA is used extensively to accelerate flower bud production in artichoke which then allows earlier harvest dates. If treatment is delayed to coincide with the appearance of flower buds, increases in head size and number have also been reported (47).

Overcoming environmental constraints on growth

GA is used to break dormancy in plants that have not received a chilling period adequate for the resumption of growth. For rhubarb crowns transplanted in the fall for forcing, GA application can substitute for the cold period normally required for bud development and subsequent petiole elongation (53). Potato tuber dormancy can also be broken by application of GA (42). This treatment is of value in the identification and screening of virus-infected tubers. In warm climates, where it is possible to plant two crops in a single year, GA treatment can break dormancy of the seed tubers from the first crop in time for a second planting.

In celery production, GA is used to increase petiole elongation under cool weather conditions, where growth in reduced.

Uses in plant breeding

GA can be used to induce precocious cone production in conifers. This may be an especially important aid to genetic improvement in silviculture. Douglas fir, for example, normally requires 20 years before seed production will occur; with $GA_{4/7}$, 6-year-old trees can be induced to produce seed (40).

GA has also been used to control flower sex expression in cucumbers and squash. GA application tends to promote maleness in these plants. When gynoecious cucumbers are treated with GA, staminate flowers are produced.

Bolting and seed stalk formation are promoted by GA in many normally biennial vegetables. This facilitates hybrid seed production for commercial purposes and also accelerates vegetable variety improvement.

ETHYLENE-RELEASING AGENTS

While the biological effects of ethylene on plant growth have been documented for some time, little practical use of ethylene in agriculture was possible due to its gaseous nature. In the early 1970s experimental formulations of compounds that decompose on or within a plant to release ethylene became available. One of the first of these compounds, ethephon, is stable at pH values less than 4, but at higher pH values, the compound decomposes to produce ethylene as well as chloride and phosphate ions. Since the cytoplasmic pH is greater than 4, once ethephon is absorbed, cleavage to ethylene inside the cell begins. Two other compounds, etacelasil (2-chloroethyl-tris(ethoxymethoxy)silane) and 2-chloroethyl-bis(phenylmethoxy)silane, also decompose to ethylene in aqueous solution. These compounds generate ethylene much more rapidly than ethephon and are less sensitive to changes in pH.

Increasing latex flow in Hevea

The amount of rubber produced in the form of coagulated latex is a function of the duration of latex flow from the tapping cut, which is made in the tree bark. Ethephon is applied to a region near the tapping cut and causes latex flow to increase in duration, resulting in an increase in the volume of latex collected. Rubber yield increases of 50-100% are common. The mechanism for increased flow of latex by ethephon is not well understood. It is believed that lutoids, non-rubber containing bodies within the latex, are disrupted by tapping and cause coagulation or plugging of the latex vessels as a result of changes either in osmotic potential or in shear forces imposed by the high flow rate through the

623

narrow pores of the vessel. Ethephon may stabilize the lutoids, making them less susceptible to disruption. Alternatively, it has been proposed that ethephon treatment leads to an increase in cell wall thickening of the vessels, making the walls less likely to contract during tapping, thereby disrupting fewer lutoids and allowing an increase in latex flow (36).

Promoting abscission

The use of mechanical harvesting devices in cherry production had once been limited because the force required to remove the mature fruit resulted in damage to the trees. Ethephon may be applied approximately 10 days before anticipated harvest to reduce the fruit removal force, thus permitting mechanical harvesting of the crop without tree injury (5) (Fig. 3).

Walnuts are also harvested mechanically after treatment with ethephon. The edible kernal of the walnut reaches maturity some 3 to 4 weeks before harvest, owing to the time required for hull dehiscence. Ethephon treatment accelerates this process. The quality of the harvested nuts is also increased, because they do not remain on the tree for long periods of time after maturation, and therefore avoid decomposition due to heat and disease (33).

Olive fruit also have a high fruit removal force at maturity. In addition, the fruit is attached to long, willowy branches that do not lend themselves to mechanical shaking. Ethephon causes fruit abscission, but also excessive leaf abscission, which reduces flowering the following

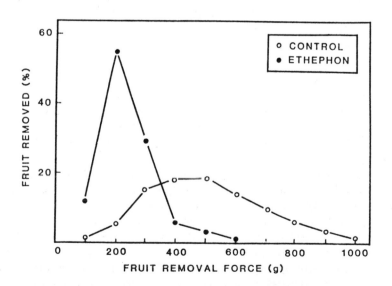

Fig. 3. The effect of ethephon on the fruit removal force in sour cherry 6 days after application. From (5)

spring. Etacelasil can be used, however, because it reduces the attachment force without defoliation. This compound releases ethylene at a much faster rate than does ethephon. It may be that leaf abscission requires elevated ethylene levels of longer duration than those needed for fruit abscission, making the silyl compounds more useful as fruit abscission agents than ethephon (21).

Ethylene-releasing agents are also being used to remove young fruit from apple and peach trees that have set a potentially excessive fruit crop. In peach, 2-chloroethylmethyl-bis(phenylmethoxy)silane has provided acceptable fruit abscission without defoliation in many areas of the southeastern US (14).

Promoting fruit ripening

The ripening process in mature fruit can be accelerated by ethephon application. Presumably the fruit are sensitive to ethylene at this stage of development, but have not produced enough endogenous ethylene to stimulate the ripening process. In apple, ethephon can be used to accelerate fruit softening and to advance fruit color production by several weeks, although additional application of compounds that will delay abscission must be used in conjunction with ethephon (15) (Table 3).

Table 3. Effects of ethephon, daminozide and NAA on abscission, firmness, and color of 'McIntosh' apples. From (15)

Treatment	Drop (%)	Firmness (kg)	Red color (%)
Control	29	6.6	51
daminozide	2	7.4	56
daminozide + ethephon	77	7.0	67
daminozide + ethephon + NAA	5	7.0	90

In tomato, ethephon is used to accelerate ripening and synchronize the maturation of the fruit for mechanical harvesting. Ethephon stimulates the production of lycopene by fruits and can therefore increase total yield of ripe fruit in the production of processing tomatoes, since this crop is harvested at one time only (13).

In grape, ethephon has been found to promote color development and decrease total fruit acidity. In some of the cooler grape-growing regions acidity is often excessive for optimum wine quality. Ethephon treatment may also be useful when natural fruit color development is poor.

Delaying flowering in fruit crops

Application of ethephon in the fall of the year prior to the spring flowering period delays bud expansion and anthesis in cherry and peach

Fig. 4. Effect of ethephon on time of flowering in peach. Center—control; side—shoots from trees sprayed with 100 ppm ethephon: left on October 12th, right on October 5th. Photo taken following April 12th. From (19)

(Fig. 4) (41,19). This treatment offers the potential to avoid spring frost damage to flowers.

Promoting leaf senescence in tobacco

Ethephon is used to promote leaf yellowing in flue-cured tobacco. Ethephon increases the number of leaves that may be harvested at one time and decreases the curing time for the leaf (49). This treatment is especially useful in the cooler tobacco-growing regions where the upper-most leaves, which are the last to be harvested, may be damaged by frost.

GROWTH RETARDANTS

The growth retardants are a diverse group of synthetic compounds that reduce stem elongation and generally increase the green color of leaves. These compounds inhibit cell division in the subapical meristem of the shoot, but have little effect on the production of leaves or on root growth. The physiological effects of the growth retardants can be reversed by application of GA, but generally no other compounds are effective. Incorporation of growth retardants into the media of cultures of *G. fujikuroi* inhibit GA production. In higher plants, the activity of the enzymes involved in kaurene synthesis and oxidation are inhibited by growth retardants. Two general classes of compounds have emerged.

Compounds such as AMO1618(4-hydroxy- 5-isopropyl-2-methylphenyl-trimethylammoniumchloride-1-pipe-ridinecarboxylate) and phosphon D (2,4-dichlorobenzyl-tributyl phophonium chloride) inhibit the enzymes involved in kaurene synthesis, whereas ancymidol (a-cyclopropyl-a-(4-methoxyphenyl)-5-pyrimidinemethanol)and paclobutrazol (2RS, 3RS-1-(4-chlorophenyl)-4,4-dimethyl-2-1,2,4-triazol-1-yl-pentan-3-ol) inhibit the kaurene oxidation sequence of reactions. In both cases the levels of GA are reduced. A number of the compounds that inhibit GA biosynthesis also inhibit sterol production in plants. There is, however, little evidence that sterols are involved in stem elongation, and it would appear therefore that the primary effect of the growth retardant on internode length is due to an inhibition in GA biosynthesis.

Controlling stem growth in greenhouse crops

The application of growth retardants to potted plants results in shorter, more rigid stems and darker green foliage, characteristics that increase the value of the crop. In chyrsanthemums, daminozide is effective as a foliar spray, and ancymidol may be used as both a foliar spray or a soil drench (28). Ancymidol treatment may, however, result in a delay in flowering.

In poinsettia, chlormequat chloride is used extensively for height control since it is less expensive than ancymidol. In Easter lily, ancymidol is used because it is the most effective compound for reducing stem height in this plant (Fig. 5). Paclobutrazol and the triazole fungicide triademefon

Fig. 5. Effect of increasing concentrations (left to right) of ancymidol on stem elongation in Easter lily.

will also control stem height, but higher concentrations are required in comparison to ancymidol.

Controlling rank growth in cotton
Under certain conditions of high fertility and favorable environmental conditions excessive vegetative growth of cotton results. Mepiquat chloride (1,1,dimethylpiperidinium chloride) applied at the time of flowering can reduce growth by 20-30% (23). Early yield of cotton is often increased by this treatment, presumably because of greater light penetration into the canopy, which allows fruit set to take place in flowers produced on the lower nodes of the plant. Reduced vegetative growth also allows greater coverage by insecticides, fungicides, and defoliants, the latter increasing the efficiency of mechanical harvesting.

Lodging control in cereals
Stem lodging is one of the most serious problems in wheat, when this crop is grown under the conditions of high fertility in Europe. The ability to use nitrogen to increase yield is limited by its adverse effect on stem growth. Chlormequat chloride can be used to reduce stem height and increase stem diameter. Yield is increased as a result of reduced stem lodging (26) (Fig. 6). In addition, in some years when lodging is not a problem, yield may still be increased because the growth retardant treatment results in a stimulation of tillering. Other cereals do not respond as well to chlormequat chloride as wheat. However, lodging control has been obtained in barley and rye with a combination of mepiquat chloride and ethephon.

Reducing growth of turf grass
Chemical control of grass growth especially on sites such as highway dividers, areas near airfields, or steep slopes that are difficult or dangerous to maintain, can be an economically feasible practice (16). Several compounds such as chlorflurenol (methyl-2-chloro-9-hydroxy-fluorene-9-carboxylate), mefluidide (N-(2,4-dimethyl-((trifluoromethyl)-sulfonyl)-amino)phenylacetamide), and paclobutrazol have been registered for use as plant growth regulators for grass control. Chlorflurenol acts by inhibiting cell division in the shoot meristem, whereas mefluidide inhibits cell elongation. Both compounds have been found to reduce the frequency of mowings required over the course of the growing season. Paclobutrazol reduces mowing frequency to an even greater extent than the other compounds. However, it does not effectively suppress seedhead formation. Combinations of paclobutrazol with either maleic hydrazide, chlorflurenol, or mefluidide should provide adequate seedhead suppression and persistence throughout the growing season for almost complete control of grass growth.

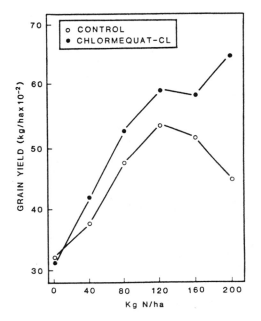

Fig. 6. Effect of chlormequat chloride and nitrogen fertilization on yield of wheat. (*Euphytica* 1, 215-18)

Increasing fruit set in grape

Application of chlormequat chloride to vinifera grapes before bloom increases fruit set of seeded berries (11). Cluster fresh weight is increased as a result of treatment. Daminozide is more effective than chlormequat chloride in increasing fruit set of the labrusca varieties (54). In addition to increasing cluster yield, vine growth is reduced by growth retardant treatment. It is not clear whether the increase in fruit set by the growth retardant is due to a direct effect on this process by decreasing GA levels (GA is used for berry thinning) or an indirect effect resulting from decreased vegetative growth. Exceedingly vigorous shoot growth is often associated with poor fruit setting in the field. Moreover, if shoot tips are removed, fruit set in grape can be increased, but the growth retardants are not capable of further increasing fruit set in detopped plants.

Advancing fruit color development

Daminozide may be used to advance anthocyanin production in the fruit skin and flesh of sweet cherry (6). The rate of color development is increased as well as the total amount of pigment synthesized. Other processes associated with fruit ripening such as fruit softening are not affected by daminozide treatment.

Daminozide will increase anthocyanin synthesis, decrease preharvest fruit drop, and reduce fruit softening during cold storage (15) in apple.

Furthermore, physiological disorders that develop at harvest or in storage have reportedly been less severe after a mid summer application of daminozide. The mechanism by which daminozide enhances color development in fruit is not clear. In apple, daminozide will inhibit ethylene production and delay the appearance of the respiratory climacteric (29). This will permit a delay in the harvest date and perhaps allow anthocyanin production to continue for a longer time before harvest. In some cases, however, daminozide will not only accelerate color production, but will also stimulate the production of greater amounts of anthocyanin in the apple skin or the flesh of cherry, thereby suggesting a more direct effect of the compound on pigment synthesis. Faust (17) has shown that anthocyanin production in apple is associated with increasing activity of the pentose phosphate pathway in the catabolism of carbohydrate. See and Foy (45) found that daminozide inhibits succinate dehydrogenase activity in isolated mitochondria. Perhaps by inhibiting Krebs cycle activity, greater carbon flow occurs in the pentose pathway, which forms the essential precursors for anthocyanin. In isolated apple skin discs, however, it was not possible to demonstrate a shift in carbon metabolism to the pentose pathway or to increase anthocyanin production in the presence of daminozide (18).

Induction of flower bud formation
    Both apple and pear trees do not generally come into full production until the trees are at least 5 years of age. Flowering can be stimulated in young trees by daminozide application. Increased bloom the year after a heavy crop on mature trees will also occur after daminozide application in apple, or chlormequat chloride treatment of pear. The growth retardants decrease shoot elongation in fruit trees, and perhaps flower bud initiation is promoted through the inhibition of vegetative growth.

Controlling tree size
    Paclobutrazol and other triazole analogs are probably the most effective compounds found to date for controlling shoot elongation in fruit trees. Controlling tree size with these compounds will be an effective way of maintaining tree height for maximum spraying and harvesting efficiency, used in conjunction with modern pruning practices, such as summer mowing of the tree canopy.

MISCELLANEOUS COMPOUNDS

Maleic hydrazide
    Maleic hydrazide has been used since the 1950s for tobacco sucker growth control, for the prevention of bud sprouting in onions and potatoes,

and for the control of turfgrass growth. At one time, maleic hydrazide accounted for almost 90% of the sales of plant growth regulators.

In tobacco production, the terminal bud is removed from the plant after a selected number of leaves have been produced. This practice, called topping, increases the size, weight, and quality of the cured leaf. Axillary buds that develop as a result of topping will reduce the effect of terminal bud removal on leaf yield and quality. Maleic hydrazide will provide excellent control of axillary bud growth when applied as a foliar spray to the upper two-thirds of the plant, after terminal bud removal (39).

Maleic hydrazide is also used to control storage sprouting of onions and potatoes. The compound is applied as a preharvest foliar spray and is rapidly translocated to the storage organs of both of these plants. Maleic hydrazide inhibits cell division in a wide range of plants, and the ability of the compound to be translocated to meristematic tissue probably accounts for its affect on axillary bud growth in tobacco and the sprouting of tubers and bulbs (35). An analog of uracil, maleic hydrezide may inhibit cell division by reducing nucleic acid biosynthesis in shoot and root meristems.

Citrus abscission agents

Several abscission agents are being developed for the mechanical harvesting of oranges intended for processing use. The compounds Release (5-chloro-3-methyl-4-nitro-1-pyrazole) and Pik-off (ethandiol dioxime) induce abscission by causing superficial injury to the rind of the fruit (25). Wound ethylene is then synthesized and is presumably responsible for the reduction in fruit removal force. Application of ethephon to trees with mature fruit will also induce abscission; however, significant defoliation often will occur with this chemical.

Sugarcane ripeners

One of the most useful compounds for increasing yield in sugarcane is glyphosine, which in Hawaii has increased sucrose yield by 10-15%. The herbicide glyphosate, an analog of glyphosine, is also effective, and lower rates of application can be used. Glyphosine will decrease terminal growth of the cane, and it has been shown that removing the upper leaves of the stalk increases sucrose translocation from lower leaves into the stem and ripening joints (22). In addition, these compounds apparently alter the partitioning of carbohydrate in the sugarcane internode. More carbohydrate goes into sucrose storage at the expense of fiber production (37).

Cotton defoliants

The organophosphate compounds DEF and Folex are used as leaf abscission agents before mechanical harvesting of cotton. The presence of

631

green leaves interferes with the harvesting procedure and can reduce the quality of the cotton fiber. Two new compounds are also being evaluated for this purpose. Dimethipin (2,3-dihydro-5,6-dimethyl-1,4-dithiin-1,1,4,4-tetraoxide) and thidiazuron (1-phenyl-3(1,2,3-thiadiazol-5-yl)urea) induce defoliation and provide superior control of regrowth vegetation after leaf abscission.

FUTURE PROSPECTS

Most of the PGRs that have been described here are used to increase crop yield or quality by modifying plant development through the hormone system-either by blocking the synthesis or action of a hormone, or by supplementing its supply at a given time. It may be possible, however, to directly affect plant metabolism and increase crop yield by modifying key physiological processes such as photosynthesis, nitrogen fixation, mineral ion uptake, and senescence (44).

Increasing photosynthetic efficiency by reducing photorespiration could potentially increase dry matter production. Inhibitors of photorespiration have been identified, but unfortunately net photosynthesis has not been increased by these compounds. Altering assimilate partitioning may be another approach to increasing the efficient use of photosynthates. Certainly hormones such as IAA and GA are known to stimulate assimilate translocation to the site of application of these compounds, although this phenomenon has yet to be put to practical use.

Nitrogen fixation requires large amounts of energy and is often limited by photosynthetic capacity. Growth regulators that improve nitrogen fixation efficiency could greatly affect crop yield as would compounds that increase the efficiency of mineral ion uptake by the root system.

Controlling plant senescence also offers a potential for yield increase. It may be possible not only to delay leaf senescence so that more of the growing season can be used for crop growth, but also to induce senescence for the remobilization of nutrients and assimilates at times when the potential for seed or fruit growth is high.

Other areas of active research include the search for chemical hybridizing agents and the discovery of herbicide antidotes. Hybridizing agents can induce male sterility in crop plants, thereby facilitating cross-pollination and the development of hybrid seeds. Herbicide antidotes reduce the toxicity of the compound to the crop plant but not the weed, allowing increased selective use of currently available herbicides in a greater variety of weed-crop situations.

References

1. Abeles, F.B. (1969). Herbicide-induced ethylene production: Role of the gas in sublethal doses of 2,4-D. Weed Sci. *16*, 498-500.
2. Abeles, F.B., Leather, G.R., Forrence, L.E., Craker, L.E. (1971). Abscission: Regulation of senescence, protein synthesis, and enzyme secretion by ethylene. HortSci. *6*, 371-376.
3. Beyer, E.M. (1973). Abscission-support for a role of ethylene modification of auxin transport. Plant. Physiol. *52*, 1-5.
4. Beyer, E.M. Jr., Quebedeaux, B. (1974). Parthenocarpy in cucumber; mechanism of action of auxin transport inhibitors. J. Amer. Soc. Hort. Sci. *99*, 385-390.
5. Bukovac, M.J., Zucconi, F., Larsen, R.P., Kesner, C.D. (1970). Chemical promotion of fruit abscission in cherries and plums with special reference to 2-chloroethylphosphonic acid. J. Amer. Soc. Hort. Sci. *94*, 226-230.
6. Chaplin, M.H., Kenworthy, A.L. (1970). The influence of succinamic acid 2,2-dimethyl hydrazide on fruit ripening of the 'Windsor' sweet cherry. J. Amer. Soc. Hort.Sci. *95*, 532-536.
7. Chen, L.G., Switzer, C.M., Fletcher, R.A. (1972). Nucleic acid and protein changes induced by auxin-like herbicides. Weed Sci. *20*, 53-55.
8. Christodoulou, A.J., Weaver, R.J., Pool, R.M. (1968). Relation of gibberellin treatment to fruit-set, berry development, and cluster compactness in *Vitis vinifera* grapes. Proc. Amer. Soc. Hort.Sci. *92*, 301-310.
9. Coggins, C.W. Jr., Hield, H.Z. (1965). Navel orange fruit response to potassium gibberellate. Proc. Amer. Soc. Hort. Sci. *81*, 227-230.
10. Coggins, C.W. Jr., Hield, H.Z. and Boswell, S.B. (1960). The influence of potassium gibberellate on Lisbon lemon trees and fruit. Proc. Amer. Soc. Hort.Sci. *76*, 199-207.
11. Coombe, B.G. (1965). Increase in fruit set of *Vitis vinifera* by treatment with growth retardants. Nature *205*, 305-306.
12. Denny, F.E. (1924). Effect of ethylene upon respiration of lemons. Bot. Gaz. *77*, 322-329.
13. Dostal, H.C., Wilcox, G.E. (1971). Chemical regulation of fruit ripening of field-grown tomatoes with (2-chloroethyl) phosphonic acid. J. Amer. Soc. Hort. Sci. *96*, 656-660.
14. Dozier, W.A. Jr., Carlton, C.C., Short, K.C., McGuire, J.A. (1984). Thinning 'Loring' peaches with CGA-15281. HortScience *16*, 56-57.
15. Edgerton, L.J., Blanpied, G.D. (1970). Interaction of succinic acid 2,2-dimethyl hydrazide, 2-chloroethylphosphonic acid and auxins on maturity, quality, and abscission of apples. J. Amer. Soc. ort. Sci. *95*, 664-666.
16. Elkins, D.M. (1983). Growth regulating chemicals for turf and other grasses. *In*: Plant Growth Regulating Chemicals Vol 2. Nickel, L.G. ed. p. 113-128.
17. Faust, M. (1965). Physiology of anthocyanin development in McIntosh apple. I. Participation of pentose phosphate pathway in anthocyanin development. Proc. Amer. Soc. Hort. Sci. *87*, 1-9.
18. Gianfagna, T.J., Berkowitz, G.A. (1986). Glucose catabolism and anthocyanin production in apple fruit. Phytochemistry (in press).
19. Gianfagna, T.J., Marini, R.P., Rachmiel, S. (1986). Effect of ethephon and GA₃ on time of flowering in peach. HortScience (in press).
20. Gustafson, F.G. (1937). Parthenocarpy induced by pollen extracts. Amer. J. Bot. *24*, 102-107.
21. Hartmann, H.T., Reed, W., Opitz, K. (1976). Promotion of olive fruit abscission with 2-chloroethyl-tris (2-methoxyethoxy)-silane J. Amer. Soc. Hort. Sci. *101*, 278-281.
22. Hartt, C.E., Kortschak, H.P., Burr, G.O. (1964). Effects of defoliation, irradiation, and darkening the blade upon translocation of ¹⁴C in sugarcane. Plant Physiol. *39*, 15-22.
23. Heilman, M.D. (1981). Interactions of nitrogen with Pix on the growth and yield of cotton. Proc. Beltwide Cotton Prod. Res. Conf. 47.
24. Hitchcock, A.E., Zimmerman, P.W. (1942). Root-inducing activity of phenoxy compounds in relation to their structure. Contrib. Boyce Thomp. Inst. *12*, 497-507.

25. Holm, R.E., Wilson, W.C. (1977). Ethylene and fruit loosening from combinations of citrus abscission chemicals. J. Amer. Soc. Hort.Sci. *102*, 576-579.
26. Humphries, E.C., Welbank, P.J., Witts, K.J. (1965). Effect of CCC (chlorocholine chloride) on growth and yield of spring wheat in the field. Ann. Appl. Biol. *56*, 351-361.
27. Kotob, M.A., Schwabe, W.W. (1971). Induction of parthenocarpic fruit in Cox's Orange Pippin apples. J. Hort.Sci. *46*, 89-93.
28. Larson, R.A., Kimmins, R.K. (1971). Response of *Chrysanthemum morifolium* Ramat to foliar and soil applications of ancymidol. Hort.Science 7, 192-193.
29. Looney, N.E. (1968). Inhibition of apple ripening by succinic acid 2-2-dimethyl hydrazide and its reversal by ethylene. Plant Physiol. 43, 1133-1137.
30. Luckwill, L.C. (1953). Studies of fruit development in relation to plant hormones. II. The effect of naphthalene acetic acid on fruit set and fruit development in apples. J. Hort.Sci *28*, 25-40.
31. Luckwill, L.C. (1953). Studies of fruit development in relation to plant hormones. I. Hormone production by the developing apple seed in relation to fruit drop. J. Hort.Sci. *28*, 14-24.
32. Magness, J.R., Overly, F.L. (1929). Relation of leaf area to size and quality of apples and pears. Proc. Amer. Soc. Hort.Sci. *26*, 160-162.
33. Martin, G.C. (1971). 2-chloroethylphosphonic acid as an aid to mechanical harvesting of English walnuts. J. Amer. Soc. Hort.Sci. *96*, 434-436.
34. Moore, P.H., Osgood, R.V., Carr, J.B., Ginoza, H.S. (1982). Sugarcane studies with gibberellin. V. Plot harvests vs. stalk harvests to assess the effect of applied GA_3 on sucrose yield. J. Plant Growth Reg. *1*, 205-210.
35. Nooden, L.D. (1969). The mode of action of maleic hydrazide: inhibition of growth. Physiol. Plant. *22*, 260-270.
36. Osborne, D.J., Sargent, J.A. (1974). A model for the mechanism of stimulation of latex flow in *Hevea brasiliensis* by ethylene. Ann. Appl. Biol. *78*, 83-88.
37. Osgood, R.V., Moore, P.M., Ginoza, H.S. (1981). Differential dry matter partitioning in sugarcane cultures treated with glyphosate Proc. Plant Growth Reg. Soc. Amer. *8*, 97.
38. Palmer, G.M. (1974). The industrial use of gibberellic acid and its scientific basis - A review. J. Inst. Brewing *80*, 13-30.
39. Petersen, E.L. (1952). Controlling tobacco sucker growth with maleic hydrazide. Agron J. *44*, 332-334.
40. Pharis, R.P., Ross, S.D., McMullan, E. (1980). Promotion of flowering in the Pinaceae by gibberellins. III. Seedlings of Douglas fir. Physiol. Plant *50*, 119-126.
41. Proebsting, E.L. Jr., Mills, H.H. (1972). Bloom delay and frost survival in ethephon treated sweet cherry. HortScience 8, 46-47.
42. Rappaport, L., Timm, H., Lippert, L.F. (1957). Sprouting, plant growth, and tuber production as affected by chemical treatment of white potato seed pieces. I. Breaking the rest period with gibberellic acid. Amer. Potato J. *34*, 254-260.
43. Rodriguez, A.G. (1932). Influence of smoke and ethylene on fruiting of pineapple (*Ananas sativus* Schult.). J. Agric. Univ. P.R. *15*, 5.
44. Scott, T.K. (1978). Plant Growth Regulation and World Agriculture. Plenum Press.
45. See, R.M., Foy, C.L. (1982). Effect of butanedioic acid mono (2, 2-dimethyl hydrazide) on the activity of membrane bound sucinate dehydrogenase. Plant Physiol. *70*, 350-352.
46. Sidahmed, D.A., Kliewer, W.M. (1980). Effects of defoliation, gibberellic acid and 4-chlorophenoxy acetic acid on growth and composition of Thompson seedless grape berries. Amer. J. Enol. Vitic. *31*, 149.
47. Snyder, M.J., Welch, N.C., Rubatzky, V.E. (1971). Influence of gibberellin on time of bud development in globe artichoke. HortScience *6*, 484-485.
48. Southwick, F.W., Demoranville, I.E., Anderson, J.F. (1953). The influence of some growth regulating substances on preharvest drop, color and maturity of apples. Proc. Amer. Soc. Hort.Sci. *59*, 155-162.

49. Steffins, G.L., Alphin, J.G., Ford, Z.T. (1970). Ripening tobacco with the ethylene releasing agent 2-chloroethylphosphonic acid. Beitr. Tobakforschung 5, 262.
50. Stewart, W.S., Hield, H.Z. (1949). Effect of 2, 4-dichlorophenoxyacetic acid and 2,4,5-trichlorophenoxyacetic acid on fruit drop, fruit production, and leaf drop of lemon trees. Proc. Amer. Soc. Hort.Sci. 55, 163-171.
51. Struckmeyer, B.E., Roberts, R.H. (1950). A possible explanation of how naphthalene acetic acid thins apples. Proc. Amer. Soc. Hort.Sci. 56, 76-78.
52. Thomas, T.H. (1982). Plant Growth Regulator Potential and Practice. BCPC Publications 14-150 London Rd, Croydon CRO 2TD U.K.
53. Tompkins, D.R. (1966). Rhubarb rest period as influenced by chilling and gibberellin. Proc. Amer. Soc. Hort.Sci. 87, 371-379.
54. Tukey, L.D., Fleming, H.K. (1968). Fruiting and vegetative effects of N-dimethyl-aminosuccinamic acid on 'Concord' grapes, Vitis labrusca L. Proc. Amer. Soc. Hort.Sci. 93, 300-310.
55. Wallerstein, I., Goren, R., Monselise, S.P. (1973). Seasonal changes in gibberellin-like substances in Shamouti orange (Citrus sinensis (L.) Osb.) trees in relation to ringing. J. Hort.Sci. 48, 75-82.
56. Walsh, C.S., Swartz, H.J., Edgerton, L.J. (1979). Ethylene evolution in apple following post-bloom thinning sprays. Hort Science 14, 704-706.
57. Westwood, M.N. (1978). Temperate Zone Pomology. W.H. Freeman and Co., San Francisco, CA.
58. Williams, M.W. (1979). Chemical Thinning of Apples. Hort. Rev. 1, 270-300.
59. Williams, M.W., Stahly, E.A. (1969). Effect of cytokinins and gibberellins on shape of 'Delicious' apple fruits. J. Amer. Soc. Hort.Sci. 94, 17-19.

E20. Genes Specifying Auxin and Cytokinin Biosynthesis in Prokaryotes

Roy O. Morris

Department of Agricultural Chemistry, Oregon State University, Corvallis, Oregon 97331, USA.

INTRODUCTION

Since the discovery and characterization of auxins and cytokinins, much time, effort and resources have been expended in order to determine the mechanism of their biosynthesis and mode of action. It would not be an understatement to say that the effort has met with only moderate success. For example, in the case of the cytokinins, the structures are known (46), their metabolism has been extensively documented (29) and the major morphological and physiological consequences of their action have been described (26). Nevertheless, there are large gaps in our understanding. Biosynthesis is a case in point. Although there have been reports of isolation of putative cytokinin biosynthetic enzymes (9,10), no enzyme has been obtained in a state of analytical homogeneity and there is no detailed understanding of the biosynthetic pathway, its expression in different plant organs and its control. The genes which encode the biosynthetic enzymes have yet to be isolated and important questions relating to the temporal and spatial expression of such genes during development remain to be answered.

Recently, light has been shed from a somewhat unexpected source on the biosynthesis of both hormones. Certain phytopathogenic bacteria, notably *Agrobacterium tumefaciens (A. tumefaciens)* and *Pseudomonas syringae pv. savastanoi, (P. savastanoi)* can incite tumors or galls on dicotyledonous plants. The tumors are hormone-autotrophic in culture and contain elevated levels of auxins and cytokinins (33). Molecular studies have now identified the bacterial genes which specify auxin (25,27,43,55) and cytokinin (1,4) biosynthesis. Both bacteria contain genes specifying the synthesis of both phytohomones. A functional auxin biosynthetic pathway exists in *P. savastanoi (21)* and the organism contains two plasmid-born genes which encode the enzymes, tryptophan-2-monooxygenase and indoleacetamide hydrolase which act in concert to produce IAA from tryptophan (55,11,12,13). Cognate genes are present in *A. tumefaciens* but are not expressed in the bacterium, only upon transfer to the plant (33,25,27,43,55). Likewise, both bacteria contain genes which

encode dimethylallyl pyrophosphate: 5'AMP transferase (DMA transferase or isopentenyl transferase) (4, 33), an enzyme responsible for
cytokinin biosynthesis.

It appears that the native expression of these genes in association
with the plant is responsible for at least part of the phenotype associated
with tumors and galls even though (as shown below) the molecular
mechanisms by which the bacteria incite such tumors are dissimilar. In
spite of the dissimilarities, corresponding genes from the two organisms
share extensive sequence homology (41, 55).

In this chapter, the molecular biology of tumorigenesis is outlined, the
evidence relating to the presence of phytohormones in tumors is presented, and details of the structure and function of the genes are
described. With the availability of genes specifying phytohormone
biosynthesis and with the recent development of vectors for transfer of
genes into plants, it may now be possible to investigate phytohormone
function in quite a novel way. The last section will consider a conceptual
framework for such studies and describe some of the experimental
opportunities.

Molecular Aspects of Plant Tumor Formation and Growth

At least three groups of phytopathogenic bacteria, A. tumefaciens,
P. savastanoi and Corynebacterium fascians, are known to produce neoplastic or hyperplastic diseases in plants (36, 37). Infection results in the
production of galls or tumors which display either completely unorganized
growth properties or incomplete organogenesis with production of abnormal shoots or roots. A. tumefaciens produces unorganized crown gall
tumors on most dicotyledonous plant species (Fig. 1A) although on some
hosts and with some strains of A. tumefaciens, shooty teratomata are produced. Such teratomata may be cloned in culture where they continue to
exhibit the shoot-bearing phenotype (Fig. 1B). A closely related organism
A. rhizogenes causes hairy root disease in which grossly abnormal roots
are produced (Fig. 1C). As described below, the Agrobacteria actually
transform the cells of the host plant by introduction of genes from an
endogenous plasmid.

The two other groups of bacteria which produce galls do so not by
transformation of the host cells but by virtue of close association with
them. P. savastanoi produces unorganized galls on a limited number of
woody hosts, notably olives, oleanders, and privet. Corynebacterium
fascians produces witches broom disease, a proliferation of shoots at the
apex of the affected plant (Fig. 1D). Although there is preliminary evidence that C. fascians contains a cytokinin biosynthetic gene (31), the
most complete evidence relates to A. tumefaciens and P. savastanoi. The
discussion will therefore be limited to these two organisms.

Fig. 1. Tumors and galls incited by phytopathogenic bacteria. (A) Undifferentiated crown gall tumor incited on *Kalanchoe* stem by *A. tumefaciens* octopine strain A6NC. Roots arise not from the tumor but from the stem below the point of infection. (B) *Nicotiana tabacum* shoot-bearing teratoma in culture. Incited by *A. tumefaciens* nopaline strain T37. (C) Hairy root tumor incited on *Kalanchoe* by *Agrobacterium rhizogenes* strain A4 (D) Witches broom disease incited on Shasta daisy by *Corynebacterium fascians* (courtesy of Don Cooksey).

Figure 2 outlines the molecular mechanism underlying tumor formation by *A. tumefaciens*; full details may be found in recent reviews (36, 37). In 1974, Zaenen et al.(56) demonstrated that *A. tumefaciens* harbors large (ca 200kb) plasmids. All virulent strains of *A. tumefaciens* were found to have these Ti or tumor-inducing plasmids. They confer virulence on the bacteria which harbor them. Thus, if a strain of *A. tumefaciens* is cured of its endogenous Ti plasmid by growth at elevated temperatures, it loses virulence. If the Ti plasmid from the same strain or from a different strain is introduced into the cured bacterium,virulence is regained. The host range of the transformant and the properties of the resulting tumors are determined primarily by the donor Ti plasmid and not by the recipient host bacterium.

If the Ti plasmids from a range of *A. tumefaciens* strains are examined for physical structure (by digestion with restriction enzymes), relatedness (by hybridization), or function; it becomes evident that there are several different classes. Each member of the class confers upon the bacterium the ability to produce a specific type of tumor on a specific host plant or set of plants (36,37). Although there are these differences, the plasmids share two common functional regions responsible for virulence and

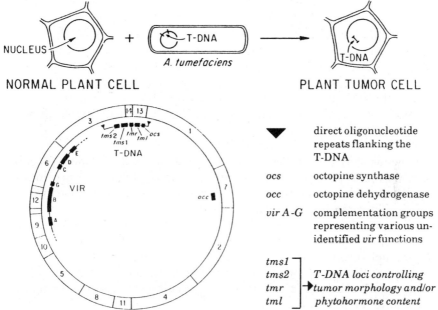

Fig. 2. Molecular mechanism of crown gall tumorigenesis and functional map of a typical octopine Ti plasmid. Top: The T-DNA of the Ti plasmid is transferred to and integrated into the nuclear genome of the host plant. Bottom: Functional map of a typical octopine Ti plasmid. Radial lines in outer circles represent restriction enzyme (HpaI) cleavage points, giving fragment sizes as indicated.

tumorigenesis, known respectively as the *vir* region and the T-DNA (Fig. 2). In a sense, the T-DNA may be regarded as the ultimate pathogenic entity. It is a region of the Ti plasmid, bounded by well-defined oligonucleotide direct repeats (54) which, during tumorigenesis, is transferred to the plant cell nucleus and integrated into the nuclear genome, where its presence and expression is ultimately responsible for maintenance of the tumorous condition.

The *vir* region is responsible for T-DNA transfer. It contains a set of at least six genes (17) which direct transfer of T-DNA to the plant but which are probably not themselves transferred. Although details of *vir* gene function are unclear, some recent experiments have shown that their expression is regulated by plant products and that this expression in turn regulates the transfer of T-DNA. Thus, there is little expression of the genes of either the T-DNA or of most of the *vir* region when the bacteria are grown alone in culture. However, when a plant extract is added, transcription of at least some of the *vir* genes is initiated. Two of the plant factors responsible for activation were characterized recently. One is a protein of undetermined function (38); the other is acetosyringone (47), a phenolic compound closely related to the intermediates involved in lignin biosynthesis. Apparently, *vir* region transcription and T-DNA transfer are not initiated unless the bacteria are in intimate contact with products excreted by the wound site.

About the transformation itself and the integration of T-DNA into the plant genome, little is known except that at least one T-DNA border sequence is required and that a circular T-DNA intermediate is probably involved (28). The exact details of the enzymes required, the process of transfer through the bacterial wall and across the nuclear membrane, and the possible requirement for specific plant factors or sequences, all remain to be determined.

Ti plasmids may be conveniently categorized according to the nature of the genes they encode for opine metabolism (36,37). Opines (octopine and nopaline among others) are condensation products of arginine and alpha-keto acids: octopine is formed by condensation of arginine with pyruvic acid; nopaline from arginine and alpha-ketoglutaric acid. Most Ti plasmids encode genes which allow the host bacteria to catabolize either octopine or nopaline. Ti plasmids which carry the octopine catabolic gene, octopine dehydrogenase (occ), confer on the bacterium the ability to use octopine but not nopaline as sole source of carbon and nitrogen. For convenience such strains are referred to as octopine strains and the plasmids as octopine Ti plasmids. Similarly, nopaline Ti plasmids carry the nopaline catabolic gene (noc).

The opine catabolic genes are located outside the T-DNA (Fig. 2) and appear to play no part in the process of tumor induction. Nevertheless, a given Ti plasmid usually carries within its T-DNA an opine biosynthetic gene corresponding to the opine catabolic gene external to the T-DNA. Thus, octopine plasmids carry the octopine synthase gene ocs, while nopaline Ti plasmids carry the nopaline synthase gene, nos. Opine synthase genes, in common with the other T-DNA genes, are not expressed in the bacterium to any appreciable extent but are expressed in the tumor where they cause the production of the corresponding opine.

Just as there are different classes of Ti plasmid, so the integration patterns of their T-DNAs are different. Generally, T-DNA from octopine plasmids is integrated as two independent segments, the T_LDNA and the T_RDNA. The two segments are integrated at different sites in the plant genome and may be present at different copy numbers per cell. The T_LDNA carries the genes necessary for transformation since the T_RDNA may be deleted without affecting tumor growth. The T_LDNA of octopine strains is usually about 12kb in length and has been shown to encode seven genes (53) (including ocs) which are transcribed only in the plant (Fig. 3). Nopaline T-DNA on the other hand is substantially larger (23kb) and encodes thirteen genes.

Hybridization studies have allowed measurement of homology between the two types of plasmid (overall about 30%) and specifically between the two T-DNAs (14). Certain regions of octopine T_LDNA exhibit high homology with the nopaline T-DNA. These common regions are

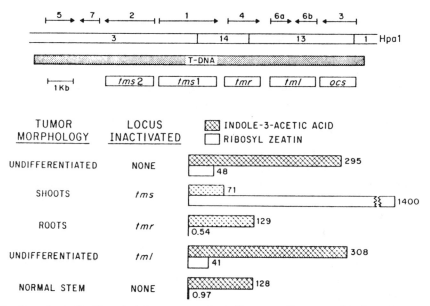

Fig. 3. Top: Loci affecting phytohormone content of crown gall tumors. Horizontal arrows represent major T-DNA transcripts from a typical octopine T-DNA. Transcripts 1 and 2 correspond to the *tms1* and *tms2* genes encoding auxin biosynthetic enzymes; transcript 4 corresponds to the cytokinin biosynthetic locus, *tmr*; and transcript 3 to the locus encoding octopine synthase (*ocs*). The boundaries of the loci determining tumor morphology are indicated by the boxes below the T-DNA. Bottom: Inactivation of *tms* gives shoot-bearing tumors, inactivation of *tmr* gives root-bearing tumors, and inactivation of *tml* gives large tumors. Horizontal bars represent the auxin and cytokinin levels present in tobacco tumors incited by strains bearing mutated loci. Amounts are expressed as ng/g tissue. Adapted from (2,15,23).

responsible for maintenance of the tumorous condition and it is within them that genes encoding auxin and cytokinin biosynthesis are found.

The overall picture to emerge then is that transfer, integration and expression of T-DNA is the primary cause of tumorigenesis and that the event represents a natural genetic engineering phenomenon exploited by the bacterium. In contrast to crown gall tumors, galls produced by *P. savastanoi* do not exhibit characteristics of permanent transformation. Galls senesce at the end of the growing season and do not resume growth in subsequent years. There is no direct experimental evidence for integration of *P. savastanoi* DNA into plant cells (T. Kosuge, personal communication).

EVIDENCE FOR INDOLEACETIC ACID PARTICIPATION IN PLANT TUMOR GROWTH

An extensive body of evidence indicates that auxin and cytokinin production, encoded by T-DNA genes, is central to the maintenance of the transformed state. The first evidence that crown gall tumors have altered phytohormone status came from the pioneering work of Braun (7). Tumors exhibit auxin-like growth effects and can be grown in axenic culture in the absence of added auxin (in contrast to untransformed tissue which cannot). He suggested that growth was supported by the endogenous production of auxins and cell division factors. Many subsequent studies (summarized in 33, 36) confirmed and extended the original work and provided evidence for altered auxin metabolism in tumors. Thus:

- Tumor IAA levels are generally elevated over those of untransformed tissues in culture. While there is some conflict between different reports, this can probably be ascribed to the use of cloned and uncloned tumor lines (containing mixtures of transformed and untransformed cells) of different species, and the use of different techiques for auxin determination.

- IAA accumulation rates of tumors are greater than those of untransformed tissues in culture. IAA levels have been reported to peak early in exponential growth, late in growth or not at all. But again, in general, levels increase more rapidly and to a greater degree than those of untransformed tissues.

- The process of conversion of tryptophan to IAA by tumor tissue differs somewhat from that of untransformed tissue. Both metabolize tryptophan to indoleacetamide, indoleacetaldehyde, and indole-3-ethanol but conversion is more efficient in crown gall and indoleacetoxime is formed only in crown gall tissue.

- Tumors incited by Ti plasmids bearing mutations in the T-DNA (usually by spontaneous transposition of a bacterial element to a locus within the T-DNA) exhibit an IAA growth requirement whereas wild type tumors do not. For example, a mutant of the octopine strain Ach5 bearing the insertion element IS60 within its T-DNA was avirulent on *Kalanchoe* and produced only small tumors on tomato (39). Tumor growth was restored by application of NAA. Other mutants (A66) exhibited a requirement for IAA for growth in culture (6).

- Tumor IAA levels are decreased significantly if the T-DNA bears certain defined mutations, as illustrated in Fig. 3. Using transposon mutagenesis (a process by which bacterial transposable elements are inserted pseudorandomly into DNA thereby inactivating any genes present) of the T-DNA prior to tumor induction, Garfinkel et al. (15) defined three T-DNA loci (*tms*, *tmr*, and *tml*) which control tumor morphology. Inactivation of the *tms* locus causes the normally undifferentiated tumors to produce shoots. Inactivation of *tmr* causes root production. Inactivation of *tml* causes an increase in tumor size. Evidence linking control of auxin levels to the *tms* locus was obtained from a study of such tumors (1). Auxin levels in shoot-bearing tumors (inactivated in the *tms* locus) were significantly lower than those of wild type tumors (Fig. 3). Because *tms* encompasses two genes (*tms1* and *tms2*) and because insertional mutagenesis of either causes reduction in IAA content, both genes must be involved in IAA metabolism.

• Some tumors incited by mutated T-DNA contain excessively high levels of the putative IAA biosynthetic intermediate, indoleacetamide (52). Levels may reach as high as 1000 times those in wild type tumors.

Auxin Biosynthetic Genes in *Agrobacterium Tumefaciens*

Definitive evidence that the *tms1* and *tms2* genes of *A. tumefaciens* actually specify auxin biosynthetic enzymes came from a series of experiments in which the genes were sequenced, found to be homologous to known auxin biosynthetic genes in *P. savastanoi* (see below) and shown to encode the appropriate enzyme activity.

The auxin biosynthetic pathway in *A. tumefaciens* has two enzymatic components (Fig. 4). The first is tryptophan-2-monooxygenase encoded by the *tms1* locus. The second is indoleacetamide hydrolase encoded by *tms2*.

Tms1 was sequenced (27) and shown to define an open reading frame capable of specifying an 84 kd protein. The deduced amino acid sequence of this open reading frame displayed significant homology with the sequence of the FAD-requiring enzyme 4-hydroxybenzoate hydroxylase from *P. fluorescens*. Specifically, the amino acid sequence known to be responsible for binding FAD in 4-hydroxybenzoate hydroxylase was present indicating that *tms1* was likely to be a flavoprotein.

Although *tms1* has yet to be cloned and expressed in *E. coli* with production of tryptophan-2-monooxygenase activity, tumor lines bearing the *tms1* gene exhibit tryptophan-2-monooxygenase activity *in vitro*, are able to convert tryptophan to indoleacetamide *in vivo* and contain excessively high levels of indoleacetamide, whereas controls do not (51).

The second enzyme of the pathway, indoleacetamide hydrolase, was identified by direct cloning. The major *tms2* open reading frame can encode a 49kd protein (44). When the complete coding region was cloned

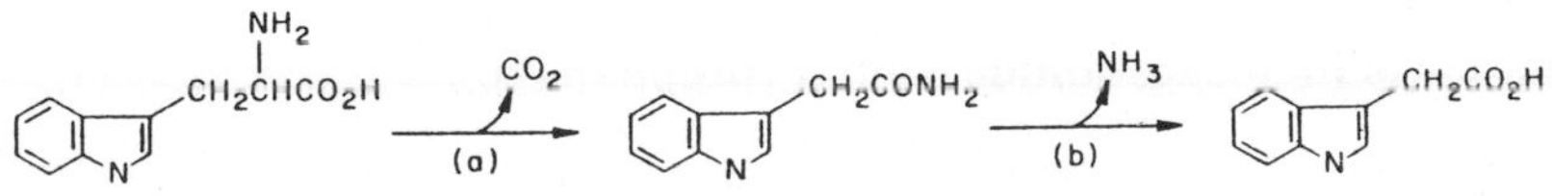

Fig. 4. Auxin biosynthetic pathway in *A. tumefaciens* and *P. savastanoi*. (a) tryptophan-2-monooxygenase encoded by *tms1* in *A. tumefaciens* and by *iaaM* in *P. savastanoi* (b) indole acetamide hydrolase encoded by *tms2* in *A. tumefaciens* and by *iaaH* in *P. savastanoi*

and expressed in *E. coli* (43), it induced the formation of a 49 kd protein and extracts from the bacteria were able to convert indoleacetamide to IAA. The enzyme was partially purified (25) and found to hydrolyse indoleacetonitrile, IAA esters, naphthaleneacetamide, phenylacetamide and, of course, indoleacetamide. It cannot hydrolyse IAA-aspartate or the IAA conjugates of alanine, glutamic acid or glycine.

Auxin Biosynthetic Genes in *Pseudomonas Savastanoi*

Early studies of *P. savastanoi* and of the galls formed on olives (37) indicated that IAA and other growth promoting substances were present. A series of elegant experiments by Kosuge and his coworkers then established (11, 12, 13, 21), that there is an active IAA biosynthetic pathway in the bacterium identical to that present in *A. tumefaciens* (Fig. 4).

Initially, both tryptophan-2-monooxygenase and indoleacetamide hydrolase activities were demonstrated in semi-purified extracts of *P. savastanoi*. Subsequently, genes encoding these activities were identifed and designated *iaaM* and *iaaH*. A clear connection between IAA production and virulence was established by examination of *P. savastanoi* strains resistant to tryptophan analogs. Some were defective in IAA production and had lost virulence, others overproduced IAA and exhibited enhanced virulence (12, 13). As in *A. tumefaciens*, the IAA biosynthetic genes of *P. savastanoi* are carried on plasmids. *P. savastanoi* strain 2009 contains four plasmids having sizes of 58, 52, 41, and 34 Kb (12). Mutants which have lost the 52 kb plasmid (pIAA1) have also lost the ability to synthesize IAA and indoleacetamide. Reintroduction of pIAA1 into IAA⁻ mutants of *P. savastanoi* causes reacquisition of tryptophan-2-mono-oxygenase, indoleacetamide hydrolase, and virulence.

Characterization of the tryptophan-2-monooxygenase and indole-acetamide hydrolase gene products was achieved by cloning and expression in *E. coli*. *IaaM* was located by transformation of *E. coli* with restriction fragments of pIAA1 DNA. Transformants bearing the appropriate fragments produced indoleacetamide but not IAA and expressed tryptophan-2-monooxygenase but not indoleacetamide hydrolase. The gene was subsequently sequenced and found to encode an open reading frame equivalent to a protein of 62 kd. The coding region contained an amino acid sequence similar to those responsible for binding FAD in 4-hydroxybenzoate hydroxylase and *tms1*. The purified enzyme was found to contain one mole FAD per mole of protein and to catalyze formation of indoleacetamide from tryptophan (20).

Similar studies (55) allowed characterization of indoleacetamide hydrolase whose gene, *iaaH*, is located adjacent to *iaaM* on pIAA1. The gene was cloned (together with *iaaM*) into *E. coli* on a 4 kb restriction fragment of pIAA1 which contained both loci. Sequence data and *in vitro* transcription/translation experiments indicated that it specifies a 47kd protein which has indoleacetamide hydrolase activity.

EVIDENCE FOR CYTOKININ PARTICIPATION IN PLANT TUMOR GROWTH

The subject of cytokinin biosynthesis has been covered in Chapter B3. The most likely pathway is that outlined by Letham (29) and illustrated in Fig. 5, namely: condensation of dimethylallylpyrophosphate (DMAPP) with 5'AMP to give isopentenyladenosine 5'-monophosphate ([9R-5P]iP). Dephosphorylation and deribosylation lead sequentially to the production of isopentenyladenosine ([9R]iP), and isopentenyladenine (iP). The condensing enzyme (a) is a prenyl transferase, dimethylallylpyrophosphate: 5'-AMP transferase (DMA transferase or isopentenyl transferase) first described in habituated tobacco callus by Chen and Melitz (10). The enzyme (b) responsible for hydroxylation, is an uncharacterized oxidase (9) which is capable, in theory, of converting any one of three substrates ([9R-5P]iP, [9R]iP, or iP) into zeatin riboside 5'-phosphate (*trans*-[9R-5'P]Z), zeatin riboside or zeatin.

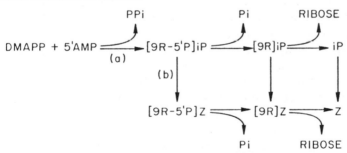

Fig. 5. Probable cytokinin biosynthetic pathway. [9R-5'P]iP = isopentenyladenosine 5'-monophosphate; [9R]iP = isopentenyladenosine; iP = isopentenyladenine; [9R-5'P]Z = zeatin riboside 5'-monophosphate; [9R]Z = zeatin riboside; Z = zeatin. Enzymes responsible for cytokinin biosynthesis are: (a) DMA transferase (isopentenyl transferase) encoded in *A. tumefaciens* by *tmr* and *tzs* and in *P. savastanoi* by *ptz*; (b) Cytokinin hydroxylase. Although expression of *tzs* and *ptz* in *E. coli* results in the production of zeatin, hydroxylase activity is probably provided by *E.coli*.

The body of evidence implicating cytokinins in crown gall growth is smaller than that related to the auxins but is somewhat more internally consistent. Differences in cytokinin levels and metabolism between untransformed and crown gall tissues are greater than those observed for the auxins. Again, the pioneering work of Braun (7) laid the foundation with the finding that crown gall tumors were autotrophic for cell division factors in culture. Subsequently, Miller showed that tumors contained high levels of zeatin and zeatin riboside (32). The evidence for altered cytokinin levels and metabolism in tumors is reviewed in detail in (33) and, in outline, is as follows:

- Tumor cytokinin levels are elevated over those of untransformed tissues. The most precise measurements have been made by Horgan and his coworkers (45 and references cited therein) using mass spectrometric techniques. In summary, they found that zeatin, zeatin riboside and many cytokinin metabolites, such as the side chain glucosides, are significantly elevated in crown gall tumors.
- Cytokinin accumulation rates are greater in tumors than in untransformed tissues in culture. Cytokinin levels were found to peak early in the exponential phase of tumor growth (16) and to reach levels greater than those of untransformed tissues.
- Biosynthesis of cytokinins is increased in tumors. Adenine is rapidly incorporated into cytokinins in *Vinca rosea* crown gall (48) and specifically into *trans*-[9R-5'P]Z (40). No incorporation occurred in cultures of untransformed *Vinca rosea*.
- In cloned octopine tumors in culture, increased cytokinin levels and elevated DMA transferase activities are associated solely with the presence of T_LDNA and not with T_RDNA (Fig. 6), indicating that the gene responsible is carried within the T_LDNA.
- Crown gall lines bearing mutations in the T_LDNA have cytokinin levels which differ from those of wild type tumors. Insertional inactivation of the *tmr* locus causes tumors which are normally undifferentiated to produce roots and, at the same time, causes a requirement for cytokinins for growth in culture (22) and a dramatic decrease in tumor cytokinin content (1) (Fig. 3). Thus, zeatin riboside levels in primary tobacco tumors decreased from 50 pmole/g in the wild type tumor to less than 1

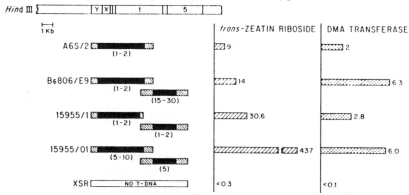

Fig. 6. Cytokinin levels and cytokinin prenyl transferase activity in cloned crown gall tumors. Cytokinin levels were determined by HPLC and RIA of extracts of cloned tumor lines grown in axenic culture. DMA transferase activity was measured *in vitro* by the procedure of Hommes *et al.*(56). Restriction enzyme cleavage sites of the T-region are shown. Horizontal bars represent the extent and ( ) the copy numbers of T_LDNA and T_RDNA segments present in tumors (determined by Kwok, Gordon and Nester; personal communication). Hatched bars represent the cytokinin and DNA transferase levels (D. Akiyoshi, Thesis, Oregon State University)

pmole/g when the *tmr* locus was inactivated. This last level was not significantly different from that present in untransformed stem tissue. Interestingly, inactivation of *tms* causes massive elevation of zeatin riboside levels to about 400 pmole/g. There is a clear relationship therefore between tumor cytokinin content and the expression of the auxin biosynthetic enzymes encoded by *tms1* and *tms2* and cytokinin production. Its nature is unknown at present.

• If *tmr* is inactivated, tumor DMA transferase activity is abolished (34).

Genes Specifying Cytokinin Biosynthesis in *Agrobacterium Tumefaciens*

Although the above data indicated that the *tmr* gene product could be a cytokinin prenyl transferase, it was not until Barry *et al.* (4) cloned and expressed *tmr* in *E. coli* that definitive evidence was obtained. The availability of the T-DNA nucleotide sequence[3] allowed identification of an open reading frame associated with *tmr* which specifies a protein of about 27kd. When *tmr* is cloned and expressed in *E. coli*, a protein of this size is produced and at the same time DMA transferase activity can be detected (4). Table 1 illustrates expression of DMA transferase activity by *tmr* cloned in both orientations behind the *lac* promoter of the multicopy *E. coli* plasmids pUC18 and pUC19. In pIP192, the *tmr* coding region is correctly aligned with the *lac* promoter of the plasmid in order to maximise transcription whereas in pIP182 it is in the opposite orientation (35). Although DMA transferase activity is expressed in both cases, it is

Table 1. Cytokinin prenyl transferase activity expressed by cloned bacterial genes.

Gene	Strain	DMA Transferase Activity fmol iP plus [9R]iP produced/hr/mg protein
tmr	HB101 (pIP192)	5.1
tmr	HB101 (pIP182)	1.5
tzs	HB101 (pTZ120)	6.5
None	HB101	not detected

higher when the gene is in the correct orientation. That the DMA transferase expressed in *E. coli* is identical to the enzyme contained in crown gall tumors is likely because Buchmann *et al.* (8) found that an antibody raised against a synthetic decapeptide, whose sequence was derived from the *tmr* coding region, crossreacted with both enzymes.

It apppears therefore that *tmr* encodes a cytokinin prenyl transferase which synthesizes [9R-5'P] iP *in vitro* and which, when expressed in crown gall tumors, results in elevated levels of many cytokinins and concomitant massive cell proliferation.

A Second Cytokinin Biosynthetic Gene is Located outside the T-DNA

The DMA transferase encoded by *A. tumefaciens* T-DNA is not the only cytokinin prenyl transferase present in this organism. There is evidence for the presence of a second gene.

The first indication of its presence came from a series of studies of cytokinin secretion by free-living *A. tumefaciens* (23, 42). The nopaline strain C58, secreted zeatin in culture but, when cured of its plasmid, lost the ability to do so. Subsequently it was shown that only wild-type nopaline *A. tumefaciens* strains or exconjugants containing nopaline Ti plasmids could secrete zeatin (42). A locus (*tzs*) in pTiC58 was therefore proposed as being responsible for zeatin production. Construction of a library of restriction fragments of pTiC58 in *E. coli* and subsequent screening of the transformants for zeatin secretion led to the suprising finding that *tzs* was active in *E. coli* and that it mapped outside the T-DNA (5). Production of zeatin by *E. coli* containing the *tzs* gene from pTiC58 is illustrated in Fig. 7.

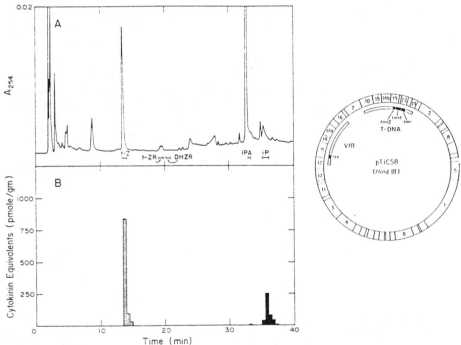

Fig. 7. Cytokinin production by *tzs* and location of *tzs* on pTiC58. Left: Cytokinin production by *tzs* cloned in *E. coli*. Production by the clone HB101(pTZ121) was measured in mid exponential phase growth by immunoaffinity chromatography followed by (A) HPLC and (B) RIA. RIA with anti-zeatin riboside antibody ▨ and anti-isopentenyladenine antibody ■. Right: Restriction map of pTiC58 showing *Hind*III cut sites and the approximate locations of *vir*, T-DNA, and the genes specifying phytohormone biosynthetic enzymes: *tmr*—cytokinin prenyl transferase (transferred from the T-DNA to the plant); *tms1*—tryptophan-2-monooxygenase; *tms2*—indole-3-acetamide hydrolase; *tzs*—cytokinin prenyl transferase (probably not transferred to plant). The position of *tzs* is known precisely but the boundaries of the the *vir* region are not.

Since expression of *tzs* in *E. coli* results in secretion of zeatin, it first appeared that the gene might encode the hydroxylase (Fig. 5b) responsible for side chain hydroxylation. However, the nucleotide sequence of *tzs* was found to exhibit extensive sequence homology with that of *tmr* which suggested that the gene might express DMA transferase activity (5). This proved to be true (Table 1) (35) and it now appears that *tzs* encodes a cytokinin prenyl transferase similar in function to that encoded by *tmr*.

The biological function of *tzs* is not known. Although it is close to the *vir* region (Fig. 7B), the *vir* region boundaries are not well defined and *tzs* may or may not be included. Octopine strains do not possess *tzs* yet are capable of inciting tumors. Interestingly, expression of *tzs* in strain C58 is enhanced significantly if the *vir* region inducer acetosyringone is included in the culture medium (Powell and Morris, unpublished data). The gene may well, therefore, play a role in the early processes of tumor induction.

Cytokinin Biosynthetic Genes in *Pseudomonas Savastanoi*

Cytokinin production by *P. savastanoi* was first observed by Surico et al. (50) who showed that culture filtrates contained iP and [9R]iP-like cytokinins. Subsequently, the presence of an unusual cytokinin, 1"-methyl-*trans*-zeatin riboside was reported (49).

Examination of several wild-type *Pseudomonas* strains reveals considerable variation in the amounts and types of cytokinins produced. In general, levels of cytokinin in culture filtrates of *P. savastanoi* are very much greater than those present in cultures of *A. tumefaciens* (30). Stationary phase cultures of *P. savastanoi* can contain as much as 10 μM cytokinin, the majority of which is zeatin and zeatin riboside. Some strains produce 1"-methyl-zeatin riboside whereas others do not. In two strains of *P. savastanoi* the existence of plasmid-born cytokinin biosynthetic genes has now been demonstrated (30). They are necesssary for pathogenicity (T. Kosuge, personal communication) and are very similar to the genes from *A. tumefaciens*.

The oleander-specific *P. savastanoi* strain 213 contains five plasmids ranging in size from 38 to 64kb; the olive-specific strain 1006 has two, 84kb and 105kb. An examination of cytokinin production by plasmid deletion mutants indicated that loss of the 42kb plasmid from strain 213 results in loss of ability to secrete zeatin and that deletion of a 40kb fragment from the 105kb plasmid of strain 1006 abolishes zeatin production (35). Cytokinin production by strain 1006 and its dependence on plasmid status is illustrated in Fig. 8. These results provide definitive evidence that at least one plasmid-born gene is directly involved with cytokinin production in *P. savastanoi*.

The isolation of that gene, named *ptz*, has now been reported (41). A library of plasmid fragments from *P. savastanoi* strain 1006 was constructed in *E. coli*, cytokinin-producing transformants were identified,

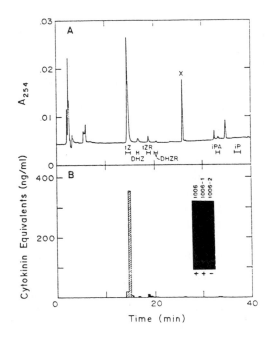

Fig 8. Cytokinin production and plasmid status of *P. savastanoi* strain 1006. Cytokinins present in one ml of mid-expenential phase culture filtrate of *P. savastanoi* strain 1006, purified by immunoaffinity chromatography and analysed by HPLC and RIA. (A) Absorbance at 254 nm. X is 1"methyl zeatin riboside. (B) RIA with anti-zeatin riboside antibody ▨ and antiisopentenyladenine antibody ■. *Inset*: Plasmid status and zeatin production: 1006: plasmids, 105 kb and 81 kb; 1006-1: plasmids, 105 kb; 1006-2: plasmids, 81 kb and 65 kb (deletion of 105 kb).

and the gene was sequenced. It contains an open reading frame of 702 nucleotides corresponding to a 26.6 kd protein. The size of the product was confirmed by coupled *in vitro* transcription/ translation and SDS gel electrophoresis. On expression in *E. coli*, zeatin is produced and cytokinin prenyl transferase can be detected.

COMPARISON OF AGROBACTERIUM AND PSEUDOMONAS GENES

Kosuge and coworkers (55) have shown that the *P. savastanoi* auxin biosynthetic genes, *iaaM* and *iaaH*, maintain substantial sequence homology with their counterparts, *tms1* and *tms2* from *A. tumefaciens*. *Tms1* is 54% homologous to *iaaM* and *tms2* is 38% homologous to *iaaH*. A comparison of deduced amino acid sequences from the coding regions of the genes specifying cytokinin biosynthesis reveals similar homology (Fig. 9) (41, 5). Overall there is 48% identity between *tzs* and *ptz*, while *tmr* and *ptz* contain perfect matches at 44% of the amino acid positions. Within the N-terminal region of the genes, the extent of homology is highest, at 66% and 58% respectively. Little similarity is found at the C-terminus, suggesting that this region plays only a minor role in the catalytic activity of the gene products.

Although substantial sequence homology is observed within open reading frames, little similarity is found in the non-coding regions. *Ptz*

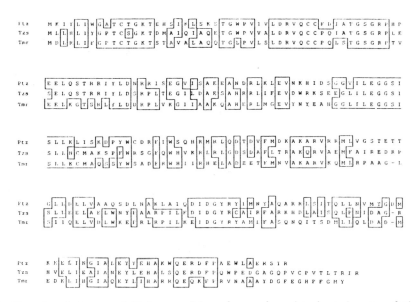

Fig. 9. Homology between cytokinin prenyl transferases from *Agrobacterium tumefaciens* and *Pseudomonas savastanoi*. The deduced amino acid sequences of the coding regions of *tmr* and *tzs* from *Agrobacterium tumefaciens* and of *ptz* from *Pseudomonas savastanoi* are illustrated. Several regions of complete identity are shown by □. Homology is especially evident at the N-termini. (A=alanine; C=cysteine; D=Aspartic acid; E=glutamic acid; F=phenylalanine; G=glycine; H−histidine; I−isoleucine; K=lysine; L=leucine; M=methionine; N=asparagine; P=proline; Q=glutamine; R=arginine; S=serine; T=threonine; V=valine; W=tryptophan; Y=tyrosine).

and *tzs* both display well-defined prokaryotic ribosome binding sites and apparent transcription termination inverted repeats external to the coding regions, although *E. coli* consensus promoter sequences are seen only in *ptz*. The *tmr* sequence bears no identifiable prokaryotic transcription or translation control signals. The nucleotide sequences encompassing the coding regions are consistent therefore with the known restriction of expression of *tmr* to the plant and *ptz* and *tzs* to the bacteria.

PROSPECTS FOR CONTROL OF PHYTOHORMONE BIOSYNTHESIS IN PLANTS

The acquisition and use of genes specifing cytokinin and auxin biosynthesis represents an interesting and unusual strategy employed by two rather different phytopathogenic bacteria, *A. tumefaciens* and *P. savastanoi*, during the process of inciting tumors on plants. Although the bacteria differ markedly in the manner in which they employ the phytohormone biosynthetic genes, the two sets of genes exhibit a startling degree of sequence and functional homology suggesting the past occurrence of some form of horizontal transfer. How such a transfer occurred and by what mechanism remains to be determined.

The relationship between the bacterial genes and those of the untransformed plant also remains to be determined. Hybridization experiments designed to detect homology between T-DNA genes and those of uninfected plants have yielded negative results. However, the techniques employed so far (hybridization under moderately stringent conditions) do not always demonstrate homology when it exists and it may be that there is a relationship between the bacterial and plant genes which remains to be discovered. Nevertheless, whatever the actual relationship, the discovery and characterization of the bacterial genes has provided a useful tool to probe phytohormone function in plants.

Many of the older studies of function had perforce to apply unacceptably large amounts of phytohormones in order to compensate for barriers to uptake. Now that the techniques for introduction of genes into plants are well established, it has become possible in principle to introduce these genes and control their expression in such a way as to cause transient or permanent increases in the phytohormone levels within specific organs or tissues.

Several methods have been developed to transfer genes to plants. Among the most interesting is the binary plasmid system first described by Schilperoort and his coworkers (19) and extended by the work of An (2). The system comprises two plasmids. The first carries, on a relatively small plasmid, defined restriction enzyme cleavage sites and a gene specifying resistance to kanamycin inserted between the T-DNA border sequences. When genes are introduced into these sites between the T-DNA borders, and the resulting chimeric plasmid is transferred to A. tumefaciens containing a "disarmed" Ti plasmid from which the T-DNA and its borders have been deleted, the desired genes and the accompanying resistance marker may then be transfered to a plant with the aid of the vir genes present on the disarmed plasmid. Plant transformants may then be selected on the basis of resistance to kanamycin or the aminoglycoside antibiotic G418 and then screened for the presence of the desired genes.

In order to introduce phytohormone biosynthetic genes into plants and arrange for their expression in desired situations, at least two criteria must be met:

• The genes must contain coding regions capable of being adequately transcribed and translated within a eukaryotic enviromnment.
• There must be stringent control of gene expression especially during intermediate stages of transfer when, for example, one desires organogenesis to occur in culture.

Fortunately, the first criterion appears to be have been met in the case of the tms1, tms2 and tmr coding regions from A. tumefaciens. The genes have evolved so that they are normally expressed adequately in plants. The same may be true for the cognate genes from P. savastanoi although it has yet to be demonstrated.

The second criterion may be harder to meet. Leaving aside for the moment the question of chromosomal position effects on expression, the problem is to select plant DNA elements which can confer on the introduced phytohormone coding regions the desired degree of tissue specific expression. Tissue specific gene expression is well documented in plants (24) but the stringency of such control and the exact mechanism by which it occurs are not yet defined. The necessity of ensuring that absolutely no transcription occurs under certain conditions may put constraints on the sequences available for use. Whether the degree of control available from existing tissue-specific promoters, or from heterologous promoters derived from heat shock or metallothionine genes will prove adequate for the task remains to be seen. If these problems can be overcome then the exciting prospect of controlled study of the influence of phytohormones on plant development will become possible and some of the questions relating, for example, to tissue sensitivity to endogenous hormone levels may be addressed.

Acknowledgements

I wish to thank my colleagues who have actively participated in some of the work described here and acknowledge helpful discussions with Gary Powell and Michel Laloue. I also wish to thank Professor Jean Guern of the CNRS for his hospitality and support during the preparation of this manuscript. The work of the author's laboratory is supported by grants from the National Science Foundation (PCM 83-03371) and the U.S. Department of Agriculture (83-CRCR-1-1249).

References

1. Akiyoshi, D.E., Morris, R.O., Hinz, R., Mischke, B.S., Kosuge, T., Garfield, D.J., Gordon, M.P., Nester, E.W. (1983). Cytokinin-auxin balance in crown gall tumors is regulated by specific loci in the T-DNA. Proc. Natl. Acad. Sci. USA *80*, 407-411.
2. An, G., Watson, B.D., Stachel, S., Gordon, M.P., Nester, E.W. (1985) New cloning vehicles for transformation of higher plants. Embo J. *4*, 277-284.
3. Barker, R.F., Idler, K.B., Thompson, D.V., Kemp, J.D. (1983). Nucleotide sequence of the T-DNA from *Agrobacterium tumefaciens* octopine Ti plasmid pTi15955. Plant Mol. Biol. *2*, 335-350.
4. Barry, G.F., Rogers, S.G., Fraley, R.T., Brand, L. (1984). Identification of a cloned cytokinin biosynthetic gene. Proc. Natl. Acad. Sci. USA *81*, 4776-4780.
5. Beaty, J.S., Powell, G.K., Lica, L., Regier, D.A., Macdonald, E.M.S., Hommes, N.G., Morris, R.O. (1986). *Tzs*, a nopaline ti plasmid gene from *Agrobacterium tumefaciens* associated with *trans*-zeatin biosynthesis. Mol. Gen. Genet. *203*, 274-280.
6. Binns, A.N., Sciaky, D., Wood, H.N. (1982) Variation in hormone autonomy and regenerative potential of cells transformed by strain A66 of *Agrobacterium tumefaciens*. Cell *31*, 605-612.
7. Braun, A.C. (1958). A physiological basis for the autonomous growth of the crown gall tumor cell. Proc. Natl. Acad. Sci. USA *44*, 344-349.
8. Buchmann, I., Marner, F.J., Schroder, G., Waffenschmidt S., Schroder, J. (1985). Tumor genes in plants: T-DNA encoded cytokinin biosynthesis. EMBO J. *4*, 853-859.
9. Chen, C.M., Leisner, S.M. (1984). Modification of cytokinins by cauliflower microsomal enzymes. Plant Physiol. *75*, 442-446.
10. Chen, C.M., Melitz, D.K. (1979). Cytokinin biosynthesis in a cell-free system from cytokinin-autotrophic tobacco tissue cultures. Febs Lett. *107*, 15-20.
11. Comai, L., Kosuge, T. (1980) Involvement of plasmid deoxyribonucleic acid in indoleacetic acid synthesis in *Pseudomonas Savanostanoi*. J. Bacteriol. *143*, 950-957.

12. Comai, L., Kosuge T. (1982) Cloning and characterization of *iaaM*, a virulence determinant of *Pseudomonas Savastanoi*. J. Bacteriol. *149*, 40- 46.
13. Comai, L., Surico, G., Kosuge, T. (1982). Relation of plasmid DNA to indoleacetic acid production in different strains of *pseudomonas syringae* pv. *savastanoi*. J. Gen. Microbiol. *128*, 2157-2163.
14. Engler, G., Depicker, A., Maenhaut, R., Villaroel-Mandiola, R., Van Montagu, M., Schell, J., Hernalsteens, J.P. (1981) Physical mapping of DNA base sequence homologies between an octopine and a nopaline Ti plasmid of *Agrobacterium tumefaciens*. J. Mol. Biol. *152*, 183-208.
15. Garfinkel, D.J., Simpson, R.B., Ream, L.W., White, F.F., Gordon, M.P., Nester, E.W. (1981). Genetic analysis of crown gall: fine structure map of the T-DNA by site-directed mutagenesis. Cell *27*, 143-153.
16. Hansen, C.E., Meins, F. Jr., Milani, A., (1985) Clonal and physiological variation in the cytokinin content of tobacco-cell lines differing in cytokinin requirement and capacity for neoplastic growth. Differentiation *29*, 1-6.
17. Hille, J., Van kan, J., Schilperoort, R. (1984) Trans-acting virulence functions of the octopine ti plasmid from *Agrobacterium tumefaciens*. J. Bact. *158*, 754-6.
18. Hommes, N.G., Akiyoshi, D.E., Morris, R.O. (1985) Assay and partial purification of the cytokinin biosynthetic enzyme dimethylallylpyrophosphate: 5'AMP transferase. Methods Enzymol. *110*, 340-347.
19. Hoekema, A., Hirsch, P.R., Hooykaas, P.J.J., Schilperoort R.A. (1983) A binary vector strategy based on separation of *vir* and T-region of the *Agrobacterium tumefaciens* Ti-plasmid . Nature *303*, 179-80.DNA-encoded auxin formation in crown-gall cells. Planta *163*, 257-262.
20. Hutcheson, S., Kosuge, T. (1985). Regulation of 3-indoleacetic acid production in *Pseudomonas syringae* pv. *savastanoi*. purification and properties of tryptophan-2-monooxygenase. J. Biol. Chem. *260*, 6281-6287.
21. Hutzinger, O., Kosuge, T. (1967) Microbial synthesis and degradation of indole-3- acetic acid. Biochim. Biophys. Acta *136*, 389-391.
22. Joos, H., Inze, D., Caplan, A., Sormann, M., Van Montagu, M., Schell, J. (1983) Genetic analysis of T-DNA transcripts in nopaline crown galls. Cell *32*, 1057-1067.
23. Kaiss-Chapman, R.W., Morris, R.O., (1977) Trans-zeatin in culture filtrates of *Agrobacterium tumefaciens*. Biochem. Biophys. Res. Comm. *76*, 453-459.
24. Kamalay, J.C., Goldberg, R.B., (1984) Organ-specific nuclear RNAs in tobacco. Proc. Natl. Acad. Sci. USA *81*, 2801-2805.
25. Kemper, E., Waffenschmidt, S., Weiler E.W., Rausch, T., Schroder, J. (1985) T-
26. Kende, H. (1971). The cytokinins. Int. Rev. Cytol. *31*, 301-338.
27. Klee, H., Montoya, A., Horodyski, F., Lichtenstein , C., Garfinkel, D. Fuller, S., Flores, C., Peschon, J., Nester, E., Gordon, M. (1984). Nucleotide sequence of the *tms* genes of the pTiA6NC octopine Ti plasmid: 2 gene products involved in plant tumorigenesis. Proc. Natl. Acad. Sci. USA *81*, 1728-1732.
28. Koukolilova-Nicola, Z., Shillito, R.D., Hohn, B., Wang, K., Van Montagu, M., Zambryski, P. (1985) Involvement of circular intermediates in the transfer of T-DNA from *Agrobacterium tumefaciens* to plant cells. Nature *313*, 191-196.
29. Letham, D.S., Palni, L.M.S. (1983). The biosynthesis and metabolism of cytokinins. Ann. Rev. Plant Physiol. *34*, 163-197.
30. Macdonald, E.M.S., Powell, G.K., Regier, D.A., Glass, L., Kosuge, T., Morris R.O. (1986) Secretion of zeatin, ribosylzeatin and ribosyl-1"-methylzeatin by *Pseudomonas savastanoi*: plasmid-coded cytokinin biosynthesis. Plant Physiol. in press.
31. Mellano, M.A., Cooksey, D.A. (1985) Detection of a cytokinin gene homologue in *Corynebacterium fascians*. Phytopathology in press.
32. Miller, C.O. (1974) Ribosyl-*trans*-zeatin. a major cytokinin produced by crown gall tumor tissue. Proc. Natl. Acad. Sci. USA *71*, 334- 338.
33. Morris, R.O. (1986). genes specifying auxin and cytokinin biosynthesis in phytopathogens. Ann. Rev. Plant Physiol. *37*, 509-538.
34. Morris, R.O., Akiyoshi, D.E., Macdonald, E.M.S., Morris, J.W., Regier, D.A., Zaerr, J.B. (1982). Cytokinin metabolism in relation to tumor induction by *Agrobacterium tumefaciens*. *In* Plant Growth Substances, P.F. Wareing, ed. Academic Press, London.
35. Morris, R.O., Powell, G.K., Beaty, J.S., Durley, R.C., Hommes, N.G. Lica, L., Macdonald, E.M.S. (1986). Cytokinin biosynthetic genes from *Agrobacterium tumefaciens* and other

plant-associated prokaryotes. *In* Plant Growth Substances, pp. 185-196. M. Bopp ed. Springer Verlag, Berlin.

36. Nester, E.W., Gordon, M.P., Amasino, R.M., Yanofsky, M.F. (1984). Crown gall: a molecular and physiological analysis. Ann. Rev. Plant Physiol. *35*, 387-413.

37. Nester, E.W., Kosuge, T. (1981) Plasmids specifying plant hyperplasias. Ann. Rev. Microbiol. *35*, 531-565.

38. Okker, R.J.H., Spaink, H., Hille, J., Van Brussel, T.A.N., Lugtenberg, B., Schilperoort, R.A. (1984) Plant-inducible virulence promoter of the *Agrobacterium tumefaciens* Ti plasmid. Nature *312*, 564-566.

39. Ooms, G., Hooykaas, P.J.J., Moolenaar, G., Schilperoort, R.A. (1981) Crown gall plant tumors of abnormal morphology induced by *Agrobacterium tumefaciens* carrying mutated octopine Ti plasmids: analysis of T-region DNA functions. Gene *14*, 33-50.

40. Palni, L.M.S., Horgan R., Darrall, N.M., Stuchbury, T., Wareing P.F. (1983) Cytokinin biosynthesis in crown gall tissue of *Vinca rosea* L. the significance of nucleotides. Planta *159*, 50-59.

41. Powell, G.K., Morris, R.O. (1986). Nucleotide sequence and expression of a *Pseudomonas savastanoi* cytokinin biosynthetic gene: homology with *Agrobacterium tumefaciens tmr* and *tzs* loci. Nucleic Acids Res. *14*, 2555-2565.

42. Regier, D.A., Morris, R.O. (1982) Secretion of *trans*-zeatin by *Agrobacterium tumefaciens*: a function determined by the nopaline Ti plasmid. Biochem. Biophys. Res. Commun. *104*, 1560-1566.

43. Schroder, G., Waffenschmidt, S., Weiler, E.W., Schroder, J. (1984) The T-region of Ti plasmids codes for an enzyme synthesizing indole-3-acetic acid. Eur.J.Biochem. *138*, 387-391.

44. Sciaky, D., Thomashow, M.F., (1984) The sequence of the *tms* transcript 2 locus of the *A. tumefaciens* plasmid pTiA6 and characterization of the mutation in pTiA66 that is responsible for auxin attenuation. Nucleic Acids Res. *12*, 1447-1461.

45. Scott, I.M., Horgan, R. (1984) Mass spectrometric quantification of cytokinin nucleotides and glycosides in tobacco crown gall tissue. Planta *161*, 345-54.

46. Skoog, F., Armstrong D.J. (1970). Cytokinins. Ann. Rev. Plant Physiol. *21*, 359-384.

47. Stachel, S.E., Messens, E., Van Montagu, M., Zambryski, P. (1985) Identification of the signal molecules produced by wounded plant cells that activate T-DNA transfer in *Agrobacterium tumefaciens* Nature *318*, 624-649.

48. Stuchbury, T., Palni, L.M., Horgan, R., Wareing, P.F. (1979) The biosynthesis of cytokinins in crown gall tissue of *Vinca rosea* L. Planta 147, 97-102.

49. Surico, G., Evidente, A., Iacobellis, N., Randazzo, G. (1985). A new cytokinin from the culture filtrate of *Pseudomonas syringae* pv. *savastanoi*. Phytochemistry *24*, 1499-1502.

50. Surico, G., Sparapano, L., Legario, P., Durbin, R.D., Iacobellis, N. (1975). Cytokinin-like activity in extracts from the culture filtrate of *Pseudomonas savastanoi*. Experientia *31*, 929-930.

51. Van Onckelen, H., Prinsen, E., Inze, D., Rudelsheim, P., Van Lijsebettens, M., Follin, A., Schell, J., Van Montagu, M., De Greef, J. (1986). *Agrobacterium* T-DNA gene 1 codes for tryptophan-2-monooxygenase activity in tobacco crown gall cells. FEBS Lett. *198* 357-360.

52. Van Onckelen, H., Rudelsheim, P., Inze, D., Follin, A., Messens, E., Hiremans, S., Schell, J., Van Montagu, M., De Greef, J. (1985). Tobacco plants transformed with the *Agrobac-terium* T-DNA gene 1 contain high amounts of indole-3-acetamide. Febs Lett. *181* 373-76.

53. Willmitzer, L., Dhaese, P., Schreier, P.H., Schmalenbach, W., Van Montagu, M., Schell, J. (1983) Size, location and polarity of T-DNA-encoded transcripts in nopaline crown gall tumors: common transcripts in octopine and nopaline tumors. Cell *32*, 1045-1056.

54. Yadav, N.S. Vanderleyden, J., Bennett, D.R., Barnes, W.M., Chilton, M.-D. (1982) Short direct repeats flank the T-DNA on a nopaline Ti plasmid. Proc. Natl. Acad. Sci. USA *79*, 6322-6326.

55. Yamada, T., Palm, C.J., Brookes, B., Kosuge, T. (1985). Nucleotide sequence of the *Pseudomonas Savastanoi* indoleacetic acid genes show homology with *Agrobacterium tumefaciens* DNA. Proc. Natl. Acad. Sci. USA *82*, 6522-6526.

56. Zaenen, I., Van Larabeke, N., Teuchy, H., Van Montagu, M., Schell, J. (1974) Super-coiled circular DNA in crown gall inducing *Agrobacterium* strains. J.Mol.Biol. *86*, 109-127.

INDEX

656

Chilling
ABA, 549
ACC, 97
bud, 545
cytokinin, 548
ethylene, 97, 549
GA, 547
growth promoting, 546
mechanism, 546
requirement, 543
seed, 545
temperature, 542
unit, 542, 543
Chlorflurenol, 628
Chlormequat chloride, 54,
447, 499, 528, 532, 533,
589, 614, 615, 627-629
2-Chloroethyl-bis(phenyl-
methoxy)silane, 623
2-Chloroethylmethyl-bis
(phenylmethoxy)silane,
625
2-Chloroethylphosphonic
acid, 259, 615
2-Chloroethyltrimethyl-am-
monium chloride, 441,
614
2-Chloroethyl-tris(ethoxymethoxy)silane, 623
4-Chloro-indole-3-acetic acid,
4, 24
p-Chloromecuribenzene
sulfonate, 487
Chloro-3-methyl-4-nitro-1-
pyrazole, 631
4-Chlorophenoxyacetic acid,
616
Chloronemata, 319, 320, 396
2RS, 3RS-1-(4-Chloro-
phenyl)-4, 4-dimethyl-2-1,
2, 4-triazol-1-yl-pentan-3-
ol, 627
Chlorophyll
breakdown, 556, 558
degradation, 555
pod removal, effect of, 564
retention, 559, 560
Chloroplasts
ABA, 414
Chlorphonium chloride, 54
Cholodny-Went theory, 32,
377-379, 383, 384
Chromatin, 196

Chromatography, 223
Chromoplasts, 580
Chrysanthemum, 271, 627
Chrysanthemum morifolium,
433, 434, 438, 447
Cichorium intybus, 443
Cinnamic acid, 282, 313
Citrus, 436
Citrus paradisi, 435
Citrus sinensis, 435
Climacteric, 274, 576, 577,
578
Cobalt, 100, 260, 559, 560
Cocklebur, 103
Coconut water(= milk), 76,
593, 594, 598, 601
Coleoptile
IAA, growth, 17, 132-144
transport, 31
Coleus, 365, 367
Columns, GC capillary, 228
Commelina communis, 412,
416-420
Compactin, 53
Compartmentation, 19
Competence, 601, 602
Concentration, plant
hormone, 18
Conditioned medium, 598
Conglycinin, 506, 507
mRNA, 510
Conifers, juvenile, 440, 441
Controlled atmosphere
storage, 585, 586, 588
Copalyl pyrophosphate, 48,
55, 326
Coreopsis grandiflora, 434
Corn-see Maize
Correlative inhibition, 395
Corynebacterium fascians,
82, 637, 638
Cosmos, 272
Cotton, 614, 628, 631
embryo, 504, 505, 508
gene expression, 510
Coumarin, 531
Cowpea, 297
Crassulacean acid
metabolism, 285
Crown gall, 7, 610, 637, 639,
641
cytokinin, 645, 646
cytokinin oxidase, 89

cytokinin/auxin ratio, 17
cytokinins, 82
immunoassay, 249
tumors, 593, 594
Vinca rosea, 84
Cruciferin, 506
Cucumber
chilling, 578
flowering, 434
GA, 65, 67, 297, 299
binding, 313
cell wall, 304
light, 299
osmotic potential, 304
polyamines, 288
sex expression, 456
squirting, 275
xylem regeneration, 367
Cucumis melo, 577, 578
Cucumis sativus-see
Cucumber
Cultures, suspension, 601
Cupressus, 388, 436
Curcurbita maxima, 45, 51
GA, 62-65
Curcurbita pepo, 354, 578
Curcurbitaceae, 268
Cuttings, leaf, 607
Cyanoalanine, 101
Cyanoformic acid, 100
Cyclamen, 272
Cyclic AMP, 196, 197
Cyclopropylamine, 100
Cycocel-see Chlormequat
chloride
Cymbidium, 595
Cytisus, 65
Cytokinin, 588
ABA, senescence, 560
abbreviations, 77
ACC synthase, 98
*Agrobacterium tume-
faciens*, 82, 532, 636-38
N-alanyl conjugates, 88
amino acid conjugates, 87
analysis, 79
antigens, 247
antobodies, 247
apical bud, 398
apical dominance, 397,
399, 532
assay, 224
auxin, 207, 560, 599

667

Guard-cell
   ABA binding, 213

Hadacidin, 398
Haploid, 596
Haptens, 241
Hazel, 547
Hedera helix, 436, 441-43
Helianthus annuus-see
   sunflower
Helianthus tuberosus, 288
Herbicide, 597, 616, 618
Heterosis, 463-471
Hevea, 273, 623
Hieracium aurantiacum, 445
High performance liquid
chromatography-see HPLC
Hordeum vulgare-see Barley
Hormone-see also Plant
   hormone
   animal, 1, 21
   binding, 194-215
      methods, 218-221
      sites, 177
   concentration, 14, 22
   definition, 1
   mammalian, 1, 14
   origin, 1, 2
   peptide, perception, 197
      transduction, 197
   perception, 195, 196
   receptor, 176
      methods, 218-221
      proteins, 215
   receptors, 194, 196
   specificity, 22
   steroid, 198
   synthesis, 1
   transduction, 195, 196
   transport, 1
Horseradish, 312
HPLC, 222-224
   detectors, 233
   fluorescence detection,
      235
   resolution, 226
   reverse phase, 225
   UV detection, 235
Hyacinth, 607
Hybrid
   production, 463
   vigor, 463-471
Hybridoma, 241

Hydraulic conductivity, 132
Hydrolase, 170
Hydroperoxide, 579
4-Hydroxybenzoate hydrox-
   ylase, 644
Hydroxycinnamic acid, 287
5-Hydroxyindole-3-acetic
   acid, 245
4'-Hydroxy-α-ionylidene
   acetaldehyde, 122
Hydroxykaurenoic acid, 51,
   328
Hydroxymethyl ABA, 124,
   125, 127, 128
3-Hydroxy-3-methyl glutaryl
   HM ABA, 126
3-Hydroxymethyloxindole,
   33, 34, 36
Hydroxyproline, 144
Hyoscyamus niger, 434, 437,
   438
IAA-see Indole-3-acetic acid
iaaH, 644, 650
iaaM, 644, 650
Immunoaffinity
   chromatography, 648, 650
   column, 251
Immunoassay, 610
   parallelism, 249
   protocols, 242
   purification, 249, 250
   quantitation problems
      with, 248
   sensitivity, 244
   validation, 249, 250
Immunofluorescence, 252
Impatiens balsamina, 446
In vitro, 600
Inbreeding depression, 465
Indole, 230
Indoleacetaldehyde, 4, 642
Indoleacetamide, 27, 245,
   642, 643, 644
Indoleacetamide hydrolase,
   636, 643, 644, 648
Indole-3-acetic acid, 4, 31-see
   also Auxin
   ABA, stomata, 419
      stomatal closure, 418
      stomatal opening, 422
   ACC synthase, 98
   accumulation, 349
   amide conjugate, 38

amide linkage, 39
amino acid conjugates, 30
   synthesis of, 38
amount, 24
antibodies, 244-246
   cross-reactivities, 245
antigens, 244
apical dominance, 407
assimilate translocation,
   632
axillary bud growth, 404,
   405, 407
binding, 200, 202
   protein, 201
   sites, 24, 203, 208
biosynthesis, 12, 24-30,
   636
   non-indolylic, 27
   Pseudomonas
      savastanoi, 27
   tryptophan, 25, 28
bound, 30
breakdown, 616
bulbs, 582
cabbage, 585, 586
calcium, 385
callus, cultures, 593
carrier, 20
catabolism, 37
CO_2, stomata, 419, 420,
   424
compartmentation, 20
concentration, response,
   17
conjugates, 12, 29, 244, 383
   385-388, 478, 561, 644
   amide, 30
   enzymatic hydrolysis,
      30
   ester, 30
   hydrolysis of, 25, 28
   synthesis, 38
   Zea mays, 28
cytokinin, stomata, 421
decarboxylation, 33, 34, 37
   dichlorophenol, 33
   monophenols, 33
derivatives, mass spectra,
   230
diffusion, 20
distribution, 383, 386, 387
   phototropism, 377
   stems, 18

675

677